System Analysis, Design, and Development

System Analysis, Design, and Development
Concepts, Principles, and Practices

Charles S. Wasson

A John Wiley & Sons, Inc., Publication

Copyright © 2006 by John Wiley & Sons, Inc. All rights reserved.

Published by John Wiley & Sons, Inc., Hoboken, New Jersey.
Published simultaneously in Canada.

No part of this publication may be reproduced, stored in a retrieval system, or transmitted in any form or by any means, electronic, mechanical, photocopying, recording, scanning, or otherwise, except as permitted under Sections 107 or 108 of the 1976 United States Copyright Act, without either the prior written permission of the Publisher, or authorization through payment of the appropriate per-copy fee to the Copyright Clearance Center, Inc., 222 Rosewood Drive, Danvers, MA 01923, 978-750-8400, fax 978-646-8600, or on the web at www.copyright.com. Requests to the Publisher for permission should be addressed to the Permissions Department, John Wiley & Sons, Inc., 111 River Street, Hoboken, NJ 07030, (201) 748-6011, fax (201) 748-6008, e-mail: permreq@wiley.com.

Limit of Liability/Disclaimer of Warranty: While the publisher and author have used their best efforts in preparing this book, they make no representations or warranties with respect to the accuracy or completeness of the contents of this book and specifically disclaim any implied warranties of merchantability or fitness for a particular purpose. No warranty may be created or extended by sales representatives or written sales materials. The advice and strategies contained herein may not be suitable for your situation. You should consult with a professional where appropriate. Neither the publisher nor author shall be liable for any loss of profit or any other commercial damages, including but not limited to special, incidental, consequential, or other damages.

For general information on our other products and services please contact our Customer Care Department within the U.S. at 877-762-2974, outside the U.S. at 317-572-3993 or fax 317-572-4002.

Wiley also publishes its books in a variety of electronic formats. Some content that appears in print, however, may not be available in electronic format.

Library of Congress Cataloging-in-Publication Data:
Wasson, Charles S., 1948–
    System analysis, design, and development : concepts, principles, and practices / by Charles S. Wasson.
        p. cm.
    "A Wiley-Interscience publication."
    Includes bibliographical references and index.
    ISBN-13 978-0-471-39333-7
    ISBN-10 0-471-39333-9 (cloth : alk. paper)
    1. System design.   2. System analysis.   I. Title.

QA76.9.S88W373 2005
004.2′1—dc22

2004061247

Printed in the United States of America

10 9 8 7 6 5 4 3 2 1

"If an error or omission is discovered, please notify the publisher with corrections in writing."

Table of Contents

Preface   ix
Acknowledgements   xii

1 Introduction   1

2 Book Organization and Conventions   3

Part I System Analysis Concepts

System Entity Concepts Series

3 What Is a System?   17

4 System Attributes, Properties, and Characteristics   27

5 System Roles and Stakeholders   39

6 System Acceptability   46

7 The System/Product Life Cycle   59

System Architecture Concepts Series

8 The Architecture of Systems   67

9 System Levels of Abstraction and Semantics   76

10 The System of Interest Architecture   86

11 The Operating Environment Architecture   97

12 System Interfaces   110

System Mission Concepts Series

13 Organizational Roles, Missions, and System Applications   122

14 Understanding the Problem, Opportunity, and Solution Spaces   135

15 System Interactions with its Operating Environment   146

16 System Mission Analysis   159

17 System Use Cases and Scenarios   167

System Operations Concepts Series

18 System Operations Model   178

19 System Phases, Modes, and States of Operation   189

20 Modeling System and Support Operations   206

System Capability Concepts Series

21 System Operational Capability Derivation and Allocation   217

22 The Anatomy of a System Capability   229

System Concept Synthesis

23 System Analysis Synthesis   241

Part II System Design and Development Practices

System Development Strategies Series

24 The System Development Workflow Strategy — 251

25 System Design, Integration, and Verification Strategy — 265

26 The SE Process Model — 275

27 System Development Models — 290

System Specification Series

28 System Specification Practices — 302

29 Understanding Specification Requirements — 315

30 Specification Analysis — 327

31 Specification Development — 340

32 Requirements Derivation, Allocation, Flow Down, and Traceability — 358

33 Requirements Statement Development — 370

System Development Series

34 Operational Utility, Suitability, and Effectiveness — 390

35 System Design To/For Objectives — 400

36 System Architecture Development — 410

37 Developing an Entity's Requirements Domain Solution — 430

38 Developing an Entity's Operations Domain Solution — 439

39 Developing an Entity's Behavioral Domain Solution — 451

40 Developing an Entity's Physical Domain Solution — 465

41 Component Selection and Development — 480

42 System Configuration Identification — 489

43 System Interface Analysis, Design, and Control — 507

44 Human–System Integration — 524

45 Engineering Standards, Frames of Reference, and Conventions — 544

46 System Design and Development Documentation — 562

Decision Support Series

47 Analytical Decision Support — 574

48 Statistical Influences on System Design — 586

49 System Performance Analysis, Budgets, and Safety Margins — 597

50 System Reliability, Availability, and Maintainability (RAM) — 615

51 System Modeling and Simulation — 651

52 Trade Study Analysis of Alternatives — 672

Verification and Validation Series

53 System Verification and Validation — 691

54 Technical Reviews — 710

| 55 | System Integration, Test, and Evaluation | 733 |
| 57 | System Operations and Support (O&S) | 773 |

System Deployment, Operations, and Support Series

| 56 | System Deployment | 758 |

Epilogue   788

Index   789

Preface

As a user, acquirer, or developer of a system, product, or service, have you ever been confronted with one of the situations listed below?

- Wondered if the people who designed a product bothered to ask potential users to simply try it before selling it to the public.
- Found that during a major program review prior to component development that someone thought a requirement was so obvious it didn't have to be written down.
- Participated in a new system development effort and discovered at Contract Award that team members were already designing circuits, coding software, and developing mechanical drawings BEFORE anyone understood WHAT system users expected the system to provide or perform?
- Procured one of those publicized "designed for assembly" products and discovered that it was not designed for maintainability?
- Interacted with a business that employed basic business tools such as desktop computers, phones, and fax machines that satisfied needs. Then, someone decided to install one of those new, interactive Web sites only to have customers and users challenged by a "new and improved" system that was too cumbersome to use, and whose performance proved to be inferior to that of the previous system?

Welcome to the domain of system analysis, design, and development or, in the case of the scenarios above, the potential effects of the lack of System Engineering (SE).

Everyday people acquire and use an array of systems, products, and services on the pretense of improving the quality of their lives; of allowing them to become more productive, effective, efficient, and profitable; or of depending on them as tools for survival. The consumer marketplace depends on organizations, and organizations depend on employees to ensure that the products they produce will:

1. Perform planned missions *efficiently* and *effectively* when called upon.
2. Leverage user skills and capabilities to accomplish tasks ranging from simple to highly complex.
3. Operate using commonly available resources.
4. Operate *safely* and *economically* in their intended environment with *minimal* risk and intrusion to the general public, property, and the environment.
5. Enable the user to complete missions and return safely.
6. Be maintained and stored until the next use for low cost.
7. Avoid any environmental, safety, and health risks to the user, the public, or the environment.

In a book entitled *Moments of Truth*, Jan Carlzon, president of an international airline, observed that every *interaction* between a customer and a business through product usage or service support is a *moment of truth*. Each customer–product/service interaction, though sometimes brief, produces and influences perceptions in the User's mind about the system, products, and services of each organization. Moment of truth interactions yield *positive* or *negative* experiences. Thus, the experiences posed by the questions above are *moments of truth* for the organizations, analysts, and engineers who develop systems.

Engineers graduate from college every year, enter the workforce, and learn system analysis, design, and development methods from the bottom up over a period of 10 to 30 years. Many spend entire careers with only limited exposure to the Users of their designs or products. As engineers are assigned increasing organizational and contract responsibilities, interactions with organizational customers also increase. Additionally, they find themselves confronted with learning *how to* integrate the efforts of other engineering disciplines beyond their field. In effect, they informally learn the rudiments of System Engineering, beginning with buzzwords, from the bottom up through observation and experience.

A story is told about an engineering manager with over 30 years of experience. The manager openly bragged about being able to bring in new college graduates, throw them into the work environment, and watch them sink or swim on their own without any assistance. Here was an individual with a wealth of knowledge and experience who was determined to let others "also spend 30 years" getting to comparable skill levels. Granted, some of this approach is fundamental to the learning experience and has to evolve naturally through personal *trials* and errors. However, does society and the engineering profession benefit from this type of philosophy?

Engineers enter the workplace from college at the lowest echelons of organizations mainly to apply their knowledge and skills in solving unique *boundary condition* problems. For many, the college dream of designing electronic circuits, software, or impressive mechanical structures is given a reality check by their new employers. Much to their chagrin, they discover that physical design is not the first step in engineering. They may be even startled to learn that their task is not to design but to find low-cost, acceptable risk solutions. These solutions come from research of the marketplace for existing products that can be easily and cost-effectively adapted to fulfill system requirements.

As these same engineers adapt to their work environment, they *implicitly* gain experience in the processes and methods required to transform a user's operational needs into a physical system, product, or service to fulfill contract or marketplace needs. Note the emphasis on *implicitly*. For many, the skills required to understand these new tasks and roles with increasing complexity and responsibility require tempering over years of experience. *If* they are fortunate, they may be employed by an organization that takes system engineering seriously and provides formal training.

After 10 years or so of experience, the demands of organizational and contract performance require engineers to assimilate and synthesize a wealth of knowledge and experience to formulate ideas about how systems operate. A key element of these demands is to communicate with their customers. Communications require open elicitation and investigative questioning, observation, and listening skills to understand the customer's operational needs and frustrations of unreliable, poorly designed systems or products that:

1. Limit their organization's ability to successfully conduct its missions.
2. Fail to start when initiated.
3. Fail during the mission, or cause harm to its operators, the general public, personal property, or the environment.

Users express their visions through operational needs for new types of systems that require application of newer, higher performance, and more reliable technologies, and present the engineer with the opportunity to innovate and create—as was the engineer's initial vision upon graduation.

Task leads and managers have a leadership obligation to equip personnel with the required processes, methods, and tools to achieve contract performance—for example, on time and within budget deliverables—and enterprise survival over the long term. They must be visionary and proactive. This means providing just-in-time (JIT) training and opportunities to these engineers when they need these skills. Instead, they defer training to technical programs on the premise that this is on-the-job (OJT) training. Every program is unique and only provides a subset of the skills that SEs need. That approach can take years!

While browsing in a bookstore, I noticed a book entitled *If I Knew Then What I Know Now* by Richard Elder. Mr. Elder's book title immediately caught my attention and appropriately captures the theme of this text.

You cannot train experience. However, you can educate and train system analysts and engineers in system analysis, design, and development. In turn, this knowledge enables them to *bridge the gap* between a user's abstract operational needs and the hardware and software developers who design systems, products, and services to meet those needs. You can do this in a manner that avoids the *quantum leaps* by local heroes that often result in systems, products, or services that culminate in poor contract program performance and products that fail to satisfy user needs.

Anecdotal evidence suggests that organizations waste vast amounts of resources by failing to educate and train engineers in the concepts, principles, processes, and practices that consume on average 80% of their workday. Based on the author's own experiences and those of many others, if new engineers entering and SEs already in the workplace were equipped with the knowledge contained herein, there would be a *remarkable* difference in:

1. System development performance
2. Organizational performance
3. Level of personal frustrations in coping with complex tasks

Imagine the collective and synergistic power of these innovative and creative minds if they could be introduced to these methods and techniques without having to make quantum leaps. Instead of learning SE methods through informal, observational osmosis, and trial and error over 30+ years, *What if* we could teach system, product, or service problem-solving/solution development as an educational experience through engineering courses or personal study?

Based on the author's experience of over 30 years working across multiple business domains, this text provides a foundation in system analysis, design, and development. It evolved from a need to fill a void in the core curriculum of engineering education and the discipline we refer to as system engineering.

Academically, some people refer to System Engineering as an *emerging* discipline. From the perspective of specific engineering disciplines, System Engineering may be emerging only in the sense that organizations are recognizing its importance, even to their own disciplines. The reality is, however, the practice of engineering systems has existed since humans first employed tools to leverage their physical capabilities. Since World War II the formal term "system engineering" has been applied to problem solving-solution development methods and techniques that many specific engineering disciplines employ. Thus, system engineering concepts, principles, and practices manifest themselves in every engineering discipline; typically without the formal label.

In the chapters ahead, I share some of the *If I Knew Then What I Knew Now* knowledge and experiences. Throughout my career I have had the good fortune and opportunities to work and learn from some of the world's best engineering application and scientific professionals. They are the professionals who advanced the twentieth century in roles such as enabling space travel to the Moon and Mars, creating new building products and approaches, developing highly complex systems, and instituting high-performance organizations and teams.

This is a practitioner's textbook. It is written for advancing the *state of the practice* in the discipline we refer to as System Engineering. My intent is to go beyond the philosophical buzzwords that many use but few understand and address the HOWS and WHYS of *system* analysis, design, and development. It is my hope that each reader will benefit from my discussions and will endeavor to expand and advance System Engineering through the application of the *concepts*, *principles*, and *practices* stated herein. Treat them as *reference guides* by which you can formulate your own approaches *derived* from and *tempered* by your own unique experiences.

Remember, every engineering situation is unique. As an engineer, you and your organization bear *sole responsibility and accountability* for the *actions* and *decisions* manifested in the systems, products, and services you design, develop, and deliver. Each user experience with those products and services will be a *moment of truth* for your organization as well as your own professional *reputation*. With every task, product, or service delivery, internally or externally, make sure the user's *moment of truth* is *positive* and *gratifying*.

ACKNOWLEDGMENTS

This work was made possible by the various contributions of the many people identified below.

My special debt of gratitude goes to Dr. Charles Cockrell, mentor, teacher, and leader; Neill B. Radke; Gerald "Jerry" Mettler; and Robert "Bob" M. Love who persevered through countless hours and iterations reviewing various sections of this work. Likewise, a special appreciation to Dr. Gregory M. Radecky for his technical counsel and commentary. Special thanks go to Sandra Hendrickson for support in revising the manuscript, to Lauren and Emily, and to Sharon Savage-Stull, and to Jean for coordinating the distribution of draft copies to reviewers. I thank members of the JPL—Brian Muirhead, Howard Eisen, David Miller, Dr. Robert Shisko, and Mary Beth Murrill—for sharing their time and experiences. Additionally, I thank Larry Riddle of the University of California, San Diego, and David Weeks for graphics submittals. Thanks also to INCOSE President-Elect Paul Robitaille and to William E. Greenwood and JoAnne Zeigler for their observations. To those true leaders who provided insightful wisdom, knowledge mentoring, training, concepts, and opportunities along my career path, I give a special word of *recognition* and *appreciation*. These include Bobby L. Hartway, Chase B. Reed, William F. Baxter, Dan T. Reed, Spencer and Ila Wasson, Ed Vandiver, and Kenneth King.

Finally, no words can describe how much I appreciate the dedication and caring of my loving wife and children who endured through the countless hours, weekends, and holidays and provided support over many years as this work evolved from concept to maturity. I couldn't have done this without you.

CHARLES S. WASSON
July, 2005

Chapter 1

Introduction

1.1 FRAMING THE NEED FOR SYSTEM ANALYSIS, DESIGN, AND DEVELOPMENT SKILLS

One of the most perplexing problems with small, medium, or large system development programs is simply being able to deliver a *system*, *product*, or *service* without latent defects on schedule, within budget, and make a profit.

In most competitive markets, changes in technology and other pressures force many organizations to *aggressively* cut realistic schedules to win contracts to sustain business operations. Many times these shortcuts violate best practices through their elimination under the premise of "selective tailoring" and economizing.

Most programs, even under near ideal conditions, are often challenged to translate User needs into *efficient* and *cost-effective* hardware and software solutions for deliverable systems, products, and services. Technical program leads, especially System Engineers (SEs), create a strategy to bridge the gap. They translate the User's *abstract* vision into a language of specifications, architectures, and designs to guide the hardware and software development activities as illustrated in Figure 1.1. When aggressive "tailoring" occurs, programs attempt to bridge the gap via a quantum leap strategy. The strategy ultimately defaults into a continuous *build–test–redesign* loop until resources such as cost and schedules are overrun and exhausted due to the extensive rework. Systems delivered by these approaches are often patched and are plagued with undiscovered latent defects.

Bridging the gap between User needs and development of systems, products, and services to satisfy those needs requires three types of technical activities: 1) system analysis, 2) system design, and 3) system development (i.e., implementation). Knowledge in these areas requires education, training, and experience. Most college graduates entering the workforce do not possess these skills; employers provide very limited, if any, training. Most knowledge in these areas varies significantly and primarily comes from personal study and experience over many years. Given this condition, programs have the potential to be staffed by personnel lacking system analysis, design, and development skills attempting to make a quantum leap from user needs to hardware and software implementation.

Technically there are solutions of dealing with this challenge. This text provides a flexible, structural framework for "bridging the gap" between Users and system developers. Throughout this text we will build on workflow to arrive at the steps and practices necessary to plan and implement system analysis, design, and development strategy without sacrificing best practices objectives.

Part II *System Design and Development Practices* presents a *framework* of practice-based strategies and activities for developing systems, products, and services. However, system development requires more than simply implementing a standard framework. You must understand the

---

System Analysis, Design, and Development, by Charles S. Wasson
Copyright © 2006 by John Wiley & Sons, Inc.

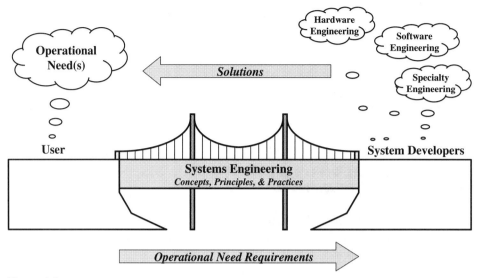

Figure 1.1 Systems Engineering—Bridging the Gap from User Needs to System Developers

foundation for the framework—HOW TO analyze systems. This requires understanding WHAT systems are; HOW the User envisions *deploying*, *operating*, *supporting*, and *disposing* of the system; under WHAT conditions and WHAT outcome(s) they are expected to achieve. Therefore, Part I addresses System Analysis Concepts as a precursor to Part 2.

This text identifies fundamental *system analysis*, *design*, and *development* practices that in the author's experiences are applicable to most organizations. The *concepts*, *principles*, and *practices* presented in Parts I and II represent a collection on topics that *condense* the *fundamentals* of key practices. Some of these topics have entire textbooks dedicated to the subject matter.

Your experiences may be different; that's okay. You and your organization are responsible and accountable for identifying the key concepts, principles, and practices unique to your line of business and programs and incorporate them into its command media—namely *policies* and *procedures*. Using this knowledge and framework, personnel at all levels of the organization are better postured to make informed decisions to bridging the gap from User needs to system, product, and service solutions to meet those needs without having to take a quantum leap.

Chapter 2

Book Organization and Conventions

2.1 HOW THIS BOOK IS ORGANIZED

There is a wealth of engineering knowledge that is well documented in textbooks targeted specifically for disciplinary and specialty engineers. In effect, these textbooks are compartmentalized bodies of knowledge unique to the discipline. The challenge is that SE requires knowledge, application, and integration of the concepts in these bodies of knowledge. The author's purpose in writing this book is not to duplicate what already exists but rather to complement and link SE and development to these bodies of knowledge as illustrated in Figure 2.1.

To accomplish these interdisciplinary linkages, the topical framework of the book is organized the way SEs think. SEs analyze, design, and develop systems. As such, the text consists of two parts: Part I *System Analysis Concepts* and Part II *System Design and Development Practices*. Each part is organized into series of chapters that address concepts or practices and include *Definitions of Key Terms* and *Guiding Principles*.

Part I: System Analysis Concepts

Part I provides the fundamentals in systems analysis and consists of a several series of topics:

- System entity concepts
- System architecture concepts
- System mission concepts
- System operations concepts
- System capability concepts

Each series within a part consists of chapters representing a specific topical discussion. Each chapter is sequentially numbered to facilitate quick location of referrals and topical discussions. The intent is to isolate topical discussions in a single location rather than a fragmented approach used in most textbooks. Due to the interdependency among topics, some overlap is unavoidable. In general, Part I provides the underlying foundation and framework of concepts that support Part II.

Unlike many textbooks, you will not find any equations, software code, or other technical exhibits in Part I. SE is a problem solving–solution development discipline that requires a fundamental understanding in HOW to think about and analyze systems—HOW systems are organized, structured, defined, bounded, and employed by the User.

System Analysis, Design, and Development, by Charles S. Wasson
Copyright © 2006 by John Wiley & Sons, Inc.

4 Chapter 2 Book Organization and Conventions

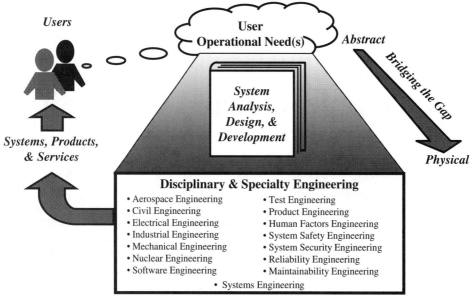

Figure 2.1 Book Scope

Part II: System Design and Development Practices

Part II builds on the *system analysis concepts* of Part I and describes the *system design and development practices* embodied by the discipline we refer to as system engineering. Part II contents consists of several *series* of practices that include:

- System development strategies
- System specification
- System design
- Decision support
- System verification and validation
- System deployment, operations, and support

Each series covers a range of topical practices required to support the series.

2.2 DEFINITIONS

SE, as is the case with most disciplines, is based on *concepts*, *principles*, *processes*, and *practices*. The author's context for each of these terms can be better understood as follows:

- **Concept**   A visionary expression of a proposed or planned action that leads to achievement of a disciplinary objective.
- **Principle**   A guiding thought based on empirical deduction of *observed behavior* or *practices* that proves to be true under most conditions over time.
- **Process**   A sequence of *serial* and/or *concurrent* operations or tasks that transform and/or add value to a set of inputs to produce a product. Processes are subject to external *controls* and *constraints* imposed by regulation and/or decision authority.

- **Operation** A collection of outcome-based tasks required to satisfy an operational objective.
- **Task** The application of methods, techniques, and tools to add value to a set of inputs—such as materials and information—to produce a work product that meets "fitness for use" standards established by formal or informal agreement.
- **Practice** A systematic approach that employs *methods* and *techniques* that have been demonstrated to provide results that are generally *predictable* and *repeatable* under various operating conditions. A practice employs processes, operations, or tools.
- **Best or Preferred Practice** A practice that has been *adopted* or *accepted* as the most suitable method for use by an organization or discipline. Some individuals rebuff the operative term "best" on the basis it is relative and has yet to be universally accepted as THE one and only practice that is above all others. Instead, they use *preferred* practice.

2.3 TEXT CONVENTIONS

This textbook consists of several types of annotations to facilitate readability. These include referrals, author's notes, guideposts, reference identifiers, and examples. To better understand the author's context of usage, let's briefly summarize each.

Referrals. SE concepts, processes, and practices are highly interdependent. Throughout the book you will find *Referrals* that suggest related chapters of the book that provide additional information on the topic.

Author's Notes. *Author's Notes* provide insights and observations based on the author's own unique experiences. Each *Author's Note* is indexed to the chapter and in sequence within the chapter.

Guideposts. *Guideposts* are provided in the text to provide the reader an understanding of WHERE you are and WHAT lies ahead in the discussion. Each *guidepost* is indexed to the chapter and sequence within the section.

Reference Identifiers. Some graphics-based discussions progress through a series of steps that require navigational aids to assist the reader, linking the text discussion to a graphic. *Reference Identifiers* such as (#) or circles with numbers are used. The navigational reference IDs are intended to facilitate classroom or training discussions and reading of detailed figures. It is easier to refer to "Item or ID 10" than to say "system development process."

Examples. *Examples* are included to illustrate *how* a particular concept, method, or practice is applied to the development of real world systems. One way SEs deal with *complexity* is through concepts such as *abstraction*, *decomposition*, and *simplification*. You do not need a Space Shuttle level of complexity example to learn a key point or concept. Therefore, the examples are intended to accomplish one objective—to *communicate*. They are not intended to insult your intelligence or impress academic egos.

References. Technical books often contain pages of references. You will find a limited number of references here. Where external references are applicable and reinforce a point, explicit call-outs are made. However, this is a *practitioner's* text intended to equip the reader with the practical knowledge required to perform system analysis, design, and development. As such, the book is intended to stimulate the reader's thought processes by introducing fresh approaches and ideas for advancing the state of the practice in System Engineering as a professional discipline, not summarizing what other authors have already published.

Naming Conventions. Some discussions throughout the book employ terms that have *generic* and *reserved* word contexts. For example, terms such as equipment, personnel, hardware, software, and facilities have a generic context. Conversely, these same terms are considered SE system elements and are treated as RESERVED words. To delineate the context of usage, we will use lowercase spellings for the generic context and all capitals for the SE unique context—such as EQUIPMENT, PERSONNEL, HARDWARE, SOFTWARE, and FACILITIES. Additionally, certain words in sentences require communication emphasis. Therefore, some words are *italicized* or CAPITALIZED for emphasis by the author as a means to enhance the readability and communicate key points.

2.4 GRAPHICAL CONVENTIONS

System analysis and *design* are graphics-intensive activities. As a result a standard set of graphical conventions is used to provide a level of continuity across a multitude of highly interdependent topics. In general, system analysis and design employ the following types of relationships:

1. Bounding WHAT IS/IS NOT part of a system.
2. Abstractions of collections of entities/objects.
3. Logical associations or relationships between entities.
4. Iterations within an entity/object.
5. Hierarchical decomposition of abstract entities/objects and integration or entities/objects that characterized by *one-to-many* and *many-to-one* entity or object relationships—for example, parent or sibling.
6. Peer-to-peer entity/object relationships.
7. Time-based, *serial* and *concurrent* sequences of *workflow*, and *interactions* between entities.
8. Identification tags assigned to an entity/object that give it a unique identity.

There are numerous graphical methods for illustrating these relationships. The Object Management Group's (OMG) *Unified Modeling Language* (UML®) provides a diverse set of graphical symbols that enable us to express many such relationships. Therefore, diagrams employing UML symbology are used in this book WHERE they enable us to better communicate key concepts. UML annotates *one-to-many* (i.e., *multiplicity*) entity relationships with "0 . . . 1," "1," "1 . . .*," and so forth. Many of the graphics contain a significant amount of information and allow us to forgo the multiplicity annotations. Remember, this text is intended to communicate concepts about system analysis, design, and development; not to make you an expert in UML. Therefore, you are encouraged to visit the UML Web site at www.omg.org for implementation specifics of the language. Currently, SE versions of UML®, SYSML, is in the process of development.

System Block Diagram (SBD) Symbology

One of the first tasks of system analysts and SE is to bound WHAT IS/IS NOT part of the system. System Block Diagrams (SBDs), by virtue of their box structure, offer a convenient way to express these relationships, as illustrated at the left side of Figure 2.2.

2.4 Graphical Conventions

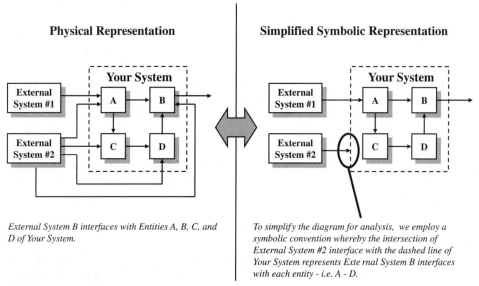

Figure 2.2 Symbolic Interface Representation Convention

If we attempt to annotate each system input/output relationship with lines, the chart would become unwieldy and difficult to read. Using the left side of Figure 2.2 as an example, External System 1 interfaces with Entity A; External System B interfaces with Entities A through D. External System 2 could be the natural environment—consisting of temperature, humidity, and the life—that affects Entities A through D within your system.

Where a system such as External System B interfaces with ALL internal entities, we simplify the graphic with a single arrow touching the outer boundary of the system—meaning your system. Therefore, any arrow that touches the boundary of an entity represents an interface with each item with the entity.

Aggregation and Composition Relationships Symbology

Object-oriented and entity relationship methods recognize that hierarchical *objects* or *entities* are comprised of lower level *sibling* objects or entities. Two types of relationships exist in these cases: *aggregation* versus *generalization*. Let's elaborate on these further.

Aggregation (Composition). *Aggregation* represents the collection of entities/objects that have direct relationships with each other. *Composition*, as a form of aggregation, characterizes relationships that represent *strong associations* between objects or entities as illustrated in Figure 2.3. Entity A consists of Entities A1, A2, A3, and A4 that have *direct* relationships via interfaces that enable them to work together to provide entity A's capabilities. Consider the following example:

EXAMPLE 2.1

An automobile ENGINE consists of PISTONS that have direct relationships via the engine's SHAFT. Therefore, the ENGINE is an aggregation of all entities/objects—such as ENGINE SHAFT and PISTONS—required to provide the ENGINE capability.

8 Chapter 2 Book Organization and Conventions

UML symbology for *aggregation or composition* is represented by a filled (black) diamond shape, referred to as an aggregation indicator, attached to the aggregated object/entity as illustrated at the left side of Figure 2.3. The diamond indicator is attached to the parent entity, System A, with linkages that connect the indicator to each object or entity that has a direct relationship.

Author's Note 2.1 *As a rule, UML only allows the aggregation indicator to be attached to the aggregated entity/object on one end of the relationship (line). You will find instances in the text whereby some abstractions of classes of entities/objects have many-to-many relationships with each other and employ the indicator on both ends of the line.*

Generalization. *Generalization* represents a collection of objects or entities that have *loose* associations with each other as illustrated in Figure 2.3. Entity B consists of sibling Entities B1, B2, B3, and B4, which have not direct relationship with each other. Consider the following example:

EXAMPLE 2.1

A VEHICLE is a generalization for classes of trucks, cars, snowmobiles, tractors, and the like, that have the capability to maneuver under their own power.

UML® symbology for *generalization* is represented by an unfilled (white) triangular shape as illustrated as the right side of Figure 2.3. The triangle indicator is attached to the parent entity with linkages that connect the indicator to each lower level object or entity that have loose associations or relationships.

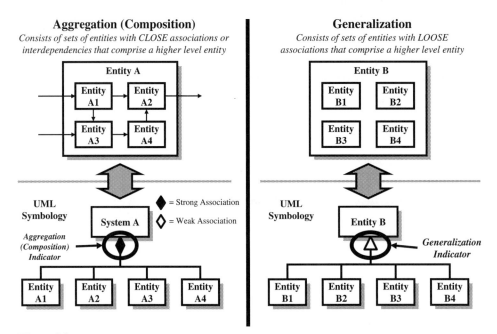

Figure 2.3 Hierarchical Aggregation and Generalization Symbology

Relationship Dependencies

In general, this text employs three types of line conventions to express entity/object relationship dependencies as illustrated in Figure 2.4.

- *Instances of a Relationship That May or May Not Exist (Panel A)*   Since there are instances that *may* or *may not* contain a specific relationship, a *dashed line* is used for all or a part of the line. Where an *aggregated* entity/object *may* or *may not* have all instances of *siblings*, the *parent* half of the line is *solid* and the *sibling* half may be dashed.
- *Electronic/Mechanical Relationships (Panel B)*   Some graphics express electronic relationships by *solid lines* and mechanical relationships by *dashed lines*. For example, a computer's electronic data communications interface with another computer is illustrated by a *solid line*. The mechanical relationship between a disc and a computer is illustrated by a *dashed line* to infer either a mechanical or a temporary connection.
- *Logical/Physical Entity Relationships (Panels C and D)*   Since entities/objects have *logical associations* or *indirect relationships*, we employ a *dashed line* to indicate the relationship.

Interaction Diagrams

UML accommodates *interactions* between entities such as people, objects, roles, and so forth, which are referred to as *actors* via *interaction* or *sequence diagrams*, as illustrated in Figure 2.5. Each *actor* (object class) consists of a vertical time-based line referred to as a *lifeline*. Each actor's *lifeline* consists of *activation boxes* that represent time-based processing. When interactions occur between *actors*, an *event* stimulates the *activation box* of the interfacing *actor*. As a result, a simple sequence of actions will represent interchanges between actors.

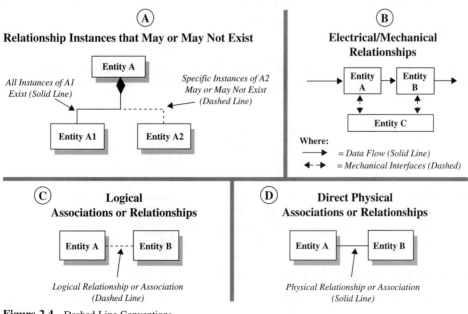

Figure 2.4   Dashed Line Conventions

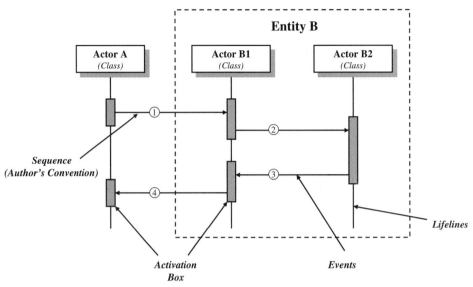

Figure 2.5 UML® Sequence Diagram Symbology

Process Activity Graphics

Systems processing consists of sequential and concurrent process flows and combinations of the two. Key UML elements for representing process flow consist of initial/final states, activities, decision blocks, and synchronization bars (forks and joins), as illustrated in Figure 2.6.

Initial and Final States. To *isolate* on specific aspects of process flow, a process requires a *beginning* referred to as an INITIAL STATE and an *ending* we refer to as a FINAL STATE. UML symbolizes the INITIAL STATE with a *filled* (black) *circle* and the FINAL STATE with a large *unfilled* (white) *circle* encompassing a *filled* (black) *circle*.

Activities. Activities consist of operations or tasks that transform and add value to one or mode inputs to produce an objective-based *outcome* within a given set of performance constraints such as resources, controls, and time. UML graphically symbolizes *activities* as having a flat top and bottom with convex arcs on the left and right sides.

Decision Blocks. Process flows inevitably have *staging* or *control points* that require a decision to be made. Therefore, UML uses a diamond shape to *symbolize* decisions that conditionally branch the process flow to other processing *activities*.

Synchronization Bars. Some entity processing requires concurrent activities that require synchronization. For these cases, *synchronization bars* are used and consist of two types: *forks* and *joins*. Forks provide a means to branch condition-based processing flow to specific activities. *Joins* synchronize and integrate multiple branches into a single process flow.

Hierarchical Decomposition Notation Conventions

Systems are composed of *parent–sibling* hierarchies of entities or objects. Each object or entity within the diagram's structural framework requires establishing a numbering convention to uniquely

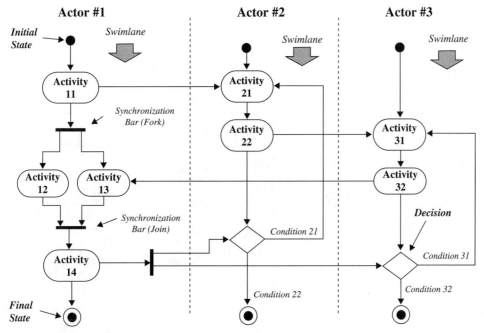

Figure 2.6 UML Activity Diagram Symbology

identify each entity. In general, there are two types of conventions used in the text: decimal based and tag based.

Decimal-Based Notation. SEs employ *decimal notation* to delineate levels of information with the most significant level being in the left most digit position as illustrated in Figure 2.7. Lower levels are identified as extensions to the previous level such as 1.0, 1.1, 1.1.1, 1.1.1.1 and so forth.

Tag-Based Notation. In lieu of the decimal system in which the decimal point can be misplaced or deleted, numerical tags are used without the decimal points as illustrated in the right side of Figure 2.7. Rather than designating these lower level entities with names such as B, C, and D, we need to explicitly identify each one based on its *root traceability* to its higher level parent. We do this by designating each one of the entities as A_1, A_2, and A_3. Thus, if entity A_2 consists of two lower level entities, we label them as A_21 and A_22. A_21 consists of A_211, A_212, and A_213. Following this convention, entity A_212 is an element of entity A_21, which is an element of entity A_2, which is an element of entity A.

2.5 EXERCISES

Most sections of the text consist of two types of exercises: *general exercises* and *organizational centric exercises*.

General Exercises

General exercises are intended to test your understanding of each chapter's topic to two types of problems:

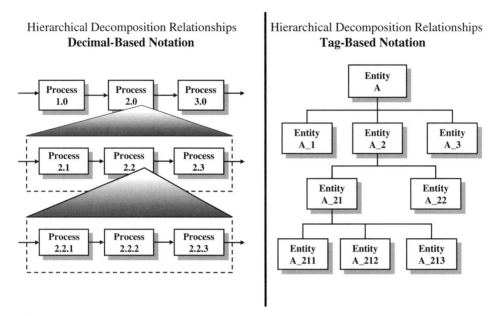

Figure 2.7 Hierarchical Decomposition Relationship Notations

1. *What You Should Learn from This Chapter* questions presented in the *Introduction* of each chapter.
2. Progressive application of knowledge to a selected system as listed is Table 2.1.

Organizational Centric Exercises

Organizational Centric Exercises are intended for organizations that may conduct internal SE training programs. SEs work within the framework of organizational command media such as *policies* and *procedures* and apply that knowledge to contract programs. Therefore, these exercises consists of two types of problems: research of organizational *command media* concerning SE topics of interest and interviewing technical leadership of contract programs to understand *how* they:

1. Approached various facets of SE on their programs.
2. What *best or preferred practices* were used?
3. What *lessons* were *learned*?

2.6 TEAM DECISION MAKING

Team decision making is all about *consensus*. Development teams such as Integrated Product Teams (IPTs) consist of personnel from different disciplines that bring knowledge and levels of experience; some senior level, some young, others in between. The context of the term *consensus* throughout this book refers to *root wisdom* decision making that stands the test of time. It's NOT about one person, one vote; seniority; dominating personalities; or compromise. It's not about showcasing IPTs to customers while continuing to do business the OLD way.

Table 2.1 Sample systems for application to General Exercises

Individual Project Suggestions	Team-Based Project Suggestions
1. Mechanical pencil	1. Exercise room treadmill
2. Desktop stapler	2. Snowmobile
3. Disposable camera	3. Automobile
4. Personal digital assistant (PDA)	4. Word processor
5. Cellular phone	5. Voice mail system
6. Desktop or laptop computer	6. Sports utility vehicle (SUV)
7. Computer mouse	7. Doctor's office
8. Computer scanner	8. Automatic car wash
9. Computer printer	9. Fast food restaurant with drive—through window
10. Computer display monitor	10. Store (video, grocery, bookstore, etc.)
11. CD/DVD player	11. Shopping mall
12. TV/CD/DVD remote control device	12. Hospital
13. Television	13. School
14. MP3 player	14. Fire department
15. Home	15. Overnight package delivery
16. Residential mailbox	16. Restaurant
17. Lawn mower	17. Garbage collection system
18. Lawn edger	18. Recyclable materials station
19. Hand-held calculator	19. Community landfill
20. Ceiling fan	20. Emergency response system (ERS)
21. Web site	21. City rapid transit system or an element
22. Fast food restaurant drive through	22. Professional sports stadium
23. Airport check-in kiosk	23. Organization within an enterprise
24. Voice mail system	24. Fighter aircraft
	25. Commercial jet aircraft
	26. NASA Space Shuttle
	27. International Space Station (ISS)

Team decision making involves *eliciting* and *integrating* team member *knowledge* and *experience* to make choices that clearly represent a *path to success* and avoid a *path to failure*. It may require smart, *informed* assessments of *risk* and *reward* decision making. The bottom line is that it's about making technical decisions *everyone* can and will *proactively* support.

2.7 WARNINGS AND CAUTIONARY DISCLAIMERS

As a *professional*, you and your organization are solely *responsible* and *accountable* for the application and implementation of the *concepts*, *principles*, *processes*, and *practices* discussed in this book, the quality of work products produced, and the impact of those actions on society, colleagues, and the environment. As a practitioner's book, the discussions reflect experiences that may or may not be relevant to you, your organization, or program.

You are advised to supplement this information with personal study, education, research, and experience to enhance your competency skills to the level of performance required and expected by your organization, contract, profession, and applicable laws and regulations. Where specialized expertise is required, employ the services of highly qualified and competent subject matter expert (SME) professionals.

Part I

System Analysis Concepts

EXECUTIVE SUMMARY

The foundation of any discipline resides in its concepts and guiding principles. Part 1 is structured around five thematic concepts that are fundamental to understanding systems—WHAT a system is; WHO its users and stakeholders are; WHY it exists and HOW it benefits its users and stakeholders; HOW it is structured; and HOW it operates, is supported, and disposed. Chapters 3–22 are grouped and presented in a sequence that supports the five concepts listed below:

- System Entity Concepts
- System Architecture Concepts
- System Mission Concepts
- System Operations Concepts
- System Capability Concepts

These basic concepts serve as the foundation for understanding Part II *System Design and Development Practices*. This foundation fills the void for people and organizations that restrict their education and training to the philosophy of SE and attempt to make a quantum leap from specifications to point design solutions due to a lack of understanding of these fundamental concepts.

To better understand what each of these concepts entails, let's explore a brief introductory synopsis of each one.

System Entity Concepts

Our first series of discussions focus on a simple concept, the system as an entity. The *System Entity Concepts* consist of Chapters 3–7. These discussions: define what a system is; identify attributes, properties, and characteristics common to most systems; address organizational systems roles and stakeholders; identify key factors that impact user acceptability of a system, and define a model for the system/product lifecycle.

Given an understanding of the *System Entity Concepts*, our next discussions shift to understanding HOW systems and their operating environments are organized and structured.

System Architecture Concepts

Most people think of a system and its operating environment in physical terms; however, some systems may be virtual such as those centered around political or cultural systems. Regardless of

---

System Analysis, Design, and Development, by Charles S. Wasson
Copyright © 2006 by John Wiley & Sons, Inc.

whether a system is physical or virtual, we can characterize them in terms of *form*, *fit*, and *function* using their structural framework as the point of reference.

Our second series, *System Architecture Concepts*, decomposes the system into its constituent parts via its architectural framework. Chapters 8–12 establish a high level analytical framework for analyzing how a system interacts with itself and its operating environment. Analyses and observations of these interactions reveal that we can characterize a SYSTEM OF INTEREST (SOI) and its operating environment via system elements—EQUIPMENT, PERSONNEL, MISSION RESOURCES, PROCEDURAL DATA, and FACILITIES. Our discussions include establishment of a semantics convention to minimize confusion relating to how developers communicate about multi-level components within a system and relate to higher-level systems.

Based on an understanding of the *System Architecture Concepts*, we are ready to explore WHY a system exists and how an organization employs it as an asset via the *System Mission Concepts*.

System Mission Concepts

Every system has a mission or a reason for its existence as envisioned by its acquirer, users, and stakeholders who expect the system to provide purposeful value and a return on investment (ROI). The *System Mission Concepts* series consisting of Chapters 13–17 describe HOW systems are employed by organizations and assigned performance-based outcome missions to fulfill specific aspects of organizational objectives.

Our discussions trace the origins of a system from an organization's operational need to HOW users envision operating and supporting the system to perform missions that satisfy that need. We investigate how systems interact with their operating environment during missions, how missions are planned, and explore how the user(s) expect the system to perform specific actions related to achieving mission objectives.

The *System Mission Concepts*, which define WHAT a system is to accomplish and HOW WELL from an organizational perspective, provide the foundation for our next topic, the *System Operations Concepts*.

System Operations Concepts

System missions require timely execution of a series of performance-based operations and tasks. The *System Operations Concepts* series consisting of Chapters 18–20 explore how users prepare and configure a system for a mission, conduct the mission, and perform most-mission follow-up. Analysis and observations of systems reveals that we can create a tailorable System Operations Model that serves as a basic construct to facilitate identification of phase-based operations common to all human-made systems. We employ the model's framework to illustrate how system interactions with its operating environment are modeled for various types of single use and multi-use applications.

The *System Operations Concepts*, which structure a system mission into specific operations and tasks, serve as a framework for identifying system capabilities to be provided by the system to accomplish mission objectives. This brings us to the final topic of Part I, *System Capability Concepts*.

System Capability Concepts

The *System Capability Concepts* series consist of Chapters 21 and 22. We explore HOW to derive and allocate mission operational capabilities to integrated sets of system elements. Whereas most people believe a capability is an end in itself, our discussions reveal that a capability has a common construct that can be universally applied to specifying and implementing all types of capabilities.

Chapter 3

What Is a System?

3.1 INTRODUCTION

Analysis, design, and development *systems*, *products*, or *services* requires answering several fundamental questions:

1. *WHAT is a system?*
2. *What is included within a system's boundaries?*
3. *WHAT role does a system perform within the User's organization?*
4. *What mission applications does the system perform?*
5. *WHAT results-oriented outcomes does the system produce?*

These fundamental questions are often difficult to answer. If you are unable to *clearly* and *concisely* delineate WHAT the system is, you have a major challenge.

Now add the element of *complexity* in bringing groups of people working on same problem to *convergence* and *consensus* on the answers. This is a common problem shared by Users, Acquirers, and System Developers, even within their own organizations.

This chapter serves as a cornerstone for this text. It answers the first question, *What is a system?* We begin by defining *what a system is* and explain the meaning of structural phrases within the definition. Based on the definition, we introduce various categories of systems and describe the differences between systems, products, and tools. We introduce the concept of *precedented* and *unprecedented* systems. Finally, we conclude by presenting an analytical and graphical representation of a system.

What You Should Learn from This Chapter

- What is a system?
- What are some examples of *types of systems*?
- What are the differences between systems, products, and tools?
- What is the difference between a *precedented* system and an *unprecedented* system?
- How do we analytically represent a system?

System Analysis, Design, and Development, by Charles S. Wasson
Copyright © 2006 by John Wiley & Sons, Inc.

3.2 DEFINITION OF A SYSTEM

The term "system" originates from the Greek term *systēma*, which means to "place together." Multiple business and engineering domains have definitions of a *system*. This text defines a system as:

- **System** An integrated set of interoperable elements, each with explicitly specified and bounded capabilities, working synergistically to perform value-added processing to enable a User to satisfy mission-oriented operational needs in a prescribed operating environment with a specified outcome and probability of success.

To help you understand the rationale for this definition, let's examine each part in detail.

System Definition Rationale

The definition above captures a number of key discussion points about systems. Let's examine the basis for each phrase in the definition.

- By "an integrated set," we mean that a system, by definition, is composed of hierarchical levels of physical elements, entities, or components.
- By "interoperable elements," we mean that elements within the system's structure must be compatible with each other in *form*, *fit*, and *function*, for example. System elements include equipment (e.g., hardware and software, personnel, facilities, operating constraints, support), maintenance, supplies, spares, training, resources, procedural data, external systems, and anything else that supports mission accomplishment.

Author's Note 3.2 *One is tempted to expand this phrase to state "interoperable and complementary." In general, system elements should have complementary missions and objectives with nonoverlapping capabilities. However, redundant systems may require duplication of capabilities across several system elements. Additionally, some systems, such as networks, have multiple instances of the same components.*

- By each element having "explicitly specified and bounded capabilities," we mean that every element should work to accomplish some higher level goal or purposeful mission. System element contributions to the overall system performance must be explicitly specified. This requires that *operational* and *functional* performance capabilities for each system element be identified and explicitly *bounded* to a level of specificity that allows the element to be analyzed, designed, developed, tested, verified, and validated—either on a stand-alone basis or as part of the integrated system.
- By "working in synergistically," we mean that the purpose of integrating the set of elements is to *leverage* the capabilities of individual element capabilities to accomplish a higher level capability that cannot be achieved as stand-alone elements.
- By "value-added processing," we mean that factors such operational cost, utility, suitability, availability, and efficiency demand that each system operation and task add *value* to its inputs availability, and produce outputs that contribute to achievement of the overall system mission outcome and performance objectives.
- By "enable a user to predictably satisfy mission-oriented operational needs," we mean that every system has a *purpose* (i.e., a reason for existence) and a *value* to the user(s). Its value may be a return on investment (ROI) relative to satisfying operational needs or to satisfy system missions and objectives.

- By "in a prescribed operating environment," we mean that for economic, outcome, and survival reasons, every system must have a prescribed—that is, *bounded*—operating environment.
- By "with a specified outcome," we mean that system stakeholders (Users, shareholders, owners, etc.) expect systems to produce results. The observed behavior, products, by-products, or services, for example, must be *outcome-oriented*, *quantifiable*, *measurable*, and *verifiable*.
- By "and probability of success," we mean that accomplishment of a specific outcome involves a degree of *uncertainty* or *risk*. Thus, the *degree of success* is determined by various *performance factors* such as reliability, dependability, availability, maintainability, sustainability, lethality, and survivability.

Author's Note 3.1 *Based on the author's experiences, you need at least four types of agreement on working level definitions of a system: 1) a personal understanding, 2) a program team consensus, 3) an organizational (e.g., System Developer) consensus, and 4) most important, a contractual consensus with your customer. Why?*

Of particular importance is that you, your program team, and your customer (i.e., a User or an Acquirer as the User's technical representative) have a mutually clear and concise understanding of the term. Organizationally you need a consensus of agreement among the System Developer team members. The intent is to establish continuity across contract and organizations as personnel transition between programs.

Other Definitions of a System

National and international standards organizations as well as different authors have their own definitions of a system. If you analyze these, you will find a diversity of viewpoints, all tempered by their personal knowledge and experiences. Moreover, achievement of a *"one size fits all"* *convergence* and *consensus* by standards organizations often results in wording that is so diluted that many believe it to be *insufficient* and *inadequate*. Examples of organizations having standard definitions include:

- International Council on Systems Engineering (INCOSE)
- Institute of Electrical and Electronic Engineers (IEEE)
- American National Standards Institute (ANSI)/Electronic Industries Alliance (EIA)
- International Standards Organization (ISO)
- US Department of Defense (DoD)
- US National Aeronautics and Space Administration (NASA)
- US Federal Aviation Administration (FAA)

You are encouraged to broaden your knowledge and explore definitions by these organizations. You should then select one that best fits your business application. Depending on your personal viewpoints and needs, the definition stated in this text should prove to be the most descriptive characterization.

Closing Point

When people develop definitions, they attempt to create *content* and *grammar* simultaneously. People typically spend a disproportionate amount of time on grammar and spend very little time on substantive content. We see this in specifications and plans, for example. Grammar is important, since it is the root of our language and communications. However, wordsmithed grammar has no value if it lacks substantive content.

You will be surprised how animated and energized people become over wording exercises. Subsequently, they throw up their hands and walk away. For highly diverse terms such as a *system*, a good definition may sometimes be simply a bulleted list of descriptors concerning what a term is or, perhaps, is not. So, if you or your team attempts to create your own definition, perform one step at a time. Obtain consensus on the key elements of substantive content. Then, structure the statement in a logical sequence and translate the structure into grammar.

3.3 LEARNING TO RECOGNIZE TYPES OF SYSTEMS

Systems occur in a number of forms and vary in composition, hierarchical structure, and behavior. Consider the next high-level examples.

EXAMPLE 3.1

- Economic systems
- Educational systems
- Financial systems
- Environmental systems
- Medical systems
- Corporate systems
- Insurance systems
- Religious systems
- Social systems
- Psychological systems
- Cultural systems
- Food distribution systems
- Transportation systems
- Communications systems
- Entertainment systems
- Government systems
    - Legislative systems
    - Judicial systems
    - Revenue systems
    - Taxation systems
    - Licensing systems
    - Military systems
    - Welfare systems
    - Public safety systems
    - Parks and recreation systems
    - Environmental systems

If we analyze these systems, we find that they produce combinations of products, by-products, or services. Further analysis reveals most of these fall into one or more classes such as individual versus organizational; formal versus informal; ground-based, sea-based, air-based, space-based, or hybrid; human-in-the-loop (HITL) systems, open loop versus closed loop; and fixed, mobile, and transportable systems.

3.4 DELINEATING SYSTEMS, PRODUCTS, AND TOOLS

People often confuse the concepts of *systems*, *products*, and *tools*. To facilitate our discussion, let's examine each of these terms in detail.

System Context

We defined the term *system* earlier in this section. A system may consist of two or more integrated elements whose combined—synergistic—purpose is to achieve mission objectives that may not be effectively or efficiently accomplished by each element on an individual basis. These systems typically include humans, products, and tools to varying degrees. In general, human-made systems require some level of human resources for planning, operation, intervention, or support.

Product Context

Some systems are created as a work *product* by other systems. Let's define the context of *product*: a *product*, as an ENABLING element of a larger system, is typically a physical device or entity that has a specific *capability—form*, *fit*, and *function*—with a specified level of performance.

Products generally lack the ability—meaning *intelligence*—to self-apply themselves without human assistance. Nor can products achieve the higher level system mission objectives without human intervention in some form. In simple terms, we often relate to equipment-based products as items you can procure from a vendor via a catalog order number. Contextually, however, a product may actually be a vendor's "system" that is integrated into a User's higher-level system. Effectively, you create a system of systems (SoS).

EXAMPLE 3.1

A hammer, as a procurable product has *form*, *fit*, and *function* but lacks the ability to apply its self to hammering or removing nails.

EXAMPLE 3.2

A jet aircraft, as a system and procurable vendor product, is integrated into an airline's system and may possess the capability, when programmed and activated by the pilot under certain conditions, to fly.

Tool Context

Some systems or products are employed as tools by higher level systems. Let's define what we mean by a tool. A *tool* is a supporting product that enables a user or *system* to leverage its own capabilities and performance to more effectively or efficiently achieve mission objectives that exceed the individual capabilities of the User or system.

EXAMPLE 3.3

A simple fulcrum and pivot, as tools, enable a human to leverage their own physical strength to displace a rock that otherwise could not be moved easily by one human.

EXAMPLE 3.4

A statistical software application, as a support tool, enables a statistician to efficiently analyze large amounts of data and variances in a short period of time.

3.5 PRECEDENTED VERSUS UNPRECEDENTED SYSTEMS

Most human-made systems evolve over time. Each new evolution of a system *extends* and *expands* the capabilities of the previous system by leveraging new or advanced technologies, methods, tools, techniques, and so forth. There are, however, instances where system operating environments or needs pose new challenges that are *unprecedented*. We refer to these as *precedented* and *unprecedented* systems. Although we tend to think in terms of the legal system and its *precedents*, there are also *precedents* in physical systems, products, and services.

22   Chapter 3   What Is a System?

3.6 ANALYTICAL REPRESENTATION OF A SYSTEM

As an abstraction we symbolically represent a system as a simple *entity* by using a rectangular box as shown in Figure 3.1. In general, inputs such as stimuli and cues are fed into a system that processes the inputs and produces an output. As a construct, this symbolism is acceptable; however, the words need to more explicitly identify WHAT the system performs. That is, the system must add value to the input in producing an output.

We refer to the transformational *processing* that adds value to inputs and produces an output as a *capability*. You will often hear people refer to this as the system's *functionality*; this is partially correct. *Functionality* only represents the ACTION to be accomplished; not HOW WELL as characterized by *performance*. This text employs *capability* as the *operative* term that encompasses both the *functionality* and *performance attributes* of a system.

The simple diagram presented in Figure 3.1 represents a system. However, from an analytical perspective, the diagram is missing critical information that relates to *how* the system operates and performs within its operating environment. Therefore, we expand the diagram to identify these missing elements. The result is shown in Figure 3.2. The attributes of the construct—which include desirable/undesirable inputs, stakeholders, and desirable/undesirable outputs—serve as a key checklist to ensure that all contributory factors are duly considered when specifying, designing, and developing a system.

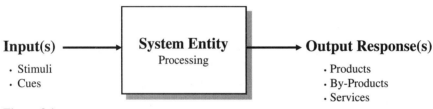

Figure 3.1   Basic System Entity Construct

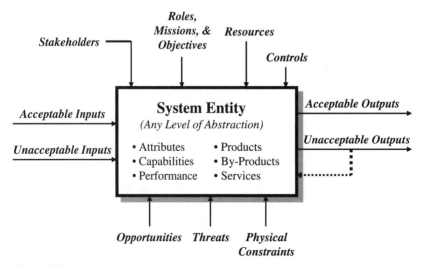

Figure 3.2   Analytical System Entity Construct

3.7 SYSTEMS THAT REQUIRE ENGINEERING

Earlier we listed examples of various types of systems. Some of these systems are *workflow*-based systems that produce systems, products, or services such as schools, hospitals, banking systems, and manufacturers. As such, they require *insightful*, *efficient*, and *effective* organizational structures, supporting assets, and collaborative interactions.

Some systems require the analysis, design, and development of specialized structures, complex interactions, and performance monitoring that may have an impact on the safety, health, and well-being of the public as well as the environment, engineering of systems may be required. As you investigate WHAT is required to analyze, design, and develop both types of systems, you will find that they both share a common set *concepts*, *principles*, and *practices*. Business systems, for example, may require application of various *analytical* and *mathematical* principles to develop business models and performance models to determine profitability and return on investment (ROI) and statistical theory for optimal waiting line or weather conditions, for example. In the case of highly complex systems, *analytical*, *mathematical*, and *scientific* principles may have to be applied. We refer to this as the *engineering of systems*, which may require a mixture of engineering disciplines such as system engineering, electrical engineering, mechanical engineering, and software engineering. These disciplines may only be required at various stages during the analysis, design, and development of a system, product, or service.

This text provides the concepts, principles, and practices that apply to the analysis, design, and development of both types of systems. On the surface these two categories imply a clear distinction between those that require engineering and those that do not. So, *how do you know when the engineering of systems is required*?

Actually these two categories represent a continuum of systems, products, or services that range from making a piece of paper, which can be complex, to developing a system as complex as an aircraft carrier or NASA's International Space Station (ISS). Perhaps the best way to address the question: *What is system engineering*?

What Is System Engineering?

Explicitly SE is the multidisciplinary engineering of systems. However, as with any definition, the response should eliminate the need for additional *clarifying* questions. Instead, the *engineering of a system* response evokes two additional questions: *What is engineering? What is a system?* Pursuing this line of thought, let's explore these questions further.

Defining Key Terms

Engineering students often graduate without being introduced to the *root* term that provides the basis for their formal education. The term, *engineering* originates from the Latin word *ingenerare*, which means "to create." Today, the Accreditation Board for Engineering and Technology (ABET), which accredits engineering schools in the United States, defines the term as follows:

- **Engineering** "[T]he profession in which knowledge of the mathematical and natural sciences gained by study, experience, and practice is applied with judgment to develop ways to utilize economically the materials and forces of nature for the benefit of mankind." (Source: Accreditation Board for Engineering and Technology [ABET])

There are a number of ways to define SE, each dependent on an individual's or organization's perspectives, experiences, and the like. *System engineering* means different things to different people. You will discover that even your own views of SE will evolve over time. So, *if you have a diver-*

sity of perspectives and definitions, what should you do? What is important is that you, program teams, or your organization:

1. Establish a consensus definition.
2. Document the definition in organizational or program command media to serve as a guide for all.

For those who prefer a brief, high-level definition that encompasses the key aspects of SE, consider the following definition:

- **System Engineering (SE)** The multidisciplinary application of analytical, mathematical, and scientific principles to formulating, selecting, and developing a solution that has acceptable risk, satisfies user operational need(s), and minimizes development and life cycle costs while balancing stakeholder interests.

This definition can be summarized in a key SE principle:

Principle 3.1 System engineering BEGINS and ENDS with the User.

SE, as we will see, is one of those terms that requires more than simply defining WHAT SE does; the definition must also identify WHO/WHAT benefits from SE. The ABET definition of engineering, for example, includes the central objective "to utilize, economically, the materials and forces of nature for the benefit of mankind."

Applying this same context to the definition of SE, the User of systems, products, and services symbolizes humankind. However, mankind's survival is very dependent on a living environment that supports sustainment of the species. Therefore, SE must have a broader perspective than simply "for the benefit of mankind." SE must also ensure a balance between humankind and the living environment without sacrificing either.

3.8 SUMMARY

This concludes our discussion of *what a system is*. We defined the term "system" and highlighted the challenges of defining the term within diverse contexts. We also explored examples of types of systems; distinguished between *precedented* and *unprecedented* systems and considered the context of systems, products, and tools.

We concluded with the identification of two categories of systems that produce other systems, products, or services. Some of these require the engineering of systems or system engineering. Therefore, we defined engineering, which in combination with the definition of a system, leads to defining system engineering.

With this basic understanding, we are now ready to investigate the key attributes, properties, and characteristics that make each system unique.

GENERAL EXERCISES

1. Answer each of the *What You Should Learn from This Chapter* questions identified in the *Introduction*.
2. Create your own definition of a system. Based on the "system" definitions provided in this chapter:
   (a) Identify your viewpoint of shortcomings in the definitions.
   (b) Provide rationale as to why you believe that your definition overcomes those shortcomings.
   (c) From an historical perspective, identify three *precedented* systems that were replaced by *unprecedented* systems.

ORGANIZATION CENTRIC EXERCISES

1. How do you and your organization define a "system"?
2. Do you and your work team have a definition for a "system"? If not, ask members to independently develop their definition of what a system is. Summarize the results and present individual viewpoints to the team. Discuss the results and formulate a consensus definition. Report the results to your class. What diversity of opinions did you observe? What concept or semantic obstacles did the team have to overcome to get to consensus?
3. Research the definitions for system engineering provided in the list below. Compare and contrast these definitions and determine which one best fits your beliefs and experiences?
   (a) AFSCM 375-1
   (b) Former FM 770-1
   (c) Former MIL-STD-499A
   (d) EIA/IS-731.1
   (e) Defense Systems Management College (DSMC)
   (f) International Council on Systems Engineering (INCOSE)
   (g) International Organization for Standardization (ISO)
4. For the system, product, or service your organization produces, identify constituent products and tools (e.g., external systems) required to create or support it.
5. Identify the paradigms you observe in your: (a) organization, (b) customers, and (c) business domain that influence system or product design. For each paradigm, what are the characteristic phrases stakeholders use that make the paradigm self-evident.
6. How does your organization view and define SE?
7. How does the author's definition of SE compare with your experiences?
8. What challenges and paradigms does your organization or program face in defining SE?

REFERENCE

Accreditation Board for Engineering and Technology (ABET). Baltimore, MD. URL: www.abet.org

ADDITIONAL READING

ANSI/EIA 632-1999 Standard. 1999. *Processes for Engineering Systems*. Electronic Industries Alliance (EIA). Arlington, VA.

BLANCHARD, B.S. 1998. *System Engineering Management*. New York: Wiley.

BLANCHARD, B.S., and W.J. FABRYCKY. 1990. *Systems Engineering and Analysis*, 2d ed. Englewood Cliffs, NJ: Prentice-Hall.

BUEDE, DENNIS M. 2000. *The Engineering Design of Systems: Models and Methods*. New York: Wiley.

FM 770-78. 1979. *System Engineering* Field Manual. Washington, DC: Headquarters of Department of the Army.

Defense Systems Management College (DSMC). 2001. *Glossary: Defense Acquisition Acronyms and Terms*, 10th ed. Defense Acquisition University Press. Ft. Belvoir, VA.

International Council on Systems Engineering (INCOSE). 1993. *Identification of Pragmatic Principles—Final Report*. SE Practice Working Group, Subgroup on Pragmatic Principles. Seattle, WA.

IEEE 1220-1998. 1998. *IEEE Standard for Application and Management of the Systems Engineering Process*. Institute of Electrical and Electronic Engineers (IEEE). New York, NY.

MIL-STD-498 (canceled). 1994. *Software Development and Documentation*. Washington, DC: Department of Defense.

MIL-STD-499B (canceled draft). *Systems Engineering*. Washington, DC: Department of Defense. 1994.

Federal Aviation Administration (FAA), ASD-100 Architecture and System Engineering. 2003. *National Air Space System—Systems Engineering Manual.* Washington, DC.

International Council on Systems Engineering (INCOSE). 2000. *System Engineering Handbook.* Version 2.0. Washington, DC.

SAGE, ANDREW P. 1995. *Systems Management for Information Technology and Software Engineering.* New York: Wiley.

Defense Systems Management College (DSMC). 2001. *Systems Engineering Fundamentals.* Defense Acquisition University Press. Ft. Belvoir, VA.

NASA SP-6105. 1995. *System Engineering Handbook.* Washington, DC: National Aeronautics and Space Administration.

Chapter 4

System Attributes, Properties, and Characteristics

4.1 INTRODUCTION

System engineering requires development of a strong foundation in understanding *how* to characterize a *system*, *product*, or *service* in terms of its attributes, properties, and performance.

This Chapter introduces *system attributes* that are common across most natural and human-made systems. Our discussions address these attributes in terms of a framework that Acquirers can use as a checklist for system specifications and System Developers/Service Providers can use to assess the adequacy of those specifications. The intent is to enable you to learn how to: 1) THINK about, 2) ORGANIZE, and 3) CHARACTERIZE systems. This knowledge equips SEs and system analysts in two ways. First, when you analyze and evaluate specifications, checklists of commonly used attributes, properties, and characteristics enable us to perform a reality check and identify any "holes" in specification requirements. Second, when we develop specifications, it provides a reference checklist for *organizing* and *specifying* key capabilities and their levels of performance.

Based on this Introduction, let's identify what you should learn from the chapter's discussions.

What You Should Learn from This Chapter

1. *What is a system attribute?*
2. *What is a system property?*
3. *What is a system characteristic?*
4. *What makes a system, product, or service unique?*
5. *Understanding categories of system, product, or service performance*
6. *What are some types of system characteristics?*
7. *What constitutes a system's state of equilibrium?*

Definition of Key Terms

- **Capability** An explicit, inherent feature activated or excited by an external stimulus to perform a function (action) at a specified level of performance until terminated by external commands, timed completion, or resource depletion.

System Analysis, Design, and Development, by Charles S. Wasson
Copyright © 2006 by John Wiley & Sons, Inc.

- **Fit** An item's compatibility to interface with another item within a prescribed set of limits with ease and without interference.
- **Form** An item's prescribed shape intended to support one or more interface boundary objectives.
- **Form, Fit, and Function** "In configuration management, that configuration comprising the physical and functional characteristics of an item as an entity, but not including any characteristics of the elements making up the item." (Source: IEEE 610.12-1990)
- **Function** An operation, activity, process, or action performed by a system element to achieve a specific objective within a prescribed set of performance limits. Functions involve work—such as to move a force through a distance, analyze and process information, transform energy or physical properties, make decisions, conduct communications, and interoperate with other OPERATING ENVIRONMENT systems.
- **Functional Attributes** "Measurable performance parameters including reliability, maintainability, and safety." (Source: ANSI/EIA-649-1998, para. 3.0, p. 5)
- **Level of Performance** An objective, measurable parameter that serves to bound the ability of a system to perform a function based on a set of scenario assumptions, initial conditions, and operating conditions. Examples include system effectiveness, PERSONNEL Element proficiency, and system efficiency.
- **Performance** "A quantitative measure characterizing a physical or functional attribute relating to the execution of an operation or function." (ANSI/IEEE 649-1998, para. 3.0, p.5)

    "Performance attributes include: quantity (how many or how much), quality (how well), coverage (how much area, how far), timeliness (how responsive, how frequent), and readiness (availability, mission/operational readiness)." (ANSI/IEEE 649-1998, para. 3.0, p. 5)
- **Physical Attributes** "Quantitative and qualitative expressions of material features, such as composition, dimensions, finishes, form, fit, and their respective tolerances." (Source: ANSI/EIA-649-1998, Section 3.0, p. 6)

4.2 OVERVIEW OF ATTRIBUTES, PROPERTIES, AND CHARACTERISTICS

You will often hear people refer to a system's *attributes*, *properties*, and *characteristics*. To the casual observer who researches the definitions of these terms, most dictionaries define these terms by referencing the other. For purposes of our discussions, we will employ the following as a means of delineating the differences between the terms.

Attributes

The term *attributes* classifies *functional* or *physical* features of a system. Examples include gender; unit cost; nationality, state, and city of residence; type of sport; organizational position manager; and fixed wing aircraft versus rotor.

Properties

The term, *properties*, refers to the *mass properties* of a system. Examples include composition; weight; density; and size such as length, width, or height.

Characteristics

The term *characteristics* refers to the behavioral and physical qualities that uniquely identify each system. *Behavioral characteristics* examples include predictability and responsivity. *Physical characteristics* examples include equipment warm-up and stabilization profiles; equipment thermal signatures; aircraft radar crosssections; vehicle acceleration to cruise speed, handling, or stopping; and whale fluke markings.

The sum of a system's *attributes*, *properties*, and *characteristics* uniquely identifies and distinguishes a system, product, or service from others of the same classification. To illustrate this uniqueness, let's explore a few aspects that are common to most systems.

4.3 EVERY SYSTEM HAS ITS OWN UNIQUE IDENTITY

All natural and human-made systems have their own attributes (traits) that uniquely characterize, for example, their roles, behavioral patterns, temperament, and appearance, even within the same species. In general, key attributes of uniqueness include the following items, which are described in Table 4.1.

4.4 UNDERSTANDING SYSTEM PERFORMANCE

In general, system performance is the main factor that determines the ultimate level of success of a system. System *functionality* is often viewed as the "qualifying criterion" for systems performance. From the user's perspective, will the system be operationally effective in accomplishing its mission and objectives? Let's begin our discussion by defining "performance."

Categories of Performance

When you investigate systems, you soon discover two basic categories of *performance*: 1) *objective performance* and 2) *subjective performance*. Let's define these terms:

- **Objective Performance** Performance that produces measurable physical evidence of system effectiveness based on pre-defined criteria. For example, the temperature of the water is 108°F.
- **Subjective Performance** Performance indicated by a subjective quality that varies by individual sensory values, interpretations, or perspectives. For example, is the water "warm or hot"?

Given these definitions, let's examine each performance category in greater detail.

Objective Performance

From an SE perspective, especially in writing specifications, a system's capabilities and expected levels of performance must be specified with clear, unambiguous, quantifiable, measurable, testable, and verifiable parameters without the influence of *subjective* interpretations. Examples of *objective performance* include:

Table 4.1 Descriptions of system attributes, properties, and characteristics

ID	Attribute	Description
1	System benefactors	Every system has at least one or more benefactors such as owners, administrators, operators, and maintainers, who benefit from its behavior, products, by-products, or services.
2	System life cycle	Every system, product, and service has a life cycle that depicts its level of maturity.
3	System operating domain	Every system has an operating domain or "sphere of influence" that bounds its area of coverage, operations, and effectiveness. Humans have learned to extend the area of coverage by employing other assets that enable a specific system to "amplify" its range. **EXAMPLE 4.1** An aircraft has a specific range under specific operating conditions such as fuel, payload, and weather. Deploying refueling sources—airborne tankers—and maintenance facilities along its mission flight path can extend the range.
4	System frame of reference	Every system at any point in time has a *frame of reference* that serves as the permanent or temporary: 1. Base of operations for its operating domain. 2. Basis for navigation. **EXAMPLE 4.2** An aircraft may be assigned to a permanent home base that serves as the center of its operations. The aircraft may be ordered to perform special (temporary) assignments from a base in Europe. The Apollo Space Program used the Kennedy Space Center (KSC) and the Earth as its frame of reference.
5	Higher order systems	Every system: 1. Operates as part of a higher order system that may govern, direct, constrain, or control its operation and performance. 2. Provides resources for missions.
6	Purpose-based role	Viewing the universe as a "system of systems (SOS)," every natural and man-made system has a beneficial role based on a reason for its existence as envisioned by its original Acquirer or System Owner.
7	System missions	Every system performs missions in fulfillment of its purpose to achieve outcome-based performance objectives established by its owner and Users.
8	Mission goals and performance objectives	Each system and mission must be characterized by a set of goals and objectives, preferably documented. *Goals* and *objectives* provide the fundamental basis for resource expenditures by the system owner and shareholders based on a planned set of multifaceted accomplishments and an expected return on investment(ROI). Each goal must be supported by one or more specific objectives that are *quantifiable*, *measurable*, *testable*, and *verifiable*.
9	System operating constraints and conditions	Every system, in execution of its mission, is subjected to a set of operating constraints and conditions controlled by higher order systems.
10	System utility	Every system must provide a physical, psychological, sociological, financial, and economic *value-added utility* to its User. System utility includes ease of use, usefulness, etc.

(continued)

Table 4.1 *continued*

ID	Attribute	Description
11	System suitability	Every system has a level of *operational suitability* to the User in terms of suiting its planned application and integration into the users organizational system.
		EXAMPLE 4.3 A gas-powered lawn edger is suitable for cutting grass around trees and flower gardens; they are not, however, suitable for mowing lawns unless you do not own a lawn mower.
12	System success criteria	Each system and mission requires a set of success criteria that the system owner and shareholders agree represent WHAT objective criteria constitute successful accomplishment of a mission via goals and results-oriented objectives. Ultimate success resides in User *acceptance* and level of *satisfaction*.
13	Mission reliability	Every system is characterized by a *probability of success* in accomplishing mission objectives for a specified mission duration and set of operating environment conditions and scenarios.
14	System effectiveness	Every system has some level of *cost* and *technical effectiveness* related to accomplishing the system's mission with an anticipated probability of success per unit of cost.
		EXAMPLE 4.4 Consider the *system effectiveness* of an educational system or a health care system. The challenge is: Effectiveness from WHAT stakeholder's perspective?
15	System efficiency	Every system has a *degree of efficiency* in processing raw materials, information, stimuli, cues, etc. As engineers, we assign an *efficiency* metric that provides a ratio of the quantity of output produced for a known quantity on input.
16	System integrity	Every system has a *level of integrity* in its ability to deliver systems, products, and services as required despite operating constraints and conditions.
17	System sustainment	To ensure success in accomplishing its mission, every system, product, or service requires resources such as personnel, funding, consumables, expendables; corrective and preventive maintenance; and support such as spares, supplies, and training.
18	System promotion	Some systems, namely businesses, promote their systems in anticipation of sales via demonstrations, advertising, etc. The promotion activities may require protection and security.
		EXAMPLE 4.5 A publisher plans to release a new book in a series on a specific day and time, promote the book via advertising, and impose sale constraints and conditions on bookstore owners. The bookstore owners must keep the book under lock and key (protection) with 24 hour surveillance (security) until the official release.
19	System threats	Every system and its missions may be threatened by competitors or adversaries in its operating environment that may exhibit friendly, benign, or hostile intentions or actions.
20	System concealment	Because of *vulnerabilities* or the need for the element of surprise, some systems require *camouflage* or *concealment* to shield or alter their identity.
21	System protection	Every system must have some *level of protection* to *minimize* its vulnerability to external threats.

(continued)

Table 4.1 *continued*

ID	Attribute	Description
22	System security	Man-made systems may maintain a *level of security* such as physical security (PHYSEC), communications security (COMSEC), operational security (OPSEC), and information security (INFOSEC).
23	System architecture	Every system consists of a multi-level, logical (functional) and physical structure or architecture that provides the framework for its *form*, *fit*, and *function*.
24	System capabilities	Every system, by definition, has inherent capabilities such as processing, strengths, transfer functions that enable it to process inputs such as raw materials, information, and stimuli and to provide a response in the form of behavior patterns, products, and by-products. System capabilities, like operating domains, can be extended using tools or other systems.
25	System concept of operations (Con Ops)	Every system has a *Concept of Operations* (*ConOps*) as envisioned by its system owner, system developer, and/or system maintainer. The ConOps provides the basis for bounding the operating space, system capabilities, interfaces, and operating environment.
26	System phases, modes, and states of operation	For each system/product life cycle phase, every system, product, or service evolves through a series of phases, modes, and state of operation that may be cyclical or nonrecurring (single use).
27	Operating norms, standards, and conventions	Every system employs a set of *operating norms*, *standards*, and *conventions* that governs its operations, morals, ethics, and tolerances.
28	System description	Every system should have a system description that characterizes the system architecture, its elements, interfaces, etc. Each of these characteristics is represented by system capabilities and engineering performance parameters that must be captured and articulated as requirements in the *System Performance Specification (SPS)*.
29	System operating constraints and conditions	Every system has *operating constraints* and *conditions* that may be physical (capabilities), imposed by higher order authority—international, governmental, environmental, social, economic, financial, psychological, etc.
30	System sensors	Every natural and human-made system possesses some form of *sensor* that enables it to detect external stimuli and cues.
31	System behavior patterns	Every system is characterized by *patterns of behavior*.
32	System responsiveness and sensitivity	Every system possesses performance-based behavioral characteristics, such as throughput, that characterize its ability to process raw materials or stimuli and provide a response. We refer to the quickness as its *responsivity*. **EXAMPLE 4.8** Accelerator boards enable computer processors to improve responsiveness.
33	System interfaces	Every system has internal and external interfaces that enable it to interact within itself and its operating environment.
34	System pedigree	Every system has a pedigree derived from predecessor system designs, technologies, and improvements to those designs to correct for flaws, defects, deficiencies, errors, etc.

(continued)

Table 4.1 *continued*

ID	Attribute	Description
35	Mission resources *(system inputs)*	Every system requires inputs such as tasking, expendables, consumables, and operator actions that can be transformed into specific actions required to stimulate motivate, maneuver, process, and output behavioral and physical responses.
36	System products, services, and by-products	Every system produces: 1. Value-added products and/or performs services that benefit its stakeholders 2. By-products that may impact system performance and/or its operating environment. **EXAMPLE 4.9** By-products include heat, waste products—trash, exhaust, thermal signatures, and colorations.
37	Procedural data	Every human-made system requires *procedural data* that describe safe operating procedures related to equipment, services, and operator interfaces and interfaces with external systems.
38	System lethality	Some defensive and offensive systems are characterized by their *lethality*—their potential to destroy or inflict damage, disable, neutralize, or otherwise cause harm to a threat or target.
39	System vulnerability	Every system has some form *vulnerability* that exposes *uncertainties* or *shortcomings* in its behavioral and physical characteristics. Vulnerability includes physical, psychological, social, economic, security, privacy, and other factors. **EXAMPLE 4.10** Military tanks have additional layers of protection to minimize the impacts of direct hits. Internet sites have vulnerabilities to computer "hackers."
40	System survivability	Every system has degrees of *fault tolerance* that enable it to perform missions and achieve mission objectives while operating at a *degraded* level of performance for a given set of internal or external induced or malfunctions.
41	System availability	The state of a system's operational readiness to perform a mission on-demand. Availability is a function of the system's reliability and maintainability.
42	System aesthetics	Every system possesses psychological or appearance characteristics that appeal to the senses or are aesthetically pleasing to its stakeholders.
43	System blemishes	Every system is unique in its development. This includes *design flaws* and *errors*, *work quality* and *material defects*, *imperfections*, etc., that may impact system performance or cosmetically diminish its value based on appearance.
44	Risk	Every system, product, or service has an element of risk related to mission operations and its operating environment that include: 1. *Probability of occurrence.* 2. *Consequence(s) of failure.*
45	System environmental, safety, and health (ES&H)	Every human-made system, at various stages of the system/product life cycle, may pose environmental, safety, or health risks to system personnel—operators and maintainers, private and public property, the environment, etc.
46	System health status	Every system has an operational health status that represents its current state of readiness to perform or support User missions.
47	System total cost of ownership	Every human-made system has a *total ownership cost* (TOC) over its life cycle that includes *nonrecurring* and *recurring* development operational costs.

- Time
- Distance
- Size
- Length
- Depth
- Thickness
- Weight
- Volume
- Density
- Physical state
- Cost
- Voltage
- Amperage
- Angle
- Displacement
- Velocity
- Acceleration
- Thrust
- Hardness
- Softness
- Horsepower
- Viscosity
- Frequency
- Intensity
- Wavelength
- Maintainability
- Reliability
- Productivity
- Effectiveness
- Efficiency
- Temperature
- Pressure
- Humidity
- Number of errors
- Field of view
- Resolution
- Defects

These are just a few examples of *objective* performance parameters. Now let's investigate the other type, subjective performance.

Subjective Performance

Subjective performance is more difficult to *characterize* and *quantify*. Interestingly, we can assign arbitrary quantities to *subjective performance* parameters that are *measurable*, *testable*, and *verifiable* via surveys and interviews, tests. However, when the survey or interview participants are asked to indicate their *degree of preference*, *agreement*, and *like/dislike* with the *measurable* statement, the response still requires interpretation, value judgment, opinion, and so on. Thus, the response may be aliased based on past experience and lessons learned. *Subjective performance* examples include:

- Quality—clarity, appearance, and color
- Affinity
- Likeability
- Opinion
- Smoothness
- Satisfaction—enjoyment and taste

4.5 SYSTEM CHARACTERISTICS

When we characterize systems, especially for marketing or analysis, there are four basic types of characteristics we consider: 1) general characteristics, 2) operating or behavioral characteristics, 3) physical characteristics, and 4) system aesthetics.

General Characteristics

The high-level features of a system are its *general characteristics*. We often see general characteristics stated in marketing brochures where key features are emphasized to capture a client or customer's interest. General characteristics often have some commonality across multiple instances or models of a system. Consider the following examples:

EXAMPLE 4.12

- *Automobile General Characteristics* Available in two-door or four-door models; convertible or sedan; air-conditioned comfort; independent suspension; tinted windows, 22 mpg city, 30 mpg highway.
- *Aircraft General Characteristics* Fanjet, 50-passenger, 2000 nautical mile range, IFR capabilities.
- *Enterprise or Organization General Characteristics* 200 employees; staff with 20 PhD, 50 Master, and 30 BS degrees; annual sales of $500 M per annum.
- *Network General Characteristics* Client-server architecture, PC and Unix platforms, firewall security, remote dial-up access, Ethernet backbone, network file structure (NFS).

Operating or Behavioral Characteristics

At a level of detail below the general characteristics, systems have *operating characteristics* that describe system features related to usability, survivability, and performance for a prescribed operating environment. Consider the following examples:

EXAMPLE 4.17

- *Automobile Operating Characteristics* Maneuverability, turn radius of 18 ft, 0 to 60 mph in 6 seconds, etc.
- *Aircraft Operating Characteristics* All-weather application, speed, etc.
- *Network Operating Characteristics* Authorization, access time, latency, etc.

Physical Characteristics

Every system is described by *physical characteristics* that relate to nonfunctional attributes such as size, weight, color, capacity, and interface attributes. Consider the following examples:

EXAMPLE 4.13

- *Automobile Physical Characteristics* 2000 lbs, curb weight 14.0 cu ft of cargo volume, 43.1 of inches (max). of front leg room, 17.1 gals fuel capacity, 240 horsepower engine at 6250 rpm, turbo, available in 10 colors.
- *Enterprise or Organization Physical Characteristics* 5000 sq ft of office space, 15 networked computers, 100,000 sq ft warehouse.
- *Network Physical Characteristics* 1.0 Mb Ethernet backbone, topography, routers, gateways.

System Aesthetic Characteristics

General, *operating*, and *physical* characteristics are *objective* performance parameters. However, what about *subjective* characteristics? We refer to these as system *aesthetic characteristics* because they relate to the "look and feel" of a system. Obviously, this includes psychological, sociological, and cultural perspectives that relate to appealing to the User's, Acquirer's, or System Owner's preferences. Thus, some buyers make independent decisions, while others are influenced by external systems (i.e., other buyers) in matters relating to community or corporate status, image, and the like.

4.6 THE SYSTEM'S STATE OF EQUILIBRIUM

Every natural and human-made system exists in a *state of equilibrium* relative to its operating environment. In general, we refer to this as the "balance of power." The state of equilibrium depends on how a system exists through its own: 1) level of dominance or 2) subordination by other systems. At any instance of time, a system is typically described by an INITIAL STATE—with conditions, statics, dynamics, strengths, weaknesses, or stabilization—and a FINAL STATE—with behavior, product, by-product, or service-oriented result controlled by the balance of power.

Prerequisite Conditions

System *stability*, *integrity*, and *consistency* of performance require that transitions between system phases, operations, and tasks have clean *unambiguous* transitions with no *ramifications*. Thus, systems are *assumed* by designers to have *pre-requisite* or *initial conditions* or *criteria* that must be accomplished prior to entering the next phase, operation, or task. By definition, since a system is composed of a set of integrated elements, this is important to ensure that all elements of the system are *synchronized* and *harmonized*.

Initial Operating Conditions and State

A system's *initial operating conditions* consist of the physical and operational *states* of the system and its surrounding operating environment at the beginning of a system mission phase, operation, or task. Since analyses often require the establishment of basic assumptions for investigating some facet of system phases, operations, or task, initial conditions serve as a "snapshot" or starting point that captures the assumptions. To illustrate this concept, consider the following example:

EXAMPLE 4.14

The aircraft took off in a crosswind of 35 knots; the early morning rush hour began as a blizzard with 30 mph windgusts moved through the area.

Statics

When we analyze systems, a key basis for the analysis is often the *physical* state of the system at a given "snapshot of time." *Statics* are used to characterize a system's current orientation, such as state vector or orientation within a larger system. From an overall system perspective, an aircraft sitting in a hanger, an automobile in a driveway, a network computer system with no message traffic, and a lighting system in the ON or OFF state, all *represent* a system in its *static* state. In contrast, lower level system components may have a static condition while the system as a whole is in a *dynamic* condition.

Mission Dynamics

Every natural and man-made system conducts missions in its operating environment in some form of dynamic, physical state. *Dynamics* are a time-based characterization of system *statics* over a defined timeframe within its operating environment. The *dynamics* may range from slow changes—rock anchored on a hillside—to moderate changes—temperature variations—to violent, sudden changes—earthquakes or volcanoes.

Dynamics occur as *inconsistencies*, *perturbations*, and *instabilities* in the *balance of power* in the local or global environment. Mankind has always been intrigued by the study of dynamics and their effect on behavior patterns—of the Earth, weather, oceans, stock market, and people—especially when its comes to predicting dynamic behavior that can have devastating economic or safety impacts. Thus, predicting the advancement of the *state of the practice* and *technology* is big business. *Why?* We need to be able to confidently predict *how* a system will behave and perform under specified—dynamic—operating conditions.

System Stabilization

All natural and man-made systems must maintain a *level of stability* to ensure their longevity. Otherwise, the system can easily become *unstable* and potentially become a threat to itself and surrounding systems. Therefore, systems should have inherent design characteristics that enable them to *stabilize* and *control* their responses to dynamic, external stimuli.

Stabilization is ultimately dependent on having some form of calibrated reference that is stable, dependent, and reliable. For man-made systems, stabilization is achieved by employing devices such as inertial navigation gyroscopes, global positioning satellites (GPS), quartz crystals for electronic watches, and reference diodes for voltage regulators. In each of these cases, the *system stabilization* is accomplished by sensing current free body dynamics; comparing them with a known, calibrated reference source; and initiating system feedback control actions to correct any variations.

The Balance of Power

Taking all of these elements into account, system *existence* and *survival* are determined by its ability to: 1) cope with the *statics* and *dynamics* of its operating environment and 2) sustain a level of capability and stabilization that harmonizes with its adjacent systems—the balance of power or state of equilibrium.

4.7 SUMMARY

Our discussion in this chapter introduced the concept of system attributes, properties, and characteristics. Topically, this information provides the foundation for: 1) characterizing an existing system's capabilities and 2) serves as an initial checklist for developing or assessing specifications.

GENERAL EXERCISES

1. Answer each of the *What You Should Learn from This Chapter* questions identified in the *Introduction*.
2. Refer to the list of systems identified in Chapter 2. Based on a selection from the preceding chapter's *General Exercises* or a new system selection, apply your knowledge derived from this chapter's topical discussions. Identify the following:
   (a) Examples of the selected system's unique identity via system mission, goals, and objectives.
   (b) Types of mission operations attributes.
   (c) General, operating, physical, and system aesthetic attributes.
3. What is the *frame of reference* and *operating domain* for your: (a) business organization and (b) systems, products, or the services it provides?
4. Using any system, product, or service your organization provides, identify the human system roles for the product.

REFERENCES

IEEE Std 610.12–1990. 1990. *IEEE Standard Glossary of Modeling and Simulation Terminology.* Institute of Electrical and Electronic Engineers (IEEE) New York, NY.

ANSI/IEEE 649-1998. 1998. *National Consensus Standard for Configuration Management.* Electronic Industries Alliance.

Chapter 5

System Roles and Stakeholders

5.1 INTRODUCTION

Every system within the universe has a purpose or mission. In other words, every system has a *reason* for its *existence* relative to other systems. Most human-made systems are directed toward contributing to and achieving the "owner" organization's roles, missions, and objectives. Each system performs a role within the owner system's overall role. This section introduces the concept of system roles and stakeholders. We explore the roles of organizations that employ *systems*, *products*, and *services* and how *physical systems*—namely assets—are acquired to perform in support of organizational roles, missions, and objectives.

The success of man-made systems in achieving success within an organization is determined by HOW WELL the system is specified, designed, developed, integrated, verified, validated, operated, and supported. This requires that humans, either directly or indirectly, that a vested interest in the *operational effectiveness* of the mission results, performance achieved, and outcome of the system's mission and objectives. We refer to these individuals as the *system stakeholders*. We conclude this chapter by identifying the primary system stakeholder roles and their contributions, sometimes positive—sometimes negative—to system mission performance and outcomes.

What You Should Learn from This Chapter

- What is a system role?
- What is a MISSION SYSTEM?
- What is a SUPPORT SYSTEM?
- Who is a stakeholder?
- Who are the various stakeholders?

Definitions of Key Terms

- **End User** An individual or organization that benefits directly from the outcome or results of a system, product, or service.
- **Stakeholder** An individual or organization that has a vested interest (e.g., friendly, competitive, or adversarial) in the outcome produced by a system in performing its assigned mission.
- **System User** An individual or organization that employs a system, product, or service or their by-products for purposes of accomplishing a mission-oriented objective or task. For example, a city transportation system employs bus drivers and buses to transport people from one location to another. The bus drivers and passengers are categorized as *users*.

System Analysis, Design, and Development, by Charles S. Wasson
Copyright © 2006 by John Wiley & Sons, Inc.

5.2 ORGANIZATIONAL SYSTEM ROLES

Organizational systems perform roles that reflect their chartered missions and objectives within their business domains. Table 5.1 provides examples of system roles and missions.

5.3 SYSTEM ROLE CONTEXT

Every man-made system performs *mission-oriented* operations and tasks intended to accomplish specific performance-based objectives and outcomes to support their customers, the Users and stakeholders. The conduct of these missions requires two types of operational roles: mission system and support system.

SYSTEM OF INTEREST (SOI) Entities

If you analyze human-made systems, you discover that EVERY system performs two *simultaneous*, contextual roles: 1) a MISSION SYSTEM role and 2) a SUPPORT SYSTEM role. Let's define each of these roles.

MISSION SYSTEM Role. MISSION SYSTEM *roles* are performed by systems that are assigned specific tasks to deliver products and services that achieve performance-based outcome objectives. Consider the following example:

EXAMPLE 5.1

NASA's Space Shuttle performs space-based missions to accomplish various scientific objectives.

Author's Note 5.1 *We should note here that the User may or may not own the MISSION SYSTEM. In some cases, they do; in other cases, they may contract to other organizations to lease systems. A trucking company, for example, may lease additional vehicles to perform organizational missions during a business surge.*

SUPPORT SYSTEM Role. SUPPORT SYSTEM *roles* are performed by systems to ensure that another MISSION SYSTEM(s) is (are) operationally ready to conduct and support assigned tasks and to perform subsequent maintenance and training. Consider the following examples:

EXAMPLE 5.2

An airline's MISSION SYSTEM consists of the aircraft and crew with an objective to safely and comfortably transport passengers and cargo between cities. The aircraft's SUPPORT SYSTEM consists of baggage handlers, mechanics, facilities, ground support equipment (GSE), and others that function in their own MISSION SYSTEM roles to prepare the aircraft for a safe flight, replenish expendables and consumables, and load/unload cargo and passenger luggage.

EXAMPLE 5.3

NASA's Space Shuttle performs a MISSION SYSTEM role to conduct space-based missions to accomplish science objectives such as payload bay experiments. As a SUPPORT SYSTEM to the International Space

Table 5.1 Example system roles and missions

System Role	Role-Based Mission
Legislative	Establish societal compliance guidance and constraints in the form of laws, statutes, regulations, ordinances, policies, etc. that govern other individuals, organizations, or or governmental entities such as cities or states.
Judicial	Adjudicate individual, organizational, or enterprise compliance with established laws, statutes, regulations, ordinances, policies, etc.
Military	Perform cooperative, emergency, peacekeeping, deterrence, and wartime roles that ensure the survival of a country and protect its constitution, security, and sovereignty.
Transportation	Provide transportation services that enable customers and products to move safely and efficiently from one location to another by land, sea, air, or space or combinations of these.
Civic	Perform public services that further the goals and objectives of an organization.
Educational	Provide educational opportunities for people to gain specialized knowledge and enhance their skills to prepare them for becoming contributing members of society.
Resource	Provide resources—time, money, fuel, electricity, etc.—commensurate with performance and risk to support the missions, goals, and objectives of an individual, organization, or enterprise with an expectation of a *return on the investment* (ROI).
Monitor and control	Monitor the performance of other systems, evaluate the performance against established standards, record objective evidence, and control the performance or the other systems.
Research and development	Investigate the research and development, productization, or application of new technologies for systems.
Producer	Produce large or mass quantities of a system design in accordance with requirements and standards for the marketplace.
Construction	Provide construction services that enable system developers to implement facilities, sites, etc., that enable Users to deploy, operate, support, train, anddispose of systems.
Agricultural	Provide nutritional food and agricultural by-products to the marketplace that are safe for human and animal consumption and safe for the environment.
Retail or wholesale business	Supply consumer *products* and *services* to the marketplace.
Medical	Provide medical consultation and treatment services.

Station (ISS), the Space Shuttle ferries astronauts and cargo back and forth between the Kennedy Space Center (KSC) and the ISS.

Referral For more detailed information about the MISSION SYSTEM and SUPPORT SYSTEM roles, refer to Chapter 13 *Organizational Roles, Missions, and System Applications*.

How Do Organizational System Roles Relate to SE?

You may be asking WHY and HOW organizational roles relate to SE and the *engineering of systems*. Physical systems, such as hardware, software, and courseware, exist because higher level human organizations, as Users, *employ* and *leverage* physical system capabilities to achieve organizational goals, missions, and objectives within budgetary cost and schedule constraints.

As an SE, you must fully *comprehend*, *understand*, and *appreciate* HOW the User *intends* to deploy and employ a system in a prescribed operating environment to achieve the organization's goals, objectives, and missions.

For example, *How does this guidance relate to an organization's Transportation Role*? If you are in the airline business, you provide or contract for reservation and ticketing services, check-in and baggage handling, aircraft, gate facilities, special services, or security. All these entities require physical systems—as well as hardware and software—to perform the organization's role. The organization may:

1. Develop a system, product, or service.
2. Procure the system, product, or service from other vendors.
3. "Outsource" (e.g., contract, lease) for the systems or services.

In any case each system *entity* is allocated goals, objectives, missions, or performance requirements that contribute to achieving the organization's transportation role.

When an organization, such as an airline, initiates operations, large numbers of humans must be an integral part of the planning, implementation, and operation activities. Each stakeholder has a *vested interest* or stake in the *outcomes* and *successes* of the organization's role and its embedded systems. This brings us to out next topic, understanding the role of system stakeholders.

5.4 UNDERSTANDING THE ROLE OF SYSTEM STAKEHOLDERS

Human-made systems, from conception through disposal, require human support, both directly and indirectly. Humans with vested interests in a system, product, or service expect to contribute to the conceptualization, funding, procurement, design, development, integration, operation, support, and retirement of every system. We refer to these people as *stakeholders*. Depending on the *size* and *complexity* of the system, including risks and importance to the User, *stakeholder roles* may be performed by an individual, an organization, or some higher level enterprise such as a corporation, government, or country.

To better understand who stakeholders are, let's scope and bound the term's application.

Who Is a System Stakeholder?

A stakeholder is anyone or an organization having a vested interest in a system and its outcomes. WHO a stakeholder is depends on their role.

System Stakeholder Roles

Every human-made system is supported by human roles with different objectives and agendas that contribute to the overall longevity and performance of the system throughout its life cycle. Consider the following examples of system roles:

EXAMPLE 5.4

- System advocate or proponent
- System shareholder
- System owner
- System user(s)
- System architect
- System acquirer
- System developer
- Services provider
- Independent Test Agency (ITA)
- System administrator
- Mission planner
- System analyst
- System support
- System maintainer
- System instructor
- System critic
- System competitor
- System adversary
- System threat

Let's introduce and define each of these roles. Table 5.2 provides a brief description of each stakeholder role.

Understanding the Multiplicity of System Stakeholder Roles

The context of stakeholder roles centers on a SYSTEM OF INTEREST (SOI). Recognize that the individual, organizations, or enterprises that perform these roles may be stakeholders in other systems. In fact, stakeholder roles may vary from system to system.

EXAMPLE 5.6

The System Advocate for your system may serve as a System Owner for numerous systems. Users may employ several other systems to achieve their own mission goals and objectives. The System Developer may be the developer of numerous systems and system maintainer of other systems.

5.5 SUMMARY

Our discussion covered the *roles of system stakeholders*. Each stakeholder has some level of interest in the *outcome* and *success* of the system, product, or service. Based on these interests, system stakeholders become key sources for system requirements. Stakeholder satisfaction with system performance ultimately determines system acceptability, our next topic.

GENERAL EXERCISES

1. Answer each of the *What You Should Learn from This Chapter* questions identified in the *Introduction*.
2. Refer to the list of systems identified in Chapter 2. Based on a selection from the preceding chapter's General *Exercises* or a new system selection, apply your knowledge derived from this chapter's topical discussions. Specifically identify the following:
   (a) What MISSION SYSTEM and SUPPORT SYSTEM roles does the system perform?
   (b) Who are the stakeholders for each system?

Table 5.2 System stakeholder role definitions

Role	Role Description
System advocate or proponent	An individual, organization, or enterprise that *champions* the system's cause, mission, or reason for existence. *System advocates* may derive *tangible* or *intangible* benefits from their support of the system, or they may simply believe the system contributes to some higher level cause that they support.
System shareholder	An individual, organization, or enterprise that "owns," either directly or indirectly, all or equity shares in the system and its development, operation, products, and by-products.
System owner	An individual, organization, or enterprise that is legally and administratively responsible and accountable for the system, its development, operation, products, by-products, and outcomes and disposal.
System user(s)	An individual, organization, or enterprise that derives direct benefits from a system and/or its products, services, or by-products. *Users* may physically operate a system or provide inputs—data, materials, raw materials, preprocessed materials, etc.—to the system and await the results of value-added processing in the form of products, services, or information. Users may directly or indirectly include the System Advocate, System Owner, or other Users.
System architect	An individual, organization, or enterprise that *visualizes*, *conceptualizes*, and *formulates* the system, system concepts, missions, goals, and objectives. Since SE is viewed as multidisciplined, the system architect role manifests itself via hardware architects, software architects, instructional architects, etc.
System acquirer	An agent (or agency) selected by the User to serve as their *technical representative* to: 1. Specify the system. 2. Select a System Developer or Services Provider. 3. Provide technical assistance. 4. Provide contractual oversight for the execution of the contract and delivery of a *verified* and *validated* system to the User.
System developer	An individual, organization, or enterprise responsible for developing a *verified* system solution based on operational capabilities and performance *bounded* and *specified* in a *System Performance Specification (SPS)*.
Services provider	An individual, organization, or enterprise chartered or contracted to provide services to operate the system or support its operation.
Independent test agent/agency (ITA)	An individual, organization, or enterprise responsible for verifying and/or *validating* that a system will meet the User's documented operational mission needs for an intended and prescribed operating environment.
System administrator	An individual, organization, or enterprise responsible for the general operation, configuration, access, and maintenance of the system.
Mission planner	An individual, organization, or enterprise that: 1. Translates mission objectives into detailed tactical implementation plans based on situational analysis; system capabilities and performance relative to *strengths*, *weaknesses*, *opportunities*, and *threats* (SWOT) 2. Develops a *course of action*, countermeasures, and required resources to achieve success of the mission and its objectives.

(continued)

Table 5.2 *continued*

Role	Role Description
System analyst	An individual or organization that applies analytical methods and techniques (scientific, mathematical, statistical, financial, political, social, cultural, etc.) to provide meaningful data to support *informed* decision making by mission planners, system operators, and system maintainers.
System support	An individual, organization, or enterprise responsible for supporting the system, its capabilities, and/or performance at a sustainment level that ensures successful achievement of the system's mission and objectives. System support includes activities such as maintenance, training, data, technical manuals, resources, and management.
System maintainer	An individual, organization, or enterprise accountable for ensuring that the EQUIPMENT System Element is properly maintained via *preventive* and *corrective maintenance*, system upgrades, etc.
System instructor	An individual or organization accountable for training all members of the PERSONNEL System Element to achieve a level of performance standard based proficiency in achieving the system mission and its objectives.
System critic	An individual, organization, or enterprise with competitive, adversarial, or hostile motivations to publicize the shortcomings of a system to fulfill its assigned missions, goals, and objectives in a cost effective, value-added manner and/or believes the system is a threat to some other system for which the *System Critic* serves as a *System Advocate*.
System competitor	An individual, organization, or enterprise whose missions, goals, and objectives compete to capture similar mission outcomes. **EXAMPLE 5.5** Examples include market share, physical space, etc.
System adversary	A *hostile* individual, organization, or enterprise whose interests, ideology, goals, and objectives are: 1. Counter to another system's missions, goals, and/or objectives. 2. Exhibits behavioral patterns and actions that appear to be threatening.
System threat	A *competitive*, *adversarial*, or *hostile* individual, organization, or enterprise *actively* planning and/or executing missions, goals, and objectives that are counter to another system's missions, goals, and/or objectives.

ORGANIZATIONAL CENTRIC EXERCISES

1. Identify a contract program within your organization. Interview the program and technical directors to identify the system's roles and stakeholders, using Table 5.2 as a checklist.
2. Identify a services group within your organization—such as communications, accounting, or contracts. Identify the role and stakeholders of the services system using Table 5.2 as a checklist.
3. Using any system, product, or service your organization provides, identify what human system roles the product supports.

Chapter 6

System Acceptability

6.1 INTRODUCTION

The *degree of success* of any human-made system and its mission(s) ultimately depends on:

1. Whether the marketplace is ready for introduction of the system—an operational need driven "window of opportunity."
2. The User's perception of the system's *operational utility*, *suitability*, and *availability*.
3. The system's ability to accomplish the User's mission—*system effectiveness*.
4. The return on investment (ROI) for the resources expended to operate and maintain the system—*cost effectiveness*.

Most people view *system acceptability* in a customer satisfaction context. Engineers often shrug it off as *something that can be measured by customer satisfaction surveys AFTER a system or product has been delivered or distributed to the marketplace*. Based on WHAT the organization learns from the postdelivery surveys, they may improve the system or product if:

1. The User awards them another contract to *correct* any *deficiencies*, assuming the deficences are not covered by the contract.
2. Longer term profit projections make internal investment worthwhile.

You should understand the User's operational needs prior to the system development rather than from postdelivery surveys. You need to understand:

1. HOW the User intends to use the system or product.
2. WHAT measures of success are to be applied?
3. The consequences *and* ramifications of User failure or degrees of success.

Author's Note 6.1 *This topic is seldom addressed by many texts and is typically one of the last concepts engineers learn. Yet, it is one of the most important concepts. If system developers do not understand the success criteria for user acceptance, the most elegant designs are worthless. Therefore, this topic is introduced as part of the* System Entity Concepts.

The four *degree of success* factors listed at the start of this chapter are seldom optimum simultaneously. Though appearing to be equal, psychologically, the *subjective* measures—namely, perception of *operational utility*, *suitability*, and *availability* (factor 2)—often obscure the *objective* measures of system success—*system* and *cost effectiveness* (factors 3 and 4).

System Analysis, Design, and Development, by Charles S. Wasson
Copyright © 2006 by John Wiley & Sons, Inc.

- *Subjective* look, feel, and perception within the peer community or by the funding source factors often obscure User acceptance that a system may only be partially successful.
- Conversely, the User for the same *subjective* reasons may reject an *objectively* successful system.

Ultimately, some event often brings a reality check. A decision to continue with the system may require an upgrade to correct deficiencies or to phase out the system and replace it with a new system. *Operational utility*, *suitability*, *availability* and *effectiveness* drive the need to ensure that SE is an integral part of overall system development from *conception* through *disposal*. These factors thrive as common buzzwords among marketers and are often very difficult to quantify for implementation. The implementation typically requires specialists that spend entire careers focused exclusively on quantifying and presenting this information in a manner that is realistic and easily understood by key decision makers.

What You Should Learn from This Chapter

- *What factors influence system acceptability to Users?*
- *Why are system introduction, feasibility, and affordability important?*
- *What is meant by the operational utility of a system?*
- *What is meant by the operational suitability of a system?*
- *What is operational availability?*
- *What is meant by the operational effectiveness of a system?*
- *What is meant by the cost effectiveness of a system?*
- *What is system verification?*
- *What is system validation?*
- *Why are system verification and validation important to Users, Acquirers, and System Developers?*

Definitions of Key Terms

- **Measures of Effectiveness (MOE)** "A qualitative or quantitative measure of the performance of a model or simulation or a characteristic that indicates the degree to which it performs the task or meets an operational objective or requirement under specified conditions." (Source: DoD 5000.59-M *Modeling and Simulation (M&S) Glossary*, Part II, Glossary A-318, p. 134)
- **Measures of Performance (MOP)** "Measures of lowest level of performance representing subsets of measures of effectiveness (MOEs). Examples are speed, payload, range, time on station, frequency, or other distinctly quantifiable performance features." (Source: DSMC— *Test & Evaluation Management Guide*, Appendix B, *Glossary of Test Terminology*)
- **Measures of Suitability (MOS)** Objective performance measures derived from subjective user criteria for assessing a system's operational suitability to the organizational and mission applications.
- **Operational Effectiveness** "An operational Test & Evaluation (OT&E) metric that measures the overall degree of mission accomplishment of a system when used by representative personnel in the environment planned or expected (e.g., natural, electronic, threat) for operational employment of the system considering organization, doctrine, tactics, survivability, vulnerability, and threat (including countermeasures, initial nuclear weapons effects, nuclear

biological, and chemical contamination threats). The operational system that is provided to users from the technical effort will be evaluated for operational effectiveness by a service OT&E agency. Also a useful metric for operational effectiveness assessments." (Source: INCOSE *Handbook*, Section 12, Appendix Glossary, p. 36)

- **Operational Suitability** "The degree to which a system can be placed satisfactorily in field use with consideration being given to availability, compatibility, transportability, interoperability, reliability, ... usage rates, maintainability, safety, human factors, manpower supportability, logistic supportability, natural environmental effects and impacts, documentation, and training requirements." (Source: Adated from DoD *Glossary*, *Defense Acquisition Acronyms and Terms*)
- **System Effectiveness** "A quantitative measure of the extent to which a system can be expected to satisfy customer needs and requirements. System effectiveness is a function of suitability, dependability, (reliability, availability, maintainability), and capability." (Source: INCOSE *Handbook*, Section 12, Appendix Glossary, p. 43)

Key Challenges in Developing Systems Acceptable to Their Users

The *success* and level of *acceptance* of a *system*, *product*, or *service* is often measured by a series of questions that are answered by analysis before a system is developed and by User feedback and *actual results* after the system is implemented in the field. The questions include:

1. *Is the timing RIGHT for the introduction of a new system? Is the marketplace "mentally and emotionally ready" ready for this system, product, or service as driven by operational needs?*
2. *Is system feasibility sufficient to warrant User/System Developer investments that may result in a return on investment (ROI) at a later date?*
3. *Does the proposed system have OPERATIONALLY UTILITY to the User relative to their organizational missions and objectives?*
4. *Is the proposed system OPERATIONALLY SUITABLE for all stakeholders relative to its intended application?*
5. *Is the proposed system OPERATIONALLY AVAILABLE when tasked to perform missions?*
6. *Is the system OPERATIONALLY EFFECTIVE, in terms of cost and technical performance, for its intended mission applications and objectives?*

The underlying foundation for all of these questions resides in User and *system stakeholder* perceptions of the system. An old adage states "perception is reality." Despite the physical realities of system success, the User and *stakeholders* must be *technically convinced* via *objective, fact-based, compelling evidence* that the system does or will meet their missions and objectives.

This chapter provides an overview of the concepts of: 1) marketplace timing, 2) system affordability, 3) system operational utility, suitability, and system effectiveness; and 4) system verification and validation.

6.2 SYSTEM INTRODUCTION, FEASIBILITY, AND AFFORDABILITY

The decision to proceed with development of new systems, products, and services involves three basic questions:

1. *Is the timing right to introduce a new system to the market/User, particularly in the commercial environment?*
2. *Is it economically feasible to develop a new system with the technologies currently available within budgetary constraints?*
3. *If development is economically feasible, will the User be able to afford the operating and maintenance costs over the planned service life of the system?*

Let's investigate each of these questions further.

System Timing to the Marketplace

History is filled with examples of systems or products that were delivered to the marketplace prematurely or too late. You can *innovate* and *develop* the best widget or electronic mouse trap. However, if the marketplace is not *mentally* or *skillfully* ready for the device or can afford it, your efforts and investments may be futile—*timing is critical* to User acceptance!

The same is true for proposing new systems or capabilities to Users. They may WANT and NEED a system yet lack sufficient funding. In other cases their funding may be placed "on hold" by decision authorities due to a lack of consensus regarding the maturity in system definition or understanding the system's requirements.

For this reason, most organizations develop a series of decision-making "gates" that qualify the *maturity* of a business opportunity and *incrementally* increase the level of commitment, such as funding. The intent is to ensure that the RIGHT system/product *solution* is introduced at the RIGHT time for the RIGHT price and is *readily accessible* when the User is ready to purchase. Therefore, organizations must do their homework and work *proactively* with the Users to ensure that system timing is right. This leads to the next point: User system/product *feasibility* and *affordability*.

System Feasibility and Affordability

If a determination is made that the timing for a system, product, or service is RIGHT, the next challenge comes in determining if the system, as currently specified, can be *feasibly* developed and produced with existing technologies within the planned development and life cycle budget at *acceptable risk* for the User or Acquirer.

System feasibility ultimately focuses on four key questions:

1. WHAT does the User WANT?
2. WHAT does the User NEED?
3. WHAT can the User AFFORD?
4. WHAT is the User WILLING to PAY?

As an SE, chances are you will be required to provide technical support to business development teams working on a new system or product acquisition. If not, you may be supporting an SE who is. From a technical perspective, the multidisciplinary SE team is *expected* to *conceptualize*, *mature*, and *propose* technical solutions to satisfy the system feasibility questions noted above.

Author's Note 6.2 *If you choose to AVOID business development support, others within your organization may potentially formulate a risky or undesirable solution or commitment that you have to live with later. Conversely, if the others solicit engineering support and you choose to IGNORE them, you may be stuck with the consequences of your own inaction. Therefore, proactively support and technically influence business development activities and decision making—It's a win–win for all stakeholders.*

6.3 OPERATIONAL UTILITY AND SUITABILITY

If you determine that the market timing is *RIGHT* and the Users can *AFFORD* the system/product, the next challenge is managing User perceptions of *operational utility* and *operational suitability*.

Operational Utility

Users *expect* systems and products to have a level of *operational utility* that enables them to accomplish the organizational missions and achieve the stated goals and objectives. These are nice words, but what does *operational utility* really mean? A system or product having *operational utility* is one that is:

1. The RIGHT *system*, *product*, or *service* for the objective to be accomplished.
2. Does not pose any *unacceptable* safety, environment, or health hazards or risks to its operators or the public.

So, if a system satisfies these operational utility criteria, how do we determine *operational suitability*?

Operational Suitability

Operational suitability characterizes HOW WELL a system or product is:

1. *Suited* to a User's specific application in a given operating environment.
2. Integrates and performs within the User's existing system.

Some systems and products may have significant *operational utility* for some applications but simply are not *operationally suited* to a specific User's intended application and operating environment. Consider the following example:

EXAMPLE 6.1

From a transportation perspective, vehicles such as cars may have *operational utility* to a User. However, if the User plans to use the vehicle off the road in a rugged, harsh environment, only specific types of vehicles may be *operationally suitable* for the application. If the User intends to carry heavy loads, only specific types of trucks may be *operationally suitable* for the application.

Operational Availability

Operational availability means that the *system*, *product*, or *service* is *ready ON DEMAND to perform* a mission WHEN tasked. *Operational availability* becomes a critical metric for assessing the *degree of readiness* to perform missions. *Availability* is a function of SYSTEM *reliability* and *maintainability*. Consider the following example:

EXAMPLE 6.2

911 calls to police, fire departments, and emergency medical responders TEST the respective organization's system *availability*—its personnel and equipment—to respond to emergencies and disasters.

6.4 OPERATIONAL EFFECTIVENESS

If the User deems a system or product to: 1) have *operational utility*, 2) be *operationally suitable* for the application, and 3) *operationally available* to perform missions, the next challenge is determining if the system is *operationally effective*. Users and organizations are chartered with specific goals, missions, and objectives. Users acquire and implement systems, products, or services specifically to support achievement of those goals, missions, and objectives. If the system, product, or service is not or is only marginally *operationally effective*, it is of limited or no value to the User.

The Elements of Operational Effectiveness

A system, product, or service must be capable of supporting User missions to a level of performance that makes it *operationally effective* in terms of accomplishing organizational goals and objectives, namely outcomes, cost, schedule, and risk.

EXAMPLE 6.3

For a military system, system effectiveness depends on environmental factors such as operator organization, doctrine, and tactics; survivability; vulnerability; and threat characteristics. (Source: DSMC, *System Engineering Fundamentals*, Chapter 14, *Measures of Effectiveness and Suitability*, p. 125)

If you analyze *operational effectiveness*, two key elements emerge:

1. HOW WELL the system accomplishes its mission objectives—operational effectiveness.
2. HOW WELL the system integrates and performs missions within the User's organizational structure and operating environment—operational suitability.

Therefore, we need to establish metrics that enable us to *analyze*, *predict*, and *measure* mission operational outcomes. We do this with two key metrics *measures of effectiveness (MOEs)* and *measures of suitability (MOSs)*.

Measures of Effectiveness (MOEs)

MOEs enable us to evaluate the first objective—*HOW WELL the system accomplished its mission objectives*. MOEs are *objective* measures that *represent* the most critical measures—performance effecters—that contribute to mission outcome success. Consider the following example:

EXAMPLE 6.4

Did we meet our financial target? How far did we miss the target? Did we meet our production goals? Did we finish the production run on schedule?

MOEs are used as the basis for deriving *measures of performance* (*MOP*) parameters stated as requirements in system performance and item development specifications.

Measures of Suitability (MOSs)

MOSs enable us to evaluate the second objective—*HOW WELL the system integrates into the User's organizational structure, mission applications, and operating environment*. MOSs represent *integrated* measures that objectively quantify issues such as supportability, human interface compatibility, and maintainability.

52 Chapter 6 System Acceptability

> **EXAMPLE 6.5**
>
> 1. What anthropometrics, skills, tools, and equipment are required to operate and maintain the system to a specific level of performance?
> 2. Do the control panels incorporate ergonomic designs and devices that *minimize* operator fatigue?

MOSs are characterized measures of performance (MOP) parameters stated as requirements in system specifications.

6.5 COST EFFECTIVENESS

The *objective measure* of *system success* ultimately depends on its *cost effectiveness*. From an organizational perspective: *Does the system produce outcome-based performance and results that provide a return on investment (ROI) that justifies continued use?*

Engineers, by virtue of their technical backgrounds, often have difficulty in relating to the concept of *cost effectiveness*; they tend to focus on *system effectiveness* instead. The reality is: *system life cycle operational costs and the profits derived from system applications DRIVE organizational decision making.* You can *innovate* the best *system, product, or service* with outstanding *system effectiveness*. However, if the recurring operating and support costs are unaffordable for its Users, the system may "dead on arrival" at system delivery, especially in commercial environments.

Cost effectiveness, as a metric, is computed from two elements: 1) life cycle cost and 2) *system effectiveness*.

System Effectiveness

As an objective factor, *system effectiveness* represents the physical reality of outcome-based performance and results.

Outcome-based performance and *results* occur in two basic forms: *planned* and *actual* performance. When system development is initiated, the System Developer is dependent on analyses, models, and simulations to provide technical insights that will reveal *how* a system is projected to perform. These data are used to:

1. *Bound* and *specify* the system or one of its items.
2. Compare *actual* versus *planned* performance or expected results.

Various techniques, such as rapid prototypes, proof of concept prototypes, and technology demonstrations, are employed to validate models and simulations as predictors of system performance ands reduce risk. The intent is to collect *objective, empirical evidence* "early on" to gain a *level of confidence* that the system, or portions thereof, perform as expected. The net result is to *verify* and *validate* the predictions.

When the actual system is ready for field-testing, *actual* performance data are collected to:

1. *Verify* accomplishment of the requirements.
2. *Validate* models that the system successfully performs according to the User's intended use.

System effectiveness requires understanding the contributory factors such as reliability, maintainability, and performance.

Quantifying System Effectiveness

Although the concepts of *cost effectiveness* and *system effectiveness* are intriguing, the challenge is being able to *translate* concepts into simple, physical reality that people can easily understand. We make the translation by establishing *metrics* that enable us to analytically and objectively *quantify* system effectiveness and the level of performance required to achieve the effectiveness. *System effectiveness* metrics include: *MOEs* and *MOSs* supported by *MOPs*. MOPs are stated as *technical performance parameters* (*TPPs*) in specifications.

MOEs and MOSs are outcome-based measures of *operational effectiveness*. Operational effectiveness is further defined as "the overall degree of a system's capability to achieve mission success considering the total operational environment" (Source: DSMC, *Systems Engineering Fundamentals*, Dec. 2000, para. 14.1, p. 125). MOEs and MOSs "identify the most critical performance requirements to meet system-level mission objectives, and will reflect key operational needs in the operational requirements document" (Source: DSMC, *Systems Engineering Fundamentals*, Dec. 2000, para. 14.1, p. 125).

6.6 REALITY CHECK

All of this discussion sounds fine, but *how does it apply to the real world*? Traditionally, organizations have developed procurement packages and employed cost models to provide a *rough order of magnitude* (ROM) estimate for system development. In effect, Acquirers solicited a *point solution* for a specified capability.

Although the cost model estimates were generally *accurate, surprises* did occur. If the cost proposals *exceeded* the original estimate that had been budgeted, the Acquirer and User were confronted with recompeting the contract. Generally, negotiations took place to arrive at a contract price for a system that was *affordable* but often with lesser capabilities.

WHAT Acquirers and Users wanted to hear was "*Give us a proposal that describes the range and mix of capabilities and levels of performance—the system/cost effectiveness—that can be procured for a specific set of system objectives or requirements.*"

In concept, this approach focuses on *cost effectiveness*—namely how *much system effectiveness can we acquire per unit of life cycle cost*. A more formal term, cost as an independent variable (CAIV), is applied to the concept. As a result, System Developer proposals may be required to submit graphical analyses that plot system performance options along the horizontal axis and cost on the vertical axis. Figure 6.1 provides an example view.

One final point. Today's society is continually *innovating* new contracting approaches to assist organizations that have sporadic funding. Due to the cost of acquiring, developing, operating, and supporting new system, organizations often shift the investment cost burden of system development to contractors. The *intent* is to procure on a *fee for service* basis to meet budgetary needs. As a result, contracts between these organizations incorporate *system availability* clauses that require the system to be *operationally available* on-demand when the User needs the system. The bottom line is that if the system is not o*perationally available*, the contractor is not paid. Therefore, *system availability* becomes a prime *measure of success criteria* for *system acceptability*—meaning User *willingness* or *desire* to employ your *systems, products*, and *services* to achieve individual or organizational goals and objectives.

Are *cost effectiveness* and *system effectiveness* analyses easy to perform? No. In fact both are often very difficult to *quantify* and *measure*. Implementation of these details varies by organization and program. Military organizations employ subject matter experts (SMEs) with specialized skills in these areas to assess survival in battlefield operations. If you *fail* to plan for system *effectiveness* and all the factors that contribute to its success, you place organizational and mission objec-

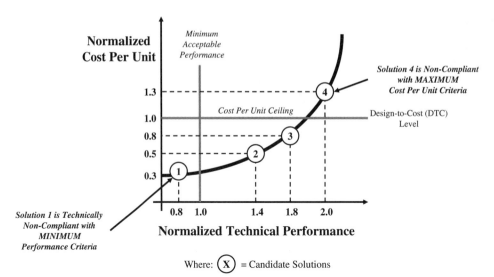

Figure 6.1 Cost as an Independent Variable (CAIV) plot for making cost-capability decisions

tives at risk. *Are these concepts important to retailer companies doing business on the World Wide Web?* You, bet! Especially from a User interface, security, and billing perspective. *Is system effectiveness important in educational and training systems?* A resounding Yes!

6.7 SYSTEM VERIFICATION AND VALIDATION (V&V)

The subject of *system acceptability* is often *abstract* and is OPEN to INTERPRETATION to different people. WHAT constitutes *acceptability* to one User may not be *acceptable* to another User. Our discussion of system *operational utility*, *suitability,* and *effectiveness* ultimately influences the User's decision as to whether the system satisfies their operational needs. The *degree of satisfaction* for a given system application establishes user perceptions concerning whether they *procured* the RIGHT system. We refer to this as *system validation*.

System validation draws yet another question: *If the Users have a vision of WHAT they WANT, HOW do they*:

1. *Translate the vision into a specification to produce the visionary system?*
2. *Have a level of confidence that the system, product, or service delivered will be that RIGHT system?*
3. *Be able to reproduce multiple copies of the system?*

The first question requires the "engineering of the system."

The second question requires building *integrity* into the evolving "engineering of the system" to ensure that the source requirements manifest themselves in the design of the deliverable *system*, *product*, or *service* and can be demonstrated at system delivery to be satisfied. We refer to this as *system verification*.

The third question requires that the "engineering of the system" establish *high-quality* documentation and process *standards* that enable skilled workers to reliably and predictably *reproduce* each copy of the system, product, or service within specified performance constraints.

Let's focus our attention briefly on the concepts of *system validation* and *system verification*. Figure 6.2 provides an illustration of system verification and system validation that helps in better understanding these concepts.

6.7 System Verification and Validation (V&V)

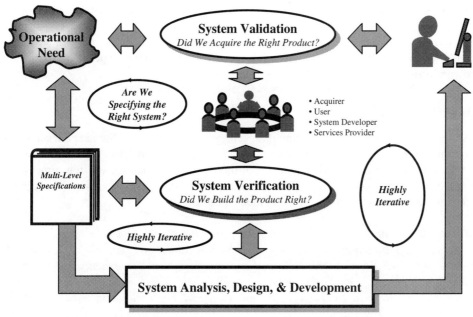

Figure 6.2 Verification & Validation (V&V) Concept Overview

Operational Need

Beginning in the upper left corner of the figure, Users have operational needs related to their assigned organizational and system missions. Later we will discuss the context of this operational need in terms of a *problem space* or and *solution space*. To satisfy this operational need, the User must *translate* the operational need requirements into a specification that is *legally sufficient* for procurement. Since the creation of the specification requires *specialized* expertise provided by subject matter expert (SME) SEs, the User's organization may lack this capability. Where this is the case, the User may employ the services of an Acquirer organization to serve as their technical and contractual representatives.

System Verification

Throughout the development of the multi-level specifications, the Acquirer and User must answer the question: *Are we specifying the RIGHT system that will satisfy the User's operational need?* This is an Acquirer-User process that is *highly iterative* until the requirements reach a *state of maturity* sufficient for initiating a procurement action. Once the procurement action begins, the formal *Request for Proposal (RFP)* solicitation process for a system provides additional insights, such as from proposals by System Developer candidates concerning the RIGHT system from an objective, technical perspective. The formal RFP solicitation process culminates in the award of a system development or procurement contract.

Once the contract is awarded, the challenge becomes: *HOW do we ensure that the system will be built RIGHT—that is, correctly — in accordance with the contract specification?* The Acquirer, User, and System Developer must answer this question mainly through a series of technical reviews, technical demonstrations, and risk assessments that occur throughout the system development contract. As a result, the teams must continually assess compliance of the *evolving* system design solution relative to the specification requirements. We refer to the *evolving* system design solution as the *Developmental Configuration* and the test activities during system development as the Developmental Test and Evaluation (DT&E).

Ultimately, when system or product development is complete, those responsible and accountable for *system acceptance* must answer the question: *Did we build the system RIGHT*—that is, in accordance with specification requirements? This question is typically answered with a System Verification Test (SVT) that documents test results for each specification requirement based on prescribed verification methods.

System Validation

Once the system has been verified as meeting its specification requirements, the next question that the Acquirer and User must answer is: *Did we procure the RIGHT system*? This is the ultimate test of HOW WELL the Acquirer, in collaboration with the User, performed the up-front process of partitioning the operational need—meaning the problem space—into a solution space that can be bounded and technically described via a procurement specification.

Author's Note 6.3 *For introductory overview purposes, the scope of the following discussion focuses on system validation of the completed system. System validation occurs throughout system development from Contract Award through formal system acceptance. Contextually, system verification and validation apply to each step along the supply chain (e.g., customer-supplier) involving the User, Acquirer, System Developer, Subcontractor, and vendor organizations as well as individuals within those organizations.*

During *system validation* the User or an Independent Test Agency (ITA) subject the system, product, or service to actual operating field conditions, using trained system operators and maintainers. *Actual devices* or *surrogate representations* of the operational need are employed to formally evaluate the overall *system responsiveness* and *effectiveness*, including operator and maintainer actions, to satisfy the operational need. We refer to this phase of system development as Operational Test and Evaluation (OT&E). Depending on contract requirements, results of OT&E are incorporated as *corrective actions* into the Developmental Configuration of the design solution.

Depending on contract requirements, on completion of the OT&E activities, the Acquirer and System Developer conduct a *functional configuration audit* (FCA) and a *physical configuration audit* (PCA). The FCA verifies that SVT results fully comply with specification requirements. The PCA verifies that the physical components of the "As Verified" and "As Validated" system identically match the "As Designed" requirements (drawings, wiring diagrams, etc.). Results of the FCA and PCA are certified at a System Verification Review (SVR) prior to system delivery and acceptance. As part of the SVR, the Product Baseline is established as a precursor for system production.

Due the expense of producing production items and the need to *avoid rework*, the Product Baseline is crucial for reproducing production quantities of the system. The Product Baseline represents the physical state of the *Developmental Configuration*. This does not mean necessarily that the design has achieved a low cost suitable for production. Additional design refinements may be required to REDUCE *recurring* production costs. Once the design solution has matured, a Production Baseline can be established.

Author's Note 6.4 *Operational needs and intended usage must be documented BEFORE system development, rather than AFTER the system is delivered for system validation. Documenting these needs AFTER system development invalidates the specification process intended to document system requirements derived from validated User operational needs. The User certainly has the prerogative to change their minds about their intended needs at any time during the system devel-*

opment process. However, this means modifying the contract requirements, which were based on the original needs, via an engineering change proposal (ECP).

One of the challenges of SE is being able to translate User operational needs into specification requirements, especially if you have a customer who says *"We don't know WHAT we WANT, but we will KNOW it WHEN we see it!"* You MUST be able to *validate, document,* and *bound* WHAT they WANT—*up front*! Otherwise, your contracts organization should work with the Acquirer to establish a cost plus contract vehicle—that allows further *exploration, definition*, and *refinement* of these needs until the User identifies WHAT they WANT.

Author's Note 6.5 *System validation may be performed formally as part of contract requirements or informally by the System Developer prior to system delivery or by the User(s) after contract system delivery. System validation is sometimes specified in contracts that may involve development of systems planned for large production quantities. As an Acquirer or User, you DO NOT want to invest large sums of money to procure systems in large quantities that DO NOT satisfy User needs. Even in smaller quantities, there may be instances where the system requires validation of human systems integration (HSI).*

6.8 GUIDING PRINCIPLES

In summary, the preceding discussions provide the basis with which to establish the guiding principles that govern system acceptability.

Principle 6.1 System acceptability is determined user satisfaction; user satisfaction is determined by five User criteria:

1. Provide value—meaning *operational utility*.
2. Fit within the user's system and mission applications—meaning *operational suitability*.
3. Be available to conduct missions—meaning *operational availability*.
4. Accomplish performance objectives—meaning *operational effectiveness*.
5. Be affordable—meaning *cost effectiveness*.

Principle 6.2 Despite the most technically *innovative* and *elegant* SE design solutions, Users' *perceptions* of a system, product, or service constitute *reality*.

6.9 SUMMARY

In our discussion of system acceptability, we have identified, defined, and provided examples of the factors that contribute to this goal: 1) system introduction, 2) system affordability, 3) system feasibility, 4) operational utility, 5) operational suitability, 6) operational availability, 7) operational effectiveness, 8) cost effectiveness, and 9) system effectiveness. Our purpose in introducing these concepts at this time is to highlight key *showstopper* drivers that determine system success but are often swept aside by most authors and SEs. Our intent is to emphasize these concepts *up front* and not as notional afterthoughts determined by business development customer satisfaction surveys AFTER a system, product, or service has been fielded.

As an SE you should visualize the *system entity* as a simple box whose performance and success are measured by these criteria. *Why?* Some engineers are notorious for immersing themselves into math and science details BEFORE they have an in-depth understanding of WHAT is expected from the system by the User. These engineers lack the fundamental knowledge of system acceptability factors that help one to better *understand* how the element of the system contributes to the overall system capabilities and performance. Your challenge as an SE is to make sure system acceptability criteria are manifested in the specifications, designs, testing, verification, and validation of each hierarchical element of the system.

GENERAL EXERCISES

1. Answer each of the *What You Should Learn from This Chapter* questions identified in the *Introduction*.
2. Refer to the list of systems identified in Chapter 2. Based on a selection from the preceding chapter's General *Exercises* or a new system, selection, apply your knowledge derived from this chapter's topical discussions. Identify acceptance criteria for the stakeholders of each system.
3. Identify three instances of systems that were deemed by Users as unacceptable, were rejected, and subsequently failed in the marketplace.
4. Identify three instances of systems that were deemed by Users as acceptable and highly successful in the marketplace.
5. Identify three instances of systems you purchased that met your specification requirements (e.g., verification) but failed to satisfy your needs (e.g., validation).
6. Identify three instances of systems you use. How would you quantify their:
   (a) System effectiveness?
   (b) Cost effectiveness?

ORGANIZATIONAL CENTRIC EXERCISES

1. What command media does your organization have that provides guidance related to system acceptability?
2. Contact a contract program in your organization. Interview the Program Director and Technical Director.
   (a) What success criteria has the user established?
   (b) How our these criteria documented and implemented in specifications?

REFERENCE

Defense Systems Management College (DSMC) Ft. Belvoir, VA. 2001. *Systems Engineering Fundamentals*.

ADDITIONAL READING

BLANCHARD, BENJAMIN S. 1998. *System Engineering Management*. New York: Wiley.

Defense Systems Management College (DSMC) FT. BELVOIR, VA. 2001. *Glossary—Defense Acquisition Acronyms and Terms*, 10th ed. Defense Acquisition University Press.

NASA SP-6105. 1995. *Systems Engineering Handbook*.

Defense Systems Management College (DSMC) FT. BELVOIR, VA. 1998. *Test and Evaluation Management Guide*, 3rd ed.

Chapter 7

The System/Product Life Cycle

7.1 INTRODUCTION

The system/product life cycle serves as the fundamental *roadmap* for understanding and communicating how natural and human-made systems *evolve* through a progression of sequential life cycle phases. For human-made systems the roadmap provides a basis for assessing *existing* system capabilities and performance relative to threats and opportunities; defining, procuring, and developing new systems to respond to the threats and opportunities; and implementing new systems to achieve mission objectives that counter or leverage the threats and opportunities.

This section introduces the concept of the system life cycle, the top-level life cycle framework for human-made systems. As the final discussion in the *System Entity Series*, we characterize the concept-to-disposal evolution of a system, product, or service.

What You Should Learn from This Chapter

- What is the system/product life cycle?
- What is the System Definition Phase; when does it start, and when does it finish?
- What is the System Procurement Phase; when does it start, and when does it finish?
- What is the System Development Phase; when does it start, and when does it finish?
- What is the System Production Phase; when does it start, and when does it finish?
- What is the System Operations and Support (O&S) Phase; when does it start, and when does it finish?
- What is the System Disposal Phase; when does it start, and when does it finish?
- What are life cycles within life cycles?
- How is a legacy system's life cycle used to establish requirements for a new system?

7.2 SYSTEM LIFE CYCLE OVERVIEW

The evolution of any system made by or known to humankind begins at the point of conception and ends at disposal. This process is referred to as the system life cycle. The system life cycle serves structurally as the foundation for system development. Human-made systems are *conceptualized*, *planned*, *organized*, *scheduled*, *estimated*, *procured*, *deployed*, *operated* and *supported*, and *disposed* of using this structure. Natural systems follow similar constructs with life phases.

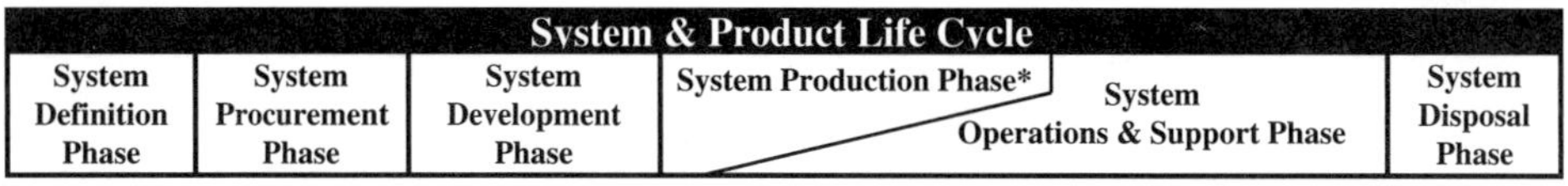

System & Product Life Cycle						
System Definition Phase	System Procurement Phase	System Development Phase	System Production Phase*		System Operations & Support Phase	System Disposal Phase

* = If applicable

Figure 7.1 The System/Product Life Cycle

The life cycle for any system, product, or service consists of a series of phases starting with system *conception* and continuing through final *disposal*. For human-made systems the beginning and ending of each phase is marked by a significant *control point* or *staging event* such as a key decision at a technical review or a field event that authorizes progression to the next phase.

Author's Note 7.1 *There are a number of ways to define a system life cycle. Ten people will have 10 different versions of this graphic. You and your organization should choose one that best reflects your organization and industry's perspective of the life cycle.*

The typical system life cycle is composed of a series of phases as shown in Figure 7.1. The phases are:

- System Definition Phase
- System Procurement Phase
- System Development Phase
- System Operations and Support (O&S) Phase
- System Production Phase
- System Disposal Phase

This chapter presents the system/product life cycle as a top-level *framework* of embedded phases required to evolve a User operational need for a system, product, or service from conceptual "vision" through disposal. Each of the phases represents a collection of activities that focus on specific program objectives and work products. As you will soon discover, some of these phases have well-defined endings marked by key milestones while other phases overlap and transition from one to another.

7.3 SYSTEM LIFE CYCLE PHASE SYNOPSIS

The System Definition Phase

The System Definition Phase begins with recognition by the User that a new system or upgrade to an existing *system*, *product*, or *service* is required to satisfy an *operational need*. The operational need may be derived from: 1) mission opportunities, 2) threats, or 3) projected system capability and performance "gaps" or deficiencies.

On determination to initiate definition of a new system, the User analyzes *existing* system operational needs and defines requirements for a new system, product, or service. In some instances the User may enlist the services of an Acquirer to procure and develop the system and serve as the User's *technical* and *contract representative*.

The Acquirer assists the User in analyzing the *opportunity* or *problem* space that created the need. The Acquirer, in collaboration with the User, bounds the *solution space* in the form of a set of system requirements to serve as the basis for a system development contract.

When the System Definition Phase has reached sufficient maturity, the Acquirer initiates the System Procurement Phase.

The System Procurement Phase

The System Procurement Phase consists of those activities required to procure the new system or upgrades to the existing system. These activities include:

1. Qualifying capable system, product, or service vendors.
2. Soliciting proposals from qualified vendors (offerors).
3. Selecting a preferred vendor (offeror).
4. Contracting with the vendor to develop the system, product, or service.

The selected vendor (offeror) becomes the System Developer or Services Provider.

The System Development Phase

The System Development Phase consists of those activities required to translate the contract system specifications into a physical system solution. Key System Development Phase activities include:

1. System Engineering Design
2. Component Procurement and Development
3. System Integration, Test, and Evaluation (SITE)
4. Authenticate System Baselines
5. Operational Test and Evaluation (OT&E)

Throughout the phase, the multi-level system design solution evolves through a progression of maturity stages. Each stage of maturity typically consists of a major technical design review with *entry* and *exit criteria* supported by analyses, prototypes, and technology demonstrations. The reviews culminate in design baselines that capture snapshots of the evolving *Developmental Configuration*. When the system engineering design is formally approved, the *Developmental Configuration* provides the basis for component acquisition and development. We refer to the initial system(s) as the *first article* of the *Developmental Configuration*.

Acquired and developed components are inspected, integrated, and *verified* against the respective design requirements and performance specifications at various levels of integration. The intent of verification is to answer the question: *Did we develop the system CORRECTLY?*—in accordance with the specification requirements. The integration culminates in a System Verification Test (SVT) that proves the system, product, or service fully complies with the contract *System Performance Specification (SPS)*. Since the System Development Phase focuses on the creation of the system, product, or service from Contract Award through SVT, we refer to this as *Developmental Test and Evaluation (DT&E)*.

When the *first article* system(s) of the *Developmental Configuration* has been *verified* as meeting the SPS requirements, one of two options may occur, depending on contract requirements. The system may deployed to:

1. Another location for *validation testing* by the User or an Independent Test Agency (ITA) representing the User's interests.
2. The User's designated field site for installation, checkout, and final acceptance.

Validation testing, which is referred to as *Operational Test and Evaluation (OT&E)*, enables Users to determine if they specified and procured the right SYSTEM to meet their operational needs. Any deficiencies are resolved in accordance with the terms and conditions (Ts&Cs) of the contract.

After an initial period of system operational use in the field to *correct deficiencies* and *defects* and collect field data to *validate* system operations, a decision is made to begin the System Production Phase, if applicable. If the User does not intend to place the system or product in production, the Acquirer and User formally accepts system delivery, thereby initiating the System Operations and Support (O&S) Phase.

The System Production Phase

The System Production Phase consists of those activities required to produce small to large quantities of the system. The initial production typically consists of a Low-Rate Initial Production (LRIP) to *verify* and *validate* that:

1. Production documentation and manufacturing processes are mature.
2. *Latent defects* such as *design errors*, *design flaws*, or poor workmanship are eliminated.

When the production process has been verified and validated based on field tests of system production samples, large-scale production, if applicable, may be initiated. Since the system and production engineering designs have already been verified, each production system is:

1. *Inspected.*
2. *Verified* against key *System Performance Specification* requirements.
3. *Deployed* to the User's designated field site(s) for implementation (i.e., operations and support).

The System Operations and Support (O&S) Phase

The System Operations and Support (O&S) Phase consists of User activities required to *operate*, *maintain*, and *support* the system including *training* for system users to perform the system's operational mission. If the system is directed to change physical or geographic locations in preparation for the next mission, the system is *redeployed*. On deployment, the system, product, or service begins *active duty*.

Throughout the system's operational life, *refinement* and *enhancement* upgrades may be procured and installed to improve system capabilities and performance in support of organizational missions. The system configuration at *initial delivery* and *acceptance* represents the *Initial Operational Capability* (IOC). System upgrades, referred to as *incremental builds*, are released and incorporated into the fielded system or product until the system reaches a planned level of maturity referred to as *Full Operational Capability* (FOC).

Although most systems have a planned *operational service life*, the expense of maintaining a system via *upgrades* and ability to upgrade the existing system with new technologies is not always cost effective. As a result, the User may be forced to procure a new system, product, or service to replace the existing system. Where this is the case, a new *system life cycle* is initiated while the existing system is still in *active duty*.

As the first articles of the new system are placed into active duty, a transition period occurs whereby the *legacy* (i.e., existing) system and the new system are in operation simultaneously in the field. Ultimately, a decision will be made to *deactivate* and *phase out* the legacy system from active duty. When system phase-out occurs, the legacy system's disposal phase begins.

The System Disposal Phase

The System Disposal Phase consists of those activities required to phase out an existing or legacy system from active duty. Each system or the lot of systems may be dispositioned for *sale, lease, storage* or *disposal*. Disposal alternatives include mothballing for future recommissioned use, disassembly, destruction, burning, and burial. System disposal may also require *environmental remediation* and *reclamation* to restore the system's field site or disposal area to its natural state.

7.4 LIFE CYCLES WITHIN LIFE CYCLES

Now that we have a basic understanding of the system life cycle we can shift our attention to understanding *how* a system's life cycle fits within an organization's context.

Understanding the Organizational Aspects of System Life Cycles

Each system, product, or service is an asset of a higher level organization (i.e., system) that also has a system life cycle. Those same systems and products may include lower level systems or products that also have system life cycles. Therefore, we have multiple levels of system life cycles within higher level system life cycles.

Suppose that a user has quantities of a product or system, including various versions in inventory. At some point in time, the User may decide to replace a specific product or a group of products.

For example, an airline might decide to replace a specific aircraft by tail number or replace an entire fleet of aircraft over a period of time. Each aircraft, which has its own system life cycle, is part of a much larger system such as a fleet of aircraft, which also has its own system life cycle. To illustrate this point, Figure 7.2 provides an example.

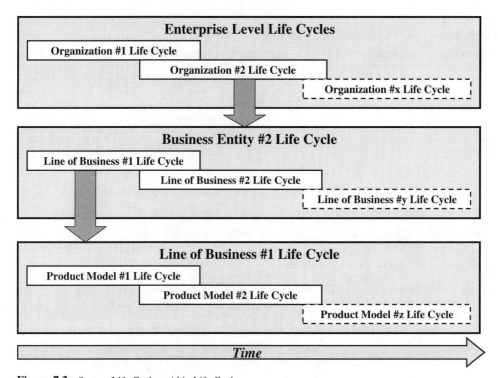

Figure 7.2 System Life Cycles within Life Cycles

Assume we have a corporation that has evolved over a number of years. Historically, we can state that the business came into existence as Organizational Entity 1. As the business entity grows, it changes its name and becomes Organizational Entity 2, and so forth.

If we examine the system life cycle of Organizational Entity 2, we might find that the organization evolves through several lines of business (LOBs): LOB 1, LOB 2, and so on. Within each LOB, the organization has a core product line that consists of Product Model 1, which evolves into Product Model 2, and so forth. Observe the overlapping of Product 1 and Product 2 life cycles. The evolution of this product line continues until the organization decides to terminate the product or LOB. *How is this concept applied to the real world?*

Application of System Life Cycles

We can apply the concept of system life cycles within system life cycles to an example such as small engine developer. The organization, which has a life cycle, may evolve through a number of organizational life cycles—as small business, corporation, and so on. During Organizational Life Cycle 2, the organization may develop several LOBs—two-cycle engines, four-cycle engines, and so on—to support marketplace opportunities such as lawn mowers, edgers, and small tractors. The organization's four-cycle LOB may evolve through Product Model 1, Product Model 2, and so forth. Each *product model* builds on its predecessor (i.e., *precedented system*) to improve capabilities and performance to meet marketplace needs.

The preceding discussion focused on a system or product suppliers system life cycles. The same analogy applies to users of system, product, and services. Their organizations *evolve* through similar life cycles. The differences occur when Product Model 1:

1. Fails.
2. Becomes to costly to maintain.
3. Is predicted to be vulnerable to system threats.
4. Lacks the specific level of capability or performance to meet predicted organizational needs.

Now, *why is this relevant to SE*? As a systems engineer, you need to understand what:

1. LOB the User is engaged in.
2. *Opportunities*, *problems*, or *issues* the User is chartered to address as part of its LOB. We refer to this as the *opportunity space*; specific targets as *targets of opportunity (TOO)*.
3. Missions the User performs to support the LOB. We refer to this as the *solution space*.
4. Capabilities are required to support *solution space* missions now and in the future.
5. Existing systems, products, or services the User employs to provide those capabilities.
6. Deficiencies—or opportunities—exist in the current system, product, or service and how you and your organization can cost effectively eliminate those deficiencies with new technologies, systems, products, or services.

Based on this knowledge and understanding, the SE's role as a *problem solver-solution developer* becomes crucial. The challenge is how do SEs work with Users and Acquirers to:

1. *Collaboratively* identify and partition the *opportunity space* into one or more *solution spaces*,
2. Technically *bound and specify* the *solution space* in terms of capability and performance requirements that are legally sufficient to procure systems, products, and services,

3. *Verify* that the new system complies with those requirements,
4. *Validate* that the system developed satisfies the User's original operational needs?

The remainder of this book is intended to answer this question.

Evolutionary Life Cycles

Although the preceding discussion appears straightforward, it can become very complicated, especially with highly complex, precedented *systems* even when the requirements are reasonably well known. The question is: *What happens when the requirements are not well known?*

In this case the system, product, or service requirements evolve over a series of *iterative* system development cycles that may include TEST markets. A *planned* product system life cycle may have a System Development Phase that may include several *prototype* or *technology demonstration* life cycles. Consider the following example.

EXAMPLE 7.1

An organization may award a series of sequential contracts to develop analyses, prototypes, and technology demonstrators of a system. The key work product of each contract may be to *mature* and *refine* the system, product, or service requirements of the preceding contract as a means of getting to a set of requirements that can be used to develop the end product. This approach, which is referred to as *spiral development*, serves to *reduce* system development *risks* when dealing with *unprecedented* or *highly complex* systems, new technologies, poorly defined requirements, and the like.

To better understand how a User acquires a new system and phases out the current system, let's explore the basic strategy.

7.5 SYSTEM TRANSITION STRATEGY AND SEQUENCING OVERVIEW

During the System Operations and Support (O&S) Phase of System 1 as shown in Figure 7.3, a decision event (1) occurs to replace System 1. The acquisition strategy is to bring the new System 2 "on-line" or into active service as noted by the *First Article Field Delivery* (5) *event*.

After a *System Transition* Period (6) for checkout and integration of System 2 into the HIGHER ORDER SYSTEM, an *Existing System Deactivation Order* (7) is issued. At that time, System 2 becomes the primary system and System 1 enters the System Disposal Phase of its life cycle. At some time period later, the disposal of System 1 is marked by the *Existing System Disposal Complete* (8) event.

So, *how do we initiate actions to get System 2 into active service by the planned new system's First Article Field Delivery (5) event without disrupting organizational operations?* Let's explore that aspect.

When the new system *Operational Need Decision* (1) *event* is made, procurement actions (2) are initiated to initiate System 2's life cycle. Thus, the System Definition Phase of System 2's life cycle begins. System 2's System Development Phase must be complete (4) and ready for field integration by the New System's First Article Field Delivery (5) event. System 2 then enters the System Operations and Support (O&S) Phase of its life cycle.

By the time the *Existing System Deactivation Order* (7) *event* is issued, System 2 must be "on-line" and in *active service*. As a result, System 1 completes its life cycle at the *Existing System Disposal Complete* (8) *event* thereby completing the transition.

Chapter 7 The System/Product Life Cycle

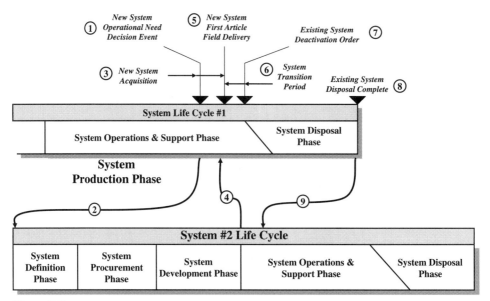

Figure 7.3 The Acquisition on New Systems and Phase-out of Legacy Systems

GENERAL EXERCISE

1. Answer each of the *What You Should Learn from This Chapter* points identified at the beginning of this chapter.

ORGANIZATIONAL CENTRIC EXERCISE

1. Research system life cycle standards of the following organizations. Summarize your findings for each and contrast them with your own system/product domain.
   (a) Department of Defense (DoD)
   (b) NASA
   (c) IEEE 1220-1998
   (d) ANSI/EIA 632-1999
   (e) International Council on Systems Engineering (INCOSE)
   (f) International Organization of Standardization (ISO)
   (g) ISO/IEC 15288
   (h) Your local organization
   (i) Your customer's life cycle

ADDITIONAL READING

ANSI/EIA 632-1999 Standard. 1999. *Processes for Engineering Systems*. Electronic Industries Alliance (EIA) Arlington, VA.

IEEE 1220-1998. 1998. *IEEE Standard for Application and Management of the Systems Engineering Process*. Institute of Electrical and Electronic Engineers (IEEE) New York, NY.

ISO/IEC 15288: 2002. *System Engineering—System Life Cycle Processes*. International Organization for Standardization (ISO). Geneva, Switzerland.

Chapter 8

The Architecture of Systems

8.1 INTRODUCTION

Every human-made and natural system is characterized by a structure and framework that supports and/or enables the integrated elements of the system to provide the system's capabilities and perform missions. We refer to this integrated framework as the system's architecture.

This chapter introduces the *System Architecture Concepts Series* and provides an overview to chapters that follow. We introduce the concept of the *system element architecture* via a construct consisting of two key entities: 1) the SYSTEM OF INTEREST (SOI) and 2) its OPERATING ENVIRONMENT.

Each of these entities is decomposed into lower tier elements. The SOI is composed of one or more MISSION SYSTEM(s) (role) and a SUPPORT SYSTEM (role). The OPERATING ENVIRONMENT consists of: 1) HIGHER ORDER SYSTEMS domain and 2) a PHYSICAL ENVIRONMENT domain. We conclude our discussion by introducing the conceptual "building blocks," referred to as the *system elements*, for each domain.

What You Should Learn from This Chapter

1. *What systems comprise the SYSTEM OF INTEREST (SOI)?*
2. *What systems comprise the OPERATING ENVIRONMENT?*
3. *Identify the MISSION SYSTEM and SUPPORT SYSTEM system elements?*
4. *How do the MISSION SYSTEM and SUPPORT SYSTEM system elements differ?*
5. *Identify the HIGHER ORDER SYSTEMS domain elements?*
6. *Identify the PHYSICAL ENVIRONMENT domain elements.*

Before we begin, let's define a few terms that are relevant to our discussion.

Definitions of Key Terms

- **Abstraction** An analytical representation of an entity for a specific purpose in which lower level details are suppressed. For example, a "family" is an abstraction that suppresses lower level entities such as father, mother, and children.
- **Entity** A *noun-based* person, place, or virtual object that represents a *logical* or *physical* system entity.
- **Entity Relationships** The hierarchical relationships and interactions between multi-level logical and physical system elements.

System Analysis, Design, and Development, by Charles S. Wasson
Copyright © 2006 by John Wiley & Sons, Inc.

- **Environment** "The natural (weather, climate, ocean conditions, terrain, vegetation, dust, etc.) and induced (electromagnetic, interference, heat, vibration, etc.) conditions that constrain the design for products and their life cycle processes." (Source: Kossiakoff and Sweet, *System Engineering*, p. 448)
- **Hierarchical Interactions** Actions between authoritarian command and control systems that lead, direct, influence, or constrain MISSION SYSTEM and SUPPORT SYSTEM actions and behavior. For analytical purposes, we aggregate these systems into a single entity abstraction referred to as the HIGHER ORDER SYSTEMS domain.
- **Logical Entity Relationship** A high-level, functional association that exists between two system entities without regard to physical implementation.
- **Object** See *entity*.
- **Physical Entity Relationship** A *physical*, *point-to-point* interface between two or more *entities* that may be unidirectional or bi-directional.
- **System Element** A label applied to classes of entities that comprise the MISSION SYSTEM/SUPPORT SYSTEM, HIGHER ORDER SYSTEMS, or PHYSICAL ENVIRONMENT domains. As a convention, specific system element names are capitalized throughout the text to facilitate identification and context of usage as illustrated in the descriptions below.
- **System of Interest (SOI)** The system consisting of a MISSION SYSTEM and its SUPPORT SYSTEM(s) assigned to perform a specific organizational mission and accomplish performance-based objective(s) within a specified time frame.
- **Taxonomy** "A classification system. Provides the basis for classifying objects for identification, retrieval and research purposes" (MORS Report, October 27, 1989. (Source: DoD 5000.59, Glossary A—Item 505, p. 163)

Based on this introduction, let's begin our discussion with the introduction of the *system architecture construct*.

8.2 INTRODUCING THE SYSTEM ARCHITECTURE CONSTRUCT

All natural and human-made systems exist within an abstraction we refer to as the system's OPERATING ENVIRONMENT. Survival, for many systems within the OPERATING ENVIRONMENT, ultimately depends on system capabilities—physical properties, characteristics, strategies, tactics, security, timing, and luck.

If we observe and analyze these systems and their patterns of behavior to understand how they adapt and survive, we soon discover that they exhibit a common construct—template—that describes a system's relationship to their OPERATING ENVIRONMENT. Figure 8.1 provides a graphical depiction of the construct. This construct establishes the foundation for all systems.

When systems interact with their OPERATING ENVIRONMENT, two types of behavior patterns emerge:

1. Systems interact with or respond to the dynamics in their OPERATING ENVIRONMENT. These interactions reflect peer-to-peer role-based behavioral patterns such as *aggressor*, *predator*, *benign*, and *defender* or combinations of these.
2. System Responses—behavior, products, by-products, or services—and internal failures sometimes result in *adverse* or *catastrophic* effects to the system—creating instability, damage, degraded performance, for example—that may place the system's mission or survival at risk.

8.2 Introducing the System Architecture Construct   69

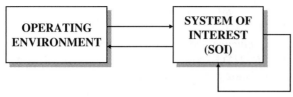

Figure 8.1   Top Level System Architecture Construct

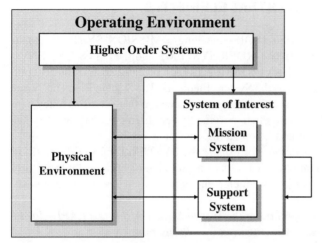

Figure 8.2   Top Level System Environment Construct

When you analyze interactions of a SYSTEM OF INTEREST (SOI) with its OPERATING ENVIRONMENT, two fundamental types of behavior emerge:

1. Hierarchical *interactions* (i.e., vertical interactions under the command and control of higher order systems).
2. Peer level interactions.

When a MISSION SYSTEM interacts with its OPERATING ENVIRONMENT, it:

1. Performs mission task assignments established by higher level, chain-of-command, decision authorities.
2. Interacts with *external systems* (i.e., human-made systems, natural environment, and its induced operating environment during mission execution environment).

We characterize SOI interactions with its OPERATING ENVIRONMENT to include two types of entities: 1) HIGHER ORDER SYSTEMS and 2) a PHYSICAL ENVIRONMENT.

The identification of OPERATING ENVIRONMENT domains enables us to expand the System Architecture construct shown in Figure 8.1. The result is Figure 8.2.

Author's Note 8.1   *Figure 8.2 features a subtlety that may not be readily apparent. Observe that the PHYSICAL ENVIRONMENT is shown on the left and the SYSTEM OF INTEREST is placed on the right-hand side.*

Most engineering environments establish standards and conventions for interpretive reading. For example, graphics read from left to right and from top to bottom. The convention is that data processing and work flow progress from left to right. If we had placed the SOI on the left side and

70 Chapter 8 The Architecture of Systems

the OPERATING ENVIRONMENT on the right side, this would have created a false perception that the SOI drives its OPERATING ENVIRONMENT.

Although the SYSTEM OF INTEREST does influence and may exercise some limited control over the OPERATING ENVIRONMENT, the OPERATING ENVIRONMENT actually stimulates the SOI to action and invokes a behavioral response. Where practical, we will employ this left-to-right convention throughout the book.

8.3 INTRODUCTION OF THE SYSTEM ELEMENTS

As *abstractions*, the SYSTEM OF INTEREST (SOI)—meaning a MISSION SYSTEM and its SUPPORT SYSTEM(s)—interact with HIGHER ORDER SYSTEMS and the PHYSICAL ENVIRONMENT within the SOI's OPERATING ENVIRONMENT. Each of these abstractions is composed of analytical *building blocks* referred to as the System Elements. The System Elements, when integrated into an architectural framework, form the System Architecture that serves as a key construct for system analysis and design. As an introduction to the System Elements, let's identify and define each *element* as it relates to the SOI and its operating environment.

To derive the system elements, we *decompose* or *expand* the SYSTEM OF INTEREST (SOI) and its OPERATING ENVIRONMENT into lower levels of abstraction or classes. Table 8.1 provides a listing. We will describe each of these system elements later in their respective sections.

Author's Note 8.2 *Due to the unique identities of the System Elements, this text CAPITALIZES all instances of each term to facilitate easy recognition in our discussions.*

Importance of the System Elements Concept

The *System Elements* concept is important for three reasons. First, the system elements enable us to organize, classify, and bound system entity abstractions and their interactions. That is, it is a way to differentiate *what is* and *what is not* included in the system. Second, the System Element Architecture establishes a common framework for developing the *logical* and *physical* system architec-

Table 8.1 Identification of system element classes by domain

Domain	System Element Classes	Apply to
SYSTEM OF INTEREST (SOI)	• PERSONNEL • FACILITIES • EQUIPMENT • MISSION RESOURCES • PROCEDURAL DATA • SYSTEM RESPONSES	• MISSION SYSTEM role • SUPPORT SYSTEM role
OPERATING ENVIRONMENT	• ROLES and MISSIONS • ORGANIZATIONAL STRUCTURE • OPERATING CONSTRAINTS • RESOURCES	HIGHER ORDER SYSTEMS
	• HUMAN-MADE SYSTEMS • NATURAL ENVIRONMENT • INDUCED ENVIRONMENT	PHYSICAL ENVIRONMENT

tures of each entity within the system hierarchy. Third, the system elements serve as an initial starting point for allocations of multi-level performance specification requirements.

Despite strong technical and analytical skills, engineers are sometimes poor organizers of information. Therein lies a fundamental problem for the *engineering of systems*. Being able to *understand* and frame/structure the problem is 50% of the solution. The organizational framework of the System Element Architecture concept provides the framework for defining the system and its boundaries.

The challenge in analyzing and solving system development and engineering problems is being able to *identify, organize, define*, and *articulate* the relevant elements of a problem (objectives, initial conditions, assumptions, etc.) in an easy-to-understand, intelligible manner that enables us to *conceptualize* and *formulate* the solution strategy. Establishing a standard analytical framework enables us to apply "plug and chug" mathematical and scientific principles, the core strength of engineering training, to the architecture of the system.

Problem Solving to Reduce Complexity

System analysis and engineering is deeply rooted in the concept of analytically decomposing large, complex problems into manageable problems that can be easily solved. Unfortunately, many engineers lack the training to be able to organize, structure, and analyze a system problem around its *system elements*.

In practical terms, you should ingrain this concept as a basis for organizing and structuring relevant parts your problem. However, a word of caution: *Avoid temptation to tailor out relevant system elements without supporting rationale that can withstand professional scrutiny*. You may pay a penalty in overlooked system design issues.

For now, we have one remaining system element concept to introduce—entity relationships (ERs).

8.4 UNDERSTANDING SYSTEM ELEMENT ENTITY RELATIONSHIPS

The architectural concept discussions that follow describe the entity relationships (ERs) between each of the System Elements identified in Table 8.1. Before we begin these discussions, let's introduce the types of relationships that exist between these elements.

System element *interactions* can be characterized by two types of relationships: *logical* and *physical*. Perhaps the best way to think of *logical* and *physical* relationships is to focus on one topic at a time and then integrate the two concepts.

Logical Entity Relationships

The *first step* in identifying *logical entity relationships* is to simply *recognize* and *acknowledge* that some form of *association* exists through deductive reasoning. You may not know the physical details of the relationship—that is, *how* they link up—but you know a relationship does or will exist. Graphically, we depict these relationships as simply a line between the two entities.

The *second step* is to characterize the *logical* relationship in terms of logical functions—that is, *what* interaction occurs between them—must be provided to enable the two entities to associate with one another. When we assemble the logical entities into a framework that graphically describes their relationships, we refer to the diagram as *logical architecture*. To illustrate, let's assume we have a simple room lighting situation as shown in Figure 8.3.

"What Logical Association Exists Between Two System Entities"

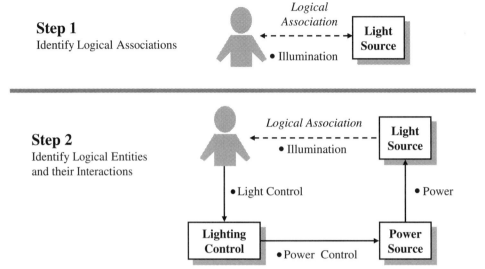

Figure 8.3 Logical Architecture Example

EXAMPLE 8.1

Room Lighting—Logical Architecture Entity Relationships The top portion of Figure 8.3 depicts a simple ROOM LIGHTING SYSTEM consisting of a PERSON (entity) desiring to control a room LIGHT SOURCE (entity). As a *logical representation*, we draw a line between the PERSON (entity) and the LIGHT SOURCE (entity) to acknowledge the relationship. Thus, we state that the PERSON (entity) has a *logical association* or *entity relationship* with the LIGHT SOURCE.

Author's Note 8.3 *Observe that we are interested in simply establishing and acknowledging the need for the logical relationship—meaning capability. The need for the capability serves as the basis for a specification requirement—meaning WHAT—and HOW much illumination is required. HOW this logical entity relationship is physically implemented via design and components becomes the basis for engineering analysis and design—with the application of mathematical and scientific principles.*

Next, we need a control mechanism for the LIGHT SOURCE (logical entity), which derives its energy from a POWER SOURCE (logical entity). We complete the *representation* by connecting the PERSON (logical entity) with the LIGHTING CONTROL (logical entity). The LIGHTING CONTROL enables the flow of current from the POWER SOURCE to the LIGHT SOURCE. When energized, the LIGHT SOURCE illuminates the room and the PERSON.

From this description you should note that we purposely avoided specifying HOW the:

1. PERSON (logical entity) *interfaced* with the LIGHTING CONTROL (logical entity).
2. LIGHTING CONTROL (logical entity) *controlled* the POWER SOURCE (logical entity).
3. POWER SOURCE (logical entity) provided current to the LIGHT SOURCE (logical entity).
4. LIGHT SOURCE (logical entity) illuminated the PERSON (logical entity).

The diagram simply documents associative relationships. Additionally, we avoided specifying what mechanisms were used for the LIGHTING CONTROL, POWER SOURCE, or LIGHT SOURCE. These decisions will be deferred to our next topic.

Based on this logical representation, let's investigate the physical implementation of the ROOM LIGHTING SYSTEM.

Physical Entity Relationships

The physical implementation of system element interfaces requires more in-depth analysis and decision making. *Why*? Typically, cost, schedule, technology, support, and risk become key drivers that must be "in balance" for the actual implementation. Since there should be a number of *viable candidate options* available for implementing an interaction, trade studies may be required to select the best selection and configuration of physical components. Graphically, we refer to the *physical* implementation of an interface as a *physical representation*.

As we select components (copper wire, light switches, lighting fixtures, etc.), we configure them into a system block diagram (SBD) and electrical schematics that depict the *physical* relationships. These diagrams become the basis for the *Physical System Architecture*. To illustrate a physical architecture depicting physical entity relationships, let's continue with our previous example.

EXAMPLE 8.2

Room Lighting—Physical Architecture Entity Relationships After some analysis we develop a physical representation or physical system architecture of the ROOM LIGHTING SYSTEM. As indicated by Figure 8.4, the system consists of the following physical entities: a POWER SOURCE, WIRE 1, WIRE 2, LIGHT SWITCH, a BUILDING STRUCTURE, a PERSON, and a LIGHT RECEPTACLE containing a LIGHT BULB. The solid black lines represent *electrical interfaces*; the dashed lines *represent* mechanical interfaces. In physical terms, the BUILDING STRUCTURE provides mechanical support for the LIGHT SWITCH, WIRE 1, WIRE 2, and LIGHT FIXTURE that holds the LIGHT BULB.

When the PERSON (physical entity) places the LIGHT SWITCH (physical entity) in the ON position, AC current (physical entity) flows from the POWER SOURCE (physical entity) through WIRE 1 (physical entity) to the LIGHT SWITCH (physical entity). The AC current (physical entity) flows from the LIGHT SWITCH (physical entity) through WIRE 2 (physical entity) to the LIGHT RECEPTACLE (physical entity) and into the LIGHT BULB. Visible light is then transmitted to the PERSON until the LIGHT SWITCH is placed in the OFF position, the LIGHT BULB burns out, or the POWER SOURCE is disconnected.

Logical and Physical Architecture Approach

The partitioning and sequencing of these discussions provides a fundamental portion of the methodology for developing systems, products, or services. If you *observe* and *analyze* human behavior, you will discover that humans characteristically have difficulty deciding WHAT decisions to make and the strategic steps required to make those decisions. We desire lots of information but are often unable to synthesize all of the data at the individual or team levels to arrive at an encompassing, multi-level design solution in a single decision. As a result, the ramifications of the decision-making process increases exponentially with the size and complexity of the system.

Given this characteristic, humans need to incrementally progress down a decision path from *simple, high-level* decisions to *lower level detail* decisions based on the higher level decisions. The flow from *logical* to *physical* entity relationships enables us to incrementally decompose complexity. Illustrations enable us to *progress* from simply acknowledging the existence of a relationship to detailed decisions regarding HOW the logical relationship can be physically implemented.

74 Chapter 8 The Architecture of Systems

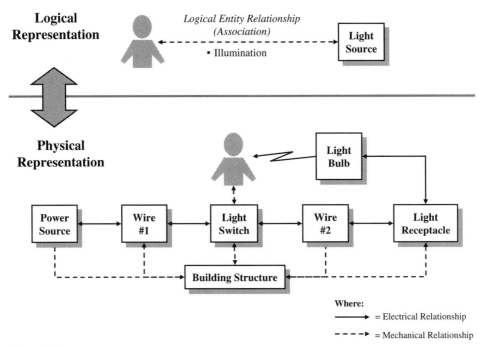

Figure 8.4 Logical-Physical Representations

As summarized in Figure 8.4, we evolve the *logical architecture representation* of the Room Lighting System from the *abstract* to the detailed physical architecture representation.

This point is important. It provides the basis for a later discussion when we introduce the concept of the *system solution domains* in Part II. The domain solutions include: 1) a Requirements Domain Solution, 2) an Operational Domain Solution, 3) a Behavioral Domain Solution, and 4) a Physical Domain Solution.

8.5 GUIDING PRINCIPLES

In summary, the preceding discussions provide the basis with which to establish the guiding principles that govern the architecture of systems.

Principle 8.1 System interact with external entities in their OPERATING ENVIRONMENT and themselves.

Principle 8.2 Every system serves at the pleasure of higher order, human and natural systems that exercise authority over the system and its operation.

Principle 8.3 Every system is part of a larger system of systems (SoS).

8.6 SUMMARY

Our discussion in this chapter introduced the fundamental concepts that form the basis for the system architecture. We introduced the concepts of the OPERATING ENVIRONMENT, SYSTEM OF INTEREST (SOI),

MISSION SYSTEM, and SUPPORT SYSTEM, and their interactions. We also decomposed each of these entities into classes or sets of *systems elements*.

The next chapter on system architecture levels of abstraction and semantics complements the discussion of this chapter by introducing the way system elements expand into lower level abstractions. In the two subsequent chapters we will discuss the SYSTEM OF INTEREST (SOI) architecture and the OPERATING ENVIRONMENT architecture, and identify system elements within their respective abstractions and describe those system elements in terms of the SOI and OPERATING ENVIRONMENT architectures.

GENERAL EXERCISES

1. Answer each of the *What You Should Learn from This Chapter* questions identified in the *Introduction*.
2. Refer to the list of systems identified in Chapter 2. Based on a selection from the preceding chapter's General *Exercises* or a new system selection, apply your knowledge derived from this chapter's topical discussions.
   (a) Identify and bound the SOI's OPERATING ENVIRONMENT
   (b) Identify and bound the HIGHER ORDER SYSTEMS, PHYSICAL SYSTEMS ENVIRONMENT, MISSION SYSTEM, and SUPPORT SYSTEM

ORGANIZATIONAL CENTRIC EXERCISES

1. Contact a system development program and investigate how their SOI interfaces with HIGHER ORDER SYSTEMS, PHYSICAL SYSTEMS ENVIRONMENT, and SUPPORT SYSTEM. How are these elements addressed in their system architecture diagrams? Report on your findings and observations.

REFERENCES

KOSSIAKOFF, ALEXANDER, and SWEET, WILLIAM N. 2003. *Systems Engineering Principles and Practice*. New York: Wiley-InterScience.

DoD 5000.59-M. 1998. *DoD Modeling and Simulation (M&S) Glossary* Washington, DC: Department of Defense.

ADDITIONAL READING

FM 770-78. 1979. *System Engineering* Field Manual. Washington, DC: Headquarters—Department of the Army.

Chapter 9

System Levels of Abstraction and Semantics

9.1 INTRODUCTION

Every natural and human-made system is part of a HIGHER ORDER SYSTEM. The universe, for example, can be viewed as a *hierarchy of systems*. Any system within that hierarchy is composed of lower level systems. As such, we refer to it as a system of systems (SoS). Systems within the hierarchy range from infinitely large, complex systems that exceed human comprehension and knowledge down to the smallest instance of physical *matter*.

When humans, especially system analysts and SEs, attempt to *communicate* about systems within the hierarchy of systems, the *context* of their *viewpoint* and *semantics* becomes a critical communications issue. Despite the breadth of the English language in terms of words, those applicable to the *engineering of systems* are finitely limited. Thus, when we attempt to apply a limited set of semantics to large numbers of system levels, confusion ultimately results.

A common comment that echoes throughout engineering development organizations is *Whose system are you referring to?* The question surfaces during conversations among Users, the Acquirer, and System Developers. Engineering organizations grapple with trying to understand the context of each person's *semantics* and *viewpoint* of the system. The problem is exacerbated by a mixture of SEs with varying degrees of semantics knowledge derived from: 1) on the job training (OJT), 2) personal study, 3) brochureware, and 4) formal training.

This section introduces the concept of *system levels of abstraction* that form the *semantics frame of reference* used in this text. We define the context of each level of abstraction. Given that system size and complexity vary from system to system, we describe how to tailor these levels to your SYSTEM OF INTEREST (SOI). We conclude with some guidelines that govern system decomposition and integration.

What You Should Learn from This Chapter

- What is an abstraction?
- For a six-level system, what are its levels of abstraction?
- How do you tailor the levels of abstraction to fit your system?
- Describe the scope and boundaries of a Level 0 or Tier 0 System?
- Describe the scope and entity relationships of a Level 1 or Tier 1 System?
- Describe the scope and entity relationships of a Level 2 or Tier 2 System?

- Describe the scope and entity relationships of a Level 3 or Tier 3 System?
- Describe the scope and entity relationships of a Level 4 or Tier 4 System?
- Describe the scope and entity relationships of a Level 5 or Tier 5 System?
- Describe the scope and entity relationships of a Level 6 or Tier 6 System?
- Describe the scope and entity relationships of a Level 7 or Tier 7 System?
- For an eight-level system, identify entity relationships between the various levels of abstraction.
- For an eight-level system, identify the entity relationships for system integration.

Definitions of Key Terms

- **Product Structure** The hierarchical structure of a physical system that represents decompositional relationships of physical entities. A top assembly drawing, specification tree, and a bill of materials (BOM) are primary documents for describing a SYSTEM's product structure.

9.2 ESTABLISHING A SEMANTICS FRAME OF REFERENCE

One of your first tasks as a system analyst or SE is to establish a semantics frame of reference for your SYSTEM OF INTEREST (SOI). When most people refer to systems, they communicate about a system from their own observer's *frame of reference* of everyday work tasks. When you listen to communications between Users, the Acquirer, and System Developers, you soon discover that one person's SYSTEM equates to another person's SUBSYSTEM, and so forth.

The System Context and Integration Points

When a system is specified and procured, one of the key issues is to understand the SYSTEM OF INTEREST (SOI) or *deliverable system's* context within the User's OPERATING ENVIRONMENT. From a hierarchical system of systems (SoS) context, a System Developer's contract *deliverable system* is an *entity* within the User's HIGHER ORDER SYSTEM abstraction. We refer to the location within the HIGHER ORDER SYSTEM where the *deliverable system* is integrated as its *Integration Point (IP)*.

Who Is the System's User?

Part of the challenge in defining the SYSTEM OF INTEREST (SOI) is identifying the User(s). Some systems have both *direct* and *indirect* Users. Consider the following example:

EXAMPLE 9.1

A computer system may have several Users. These include:

1. The day-to-day operator of the computer system.
2. Maintenance personnel.
3. Personnel who receive *work products* generated by the computer system.
4. Trainers who provide hands-on instruction to the operators.
5. Electronic mail recipients.

So, *when you say you are developing a system for the "User," to which user are you referring*?

Solving the Semantics Communication Problem

One of the ways to alleviate this problem is to establish a *standard set of semantics* that enable SEs from all disciplines to communicate intelligibly using a common contextual language. We need to establish a *standard semantics convention*. Once the convention is established, update the appropriate command media—meaning the organizational policies and procedures for use in training organizational personnel.

9.3 UNDERSTANDING SYSTEM LEVELS OF ABSTRACTION

When we address the *deliverable system* context, we need a standard way of communicating the embedded levels of abstraction. Since all systems are hierarchical, the User's system may be a supporting element of HIGHER ORDER SYSTEM. Given the large number of direct and indirect Users of the system, how can we establish a simple method of communicating these levels of abstraction? First, let's define the term:

Author's Note 9.1 *On the surface, you may view the discussion in this section as academic and esoteric. The reality is you, your customer (the User, Acquirer, etc.), and vendors must reach a common consensus as to WHAT IS and WHAT IS NOT part of the SOI or deliverable system. Does your contract or agreement explicitly delineate boundaries of the deliverable system? Ultimately, when the system is verified and validated, you do not want any contract conflicts as to WHAT IS or IS NOT included in the deliverable system.*

In the contracts world, objective evidence such as a Statement of Objectives (SOOs), System Performance Specification (SPS), Contract Work Breakdown Structures (CWBSs), and Terms and Conditions (Ts&Cs) serve as mechanisms for documenting the understanding between both parties regarding the system's boundaries and work to be accomplished relative to those boundaries. As such, they provide the frame of reference for settling programmatic and technical issues. In general, undocumented intentions and assumptions about a system's boundaries are unacceptable.

How the term *abstraction* applies to systems is illustrated in Figure 9.1. Beginning in the upper left corner, a system, product, or service consists of an initial set of loosely coupled entities such as ideas, objectives, concepts, and parts (i.e., items *A* through *N*).

If we analyze these *entities* or *objects*, we may determine that various groupings may share a common set of objectives, characteristics, outcomes, etc. as illustrated in the lower left portion of the figure. We identify several groupings of items:

1. Entity 10 consists of Entities *A* and *E*.
2. Entity 20 consists of Entities *C*, *F*, and *I*.
3. Entity 30 consists of Entities *D*, *J*, *H*, and *M*.
4. Entity 40 consists of Entities *B*, *K*, *L*, *N*, and *O*.

Each e*ntity* represents a *class* of object or *abstraction*. *Abstractions* or classes of objects are actually hierarchical groupings that suppress lower level details as illustrated by the right side of the Figure 9.1. Here we have a structure that *represents* the hierarchical structure, or taxonomy, of a system and each of its levels of abstraction.

The concept of generic levels of abstraction is useful information for simple systems. However, large, complex systems involve multiple levels of detail or abstraction. Where this is the case, *How do system analysts and SEs delineate one level of abstraction from another*? They do this by estab-

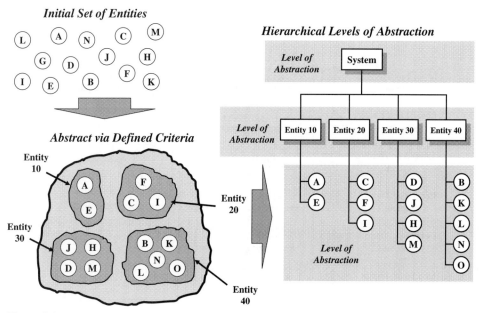

Figure 9.1 Definition of Abstraction

lishing an observer's *frame of reference* convention. In a contractual context, the contract establishes the basis.

Establishing a Frame of Reference Convention

Some organizations establish a contextual frame of reference convention for a system to facilitate communications about specific entities within the system. The intent is to designate a reference point for the *deliverable system*. One example convention employs Level 0, Level 1, Level 2, and so forth semantics as depicted in Figure 9.2. Another convention employs Tier 0, Tier 1, and so forth semantics. However, Levels or Tiers 0 through X are simply identifiers that need more explicit nomenclature labels. Table 9.1 provides an example nomenclature naming convention and a brief, scoping definition.

The key point of this discussion is for you and your colleagues to establish a semantics convention (Level 0/Tier 0, Level 1/Tier 1, etc.) that enables the team to communicate with a common *frame of reference* relative to the User's system—i.e. Level 0 or Tier 0. Based on an understanding of the *system levels of abstraction*, let's explore *how* we define and depict these graphically.

Relating System Levels of Abstraction and Semantics to System Architecture

The preceding discussion describes each level and entity in terms of its hierarchical and peer level entity relationships. These relationships provide the basic framework for defining the system *logical* and *physical architectures* discussed later in Part II on *System Design and Development Practices*.

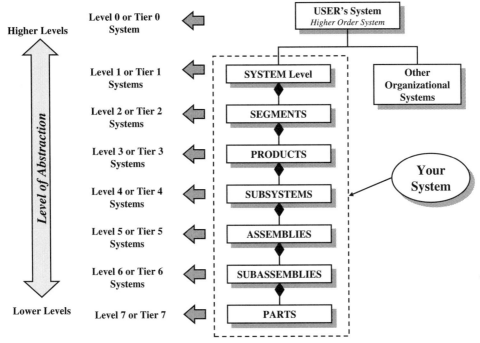

Figure 9.2 System Levels of Abstraction

Tailoring Levels of Abstraction for Your System's Application

The preceding discussion also introduced a set of *semantics* for application to large, complex systems. You and your organization may or may not have an eight-level system. Tailor the number of *system levels of abstraction* to match your system's application.

To illustrate this point, Figure 9.3 presents a tailored application of the standard system levels. The left side of the figure represents the standard system levels; the right side represents an organization's tailoring of the standard system levels. In this case the organization has adopted the following semantics: User system, SYSTEM, SUBSYSTEM, ASSEMBLY, and PART levels. As a result reference level numbers. (Level 1, Level 2, etc.) have been sequentially applied to match the tailoring.

9.4 SYSTEM DECOMPOSITION AND INTEGRATION DESIGN GUIDELINES

System structures are viewed from two SE perspectives:

1. Analytically, as a top-down, hierarchical decomposition or expansion.
2. Physically, as bottom-up, vertically integrated sets of entities.

System composition entity relationships (ERs) enable us to analytically decompose hierarchical systems into manageable design levels of complexity. Figure 9.4 provides a framework for the rules stated in Table 9.2.

Table 9.1 System levels of abstraction descriptions

Level or Tier	Nomenclature (Optional)	Scoping Definition
0	User's SYSTEM Level	A *level of abstraction* that represents the User's business environment with your SYTEM OF INTEREST (SOI) as an embedded element. We refer to this as the Level 0 or Tier 0 system.
1	SYSTEM Level	A *level of abstraction* that describes on the top-level *representation* of your SOI from the organization's (observer's) frame of reference. *Architectural representations* (i.e., an architectural block diagram, ABD) of the SYSTEM level include the SOI boundaries, interfaces to external systems in the operating environment, and embedded SEGMENT level entities (if applicable) or PRODUCT level entities and their interfaces. This level of abstraction is referred to as a Level 1 or Tier 1 system.
2	SEGMENT Level (optional)	Refers to system entities at the first level of decomposition below the SYSTEM level. Each instance of a SEGMENT level entity is referred to as a Level 2 or Tier 2 system. *An architectural representation* of a SEGMENT level entity includes: 1. Level 3 or Tier 3 PRODUCTS and lower level entities. 2. Their internal PRODUCT level relationships. 3. The SEGMENT's relationships with external entities within the SYSTEM (i.e., with other SEGMENTS) and external systems beyond the SYSTEM's boundaries, as applicable. Consider the following example: **EXAMPLE 9.2** A communications SYSTEM might consist of land-based, sea-based, air-based, and space-based SEGMENTS.
3	PRODUCT Level (optional)	Refers to system entities at the first level of decomposition below the SEGMENT level. Each instance of a PRODUCT Level entity is referred to as a Level 3 or Tier 3 system. *An architectural representation* of a PRODUCT level entity includes: 1. Level 4 or Tier 4 SUBSYSTEMS and lower level entities. 2. Their internal SUBSYSTEM level entity relationships. 3. The PRODUCT's relationships with external entities within the SYSTEM (i.e., with other PRODUCTS) and external systems beyond the SYSTEM's boundaries, as applicable.
4	SUBSYSTEM Level	Refers to system entities at the first level of decomposition below the PRODUCT Level. Each instance of a SUBSYSTEM level entity is referred to as a Level 4 or Tier 4 system. *An architectural representation* of a SUBSYSTEM level entity includes: 1. Level 5 or Tier 5 ASSEMBLIES and lower level entities. 2. Their internal ASSEMBLY level entity relationships. 3. The SUBSYSTEM's relationships with *external entities* within the SYSTEM (i.e., with other SUBSYSTEMS) or *external systems* beyond the SYSTEM's boundaries, as applicable.
5	ASSEMBLY Level	Refers to system entities at the first level of decomposition below the SUBSYSTEM level. Each instance of an ASSEMBLY level entity is referred to as a Level 5 or Tier 5 system. *An architectural representation* of an ASSEMBLY level entity includes:

(continued)

Table 9.1 *continued*

Level or Tier	Nomenclature (Optional)	Scoping Definition
		1. Level 6 or Tier 6 COMPONENTS and lower level entities. 2. Their internal SUBASSEMBLY level entity relationships. 3. The ASSEMBLY's relationships with *external entities* within the SYSTEM (i.e., with other ASSEMBLIES) or *external systems* beyond the SYSTEM's boundaries, as applicable.
6	SUBASSEMBLY Level	Refers to system entities at the first level of decomposition below the ASSEMBLY level. Each instance of a SUBASSEMBLY level entity is referred to as a Level 6 or Tier 6 system. *Architectural representation* of a SUBASSEMBLY Level entity includes: 1. Level 7 or Tier 7 PARTS. 2. Their internal PART level entity relationships. 3. The SUBASSEMBLY's relationships with *external entities* within the SYSTEM (i.e., with other SUBASSEMBLIES) or *external systems* beyond the SYSTEM's boundaries, as applicable.
7	PART Level	Refers to the lowest level decompositional element of a system. An *architectural representation* of a PART includes form factor envelope drawings, schematics, and models. **Author's Note 9.2** *Engineering drawings, which are produced at all system levels of abstraction, consist of two basic types:* 1. *Dimensional drawings or schematics for internally developed or externally procured and modified internally items.* 2. *Source control drawings that bound part parameters and characteristics for externally procured parts.* For software systems, the PART level equates to a *source line of code* (SLOC).

Author's Note 9.3 *Hierarchical decomposition follows the same rules as outlining a report. Avoid having a single item subordinated to a higher level item. Good design practice suggests that you should always have at least two or more entities at a subordinated level. In the case of systems this does not mean that the subordinate entities must be at the same level of abstraction.*

For example, the Bill of Materials (BOM)—meaning a parts list—for a top level assembly within a product structure may consist of at least one or more SUBSYSTEMS, one or more ASSEMBLIES, and at least one or more PARTS kit (e.g., nuts and bolts). If we depict the BOM as an indentured list, the PARTS kit used to mechanically and electrically connect the SUBSYSTEMS and ASSEMBLIES into an INTEGRATED SYSTEM may be at the same level of abstraction as the SUBSYSTEMS and ASSEMBLIES.

As a final note, this text treats entities at any level of abstraction as *components* of the system.

9.5 SUMMARY

Our discussion in this chapter has introduced the hierarchical concept of system levels of abstraction and semantics. This hierarchical framework enables SEs to standardize analysis and communications about their SYSTEM OF INTEREST (SOI). The intent of a *semantics convention* is to synchronize members of a System Developer's team, the Acquirer, and the User on a common set of terms to use in communicating complex hierarchies.

9.5 Summary   83

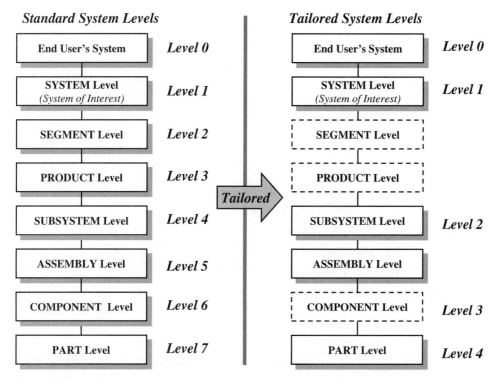

Figure 9.3 System Levels of Abstraction and Tailoring Example

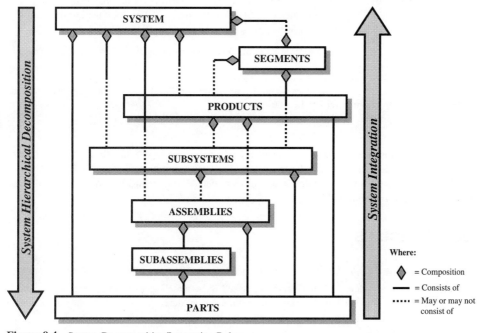

Figure 9.4 System Decomposition/Integration Rules

Table 9.2 System entity decomposition and integration rules

Level or Tier	Entity	Entity Decomposition/Integration rules
0	User Level	The User's SYSTEM is bounded by its *organizational mission* and consists of the system element assets required to accomplish that mission within its OPERATING ENVIRONMENT.
1	SYSTEM Level	Each instance of a SYSTEM consists of at least two or more instances of SEGMENT, PRODUCT, SUBSYSTEM, ASSEMBLY, SUBASSEMBLY, or PART level entities or combinations thereof.
2	SEGMENT Level	If the SEGMENT level of *abstraction* or *class* is applicable, each SEGMENT level entity consists of at least two or more instances of PRODUCT, SUBSYSTEM, ASSEMBLY, SUBASSEMBLY, or PART level entities or combinations thereof.
3	PRODUCT Level	If the PRODUCT level of *abstraction* or *class* is applicable, each instance of a PRODUCT level entity consists of at least two or more instances of SUBSYSTEM, ASSEMBLY, SUBASSEMBLY, or PART level entities or combinations thereof.
4	SUBSYSTEM Level	If the SUBSYSTEM level of *abstraction* or *class* is applicable, each instance of a SUBSYSTEM level entity consists of at least two or more instances of ASSEMBLY, SUBASSEMBLY, or PART level entities or combinations thereof.
5	ASSEMBLY Level	If the ASSEMBLY level of *abstraction* or *class* is applicable, each instance of an ASSEMBLY level entity consists of at least two or more instances of SUBASSEMBLY or PART level entities or combinations thereof.
6	SUBASSEMBLY Level	If the SUBASSEMBLY level of *abstraction* or *class* is applicable, each instance of a SUBASSEMBLY level entity must consist of at least two or more instances of PART level entities.
7	PART Level	The PART level is the lowest decompositional element of a system.

Note: For *hierarchical decomposition*, read the table top-down from Level 0 to Level 7 order; for *integration*, read the table bottom-up in *reverse order*—from Level 7 to Level 0.

As part of our discussion, we provided a mechanism for System Developers that allows them to benchmark their system in the context of the User's larger system. This was accomplished using the Level 0 or Tier 0 convention. This convention, or whatever you and your team decide to use, must answer the central question, *Whose system are you referring to?*

GENERAL EXERCISES

1. Answer each of the *What You Should Learn from This Chapter* questions identified in the *Introduction*.
2. Refer to the list of systems identified in Chapter 2. Based on a selection from the preceding chapter's General *Exercises* or a new selection, apply your knowledge derived from this chapter's topical discussions.
   (a) Equate *system levels of abstraction* to multi-level entities of the physical system selected.
   (b) Develop an entity relationships diagram that identifies physical entity relationships.

ORGANIZATION CENTRIC EXERCISES

1. Contact a program organization and investigate what levels of abstraction and semantics are used.
2. Research your organization's command media for guidance in identifying levels of abstraction.
   (a) What guidance, if any, is provided?
   (b) How does each program establish its own guidance?

Chapter 10

The System of Interest Architecture

10.1 INTRODUCTION

The central focal point for the development of any system is the SYSTEM OF INTEREST (SOI) consisting of the MISSION SYSTEM(s) and its SUPPORT SYSTEM(s). Since every system within the supply chain performs two roles—MISSION SYSTEM and SUPPORT SYSTEM—analysis reveals that each role consists of seven classes of *system elements* that form its architectural framework.

This section introduces the concept of SOI system elements. We identify the *system elements* and describe the System Element Architecture that forms the analytical basis for analyzing systems and their interactions with their OPERATING ENVIRONMENT. We describe the scope and bound the contents of each element.

What You Should Learn from This Chapter

- *What is an SOI?*
- *What are the key elements of an SOI?*
- *Graphically depict the generalized system architecture of an SOI including its external interfaces.*
- *What are the key elements of a MISSION SYSTEM?*
- *Graphically depict the generalized system architecture of a MISSION SYSTEM including its external interfaces.*
- *What is the scope of the PERSONNEL, EQUIPMENT, MISSION RESOURCES, PROCEDURAL DATA, FACILITIES, and SYSTEM RESPONSES elements?*
- *What are the key elements of the EQUIPMENT Element?*
- *Why is the SOFTWARE Element separate from the EQUIPMENT Element?*
- *What are the key elements of the SUPPORT SYSTEM?*
- *Graphically depict the generalized system architecture of a SUPPORT SYSTEM including its external interfaces.*
- *How does the MISSION SYSTEM architecture differ from the SUPPORT SYSTEM architecture?*
- *What are the key elements of a SUPPORT SYSTEM's EQUIPMENT Element?*
- *What are CSE and PSE?*

System Analysis, Design, and Development, by Charles S. Wasson
Copyright © 2006 by John Wiley & Sons, Inc.

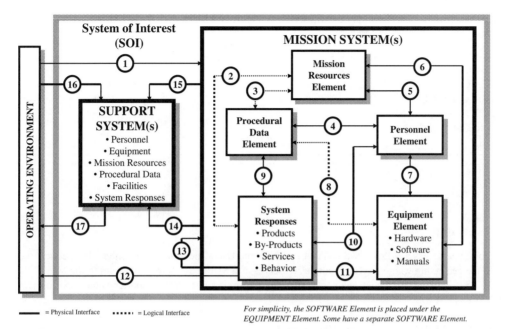

Figure 10.1 The System of Interest (SOI) Architecture and its System Elements

10.2 THE SYSTEM ELEMENT ARCHITECTURE CONSTRUCT

Every human-made system consists of a generalized framework we refer to as the *System Element Architecture (SEA)*. The SEA represents a logical arrangement of *system elements* that serve as generalized construct or template for systems design and analysis. Figure 10.1 provides a graphical representation of the SEA. To promote readability and simplicity in the figure, the OPERATING ENVIRONMENT is abstracted as a single entity.

Every system performs MISSION SYSTEM and SUPPORT SYSTEM roles as part of its tasking from HIGHER ORDER SYSTEMS. Regardless of the system role, each system consists of combinations of *system elements* with specific system and mission objectives.

Mission System and Support System Compositional Elements

Each MISSION SYSTEM and SUPPORT SYSTEM consists of a unique set of integrated *system elements* that enable the system to accomplish its mission and objectives. The mission/support *system elements* are PERSONNEL, EQUIPMENT, SOFTWARE, MISSION RESOURCES, PROCEDURAL DATA, and SYSTEM RESPONSES. Table 10.1 relates each of these elements to the MISSION SYSTEM and SUPPORT SYSTEM roles.

Author's Note 10.1 *MISSION SYSTEMS such as aircraft, vehicles, etc. do not have a FACILITIES Element. In contrast, a home or an office owned by the occupants, as a MISSION SYSTEM, does have a FACILITIES Element.*

SOI Architectural System Elements

The System Element Architecture consists of analytical abstractions that *represent* the SYSTEM OF INTEREST (SOI) *interactions* with its OPERATING ENVIRONMENT. The SOI consists of the MISSION SYSTEM as an abstraction with interfaces to the SUPPORT SYSTEM.

Table 10.1 System elements common to MISSION SYSTEM and SUPPORT SYSTEM roles

System Element	MISSION SYSTEM Role	SUPPORT SYSTEM Role
PERSONNEL	•	•
EQUIPMENT	•	•
MISSION RESOURCES	•	•
PROCEDURAL DATA	•	•
SYSTEM RESPONSES	•	•
FACILITIES		•

EXAMPLE 10.1

A new automobile serves a MISSION SYSTEM (role) for the User. The transporter vehicle that delivers the new car to the dealership where the car is purchased serves as a SUPPORT SYSTEM (role) to the MISSION SYSTEM. To the organization that *owns* and *operates* the transporter, the transporter vehicle performs a MISSION SYSTEM role.

The PERSONNEL System Element

The PERSONNEL element 1) consists of all human roles required to perform the system mission operations in accordance with safe operating practices and procedures, and 2) has overall accountability for accomplishing mission objectives assigned by HIGHER ORDER SYSTEMS.

- **MISSION SYSTEM PERSONNEL Roles** Include all personnel directly required to operate the MISSION SYSTEM and accomplish its objectives. In general, these personnel are typically referred to as *System Operators*.
- **SUPPORT SYSTEM PERSONNEL Roles** Include personnel who support the MISSION SYSTEM through maintenance, supply support, training, publications, security, and other activities.

The EQUIPMENT System Element

The EQUIPMENT system element consists of any physical, multi-level, electromechanical optical device that represents an integration of the HARDWARE Element and the SOFTWARE Element, if applicable. This integration of elements is:

1. Developed and/or procured to satisfy a system entity capability and performance requirement.
2. Used to operate and maintain the system.
3. Used to generate or store energy required by the system.
4. Used to dispose of a system.

The ultimate success of the MISSION SYSTEM requires that the EQUIPMENT Element be *operationally available* and fully capable of supporting the system missions and the safety of its PERSONNEL to ensure a *level of success*. As a result specialty engineering (reliability, availability, maintainability, vulnerability, survivability, safety, human factors, etc.) becomes a key focus of the EQUIPMENT Element. Depending on the application, EQUIPMENT may be *fixed*, *transportable*, or *mobile*.

The HARDWARE System Element

The Hardware Element represents the integrated set of multi-level, physical components—mechanical, electrical/electronic, or optical *less* software—configured in accordance with the system architecture. Whereas the Hardware Element is common to both the MISSION SYSTEM and the SUPPORT SYSTEM, there are difference in the classes of components. MISSION SYSTEM hardware components are physically integrated to provide the capabilities required to accomplish mission objectives. SUPPORT SYSTEM hardware components consist of tools required to maintain and support the MISSION SYSTEM. These tools are categorized as: 1) *Common Support Equipment (CSE)* and 2) *Peculiar Support Equipment (PSE)*.

Common Support EQUIPMENT (CSE). Common support equipment (CSE) consists of the items required to support and maintain the system or portions of the system while not directly engaged in the performance of its mission. CSE items are in the organization's inventory for support of other systems. CSE excludes:

- Overall planning, management and task analysis functions inherent in the work breakdown structure element, Systems Engineering/Program Management.
- Common support equipment, presently in the inventory or commercially available, bought by the User, not by the Acquirer.

Peculiar Support EQUIPMENT (PSE). MIL-HDBK-881 characterizes peculiar support equipment (PSE) as follows:

> ... the design, development, and production of those deliverable items and associated software required to support and maintain the system or portions of the system while the system is not directly engaged in the performance of its mission, and which are not common support equipment (CSE).
>
> (Source: Mil-HDBK-881, Appendix H, para. 3.6).

PSE includes:

- *Vehicles, equipment, tools, etc., used to fuel, service, transport, hoist, repair, overhaul, assemble, disassemble, test, inspect, or otherwise maintain mission equipment.*
- *Any production of duplicate or modified factory test or tooling equipment delivered to the (Acquirer) for use in maintaining the system. (Factory test and tooling equipment initially used by the contractor in the production process but subsequently delivered to the (Acquirer) will be included as cost of the item produced.)*
- *Any additional equipment or software required to maintain or modify the software portions of the system.*

> (Source: Mil-HDBK-881, Appendix H, para. 3.6).

CSE and PCE Components

Common support equipment (CSE) and peculiar support equipment (PSE) each employ two categories of equipment that are common to both types: 1) test, measurement, and diagnostics equipment (TMDE) and 2) support and handling equipment.

Test, Measurement, and Diagnostics Equipment (TMDE). MIL-HDBK-881 characterizes test, measurement, and diagnostics equipment (TMDE) as follows:

> ... consists of the peculiar or unique testing and measurement equipment which allows an operator or maintenance function to evaluate operational conditions of a system or equipment by performing

specific diagnostics, screening or quality assurance effort at an organizational, intermediate, or depot level of equipment support.
  TMDE, for example, includes:

- Test measurement and diagnostic equipment, precision measuring equipment, automatic test equipment, manual test equipment, automatic test systems, test program sets, appropriate interconnect devices, automated load modules, taps, and related software, firmware and support hardware (power supply equipment, etc.) used at all levels of maintenance.
- Packages which enable line or shop replaceable units, printed circuit boards, or similar items to be diagnosed using automatic test equipment.

(Source: Mil-HDBK-881, Appendix H, para. 3.7.1)

Support and Handling EQUIPMENT. Support and handling EQUIPMENT consists of the deliverable tools and handling equipment used for support of the MISSION SYSTEM. This includes "... *ground support equipment (GSE), vehicular support equipment, powered support equipment, unpowered support equipment, munitions material handling equipment, materiel-handling equipment, and software support equipment (hardware and software).*" (Source: Mil-HDBK-881, Appendix H, para. 3.7.2)

The SOFTWARE System Element

The SOFTWARE element consists of all software code (source, object, etc.) and documentation required for installation, execution, and maintenance of the EQUIPMENT Element. You may ask *why* some organizations separate the SOFTWARE Element from its EQUIPMENT element. There are several reasons:

1. EQUIPMENT and SOFTWARE may be developed separately or procured from different vendors.
2. SOFTWARE may provide the flexibility to alter system capabilities and performance (decision-making, behavior, etc.) *without* having to *physically modify* the EQUIPMENT, assuming the current EQUIPMENT design is adequate.

Author's Note 10.2 *The EQUIPMENT element consists of integrated HARDWARE and SOFTWARE as subordinated and supporting elements. Engineers become prematurely focused with hardware and software details long before higher level SOI decisions have been made—namely EQUIPMENT requirements. EQUIPMENT decisions lead to lower level HARDWARE and SOFTWARE use cases and decisions. These decisions subsequently lead to the question: What capabilities should be implemented in HARDWARE versus those implemented in SOFTWARE?*

SE logic says that HARDWARE and SOFTWARE may be separately procurable items. The underlying philosophy is SOFTWARE, as a *system element*, should be isolated to accommodate modification without necessarily having to modify the HARDWARE.

Application specific SOFTWARE can be procured as a separate item, regardless of its position within the system structure as long as a controlled *software requirements specification* (SRS) exists for the item's development. As new versions of application specific SOFTWARE are released, the User can procure the item without modifying the EQUIPMENT. There may be exceptions, however, where SOFTWARE requirements and priorities force the User to upgrade the computer HARDWARE capabilities and performance.

10.2 The System Element Architecture Construct

From \ To — System Element	Personnel	Equipment	Procedural Data	Mission Resources	System Responses	Facilities
Personnel	①	②	③	④	⑤	⑥
Equipment	⑦	⑧	⑨	⑩	⑪	⑫
Procedural Data	⑬	⑭	⑮	⑯	⑰	⑱
Mission Resources	⑲	⑳	㉑	㉒	㉓	㉔
System Responses	㉕	㉖	㉗	㉘	㉙	㉚
Facilities	㉛	㉜	㉝	㉞	㉟	㊱

Where: (X) = entity relationships and associated capabilities

Figure 10.2 System Element Entity Relationships Matrix

The PROCEDURAL DATA System Element

The PROCEDURAL DATA element consists of all documentation that specifies HOW to safely *operate*, *maintain*, *deploy*, and *store* the EQUIPMENT Element. In general, the PROCEDURAL DATA Element is based on operating procedures. *Operating procedures* document sequences of PERSONNEL actions required to ensure the proper and safe operation of the system to achieve its intended level of performance under specified operating conditions. The PROCEDURAL DATA Element includes items such as reference manuals, operator guides, standard operating practices and procedures (SOPPs), and checklists.

Author's Note 10.3 *Unfortunately, many people view checklists as bureaucratic nonsense, especially for organizational processes. Remember, checklists incorporate lessons learned and best practices that keep you out of trouble. Checklists are a state of mind—you can view them as "forcing" you to do something or as a "reminder" to "think about what you may have overlooked." As a colleague notes, when landing an aircraft, if the checklist says to "place the landing gear in the deployed and locked position," you may want to consider putting the landing gear down before you land! Bureaucratic or not, the consequences for a lack of compliance can be catastrophic!*

The MISSION RESOURCES System Element

The MISSION RESOURCES element includes all *data*, *consumables*, and *expendables* required on-board to support the system mission. This element consists of:

1. Data to enable the EQUIPMENT and the PERSONNEL Elements to successfully plan and conduct the mission based on "informed" decisions.
2. *Consumables* such as fuel and water to support the EQUIPMENT and PERSONNEL Elements during the mission.

3. *Expendables* such as personal and defensive systems and products to ensure a safe and secure mission.

4. Recorded data for postmission performance analysis and assessment.

Let's scope each of the four types of MISSION RESOURCES in detail.

- *Mission Data Resources* Consist of all real-time and non-real-time data resources such as tactical plans, mission event timelines (METs), navigational data, weather conditions and data, situational assessments, telecommunications, telemetry, and synchronized time that are necessary for performing a mission with a specified level of success.
- *Expendable Resources* Consist of physical entities that are used and discarded, deployed, lost, or destroyed during the course of mission operations.
- *Consumable Resources* Consist of physical entities that are ingested, devoured, or input into a processor that transforms or converts the entity into energy to power the system and may involve physical state changes.

The SYSTEM RESPONSES Element

Every natural and human-made system, as a stimulus-response mechanism, responds internally or externally to stimuli in its OPERATING ENVIRONMENT. The responses may be *explicit* (reports, communications, altered behavior, etc.) or *implicit* (mental thoughts, strategies, lessons learned, behavioral patterns, etc.).

SYSTEM RESPONSES occur in a variety of forms that we *characterize* as behavioral patterns, products, services, and by-products throughout the system's *pre-mission*, *mission*, and *post-mission* phases of operation. So, what do we mean by system behavior, products, services, and by-products?

- *System Behavior* Consists of SYSTEM RESPONSES based on a plan of action or physical stimuli and audiovisual cues such as threats or opportunities. The stimuli and cues invoke system behavioral patterns or actions that may be categorized as *aggressive*, *benign*, *defensive*, and everywhere in between. Behavioral actions include strategic and tactical tactics and countermeasures.
- *System Products* Include any type of *physical outputs*, *characteristics*, or *behavioral responses* to planned and unplanned events, external cues, or stimuli.
- *System By-products* Include any type of physical system output or behavioral that is not deemed to be a system, product, or service.
- *System Services* Any type of physical system behavior, excluding physical products, that *assist* another entity in the conduct of its mission.

Author's Note 10.4 *It is important to note here that abstracting "behavioral responses" via a "box" called SYSTEM RESPONSES is simply an analytical convenience—not reality. Most system element descriptions fail to recognize this box for what it is, expected outcome-based system performance such as operational utility, suitability, availability, and effectiveness.*

The SUPPORT SYSTEM Element

The SUPPORT SYSTEM consists of an integrated set of *system elements* depicted in Table 10.1 required to support the MISSION SYSTEM. The SUPPORT SYSTEM and its *system elements* perform *mission system support* operations that consist of the following:

10.2 The System Element Architecture Construct

- Decision support operations
- System maintenance operations
- Manpower and personnel operations
- Supply support operations
- Training and training support operations
- Technical data operations
- Computer resources support operations
- Facilities operations
- Packaging, handling, storage, and transportation (PHST) operations
- Publications support operations

Let's briefly describe each of these types of operations.

Decision Support Operations. Mission operations often require *critical* and *timely* decisions based on massive amounts of information that exceed the mental capabilities of the human decision maker. Decision support operations ensure that the decision maker always has immediate access to the most current, processed system mission information to support an informed decision.

System Maintenance Operations. System maintenance operations include system maintenance support concepts and requirements for manpower and personnel; supply support; support and test equipment; technical data; training and support; facilities; and packaging, handling, storage, transportation; computer resources support required insightful planning. *Maintenance support operations* establish support for both general system operations and mission specific operations throughout the System O&S Phase. Examples include software maintenance concepts, pre-planned product improvements (P^3I), outsourcing, and transition planning.

Between operational missions, maintenance personnel normally perform *corrective* and *preventative* maintenance on the system. Maintenance activities may be conducted in the field, at a depot or maintenance facility at some central location, or at the manufacturer's facility.

Generally, a system may be temporarily removed from active duty during maintenance. Types of maintenance include:

1. System upgrades, enhancements, and refinements.
2. Replenishment and refurbishment of system resources (fuel, training, etc.).
3. Diagnostic readiness tests.
4. Backup of system information for archival purposes.

Maintenance may also include investigations of mission or system anomalies through duplication of technical problem or investigations of system health and status resulting from abnormal operating conditions before, during, or after a mission. On completion of the maintenance activities, the system may be returned to a state of *operational readiness* and *active duty* or scheduled for system operator or readiness training.

Manpower and Personnel Operations. *Manpower and Personnel operations* consist of all personnel required to manage, operate, and support the system throughout its operational life. Manpower and personnel considerations include in-house versus contractor tasks, skill mixes, and human-system interfaces. When designing systems, systems engineering must factor in considerations of the skill levels of the available personnel, training costs, labor costs of operate, and leverage technology such as built-in diagnostics, information technology, and expert systems to provide the best life cycle cost trade-offs—given the skill levels.

Supply Support Operations. Supply support operations ensure that all equipment and spares required to support the primary system, system support, and test equipment and the "supply lines" for those items are established to support pre-mission, mission, and post-mission operations. Some organizations refer to the initial support as *provisioning* and follow-on requirements as *routine replenishment*. A key supply support objective is to *minimize* the number of different PARTS and promote *standardization* of parts selection and usage. Key issues to be addressed include communication transfer media, inventory management, configuration management, storage, security, and licensing.

Training and Training Support Operations. Training and training support operations ensure that all personnel required to support the pre-mission, mission, and post-mission operations are fully trained and supported. Personnel training consists of those activities required to prepare system personnel such as operators, maintenance personnel, and other support personnel to conduct or support pre-mission, mission, and postmission operations. Personnel must be trained with the appropriate skills to perform their assigned mission objectives and tasks on specific equipment to achieve the required level of performance and expected results.

Technical Data Operations. Pre-mission, mission, and post-mission operations require accurate, precise, and timely technical data. Technical data operations support personnel and computer equipment decision making to ensure that all decisions are made when scheduled or as appropriate to achieve the mission objectives, level of performance, and expected results.

Computer Resources Support Operations. Many systems are highly dependent on technology such as computer resources to provide information and processing of data to support pre-mission, mission, and post-mission operations. Computer resources support operations include: planning, procurement, upgrade, and maintenance support to ensure that system reliability, availability, and maintainability (RAM) requirements will be achieved in a cost-effective manner.

Packaging, Handling, Storage, and Transportation (PHST) Operations. During system deployment, redeployment, and storage, system support activities must provide the capability to *safely* and *securely* transport and store the system and its components. Shipment and storage must be accomplished to avoid damage or decomposition. PHST activities must have a clear understanding of the environmental conditions and characteristics that pose risk to the system while in transit as well as long-term shelf-life effects on materials.

Publications Support Operations. System maintenance and support personnel require immediate access to the most current information regarding system and system support equipment to ensure proper maintenance and usage. Publication support activities, sometimes referred to as Tech Pubs, produce technical manuals, reference guides, and the like, that enable SUPPORT SYSTEM personnel training activities, maintenance activities, and general operations in support of pre-mission, mission, and post-mission operations.

The FACILITIES System Element

The FACILITIES element includes all the system entities required to support and sustain the MISSION SYSTEM Elements and the SUPPORT SYSTEM Elements.

System support for *pre-mission*, *mission*, and *post-mission* operations requires FACILITIES to enable operator and support PERSONNEL to accomplish their assigned mission tasks and objectives in a reasonable work environment. Depending on the type of system, these FACILITIES support the following types of tasks:

1. Plan, conduct, and control the mission.
2. Provide decision support.
3. Brief and debrief personnel.
4. Store the physical system or product between missions.
5. Configure, repair, maintain, refurbish, and replenish system capabilities and resources.
6. Analyze post-mission data.

Where practical, the FACILITIES element should provide all of the *necessary* and *sufficient* capabilities and resources required to support MISSION SYSTEM and human activities during pre-mission, mission, and post-mission operations objectives. FACILITIES may be owned, leased, rented, or be provided on a limited one-time use agreement.

In general, people tend to think of FACILITIES as enclosures that provide shelter, warmth, and protection. This thought process is driven by human concerns for creature comfort rather than the mission system. Rhetorically speaking, *Does an aircraft ever know that it is sitting in the rain?* No, but members of the SUPPORT SYSTEM'S PERSONNEL Element are aware of the conditions.

Perhaps the best way to think of the FACILITY Element is to ask the question: *What type of interface is required to support the following MISSION SYSTEM Elements—namely EQUIPMENT Element, PERSONNEL Element, MISSION DATA Element, and RESOURCES Element—during all phases of the mission.*

The FACILITIES Element fulfills a portion of the SUPPORT SYSTEM role. Depending on the MISSION SYSTEM and its application, the FACILITIES Element may or may not be used in that context. For example, an aircraft—which is a MISSION SYSTEM—does not require a FACILITY to perform its primary mission. However, between primary missions, FACILITIES are required to maintain and prepare the aircraft for its next mission.

10.3 SYSTEM ELEMENT INTERACTIONS

Figure 10.1 illustrates the System Element Architecture (SEA) for a SOI. When you define your system and identify physical instances of each *system element*, the next step is to characterize the levels of *interactions* that occur between each interface. The Contract Work Breakdown Structure (CWBS) should include a *CWBS Dictionary* that *scopes* and *documents* the physical items included in each CWBS element such as EQUIPMENT.

One approach to ensuring *system element* interactions are identified and scoped is to create a simple matrix such as the one shown in Figure 10.2. For illustration purposes, each cell of the system element matrix represents interactions between the row and column elements. For your system, employ such a scheme and document the interactions in a description of the system architecture. Then, baseline and release this document to promote communications among team members developing and making *system element* decisions.

10.4 SUMMARY

In our discussion of the SYSTEM OF INTEREST (SOI) architecture we introduced the concept of the system elements, which are common to all human-made systems be they friendly, benign, or adversarial. The system elements common to the MISSION SYSTEM and SUPPORT SYSTEM consist of the following:

- PERSONNEL Element
- EQUIPMENT Element
- MISSION RESOURCES Element
- PROCEDURAL DATA Element
- SYSTEM RESPONSES Element
- FACILITIES Element

The FACILITIES Element is typically unique to the SUPPORT SYSTEM.

Our discussion included the System Element Architecture and illustrated *how* the system elements are integrated to establish the basic architectural framework for a MISSION SYSTEM or a SUPPORT SYSTEM. We are now ready to examine how the system interfaces with its OPERATING ENVIRONMENT and its elements.

GENERAL EXERCISES

1. Answer each of the *What You Should Learn from This Chapter* questions identified in the *Introduction*.
2. Refer to the list of systems identified in Chapter 2. Based on a selection from the preceding chapter's General *Exercises* or a new system selection, apply your knowledge derived from this chapter's topical discussions.
   (a) Equate system elements to real world components for each system.
   (b) Prepare a paper entitled "An Architectural Description of the [fill in from list in Chapter 2] System."

ORGANIZATIONAL CENTRIC EXERCISES

1. Contact a system development program within your organization. Research how the program documented and specified the SYSTEM OF INTEREST (SOI) architecture in the following *System Element* areas:
   (a) MISSION SYSTEM and SUPPORT SYSTEMs
   (b) PERSONNEL
   (c) MISSION RESOURCES
   (d) PROCEDURAL DATA
   (e) EQUIPMENT
   (f) SYSTEM RESPONSES
   (g) FACILITIES
2. For the system development program above, what lessons learned did technical management identify that were overlooked.

REFERENCES

DSMC. 1997. *Acquisition Logistics Guide*, 3rd ed. Ft. Belvoir, VA.

MIL-HDBK-881. 1998. *Work Breakdown Structure*. Department of Defense (DoD) Handbook. Washington, DC.

Chapter 11

The Operating Environment Architecture

11.1 INTRODUCTION

The success of any system is ultimately determined by its ability to:

1. Conduct its planned missions and achieve performance mission objectives.
2. Cope with threats—namely, *vulnerability* and *survivability*—within its prescribed OPERATING ENVIRONMENT.

For a system to accomplish these objectives, SE's face several major challenges.

1. Understand WHAT missions the User plans for the system to accomplish.
2. Based on the missions, specifically bound the system's OPERATING ENVIRONMENT.
3. Understand the OPERATING ENVIRONMENT *opportunities* and *threats* related to those missions.
4. Identify the *most likely* or *probable* OPERATING ENVIRONMENT *opportunity* and *threat* scenarios that influence/impact system missions.

The OPERATING ENVIRONMENT *represents* the totality of natural and human-made entities that a system must be prepare to cope with during missions and throughout its lifetime. As one of a system's key life expectancy dependencies, the system's ability to: 1) prepare for, 2) conduct, and 3) complete missions successfully is influenced by its OPERATING ENVIRONMENT.

What You Should Learn from This Chapter

1. *What are the two classes of OPERATING ENVIRONMENT domains?*
2. *What are the four classes of system elements of a system's HIGHER ORDER SYSTEMS domain?*
3. *What are the three classes of system elements of a system's PHYSICAL ENVIRONMENT domain?*
4. *What are the four classes of systems that comprise the NATURAL ENVIRONMENT domain?*
5. *How do you graphically depict the OPERATING ENVIRONMENT's architecture that includes detail interactions of the HIGHER ORDER SYSTEMS and PHYSICAL ENVIRONMENT domains?*

System Analysis, Design, and Development, by Charles S. Wasson
Copyright © 2006 by John Wiley & Sons, Inc.

11.2 APPLICATION OF THIS CONCEPT

A significant aspect of systems engineering (SE) is recognizing the need to fully *understand*, *analyze*, and *bound* relevant portions a system's OPERATING ENVIRONMENT. *Why*? System analysts and SEs must be capable of:

1. *Fully understanding* the User's organizational system mission, objectives, and most probable *use case* applications.
2. Analytically *organizing*, *extracting*, and *decomposing* the OPERATING ENVIRONMENT relative to the mission into manageable *problem space(s)* and *solution space(s)*.
3. Effectively developing a system solution that ensures mission success within the constraints of cost, schedule, technology, support, and risk factors.

Engineers and analysts can academically analyze a system's OPERATING ENVIRONMENT's problem space forever—a condition referred to as "analysis paralysis." However, success ultimately depends on making *informed* decisions based on the key facts, working within the reality of limited resources, drawing on seasoned system design experience, and exercising good judgment. The challenge is that *most large complex problems require large numbers of disciplinary specialists to solve these problems*. This set of people often has *diverging* rather than *converging* viewpoints of the OPERATING ENVIRONMENT.

As a system analyst or SE, your job is to facilitate a *convergence* and *consensus* of viewpoints concerning the definition of the system's OPERATING ENVIRONMENT. Convergence and consensus must occur in three key areas:

1. WHAT IS/IS NOT relevant to the mission in the OPERATING ENVIRONMENT?
2. WHAT is the degree of importance, significance, or influence?
3. WHAT is the probability of occurrence of those items of significance?

So how do you facilitate a convergence of viewpoints to arrive at a consensus decision?

As leaders, the system analyst and the SE must have a strategy and approach for quickly organizing and leading the key stakeholders to a *convergence* and *consensus* about the OPERATING ENVIRONMENT. Without a strategy, *chaos* and *indecision* can prevail. You need analytical skills that enable you to establish a framework for OPERATING ENVIRONMENT definition decision making.

As a professional, you have a moral and ethical obligation to yourself, your organization, and society to ensure the safety, health, and well-being of the User and the public when it comes to system operations. If you and your team OVERLOOK or choose to IGNORE a key attribute of the OPERATING ENVIRONMENT that impacts human life and property, there may be SEVERE consequences and penalties. Therefore, establish common analytical models for you and your team to use in your business applications to ensure you have thoroughly identified and considered all OPERATING ENVIRONMENT entities and elements that impact system capabilities and performance.

11.3 OPERATING ENVIRONMENT OVERVIEW

Analytically, the OPERATING ENVIRONMENT that influences and impacts a system's missions can be abstracted several different ways. For discussion purposes the OPERATING ENVIRONMENT can be considered as consisting of two high-level domains: 1) HIGHER ORDER SYSTEMS and 2) the PHYSICAL ENVIRONMENT as shown in Figure 11.1. Let's define each of these system elements.

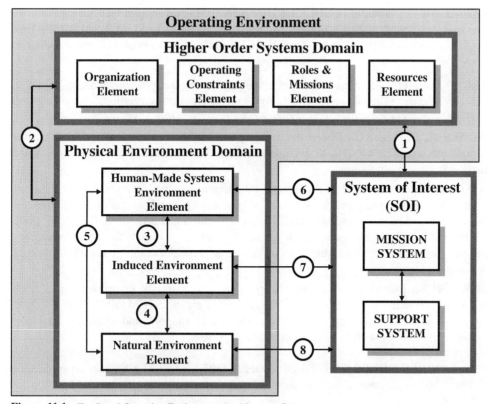

Figure 11.1 Top Level Operating Environment Architecture Construct

HIGH ORDER SYSTEMS Domain

All natural and human-made systems function as individual SYSTEMS OF INTEREST (SOIs) within a hierarchical *system of systems*. Each higher level abstraction serves as a HIGHER ORDER SYSTEM within the system of systems hierarchy that has its own *scope of authority* and operational boundaries. HIGHER ORDER SYSTEMS are characterized by:

1. Organizational purpose or mission.
2. Organizational objectives.
3. An organizational structure.
4. Command media such as rules, policies, and procedures of operation.
5. Resource allocations.
6. Operating constraints imposed on embedded system entities.
7. Accountability and objective evidence of valued-added tasks performed.
8. Delivery of systems, products, and services.

For most human-made systems, we refer to the vertical HIGHER ORDER SYSTEM–to–SYSTEM OF INTEREST (SOI) interaction as C^3I—meaning *command, control, communications, and intelligence*. The Information Age has added a fourth item—computers—thereby changing the acronym to C^4I—*command, control, computers, communications, and intelligence*.

If we observe the behavior of HIGHER ORDER SYSTEMS and analyze their interactions, we can derive four *classes* of *system elements*: 1) ORGANIZATION, 2) ROLES and MISSIONS, 3) OPERATING CONSTRAINTS, and 4) RESOURCES. Let's define each of these system element classes.

- **ORGANIZATION Element** The hierarchical command and control reporting structure, authority, and its assigned accountability for organizational roles, missions, and objectives.
- **ROLES AND MISSIONS Element** The various roles allocated to and performed by HIGHER ORDER SYSTEMS and the missions associated with these roles and objectives to fulfill the organization's vision. Examples include: strategic and tactical plans, roles, and mission goals and objectives.
- **OPERATING CONSTRAINTS Element** International, federal, state, and local statutory, regulatory, policies, and procedures as well as physical laws and principles that *govern* and *constrain* PHYSICAL ENVIRONMENT systems and SYSTEM OF INTEREST (SOI) actions and behavior. Examples include: assets, capabilities, consumables and expendables; weather conditions; doctrine, ethical, social and cultural considerations; and moral, spiritual, philosophical.
- **RESOURCES Element** The natural and physical raw materials, investments, and assets that are allocated to the PHYSICAL ENVIRONMENT and SYSTEM OF INTEREST (SOI) to *sustain* missions—namely deployment, operations, support, and disposal. Examples include commodities such as time, money, and expertise.

Contexts of HIGHER ORDER SYSTEMS. HIGHER ORDER SYSTEMS have two application contexts: 1) human-made systems, such as command and control or social structure, and 2) *physical or natural laws*.

- **Human-made Systems Context** Organizations and governments, exercise hierarchical authority, command, and control over lower tier systems via organizational "chain of command" structures, policies, and procedures, and mission tasking; constitutions, laws, and regulations, public acceptance and opinion, and so on.
- **Physical or Natural Laws Context** All systems, human-made and natural, are governed by natural and physical laws such as life science, physical science, physics, and chemistry.

Given this structural framework of the HIGHER ORDER SYSTEMS domain, let's define its counterpart, the PHYSICAL ENVIRONMENT.

The PHYSICAL ENVIRONMENT Domain

Human-made systems have some level of interaction with *external systems* within the PHYSICAL ENVIRONMENT. In general, we characterize these interactions as *friendly*, *cooperative*, *benign*, *adversarial*, or *hostile*.

If we *observe* the PHYSICAL ENVIRONMENT and *analyze* its interactions with our system, we can identify classes of constituent *system elements*: 1) NATURAL ENVIRONMENT, 2) HUMAN-MADE SYSTEMS, and 3) the INDUCED ENVIRONMENT as shown in Figure 11.1. Let's briefly define each of these:

- **NATURAL ENVIRONMENT Element** All nonhuman, living, atmospheric, and geophysical entities that comprise the Earth and celestial bodies.
- **HUMAN-MADE SYSTEMS Element** External organizational or fabricated systems created by humans that interact with your system entity at all times, during pre-mission, mission, and post-mission.

- **INDUCED ENVIRONMENT Element** *Discontinuities, perturbations,* or *disturbances* created when natural phenomenon occur or HUMAN-MADE SYSTEMS interact with the NATURAL ENVIRONMENT. Examples include thunderstorms, wars, and oil spills.

Based on this high-level introduction and identification of the OPERATING ENVIRONMENT *elements* definitions, we are now ready to establish the architecture of the OPERATING ENVIRONMENT. Let's begin with the Physical Environment Domain.

PHYSICAL ENVIRONMENT Domain Levels of Abstraction

The PHYSICAL ENVIRONMENT consists of three levels of analytical abstractions: 1) *local environment*, 2) *global environment*, and 3) *cosmospheric environment*. Figure 11.2 uses an entity relationship diagram (ERD) to illustrate the composition of the PHYSICAL ENVIRONMENT. At a high level of abstraction, the PHYSICAL ENVIRONMENT consists of the three classes of system elements—HUMAN-MADE, INDUCED, and NATURAL environment elements. Each type of environment consists of three levels of abstraction-Cosmospheric, Global, or Local.

Author's Note 11.1 *You may determine that some other number of system levels of abstraction is more applicable to your line of business. That's okay. What is IMPORTANT is that you and your team have a simple approach for abstracting the complexity of the PHYSICAL ENVIRONMENT domain into manageable pieces. The "pieces" must support meaningful analysis and ensure coverage of all relevant aspects that relate to your problem and solution spaces. Remember, bounding abstractions for analysis is analogous to cutting a pie into 6, 8, or 10 pieces. As long as you account for the TOTALITY, you can create as many abstractions as are REASONABLE and PRACTICAL; however, KEEP IT SIMPLE.*

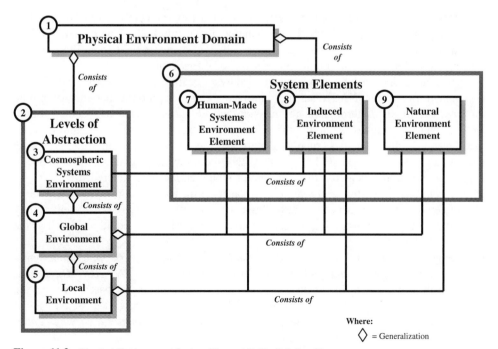

Figure 11.2 Physical Environment System Element Entity Relationships

11.4 PHYSICAL ENVIRONMENT DOMAIN SYSTEM ELEMENTS

A system's PHYSICAL ENVIRONMENT consists of three classes of *system elements*. These include: 1) the HUMAN-MADE SYSTEMS element, 2) the INDUCED ENVIRONMENT element, and 3) the NATURAL ENVIRONMENT element. Let's scope and define each of these elements.

The NATURAL ENVIRONMENT System Element

The NATURAL ENVIRONMENT System Element includes all *naturally* occurring entities that are not human-made. These entities are actually environmental "systems" that coexist within a precarious balance of power. In general, these systems represent geophysical and life form classes of objects.

Cosmospheric Environment Level of Abstraction. The Cosmospheric Environment is an abstraction that represents the totality of the Cosmos—the universe—as humankind understands it. Analytically, the Cosmospheric Environment consists of an infinite number of Global Environments.

You may ask *why* the Cosmospheric Environment is relevant to SE. Consider space probes that have flown beyond the boundaries of our solar system. The SEs and physicists who bounded the OPERATING ENVIRONMENT had to identify all of the global entities that had a potential impact on the space probe's mission. Obviously, these global entities did not move out of the way of the probe's mission path. The SEs and physicists had to understand:

1. WHAT entities they might encounter during the missions.
2. WHAT each global entity's performance characteristics are.
3. HOW to navigate and maneuver among those entities throughout the mission without an adverse or catastrophic impact.

Global Environment Level of Abstraction. The Global Environment is an abstraction that represents the PHYSICAL ENVIRONMENT surrounding a heavenly body. This includes entities such as stars, planets, and satellites of planets. Analytically, we state that the Global Environment of any heavenly body consists of an infinite number of Local Environments.

If you operate an airline, each aircraft is surrounded by a Local Environment within the Earth's *global environment*. In contrast, NASA launches interplanetary space probes experience an infinite number of *local environments* throughout the mission and *global environments*—namely Earth, planets, and the planetary moons. Although the *global environments* may share a common set of physical attributes and characteristics, such as gravity and the atmosphere, their values can vary significantly. As a result, OPERATING ENVIRONMENT requirements may require SEs to bound ranges of worst-case parameters across the spectrum of global entities encountered.

As humans, our *observer's frame of reference* is planet Earth. Therefore, we typically relate to the global entity environment as that of the Earth and its gaseous atmosphere—some people simply call it the Earth's environment. Applying the same convention to Mars, we could refer to it as the Mars's global environment.

The challenge question is: *Where do SEs draw the line to delineate overlapping global environment influences. How do we bound the Earth's and Moon's OPERATING ENVIRONMENT influences?* Do we arbitrarily draw a line at the midpoint? Obviously, this is not based on scientific principles. Alternatively, *do we bound the Earth's and Moon's environments as a region encompassing both bodies with a "no man's" zone in between? How do we "bound" a varying continuum such as the electromagnetic field, the atmosphere, or gravity?*

From an SE perspective, one approach is to delineate the two environments by asking the question, *What characteristics below some arbitrary threshold are unique or "native" to the "global" SYSTEM OF INTEREST (SOI)*? If the boundary conditions of two or more entities overlap, the objective would be to determine criteria for the boundary conditions.

As a systems analyst or SE, you represent the technical side of the program-technical boundary for the discussions that follow. Thus, you have a *professional, technical* OBLIGATION to ensure that the INTEGRITY of these decisions is supported by:

1. Objective, factual data to the extent practical and available.
2. Valid assumptions that withstand peer and stakeholder scrutiny.
3. *Most likely* or *probable* operational scenarios and conditions.

Remember, the *analytical relevance* of the decisions you make here may have a *major* impact on your system's design, cost, reliability, maintainability, vulnerability, survivability, safety, and risk considerations. These decisions may have adverse impacts on human life, property, and environment.

Analytically, we can partition the NATURAL ENVIRONMENT Global and Local levels of abstraction into five subelements:

- Atmospheric systems environment
- Geospheric systems environment
- Hydrospheric systems environment
- Biospheric systems environment

Author's Note 11.2 *A word of caution: The point of our discussion here is not to present a view of the physical science. Our intent is to illustrate HOW SEs might approach analysis of the NATURAL ENVIRONMENT. Ultimately, you will have to bound and specify applicable portions of this environment that are applicable to your SYSTEM OF INTEREST (SOI). The approach you and your team choose to use should be ACCURATELY and PRECISELY representative of your system's operating domain and the relevant factors that drive system capabilities and levels of performance. However, you should always consult subject matter experts (SMEs), who can assist you and your team in abstracting the correct environment.*

Atmospheric Systems Environment. The *Atmospheric Systems Environment* is an abstraction of the gaseous layer that extends from the surface of a planetary body outward to some predefined altitude.

Geospheric Systems Environment. The *Geospheric Systems Environment* consists of the totality of the physical landmass of a star, moon, or planet. From an Earth sciences perspective, the Earth's *Geospheric Systems Environment* includes the *Lithospheric Systems Environment*, the rigid or outer crust layer of the Earth. In general, the lithosphere includes the continents, islands, mountains, and hills that appear predominantly at the top layer of the Earth.

Hydrospheric Systems Environment. The *Hydrospheric Systems Environment* consists of all liquid and solid water systems, such as lakes, ponds, rivers, streams, waterfalls, underground aquifers, oceans, tidal pools, and ice packs, that are not part of the gaseous atmosphere. In general, the key abstractions within the *Hydrospheric Systems Environment* include: rainwater, soil waters, seawater, surface brines, subsurface waters, and ice.

Biospheric System Environment. The *Biospheric Systems Environment* is defined as the environment comprising all living *organisms* on the surface of the Earth. In general, the Earth's biosphere consists of all environments that are capable of supporting life above, on, and beneath the Earth's surface as well as the oceans. Thus, the biosphere overlaps a portion of the atmosphere, a large amount of the hydrosphere, and portions of the lithosphere. Examples include: botanical, entomological, ornithological, amphibian, and mammalian systems. In general, biospheric systems function in terms of two metabolic processes: photosynthesis and respiration.

Local Environment Level of Abstraction. The *Local Environment* is an abstraction that represents the PHYSICAL ENVIRONMENT encompassing a system's current location. For example, if you are driving your car, the Local Environment consists of the OPERATING ENVIRONMENT conditions surrounding the vehicle. Therefore, the Local Environment includes other vehicles and drivers, road hazards, weather, and any other conditions on the roads that surround you and your vehicle at any instant in time. As an SE, your challenge is leading system developers to consensus on:

1. WHAT is the Local Environment's frame of reference?
2. WHAT are the local environment's bounds (initial conditions, entities, etc.) at any point in time?

Author's Note 11.3 *If you are a chemist or physicist, you may want to consider adding a fourth level molecular environment assuming it is germane to your SOI.*

Author's Note 11.4 *Your "takeaway" from the preceding discussion is not to create five types of NATURAL ENVIRONMENT abstractions and document each in detail for every system you analyze. Instead, you should view these elements as a checklist to prompt mental consideration for relevance and identify those PHYSICAL ENVIRONMENT entities that have relevance to your system. Then, bound and specify those entities.*

HUMAN-MADE SYSTEMS Environment Element

Our discussions of the NATURAL ENVIRONMENT omitted a key entity, humankind. Since SE focuses on benefiting society through the development of systems, products, or services, we can analytically isolate and abstract humans into a category referred to as the HUMAN-MADE SYSTEMS Environment element. HUMAN-MADE SYSTEMS include subelements that influence and control human decision making and actions that affect the balance of power on the planet.

If we observe HUMAN-MADE SYSTEMS and analyze how these systems are organized, we can identify seven types of subelements: 1) historical or heritage systems, 2) cultural systems, 3) urban systems, 4) business systems, 5) educational systems, 6) transportation systems, and 7) governmental systems. Let's explore each of these further.

Historical or Heritage Systems. *Historical or heritage systems* in abstraction include all artifacts, relics, traditions, and locations relevant to past human existence such as folklore and historical records.

You may ask why the historical or heritage systems are relevant to an SE. Consider the following example:

EXAMPLE 11.1

Let's assume that we are developing a system such as a building that has a placement or impact on a land use area that has HISTORICAL significance. The building construction may disturb *artifacts* and *relics* from the legacy culture, and this may influence technical decisions relating to the physical location of the system. One perspective may be to provide physical space or a buffer area between the historical area and our system. Conversely, if our system is a museum relevant to an archeological discovery, the location may be an INTEGRAL element of the building design.

From another perspective, military tactical planners may be confronted with planning a mission and have an objective to avoid an area that has historical significance.

Urban Systems. *Urban systems* include all entities that relate to *how* humans cluster or group themselves into communities and interact at various levels of organization—neighborhood, city, state, national, and international.

EXAMPLE 11.2

Urban systems include the infrastructure that supports the *business systems environment*, such as transportation systems, public utilities, shopping, distribution systems, medical, telecommunications, and educational institutions.

Systems engineering, from an *urban systems* perspective, raises key issues that effectively require some form of prediction. *How do systems engineers plan and design a road system within resource constraints that provide some level of insights into the future to facilitate growth and expansion?*

Cultural Systems. *Cultural systems* include multi-faceted attributes that describe various identities of humans and communal traits, and *how* humans *interact, consume, reproduce,* and *survive*. Examples include: the performing arts of music and other entertainment, civic endeavors, and patterns of behavior. As the commercial marketplace can attest, cultural systems have a major impact on society's acceptance of systems.

Business Systems. *Business systems* include of all entities related to *how* humans organize into economic-based enterprises and commerce to produce products and services to sustain a livelihood. These include research and development, manufacturing, products and services for use in the marketplace.

Educational Systems. *Educational systems* include all institutions dedicated to educating and improving society through formal and informal institutions of learning.

Financial Systems. *Financial systems* include banks and investment entities that support personal, commercial, and government financial transactions.

Government Systems. *Government systems* include all entities related to governing humans as a society—international, federal, state, county, municipality, etc.

Medical Systems. *Medical systems* include hospitals, doctors, and therapeutic entities that administer to the healthcare needs of the public.

Transportation Systems. *Transportation systems* include land, sea, air, and space transportation systems that enable humans to travel *safely, economically,* and *efficiently* from one destination to another.

The INDUCED ENVIRONMENT Element

The preceding discussions isolated the NATURAL ENVIRONMENT and HUMAN-MADE SYSTEMS as element abstractions. While these two elements enable us to analytically *organize* system OPERATING ENVIRONMENT entities, they are physically *interactive*. In fact, HUMAN-MADE SYSTEMS and the NATURAL ENVIRONMENT each create *intrusions*, *disruptions*, *perturbations*, and *discontinuities* on the other.

Analysis of these interactions can become very complex. We can alleviate some of the *complexity* by creating the third PHYSICAL ENVIRONMENT element, the INDUCED ENVIRONMENT. The INDUCED ENVIRONMENT enables us to isolate entities that represent the *intrusions*, *disruptions*, *perturbations*, and *discontinuities* until they diminish or are no longer significant or relevant to system operation. The degree of significance of INDUCED ENVIRONMENT entities may be *temporary*, *permanent*, or *dampen over time*.

11.5 OPERATING ENVIRONMENT DECISION-MAKING METHODOLOGY

The OPERATING ENVIRONMENT imposes various factors and constraints on the capabilities and levels of performance of a SYSTEM OF INTEREST (SOI), thereby impacting missions and survival over the planned life span. As a system analyst or SE, your responsibility is to:

1. *Identify* and *delineate* all of the critical OPERATING ENVIRONMENT conditions.
2. *Bound* and *describe* technical parameters that characterize the OPERATING ENVIRONMENT.
3. *Ensure* those descriptions are incorporated into the *System Performance Specification (SPS)* used to procure the SOI.

The process of identifying the SOI's OPERATING ENVIRONMENT requirements employs a simple methodology as depicted in Figure 11.3. In general, the methodology implements the logic reflected in the PHYSICAL ENVIRONMENT levels of abstraction and classes of environments previously described.

The methodology consists of three iterative loops:

Loop 1: Cosmospheric level requirements (1).
Loop 2: Global entity level requirements (2).
Loop 3: Local level requirements (3).

When each of these iterations is applicable to a SOI, the logic branches out to a fourth loop that investigates which of three *classes of environments*—HUMAN-MADE SYSTEMS (4), NATURAL ENVIRONMENT (6), or INDUCED ENVIRONMENT (8) is applicable. For each type of environment that is applicable, requirements associated with that type are identified—HUMAN-MADE SYSTEMS (5), NATURAL (7), and INDUCED (9). When the third loop completes its types of environment decision-making process, it returns to the appropriate level of abstraction and finishes with the local level requirements (3).

11.6 ENTITY OPERATING ENVIRONMENT FRAME OF REFERENCE

The preceding discussions address the OPERATING ENVIRONMENT from the SYSTEM's *frame of reference*. However, lower levels of abstraction within a SYSTEM are, by definition, self-

11.6 Entity Operating Environment Frame of Reference

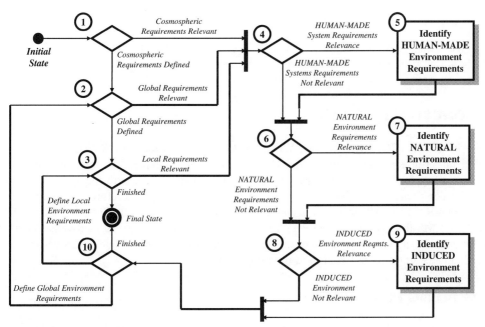

Figure 11.3 Environment Requirements Identification

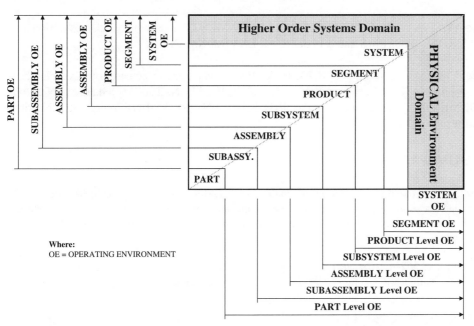

Figure 11.4 OPERATING ENVIRONMENT Relative to Observer's Levels of Abstraction

contained systems of integrated components. The OPERATING ENVIRONMENT contextually must be established from the entity's frame of reference as illustrated in Figure 11.4. So, *WHAT constitutes an ASSEMBLY's OPERATING ENVIRONMENT?* This can be anything external to the ASSEMBLY's boundary, such as other ASSEMBLIES, SUBSYSTEMS, or PRODUCTS. Consider the following example:

EXAMPLE 11.3

A processor board within a desktop computer chassis has an OPERATING ENVIRONMENT that consists of the motherboard; other boards it interfaces directly with; electromagnetic radiation from power supplies; switching devices, and so on.

11.7 CONCLUDING POINT

You may ask isn't the HIGHER ORDER SYSTEMS domain part of the PHYSICAL ENVIRONMENT domain? You could argue this point. However, Figure 11.1 represents an analytical perspective with a key focus on DIRECT, peer-to-peer and command and control (C^2) interactions. From an SOI perspective, it responds to human managerial authority. Therefore, we depict the HIGHER ORDER SYSTEMS in terms of human supervisory control of the SOI. Are HIGHER ORDER SYSTEMS (human) above the PHYSICAL ENVIRONMENT? No, in fact we are physically subordinated to it. Yet, as we see in phenomena such as global warming, our collective actions can have an adverse impact on it, and in turn on our lives.

11.8 GUIDING PRINCIPLES

In summary, the preceding discussions provide the basis with which to establish the guiding principles that govern the architecture of a system's operating environment.

Principle 11.1 A system's OPERATING ENVIRONMENT consists of two classes of domains: HIGHER ORDER SYSTEMS and a PHYSICAL ENVIRONMENT.

Principle 11.2 A system's PHYSICAL ENVIRONMENT domain consists of three classes of *system elements*: NATURAL, HUMAN-MADE, and INDUCED. NATURAL and HUMAN-MADE ENVIRONMENTs systems interact; the INDUCED ENVIRONMENT represents the time-dependent result of that *interaction*.

11.9 SUMMARY

Our discussion in this section provides an orientation of the OPERATING ENVIRONMENT, its levels of abstraction, and classes of environments. Based on identification of these OPERATING ENVIRONMENT elements, we introduced Figure 11.1 to depict relationships among PHYSICAL ENVIRONMENT levels of abstraction and its *system elements*.

Next we introduced the concept of the OPERATING ENVIRONMENT architecture as a framework for linking the OPERATING ENVIRONMENT system elements. The architectural framework of interactions provided a basis for us to define several *system element* interaction principles.

GENERAL EXERCISES

1. Answer each of the *What You Should Learn from This Chapter* questions identified in the *Introduction*.
2. Refer to the list of systems identified in Chapter 2. Based on a selection from the preceding chapter's *General Exercises* or a new system selection, apply your knowledge derived from this chapter's topical discussions. Identify the following:
   (a) HIGHER ORDER SYSTEMS domain and its *system elements*
   (b) PHYSICAL ENVIRONMENT *domain*, its *system elements*, and levels of abstraction.

ORGANIZATIONAL CENTRIC EXERCISES

1. Contact a system development program in your organization. Research how they analyzed their SYSTEM OF INTEREST (SOI), its OPERATING ENVIRONMENT, and their respective system elements. How was this analysis reflected in the SOI architecture?

ADDITIONAL READING

KOSSIAKOFF, ALEXANDER, and SWEET, WILLIAM N. 2003. *Systems Engineering Principles and Practice*. New York: Wiley-InterScience.

MIL-HDBK-1908B. 1999. *DoD Definitions of Human Factors Terms*. Washington, DC: Department of Defense (DoD).

Chapter 12

System Interfaces

12.1 INTRODUCTION

One of the crucial factors of system success is determined by what happens at its internal and external interfaces. You can engineer the most elegant algorithms, equations, and decision logic, but if the system does not perform at its interfaces, the elegance is of no value. System interface characterizations range from cooperative interoperability with external friendly systems to layers of protection to minimize vulnerability to external threats (environment, hostile adversary actions, etc.) and structural integrity to ensure survivability.

This chapter introduces the *context* of system interfaces, their purpose, objectives, attributes, and how they are implemented. Our discussions explore the various types of interfaces and factors that delineate success from failure. This information provides the basis for the next chapter, which addresses interface design and control.

What You Should Learn from This Chapter

1. *What is an interface?*
2. *What is the purpose of an interface?*
3. *What are the types of interfaces?*
4. *What is a point-to-point interface?*
5. *What is a logical interface?*
6. *What is a physical interface?*
7. *How do logical and physical interfaces interrelate?*
8. *Identify seven types of physical interfaces?*
9. *What are the steps of the interface definition methodology?*
10. *What is a generalized interface solution?*
11. *What is a specialized interface solution?*
12. *How does a generalized solution relate to a specialized solution?*
13. *How do generalized and specialized solutions relate to logical and physical interfaces?*
14. *What are some methods of limiting access to interfaces?*
15. *What constitutes an interface failure?*
16. *What are some examples of interface failures?*

System Analysis, Design, and Development, by Charles S. Wasson
Copyright © 2006 by John Wiley & Sons, Inc.

Definitions of Key Terms

- **Human-Machine Interface** "The actions, reactions, and interactions between humans and other system components. This also applies to a multi-station, multi-person configuration or system. Term also defines the properties of the hardware, software or equipment which constitute conditions for interactions." (Source: MIL-HDBK-1908, p. 21)
- **Interchangeability** "The ability to interchange, without restriction, like equipments or portions thereof in manufacture, maintenance, or operation. Like products are two or more items that possess such functional and physical characteristics as to be equivalent in performance and durability, and are capable of being exchanged one for the other without alteration of the items themselves or of adjoining items, except for adjustment, and without selection for fit and performance." (Source: MIL-HDBK-470A, Appendix G, Glossary, p. G-7)
- **Interface** "The functional/logical relationships and physical characteristics required to exist at a SYSTEM or entity boundary with its OPERATING ENVIRONMENT that enable the entity to provide a mission capability." (Source: Adapted from DSMC—Glossary of Terms)
- **Interface Control** "The process of: (1) identifying all functional and physical characteristics relevant to the interfacing of two or more items provided by one or more organizations; and (2) ensuring that proposed changes to these characteristics are evaluated and approved prior to implementation." (Source: Former MIL-STD-480B, para. 3.1.43)
- **Interface Device** "An item which provides mechanical and electrical connections and any signal conditioning required between the automatic test equipment (ATE) and the unit under test (UUT); also known as an interface test adapter or interface adapter unit." (Source: MIL-HDBK-470A, Appendix G, Glossary, p. G7)
- **Interface Ownership** The assignment of accountability to an individual, team, or organization regarding the definition, specification, development, control, operation, and support of an interface.
- **Interoperability** "The ability of two or more systems or components to exchange information and to use the information that has been exchanged." (Source: IEEE 610.12-1990)
- **Peer level Interactions** SYSTEM OF INTEREST (SOI) interactions—namely MISSION SYSTEM and SUPPORT SYSTEM—with external systems in the OPERATING ENVIRONMENT. For analytical purposes, we aggregate these systems into a single entity abstraction referred to as the PHYSICAL ENVIRONMENT SYSTEMS domain.
- **Point-to-Point** An interface configuration that characterizes the physical connectivity between two points, typically accomplished via dedicated, direct line. For example, a light switch connection to a room light.

Based on this introduction, let's begin our discussion by exploring what an interface is.

12.2 WHAT IS AN INTERFACE?

System engineering efforts often focus on:

1. The *composition* of the system architecture.
2. How those elements *interact* with each other and their OPERATING ENVIRONMENT.

You can develop the most innovative devices, computers, and algorithms. Yet, if those innovations are unable to reliably *interact* and *interoperate* with their OPERATING ENVIRONMENT when required, they may be of limited or no value to the entity or SYSTEM.

Interfaces occur between combinations of two or more system elements—such as EQUIPMENT, PERSONNEL, and FACILITIES—or between entities within system element levels of abstractions. However, *what is the purpose of an interface*?

Interface Purposes

The purpose of an interface is to associate or physically connect a SYSTEM, PRODUCT, SUBSYSTEM, ASSEMBLY, SUBASSEMBLY, or PART level component to other components within its OPERATING ENVIRONMENT. A component may *associate* or *connect* to several components; however, each linkage represents a *single* interface. If a component has multiple interfaces, the performance of that interface may have an influence or impact on the others.

The purpose stated above is a very broad description of WHY an interface exists. The question is: *HOW does an interface accomplish this*? An interface has at least one or more objectives, depending on the component's application. Typical interface objectives include the following:

Objective 1: Physically link or bind two or more system elements or entities.

Objective 2: Adapt one or more incompatible system elements or entities.

Objective 3: Buffer the effects of incompatible system elements.

Objective 4: Leverage human capabilities.

Objective 5: Restrain system element or its usage.

Let's explore each of the objectives further.

Objective 1: Physically Link or Bind Two or More System Elements or Entities

Some systems link or bind two or more compatible *system elements* or element components to anchor, extend, support, or connect the adjoining interface.

EXAMPLE 12.1

A communications tower has cables at critical attach points to anchor the tower to the ground for stability.

Objective 2: Adapt One or More Incompatible System Elements or Entities

Some *system elements*—such as EQUIPMENT and PERSONNEL—or entities may not have compatible or interoperable interfaces. However, they can be *adapted* to become compatible. Figure 12.1 illustrates how NASA developed an adapter for the Apollo-Soyuz Program. Software applications that employ reusable models may create a "wrapper" around the model to enable the model to communicate with an external application, and vice versa.

Objective 3: Buffer the Effects of Incompatible System Elements

Some systems such as automobiles are generally not intended to interact with each other. Where the unintended interactions occur, the effects of the interaction must be minimized in the interest of the safety and health of the Users. Consider the following two cases:

Case 1: An automobile's impact on another can be lessened with a shock absorber bumper and body crumple zones.

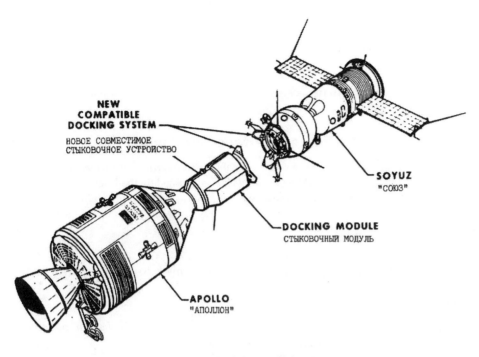

Figure 12.1 NASA Apollo-Soyuz Spacecraft Docking System
Source: NASA History Office Web Site—www.hq.nasa.gov/office/pao/History/diagrams/astp/pk69.htm

Case 2: System A is required to transmit data to System B. Because of the limited speed of the interfacing components, System A includes a buffer area for storing data for communication to free up the processor to perform other tasks. On the other side of the interface, System B may be unable to process all of the incoming data immediately. To avoid this scenario, a buffer area is created to store the incoming data until the processor can process the data.

For this objective, SEs analyze the interface and take reasonable measures to create a "boundary layer" or buffer between *system elements* or entities. Thus each is buffered to minimize the effects of the impact to the system or environment and, if applicable, safety to the operators or the public.

Objective 4: Leverage Human Capabilities

Humans employ interface capabilities to leverage our own skills and capabilities. Early humans recognized that various tools—namely simple machines—could serve as interface devices to expand or leverage our own physical capabilities in accomplishing difficult tasks.

Objective 5: Restrain SYSTEM Element or Its Usage

Some interfaces serve as *restraints* to ensure a level of safety for *system elements*. Consider the following example:

114 Chapter 12 System Interfaces

> **EXAMPLE 12.2**
>
> A safety chain is added to a trailer hitch used by an automobile to tow a trailer. A lock is added to a power distribution box to prevent opening by unauthorized individuals.

Each of these objectives illustrates HOW an interface is implemented to achieve a purposeful action. Depending on a systems application, other objectives may be required.

Interoperability—The Ultimate Interface Challenge

The ultimate success of any interface resides in its capability to *interact* with *friendly* and hostile systems in its intended OPERATING ENVIRONMENT as envisioned by the User, specified by the Acquirer, and designed by the System Developer. We refer to this as *interoperability*.

12.3 INTERFACE TYPES

Interfaces exhibit three types of operation: *active*, *passive*, or *active/passive*.

Active Interfaces

Active interfaces interact with external systems or components in a *friendly*, *benign*, or *cooperative* manner. Consider the following example:

> **EXAMPLE 12.3**
>
> Radio stations, as active "on the air" systems, radiate signals at a designated frequency via patterns to specific areas for coverage.

Passive Interfaces

Passive interface interactions with external components simply *receive* or *accept* data without responding. Consider the following example:

> **EXAMPLE 12.4**
>
> A car radio, when powered ON and viewed by a radio station system, PASSIVELY receives signals over a tuned frequency. The radio processes the information and provides an ACTIVE audio interface for the occupants in the car.

Combination Active/Passive Interfaces

Active/passive interfaces perform under the control of transmitters or receivers. Consider the following example:

> **EXAMPLE 12.5**
>
> A two-way walkie-talkie radio has an ACTIVE *interface* when the User presses the "Push to Talk" button to broadcast audio information to others listening on the same frequency within a specified transmission range and conditions. When the "Push to Talk" button is OFF, the device has a PASSIVE interface that monitors incoming radio signals for processing and audio amplification as controlled by the User.

12.4 UNDERSTANDING LOGICAL AND PHYSICAL INTERFACES

One of the recurring themes of this book is the need to *decompose complexity* into one or more manageable levels. We do this by first identifying *logical/functional* interfaces that ask several questions:

1. WHO interacts with WHOM?
2. WHAT is transferred and translated?
3. WHEN does the transference or translation occur?
4. Under WHAT conditions?

Then, we translate the *logical/functional* connectivity into a *physical* interface that represents HOW and WHERE the interface will be implemented.

Analytically, system interfaces provide the mechanism for point-to-point connectivity. We characterize interface connectivity at two levels: 1) *logical* and 2) *physical*.

- *Logical interfaces* Represent a *direct* or *indirect* association or relationship between two entities. Logical interfaces establish:
  1. WHO—Point A—communicates with WHOM—Point B.
  2. Under WHAT scenarios and conditions the communications occur.
  3. WHEN the communications occur.
  Logical interfaces are referred to as "generalized" interfaces.

- *Physical interfaces* Represent *physical interactions* between two interfacing entities. Physical interfaces express HOW devices or components (boxes, wires, etc.) will be configured to enable Point A to communicate with Point B. Physical interfaces are referred to as "specialized" interfaces because of their dependency on specific mechanisms (electronics, optics, etc.) to realize the interface.

Consider the following example:

EXAMPLE 12.6

The Internet provides a mechanism for a User with a device such as a computer equipped with the appropriate hardware and software to communicate with an external Web site. In this context a *logical interface* or *association* exists between the user and the Web site. When the User connects with the Web site, a *physical interface* is established.

The preceding discussion highlighted two "levels" of connectivity. This is an important point, especially from a system design perspective involving humans. Engineers have a strong tendency to jump to defining the *physical interface* BEFORE anyone has decided WHAT the interface is to accomplish. Therefore, you must:

1. Identify which *system elements* or entities must *associate* or interact.
2. Understand WHY they "need to connect."
3. Determine WHEN the associations or interactions occur.

12.5 INTERFACE DEFINITION METHODOLOGY

The preceding discussions enable SEs to establish a basic methodology for identifying and characterizing interfaces.

Step 1: Identify logical interfaces. When systems are designed, *logical* interfaces enable us to acknowledge that an *association* or *relationship* exists. Therefore, the interface becomes a logical means of expression that enables us to characterize WHAT the interface is required to accomplish.

Step 2: Identify and define physical interfaces. The physical implementation of a *logical* or *generalized* interface requires selection from a range of candidate solutions subject to technical, technology, cost, schedule, and risk constraints. This being the case, SEs typically conduct one or more trade studies to select the most appropriate implementation.

12.6 PHYSICAL INTERFACE TYPES

If we analyze how interfaces are implemented in the physical domain, our analysis will reveal that interfaces occur in *mechanical, electrical, optical, acoustical, natural environment, chemical*, and *biological* forms, and as combinations of these forms. For all these types of physical interfaces there are *specialized solutions*. To further understand the specialized nature of these interface solutions, let's explore each one.

Mechanical Interfaces

Mechanical interfaces consist of boundaries that exist between two physical objects and include characterizations such as *function, form*, and *fit*. Characterizations include:

1. *Material Properties* Composition.
2. *Dimensional Properties* Length, width, and depth; mass properties such as weight, density, and shape.
3. *Structural Integrity Properties* Shock, vibration, etc.
4. *Aerodynamic Properties* Drag, fluid flow, etc.

Electrical Interfaces

Electrical interfaces consist of direct electrical or electronic connections as well as electromagnetic transmission in free air space. Attributes and properties include voltages, current, resistance, inductance, capacitance, grounding, shielding, attenuation, and transmission delays.

Optical Interfaces

Optical interfaces consist of the transmission and/or receipt of visible and invisible wavelengths of light. Attributes and properties include intensity, frequency, special ranges, resolution, distortion, contrast, reflection, refraction, filtering, modulation, attenuation, and polarization.

Acoustical Interfaces

Acoustical interfaces consist of the creation, transmission, and receipt of frequencies that may be audible or inaudible to humans. Attributes and properties include volume, frequency, modulation, and attenuation.

Natural Environment Interfaces

Natural environment interfaces consist of those elements that are natural occurrences of nature. Attributes and properties include temperature, humidity, barometric pressure, altitude, wind, rain, snow, and ice.

Chemical Interfaces

Chemical interfaces consist of *interactions* that occur when chemical substances are purposefully introduced or mixed with other chemicals or other types of interfaces. Attributes and properties include heat, cold, explosive, toxicity, and physical state changes.

Biological Interfaces

Biological interfaces consist of those interfaces between living organisms or other types of interfaces. Attributes and properties include touch, feel, smell, hearing, and sight.

12.7 STANDARDIZED VERSUS DEDICATED INTERFACES

Interfaces allow us to establish *logical* or *physical* relationships between *system elements* via a common, compatible, and interoperable boundary. If you analyze the most common types of interfaces, you will discover two basic categories: 1) standard, modular interfaces and 2) unique, dedicated interfaces. Let's define the context of each type.

- *Standard, Modular Interfaces* System developers typically agree to employ a modular, interchangeable interface approach that complies with a "standard" such as RS-232, Mil-Std-1553, Ethernet, and USB (Universal Serial Bus).
- *Unique, Dedicated Interfaces* Where standard interfaces may not be available or adequate due to the uniqueness of the interface, SE designers may elect to create a unique, dedicated interface design for the sole purpose limiting compatibility with other *system elements* or entities. Examples include special form factors and encryption that make the interface unique.

12.8 ELECTRONIC DATA INTERFACES

When the User's *logical* interfaces are identified in the SYSTEM or entity architecture, one of the first decisions is to determine HOW the interface is to be implemented. Key questions include:

1. Does each interface require discrete inputs and outputs?
2. WHAT function does the interface perform (data entry/output, event driven interrupt, etc.)?
3. Are the data periodic (i.e., synchronous or asynchronous)?
4. WHAT is the quantity of data to be transmitted or received?
5. WHAT are the time constraints for transmitting or receiving the data?

Electronic data communications mechanisms employ *analog* or *digital* techniques to communicate information.

Analog Data Communications

Analog mechanisms include amplitude modulated (AM) microphone and speaker based I/O devices such as telephones and modems.

Digital Data Communications

Digital mechanisms include synchronous and asynchronous bipolar signals that employ specific transmission protocols to encapsulate encoded information content. Digital data communications consists of three basic types of data formats: *discrete*, *serial*, or *parallel*.

Discrete Data Communications. Discrete data consist of dedicated, independent instances of static ON or OFF data that enable a device such as a computer to monitor the *state* or *status* condition(s) or initiate actions by remote devices. *Digital discrete data* represent electronic representations for various conditions or states—such as ON/OFF, INITIATED, COMPLETE, and OPEN/CLOSED.

Discrete data communications also include *event driven interrupts*. In these applications a unique, dedicated signal line is connected to an external device that senses specific conditions. When the condition is detected, the device *toggles* the discrete signal line to notify the receiving device that a conditional event has occurred.

Parallel Data Communications. Some systems require high-speed communications between electrical devices. Where this is the case, parallel data communications mechanisms may be employed to improve SYSTEM performance by simultaneously transmitting synchronous data simultaneously over discrete lines. Consider the following example:

EXAMPLE 12.7

An output device may configured to set any one or all 8 bits of discrete binary data to turn ON/OFF individual, external devices.

Parallel data communications mechanisms may increase hardware component counts, development and unit costs, and risks. In these cases, performance must be traded off against cost and risk. Parallel data communications may be *synchronous*—meaning periodic—or *asynchronous*, depending on the application.

Serial Data Communications. Some systems require the transmission of data to and from *external systems* at rates that can be accomplished using serial data communications bandwidths. Where applicable, serial data communication approaches *minimize* parts counts, thereby affecting PC board layouts, weight, or complexity.

Serial data communications mechanisms may be *synchronous*—meaning periodic—or *asynchronous*, depending on the application. Serial data communications typically conform to a number of standards such as RS-232, RS-422 and Ethernet.

12.9 LIMITING ACCESS TO SYSTEM INTERFACES

Some interfaces require restricted access to only those devices accessible by *authorized* Users. In general, these interfaces consist of those applications whereby the User pays a fee for access, data security on a NEED TO KNOW basis, posting of configuration data for decision making, and so on. *How is this accomplished?*

Limited access can be implemented via several mechanisms such as: 1) authorized log-on accounts, 2) data encryption devices and methods, 3) floating access keys, 4) personal ID cards, 5) personal ID scanners, and 6) levels of need to know access.

- *Authorized User Accounts* Employed by Web sites or internal computer systems and require a UserID and password. If the user forgets the password, some systems allow the user to post a question related to the password that will serve as memory jogger. These accounts also must make provisions to reset the password as a contingency provided the user can authenticate themselves to the computers system via personal ID information.
- *Data Encryption Methods and Techniques* Employed to encrypt/decrypt data during transmission to prevent unauthorized disclosure. These devices employ data "keys" to limit access. Encryption applications range from desktop computer communications to highly sophisticated banking and military implementations.
- *Floating Access Key* Software applications on a network that allow simultaneous usage by a subset of the total number of personnel at any given point in time. Since organizations do not want to pay for unused licenses, floating licenses are procured based on projected peak demand. When a user logs onto the application, one of the license keys is locked until the user logs out. Since some users tend to forget to log out and thereby locking other users from using the key, systems may incorporate timeout features that *automatically* log out a user and make the key accessible to others.
- *Personal ID Cards* Magnetically striped ID badges or credit cards assigned to personnel that allow access to facilities via security guards or access to closed facilities via magnetic card readers and passwords.
- *Personal Authentication Scanners* Systems enabling limited access by authenticating the individual via optical scanners that scan the retina of an eye or thumbprint and match the scanned image against previously stored images of the actual person.
- *Levels of Need to Know Access* Restricted access based on the individual's need to know. Where this is the case, additional authentication may be required. This may require compartmentalizing data into levels of access.

12.10 UNDERSTANDING INTERFACE PERFORMANCE AND INTEGRITY

Interfaces, as an entry point or portal into a system, are *vulnerable* to *threats* and *failures*, both internally and externally. Depending on the extent of the physical interface interaction and resulting damage or failure, the interface capability or performance may be limited or terminated. Our discussion here focuses on understanding interface design performance and integrity. Let's begin by first defining the context of an *interface failure*.

What Constitutes an Interface Failure?

There are differing contexts regarding WHAT constitutes an interface failure including *degrees of failure*. An interface might be considered *failed* if it ceases to provide the required capability at a specified level of performance when required as part of an overall system mission. *Interface failures* may or may not jeopardize a system mission.

Consequences of an Interface Failure

Interface failures can result in the LOSS of system control and/or data; physical damage to system operators, equipment, property or the environment. As a result, system interfaces that may potentially impact mission success, cause damage to the system, the public, or environment; as well as loss of life should be *thoroughly* analyzed. This includes understanding HOW the internal or external interface may fail and its impact on other components. Let's examine how interface failures may occur.

Interface Failures

Interfaces fail in a number of ways. In general, physical interfaces can fail in at least four types of scenarios: 1) *disruption*, 2) *intrusion*, 3) *stress loading*, and 4) *physical destruction*.

- *Disruptions* can be created by acts of nature, component reliability, poor quality work, animals, lack of proper maintenance, and sabotage. Examples include: 1) failed components, 2) cable disconnects; 3) loss of power, 4) poor data transmission; 5) lack of security; 6) mechanical wear, compression, tension, friction, shock, and vibration; 7) optical attenuation and scattering; and 8) signal blocking.
- *Intrusion* examples include: 1) unauthorized electromagnetic environment effects (E^3); 2) data capture through monitoring, tapping, or listening; and 3) injection of spurious signals. Intrusion sources include electrical storms and espionage. Intrusion presentation solutions include proper shielding, grounding, and encryption.
- *Stress Loading* includes the installation of devices that "load," impede, or degrade the quality or performance of an interface.
- *Physical Attack* includes physical threat contact by *accident* or *purposeful* action by an external entity on the system to inflict *physical harm, damage*, or *destruction* to a SYSTEM, entity, or one of their capabilities.

Interface Vulnerabilities

Interface *integrity* can be *compromised* through inherent *design defects, errors, flaws*, or *vulnerabilities*. Interface *integrity* and *vulnerability* issues encompass electrical, mechanical, chemical, optical, and environmental aspects of interface design. Today most awareness to interface vulnerability tends to focus on *secure* voice and data transmissions, and network firewalls. Vulnerability solutions include secure voice and data encryption; special, shielded facilities; armor plating; compartmentalization of tanks; cable routing and physical proximity; and operational tactics.

Interface Latency

Interface latency is a *critical issue* for some systems, especially if one interfacing element requires a response within a specified timeframe. As an SE, you will be expected to lead the effort that determines and specifies time constraints that must be placed on interface responses. If time constraints are critical, what is the allowable time budget that ensures the overall system can meet its own time constraints.

Interface Failure Mitigation and Prevention

When you design system interfaces, there are a number of approaches to *mitigate* the occurrence of interface failures or results. In general, the set of solutions have a broad range of costs. SEs often focus exclusively on the hardware and software aspects of the interface design. As a natural starting point, hardware and software *reliability, availability, and maintainability (RAM)*, in combination with *failure modes and effects analysis (FMEA)*, should be investigated.

System operation involves all of the *system elements*: PERSONNEL, EQUIPMENT, and SUPPORT. The point is that there may be combinations of system element actions or tactics that allow you to optimize system performance while reducing system cost.

EXAMPLE 12.9

Mirrors on vehicles provide a critical interface for the vehicle operator. The mirrors "bend" the operator's line of sight, thereby enabling the operator to maneuver the vehicle in close proximity to other vehicles or structures. Vehicles, such as trucks, with side mirrors that extend away from the cab are especially vulnerable to being damaged or destroyed.

Hypothetically, one solution is to design outside mirrors with deflectors that protect the mirrors from damage. However, since vehicles move forward and backward, deflectors on the front side of the mirror might limit the operator's line of sight. Additionally the cost to implement a design of this type would be expensive. A low-cost solution is to simply train the operators on the importance of *operational safety*. Thus, we preserve SYSTEM performance while minimizing EQUIPMENT costs.

12.11 SUMMARY

During our discussion of system interface practices we identified the key objectives of interfaces, identified various types, and emphasized the importance of system interface integrity.

Our discussion of logical and physical entity relationships should enable you to describe how the system elements interact at various levels of detail. The entity relationships, in turn, should enable you to assimilate the relationships into an architectural framework. To facilitate the decision-making process, we established the logical entity relationships or associations between the system elements via a logical architectural representation. Once the logical architecture was established, we progressed the decision-making process to the physical architecture representation or physical architecture.

GENERAL EXERCISES

1. Answer each of the *What You Should Learn from This Chapter* questions identified in the *Introduction*.
2. Refer to the list of systems identified in Chapter 2. Based on a selection from the preceding chapter's General *Exercises* or a new system selection, apply your knowledge derived from this chapter's topical discussions. Identify the following:
   (a) Types of interfaces.
   (b) Specific interfaces and objectives.
   (c) Examples of *generalized* and *specialized* solutions for each types of interface.

REFERENCES

Glossary: Defense Acquisition Acronyms and Terms, 10th ed. Defense Systems Management College (DSMC) 2001. Defense Acquisition University Press Ft. Belvoir, VA.

IEEE Std 610.12-1990. 1990. *IEEE Standard Glossary of Modeling and Simulation Terminology.* Institute of Electrical and Electronic Engineers (IEEE) New York, NY.

MIL-HDBK-470A. 1997. *Designing and Developing Maintainable Products and Systems.* Washington, DC: Department of Defense (DoD).

MIL-STD-480B (cancelled). 1988. *Configuration Control—Engineering Changes, Deviations, and Waivers.* Washington, DC: Department of Defense (DoD).

MIL-HDBK-1908B. 1999. *DoD Definitions of Human Factors Terms.* Washington, DC: Department of Defense (DoD).

Chapter 13

Organizational Roles, Missions, and System Applications

13.1 INTRODUCTION

System Owner and User organizational roles, missions, and objectives establish the driving need for system and mission capabilities and performance requirements. Each role, mission, and objective serves as the benchmark frame of reference for scoping and bounding what *is* and *is not* relevant as an organization's mission space.

Understanding the *problem, issue, or objective* the User is attempting to *solve, resolve*, or *achieve* is key to understanding WHY a system *exists* and WHAT *purpose* it serves within the System Owner's organization. HIGHER ORDER SYSTEMs, such as corporate enterprise management, shareholders, and the general public, have an expectation that *short-* and *long-term* benefits (survival, profits, return on investment (ROI), etc.) are derived by establishing an organization that fulfills marketplace needs.

Our discussion introduces the closed loop system of organizational entity relationships that define requirements for a system in terms of the organization's *roles, missions*, and *objectives*. The description centers on *technical accountability* and discusses how system capability and performance requirements are derived to address the organization's assigned OPERATING ENVIRONMENT *problem* and *solution spaces*.

Definitions of Key Terms

- **Paradigm** An in-grained mindset or model that filters or rejects considerations to adopt or employ new innovations and ideas that may impact the status quo.
- **Situational Assessment** An objective evaluation of current *strengths, weaknesses, opportunities*, and *threats* (SWOTs) of a SYSTEM OF INTEREST (SOI) relative to operating conditions and outcome-based objectives. Results of a situational assessment document the prioritized mission operational needs for the organization.
- **Strategic Plan** An *outcome*-based, global or business domain document that expresses the organizational vision, missions, and objectives of WHERE it wants to be at some point in time and *what* it wants to accomplish in the long term, typically five years or more hence. The challenge for most organizations is: *WHAT business or line of business (LOB) are your currently in versus WHAT do you WANT to be in five years from now versus WHAT LOB you should be in*?

System Analysis, Design, and Development, by Charles S. Wasson
Copyright © 2006 by John Wiley & Sons, Inc.

- **Tactical Plan** A near-term, mission-specific plan that expresses *how* the organization's leadership plans to deploy, operate, and support existing assets—such as people, products, processes, and tools—to achieve organizational objectives allocated from the *strategic plan* within time frame and resource constraints, typically one year or less.

13.2 ORGANIZATIONAL MISSIONS WITHIN THE ENTERPRISE

While many people have the perception that organizations procure systems based on a whim or simple need, we need to first *understand why* organizations *acquire* systems to support strategic and tactical objectives. The *operational need* is deeply rooted in understanding the organization's *vision* and *mission*.

Organizationally we model the strategic and tactical planning process as illustrated in Figure 13.1. In general, the process consists of a Strategic Planning Loop (1) and a Tactical Planning Loop (7). These two loops provide the basis for our discussion.

The Strategic Planning Loop (1)

The seed for long-term organizational growth and survival begins with an organizational *vision*. Without a vision and results-oriented plan for action, the organization's founders would be challenged to *initially* or *continually* attract and keep investors, investment capital, and the like.

The directional heading of most organizations begins with a domain analysis of the OPERATING ENVIRONMENT consisting of targets of opportunity (TOOs) and threat environment. The analysis task, which is scoped by the organizational vision, produces a *Market and Threat Assessment Report*. The report, coupled with the long-term organizational *vision* of what is to be accomplished, provides the basis for developing the organization's strategic plan.

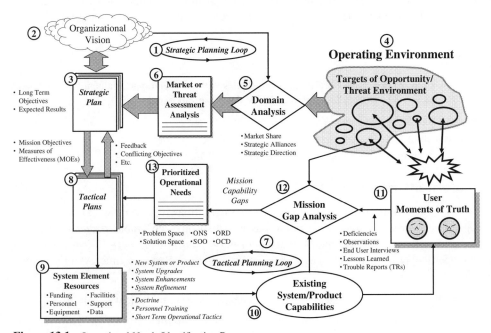

Figure 13.1 Operational Needs Identification Process

As the organization's capstone planning document, the strategic plan defines where the organization expects to be five years or more from now. The plan identifies a set of long-term objectives, each of which should be realistic, measurable, and achievable. As with any system, strategic planning objectives are characterized by performance-based metrics that serve as benchmarks for assessing *planned* versus *actual* progress.

Author's Note 13.1 *It is important to underscore the global nature of the strategic plan. The approved document forms the frame of reference for initiating tactical plans that focus on organizational LOB specific missions and objectives for their systems, products, or services.*

The Tactical Planning Loop (7)

Once the strategic plan is established, the key question is: *HOW do we get from WHERE we are NOW to five years from now?* The answer resides in creating and maintaining incremental, short-range tactical plans that elaborate near-term (e.g., one year) objectives and actions required to achieve strategic planning objectives.

Executive management decomposes *strategic* objectives into *tactical* objectives and assigns the objectives to various organizational elements. Performance-based metrics or *measures of performance* (MOPs) benchmark the required performance of each *tactical* objective. The MOPs serve as benchmarks for assessing *planned* versus *actual* progress in achieving the objectives.

Tactical Plans (8)

In response to the *tactical objectives*, each organizational element develops a *tactical plan* that describes HOW the each organization's leadership plans to achieve the objectives relative to the MOP benchmarked. In terms of HOW, the *tactical plan* describes the ways the Organizational System Elements (OSEs)—such as PERSONNEL, FACILITES, EQUIPMENT, PROCEDURAL DATA, and MISSION RESOURCES—will be *deployed*, *operated*, and *supported*. Thus, each OSE requires a specified level of capabilities and performance to support accomplishment of the tactical objectives. To illustrate an aspect of a *tactical plan*, consider the following example:

EXAMPLE 13.1

Assume an organization requires a fleet of 10 delivery vehicles with an operational availability of 0.95 and you only have 6 vehicles, the *tactical plan* describes HOW the organization plans to:

1. *Acquire* at least four additional vehicles.
2. *Operate* the vehicles to achieve organizational objectives.
3. *Maintain* the vehicles to achieve an operational availability of 0.95.

System Element Resources (9)

System element resources in inventory and their current conditions represent the existing system/product capabilities. Assuming the organization has a realistically achievable *strategic plan* and supporting tactical plans, these documents and supporting *organizational system elements* have a *shelf life*. Competitive or hostile threats, as well as opportunities, evolve over time. As a result, two types of situations occur:

1. Threat capabilities begin to *exceed* your organization's system element capabilities.
2. As new opportunities arise, each *system element* asset requires a projected level of performance to *defend* against threats or *capitalize* on the opportunities.

Given that the organizational system elements have a shelf life, your organization may find itself with gaps between *WHAT you currently have* and *WHAT you operationally NEED to survive organizational threats or capitalize on opportunities.*

As the tactical plans for achieving specific missions mature and are approved, *system elements*—namely PERSONNEL, FACILITIES, DATA, and EQUIPMENT—are fully funded to achieve the mission objectives. The updates may include:

1. Deployment of a new system, product, or service.
2. Upgrades, enhancements, and refinements to existing systems, products, or services.
3. Updates to organizational doctrine and command media revisions.
4. Personnel training and skills enhancement.
5. Revisions to operational tactics.

Depending on the *level of urgency* for these capabilities, a tactical plan may require several days, weeks, months, or a year to implement and bring the organizational capabilities up to required level of performance. Consider the following example:

EXAMPLE 13.2

An organization fields a system or product with an Initial Operational Capability (IOC) and incrementally upgrading via "builds" until a Full Operational Capability (FOC) is achieved.

Additionally, until the required capabilities are firmly established, interim *operational tactics* may be employed to project a perception to *adversarial* or *competitive* threats to a capability that may only exist in virtual space. History is filled with examples of decoy systems or products that influence competitor, adversary, or customer perceptions of reality until the actual system, product, or service capability is fielded.

Guidepost 13.1 *At this point, recognize that the organization was established to capitalize on targets of opportunity (TOOs) in the marketplace. As the organization delivers or employs those products or services in the OPERATING ENVIRONMENT, the organization must continually assess a system's operational utility, suitability, availability and effectiveness via mission gap analysis. The analysis collects and analyses data from User interviews, observations, lessons learned, trouble reports (TRs), and deficiencies, for example—comprising the mission capability gaps. The bottom line is:*

1. Here's WHAT we set out to accomplish with our products and services.
2. Here's HOW they performed in the marketplace.
3. Here's WHAT our customers told us about their PERCEPTIONS and level of SATISFACTION.
4. Here's the SCORECARD on performance results.

Existing System/Product Capabilities (10)

Effective mission gap analysis requires a realistic, introspective assessment of the existing system/product capabilities. Organizations, by nature and "spin control" media relations, project a positive image outside the organization, thereby creating a "perception" that the organization may appear to be much stronger than the existing capabilities indicate. Depending on the context, serious business ethics may be at issue with specific consequences.

For internal assessment purposes, the path to survival *demands* objective, unbiased, realistic assessments of system, product, or service capabilities. Otherwise, the organization places itself and its missions at risk by believing their own rhetoric. This *paradigm* includes a concept referred to as *group think*, whereby the organization members *synergize* their thought processes to a level of belief that *defies* fact-based reality.

User Moments of Truth (11)

Real world field data based on User *moments of truth* represent a level of risk that does have an impact on the organization, physical assets, or human life. The physical interactions may be *aggressive*, *defensive*, or *benign* encounters as part of the normal course of day-to-day business operations.

Field engineers, User interviews and feedback, or broader-based User community surveys serve as key collection points for *moments of truth* data. The interviews accumulate User experiences, lessons learned, and best practices based on direct physical interactions with the TOOs or threat environment.

Author's Note 13.2 *A common challenge many organizations face is having decision makers who emphatically insist that they "know best" what the User wants. Although this may be true in a few instances, objective evidence should reasonably substantiate the claim. Otherwise, the claim may be nothing more than pompous chest beating, unsubstantiated observation, or assertion. Competent professionals recognize this trap and avoid its pitfalls.*

Mission Gap Analysis (12)

Mission gap analysis focuses more in-depth on a strengths, weaknesses, threats, and opportunities (SWOT) or gap analysis between the organization's existing system, product, or service capabilities, operational state of readiness and targets of opportunity (TOOs) or threats. The analysis includes conducting *what if* scenarios; assessment of operational strengths, definition of measures of effectiveness (MOEs) and measures of suitability (MOSs); and leveraged capabilities. Based on the mission gap analysis results, prioritized operational needs are documented and serve as inputs into tactical plans.

Referral For more information about MOEs and MOSs, refer to Chapter 34 on *Operational Utility, Suitability, and Effectiveness Practices* in Part II.

It is important to note here that the gap analysis may have two types of information:

1. The paper analysis comparison.
2. Real world field data based on actual physical interactions between the existing system or product and the TOOs or threat environment.

The *paper analysis* is simply an abstract analysis and comparison exercise based on documented evidence such as "brochureware" in trade journals, customer feedback, surveys, intelligence, and problem reports. The analysis may be supported by various validated models and simulations that can project the virtual effects of interactions between the existing system, product, or service capabilities and TOOs or threats. Though potentially lacking in physical substance and validation, the paper analysis approach should convey a level of risk that may have an impact on the organization, physical assets, human life, property, or the environment.

Prioritized Operational Needs (13)

The results of the mission gap analysis are documented as prioritized operational needs. Documentation mechanisms include *Mission Needs Statements (MNS)*, *Statement of Objectives (SOOs)*, *Operational Requirements Documents (ORDs)*, and *Operational Concept Descriptions (OCDs)*.

Obviously the nature of these documents focuses on technical capability requirements rather than resources to implement the capabilities. Various criteria and labels are assigned to indicate the PRIORITY level (numerical, phrases, etc.) such as absolutely mandatory, desired, and nice-to-have.

Guidepost 13.2 *The preceding discussion provides the backdrop as to HOW organizations identify marketplace system, product, and service needs. Let's shift our focus to understanding the technical aspect of HOW organizations employ organizational roles and missions to identify system capability requirements.*

13.3 THE SYSTEM AND MISSION CAPABILITIES FOUNDATION

Before a system is *acquired* or *developed*, SEs analyze the OPERATING ENVIRONMENT and *opportunity/problem space* to support system acquisition or development decision making. The results of these analyses ultimately *bound* the *solution space*(s) within which the MISSION SYSTEM is intended to operate. In performing this role, SEs support business development organizations attempting to bound the operating environment and define the types of missions the system must be capable of conducting.

System Objectives

Every system, product, or service is characterized by a set of operational need-based objectives that drive the system's missions and applications. Consider the following example:

EXAMPLE 13.3

An automobile manufacturer may conduct marketing analysis and identify an operational need(s) for a family vehicle that can comfortably provide general transportation for up to four adults. Further analysis reveals that families searching for this type of vehicle are sports oriented and need to carry bicycles, kayaks, or storage containers over rough roads in mountainous territory in all weather conditions. Thus, a general *system objective* is established for a class of vehicle.

Mission Objectives

Most families, however, do not have the luxury of performing sports all week and have other family-based *missions* (shopping, school carpools, etc.). Therefore, the vehicle's *system objectives* must be capable of supporting *multiple mission objectives* that are User-oriented and support all *system stakeholders*.

How do system and mission objectives relate to a system's purpose? Consider the following example.

EXAMPLE 13.4

A retail store may target a specific market niche based on demographics, products and services, and market environment—coupled with financial performance to form *organizational objectives*. Accomplishment of the *organizational objectives* requires supporting *system elements* of the business—such as PERSONNEL EQUIPMENT, and FACILITIES—on specific mission objectives—such as increasing customer awareness, boosting sales, improving profitability, and introducing new merchandise.

This leads to two key questions:

1. WHAT is the importance of delineating organizational mission and system objectives?
2. WHY do we do it?

The reason is the System Owner and/or the User has organization *roles* and *missions* to fulfill. *Roles* and *missions* are accomplished by the integrated set of organizational systems or assets such the PERSONNEL, EQUIPMENT, and PROCEDURAL DATA elements.

Each of these systems or assets provides *system capabilities* that fulfill a bounded *solution space* subject to affordability and other constraints. *System capabilities*, which vary by phase of operation, are identified, analyzed, and translated into *system or item specification requirements*.

Organizational Efficiency and Accountability

All multi-level elements of a system must be traceable back to *system objectives* that define the system's *purpose* or *reason for its existence*. The typical profit-centric business view is that each capability must:

1. Enable to organization to achieve both strategic and/or tactical planning objectives.
2. *Efficiently* and *effectively* provide streamlined, value-added benefits—such as return on investment (ROI)—to the organization.

Therefore, from an SE perspective, there must be some form of *system accountability*.

System accountability means that all system capabilities, requirements, and "end game" performance must be:

1. Assigned to organizational owners—Users, Acquirer, System Developer personnel, Subcontractors, etc.—or *organizational accountability*.
2. Traceable back to higher level *system objectives* that identify and bound the opportunity space or solution space—or *technical accountability*.

Organizations establish *technical accountability* by a "linking" process referred to as requirements TRACEABILITY. To better understand the *requirements traceability* process, let's examine the *key entities of technical accountability*.

13.4 KEY ENTITIES OF TECHNICAL ACCOUNTABILITY

The key entities of technical accountability include organizational objectives, system objectives, and mission objectives. To see how these topics interrelate from a technical accountability perspective, refer to Figure 13.2.

Operating Environment (1)

The organization characterizes its OPERATING ENVIRONMENT in terms of problem/opportunity spaces as illustrated in Figure 13.2. From an organizational perspective, each *problem/opportunity space* includes a number of *validated* operational needs that can be partitioned into *solution spaces*. Depending on the type of organization—whether military or business—the *operational need* may be driven by the marketplace or by threats. In either case, *targets of opportunity* (TOO) may emerge.

You may ask, how does an organization respond to its OPERATING ENVIRONMENT? Corporate management assigns responsibilities to most organizations to provide *systems*, *products*, and *services* to customers. To fulfill the corporate charter, each organization performs various *roles* and conducts *MISSIONS* to achieve goals, be they financial, or otherwise.

13.4 Key Entities of Technical Accountability

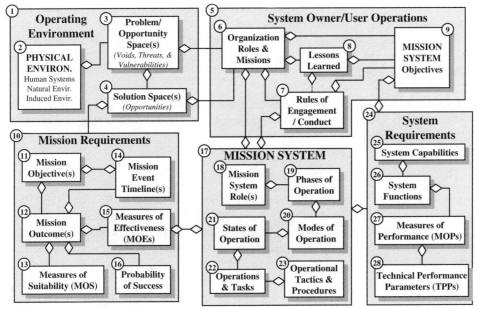

Figure 13.2 Understanding the Entity Relationships of Organizational & System Roles

At the highest level, the organization assesses the OPERATING ENVIRONMENT for threats and targets of opportunity (TOOs), whichever is applicable. Results of the assessment and derived strategy are documented in the organization's strategic plan and implemented via a series of mission-oriented tactical plans.

In performing this corporate mission, each organization must use existing resources to procure, implement, and upgrade a MISSION SYSTEM (17) that can perform the missions and achieve the objectives. This, in turn, requires specifying system "solutions" that can achieve specific mission objectives.

To satisfy the solution space(s), System Owner/User Operations (5) and Mission Requirements (10) are identified.

Mission Requirements (10)

Fulfillment of a *solution space* may require several types of missions to satisfy, correct, or fill the void. Each type of mission is documented by a *strategic plan* and supporting *tactical plans*. These plans identify, define, and document specific mission requirements.

Author's Note 13.3 *In the interest of space and simplicity, we have arbitrarily abstracted the MISSION REQUIREMENT elements as shown. Entities such as resources should also be considered. The six entities presented are intended to illustrate the technical "thought process."*

Based on an understanding of the *mission requirements*, let's shift to understand HOW a MISSION SYSTEM (17) operates to conduct its missions.

Mission System (17)

When tasked by HIGHER ORDER SYSTEMS—namely by System Owners/Users—each MISSION SYSTEM (17) must be *physically configured* to meet the specific mission requirements.

These requirements include mission objectives, mission outcomes, and Mission Event Timelines—as part of the organizational roles and missions. Additionally, the MISSION SYSTEM (17) is selected as the "system of choice" and is linked to the Mission Requirements (10). For example, Phases of Operation (19) and Modes of Operation (20), Mission Outcome(s) (12), and Mission Event Timelines (14) must be linked.

Guidepost 13.3 *At this point we have established the mission requirements and identified the MISSION SYSTEM (17). The final part of our discussion focuses on the System Requirements (24) of the selected MISSION SYSTEM.*

System Requirements (24)

MISSION SYSTEM objectives are achieved by procuring and implementing MISSION SYSTEMS (17) designed specifically for satisfy Mission Requirements (10). In general, systems are procured to handle a diversity of missions ranging from highly specialized to very general. The mechanism for documenting the total set of mission requirements is the system specification or *System Performance Specification (SPS)*.

Author's Note 13.4 *Under System Requirements (24), there should also be entity relationships for nonfunctional requirements. Because of space limitations, Figure 13.2 only depicts the functional requirements.*

As Figure 13.2 shows, System Requirements (24) *specify* and *bound* System Capabilities (25), which consist of System Functions (26). Each system function is bounded by at least one or more *Measures of Performance* (MOPs) (27). *Technical Performance Parameters* (TPPs) (28), which represent areas of performance that are critical to the mission and system, are selected for special tracking from the set of *Measures of Performance* (MOPs).

Referral For more information about traceability of system capabilities, please refer to Chapter *System Operational Capability Derivation and Allocation*.

13.5 UNDERSTANDING ORGANIZATIONAL ROLES AND MISSIONS

MISSION SYSTEM and SUPPORT SYSTEM roles are determined by their role-based contexts within the System of Systems (SoS). Human-made systems consist of integrated *supply chains* of systems in which an organization produces systems, products, and services to support another "downstream" system. Two key points:

- In the system-centric context, a MISSION SYSTEM performs specific objectives defined by contract, tasking, or personal motivation.
- In the *supply chain* context, the same system serves as a SUPPORT SYSTEM to another via the delivery of systems, products, or services.

As a result, every system in the SoS supply chain serves in two contextual roles as:

1. MISSION SYSTEM
2. SUPPORT SYSTEM

13.5 Understanding Organizational Roles and Missions

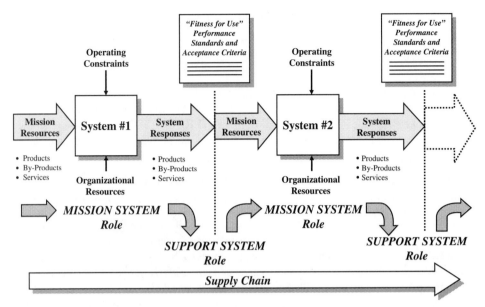

Figure 13.3 Understanding the MISSION SYSTEM and SUPPORT SYSTEM Roles Supply Chain

If we investigate the MISSION SYSTEM and the SUPPORT SYSTEM roles and their interactions with the SOI's OPERATING ENVIRONMENT, every system performs a *function*, by definition, and produces *value-added products*, *by-products*, and *services* that are used by other systems. In effect we can establish a *customer–supplier relationship* boundary between the system and external systems within its OPERATING ENVIRONMENT.

We can describe the SOI's value-added processing graphically as shown in Figure 13.3. System 1 performs *value-added processing* to transform MISSION RESOURCES inputs—namely data and raw materials—into SYSTEM RESPONSES outputs—namely behavior, products, by-products, and services. The success of MISSION SYSTEM operations is subject to HIGHER ORDER Systems:

- *Operating Constraints*—Command-and-control mission tasking, policies and procedures, contracts, etc.
- *Organizational Resources Constraints*—Processes and methods, facilities, training, tools, personnel, etc.

"Fitness-for-Use" Criteria

Customers or Acquirers of systems, by virtue of *validated* operational needs, have *minimum* requirement *thresholds* and *expectations* that must be met to ensure the *products*, *by-products*, and *services* they acquire are acceptable for use by their technical capability, quality, and safety and are not detrimental to the environment and human health. Analytically, we refer to these expectations and requirements as *fitness-for-use criteria*. Contractually, fitness-for-use criteria establish a basis for:

1. Acquirer (role) *specification of* system/entity requirements
2. Acceptance of contract deliverables.

System Interfaces and PRODUCER–SUPPLIER Relationships

The construct depicted in Figure 13.3 represents linkages within an overall chain of SUPPLIER (role) to CUSTOMER (role) exchanges. We refer to the "linkages" as supplier–customer role relationships.

If we analyze each system within the supplier–customer *chain*, we discover that every system has a *mission and objectives* to achieve. The missions and objectives focus on *products*, *by-products*, and *services* provided by a SYSTEM OF INTEREST (SOI) to satisfy customer operational needs and provide a *return on investment* (ROI) to the supplier's organizational stakeholders. Therefore, a system fulfills two roles:

1. A *Mission Role* Produce value-added products, by-products, and services.
2. A *Support Role* Deliver those products, by-products, and services to other systems.

Each system in the *supplier–customer* chain performs MISSION SYSTEM and SUPPORT SYSTEM roles relative the *predecessor* and *successor* interface systems.

As illustrated by the figure, System 1 performs a MISSION SYSTEM role as a producer of *value-added products*, *by-products*, and *services* to meet *consumer marketplace* or *contract* requirements. The marketplace or contract establishes *fitness-for-use standards* and *acceptance criteria* for use by System 2. As a *supplier* of products, by-products, and services, System 1 performs a SUPPORT SYSTEM role to System 2.

To illustrate HOW organizational elements perform in their MISSION SYSTEM and SUPPORT SYSTEM roles, let's explore the role-based interaction between a system development program and the accounting department.

EXAMPLE 13.5

A system development program, as a MISSION SYSTEM, designs and develops a *product* and provides *services* to fulfill the requirements of a contract. The program organization, functioning in a SUPPORT SYSTEM role to the accounting organization, provides weekly labor charges—namely *mission support* data.

The accounting organization, functioning in its MISSION SYSTEM role, is tasked to collect, analyze, and report financial data via accounting reports (products) and services. The accounting organization, functioning in its SUPPORT SYSTEM role, provides timely *mission support* data and services to system development programs to assess earned value against plans that enable them to make corrective actions, if necessary, to meet their contract performance obligations.

EXAMPLE 13.6

During NASA Space Shuttle missions, organizations such as the Kennedy Space Center (KSC) in Florida, Mission Control at the Johnson Space Center (JSC) in Houston, the Marshall Space Flight Center (MSFC) in Huntsville, Alabama, other NASA centers, and Vandenburg Air Force Base (VAFB) in California perform SUPPORT SYSTEM roles that include *pre-mission*, *mission*, and/or *postmission* operations support.

Let's delineate the context of the MISSION SYSTEM and SUPPORT SYSTEM roles further.

- While in its MISSION SYSTEM role, a SOI conducts assigned missions to accomplish mission objectives.
- While performing its MISSION SYSTEM role, a SOI serves as a SUPPORT SYSTEM to the HIGHER ORDER SYSTEM that assigned the task (vertical entity relationships). In so doing, the SOI may support other peer level systems in their respective MISSION SYSTEM roles (horizontal entity relationships).

Table 13.1 Organizational roles and system application rules

Rule	Title	Role-Based System Rules
13.1	Organizational vision, mission, and objectives	Establish your organization's vision, mission, and objectives in a *strategic plan*.
13.2	Tactical planning	Based on your organization's strategic plan's vision, mission and objectives, establish mission driven tactical plans.
13.3	Capability gap analysis	Conduct a gap analysis of organizational and system *strengths*, *weaknesses*, *opportunities*, and *threats* (SWOTs) capabilities required to achieve the *strategic* and *tactical plans*.
13.4	Targets of opportunity and threat environments	Based on your organizational vision, conduct a gap analysis of the global marketplace's targets of opportunity (TOOs) and threat environments.
13.3	Strategic planning	Establish an organizational strategic plan that defines the vision for where the organization intends to be in 5 years or longer.
13.4	Producer and supplier roles	Every system performs two types of supply chain roles: 1. As a MISSION SYSTEM, the system is a *producer* of *products*, *by-products*, or *services* based on marketplace needs or contract requirements. 2. As a SUPPORT SYSTEM, the system is a *supplier* of *products*, *by-products*, or *services* to satisfy Acquirer system needs.
13.5	Value-added producer-supplier	In its MISSION SYSTEM role, every system: 1. Consumes MISSION RESOURCES inputs—namely expendable and consumables such as raw or processed materials and data. 2. Transforms, converts, or processes mission resources inputs to add value. 3. Supplies value-added *products*, *by-products*, and *services* that are used by other systems.
13.6	Acquirer fitness for use criteria	Each instance of system *products*, *by-products*, and *services* supplied to an Acquirer must meet *pre-defined* and *agreed to* fitness for use standards and performance criteria.

Organizational Roles and System Application Rules

Based on the preceding discussion, we can establish several *organizational roles* and *system application* development rules. Table 13.1 provides a summary of these rules.

13.6 GUIDING PRINCIPLES

In summary, the preceding discussions provide the basis with which to establish the guiding principles that govern organizational roles, missions, and system applications concepts.

Principle 13.1 Every system performs two contextual roles: a MISSION SYSTEM role and a SUPPORT SYSTEM role.

Principle 13.2 As MISSION SYSTEM, a system performs task-based missions to achieve specific outcomes.

Principle 13.3 As a SUPPORT SYSTEM, a system delivers products, by-products, or services that serve as the MISSION RESOURCES element to another system's MISSION SYSTEM role.

Principle 13.4 A system lacking purpose and missions reflects organizational neglect, product obsolescence, or both.

Principle 13.5 For a supply chain to be successful, every system input and output must meet pre-defined sets of fitness for use performance criteria established with Users.

13.7 SUMMARY

During our discussion of organizational roles and system applications, we employed entity relationships to illustrate *how* the OPERATING ENVIRONMENT's opportunity and solution spaces are linked to mission requirements, User operations, mission systems, and subsequently system requirements.

We addressed the highly iterative strategic planning loop and tactical planning loop. Each of these loops provide situational assessments of existing organizational systems, products, and services relative to TOOs and threats in the operating domain and the organization's strategic and tactical plans. When the level or urgency triggers the need to procure a new system, product, capability, or service, work products (ONS, SOOs, ORD, etc.) are generated to serve as inputs to the *system procurement phase*.

GENERAL EXERCISES

1. Answer each of the *What You Should Learn from This Chapter* questions identified in the *Introduction*.
2. Refer to the list of systems identified in Chapter 2. Based on a selection from the preceding chapter's General *Exercises* or a new system selection, apply your knowledge derived from this chapter's topical discussions. Specifically, use Figure 13.2 as a topical framework to create a presentation for the selected system or product. Describe how its capabilities support the organizational mission.

ORGANIZATIONAL CENTRIC EXERCISES

1. Depending on the nature of your business, use Figure 13.2 as a framework to describe your organization's or User's organizational roles, missions, and system requirements.

ADDITIONAL READING

CARLZON, JAN. 1987. *Moments of Truth: New Strategies for Today's Customer-Driven Economy.* New York: Harper and Row.

GOODSTEIN, LEONARD, NOLAN, TIMOTHY, and PFEIFFER, J. WILLIAM. 1993. *Applied Strategic Planning: How to Develop a Plan that Really Works.* New York: McGraw-Hill.

MORRISEY, GEORGE L. 1996. *A Guide to Strategic Thinking: Building Your Planning Foundation.* San Francisco: Jossey-Bass.

Chapter 14

Understanding the System's Problem, Opportunity, and Solution Spaces

14.1 INTRODUCTION

The concept of identifying and bounding the *problem space* and the *solution space* is one that you occasionally hear in buzzword vocabularies. Sounds great! Impresses bystanders! However, if you ask the same people to differentiate the *problem space* from the *solution space(s)*, you either get a blank stare or a lot of animated arm-waving rhetoric.

Most successful missions begin with a thorough identification and understanding of the *problem, opportunity*, and the *solution space(s)*. One system's *problem space* may be an *opportunity space* for another that desires to capitalize on the weakness.

This chapter introduces the concept of the *problem/opportunity space* and its *solution space(s)*. Once you understand the organization's roles and mission, the first step is to understand WHAT *problem* the User is trying to solve. People can become experts in analyzing system requirements; however, system analysis, design, and development *success* begins with understanding the User's *problem/opportunity* and *solution spaces*. A *worst-case scenario* is writing perfectly worded requirements for the *wrong* problem.

Our discussions begin with an overview of the *problem space* and formulation of the *problem statement*. The discussion continues with an overview of the *solution space* and its relationship to all or a portion of the *problem space*. We contrast common *misperceptions* of the *solution space* using a geographical context. Although the *solution space* is capability based, we illustrate through examples how a lesser capability can leverage the capabilities of other systems to greatly project or expand, as a force multiplier, its *sphere of influence*.

What You Should Learn from This Chapter

1. What is a *problem space*?
2. What is an *opportunity space*?
3. What is the relationship between a *problem space* and an *opportunity space*?
4. What is a *solution space*?
5. What is the relationship between the *problem/opportunity space* and a *solution space*?
6. How do you write a *problem statement*?

System Analysis, Design, and Development, by Charles S. Wasson
Copyright © 2006 by John Wiley & Sons, Inc.

7. Identify three rules for writing *problem statements*.
8. How do you forecast the *problem/opportunity spaces*?
9. How does an organization resolve gaps between a *problem space* and its *solution space(s)*?

Definitions of Key Terms

- **Opportunity Space** A *gap* or *vulnerability* in a system, product, or service capability that represents an opportunity for: 1) a competitor or adversary to exploit or 2) supplier to offer solutions.
- **Problem Space** An abstraction within a system's OPERATING ENVIRONMENT or mission space that represents an *actual, perceived,* or *evolving* gap, hazard, or threat to an existing capability. The potential threat is *perceived* either to pose some level of financial, security, safety, health, or emotional, risk to the User or to have already had an adverse impact on the organization and its success. One or more lower level solution space systems, products, or services resolve the problem space.
- **Problem Statement** A brief, concise *statement of fact* that clearly describes an *undesirable* state or condition without identifying the source or actions required to solve the problem.
- **Situational Assessment** Refer to the definition in Chapter 13 on *Organizational Roles, Missions, and System Application*.
- **Solution Space** A bounded abstraction that represents a capability and level of performance that, when implemented, is intended to satisfy all or a portion of a higher level *problem space*.

14.2 UNDERSTANDING OPERATING ENVIRONMENT OPPORTUNITIES

HUMAN-MADE SYSTEMS, from an organizational viewpoint, exploit *opportunities* and respond to *threats*. Organizations assign *missions* and *performance objectives*—financial, market share, and medical—that support the founder's or system owner's vision of *capitalizing* on opportunities or *neutralizing* threats.

Survivalist *opportunity* and *threat* motives are common throughout the OPERATING ENVIRONMENT. Depending on one's perspective, some refer to this as the "natural ordering of systems." The animal kingdom that exists on the plains of Africa is an illustrative example. One animal's prey—or *opportunity*—may be viewed by the prey as a *threat*. Consider the following example:

EXAMPLE 14.1

An aggressor organization views a fledgling business as an *opportunity* to exploit or capitalize on the fledgling's weaknesses. In turn, the larger organization may be viewed by the fledgling business as the *threat* to its survival.

Types of System Opportunities

Opportunities generally are of two basic types:

1. *Time-based*. Waiting for the right time.
2. *Location-based*. Waiting for a lease to expire.

Let's explore both of these further.

Time-Based Opportunities. *Time-based opportunities* can occur *randomly* or *predictably*. *Random opportunities* are sometimes viewed as "luck." *Predictable opportunities* are dependent on periodic or repeatable behavioral patterns (i.e., knowledge applied to practice) that enable an aggressor system to capitalize on a situational weakness.

Location-Based Opportunities. *Location-based opportunities*, as the name implies, relate to being in the *right* place at the *right* time. In the business world success is often said to be driven by "*Location! Location! Location!*" Obviously, a good location alone does not make a business successful. However, the location positions the business for mission success.

14.3 UNDERSTANDING THE PROBLEM SPACE

One of the first steps in SE is to understand WHAT problem the User is attempting to solve. The term *problem space* is relativistic. Consider competition in the commercial marketplace or military adversaries. An organization may view a competitor or adversary and their *operating domain* as a *problem space*. Hypothetically, if you were to ask the competitor or adversary if they were a "problem," to the other organization, they may state unequivocally "yes!" or emphatically "no!" Therefore, the context of a *problem space* resides in the eyes and minds of those who *perceive* the situation. Sometimes there is little doubt as evidenced by *acts of aggression* or *hostility* such as invasion of a country's air space or hostile business takeovers.

The term *problem* has two contexts: 1) a User-Acquirer's perspective and 2) a System Developer perspective. A *problem space* for the User-Acquirer represents an *opportunity—solution space* for the System Developer. In turn a System Developer's *problem space* of finding a design solution becomes an *opportunity space* for subcontractors, vendors, and consultants to offer solutions.

OPPORTUNITY Versus PROBLEM Semantics

Technically a problem does not exist until the *hazard* that poses a potential *risk* occurs. Then you actually have a problem! The infamous "*OK, Houston, we have a problem...*" communicated by *Apollo 13* Commander Jim Lovell is one of the best illustrations of the context used here.

In a highly competitive marketplace and adversarial, hostile world, survival for many organizations requires proactive *minimization* of system *vulnerability*. Organizations that are proactive in recognizing *opportunities* initiate risk mitigation actions to prevent hazards from occurring and becoming problems—or tomorrow's corporate headlines. In contrast, procrastinators deal with problems by becoming *reactionary* "firefighters," assuming that they were aware of the potential *hazard* and did not *mitigate* it; they seem to never get ahead. Since the term *problem space* is commonly used and SE focuses on *problem solving*, this text uses the term *problem space*.

Problem Solving or Symptom Solving?

Organizations often convince themselves and their executive management they are *problem solving*. In many cases the so-called *problem solving* is actually *symptom solving*. This question leads to critical question for the User, Acquirer, and System Developers: *Is this the RIGHT problem to solve or a downstream symptom of the problem*?

Dynamics of the Problem Space

For most organizational systems, *problem spaces* are *dynamic* and *evolutionary*. They evolve over time in a number of ways. Some occur as instantaneous, catastrophic events, while others emerge over several years—such as the hole in the ozone layer of Earth's atmosphere. The *root causes* for *problem spaces* in system capabilities and performance originates from several potential sources such as:

1. System neglect.
2. Improper oversight or maintenance.
3. Ineffective training of the User Operators.
4. Budgetary constraints.

Organizationally, managers have an obligation to track *problem spaces*. The problem is that some managers are *reluctant* to surface the *problem spaces* until it is too late. In other cases management *fails* to provide the proper *visibility* and *priority* to seemingly trivial issues until they become full-fledged *problem spaces*—the proverbial "head in the sand." When this happens, four potential outcomes can occur:

1. Operationally, the source of the *problem space* goes away.
2. The organization's management becomes distracted or enamored by other priorities.
3. Organizational objectives change.
4. Catastrophic (worst-case) events heighten awareness and sensitivity.

In general, people tend to think of the *problem space* as *static*. In fact, the primary issue with the *problem space* is its continual *dynamics*, especially in trying to bound it.

Forecasting the Problem Space

The challenge for most organizations is *How do we forecast a problem space in system capabilities with some level of confidence?* The answer resides in the organizational and system level *strategic* and *tactical plans*, system missions, and objectives.

Organizational strategic and tactical plans establish the reference framework for evaluating *current* capabilities versus *planned* capabilities. Using these objectives as the basis, *situational assessments* and *gap analysis* are employed as tools to compare the state of existing to *projected* system capabilities and performance against projected capabilities and performance of *competitors* or *adversaries*. The results may indicate a potential GAP in capabilities and/or levels of performance. The identification of the gap establishes the basis for organizational problem spaces. Dynamically, *gaps* may occur rapidly or evolve slowly over time.

When Does a "Gap" Become a Problem? This is perhaps the toughest question, especially in a forecasting sense. Obviously, you know you have a *problem* when it occurs such as *malfunction*, *emergency*, or *catastrophic events*. One approach may be to determine whether a potential hazard with a level of risk has outcome-based, consequences that are *unacceptable*. In effect, you need to establish levels or thresholds for assessing the *degree of problem significance*.

Establishing Problem Space Boundaries. Conceptually, we illustrate a *problem space* with solid lines to symbolically represent its perimeter. For some systems such as a lawn, the property boundaries are clearly defined for ownership accountability. In other cases, the boundaries are *elusive* and *vague*. Consider civil unrest and wars in countries where people take sides but look, dress, and communicate similarly. *How does one differentiate friend versus foe? On WHICH day of the week?*

Lines drawn around abstractions such as ideology, politics, and religion are often blurry, vague, and ill defined. To see this, consider the graphic shown in Figure 14.1. The figure illustrates problem space boundaries by gray edges. The "center of mass" is indicated by the dark area whose edges, however, are blurry and indistinct. In some cases the problem space has tentacles that connect to

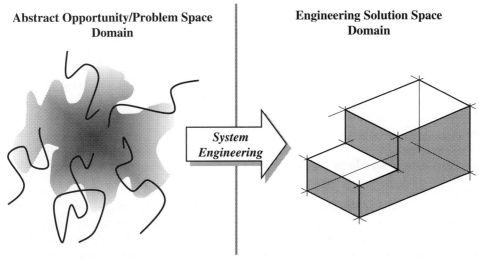

Figure 14.1 SE—Translating the Abstract Opportunity/Problem Space into an SE Solution Space

other problem spaces, each having an effect on the other. The blurriness may not be *static* but a continuum of *dynamic*, evolving changes as with clouds or a thunderstorm.

Controlling the Problem Space

Depending on the *source* or *root cause* of the problem and the *degree of risk* to your system or its objectives, the natural tendency of most organizations is to *eliminate the problem*—namely the problems within their control, resources, or *sphere of influence*. However, the reality is you may not be able to *eliminate* the *problem space*. At best, you may only be able to *manage* and *control* it—that is, keep it in balance and check. Consider the following example:

EXAMPLE 14.2

Weeds are a continual problem for lawns and reappear annually for a variety of reasons. There are a number of approaches to getting rid of the weeds, some more desirable than others in terms of environmental, health, application, and cost considerations. Since you have limited control over how the weeds get transplanted (winds, water, birds, etc.), you have a choice: 1) coexist and control the weeds or 2) "pay the price" to hire a contractor to eliminate the weeds.

Defining the Problem Statement

Our discussion to this point focuses on the *problem space* in an abstract sense. The rhetorical question that requires specificity is: *WHAT problem is the User attempting to solve?* Before any *solution* analysis can proceed, it is crucial for you and your development team, preferably in collaboration with the User, to simply document WHAT problem the User is attempting to solve. You need to define a *problem statement*. Ultimately this leads to the question: *HOW should a problem statement be written?* Although there are a number of ways of developing a *problem statement*, there are some general guidelines to apply:

1. Avoid identifying the *source* or *root cause* of a problem.
2. Identify the operational scenario or operating conditions under which the problem occurs.
3. Avoid stating any explicit or implicit solutions.

Consider the following example:

EXAMPLE 14.3

Viruses are corrupting desktop computers on our network.

Note that the example does not specify: WHERE the viruses originate, the source or root cause of the problem, WHAT the impact is, or HOW to solve the problem.

Partitioning the Problem Space

As your understanding of the *problem space* matures, the next step is to partition it into one or more *solution spaces*. Work with system stakeholders to partition the complex *problem space* into more manageable *solution spaces*. The identification of one or more *solution spaces* requires highly iterative *collaboration*, *analysis*, and *decision making*. Consider the challenge of attempting to partition the ambiguous *problem space* shown at left side of Figure 14.1. Through partitioning, our objective is to isolate key *properties* and *characteristics* of the problem as abstractions that enable us to ultimately develop solutions.

Author's Note 14.1 *Problem solving requires establishing a conceptual solution as a starting point. This solution may evolve throughout the process and may not be recognizable at completion. Some people are very ineffective at conceptualizing the starting point. One approach is to gather the facts about a problem space and create DRAFT solution space boundaries. Then, as the analysis progresses, adjust the boundaries until decisions about the solution space boundaries mature or stabilize. The key point is: some problem spaces are described as "wicked" due their dynamic nature and are effectively unsolvable. In general, you have a choice:*

1. *Flounder in the abstractness, or*
2. *Make a decision, move on to the next decision, and then revisit and revise the original decision when necessary.*

Problem Space Degree of Urgency

The *level of risk* and *degree of urgency* of the *problem space* as a whole or portions thereof may influence or drive *solution space* decisions, especially, where budgets or technology are constrained. The net result may be a decision to prioritize *solution spaces* and levels of capability within *solution spaces*.

We meet this challenge by establishing an Initial Operational Capability (IOC) at system delivery and acceptance. Then, as budgets or technologies permit, IOC is followed by a series of *incremental builds* that enhance the overall capability. Finally, as the system matures with the integration of the "builds," an overall capability, referred to as a Full Operational Capability (FOC), is achieved.

To illustrate the partitioning of the *problem space* into *solution spaces*, consider the graphic in Figure 14.2. Symbolically, we begin with a problem space represented by a large box. Next, we partition the box into five solution spaces, each focused on satisfying a set of problem space capability and performance requirements allocated to the solution space. Initially we could have started with four or six solution spaces. Through analysis, we ultimately decide there should be five solution spaces. So, *HOW does this relate to system development*? The large box symbolizes the total system solution. We partition the complexity of the system solution into PRODUCT level or SUBSYSTEM level solution spaces. Let's explore this point further.

14.4 Understanding the Solution Space(s)

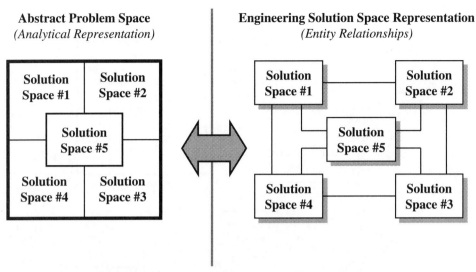

Figure 14.2 Partitioning the Problem Space into an SE Solution Space Representation

Problem Space Partitioning and Decomposition

When SEs deal with problem spaces, the exercise is accomplished through a multi-level decomposition or narrowing process as shown in Figure 14.3. We begin with the high-level *problem space* shown in the upper left-hand corner of the diagram. We partition the problem space into four *solution spaces*, 1.0 through 4.0. Solution space 3.0 becomes problem space 3.0 for the next lower level and is partitioned into solution spaces 3.1 through 3.4. The partitioning process continues to the lowest level. The net result of the narrowing process is shown in the upper right-hand corner of the figure.

Guidepost 14.1 *Given a fundamental understanding of the problem space and its relationship to the solution space, we are now ready to explore the development of the solution space.*

14.4 UNDERSTANDING THE SOLUTION SPACE(S)

Solution spaces are characterized by a variety of boundary conditions:

1. Distinct, rigid boundaries.
2. Fuzzy, blurry boundaries.
3. Overlapping or conflicting boundaries.

The degree to which the solution space is filled is determined by the capabilities required, priorities assigned to, and resources allocated to the problem space "zone." HIGHER ORDER SYSTEMS contractually and organizationally impose RESOURCE and OPERATING CONSTRAINTS system elements that may ultimately limit the degree of solution coverage. Consider the following examples:

EXAMPLE 14.4

Because of reduced budgets and funding, home refuse pick up and disposal service may be reduced from two days per week to one day per week.

142 Chapter 14 Understanding the System's Problem, Opportunity, and Solution Spaces

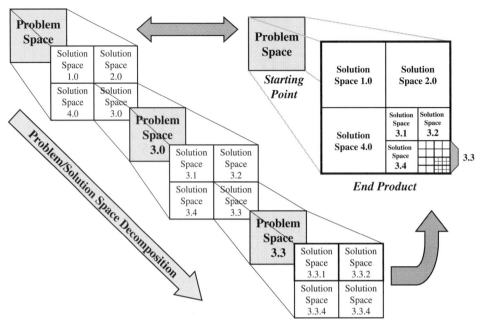

Figure 14.3 Decomposing the Problem Space into Manageable Pieces

EXAMPLE 14.5

A solution space normally satisfied by airline flights between two cities will be reduced from three flights per week to one flight per week during the non-tourist season.

Eliminating the Problem–Solution Space Gap

When a gap is identified in a system or product capability, it generally takes a finite amount of time to fill, especially if system development is involved. If the gap is of a *defensive* nature, the system or product may be *vulnerable* or *susceptible* to *acts of aggression* and hostilities from competitors or adversaries. If the gap represents a deficiency in an *offensive* capability, work must be performed to eliminate the gap by upgrading system capabilities and performance.

Depending on the system or product's application, *operational tactics* such as decoys, camouflage, and operational patterns may be employed to supplement the gap until a new system, product, or service is available.

Solution Capability Force Multipliers

Most people tend to think of the *solution space* in a geographical boundary context. Our portrayal in Figure 14.3 resembles real-estate plots. Remember, the *problem/solution spaces* are *capability* based. So, *what does this mean*?

Capability-based solutions represent *abstractions* of the *strength*, *capacity*, and *reliability* required to accomplish mission objectives.

1. *Strength*. Power to accept a specific type of mission challenge.
2. *Capacity to Project*. Multiply that power over a defined range.
3. *Reliability*. A probability of completing a mission of a given duration in a specified OPERATING ENVIRONMENT.

You can expand the solution capability of a system by synergistically leveraging capabilities of other systems to dramatically change their *sphere of influence*. Consider the following example:

Author's Note 14.2 *Recall the shadow game children play by standing in front of a ground-based floodlight pointed toward a nearby wall? The floodlight, in combination with body movements, projects a greater than life size silhouette onto the wall, thereby creating for the child the perception of a much larger person. So we can leverage the geometry of object position and light to create perceptions of capabilities beyond our individual capability. The same is true for the organizational or operational tactics for creating perceptions that are virtual but not reality.*

The key point is you don't have to build an expensive aircraft to cover X square miles. What you need to do is figure out *how* to build a reasonably priced aircraft and be able to "project" that capability into a $10X$ square mile domain by leveraging other systems and their capabilities. Consider the following example:

EXAMPLE 14.6

A fighter aircraft has a limited engagement range due to fuel capacity. However, by leveraging the capabilities of air-to-air refueling, the fighter can command a much larger area of coverage.

Author's Note 14.3 *You may recall from our discussion in Chapter 4 on system attributes, properties, and characteristics that systems have a frame of reference and an operating domain. For example, an aircraft has a home base—or frame of reference—and has an operating range—or domain—that is limited by fuel consumption and maintenance considerations. Employing "tanker" aircraft to replenish fuel in-flight can expand the aircraft's effective operating range.*

Selecting Candidate Solutions for the Solution Space

Each *solution space* is bounded by technical, technology, support, cost, and schedule constraints. The challenge is to identify and evaluate several *viable* candidate solutions that satisfy the technical requirements and then recommend the preferred solution. The selection requires establishing *pre-defined*, *objective criteria* and then performing a trade study to select the recommended solution. We will elaborate on this point further in the following discussions.

14.5 EXPOSURE TO PROBLEM–SOLUTION SPACES

Every person in your organization should have exposure to and an understanding of your organization's *opportunity-solution spaces*. Unfortunately, travel budgets and the User's desire and ability to accommodate throngs of people prevents first hand observations of HOW the system they develop will be deployed, operated, and supported. As with any scientific field, *observation* is a critical skill of SEs. *Seeing*, *touching*, *feeling*, *operating*, *hearing*, and studying existing systems in action has a profound influence on SEs throughout system development. Consider the following example:

EXAMPLE 14.7

A company is contracted to design a large piece of computer equipment. Since one of the considerations is always getting equipment through doorways, business development personnel describe the doorway and hallway leading to the room entrance. Engineering personnel were denied the opportunity to conduct a site survey of the facility.

Convinced that they could develop a cabinet to fit through a narrow doorway, engineering proceeded with the design. When the system was delivered, the installation team encountered major problems. They discovered that the cabinet could be moved down the narrow hallway to the door but could not make a 90-degree turn through the door. Lesson learned: ALWAYS send an SE to accompany business development personnel during on-site visits to understand, analyze, and document the opportunity/problem and solution spaces.

Final Thoughts

Understanding problem–solution spaces requires continual assessments due to the dynamics of the OPERATING ENVIRONMENT. SEs often erroneously believe that filling the solution space with development of a new system, product, or service is an *end all* answer. As is the case of scientific law, every *action* by the User can be expected to have an *equal* and *opposite reaction* by *competitors* and *adversaries*. So, when you bound the solution space, the *bounding* process must also consider:

1. The potential *reactions* of competitors and adversaries.
2. HOW the new system, product, or service *minimizes* susceptibility and vulnerability to those threats, at least for a reasonable period of time.

14.6 GUIDING PRINCIPLES

In summary, the preceding discussions provide the basis with which to establish the guiding principles that govern our understanding of a system's problem, opportunity, and solution spaces.

Principle 14.1 System analysis requires recognition and validation of three types of User operational needs: *real*, *perceived*, or *projected*.

Principle 14.2 One system's *problem space* is an *opportunity space* for a competitor or adversarial system.

Principle 14.3 Manage *problem space* complexity by decomposing it into one or more *solution spaces* that are solvable.

Principle 14.4 When bounding a *solution space*, anticipate competitor or adversarial reactions to counter the *solution space* capabilities.

Principle 14.5 *There are two types of solution development activities: problem solving and symptom solving.* Recognize the difference.

14.7 SUMMARY

Our discussion of *problem* and *solution spaces* serves as an SE concept for a number of reasons.
- Problem solving begins with bounding and understanding the *problem space* and its constraints.
- The fundamental concepts discussed here serve as key element of the SE Process Model.

Referral More information on the SE Process Model is provided in Part II in Chapter 26.

- The investigation of alternative concepts and solutions that satisfy the *solution space* provide the foundation for developing candidate system solutions for evaluation and decision.

Referral More information on system solution development is provided in Chapter 23 *System Analysis Synthesis* and Chapters 37–40 on system solution development.

Now that we have an understanding of the problem space and the solution space concept, we are ready to investigate HOW the User employs systems, products, and services to perform organizational and system missions. This brings us to our next topic, *system interactions with its operating environment*.

GENERAL EXERCISES

1. Answer each of the *What You Should Learn from This Chapter* questions identified in the *Introduction*.
2. Refer to the list of systems identified in Chapter 2. Based on a selection from the preceding chapter's *General Exercises* or a new system selection, apply your knowledge derived from this chapter's topical discussions. Specifically identify the following:
   (a) Problem space(s)
   (b) Opportunity space(s)
   (c) Solution space(s)
3. Cite various *solution space* tools that enable a homeowner to leverage their time, resources, and skills to maintain their lawn.
4. Cite two examples of human-made systems and two examples of natural systems that project or expand their *sphere of influence* by leveraging the capabilities of other systems.

ORGANIZATIONAL CENTRIC EXERCISES

1. For your organization, identify the following related to the systems, product, or services it provides:
   (a) Problem space(s)
   (b) Opportunity space(s)
   (c) Solution space(s)
2. Pick a contract program within your organization. From the User's organizational mission perspective, equate the following for the system, product, or service deliverables the contract provides:
   (a) Problem space(s)
   (b) Opportunity space(s)
   (c) Solution space(s)

ADDITIONAL READING

Chang, Richard Y., and Kelly, P. Keith. *Step-by-Step Problem Solving*. Publications Division, Richard P. Chang & Associates, 1993, Irvime, CA.

Kelley, Tom. The Art of Innovation, Doubleday, 2001, New York, NY.

ASD-100, Architecture and System Engineering. 2003. *National Air Space System—Systems Engineering Manual*. Washington, DC: Federal Aviation Administration (FAA).

Chapter 15

System Interactions with Its Operating Environment

15.1 INTRODUCTION

Every natural and human-made system exhibits a fundamental *stimulus–response* behavior pattern. For example, systems may respond *positively* to good news. Conversely, a system may respond *negatively* to threats and employ *defensive* tactics, pre-emptive, or retaliatory strikes. The response ultimately depends on *how* your system is designed and trained to respond to various types of inputs—*stimuli* and information, under specified types of operating conditions and constraints.

This section builds on the *system architecture concepts* discussions. Each SYSTEM OF INTEREST (SOI) *coexists* and *interacts* with external systems that comprise its OPERATING ENVIRONMENT—namely HUMAN-MADE SYSTEMS, the PHYSICAL ENVIRONMENT, and the INDUCED ENVIRONMENT. System analysts, designers, and developers need an understanding in HOW those systems INTERACT and respond to *stimuli* and *cues* in their OPERATING ENVIRONMENT.

Our discussion begins with the fundamentals of *system behavior*. We establish a basic *system behavioral model* that depicts HOW a SOI *interacts with* and *responds to* its OPERATING ENVIRONMENT. The discussion highlights key concepts such as system stimuli, transfer function, response time, and feedback control loops.

We introduce the concepts of *strategic* and *tactical interactions*, and *system adaptation* to its operating environment. From a SE perspective, SOI interactions with its OPERATING ENVIRONMENT require an *analytical* understanding of each *interaction*. To establish this understanding, we introduce an approach for analyzing system interfaces and their outcomes. As we analyze system interactions and responses, we investigate HOW some SOIs employ tactics, countermeasures, and counter-countermeasures (CCMs).

The final part of this section concludes with a discussion of the system's *degree of compliance* with its OPERATING ENVIRONMENT. All systems, whether desired or not, operate and interact within the balance of power that exists in their respective domains of operation. The level of society's acceptance of a system and a key factor in its success is determined by its ability to *function* and *comply* within its prescribed OPERATING ENVIRONMENT.

What You Should Learn from This Chapter

1. What is the purpose of the *system behavioral response model*?
2. What are the key *elements* and *interfaces* of the model?

System Analysis, Design, and Development, by Charles S. Wasson
Copyright © 2006 by John Wiley & Sons, Inc.

3. What is a *compatible* interface?
4. What is an *interoperable* interface?
5. What are examples of HUMAN-MADE SYSTEMS ENVIRONMENT threat sources?
6. What are examples of NATURAL ENVIRONMENT threat sources?
7. What are examples of INDUCED ENVIRONMENT threat sources?
8. What are *system countermeasures*?
9. What are *system counter-countermeasures*?
10. What is meant by a system's *compliance* with its OPERATING ENVIRONMENT?
11. What are the consequences of a system's noncompliance with its OPERATING ENVIRONMENT?
12. Identify two levels of system interactions.
13. Identify and describe six types of *system interactions* with its OPERATING ENVIRONMENT.

Definitions of Key Terms

- **Compatibility** "The ability of two or more systems or components to perform their required functions while sharing the same hardware or software environment." (Source: IEEE 610.12-1990)
- **Comply** To obey in strict accordance with the provisions of a contract or agreement's terms and conditions. (Ts & Cs).
- **Conform** To adapt or tailor an organization's standard methods or processes to another organizations request or instructions.
- **Countermeasure** An operational capability or tactic employed by a system to camouflage its identity, deceive or defeat adversarial or hostile system's capabilities, or minimize vulnerability by protecting itself from unauthorized access.
- **Counter-Countermeasure (CCM)** An operational capability or tactic employed by a system to neutralize another system's countermeasures.
- **Engagement** A single instance of a *friendly*, *cooperative*, *benign*, *competitive*, *adversarial*, or *hostile* interaction between two systems.
- **Strategic Threats** Entities that have long-term plans to exploit opportunities that leverage or enhance the organization's reputation or equity to achieve a long-term vision and upset the "balance of power." For example, an organization has a long-term vision to predominate a software market.
- **System Adaptation** The ability of a system to acclimate physically and functionally to a new OPERATING ENVIRONMENT with a minimal degree of degradation to capability performance.
- **System Threat** Any type of entity that has the potential to cause or inflict varying degrees of harm on another entity and its mission, capabilities, or performance. A *system threat* is any interaction by an external system that is hostile or impedes the operation and performance of your system in accomplishing its intended mission.

148   Chapter 15   System Interactions with Its Operating Environment

- **Tactical Threats**   Entities that pose a potential, short-term hazard to another organization or system and its mission. For example, counter a competitor's advertising campaign.
- **Transfer Function**   A mathematical expression used to model the relationship between a system's behavioral response to a range of inputs and constraints.

15.2 SYSTEM BEHAVIORAL RESPONSE MODEL

During our discussion of the OPERATING ENVIRONMENT architecture, Figure 11.1 served as a high-level model to illustrate a SYSTEM OF INTEREST's interactions with its OPERATING ENVIRONMENT. To see *how* this interaction occurs, let's investigate a simple behavioral response model.

Modeling the SOI's OPERATING ENVIRONMENT Interfaces

If we expand the SYSTEM OF INTEREST (SOI) aspect of Figure 11.1 to the next level of detail, the top-level system shown in Figure 15.1 emerges. The top-level system consists of the PHYSICAL ENVIRONMENT (1) and the SYSTEM OF INTEREST (2), both of which are controlled by a HIGHER ORDER SYSTEM (3).

The HIGHER ORDER SYSTEM (3) provides an ORGANIZATION (4), allocates ROLES and MISSIONS (5), imposes OPERATING CONSTRAINTS (6), and provides RESOURCES (7) to the SOI.

Elements of the PHYSICAL ENVIRONMENT (1)—such as HUMAN-MADE SYSTEMS (8), INDUCED ENVIRONMENT (9), and NATURAL ENVIRONMENT (10)—provide input stimuli into the SOI as well as affect its operating capabilities and performance.

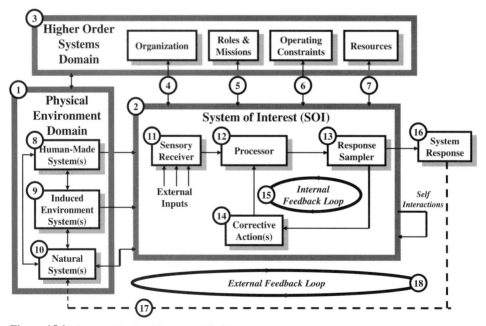

Figure 15.1   System Behavioral Responses Model

Modeling the SOI's Behavior

Stimuli serve as inputs to the Sensory Receiver (11) capability of the SOI (2). These can be cues, information, data, interrupts, and actions. The Sensory Receiver (11) decodes the stimuli and information as inputs to the Processor (12). The processor adds value to the data within the boundaries of OPERATING CONSTRAINTS (6) and RESOURCES (7) that are levied by HIGHER ORDER SYSTEM (3).

The Response Sampler (13) samples the results of the value-added processing and compares those results to OPERATING CONSTRAINTS (6)—namely mission tasking—established by a HIGHER ORDER SYSTEM (3). Based on the results of the Comparison, Corrective Actions (14) are initiated as feedback to the Processor (12).

The clockwise workflow of steps (12) through (14) form an Internal Feedback Loop (15). When the processing is deemed acceptable relative to the OPERATING CONSTRAINTS (6), the System Response (16) is produced. The System Response (16) is then fed back (17) to the OPERATING ENVIRONMENT (1) thereby completing the External Feedback Loop (18).

Understanding the Model's Behavioral Transfer Function

In terms of the system control model introduced in Figure 15.1, the SOI's *transfer or response function* is dependent on the planned behavior of the SYSTEM OF INTEREST (SOI) relative to its OPERATING ENVIRONMENT. Consider HOW this model relates to an organization.

EXAMPLE 15.1

An organization's executive management formulates a strategic plan based on its vision and analysis of the OPERATING ENVIRONMENT—in *threats* and *opportunities*—as well as *tactical plans* (refer to Figure 13.1). The vision, philosophy, missions, and mission objectives conveyed in these documents, as well as its command media—its *policies and procedures*—establishes *how* the organization and elements of the organization are to respond to the PHYSICAL ENVIRONMENT.

Several Key Points

Figure 15.1 illustrates several key points regarding *system interactions* with its OPERATING ENVIRONMENT.

- *System Interactions* Stimuli or data (8)(9)(10)—composed of cues, information, and behavior—as well as the system response (16) to its OPERATING ENVIRONMENT (1) form a closed loop (18) of system interactions.
- *Input Data Occurrence* External stimuli or data (8)(9)(10) may occur as a *"triggering event"*—as communications data, an observation, or transfer of information—or as *"trend data"* over time.
- *Measured or Conditioned System Response* Steps (11) through (14) form an internal control loop (15) that results in *measured response* appropriate for the stimuli and information (8).
- *System Transfer Function* Steps (11), (12), (13), and (14) collectively form a system *transfer function* that shapes the system response (16).
- *System Responsiveness* The time required from the SYSTEM OF INTEREST (2) to respond to external stimuli (8), (9), and (10) and information until a system response (16) is produced is referred to as system responsiveness, system response time, or system throughput.

15.3 SYSTEM BEHAVIORAL RESPONSES

Most people tend to think of a system's responses to its OPERATING ENVIRONMENT in terms of *products*, *by-products*, and *services*. However, system behavioral responses such as body language, communications (degree of bluntness, etc.) *intentionally* or *unintentionally* communicate the true message that may or may not correlate with the *verbal* message.

Systems generally respond to external stimuli and information as aggressors, neutral, or as defenders. The system response may be *aggressive* (i.e., proactive) or *defensive* (i.e., reactive). Let's consider some examples of system behavior.

- *Police* In a potentially hostile protest demonstration, strategically placed police in riot gear stand prepared to respond on command to acts of violence or public disturbances.
- *Paramedics* On receipt of an emergency call, paramedics respond with medical attention.
- *Education* Corrective action based on test results is taken to eliminate a deficiency in student skill levels in mathematics.

System Interaction Compatibility and Interoperability

When two or more systems interact, we refer to the interaction as an *engagement or encounter*. *Engagements* can be characterized with a number of terms. Examples include *friendly*, *cooperative*, *neutral*, *adversarial*, and *hostile*. The effects or results of the engagement can be described as *positive*, *benign*, *negative*, *damaging*, or *catastrophic*, depending on the system roles, missions, and objectives. Generally, the *effects* or *outcomes* can be condensed into a key question. *Was the engagement compatible and interoperable from each system's perspective?* Let's explore both the context of both of these terms.

Differentiating Compatibility and Interoperability

Compatibility often has different contextual meanings. We use the term in the context of physical *form*, *fit*, and *function* capability. Notice that we used the operative term *capability*. Having the *capability* does not mean the engagement or interface is *interoperable* or enabled.

To illustrate the application of the terms *compatibility* and *interoperability*, consider the following examples.

EXAMPLE 15.2

Two people from different countries speaking different languages may attempt to communicate—an *interaction* or *engagement* between system entities. We could say their voice communications are *compatible*—transmitting and receiving. However, they are unable to decode, process, assimilate, or "connect" *what* information is being communicated—*interoperability*.

EXAMPLE 15.3

You can have an RS-232 data communications interface between two systems that use a standard cable and connectors for transmitting and receiving data. Thus, the interface is *physically* compatible. However, the data port may not be *enabled* or the receiving system's software capable of *decoding* and *interpreting* the information—*interoperability*.

15.4 UNDERSTANDING THE SYSTEM THREAT ENVIRONMENT

The exploitation of *opportunities* may be viewed by some organizations as threatening to the sustainment and survival of the organization. Whether the *scenario* involves *increasing* market share, defending national borders, or developing a secure Internet Web site, you must ensure that your system is capable of sustaining itself and its long-term survival.

Long-term survival hinges on having a *thorough* and *complete understanding* of the potential threat environment and having the *system capabilities* to counter the threat that serve as obstacles to mission success. So, *how does this relate to SE*? When you specify requirements for your system, system requirements must include considerations of what capabilities and levels of performance are required to counter threat actions.

Sources of Threats

System threats range from the known to the unknown. One approach to identifying potential system threats can be derived from the PHYSICAL ENVIRONMENT elements—NATURAL ENVIRONMENT, HUMAN-MADE SYSTEMS, and the INDUCED ENVIRONMENT.

NATURAL ENVIRONMENT Threat Sources. NATURAL ENVIRONMENT threat sources, depending on perspective, include lightning, hail, wind, rodents, and disease.

HUMAN-MADE SYSTEMS Threat Sources. External HUMAN-MADE SYSTEMS threat sources include primarily PERSONNEL and EQUIPMENT elements. The motives and actions of the external systems delineate friendly, competitive, adversarial, or hostile intent.

INDUCED ENVIRONMENT Threat Sources. INDUCED ENVIRONMENT threat sources include contaminated landfills, electromagnetic interference (EMI), space debris, ship wakes, and aircraft vortices.

Types of System Threats

System threats occur in a number of forms, depending on the environment—HUMAN-MADE SYSTEMS, INDUCED ENVIRONMENT, and NATURAL ENVIRONMENTs.

HUMAN-MADE SYSTEMS and NATURAL SYSTEMS ENVIRONMENT Aggressor Threats. Generally, most HUMAN-MADE SYSTEMS and NATURAL ENVIRONMENT aggressor threats fall into the categories of *strategic threats* or *tactical threats* related to the *balance of power*, motives, and objectives.

Other NATURAL ENVIRONMENT Threats. *Other natural environment threats* are attributes of the NATURAL ENVIRONMENT that impact a system's inherent capabilities and performance. Examples include temperature, humidity, wind, lightning, light rays, and rodents. Although these entities do not reflect premeditated aggressor characteristics, their mere *existence* in the environment, seemingly benign or otherwise, can *adversely impact* system capabilities and performance.

Threat Alliances. Sometimes threats emerge from a variety of sources that form strategic alliances. Examples include businesses, nations, and individuals.

Threat Behavioral Characteristics, Actions, and Reactions

Threats exhibit characteristics and actions that may be described as *adversarial, competitive, hostile,* and *benign*. Some threats may be viewed as *aggressors*. In other cases, threats are generally *benign* and only take action when someone *"gets into their space."* For example, an *unauthorized* aircraft *purposefully* or *unintentionally* intrudes on another country's airspace creating a provocation of tensions. The defending system response course of action may be based on protocol or a measured retaliatory physical or verbal warning.

Threats, in general, typically exhibit three types of behavior patterns or combinations thereof: *aggressive, concealed,* and *benign*. Threat patterns often depend on the circumstances. For example, *aggressors* exhibit *acts* of aggression. *Benign threats* may "tend to their business" unless provoked. *Concealed threats* may appear to be *benign* or *disguised* and strike targets of opportunity (TOO) unexpectedly.

Threat Environment Constraints. The *threat environment* is characterized by boundaries that have various attributes that are confined by constraints such as resources and physical constraints.

Ideology, Doctrine, and Training Constraints. Ideology, doctrine, and training are often key factors in threat actions.

Threat Encounters

When systems interact with known threats, the interactions—such as encounters or engagement—should be documented and characterized for later use by other systems in similar encounters. Threat encounters intended to *probe* another system's defenses can be described with a number of descriptors: *aggressive, hostile, cooperative, inquisitive, investigative, bump and run,* and *cat and mouse*.

When threat encounters turn hostile and defensive action must be taken, systems resort to various *tactics* and *countermeasures* to ensure their survival.

System Tactics. When systems interact with their OPERATING ENVIRONMENT, they often ENGAGE *threats* or *opportunities*. Systems and system threats often employ or exhibit a series of *evasive actions* intended to *conceal, deceive,* or *camouflage* the *target of opportunity (TOO)*. Generally, when *evasive tactics* do not work, systems deploy *countermeasures* to disrupt or distract hostile actions. Let's examine this topic further.

Threat Countermeasures. To counter the impact or effects of threats on a system, systems often employ *threat countermeasures*. *Threat countermeasures* are any physical action performed by a system to *deter* a threatening action or counter the impact of a threat—i.e., survivability. Sometimes adversarial systems acquire or develop the technology to counter the TOO's system *countermeasures*.

Threat Counter-Countermeasures (CCM). Sometimes system threats compromise the established security mechanisms by deploying *counter-countermeasures (CCM)* to offset the effects of a TOO's countermeasures.

Concluding Thoughts

This concludes our overview of system *threats* and *opportunities*. You should emerge from this discussion with an *awareness* of *how* the system you are using or developing must be capable of interacting with threats and opportunities in the OPERATING ENVIRONMENT. More explicitly, you will be expected to develop a level of technical knowledge and understanding to enable your team to specify or oversee the specification of system capabilities and levels of performance related to threats and opportunities.

Now that we have an understanding of the opportunistic and potentially hostile entities, we are now ready to investigate *how* a system interacts with its OPERATING ENVIRONMENT and the balance of power to perform missions.

15.5 EXAMPLE CONSTRUCTS OF SYSTEM BEHAVIORAL INTERACTIONS

If we *observe* and *analyze* the *pattern of interactions* between human-made systems, we can identify some of the primary interaction constructs. In general, examples of common interactions of most friendly systems include the following:

- Open loop command interactions (Figure 15.2)
- Closed loop command and control (C^2) interactions (Figure 15.2)
- Peer data exchange system interactions (Figure 15.3)
- Status and health broadcast system interactions (Figure 15.4)
- Issue arbitration/resolution system interactions (Figure 15.5)
- Hostile encounter interactions (Figure 15.6)

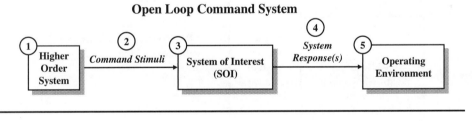

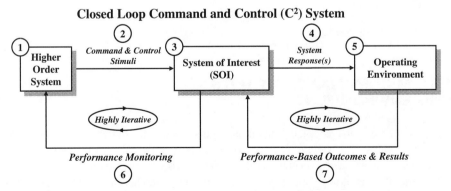

Figure 15.2 Open Loop and Closed Loop Command and Control (C^2) System Examples

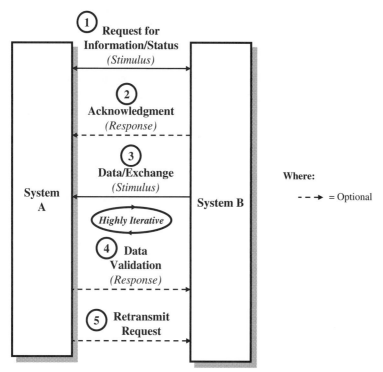

Figure 15.3 Peer Data Exchange System Interactions Example

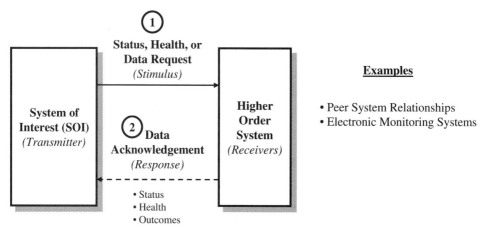

Figure 15.4 Status & Health Broadcast System Interactions Example

15.6 SYSTEM COMPLIANCE WITH ITS OPERATING ENVIRONMENT

The balance of power of systems, coupled with humankind's general desire for peace and harmony, requires systems to comply with *standards* imposed by society. Standards in this context refer to *explicit* and *implicit*, self-imposed expectations by society such as laws, regulations, ordnances, codes of conduct, morals, and ethics. Thus, system survival, peace, and harmony are often driven by a system's compliance to these standards. System adherence to these standards involves two terms that are often interchanged and require definition. The terms are *compliance* and *conformance*.

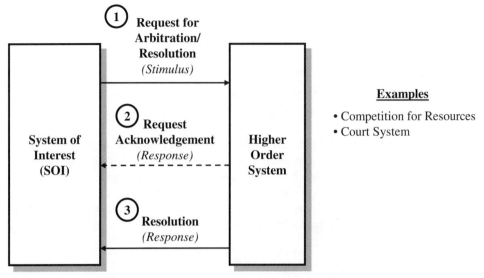

Figure 15.5 Issue Arbitration/Resolution System Interactions Example

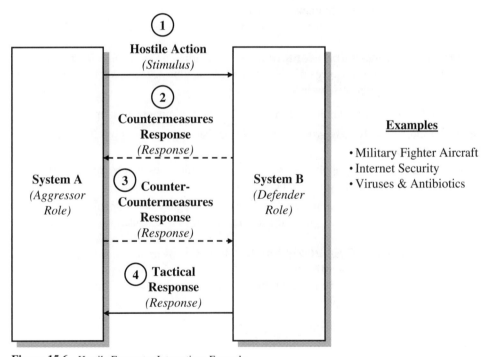

Figure 15.6 Hostile Encounter Interactions Example

The Consequences of Noncompliance

When a system fails to *adhere* to established standards, it places itself at risk with society. Society's response to a *lack of compliance* generally involves formal or informal notification, establishment that *noncompliance* occurred, adjudication of the degree or noncompliance, and sentencing in accordance with prescribed *consequences* or *penalties*. In some cases the system may *voluntarily* elect to bring itself into compliance or be *mandated* to be compliant. In other cases, society may *ostracize* or *punish* the instigators.

For systems such as ships, aircraft, and automobiles intentional or unintentional noncompliance with the NATURAL ENVIRONMENT, HUMAN-MADE SYSTEMS, and INDUCED ENVIRONMENT can be very unforgiving or even worse, *catastrophic*.

Levels of System Interactions

System interactions with its OPERATING ENVIRONMENT occur at two levels: *strategic interactions* and *tactical interactions*. Let's explore each of these in detail.

Strategic Interactions. HUMAN-MADE SYSTEMS exhibit a higher level of behavior that reflects a desire to advance our current condition as a means of achieving higher level vision. To achieve the higher-level vision, humans must implement a well-defined strategy, typically long term, based on stimuli and information extracted from out operating environment. We refer to implementation of this long-term strategy as *strategic interactions*. These *strategic interactions* are actually implemented via a series of premeditated missions—tactical interactions—with specific mission objectives.

Tactical Interactions. All life forms exhibit various types of tactics that enable the system to survive, reproduce, and sustain itself. We refer to a system's implementation of these tactics within the confines of its operating environment as *tactical interactions*. In general, this response mechanism focuses all existing survival needs in the short term—obtaining the next meal.

System Interaction Analysis and Methodology. Depending on the *compatibility* and *interoperability* of an interface, consequences of the engagement may be *positive*, *neutral*, or *negative*. As a system analyst or SE, your mission is to:

1. Develop a thorough understanding of the engagement participants (systems).
2. Define the most probable *use cases* and *scenarios* that characterize how the User intends to use the system.
3. Analyze the *use cases* by applying natural and scientific laws of physics to thoroughly understand the potential outcomes and consequences.
4. Specify system interface requirements that ensure engagement compatibility and interoperability success within cost, schedule, and technology constraints.

Adapting to the OPERATING ENVIRONMENT. Most systems are designed to perform in a prescribed OPERATING ENVIRONMENT. There are situations whereby a system is transferred to a new location. The net result is the need for the system to *adapt* to its new operating environment. Consider the following examples:

EXAMPLE 15.4

Military troops deploy to arid or snow regions. Depending on the initial conditions—namely acclimation to the previous environment—the troops must learn to adapt to a new operating environment.

EXAMPLE 15.5

As part of a strategy for climbing a high mountain, mountain climbers travel to a series of base camps to satisfy logistics requirements and allow their bodies time to acclimate to the thin air environment over a period of several days.

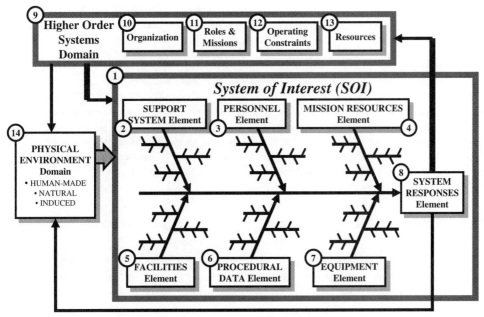

Figure 15.7 System Element Contributions to Overall System Performance

System Interactions Synthesis

As an SE, you must learn to synthesize these interactions in terms of an overall system solution. Figure 15.7 provides an illustration.

Here we have a diagram that captures the high-level interactions between the SYSTEM OF INTEREST (SOI) (1), HIGHER ORDER SYSTEM (9) and the OPERATING ENVIRONMENT (14). The SOI is illustrated via a "fishbone" diagram. We include in the diagram the system elements that are performance AFFECTER factors that must integrate harmoniously to achieve the mission objectives. In combination the SOI elements produce the SYSTEM RESPONSES (8) element, which consists of behavior, products, by-products, and services.

In operation, the SOI (1) responds to command and control guidance and direction from the HIGHER ORDER systems element that consists of ORGANIZATION (10), ROLES AND MISSIONS (11), OPERATING CONSTRAINTS (12), and RESOURCES (13) system elements. Based on this direction, the SOI system elements interact with the OPERATING ENVIRONMENT and provide SYSTEM RESPONSES (8) back to the OPERATING ENVIRONMENT and the HIGHER ORDER SYSTEMS element.

15.7 GUIDING PRINCIPLES

In summary, the preceding discussions provide the basis with which to establish the guiding principles that govern system interactions with its OPERATING ENVIRONMENT.

Principle 15.1 System interactions with its OPERATING ENVIRONMENT during an engagement may be cooperative, friendly, benign, competitive, adversarial, hostile, or combination of these.

Principle 15.2 Every system responds to stimuli and cues in its OPERATING ENVIRONMENT with behavioral actions, products, by-products, services, or combinations thereof.

15.8 SUMMARY

During our discussion of *system interactions with its operating environment*, we described a system's interactions via the *Behavioral Responses Model*. A system's responses are driven by strategic and tactical interactions related to opportunities and threats in the environment. Systems generally interact with *cooperative*, *benign*, *competitive*, or *aggressor* systems. Based on those responses, we indicated how a system might employ countermeasures and counter-countermeasures to *distract*, *confuse*, defend or *interact* with other systems. We concluded our discussion by highlighting the context of the OPERATING ENVIRONMENT based on the SYSTEM OF INTEREST perspective.

GENERAL EXERCISES

1. Answer each of the *What You Should Learn from This Chapter* questions identified in the *Introduction*.
2. Refer to the list of systems identified in Chapter 2. Based on a selection from the preceding chapter's General *Exercises* or a new system selection, apply your knowledge derived from this chapter's topical discussions.
   (a) If applicable, identify whether the system operates by a closed loop or an open loop.
   (b) If by a closed loop, how does the system process stimuli and cues and provide measured responses.
3. Identify external systems that interface with your product or service. Characterize them in terms of cooperative, benign, or adversarial.
4. What *vulnerabilities* or *susceptibilities* does your system, product, or service have to threats in its operating environment? What capabilities, tactics, or procedures have been added to the product to minimize vulnerability or susceptibility?

Chapter 16

System Mission Analysis

16.1 INTRODUCTION

The primary purpose of any system is to satisfy individual or organizational objectives with an expected *tangible* or *intangible* return on investment (ROI). These objectives may range from the quality of life such as happiness, entertainment, education, and health to the basic necessities of life—organizational survival, profitability, food, and shelter. The act of striving to accomplish these objectives can be summarized in one operative term, *mission*.

The accomplishment of individual and organizational missions requires the employment of *systems*, *products*, and *services* that leverage human capabilities. Selection or acquisition of those systems begins with understanding the WHO, WHAT, WHEN, WHERE, and HOW system User(s) plan to accomplish the mission(s). We refer to activities required to develop this understanding as a *mission analysis*.

This chapter introduces the key elements of the mission analysis and provides the foundation for deriving system capabilities and requirements. Our discussions focus on the key attributes of a mission.

What You Should Learn from This Chapter

1. What are the key tasks required to define a system mission?
2. What is a *Mission Event Timeline (MET)*?
3. What is *mission task analysis*?
4. What are the primary mission *phases of operation*?
5. What system related *operations and decisions* are performed during the pre-mission phase?
6. What system related *operations and decisions* are performed during the mission phase?
7. What system related *operations and de*cisions are performed during the postmission?
8. What are the *key decisions* that occur within mission phases and trigger the next phase?

Definitions of Key Terms

- **Mission** A pre-planned exercise that integrates a series of sequential or concurrent operations or tasks with an expectation of achieving outcome-based success criteria with quantifiable objectives.
- **Mission Critical System** "A system whose operational effectiveness and operational suitability are essential to successful completion or to aggregate residual (mission) capability. If

---

System Analysis, Design, and Development, by Charles S. Wasson
Copyright © 2006 by John Wiley & Sons, Inc.

this system fails, the mission likely will not be completed. Such a system can be an auxiliary or supporting system, as well as a primary mission system." (Source: DSMC—adapted from *Glossary: Defense Acquisition Acronyms and Terms*)

- **Mission Needs Statement (MNS)** "A nonsystem specific statement that identifies an organizational operational capability need." (Source: Adapted from *DSMC T&E Mgt. Guide*, Appendix B, *DoD Glossary of Test Terminology*, p. B-20-21)
- **Mission Reliability** "The probability that a system will perform its required mission critical functions for the duration of a specified mission under conditions stated in the mission profile." (Source: *Glossary: Defense Acquisition Acronyms and Terms*)
- **Operational Constraints** "Initially identified in the Mission Need Statement (MNS). As a minimum, these constraints will consider the expected threat and natural environments, the possible modes of transportation into and within expected areas of operation, the expected (operating) environment, operational manning limitations, and existing infrastructure support capabilities." (Source: Adapted from DSMC—*Glossary: Defense Acquisition Acronyms and Terms*)
- **Phase of Operation** A high-level, objective-based abstraction *representing* a collection of SYSTEM OF INTEREST (SOI) operations required to support accomplishment of a system's mission. For example, a system has pre-mission, mission, and postmission phases.
- **Point of Delivery** A waypoint or one of several waypoints designated for delivery of mission products, by-products, or services.
- **Point of Origination or Departure** The initial starting point of a mission.
- **Point of Termination or Destination** The final destination of a mission.
- **Task Order** A document that: 1) serves as triggering event to initiate a mission and 2) defines mission objectives and performance-based outcomes.
- **Time Requirements** "Required functional capabilities dependent on accomplishing an action within an opportunity window (e.g., a target is vulnerable for a certain time period). Frequently defined for mission success, safety, system resource availability, and production and manufacturing capabilities." (Source: Former MIL-STD-499B Draft)
- **Timeline Analysis** "Analytical task conducted to determine the time sequencing between two or more events and to define any resulting time requirements. Can include task/timeline analysis. Examples include:
  a. A schedule line showing key dates and planned events.
  b. An engagement profile detailing time based position changes between a weapon and its target.
  c. The interaction of a crewmember with one or more subsystems." (Source: Former MIL-STD-499B Draft)
- **Waypoint** A geographical or objective-based point of reference along a planned roadmap to mark progress and measure performance.

16.2 MISSION DEFINITION METHODOLOGY

Organizational and system missions range from simple tasks such as writing a letter to performing highly complex International Space Station (ISS) operations, managing a government. Regardless of application, mission analysis requires consideration of the steps specified below:

Step 1: Define the primary and secondary mission objective(s).
Step 2: Develop a mission strategy.
Step 3: Define phase-based operations and tasks.
Step 4: Create a Mission Event Timeline (MET).
Step 5: Bound and specify the mission OPERATING ENVIRONMENT interactions.
Step 6: Identify outcome-based system responses to be delivered.
Step 7: Identify mission resources and sustainment methods.
Step 8: Perform a mission task analysis.
Step 9: Assess and mitigate mission and system risk.

Let's explore each of these steps in more detail.

Step 1: Define the Primary Mission Objective(s)

People often mistakenly believe that missions begin with the assignment of the "mission" to be accomplished. However, *missions* are action-based applications of systems, products, or services to *solution spaces* for the purposes of *resolving* or *eliminating* all or a portion of an *operational need*—meaning a *problem* or *opportunity space*. These actions may be oriented toward a single *event* or occur via one or more *reusable* missions over a period of time. Consider the following examples:

EXAMPLE 16.1

The NASA Space Shuttle's external tank (ET), which is expendable, represents a *single event* mission application. On completion of its mission, the ET is jettisoned and burns up in the atmosphere.

EXAMPLE 16.2

NASA's Space Shuttle Orbiter vehicle performs via a series of mission applications to ferry components of the International Space Station (ISS) for integration and support of science.

As with any system, the initial step in performing mission analysis is to understand the underlying motivation and primary/secondary objectives to be accomplished. Mission objectives are characterized by several attributes. For our discussion the two primary attributes are:

1. Outcome-based results to be achieved.
2. Mission reliability required to achieve those results.

Identify the Outcome-Based Results. When you define a mission objective, the first step is to define WHAT results are expected to be produced. The results should be:

1. Preferably tangible as well as *measurable*, *testable*, and *verifiable*.
2. Contribute to accomplishment of HIGHER ORDER SYSTEM tasking.

Determine the Mission Reliability. Human systems, despite careful planning and execution, are not infallible. The question is: *Given resource constraints, WHAT is the minimum level of level of success you are willing to accept to provide a specified return on investment (ROI)*. From an SE

162 Chapter 16 System Mission Analysis

point of view, we refer to the *level of success* as *mission reliability*. *Mission reliability* is influenced by internal EQUIPMENT element failures or over/undertolerance conditions, human operator performance (judgment, errors, fatigue etc.) and interactions with OPERATING ENVIRONMENT entities and threats.

Mission reliability is the probability that a system will successfully accomplish a mission of a specific duration in a prescribed OPERATING ENVIRONMENT and accomplish objectives without a *failure event*. Depending on the system application, 100% mission reliability may be prohibitively *expensive*, but a 90% mission reliability may be *affordable*.

Authors Note 16.1 *Since reliability ultimately has a cost, establish an initial reliability estimate as simply a starting point and compute the cost. Some Acquirers may request a Cost-as-an-Independent-Variable (CAIV) plot of cost as a function of capability or reliability to determine what level of capability or reliability is affordable within their budgetary constraints.*

Specify and Bound the Required Level of Performance. Once the *mission reliability* is established, system designers can proceed with identifying the level of performance required of the *system elements*, such as EQUIPMENT and PERSONNEL, subject to cost, schedule, and risk constraints.

Once we establish the primary mission objectives, the next step of mission analysis is to define the mission profile.

Step 2: Develop a Mission Strategy

A mission begins with a *point of origination* and terminates at a *point of destination*. As end-to-end boundary constraints, the challenge question is: *HOW do we get from the point of ORIGINATION to the point of DESTINATION?*

We begin by establishing a strategy that leads to a *mission* profile or a roadmap that charts progress through one or more *staging* and *control* points, or *waypoints*. A *waypoint* represents a geographical location or position, a point in time, or objective to be accomplished as an *interim* step toward the destination, as illustrated in Figure 16.1. Each phase of operation is *decomposed* into one or more objectives focused on the pathway to successful completion of the mission. Consider the following examples:

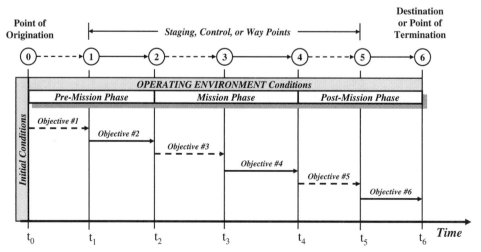

Figure 16.1 Operational Concept Timeline Example

EXAMPLE 16.3

A ship cruise line has several ports of call or waypoints on its scheduled item erary for a 7-day voyage. A package delivery service has performance-based deliveries or waypoints for a delivery route.

Step 3: Define Phase-Based Operations and Tasks

Human-made systems, especially cyclical systems, sequence through three sets of objective-based actions to accomplish a mission: 1) prepare for the mission, 2) conduct the mission, and 3) perform post-mission actions and processing. We characterize these objectives as the pre-mission, mission, and post-mission phases of operation. For those systems to be placed in storage following the mission, an interim phase—storage—may be added.

When implemented, the SYSTEM OF INTEREST (SOI) consisting of the MISSION SYSTEM and SUPPORT SYSTEM must provide capabilities and levels of performance to support these phases of operation. Each phase consists of *use case* based *operations and tasks*, all focused on accomplishing the phase outcome-based performance objective(s).

Step 4: Create a Mission Event Timeline (MET)

Once we establish the *waypoints*, the next task is to determine *waypoint* time constraints. We refer to these time *constraints* as milestone requirements derived from the *mission event timeline (MET)*. The MET can be presented as a simple, high-level schedule down to a highly detailed, multi-level, networked schedule.

Guidepost 16.1 *Mission analysis up to this point has focused on the "ideal" mission—namely what we intend to accomplish. However, to accomplish a mission, the MISSION SYSTEM must interact with the OPERATING ENVIRONMENT and its elements, consisting of HUMAN-MADE systems, the NATURAL ENVIRONMENT, and the INDUCED ENVIRONMENT. This brings us to our next mission analysis task: bound and specify the OPERATING ENVIRONMENT.*

Step 5: Bound and Specify the Mission OPERATING ENVIRONMENT Interactions

Once the basic mission is defined, the next step is to *bound and specify* its OPERATING ENVIRONMENT. Throughout the pre-mission, mission, and postmission phases, the SOI interacts with external systems within its OPERATING ENVIRONMENT. These systems may include friendly systems, benign systems, or hostile threats and harsh environmental conditions.

Collectively the mission analysis *identifies* and *analyses* these systems, their roles relative to the mission, and what impacts they may have on performing the mission and accomplishment of its performance objectives. For example, what systems does the MISSION SYSTEM need to: 1) communicate with, 2) perform deliveries and transfers to, and 3) interact with on an encounter/engagement basis along with the mission profile.

Guidepost 16.2 *Our earlier discussion emphasized the need to identify the outcome-based results of the mission. The question is: WHAT products, by-products, or services is the system required to PRODUCE or AVOID to achieve the OUTCOME-based results. This brings us to the next task: identify system responses.*

Step 6: Identify Outcome-Based System Responses

Throughout all phases of the mission, an SOI produces a series of *behaviors*, *products*, *by-products*, and *services* to satisfy internal and external requirements. *Internal requirements* include performance monitoring, resource consumption, and payload/cargo manifests. External requirements include the examples listed in Table 16.1.

Step 7: Identify Mission Resources and Sustainment Methods

Human-made systems have finite resource capacities that require *replenishment* and *refurbishment*. Depending on the mission operating range of the system relative to its current mission application, mission analysis must consider HOW the system's *expendables* and *consumables* will be *resupplied* and *replenished*. Operationally, the question is: *How will the organization sustain and maintain the mission from beginning to end?*

Step 8: Perform a Mission Task Analysis

Throughout the pre-mission, mission, post-mission phases, specific operational tasks must be performed to accomplish the phase-based mission objectives. These tasks ultimately provide the basis for capabilities the SOI must provide to accomplish the mission. Therefore the *mission analysis* should:

1. Identify the high-level outcome-based mission tasks to be performed.
2. Synchronize those tasks to the Mission Event Timeline (MET).
3. Identify the task performance-based objectives.

Step 9: Assess and Mitigate Mission and System Risk

Some systems are required to perform missions in harsh OPERATING ENVIRONMENTs that may place the system at risk to threats, not only in completing its mission but also in returning safely to its home base. Consider the following example:

EXAMPLE 16.4

Loose, hidden objects on a lawn can cause injury to people and damage a lawnmower blade and engine. Birds, ducks, and geese pose threats to airports and aircraft in flight. Loose objects and debris thrown into the air by vehicles on the road can cause injury to others and damage to vehicles. Unprotected computer systems are vulnerable to viruses.

Table 16.1 Examples of mission requirements derived from analysis of external systems

Type of External System	Example Sources of Requirements
Friendly or *cooperative* systems	Communications Deliverable items, etc.
Benign systems	Communications Detection and avoidance Evasive tactics, etc.
Hostile or *Adversarial* systems	Rules of engagement Detection and avoidance Countermeasures/counter-counter measures Aggressive/defensive actions, etc.

Risks assessments include considerations of system *vulnerability*, *susceptibility*, *survivability*, and *maintainability*. Most people tend to think in terms of external *benign*, *adversarial*, or *hostile* systems that may be threats to the system. However, since a system interacts with itself, it can also be a threat to itself. Recall from our discussion of the architecture of systems in Chapter 8 how a system interacts with: 1) its OPERATING ENVIRONMENT and 2) itself. Analytically, the sources of these threats begin with the MISSION SYSTEM and SUPPORT SYSTEM elements—comprised of PERSONNEL, EQUIPMENT, MISSION RESOURCES, PROCEDURAL DATA, SYSTEM RESPONSES, and FACILITIES. Consider the following EQUIPMENT system element examples.

EXAMPLE 16.5

Failed automobile components, such as a blown tire, can cause a driver to loose control of the vehicle while driving. Failures in an aircraft's flight control system or broken blades in a jet engine fan can force and emergency landing or have catastrophic consequences.

Internal failures or *degraded performance* also has *negative* impacts on system performance that ultimately translates into mission failure or degree of success. Perhaps one of the most notable examples is the Apollo 13 catastrophe. *Mission analysis* should identify those areas in system capabilities that may be *vulnerable* or *susceptible* to external internal threats, especially for *mission critical* components.

Author's Note 16.1 *Internal threat analysis is typically performed via failure modes and effects analysis (FMEA). For mission critical components, the FMEA may be expanded into a failure modes, effects, and criticality analysis (FMECA) that assesses the degree of criticality.*

16.3 GUIDING PRINCIPLES

In summary, the preceding discussions provide the basis with which to establish the guiding principles that govern mission analysis.

Principle 16.1 Every system mission must accomplish one or more organizational performance objectives.

Principle 16.2 Human-made systems have three primary phases of operation: pre-mission, mission, and post-mission; an interim phase may be further required for some systems.

Principle 16.3 Every system phase of operation must satisfy one or more outcome-based performance objectives associated with accomplishment of an overall system mission and its performance objectives.

Principle 16.4 Mission success requires five key elements: a purpose, resources, a reasonably achievable outcome-based performance objective(s), a Mission Event Timeline (MET), and a willingness to perform. Where there is no willingness to act, the other elements are meaningless.

16.4 SUMMARY

Our discussion of mission analysis highlighted several key points that require emphasis:
- Every system concept consists of interactions of abstract entities derived from the System Element Architecture: 1) SYSTEM OF INTEREST (SOI) consisting of the MISSION SYSTEM and the SUPPORT SYSTEM and 2) the OPERATING ENVIRONMENT.

- Each mission *begins* with a *point of origination* and *concludes* with a *destination* or *point of termination* with intervening *staging*, *control*, or *waypoints* based on specific objectives and Mission Event Timeline (MET) *events*.
- Between the *point of origination and point of termination*, some missions may require interim *waypoints* or *delivery points* that satisfy specific mission objectives.
- Every mission must be founded on an operational strategy that defines HOW the *system elements*—namely PERSONNEL, EQUIPMENT, and FACILITIES—will be *deployed* and *employed* at critical staging events to accomplish mission objectives constrained by a *Mission Event Timeline (MET)*.
- Every mission is characterized by at least three mission phases of operation: 1) *pre-mission*, 2) *mission*, and 3) *post-mission*.
- During each phase of operation, *system element* interactions must be *orchestrated* and *synchronized* in accordance with mission objectives and a Mission Event Timeline (MET).
- Every mission requires *mission critical* capabilities with a *minimum* level of performance to be provided by the system elements—namely PERSONNEL, EQUIPMENT, and MISSION RESOURCES—to achieve a specified outcome.

GENERAL EXERCISES

1. Answer each of the *What You Should Learn from This Chapter* questions identified in the *Introduction*.
2. Refer to the list of systems identified in Chapter 2. Based on a selection from the preceding chapter's General *Exercises* or a new system selection, apply your knowledge derived from this chapter's topical discussions. Identify the following:
   (a) How many different missions does the system perform?
   (b) What are those missions?
   (c) Pick two missions and perform a mission analysis using the methodology described in this section.

ORGANIZATIONAL CENTRIC EXERCISES

1. Select a contract program within your organization. Interview program personnel to understand what form of *mission analysis* was performed on the program.
   (a) Was the mission analysis required as a contract deliverable? If so, when was it required to be delivered? Were subsequent updates required? Was there a required outline format?
   (b) Who performed the mission analysis?
   (c) How was the mission analysis documented?
   (d) What was the most difficult parts of the analysis to accomplish?
   (e) In what ways do program personnel believe the analysis benefit the program?
   (f) What would you do differently next time?
   (g) How did program and executive management view the importance of the analysis?
   (h) Were the right amount of resources and expertise applied to accomplish the task?
   (i) What were the shortcomings, if any, of the end product?

REFERENCES

Defense Systems Management College (DSMC). 1998. DSMC *Test and Evaluation Management Guide*, 3rd ed. Defense Acquisition Press Ft. Belvoir, VA.

Defense Systems Management College (DSMC). *Glossary: Defense Acquisition Acronyms and Terms*, 10th ed. Defense Acquisition University Press Ft. Belvoir, VA. 2001.

MIL-STD-499B (cancelled draft). 1994. *Systems Engineering*. Washington, DC: Department of Defense (DoD).

Chapter 17

System Use Cases and Scenarios

17.1 INTRODUCTION

From an SE perspective the challenge in developing systems is being able to *translate* mission objectives, operations, and tasks into a set of capability requirements that can be transformed into a physical design solution. Most organizations and individuals attempt to make a quantum leap from the mission objectives to writing text requirements into the *System Performance Specification (SPS)*. This is typically accomplished without understanding:

1. What *problem space* the User is attempting to solve.
2. How they intend to deploy and employ the *solution space* system to perform missions to address all or a portion of the *problem space*.

As systems become more complex, they require a solution development methodology that is easily understood by Acquirers, Users, and System Developers. One method is to employ system use cases and scenarios to bridge the gap between mission objectives and specification requirements.

What You Should Learn from This Chapter

- What is a system *use case*?
- What are the attributes of a *use case*?
- What is a *use case analysis* and how do you perform one?
- How do *use cases* relate to *system capability requirements*?

Definitions of Key Terms

- **Actor** "A role of object or objects outside of a system that interacts directly with it as part of a coherent work unit (a use case). An Actor element characterizes the role played by an outside object; one physical object may play several roles and therefore be modeled by several actors." (*UML Notation Guide*, para. 6.1.2, p. 75)
- **Operational Scenario** A hypothesized narrative that describes system entity interactions, assumptions, conditions, activities, and events that have a likelihood or probability of actually occurring under prescribed or worst-case conditions.
- **Sequence Diagram** "A diagram that represents an interaction, which is a set of messages exchanged among objects within a collaboration to effect a desired operation or result." (*UML Notation Guide*, para. 7.2.1, p. 80)

System Analysis, Design, and Development, by Charles S. Wasson
Copyright © 2006 by John Wiley & Sons, Inc.

168 Chapter 17   System Use Cases and Scenarios

- **Use Case** A statement that expresses *how* the User envisions *deploying, operating, supporting, or disposing* of a system, product, or service to achieve a desired performance-based outcome.
- **Use Case Diagram** "A graph of actors, a set of use cases enclosed by a system boundary, communication (participation) associations between the actors and the use cases, and generalizations among the use cases." (*UML Notation Guide*, para. 6.1.2, p. 75)
- **Use Case Scenario** A set of conditions a MISSION SYSTEM use case may encounter in its OPERATING ENVIRONMENT that requires a unique set of capabilities to produce a desired result or outcome. Scenarios include considerations of HOW a User or threat might apply, misapply, use, misuse, or abuse a system, product, or service.

17.2 UNDERSTANDING THE ANALYTICAL CONTEXT OF USE CASES AND SCENARIOS

A reference framework that illustrates the context of use cases and scenarios and their importance is provided in Figure 17.1. Note that the *entity relationships* in the figure are partitioned into three domains: an Operations Domain, an Analysis Domain, and an Engineering Domain. Based on this framework, let's characterize these relationships.

1. Each mission is assigned at least one or more mission objectives.
2. Each mission objective is accomplished by an integrated set of mission operations performed by the MISSION SYSTEM and SUPPORT SYSTEM.
3. Each MISSION SYSTEM and SUPPORT SYSTEM operation is decomposed into hierarchical chains of sequential and concurrent tasks.
4. Each task is performed and measured against one or more performance standards and implemented via at least one or more use cases.

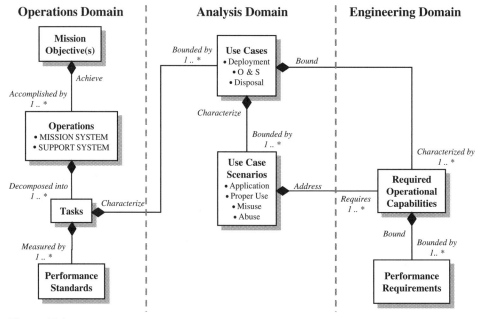

Figure 17.1 System/Product Use Cases and Scenarios Entity Relationships

5. Each use case is *bounded* by at least one or more use case scenarios.
6. Each use case scenario represents HOW the use case can be *applied*, *used*, *misused*, *abused*, etc. and is *bounded* and *specified* by at least one or more required operational capabilities.
7. Each required operational capability identifies WHAT the system, product, or service must perform to fulfill each use case and accommodates various use case scenarios and is *quantified* by at least one or more performance requirements.

This introductory framework establishes the backdrop for our discussion of *system use cases and scenarios*.

17.3 WHAT ARE USE CASES?

A *use case* is characterized by a set of attributes that describe HOW the User might *deploy*, *operate*, *support*, or *dispose* of the system. The *attributes*, which serve as a checklist for developing use cases, include:

- Unique identifier
- Objective
- Outcome-based results
- Assumptions
    - Initial state
    - Final state
    - Environmental conditions
    - Preceding circumstances (optional)
    - Operating constraints
    - External inputs
    - Resources
    - Event-based timeline
    - Frequency of occurrence and utility priorities
- Processing capabilities
- Scenarios and consequences
    - Probability of occurrence
    - Use case scenario actors
    - Stimuli and cues
    - Consequences
    - Compensating/mitigating actions

Given this list, let's briefly describe each one and its contribution to the characterization.

Attribute 1: Unique Identifier

Each *use case* should have its own unique identity and not *overlap*, *conflict*, or *duplicate* other *use cases*. Therefore each *use case* should be tagged with its own unique identifier and title.

Attribute 2: Objective

Each *use case* must have at least one or more *outcome-based* performance objectives.

Attribute 3: Outcome-Based Results

Use cases should produce at least one or more system/entity *behaviors*, *products*, *by-products*, or *services* that may be *tangible* or *intangible* to achieve the desired *outcome-based* objective.

Attribute 4: Assumptions

The formulation of *use cases* requires that SEs make *assumptions* that characterize the *initial conditions* that form the basis for initiating a *use case*. Assumptions include the following:

Initial State. The INITIAL state of a *use case* represents the *assumed* physical operational state of the *system*, *product*, or *service* when the use case is initiated.

Final State. The FINAL state represents the desired *physical* or *operational* state of the system when the desired outcome has been achieved.

Environmental Conditions. The current environmental conditions specify and bound the OPERATING ENVIRONMENT *conditions* that exist when a *system*, *product*, or *service use case* is initiated.

Preceding Circumstances (Optional). For some applications, the *circumstance* or *sequence* of events leading up to the initiation of a *use case* need to be identified. *Preceding circumstances* provide a basis for documenting this assumption.

Operating Constraints. For some *use cases* the system, product, or service may have operational constraints such as organizational policies, procedures, task orders; local, federal, state, and international regulations or statutory laws; or public opinion; or a *Mission Event Timeline (MET)*. Operational constraints thus serve to *bound* or *restrict* the acceptable set of corporate, moral, ethical, or spiritual actions allowed for a *use case*.

External Inputs. Every system, product, or service processes external inputs to add value to achieve the specified outcome.

Resources. Every system, product, or service requires MISSION RESOURCES to perform its mission. MISSION RESOURCES are typically finite and are therefore constrained. The *resources* attribute documents what types of resources (i.e. *expendables* or *consumables*) are required to sustain system/entity operations.

Event-Based Timeline. *Use cases* may require a Mission Event Timeline (MET) to synchronize the planned actions or intervention of human operators or expected responses from the system or its operators.

Frequency of Occurrence and Utility Priorities. Every *use case* has a cost and schedule for development, training, implementation, and maintenance. The realities of budgetary cost and schedule *limit* the number of use cases that can be practically implemented. Therefore, prioritize *use cases* and implement those that maximize application and safety utility to the User.

Note that we said, *to maximize application and safety utility*. If you prioritize use cases, emergency capabilities and procedures should have a very remote frequency of occurrence. Nevertheless, they can be the most critical. As is the case in trade study evaluation criteria, you may need to analytically express utility in terms of a multiplicative factor. Instead of assigning a 1 (low) to

5 (high) weighting factor priority to each *use case*, multiply the factor by a *level of criticality* from 1 (low) to 5 (high) to ensure the proper visibility from a safety perspective.

Commercial organizations produce products and services to sell in the marketplace at a profit and sustain the business operations for the long term. Organizational products and services MUST be SAFE for the Users to deploy, operate, and support. Hypothetically, you could focus all resources on safety features and produce a product that is so burdened with safety features that it has no *application utility* to the User.

Although our discussion focuses on the development of a product or service, remember that the system has other elements than just EQUIPMENT (PERSONNEL, PROCEDURAL DATA, etc.). So, when confronted with increasing design costs, there may be *equally effective* alternatives for *improving* safety such as operator certification, training and periodic refresher training, cautionary warning labels, and supervision that may not require product implementation.

Attribute 5: Processing Capabilities

The heart of a *use case* centers on stimulus-response processing for a specified set of conditions to produce the *desired* or *required outcome*. Some domains refer to this as a *transfer* or *response function*. Consider the following example:

EXAMPLE 17.1

A photovoltaic or solar cell *transforms* sunlight into electrical energy.

Author's Note 17.1 *Remember, the context here is a simple box representing a system or item at any level of abstraction with input(s) and output(s). Focus on simplification of the solution space. Avoid attempting to define processing for multiple levels simultaneously.*

Attribute 6: Scenarios and Consequences

As humans, we tend to be optimistic and ideally *believe* that everything will be *successful*. While this is true most of the time, *uncertainties* do occur that create conditions we have not planned for operationally or in terms of system, product, or service capability. Once a *use case* is identified, ask the question: *WHEN the User employs this use case:*

1. WHAT can go *wrong* that we haven't anticipated?
2. WHAT are the *consequences* of failure and how do we mitigate them?

We refer to each of these instances as *use case scenarios and consequences*? Consider the following example:

EXAMPLE 17.2

Suppose that we are designing a compact disk (CD) or digital video disk (DVD) player. Ideally, the high-level CD/DVD use case describes a User inserting the CD/DVD into the player—and magic happens! The User gets the result, be it music or a movie. The User has thus a *use case scenario* with a POSITIVE outcome. Now, *what happens if the User inserts the CD/DVD upside down*? The User has a use case scenario with a NEGATIVE outcome. This leads to the question: *When we design the CD/DVD device, how should the User be advised of this situation*? If we add notification capability to the device, development costs increase. In contrast, the LOW-COST solution may be to simply inform the User via the product manual about the convention of ALWAYS inserting the CD/DVD so that the title faces upward.

172 Chapter 17 System Use Cases and Scenarios

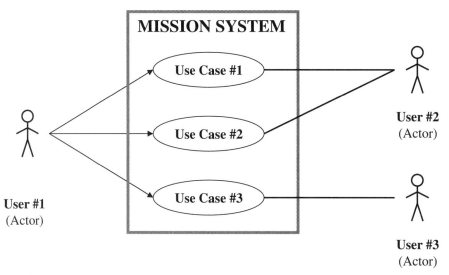

Where: UML® = Unified Modeling Language

Figure 17.2 UML® Use Case Diagram

Probability of Occurrence. Once *use case scenarios* are identified, we need to determine the *probability of occurrence* of each one. As in the earlier discussion of use case priorities, scenarios have a *probability of occurrence*. Since additional design features have a cost, prioritize scenarios with User safety as a predominant consideration.

Use Case Scenario Actors. Our discussion up to this point has focused on the WHAT is *most likely* or *probable* to occur: *use cases* or *scenarios*. The key question is: WHO or WHAT are the *interacting* entities during *use cases* and *scenarios*. The Unified Modeling Language (UML®) characterizes these entities as "actors."

Actors can be persons, places, real or virtual objects, or events. UML® represents actors as stick figures and use cases as ellipses, as shown in Figure 17.2.

- User 1 (actor) such as a system administrator/maintainer interacts with use cases 1 through 3.
- User 2 (actor) interacts with use cases 1 and 2 (capabilities).
- User 3 (actor) interacts with use case 3 (capabilities).

In our previous example, the actors include User, CD/DVD, and CD/DVD player.

Stimuli and Cues. Use cases are initiated based on a set of *actions* triggered by system operator(s), external systems, or the system. Consider the following stimulus–response actions:

- The User or an external system initiates one or more actions that cause the system to respond behaviorally within a specified time period.
- The system notifies the User to perform an action—to make a decision or input data.
- The User *intervenes* or *interrupts* ongoing actions by the system.

Each of these examples represents instances whereby the User or System stimulates the other to action. Figure 17.3 illustrates such a sequence of actions using a UML sequence diagram.

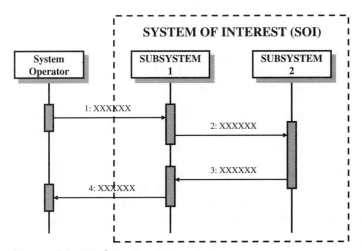

Figure 17.3 UML® Use Case Sequence Diagram

Scenario Consequences. Each *use case* and *scenario* produces an *outcome* that may have *consequences*. Consider the following example:

EXAMPLE 17.3

If scenario X occurs and the operator or system responds in a specified manner, *instabilities* and *perturbations* may be induced into the system that may have NEGATIVE *consequences*. Therefore each use case and scenario should identify the potential consequences of proper use/misuse, application/misapplication, and abuse.

Compensating/Mitigating Actions. Given the set of *consequences* identified for *use cases* and *use case scenarios*, we need to identify what *compensating/mitigating actions* should be incorporated into the system, product, or service to *eliminate* or *minimize* the effects of a NEGATIVE outcome consequences. Consider the following example:

EXAMPLE 17.4

Suppose that we are to design a car. Since a car can collide with other vehicles, walls, or trees, a generalized interface solution of the car body-to-external system is insufficient. An analysis of use cases and use case scenarios suggests that passengers can lose their lives or sustain injuries in a collision. So a specialized interface consisting of a bumper is added to the car frame as a *compensating/mitigating action*. However, impact tests reveal that the bumper is inadequate and requires yet a more *specialized* solution including the following sequences of design actions:

Design action 1: Incorporate shock absorbers into the vehicle's bumpers.
Design action 2: Install and require use of seat belts.
Design action 3: Install an air bag system.
Design action 4: Install an anti-lock braking system (ABS).
Design action 5: Specify proper vehicle operating procedures.
Design action 6: Increase driver awareness to drive safely and defensively.

17.4 USE CASE ANALYSIS

Each *use case* and its *most likely* or *probable* scenarios represent a series of anticipated interactions among the actors. Once the scenarios and actors are identified, system analysts need to understand the *most likely* or *probable* interactions between the system or entity of interest and external systems within its OPERATING ENVIRONMENT.

UML® tools are useful in understanding the *stimuli, cues,* and *behavioral responses* between interacting systems. *Sequence diagrams* serve as a key tool. Sequence diagrams consist of actors and lifelines as illustrated in Figures 2.5 and 17.3.

- **Actors** Consist of entities at a given level of abstraction—such as SYSTEM, PRODUCT, and SUBSYSTEM—and external systems within the abstraction's OPERATING ENVIRONMENT. For example, a SUBSYSTEM use case might depict its operator(s), if applicable, and external systems such as other SUBSYSTEMS, and PHYSICAL ENVIRONMENT conditions.
- **Lifeline** Consists of a vertical line to represent time relative processing. Bars are placed along the lifeline to represent entity activities or processing of external inputs, stimuli, or cues and behavioral responses.
- **Swim Lanes** Consist of the regions between the actor lifelines for illustrating sequential operations, tasks, products, by-products, or services *interactions* and *exchanges* and between each actor.

To illustrate HOW these are employed, consider the following example:

EXAMPLE 17.5

Let's suppose a User (actor) interacts with a calculator (actor) to accomplish a task to perform a mathematical calculation and communicate the results. Figure 17.4 provides a simple illustration of the interaction. Observe that Figure 17.4 is structurally similar to and expands the level of detail of Figure 17.3. To keep the example simple, assume the calculator consists of two SUBSYSTEMS, 1 and 2.

The User and each of the SUBSYSTEMS have an INITIAL State and FINAL State and conditional loops that cycle until specific decision criteria are met to terminate operation. We assume each of the SUBSYSTEM activities include wait states for inputs. When inputs arrive, processing is performed, and control is passed to the next activity. Here's how a potential *use case scenario* might be described.

- SUBSYSTEM 1 performs Activity 20 to await user inputs via the keyboard.
- When the System Operator Activity 10 enters data as Output 10, Activity 20 accepts the operator keyboard entries and converts the information into machine-readable code as Output 20.
- SUBSYSTEM 2 performs: 1) Activity 30 to await data inputs and 2) Activity 31 to perform the required computation and output the mathematical results as Output 31.
- In the interim SUBSYSTEM 1 Activity 21: 1) awaits Output 31 results, 2) converts the results into meaningful operator information, and 3) displays the results as Output 21.
- Following data entry (i.e., Activity 10), the System Operator performs Activity 11 to: 1) await Output 21 results, 2) record the results, and 3) communicate the results as Output 11.

These cycles continue until the calculator is turned off, which is the FINAL State.

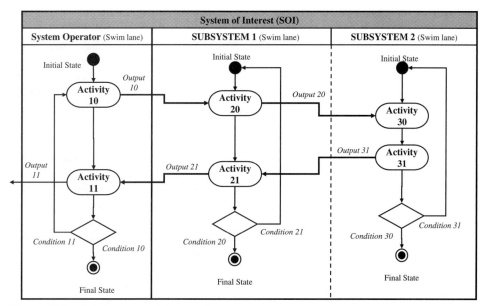

Figure 17.4 UML Symbology Swim Lanes Example

How Many Use Cases?

A key question people often ask is: *How many use cases are required for a system*? There are no magic answers; 10 through 30 use cases might be average. Some highly complex systems may just have 5 or 6; others, 10 to 20. All depends on the individuals and organizations involved. Some want simplicity to keep the number small; others want detailed lists. Keep in mind that a system may have 5 to 8 primary use cases; the remainder may be *secondary* or *subordinate* use cases that support the primary use cases.

17.5 RELATING USE CASES TO SPECIFICATION REQUIREMENTS

You may note that the application of *use cases* is fine for system design, but WHY are we addressing them here? There are several reasons:

1. *Use cases* serve as a valuable tool for Acquirer SEs to work with Users to understand their needs and translate their visions of HOW they intend to deploy, operate, and support the system into a more technical description. Use cases *isolate* on specific system features and associated capabilities that the User can easily understand. This avoids the need to write abstract system requirements language that may or may not have meaning or interest to the User.

2. Once the *use cases* and *scenarios* are identified and prioritized, they provide a basis for translation into specification requirements suitable for system, product, or service acquisition.

3. *Use cases*, which can be derived from Acquirer specifications, provide a mechanism for System Developers to formulate system operational concepts and design solutions.

17.6 FINAL THOUGHTS

From a SE perspective, *use case* analysis should be a key tool of any system development effort. However, engineers often view this activity as non–value-added paperwork to the User and product, and believe their time is better spent contemplating creation of elegant designs. The reality is the most *elegant* designs are *useless* unless the User can *easily* and *understandably* implement them with their current skill set to perform missions. This is WHY "just in time" training for system operators must take place *prior to* system acceptance and delivery.

Keep in mind that the people who give the bureaucratic argument are the same people who, after a system fails during integration and test, capitulate and remark, HOW was I to know WHAT the User wanted? I'm only human. . . . Besides they couldn't decide what they wanted. Documenting *use cases* is a simple matter. It requires professional discipline, something that tends to get lost in modern-day engineering efforts. If you doubt this, ask yourself how many products have disappointed you and made you wonder . . . why no one within the System Developer's organization bothered to consult the Users. If they had, they would have easily learned that this step is critical to the User's success and acceptance of the system, product, or service.

17.7 GUIDING PRINCIPLES

In summary, the preceding discussions provide the basis with which to establish the guiding principles that govern system use cases and scenarios.

Principle 17.1 Every use case has at least one or more most likely or probable scenarios, some with positive outcomes, others with negative outcomes: your mission as an SE is to mitigate risk and maximize positive outcomes.

Principle 17.2 Every use case, scenario, and requirement has a value to the User, a cost to implement, and a level of acceptable operational risk.

Principle 17.3 Every system mission consists of one or more use case based capabilities.

17.8 SUMMARY

Our discussion of system use cases and scenarios highlighted the need to employ use cases as a means of avoiding quantum leaps between User's visionary requirements and system design. We also showed that use cases and scenarios provide a powerful tool that Users, Acquirers, and System Developers. They can be used to improve communications and to understand how the envisioned system is to be deployed, operated, and supported.

- *Use cases* provide a means of *identifying* and *prioritizing* key User requirements for implementation.
- *Use cases* must be *prioritized*, based on *most likely* or *probable* occurrences for development subject to program technical, cost, and schedule constraints.
- *Use case scenarios* provide a basis for understanding not only *how* the User might use a system, product, or service. Also how the *misuse* or *abuse* might result in risks with consequences that require design *compensating* or *mitigating actions*.
- *Use case scenarios* must be *prioritized* within use case technical, cost, and schedule constraints.
- *Use case attributes* provide a standard framework to uniformly and consistently characterize each *use case*.

- UML *interaction diagrams* serve as a useful tool for understanding the sequencing of actor interactions and behavioral responses.
- *Each use case* and its *attributes* should be documented and placed under baseline management control for decision making.

NOTE

The *Unified Modeling Language* (UML®) is a registered trademark of the Object Management Group (OMG).

GENERAL EXERCISES

1. Answer each of the *What You Should Learn from This Chapter* questions identified in the *Introduction*.
2. Refer to the list of systems identified in Chapter 2. Based on a selection from the preceding chapter's *General Exercises* or a new system selection, apply your knowledge derived from this chapter's topical discussions. Identify the following:
   (a) System actors
   (b) System use cases
   (c) System use case diagrams
   (d) Use case sequence diagrams

ORGANIZATION CENTRIC EXERCISES

1. Check with your organization to see if any programs employ use cases and scenarios for deriving system capabilities and requirements.
   (a) Were these required by contract or a program decision?
   (b) What were the programs experiences?
   (c) Were the teams properly trained in applying use cases and scenarios?
   (d) What tools were used to perform use case analysis?
   (e) How were the use cases and scenarios documented?
   (f) How did the program link use cases and scenarios to specification requirements?

REFERENCE

Object Management Group (OMG). 1997. *Unified Modeling Language (UML®) Notation Guide*, Version 1.1, Needham, MA.

Chapter 18

System Operations Model

18.1 INTRODUCTION

As the mission analysis identifies a system's *use cases* and *scenarios*, preferably in *collaboration* with the User, the next challenging concept is working with the User to conceptualize *how* they *intend* to deploy, operate, support, and dispose of a system. One of the mechanisms for documenting the conceptualization is the *system concept of operations*, or ConOps. This section introduces the *System Operations Model* that provides the structural framework for developing the ConOps.

The *System Operations Model* provides a high-level operational workflow that *characterizes* HOW a system: 1) is configured for a mission, 2) conducts the mission, and 3) is supported following a mission. The structure of the model consists of operations and tasks that can be translated into specification requirements or as workflow for the system engineering design solution.

Our discussions provide insights regarding how the model's operational capabilities are *allocated* and *flowed down* to the system elements—such as EQUIPMENT, PERSONNEL, and FACILITIES. As a result, these discussions provide the foundation for the topic that follows, *system mission and support operations*.

What You Should Learn from This Chapter

1. What is the *System Operations Model*?
2. What is a *Concept of Operations (ConOps)*?
3. What is the *purpose* of the System Operations Model?
4. Graphically illustrate the System Operations Model.
5. Describe each of the model's operations or tasks.
6. Delineate the differences in the model from its robust version.
7. What is a System Operations Dictionary?

Definitions of Key Terms

- **Concept of Operations (ConOps)** A description of the workflow of a system's sequential and/or concurrent operations required to achieve pre-mission, mission, and postmission phase outcome-based performance objectives.
- **Control or Staging Point** A major decision point that limits advancement of workflow progress to the next set of objective-based operations until a set of *go–no go* decision criteria are accomplished.

System Analysis, Design, and Development, by Charles S. Wasson
Copyright © 2006 by John Wiley & Sons, Inc.

- **System Operations** A set of multi-level, interdependent activities or tasks that collectively contribute to satisfying a pre-mission, mission, or postmission phase objective.
- **System Operations Dictionary** A document that *scopes* and *describes* system entity *operational relationships* and *interactions* required to support a specific phase and mode of operation. The operational relationships and interactions are analyzed and translated into a set of required operational capabilities, which are then transformed into system performance requirements to support each mode of operation.
- **System Operations Model** A generalization of system operations that can be employed as an initial starting point for identifying the workflow and operations for most systems.

18.2 THE SYSTEM CONCEPT OF OPERATIONS (ConOps)

Once a system's *problem space* and *solution spaces* are bounded, the next step is to understand HOW the User *intends* to use a solution space system. Most systems are *precedented* and simply employ new technologies to build on the existing infrastructure of operations, facilities, and skills. This does not mean, however, that *unprecedented* systems do not occur.

Referral For more information about *precedented* and *unprecedented* systems, refer to Chapter 3 on the definition of these systems.

If you expand the *problem–solution space* concept, our analysis reveals that the *solution space time-based interactions*—namely entity relationships—can be characterized by a set of operations that can be generalized into the *System Operations Model*. In turn, the model provides a framework for developing the ConOps, which describes the top-level sequential and concurrent operations required to accomplish the system's mission.

EXAMPLE 18.1

A system *concept of operations (ConOps)* for a system such as the Space Shuttle system describes the operational sequences required to deliver a payload into outer space, deploy the payload and conduct experiments, and return the cargo and astronauts safely to Earth.

We refer to Example 18.1 as *cyclical* operations within the system/product life cycle. Living organisms, such as humans, exhibit this *cyclical* characterization as evidenced by our daily need for food, water, rest, and sleep.

We can generalize a ConOps in terms of a common set of objectives that reflect *how* the User plans to use the system. These objectives include:

1. Deploy the system.
2. Configure the system for deployment and operational use.
3. Check the system's readiness to conduct pre-mission, mission, and postmission operations.
4. Employ the system asset.
5. Clean it up and store it for the next use.
6. Discard the system, when appropriate.

To better understand HOW these objectives can be integrated into a total system operations solution, let's explore a description of the *System Operations Model*.

180 Chapter 18 System Operations Model

18.3 THE SYSTEM OPERATIONS MODEL

The System Operations Model depicted in Figure 18.1 provides a *construct* that can be applied to HUMAN-MADE systems. Although there are a number of variants on this graphic, let's explore this model to better understand how it operates. First, a word about the contents of the graphic.

Each box in Figure 18.1 *represents* an integrated, multi-level collection of system use case-based capabilities and activities required to achieve an overall *mission* objective. We can expand or decompose each of these capabilities into lower level operational capabilities. Ultimately these capabilities and their respective levels of performance are allocated to one or more of the system elements—such as PERSONNEL, EQUIPMENT, and FACILITIES.

Each decision block (diamond) is referred to as a *control or staging point* and requires a *go–no go* decision from a decision authority based on a predefined set of *exit* or *entrance criteria*. Each operation and *control point* is *tagged* with a unique identifier. The identifier is used to *map* to a specific requirements section in the *System Performance Specification (SPS)* or to a detailed narrative in an *operational concept description (OCD)*.

Referral For more information about linking ConOps operations and capabilities to the SPS, refer to Chapter 28 *System Specification Practices*.

Author's Note 18.1 *Observe the usage of ConOps and OCD. Various organizations use one or the other term. ConOps, to some people, infers a summary discussion of how a system will operate while OCD infers supporting detail to a ConOps. Pick one term or the other and apply it consistently across programs.*

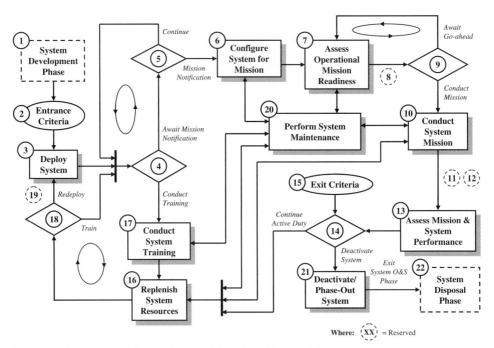

Figure 18.1 Generalized System Concept of Operations (ConOps) Model

18.4 SYSTEM OPERATIONS MODEL DESCRIPTION

Figure 18.1 depicts the *System Operations Model* that applies to most HUMAN-MADE systems. Entry into the model begins when a system is transitioned from its System Development Phase (1) of the system/product life cycle. *Entrance criteria* (2) are evaluated to assess system readiness to begin *active duty*. Let's explore the field operations that follow.

Operation 3.0: Deploy System

Operation 3.0 to deploy the system addresses system capabilities and activities required to deliver and install the system at the User's required destination. As each system rolls off the production line and is *verified* against performance requirements, the system is packed and shipped for deployment or distribution to the User. Activities include: transportation; load/unloading; crate/uncrating; initial setup, installation, and assembly; system checkout; verification; integration into higher level systems; and verification of interoperability at that level.

On completion of all planned activities, the Operation 4.0 Conduct System/Mission Training decision is made.

Operation 4.0: Conduct System/Mission Training Decision

Operation 4.0 Conduct System/Mission Training, a decision *control point*, determines if the system is to be placed immediately into *active duty* or reserved for operator training or demonstrations.

- If the system/mission training decision is *Yes* or TRUE, workflow progresses to Operation 17.0 Conduct System Training.
- If the system/mission training decision is *No* or FALSE, workflow progresses to Operation 5.0 Await Mission Notification decision.

Operation 5.0: Mission Notification Decision

Operation 5.0 Mission Notification, a decision *control point*, must await notification to prepare to conduct a mission. Depending on the system and application, Operations 4.0 and 5.0 are each effectively a cyclical WAIT STATE for the system that loops until a higher level authority issues an order to conduct the mission.

- If the mission notification decision is *Yes* or TRUE, workflow progresses to Operation 6.0 Configure System for Mission.
- If the mission notification decision is *No* or FALSE, workflow cycles back to Operation 4.0 Conduct System/Mission Training decision.

Operation 6.0: Configure System for Mission

Operation 6.0 Configure System for Mission includes system capabilities and activities required to prepare and configure the system for the required mission. On receipt of mission orders, the system is configured and supplied for the mission. Operational activities include pre-mission planning, physical hardware and software changes, personnel training, and refueling. System configuration/reconfiguration activities include the *synchronized orchestration* of the system elements such as:

1. PERSONNEL Operators, administrators, etc.
2. PROCEDURAL DATA Operating procedures, media, etc.

3. Interfaces with HUMAN-MADE SYSTEMS Friendly, cooperative systems.
4. SUPPORT SYSTEM Media, instructors, supply, maintainers, etc., into a planned, coherent operation focused on achieving the allocated mission tasks and objectives.

On completion of the activities, system verification is performed to ensure that the system is properly configured for the mission.

- If the system verification check is successful, workflow progresses to Operation 7.0, Assess Operational Mission Readiness.
- If system *defects* or *deficiencies* are discovered, workflow progresses to Operation 20.0 Perform System Maintenance.

Operation 7.0: Assess Operational Mission Readiness

Operation 7.0 Assess Operational Mission Readiness includes system capabilities and activities required to review the overall readiness to conduct the assigned mission. After the system has been configured for the mission and all system element resources are fully integrated and operational, mission operational readiness is assessed. The assessment evaluates the *readiness posture* of the integrated set of system elements—such as EQUIPMENT, PERSONNEL, and FACILITIES—to perform their assigned mission.

If the readiness assessment is *No*, the system is *tagged* as operationally *deficient* with a RED or YELLOW tag. A mission impact risk assessment decision is made to determine if the *deficiency* warrants cancellation of the mission or replacement of the system/element with a backup system.

- If the system requires maintenance, workflow progresses to Operation 20.0 Perform System Maintenance.
- If the system is determined to provide the capabilities required to support the mission, workflow progress to Operation 9.0 Await Mission Go-ahead Decision.

Author's Note 18.2 *To facilitate a later discussion in this chapter, Operations 8.0, 11.0, 12.0, and 19.0 are unused in Figure 18.1 and reserved for our follow-on topical discussion.*

Operation 9.0: Mission Go-Ahead Decision

Operation 9.0 Mission Go-ahead, a decision *control point*, determines if tasking orders to conduct the mission have been issued.

- If Operation 9.0 Await Mission Go-ahead Decision is *Yes* or TRUE, workflow proceeds to Operation 10.0 Conduct Mission.
- If the Operation 9.0 Await Mission Go-ahead Decision is *No* or FALSE, system readiness is periodically checked by *cycling* back to Operation 7.0 Assess Operational Mission Readiness.

Operation 10.0: Conduct System Mission

Operation 10.0 Conduct System Mission includes system capabilities and activities required to conduct the system's primary and secondary mission(s). During this operation the system may encounter and engage threats and opportunities as it performs the *primary* and *secondary* mission objectives.

If the system requires maintenance during the conduct of the mission, Operation 16.0 Replenish System Resources or Operation 20.0 Perform System Maintenance may be performed, if practical.

EXAMPLE 18.2

A fighter aircraft may require refueling during the conduct of an operational mission.

On completion of the mission, workflow proceeds to Operation 13.0 Assess Mission and System Performance.

Operation 13.0: Assess Mission and System Performance

Operation 13.0 Assess Mission and System Performance includes system capabilities and activities required to review the level of mission success based on mission primary and secondary objectives and system performance contributions to that success. *Activities* include postmission data reduction, target impact assessment, strengths, and weaknesses; threats; mission debrief observations and lessons learned; and mission success. These operations also provide the opportunity to review human performance, strengths, and weaknesses in the conduct of the mission. On completion of Operation 13.0 Assess Mission System Performance, workflow progresses to Operation 14.0, a Deactivate/Phase-out System Decision.

Operation 14.0: Deactivate/Phase-out System Decision

Operation 14.0 Deactivate/Phase-out System, a decision *control point*, determines if the system is to continue current operations, be upgraded, or be decommissioned or phased out of *active duty*. The decision is based on *exit criteria* (15) that were established for the system.

- If the deactivate/phase-out system decision is *Yes* or TRUE, workflow progresses to Operation 21.0 Deactivate/Phase-out System.
- If the decision is *No* or FALSE, workflow proceeds to Operation 16.0 Replenish System Resources.

Operation 16.0: Replenish System Resources

Operation 16.0 Replenish System Resources includes system capabilities and activities required to restock or *replenish* system resources such as personnel, fuel, and supplies. If deficiencies are found in the system, the system is sent to Operation 20.0 Perform System Maintenance. On completion of Operation 16.0 Replenish System Resources, workflow progresses to Operation 18.0 Redeploy System decision.

Operation 17.0: Conduct System/Mission Training

Operation 17.0 Conduct System Training includes capabilities and activities required to train Users or system operators in how to properly operate the system. For larger, more complex systems, initial operator training is sometimes performed at the System Developer's factory prior to system deployment to the field. Remedial and skills enhancement training occurs after the system is already in field service.

During Operation 17.0 Conduct System Training, new system operators are instructed in the safe and proper use of the system to develop basic skills. Experienced operators may also receive *remedial*, *proficiency*, or *skills enhancement* training based on lessons learned from previous missions or new tactics employed by adversarial or competitive threats.

On completion of a training session, workflow progresses to Operation 16.0 Replenish System Resources. If the system requires maintenance during training, Operation 20.0 Perform System Maintenance is activated.

Operation 18.0: Redeploy System Decision

Operation 18.0 to redeploy the system, a decision *control point*, determines if the physical system is to be repeployed to a new User location to support organizational mission objectives.

- If Operation 18.0 Redeploy System decision is *Yes* or TRUE, workflow progresses to Operation 3.0, which is to deploy the system.
- If Operation 18.0 Redeploy System decision is *no* or FALSE, workflow proceeds to Operation 4.0 Conduct System/Mission Training decision and the cycle repeats back to Operation 18.0 Deploy System Decision.

Operation 20.0: Perform System Maintenance

Operation 20.0 Perform System Maintenance includes system capabilities and activities required to upgrade system capabilities or correct system deficiencies through *preventive* or *corrective maintenance*. Systems are tagged with easily *recognizable* color identifiers such as RED or YELLOW to represent *corrective* or *preventive maintenance* actions required to correct any *defects* or *deficiencies* that may impact mission success.

On successful completion of system maintenance, the system is returned to active duty via the next operation—be it Operation 6.0 Configure System Mission, Operation 7.0 Assess Operational Mission Readiness, Operation 10.0 Conduct Mission, Operation 16.0 Replenish System Resources, or Operation 17.0 Conduct System/Mission Training—of the requested need for maintenance.

Operation 21.0: Deactivate/Phase-out System

Operation 21.0 Deactivate/Phase-out System includes system capabilities and activities required to disengage and remove the system from *active duty*, store, warehouse, "mothball," or disassemble the system and properly dispose of all its components and elements. Some systems may be stored or "mothballed" until needed in the future to support surges in mission operations that cannot be supported by existing systems. On completion of the deactivation, the system proceeds to the System Disposal Phase (22) of its system/product life cycle.

System Operations Dictionary

Obtaining team agreement on the graphical depiction of the concept of operations is only the first step. When working with larger, complex systems and development teams, diagrams at this level require scoping definitions for each capability to ensure proper understanding among team members. For example, you and your team may define a specific operational capability differently from a team operating in another business domain, depending on the system's application.

One solution is to create a *System Operations Dictionary*. The *dictionary*, which defines and scopes each capability similar to the previous System Operations Model descriptions, should be maintained throughout the life of the system.

Final Thoughts

The System Operations Model is used to define systems, products, organizations, services, etc. We can apply this model as an initial starting point for most, if not all, HUMAN-MADE SYSTEMS, such as automobiles, the Space Shuttle, airlines, hospitals, businesses, fire and ambulance services.

Collectively and individually, each of the model's operations *represents* a generalized construct applicable to most systems. As an SE, collaborate with the User(s) to tailor the System Operations Model to reflect their needs within the constraints of contractual, statutory, and regulatory requirements. Each operation should be *scoped* and *bounded* via a *System Operations Dictionary* to ensure all members of the Acquirer, User, and system development teams clearly understand *what is/is not* included in specific operations—with no surprises!

From an individual's perspective, the System Operations Model may appear to be very simple. However, on closer examination, even simple systems often require forethought to adequately define the operational sequences. If you challenge the validity of this statement, consider the following:

- Develop the System Operations Model for a car and driver.
- Conduct a similar exercise with each of three colleagues who are unfamiliar with the Model. The diversity of colleague opinions may be enlightening.
- Repeat the exercise as a team focused on achieving a single, collaborative consensus for the final diagram.

Now consider the case where the System Operations Model involves the definition of a more complex system with a larger stakeholder community. If you contemplated the previous car and driver exercise, you should appreciate the challenges of getting a diverse group of people from various disciplines, political factions, and organizations to arrive at a consensus on a System Operations Model for a specific system.

You will discover that engineers often refer to the System Operations Model as "textbook stuff." They:

1. Act offended to spend time addressing this concept.
2. Have a natural tendency to focus immediately on physical hardware and software design, such as resistors, capacitors, data rates, C++ language, and operating systems.
3. *Whine* to their management about the need to focus their time on resistors, coding, etc.

Beware If your program, customer, and User community have not agreed to this top-level concept and its lower level decomposition, system development problems further downstream historically can be traced back to this fundamental concept. Even worse, fielding a system that does not pass customer *validation* for intended usage presents even greater challenges, not only technically but also for your organization's reputation.

- Obtain Acquirer and User community consensus and "buy in" prior to committing resources for development of the system. Investigate *how* the User envisions operating the planned system to achieve organizational mission objectives. *Avoid* premature hardware and software development efforts until these decisions are approved and flowed down and allocated to hardware and software specifications.
- Use the System Operations Model as an infrastructure for *identifying* and *specifying* operational capabilities that can be translated into *System Performance Specification* (SPS) requirements.
- When reviewing and analyzing specifications prepared by others, use the System Operations Model to assess top-level system performance requirements for completeness for system operations.

18.5 DEVELOPING A MORE ROBUST SYSTEM OPERATIONS MODEL

The preceding System Operations Model provided a fundamental understanding of *how* a system might be employed by the User. As a high level model, it serves a useful tutorial purpose. The model, however, has some areas that need to be strengthened to accommodate a broader range of

186 Chapter 18 System Operations Model

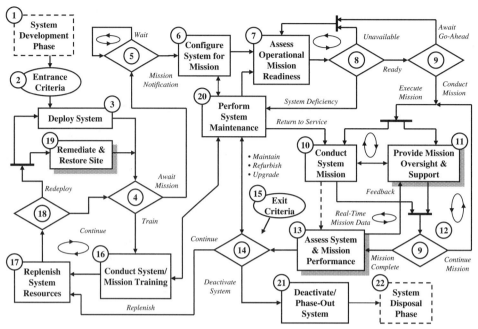

Figure 18.2 Robust System Concept of Operations (ConOps) Model

system applications. Figure 18.2 provides an expanded System Operations Model. To maintain continuity with the previous model, we have preserved the original numbering convention and simply added the following operations:

Operation 8.0: Mission Ready Decision
Operation 11.0: Provide Mission Oversight and Support
Operation 12.0: Mission Complete Decision
Operation 19.0: Remediate and Restore Site

18.6 THE IMPORTANCE OF THE GENERALIZED SYSTEM OPERATIONS MODEL

The System Operations Model serves as a high-level framework that orchestrates the *totality* of system synchronized to a time-based schedule. Operations in the model are performed by one of more of the system elements (EQUIPMENT, PERSONNEL, FACILITIES, etc.). The allocation of these operations to the system elements is important from several perspectives.

Specification Developer's Perspective

From a specification developer's perspective, the System Operations Model *construct* provides the *infrastructure* for working with customers and Users to *capture*, *organize*, and *specify* system requirements. Operational capabilities and performance decomposed and derived from this infrastructure can be translated into text requirements for system or lower level specifications.

Referral For more information about translating operational capabilities into specification requirements, please refer to Chapter 32 *Specification Development Practices*.

Specification Analyst Perspective

From a system analyst's perspective, the System Operations Model construct can be used to as an infrastructure to assign existing specification requirements for specific operations. If each System Operations Model operation is decomposed into hierarchical levels of sub operations, the system analyst can easily find the *holes* representing *missing* requirements or the need for clarification.

Author's Note 18.3 *Properly trained system engineers and others who develop systems (products, organizations, services, etc.) understand and appreciate the importance of capturing Acquirer and User community expectations of system capabilities, behavior, and performance based on how the system will be used. The System Operations Model, as a high level of system abstraction, serves as the top-level infrastructure to define how the system is to be operated. Using the model as a framework, system operational capabilities and performance can be easily specified. System operational analysis enables us to decompose each of the system life cycle operations into successively lower level tasks and activities, each of which is characterized by a specific system capability, behavior, and performance.*

Author's Note 18.4 *Many untrained specification writers focus exclusively on Operation 10.0 Conduct Mission. Even worse, they employ the feature-based approach by specifying features of the system for Operation 10.0. As human products of electrical, mechanical, and software disciplines, engineers of this type immediately focus on their "comfort zone," physical system hardware and software requirements and solutions. As a result the specifications often fall short of complete system requirements coverage as noted by the absence of mission requirements for Operations 3.0 through 13.0 and 16.0 through 19.0. Even within Operation 10.0 Conduct Mission, these writers focus only on specific physical features without consideration for system phases, modes, and states when using cases and scenarios, and so on. As a result, many requirements are missed or misplaced.*

1. *Despite the shortcoming noted in the previous points, standard system specification outlines such as the former MIL-STD-490A tend to force the specification developers to at least partially consider these missing steps (Operations 3.0–13.0 and 16.0–19.0) in areas such as Design and Construction Constraints.*
2. *Based on the author's experience, competent systems engineers begin their systems analysis work with the System Operations Model or some version tailored specifically for their system application and User needs. This statement serves as a key indicator of the training and maturity level of system engineers. Application of the System Operations Model enables you to sort out the true system engineers from the "wannabes" and the level of risk associated with their position on the program.*

18.7 GUIDING PRINCIPLES

In summary, the preceding discussions provide the basis with which to establish the guiding principles that govern development of the System Operations Model.

Principle 18.1 Every human-made system has its own unique System Operations Model that represents HOW the User deploys, operates, supports, and phases out the MISSION SYSTEM.

Principle 18.2 Each MISSION SYSTEM operation or task and its supporting system elements either adds value and contributes to achieving a phase-based performance objective or not; if not, eliminate it!

Principle 18.3 Synchronize every System Operation Model operation or task to a performance-based Mission Event Timeline (MET).

Principle 18.4 Every System Operation Model operation or task represents a system use case; each use case represents a capability that must produce a defined outcome while coping with one or more most likely or probable OPERATING ENVIRONMENT scenarios.

18.8 SUMMARY

The preceding discussions represent the embryonic, conceptual views of *how* the User intends to use the system. The operations can be translated into explicit system level capabilities and performance requirements. These requirements should ultimately be allocated to the *system elements*, the EQUIPMENT, PERSONNEL, FACILITIES, and so on. Our next discussion will decompose these operations into greater levels of detail via *system phases, modes, and states of operation*. We will thus begin to *narrow*, *bound*, and *specify* system capabilities and performance required of each of the *system elements* to support the ConOps operational tasks.

Author's Note 18.5 *As we stated earlier, it is impractical to illustrate all conceivable applications of systems. Our discussion over the past few pages has been intended to provide a basic orientation and awareness that will stimulate your thought processes and enable you to translate these approaches into your own business domain systems.*

GENERAL EXERCISES

1. Answer each of the *What You Should Learn from This Chapter* questions identified in the *Introduction*.
2. Refer to the list of systems identified in Chapter 2. Based on a selection from the preceding chapter's General *Exercises* or a new system selection, apply your knowledge derived from this chapter's topical discussions.
   (a) Apply the System Operations Model to the selected system to determine which operations apply.
   (b) Does the system have operations that are not identified in the model?
   (c) Using the System Operations Model as the initial starting point, tailor the model to the system and its application.
   (d) As a sanity check, map each use case to the System Operations Model operations. For those operations that are not addressed by a use case, is the set of use cases deficient? If the User did not have resource constraints, how would you address the deficiencies with the User?

ORGANIZATIONAL CENTRIC EXERCISES

1. Research your organization's command media for direction and guidance in the development of the *system Concept of Operations (ConOps)* or *operational concept description (OCD)*.
   (a) What requirements are levied on development of the ConOps or OCD?
   (b) When does the media required development of either of these documents?
2. Contact a small, medium and large contract program within your organization.
   (a) Does the contract require development of a ConOps or OCD?
   (b) Does the program have a ConOps or OCD? If not, why not?
   (c) If so, how did the ConOps or OCD benefit the program?
   (d) Based on development of the ConOps or OCD, what would the program do differently next time?

Chapter 19

System Phases, Modes, and States of Operation

19.1 INTRODUCTION

Our discussion of the System Operations Model provides a workflow that illustrates HOW the User might *deploy*, *operate*, and *support* a system or product to perform organizational missions. In general, the workflow consists of objective-based *sequential* and *concurrent* operations, each requiring two or more tasks to be accomplished. Tasks in turn are subdivided into subtasks, and so on.

As we probe deeper into the System Operations Model, our analysis reveals that it consists of three types of SYSTEM OF INTEREST (SOI) operations. These operations are required to: 1) *prepare* for a mission, 2) *conduct* and *support* a mission, and 3) *follow-up* after the mission. These SOI operations are performed by the integrated efforts of the MISSION SYSTEM (s) and the SUPPORT SYSTEM.

When a system is fielded, Users learn the basics of system operations that include HOW to employ a system, product, or service during these *phases* of operation via user's guides, reference manuals, and checklist procedures. Each *phase of operation*, which is assigned objectives to be accomplished, consists of embedded *modes of operation*. Each *mode of operation* represents User selectable *options* available to perform specific mission operations and tasks. Users also learn about EQUIPMENT capabilities, safe operating procedures, and performance limitations available to support these operations and tasks. WHAT the User sees are the results of system development; however, they do not reflect the *highly iterative*, time-consuming analysis and decision making that the SE design process requires to produce these results.

Our discussion in this chapter introduces the concept of system *phases*, *modes*, and *states* of operation. We build on the foundation of *use cases and use case scenarios* and System Operations Model to illustrate how SEs:

1. Establish phases of operation.
2. Derive modes of operation from use cases.
3. Derive system architectural configurations and interfaces that represent the system's state of operation.

Given a foundation in HOW a system is organized, we explore how *modal transitions* occur within and between system phases of operation. Finally, we illustrate how modal capabilities are accumulated and integrated as physical configurations or *states* of the architecture to support User phase-based objectives.

System Analysis, Design, and Development, by Charles S. Wasson
Copyright © 2006 by John Wiley & Sons, Inc.

190 Chapter 19 System Phases, Modes, and States of Operation

What You Should Learn from This Chapter

- What is a *phase of operation*?
- What is the *objective* of the *pre-mission phase* of operation?
- What is the *objective* of the *mission phase* of operation?
- What is the *objective* of the *postmission phase* of operation?
- What is a *mode of operation*?
- What is a *state of operation*?
- What is the difference between an *operational state* and a *physical state*?
- What are the *relationships* among phases, modes, and states of operation?
- How do *use cases and scenarios* relate to modes of operations?
- What is a *modal triggering event*?

Definitions of Key Terms

- **Mode of Operation** An abstract label applied to a collection of system operational capabilities and activities focused on satisfying a specific phase objective.
- **Phase of Operation** Refer to definition provided in Chapter 16 *System Mission Analysis*.
- **State of Operation** The *operational* or *operating condition* of a SYSTEM OF INTEREST (SOI) required to safely conduct or continue its mission. For example, the *operational state* of an aircraft during take-off includes architectural configuration settings such as wing flap positions, landing gear down, and landing light activation.
- **State** "A condition or mode of existence that a system, component, or simulation may be in; for example, the pre-flight state of an aircraft navigation program or the input state of given channel." (Source: IEEE 610.12-1990)
- **State Diagram** "A diagram that depicts the states that a system or component can assume, and shows the events or circumstances that cause or result from a change from one state to another." (Source: IEEE 610.12-1990) State diagrams are also called *state transition diagrams*.
- **State Machine** A device that employs a given configuration state to perform operations or tasks until *conditions* or an external *triggering event* causes it to transition to another configuration state.
- **Triggering Event** An external OPERATING ENVIRONMENT stimuli or cue that causes a system to initiate behavioral response actions that shift from a current mode to a new mode of operation.

19.2 SYSTEM PHASES, MODES, AND STATES RELATIONSHIPS

To facilitate your understanding of system *phases*, *modes*, and *states* of operation, let's establish their context using the entity relationship framework shown in Figure 19.1.

Author's Note 19.1 *The following description depicts the results of a highly iterative analysis of system phases, modes, and states that may be very time-consuming, depending on system com-*

19.2 System Phases, Modes, and States Relationships   191

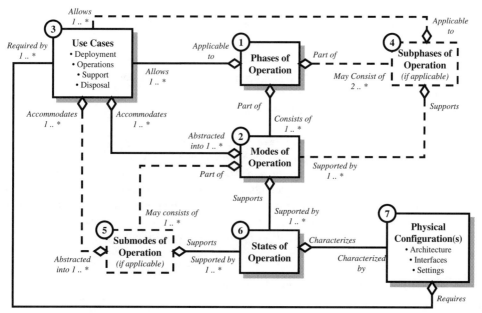

Figure 19.1 Relationships Between System Use Cases and Phases, Modes, and States of Operation

plexity. In the list there are included subphases and submodes of operation, although most systems do not employ these features. They are provided here for illustration purposes for those systems that do employ those terms.

1. Each *phase of operation* (1):
   a. Consists of at least one or more *modes of operation* (2).
   b. *Allows* application of at least one or more *use cases* (3).
   c. May consist of at least two or more *subphases of operation* (4).
2. Each *subphase of operation* (4) (if applicable) is:
   a. *An element of* a higher level *phase of operation* (1).
   b. *Accommodates* at least one or more *use cases* (3).
   c. *Supported by* at least one or more *modes of operation* (2).
3. Each *use case* (3) is:
   a. *Applicable to* at least one or more *phases of operation* (1).
   b. Analytically *abstracted* into at least one or more higher level *modes of operation* (2) or into *submodes of operation* (5).
   c. May *require* one or more *physical configurations* (7).
4. Each *mode of operation* (2):
   a. Is *unique to* one and only one *phase of operation* (1).
   b. *Accommodates* at least one or more *use cases* (3).
   c. *Supported by* at least one or more *physical configurations* (7).
5. Each *submode of operation* (if applicable) is:
   a. *Unique to* one and only one *mode of operation* (2).
   b. *Accommodates* at least one or more *use cases* (3).
   c. *Supported by* at least one or more *states of operation* (6).

192 Chapter 19 System Phases, Modes, and States of Operation

6. Each *state of operation* (6):
   a. *Supports* at least one or more *modes of operation* (2) or *submodes of operation* (5).
   b. Consists of at least one or more *physical configurations* (7).
7. Each *physical configuration* (7) is:
   a. *Characterized by* the system/item architecture, interfaces, and settings.
   b. *Unique to* a *state of operation* (6).
   c. Employed by at least one or more *use cases* (3).

Author's Note 19.2 *Since our focus here is on general relationships of system phases, modes, and states of operation, Figure 19.1 is presented with those key elements. As we will see in Chapter 21 System Operational Capability Derivation and Allocation, the linkages between use cases, modes and states of operation, and physical configuration states (i.e., the physical design solution) are accomplished via required operational capabilities that lead to performance requirements. The subject of phases, modes, and states is often confusing; this is because it is complicated by people who often misapply the terms. Therefore, we defer the required operational capabilities dimension until later.*

Given this framework of entity relationships, let's begin our discussion with *system phases of operation*.

19.3 UNDERSTANDING SYSTEM PHASES OF OPERATION

Our discussion of the System Operations Model introduced a key concept in understanding *how* human-made systems typically operate. The operations presented in Figures 18.1 and 18.2 provide an *initial framework* for organizing and collecting system capability requirements as well as developing the initial system engineering design. Let's explore the relationship of phases and operations.

Operational Phase Objectives

If we analyze and assimilate the set of system operations and their objectives, we can partition the operations into three distinct classes of abstraction: *pre-mission*, *mission*, and *postmission* operations. Analytically we refer to these abstractions as phases of operation.

Author's Note 19.3 *All human-made systems have at least three phases of operation. Although some systems may be placed in storage, keep in mind that the operative term is "operation." Since the system does not perform an action, storage represents an action performed on a system and is therefore not considered an operation.*

Pre-mission Phase Objective. The *objective* of the pre-mission phase of operations, at a *minimum*, is to ensure that the SYSTEM OF INTEREST (SOI) (i.e., MISSION SYSTEM and SUPPORT SYSTEM) is fully prepared, configured, operationally available and ready to conduct its organizational mission when directed.

Mission Phase Objective. The *objective* of the mission phase of operations, at a *minimum*, is to conduct the mission SYSTEM OF INTEREST (SOI). Besides achieving the SOI's mission objectives, one must mitigate mission risks and ensure the system's safe operation and return.

Post-mission Phase Objective. The *objectives* of the postmission phase of operations, at a *minimum*, are to:

1. Analyze mission outcome(s) and performance objective results.
2. *Replenish* system *consumables* and *expendables*, as applicable.
3. Refurbish the system.
4. *Capture* lessons learned.
5. *Analyze* and *debrief* mission results.
6. *Improve* future system and mission performance.

To see how phases of operation may apply to a system, consider the example of an automobile trip.

EXAMPLE 19.1

During the *pre-mission phase* prior to driving an automobile on a trip, the driver:

1. Services the vehicle (oil and filter change, new tires, repairs, etc.).
2. Fills the tank with gasoline.
3. Checks the tire pressure.
4. Inspects the vehicle.
5. Loads the vehicle with personal effects (suitcases, coats, etc.).

During the *mission phase* following a successful pre-mission checkout, the driver:

1. Departs on the trip from the *point of origination*.
2. Drives defensively in accordance with vehicle safe operating procedures.
3. Obeys vehicular laws.
4. Navigates to the *destination*.
5. Periodically checks and replenishes the fuel and coolant supply enroute.
6. Arrives at the *destination*.

During the *post-mission phase* on arrival at the *point of destination*, the driver:

1. Parks the vehicle in a permissible space.
2. Unloads the vehicle.
3. Safely secures the vehicle until it is needed again.

Subphases of Operation

Some complex systems may employ *subphases of operation*. Airborne systems such as aircraft and missiles have *phases of flight* within the mission phase. *Phases of flight* for an aircraft system might include: 1) push back, 2) taxi, 3) take-off, 4) climb, 5) cruise, 6) descend, 7) land, 8) taxi, and 9) park. Using the phases of operation as the frame of reference, the *phases of flight* would be equated to subphases of the *mission phase of operation*. Each of the subphases focuses on specific aspects and objectives that contribute to the overall aircraft system objective: transport passengers safely and securely from a *Point of Origination* or *Departure to a Point of Termination or Destination*.

Guidepost 19.1 *Once the systems phases of operation are established, we can proceed with aligning the use cases with the respective phases.*

Aligning Use Cases with System Phases of Operation

Once the system's use cases have been identified, SEs align each use case with a specific phase of operation as shown in Table 19.1. Each use case in the table is associated with a phase of operation.

Table 19.1 Assignment of use cases to specific system phases of operation

Use Case (UC)	Use Case Description	Phase of Operation		
		Pre-mission Phase	Mission Phase	Postmission Phase
UC 1	(Description)	•		
UC 2	(Description)	•		
UC 3	(Description)	•		
UC 4	(Description)		•	
UC 5	(Description)		•	
UC 6	(Description)		•	
UC 7	(Description)		•	
UC 8	(Description)		•	
UC 9	(Description)		•	
UC 10	(Description)			•
UC 11	(Description)			•
UC 12	(Description)		•	
UC 13	(Description)		•	
UC 14	(Description)			•

19.4 UNDERSTANDING SYSTEM MODES OF OPERATION

Operationally, modes of operation *represent* options available for selection at the User's discretion, assuming certain *conditions* and *criteria* are met. Consider the following example:

EXAMPLE 19.2

An automobile driver, assuming certain *conditions* and *criteria* are met, has the following modes of operation available for discretionary selection: PARK, REVERSE, NEUTRAL, DRIVE, and LOW. While the vehicle is in the DRIVE mode, the driver is permitted by the vehicle's design to shift to the REVERSE mode subject to satisfying the following *conditions* and *criteria*:

1. The vehicle is at a safe location conducive to the action permissible under law and safe driving rules.
2. The vehicle is safely stopped and the brake pedal is depressed.
3. The driver can view on-coming traffic from all directions.
4. The action can be safely completed before other traffic arrives.

Deriving Modes of Operation

When we analyze use cases aligned with specific phases of operation, we soon discover that some use cases share an *objective* or an *outcome*. Where there is sufficient commonality in these sets or clusters of use cases that share an objective, we abstract them into higher level *modes of operation*. Figure 19.2 illustrates *how* use cases (UCs) are *abstracted* into modes of operation.

Author's Note 19.4 *You may discover that some modes can be further abstracted into higher level modes. Where this is the case, we establish a modal hierarchy and designate submodes within a mode of operation. For simplicity, we assume that all use cases reside at the same level in the Figure 19.2 illustration.*

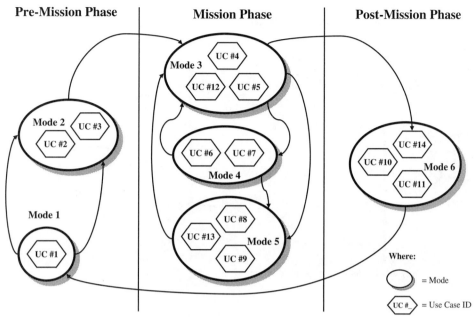

Figure 19.2 Use Cases Mapped to System Phases of Operation

Author's Note 19.5 *In some cases there are no specific guidelines for identifying modes of operation. Independent teams of equally capable SE analysts and developers can then hypothetically design and produce a system or product that complies with a set of User capability and performance requirements. Each team may nevertheless have variations of system modes of operation. The point is to learn to recognize, understand, and establish a team-based consensus concerning system phases and modes of operation. Then, together, the team can apply common sense in abstracting operations into modes of operation.*

Understanding the Modes of Operation Construct

Modes of operation can be depicted graphically or in tabular form. Since SEs communicate with graphics, we will employ the basic construct shown in Figure 19.3.

The construct is divided into pre-mission, mission, and postmission phases of operation to facilitate a left to right control flow. Although each of these *phases of operation* may consist of one or more *modes of operation*, only one mode is shown in each phase for simplicity.

Other than a general left-to-right cyclical workflow (pre-mission to mission to postmission), *modes of operation* are both *time dependent* and *time independent*. Many systems establish Mission Event Timelines (METs) that constrain: 1) the pre-mission-to-mission transition, 2) mission-to-post-mission transition, and 3) post-mission-back to pre-mission transition during the system turn-around. Within each *mode of operation* the MET event constraints may be further subdivided.

Understanding System Modal Transitions

The preceding discussion highlights the *identification* of system modes of operation by phase and the transition diagram used to modal interactions. Based on this understanding, we are now ready to investigate the *stimulus* or *triggering events* and *conditions* that initiate transition from one mode to another.

196 Chapter 19 System Phases, Modes, and States of Operation

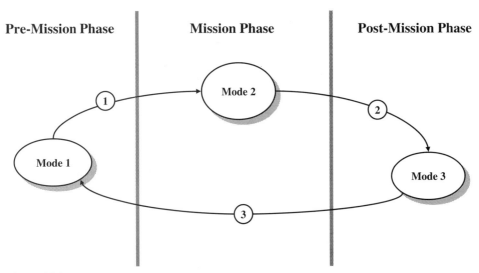

Figure 19.3 Basic Phase and Modes of Operation Construct

Note that each mode in Figure 19.3 is interconnected via curved lines with arrows that *represent* transitions from one mode to another. These transitions are initiated by pre-defined *triggering events* or conditions.

Our discussion here focuses on the *entity* relationships of phases, modes, use cases, use case scenarios, and required operational capabilities. The question is: *What causes a system, product, or service to transition from one phase, mode, and use case to another during a mission?* This brings us to a discussion of *triggering events*.

Triggering Event-Based Transitions. Most systems, as *state machines*, are designed to *cycle* within a given *mode of operation* until some external stimulus such as an operator initiates a transition to a new *mode of operation*. The occurrence of the external stimulus is marked as a *triggering event*. As discussed earlier, the *triggering event* may be *data or interrupt driven* whereby:

1. The system receives data from an external system.

2. A system User enters/inputs data into the system in accordance with prescribed *Standard Operating Practices and Procedures (SOPPs)* or *rules of engagement* to transition to the next phase or mode of operation.

Data-driven *triggering events* may be *synchronous* (i.e., periodic) or *asynchronous* (i.e., random) occurrences. When making the transition from one mode to another, the User may impose specific time requirements and constraints.

Figure 19.4 illustrates a simple, two-mode system that transitions from Mode 1 to Mode 2 when triggering Event 1 occurs and from Mode 2 back to Mode 1 on triggering Event 2. Triggering event transitions from Mode 1 to Mode 2 and back to Mode 1 require different sets of *assumptions* and *conditions*.

The graphic depicts the *bidirectional transition* as two separate transitions, T1 and T2. For T1, some external triggering Event t_1 *initiates* the transition from Mode 1 to Mode 2; transition to Mode 2 is completed at Event t_2. Some time later, another stimulus triggers Event t_3, which initiates a transition from Mode 2 to back to Mode 1; transition to Mode 1 is completed by Event t_4.

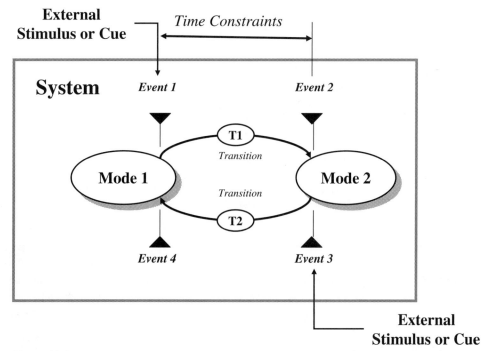

Figure 19.4 Modal Transition Loop Construct

Author's Note 19.6 *The system development team must establish the convention for transitioning from one mode of operation to another. For example, do you establish an entry criterion for the next mode or an exit criterion for the current mode? Typically a mode does not have both entry and exit criteria. Why? The transition from the current mode, n, to the next mode, n + 1, is a single transition. You do not specify the exit criterion for the current mode and then specify that same condition as the entry criterion for the other side of the modal interface. In most applications, the best approach is to define only the exit criterion for the current mode of operation.*

Describing Modal Transitions. Once you establish a conceptual view of the system phases and modes, the next step is to characterize the *triggering events* and *conditions* for initiating the *modal transitions*. One mechanism for accomplishing this is the mode transition table shown in Table 19.2.

In general, Table 19.2 depicts *how* a system transitions from its current mode: leftmost column to the next mode, which is the rightmost column. Interior columns define information relating to the *transition conditions*. Each row in the table represents a single *mode of operation* and includes a numerical transition ID. Transition information includes: identification of the *triggering source* or *event*, the *type of event*—asynchronous or synchronous—and any *transition resource* or *time constraints* for completion of the transition. Consider the following example from Table 19.2.

EXAMPLE 19.3

If the system is in the OFF Mode, placement of the Power Switch in the ON position (e.g. ASYNChronous triggering event) initiates Transition ID T1. Transition from the OFF MODE to the POWER-UP INITIALIZATION Mode is constrained to 30 seconds maximum. This leads to various power system stabilization capability and performance requirements.

Table 19.2 Modal transition table example

Current Mode n	Triggering Event	Triggering Event Condition	Event Type	Transition Constraints (seconds)	Next Mode $n + 1$
OFF	T1	Power Switched ON	ASYNC	30 Sec.	POWER-UP/ INITIALIZATION
ACTIVATE	T2	I/O Device	ASYNC	3 Sec.	CONFIGURE
ACTIVATE	T6	I/O Device	ASYNC	3 Sec.	NORMAL OPERATIONS
CONFIGURE	T4	I/O Device	ASYNC	3 Sec.	POWER-UP INITIALIZATION
CONFIGURE	T3	I/O Device	ASYNC	3 Sec.	CALIBRATE
CONFIGURE	T5	I/O Device	ASYNC	3 Sec.	POWER-UP INITIALIZATION
Etc.					

Author's Note 19.7 *Note that the process of identifying system modes of operation and modal transitions is highly iterative and evolves to maturity. Since subsequent technical decisions are dependent on the system modes decisions, your role as an SE is to facilitate and expedite team convergence.*

Analyzing Mode Transitions. Modal transitions represent interfaces whereby control flow is passed from one mode to another. While the modes characterize SYSTEM level operations, there may be instances whereby different PRODUCTs or SUBSYSTEMs provide the mode's primary capabilities. If separate system development teams are addressing the same modes, make sure that both teams operate with the same set of *assumptions* and *decisions*. Otherwise, *incompatibilities* will be created that will not surface until system integration and test. If left *undiscovered* and *untested*, a potential hazard will exist until field system *failures* occur, sometimes catastrophically.

The challenge for SEs is to ensure compatibility between any two modes such that initializations and conditions established for one mode are in place for the successor mode processing. The intent is to ensure that modal transitions are seamless. *How do you ensure consistency?* Document the modes of operation, Mission Event Timeline (MET), and modal transitions in the system's ConOps or *operational concept description (OCD)* document.

Final Thoughts about Modal Transitions. In our discussion of system operations and applications, we highlighted various types of system applications—*single use*, *reusable*, *recyclable*, and so forth. Most *reusable* systems are characterized by *cyclical* operations. *Cyclical systems* have a feedback loop that typically returns *workflow* or *control flow back* to the pre-mission phase of operation, either powered down or prepared for the next mission.

Now consider a system such as the Space Shuttle's External Tank (ET). From a mission perspective, the ET's fuel resource is a *consumable* item and the ET's entity is an *expendable* item. At a specific phase of flight and MET event, the ET is jettisoned from the Orbiter Vehicle, tumbles back toward Earth, and burns up on reentry into the atmosphere.

In the case of an expendable system such as the ET, one might expect the modes of operation to be sequential without any loop backs to previous modes. However, from an SE design perspective, ET operations involve systems that may require cycling back to an initial mode due to "scrubbed" launches. Therefore, *expendable* systems require modes of operation that ultimately transition to a *point of termination mode*—such as REENTRY.

Generalized Modes of Operation for Large, Complex Systems

The preceding discussions addressed a simple product example as a means of introducing and illustrating system *modes of operation*. When you investigate the modes of operation for large, complex, multipurpose, reusable systems, additional factors must be considered.

EXAMPLE 19.3

Example considerations include *reconfiguration*, *calibration*, *alignment*, replenishment of *expendables*, and *consumables*.

As in the case of the System Operations Model, we can define a *template* that can be used as an initial starting point for identifying classes of modes.

Generalized Modes of Operation Template

Theoretically we can spend a lot of time analyzing and abstracting use cases into a set of modes by phase of operation. Once you develop a number of systems, you soon discover that system modes are similar across numerous systems and you begin to see common patterns emerge. This leads to the question: *Rather than reinvent via modal use case analysis the modes for every system, WHY NOT explore the possibility of starting with a standard set of modes and tailoring them to the specific system application*? If this is true what are the common set of modes?

Table 19.3 provides a listing of common phase-based modes of operation. Further analysis reveals that these modes have interdependent relationships. We illustrate these relationships in Figure 19.5. Is this template applicable to every system as a starting point? Generally, so. Recognize this is simply a starting point, not an end result.

19.5 UNDERSTANDING SYSTEM STATES OF OPERATION

Scientifically, the term *state* refers to the *form* of *physical matter*, such as solid, liquid, or gas. In this context a *state* relates to the structure—meaning a configuration—and the level of activity present within the structure. Therefore, a *state*:

Table 19.3 Generalized modes of operations by phase

Phase of Operation	Potential Mode of Operation
Pre-mission Phase*	• OFF mode (coincides with state) • POWER-UP /INITIALIZATION mode • CONFIGURE mode • CALIBRATE/ALIGNMENT mode • TRAINING mode • DEACTIVATE mode
Mission Phase*	• NORMAL operations mode • ABNORMAL operations mode
Postmission Phase*	• SAFING mode • ANALYSIS mode • MAINTENANCE mode

* Some systems may permit MAINTENANCE mode(s) in all three phases.

200 Chapter 19 System Phases, Modes, and States of Operation

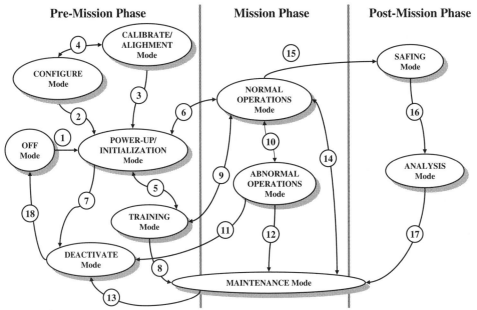

Figure 19.5 Fundamental System Phase and Modes of Operation Construct

1. Should be observable, testable, measurable, predictable, and verifiable.
2. May be STATIC or DYNAMIC—having an *infinite* number of time-variant states.

Author's Note 19.8 *The term "infinite states" refers to situations, such consumables, whereby the amount of the resource remaining represents a physical state. Examples include fluid levels, brake pad wear, and tire wear.* SE, as a discipline, builds on the *physical configuration* theme to define the *STATE* of a system, product, or service. *Therefore, STATE encompasses the system's physical architecture, such as interfaces and configuration settings*, and a range of acceptable tolerances of physical components at a given instance in time or time period.

Operational versus Physical States

The concept of states of operation is often confusing because of two contexts: *operational* versus *physical*. Let's describe each of these.

Operational States. We describe the *state* of a system or product in simple terms such as ON (*operating*) and OFF (nonoperating). Organizations often use the term *state* to represent the *state of readiness* or *operational readiness* of the integrated set of system elements—such as PERSONNEL or EQUIPMENT. By inference, this means that the integrated system and each system element is:

- *Physically Configured* Architectures and interfaces with a *capability* necessary and sufficient to conduct specific types of mission(s).
- *In acceptable operating condition* or *health* at a given point in time sufficient to *safely* and *reliably* accomplish the mission and its objectives.

Operating states are generally identified by terms with "-ed" or "-ing" suffixes. Consider the following examples:

EXAMPLE 19.4

The *operational states* of an aircraft (EQUIPMENT, crew, etc.) include parked, taxiing, taking off, ascending, cruising, descending, and landing.

EXAMPLE 19.5

A lawnmower having an adequate supply of fuel and oil and tuned for efficient operation can be described in the following *operational states*:

1. STOPPED/READY In a state of readiness for the homeowner (the User) to mow the lawn or store the mower.
2. IDLING Engine operating with blade disengaged; mower standing still.
3. MOWING Engine operating with blade engaged; mower moving across lawn mowing grass.
4. STORAGE Stored with the fuel drained from the tank, etc.

EXAMPLE 19.6

Consider an automobile operating on a flat road surface. When the driver configures the vehicle in the DRIVE (mode), the vehicle's physical design responds to the driver's inputs. In response, the vehicle configures its subsystems to respond to the stimulus accordingly. The drive train—meaning the engine and transmission—are designed to allow the driver to control the following vehicle *operational states*: PARKED, STOPPED, ACCELERATING, CRUISING, DECELERATING, and BRAKING.

Physical Configuration States. The OPERATIONAL *state* of a system or product may consist a number of allowable PHYSICAL configuration states. Building on our previous discussion, consider the following example:

EXAMPLE 19.7

An automobile's physical system design consists of several subsystems such as drive train, electrical, environmental, fuel, and entertainment. Subject to some prerequisite conditions, the automobile's design allows each of these subsystems to operate simultaneously and independently. Thus, when the vehicle is in an *operational state*—such as STOPPED or CRUISING—the driver can choose to:

1. Turn the sound system ON or OFF.
2. Turn the windshield wipers ON, OFF, or change their wiping interval.
3. Adjust the heating/cooling temperature level.
4. Adjust seat positions.
5. Apply force to or release the accelerator.
6. Apply braking.

Each of these physical adjustments ranges from finite to an infinite number of *physical configuration states*.

The preceding example illustrates WHY a system's *operational state* includes one or more *physical configuration states*. This leads to the question: *What is the relationship between system phases, modes, operational states, and physical configuration states of operation?* As illustrated in Figure 19.6, we can define the entity relationships via a set of design rules listed in Table 19.4.

To see how these entity relationships are applied, consider the following example:

202 Chapter 19 System Phases, Modes, and States of Operation

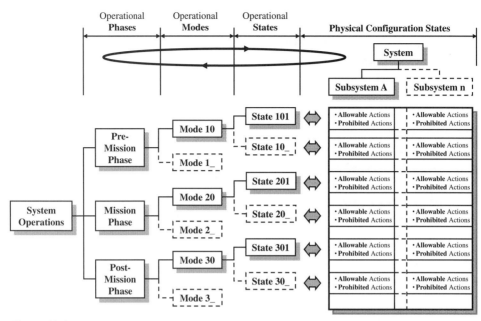

Figure 19.6 How Operational Phases, Modes, and States Influence Physical System Design and Vice Versa

Table 19.4 System operations, phases, modes, operational states, and physical configuration states entity relationship rules

ID	Entity	Entity Relationship Rules
19.1	System operations	System operations consists of at least three *phases of operation*.
19.2	Phases of operation	Each *phase of operation* consists of at least one or more use case-based *modes of operation*.
19.3	Modes of operation	Each *mode of operation* controls of at least one or more system *operational states*.
19.4	Operational states	Each system *operational state* is supported by at least one or more *physical states of operation*.
19.5	Physical configuration states	Each *physical configuration state of operation* is characterized by at least one or more *allowable actions* and at least one or more *prohibited actions*.

EXAMPLE 19.8

The driver of an automobile places the vehicle in DRIVE (mode) and starts moving forward (operational state), both *allowable* actions. As the vehicle moves forward (operational state) at a threshold speed, the vehicle's computer system automatically *locks* (*allowable* action) the doors to prevent passengers from inadvertently opening the doors and falling from the vehicle (*prohibited* action).

19.6 EQUATING SYSTEM CAPABILITIES TO MODES

The preceding discussions enable us to link system operations, phases, modes, and states. Ultimately these linkages manifest themselves in the deliverable system that has *allowable* and *prohibited actions* as documented in the User's manual for *safe* and *proper* operating practices. *Allowable actions* represent physical capabilities that the system provides subject to contract cost, schedule, technology, and risk constraints.

In our discussion of the system *modes of operation*, we considered the need to *organize* and *capture* capabilities for a given *system element*, OPERATING ENVIRONMENT, and design construction and constraints. For each type of capability, we described HOW operational capabilities are:

1. Represented by aggregations of integrated sets of requirements.
2. Documented in the SPS.
3. *Allocated* and *flowed down to* multiple system levels of abstraction.

If we analyze the System Capabilities Matrix from a *mode of operation* perspective, the matrix reveals combinations of *system elements* that must be integrated to provide the required operational capabilities for each *mode of operation*. Each of these combinations of elements and interactions are referred to as the system's *physical configuration* state or architecture. Thus each *mode of operation* is accomplished by architectural configurations of system elements. Let's explore this point further by using Figure 19.7.

The icon in the upper left side of the chart symbolically represents the System Capabilities Matrix. Note that an arrow links each of the horizontal rows of capabilities for each *mode of operation* to a specific Physical (Architecture) State. Thus, for the POWER OFF mode, there is a unique

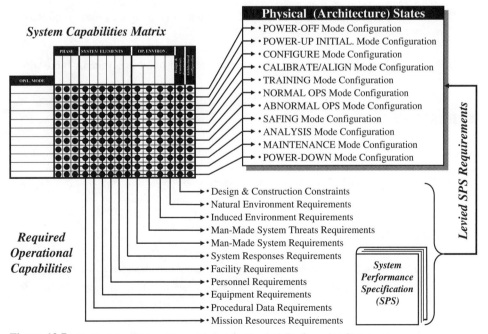

Figure 19.7 Synthesizing System Phase, Modes, & Physical Configuration States

physical architecture required to support the mode. Similarly, for the NORMAL OPERATIONS mode, each system entity within each *level of abstraction* has a unique architectural configuration that is integrated to form the Normal Operations mode configuration.

Concept Synthesis

In summary, Figure 19.7 illustrates how required operational capabilities flow from the System Capabilities Matrix into the SPS and subsequently are allocated to and provided by the system architecture *physical* states of operation for each *mode of operation*.

19.7 GUIDING PRINCIPLES

In summary, the preceding discussions provide the basis with which to establish guiding principles that apply to development of system phases, modes, and states of operation.

Principle 19.1 Every system *phase of operation* consists of at least two or more User-selectable *modes of operation*.

Principle 19.2 Every system *mode of operation* represents a capability to accommodate one or more *use cases*, each with one or more *probable scenarios*.

Principle 19.3 Each system *mode of operation* requires a *pre-defined set of condition-based triggering event criteria* for transitioning to another mode of operation.

Principle 19.4 Every system *mode of operation* requires at least one or more capability-based *operational states* to accomplish its performance-based objective.

Principle 19.5 Each *operational state* is supported by at least one or more *physical configuration states of operation*.

Principle 19.6 Each *physical state of operation* is characterized by at least one or more *allowable actions* and at least one or more *prohibited actions*.

19.8 SUMMARY

We have seen *how* system modes of operation are supported by the system element capabilities. Capabilities, in turn, are documented as a set of capabilities with bounded levels of performance in the *System Performance Specification (SPS)*, *flowed down* and *allocated* to system elements *levels of abstraction* within the various levels of its abstraction.

GENERAL EXERCISES

1. Answer each of the What You Should Learn from This Chapter questions identified in the Introduction.
2. Refer to the list of systems identified in Chapter 2. Based on a selection from the preceding chapter's *General Exercises* or a new system selection, apply your knowledge derived from this chapter's topical discussions.
3. For each of the following types of systems, identify the various states of operation.
   (a) Data communications

(b) Aircraft operations
(c) Computer system states
(d) Package delivery service
(e) Organization operations
(f) Building environmental control
(g) Food service operations
(h) Natural environment
(i) Vehicle operations

REFERENCE

IEEE Std 610.12-1990, 1990. *IEEE Standard Glossary of Modeling and Simulation Terminology*. Institute of Electrical and Electronic Engineers (IEEE) New York, NY.

Chapter 20

Modeling System and Support Operations

20.1 INTRODUCTION

Our previous discussion introduced the System Operations Model as the foundation for identifying, organizing, and conceptualizing the capabilities of the SYSTEM OF INTEREST (SOI). The results of this exercise can be translated into *System Performance Specification (SPS)* by the Acquirer or employed by a System Developer to analyze the SPS.

Referral   For more information about analyzing specifications, refer to the discussion on *Specification Analysis* practices in Chapter 28.

Once the SOI's capabilities are identified, the key question is: *WHO is accountable for performing those capabilities and achieving the associated level of performance?* More specifically, *are they performed by the MISSION SYSTEM or SUPPORT SYSTEM or both?*

Once we understand HOW the MISSION SYSTEM and SUPPORT SYSTEM interact, we can allocate the interactions as constraints down to each entity's *system elements* (PERSONNEL, EQUIPMENT, etc.). Ultimately *system element* interactions will be flowed down as constraints to the EQUIPMENT and its supporting HARDWARE and SOFTWARE elements.

This section explores the MISSION SYSTEM and SUPPORT SYSTEM interactions. Our discussions employ a matrix mapping technique to illustrate how system operations within the system Concept of Operations (ConOps) are allocated to pre-mission, mission, and postmission phases. Since MISSION SYSTEM and SUPPORT SYSTEM operations are application dependent, we introduce a method for modeling the operations. The discussion employs several examples to illustrate application of the method.

Author's Note 20.1   *The discussions in this chapter establish a base for understanding how the system will be applied to perform missions and interactions. You will find people who believe this is a case of analysis paralysis. Granted, you have to strike a BALANCE between LEAF level analysis details and being able to "see the forest." Those who diminish the value of defining system interactions are also the ones who:*

1. Carry the information around in their heads.
2. Fail to document anything.

System Analysis, Design, and Development, by Charles S. Wasson
Copyright © 2006 by John Wiley & Sons, Inc.

3. Do not communicate their understanding to others.
4. Complain that everyone else should know what they know about the system.

Remember, as an SE, one of your many roles is simplification that leads to decision making and immediate communications about those decisions. Complex system designs are simplified by decomposition and analysis; documentation is the preferred method for capturing engineering decision making for communications.

What You Should Learn from This Chapter

1. What are some *types* of system applications?
2. What are *dedicated use operations*?
3. What are *reusable/recyclable systems*?
4. What is means by a system's *control flow*?
5. What is means by a system's *data flow*?
6. How are *control flow* and *data flow related*?
7. What are *system operations cycles*?

20.2 MISSION PHASES AND OPERATIONS

Our previous discussion of the *generalized* System Operations Model provided conceptual insights into the system Concept of Operations (ConOps). The key question is: *WHAT system elements are accountable for providing the capabilities required to perform these operations?*

Referral For more information about the *generalized* system operations model, refer to the discussion of *System Concept of Operations* in Chapter 18.

Recall the ConOps tasks depicted in Figures 18.1 and 18.2. We observed that the tasks can be ORGANIZED into three types of system phases and operations: 1) *pre-mission operations*, 2) *mission operations*, and 3) *post-mission operations*. Each of these mission phase operations includes concurrent *MISSION SYSTEM Operations* (e.g., aircraft) and *SUPPORT SYSTEM Operations* (e.g., aircraft ground-based flight controllers).

To better see *how* the operational tasks in Figures 18.1 and 18.2 relate to the operations categories, let's employ a matrix mapping technique that enables us to relate System Operations Model operations to the MISSION SYSTEM and SUPPORT SYSTEM and their respective phases of operation.

Equating System Operations to MISSION SYSTEM and SUPPORT SYSTEM

Figure 20.1 shows the numerous *concurrent* MISSION SYSTEM and SUPPORT SYSTEM operations that occur throughout the pre-mission, mission, and post-mission phases. *Concurrent operations* require *synchronized, simultaneous interactions* between the MISSION SYSTEM (e.g., the Space Shuttle) and the SUPPORT SYSTEM (the ground/flight controllers, communications systems, etc.). Interoperability between the MISSION SYSTEM and SUPPORT SYSTEM is crucial to mission success. Recognize that Figure 20.1 is simply an analytical tool to understand which operational tasks are performed by the MISSION SYSTEM and SUPPORT SYSTEM by operational phase.

208 Chapter 20 Modeling System and Support Operations

	System Phase of Operation					
	Pre-Mission		Mission		Post Mission	
Operational Task	MISSION SYSTEM Elements	SUPPORT SYSTEM Elements	MISSION SYSTEM Elements	SUPPORT SYSTEM Elements	MISSION SYSTEM Elements	SUPPORT SYSTEM Elements
3.0 Deploy System						
4.0 Conduct Training Decision						
5.0 Mission Notification Decision						
6.0 Configure System for Mission						
7.0 Assess Mission Readiness						
9.0 Mission Go-Ahead Decision						
10.0 Conduct System Mission						
13.0 Assess Mission & System Performance						
14.0 Deactivate System Decision						
16.0 Replenish System Decision						
17.0 Conduct System Training						
18.0 Redeploy System Decision						
20.0 Perform System Maintenance						
21.0 Deactivate / Phase-Out System						

(#) = Reference to description
▓ = Not Applicable

Figure 20.1 Mapping Operational Tasks to MISSION SYSTEM and SUPPORT SYSTEM Elements as a Function of System Phases of Operations

Modeling SYSTEM OF INTEREST (SOI) Operations

Once the *logical* entity relationships are identified between *system elements*, the next step is to understand *how* the relationship interactions occur. We can do this by modeling the *system element* interactions.

Figure 20.2 illustrates how a SYSTEM OF INTEREST (SOI) might operate relative its OPERATING ENVIRONMENT. To keep the diagram simple, we depict Operation 1.0 (pre-mission operations) and Operation 3.0 (post-mission operations) as high-level *abstractions*. Operation 2.0 on mission operations is expanded to a lower level of detail to depict Operation 2.1 on MISSION SYSTEM operations and Operation 2.2 on SUPPORT SYSTEM operations.

Author's Note 20.2 *Based on the convention previously established, the OPERATING ENVIRONMENT interfaces (arrows) that touch the Operation 2.0 dashed boundary represent interfaces or interactions with all embedded operations—operations 2.1 and 2.2, operation 4.1 (HIGHER ORDER SYSTEMS interfaces) with Operations 2.1 and 2.2, and so forth.*

This diagram illustrates two key points concurrently:

1. SOI operations *cycle* or *loop* sequentially from pre-mission to mission to post-mission *phases of operation*.
2. The system interacts with its OPERATING ENVIRONMENT while performing these operations.

Regarding the second point, during the SOI's mission phase the MISSION SYSTEM and the SUPPORT SYSTEM *interact continuously* via physical contact, informational data exchanges, or visually as indicated by the interconnecting, bidirectional arrow. To better understand these two key points, let's investigate this flow in detail.

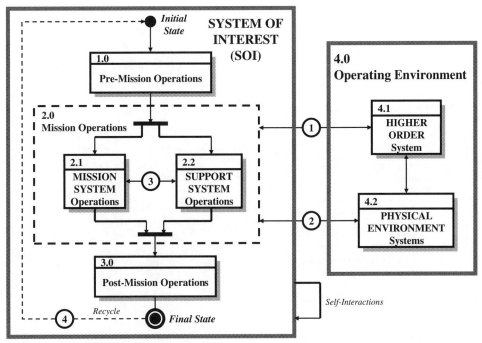

Figure 20.2 Concurrent Mission Operations

Operational Control Flow and Data Flow

When you analyze systems, two types of *flows* occur: 1) *control flow* or *work flow* and 2) *data flow*. *Control flow or workflow* enables us to understand *how* system operations are sequenced. *Data flows* enable us to understand *how* information—such as electrical, optical, or mechanical—is *transferred* or *exchanged* between the *system entities*. To illustrate these points, let's investigate the *control flow* and *data flow* aspects further.

Figure 20.3 depicts intersecting arrows on the left hand side and the SOI operations on the right side. The vertical arrow symbolizes *control flow*; the horizontal arrow symbolizes *data flow*. By convention, *control flow* or *workflow* sequences *down* the page from one operation to another; *data flow* moves *laterally* back and forth across the page between system entities.

By this convention, *control flow* or *work flow*:

1. Proceeds from Initial/State to Operation 1.0, Pre-Mission Operations
2. Proceeds to Operation 2.0, Mission Operations
3. Proceeds to Operation 3.0, Post-Mission Operations
4. Proceeds to Final/State.

Cyclical systems may employ a *feedback loop* as indicated by the line from the Final/State decision block back to START. From a *data flow* perspective, information is exchanged between Operation 2.1 (MISSION SYSTEM Operations) and Operation 2.2 (MISSION SUPPORT SYSTEM Operations) throughout Operation 2.0 (Mission Operations).

If we expand the MISSION SYSTEM and SUPPORT SYSTEM operational interactions to include all *phases of operation*—*pre-mission*, *mission*, and *post-mission*, Figure 20.4 results. Each of the *phases of operation* is expanded into concurrent MISSION SYSTEM and SUPPORT SYSTEM operations.

210 Chapter 20 Modeling System and Support Operations

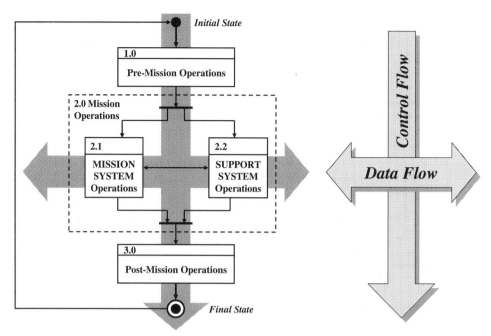

Figure 20.3 Concurrent Mission Operations Construct

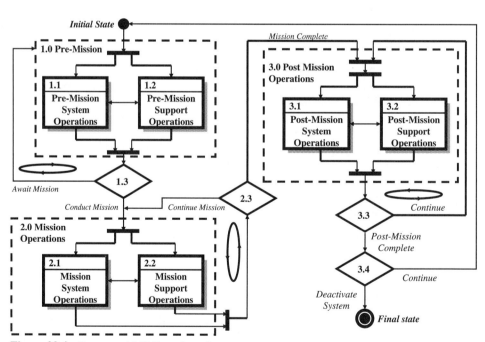

Figure 20.4 Concurrent Multi-Phase Operations

Transitions between the pre-mission, mission, and post-mission phases are illustrated with *decision control points*: 1) Operation 1.3 (Conduct Mission?), 2) Operation 2.3 (Mission Complete?), and 3) Operation 3.3 (Post-Mission Complete?) to *represent* cyclical loop back decision points. A fourth *decision control point*, Operation 3.4 (Deactivate System?), is also included. If a system is to continue *active duty service*, *control flow* or *workflow* passes back to Inital/State.

20.3 UNDERSTANDING SYSTEM OPERATIONAL APPLICATIONS

When analyzing and designing systems, SEs need to understand *how* the User intends to operate the system. Although every system exhibits areas of commonality at high levels of abstraction, uniqueness occurs at lower levels based on User applications. These graphical illustrations represent only a few unique applications. You should analyze the types of products and services your organization provides to identify how this discussion applies to your business domain.

Types of User Applications

User system applications and operations are categorized a number of ways. Examples include:

1. *General* use or *multipurpose* applications.
2. *Dedicated use* applications.
3. *Types of stakeholders*.

To better illustrate these categories, let's explore some examples.

General Use or Multipurpose System Applications. *General use or multipurpose system applications* refer to a MISSION SYSTEM that is developed for general application across a broad spectrum of User missions and mission applications. Consider the following example:

EXAMPLE 20.1

General use or multipurpose system applications include desktop computers, automobiles, tables, and chairs, and so on. Figure 20.5 provides a high-level model of an automobile's operations.

Some systems, such as aircraft and computers, may be *multipurpose* and capable of being configured for specific mission applications. Consider the following example:

EXAMPLE 20.2

A military aircraft platform is designed with *standardized* mechanical, electrical, and communications interfaces to accommodate a variety of payloads that enable the aircraft system aircraft and pilot—to perform reconnaissance, air-to-air, air-to-ground, damage assessment, among the missions.

Dedicated Use System Application. When a system is configured for a specific type of mission, we refer to the configuration as a *dedicated use application*. Consider the following example:

EXAMPLE 20.3

A commercial aircraft may be configured for dedicated *use* in transporting ticketed passengers on charter flights or have the seats removed for *multi-use* transportation of cargo.

Based on this discussion and examples, some SOIs are *multi-use* systems capable of being configured for dedicated *use* applications. The act of *reconfiguring* or *replenishing* MISSION RESOURCES for the system after each mission for new applications is referred to as *cyclical* operations.

212 Chapter 20 Modeling System and Support Operations

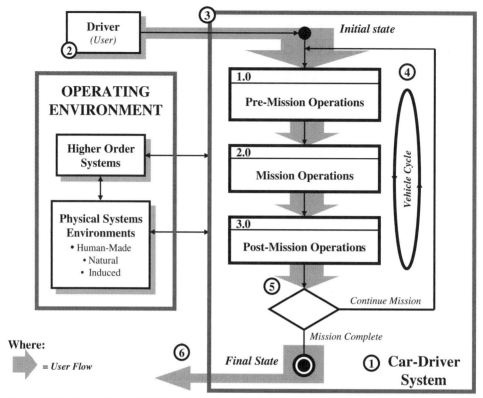

Figure 20.5 Single User—Multi-Phase Operations

Single-Use or Reusable System Applications. In general, most systems are either *single use* or *reusable*. *Single-use* applications are those such as missile systems that have a *single* concept-to-disposal operational life cycle. *Reusable or recyclable applications* refer to *cyclical* systems that are used repeatedly with a defined life expectancy, such as the automobile. Consider the following examples:

EXAMPLE 20.4

Single-use system applications include the Space Shuttle's External Tank (ET), missiles and munitions, commercial product packaging, and fireworks.

EXAMPLE 20.5

Reusable applications include the Space Shuttle's Orbiter Vehicle (OV) and Solid Rocket Boosters (SRBs), automobiles, and disposable cameras (i.e., from the consumer's perspective).

Reusable systems typically require some level of *replenishment* and/or *refurbishment* after each application or operational cycle to restore it to a *state of readiness* for the next mission. *Reusable* systems must be designed to include technical considerations such as reliability, availability, maintainability, and safety.

20.3 Understanding System Operational Applications

Types of Stakeholders. Types of Users influence system operational capabilities. Thus, a MISSION SYSTEM may require capabilities to: 1) conduct actual missions and 2) serve as training mechanisms when not conducting missions. Consider the following example:

EXAMPLE 20.6

Some systems such as aircraft used for training pilots may require capabilities to accommodate flight instructors and controls for the instructors. Automobile driving schools employ special vehicles that include an instructor brake.

Modeling Business Operations

The preceding discussions focused exclusively on a SYSTEM OF INTEREST (SOI)—namely the car and driver. Now, what about businesses that perform MISSION SYSTEM and SUPPORT SYSTEM roles such as restaurants and car dealerships? From the business owner's perspective, the organization performs as a MISSION SYSTEM. In performing that role, the business must provide systems, products, or services to customers in a SUPPORT SYSTEM role.

We can model these dual MISSION SYSTEM/SUPPORT SYSTEM roles by the simple model shown in Figure 20.6 via operations *A* through *n*. In the model a customer enters the business via Operation *A*, progresses through the workflow, and exits the shopping and dining experience via Operation *n*. From the business perspective operations *A* through *n* represent concurrent operations such as clothing, sporting goods, and lawn and garden tools. As a recyclable system, daily business operations begin with the Initial State and conclude with Final State representing termination of the day's operations. Figure 20.6 illustrates a basic business operations model such as a restaurant, automobile dealership, amusement park, or photo processing LAB.

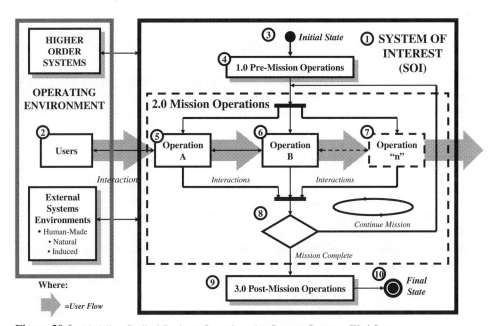

Figure 20.6 Modeling Cyclical Business Operations that Support Customer Workflow

Points to Ponder

The preceding paragraphs provided examples of *general use*, *multipurpose*, and *dedicated use* system applications; *single-use*, *multi-use*, and *reusable applications*; and types of *stakeholders*. As an SE, you may not always be able to delineate these types of systems easily. Your job will be to make sure that all stakeholders have a clear and concise understanding about the semantics of usage and the applications. To illustrate how these terms are stakeholder dependent, consider the following example.

EXAMPLE 20.7

Disposable cameras illustrate a *reusable* MISSION SYSTEM that can be viewed from a number of different perspectives. A customer purchases a disposable camera, makes pictures, returns the camera to a processing laboratory, and receives a set of prints. From the customer's perspective, the camera—or MISSION SYSTEM—is:

1. Single use with one operational cycle—namely one roll of film limited by frame count.
2. Disposable—namely returned to the processing laboratory.

From the processing laboratory perspective, the camera is viewed as:

1. *Dedicated use*—namely designed for specific film speed, aperture settings, etc.
2. *Recyclable*—namely replenished with film after each use for another consumer application, assuming that the camera meets a set of technical "fitness for use" criteria.

Reusable or multi-use mission operations, maintenance, and training activities continue until a deactivation order is received to transition the existing system from active duty service. When this occurs, the existing system or capability is phased out over time, simultaneously with a phase-in of a new system or capability. Typically, the new system operates in a concurrent, "shadow," or "pilot" mode to verify and validate interoperability with external systems until replacement of the existing system or capability begins.

20.4 OPERATIONAL CYCLES WITHIN CYCLES

While businesses have daily cyclical operations at the enterprise level, entities within the business may have iterative cycles. If you investigate the context of the MISSION SYSTEM's application, analysis reveals several *embedded* or *nested* operational cycles. Let's explore an example of this type.

City Bus Transportation System

Let's assume we have a city bus transportation system that makes a loop around the city several times a day, every day of the week, seven days a week. As part of each route, the bus makes scheduled stops at designated passengers pickup/drop-off points, passengers board, pay a token, ride to their destination, and exit the vehicle.

At the end of each route, the vehicle is returned to a maintenance facility for routine *preventive maintenance*. If additional *corrective maintenance* is required, the vehicle is removed from active service until the maintenance action is performed. During maintenance, an assessment determines if the vehicle is scheduled for replacement. If repairable, the vehicle is returned to *active service*. If the vehicle is to be replaced, a new vehicle is acquired. The current vehicle remains in

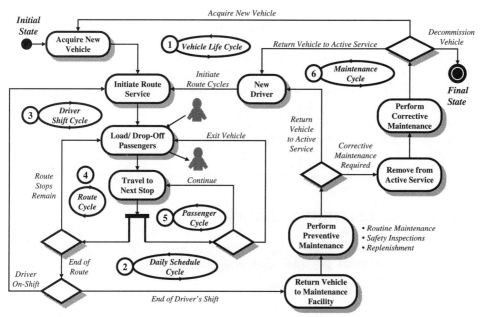

Figure 20.7 Bushiness Operational Cycles within Cycles

active service until it is decommissioned, which may or may not be linked to the new vehicle entering service, depending on business needs.

Figure 20.7 provides an operational model for this example to illustrate nested operational cycle within cycles. The six operational cycles, which are assigned reference identifiers (1) through (6), include: a Vehicle Life Cycle (1), a Daily Schedule Cycle (2), a Driver Shift Cycle (3), a Route Cycle (4), a Passenger Cycle (5), and a Maintenance Cycle (6). Depending on the structure of the business a seventh, Vehicle Fleet Cycle may be applicable. Such is the case with aircraft, delivery vehicles, rental cars, police cars, and so forth. While there are numerous ways of creating this graphic, the primary message here is LEARN to RECOGNIZE embedded operational cycles that include the integration of the MISSION SYSTEM with the SUPPORT SYSTEM during various cycles.

20.5 SUMMARY

The preceding discussions describe high-level, conceptual views of *concurrent, dedicated use, recyclable* systems and *how* the User intends to use the MISSION SYSTEM. The operations *represent*, and can be translated into, explicit system-level capabilities and performance requirements. These requirements will ultimately be allocated to the system elements (EQUIPMENT, PERSONNEL, FACILITIES, etc.). However, requirements allocations often require support from modeling and simulation of operations.

GENERAL EXERCISES

1. Answer each of the *What You Should Learn from This Chapter* questions identified in the *Introduction*.
2. Refer to the list of systems identified in *Chapter 2*. Based on a selection from the preceding chapter's *General Exercises* or a new system selection, apply your knowledge derived from this chapter's topical discussions on that system.

216 Chapter 20 Modeling System and Support Operations

3. NASA's Space Shuttle employs a launch configuration that integrates the Orbiter Vehicle (OV), Solid Rocket Boosters (SRBs), and External Tank (ET). During the launch the SRBs are jettisoned. Later the ET is jettisoned and the Orbiter Vehicle continues in flight. If you were to model the system from the SRB or ET perspective, using Figure 20.5 as a reference, how would you depict the time-based control flow operations?

4. Develop a multi-phase operations model that includes control flow and data flow operations for the following systems. Use Figure 20.6 as a reference construct.
   (a) Automobile dealership and customers.
   (b) Retail business such as a movie theatre, fast food restaurant, or baseball stadium.
   (c) K-12 public school.

5. What embedded cyclical operations occur in each of the systems in exercise 4.

ORGANIZATIONAL CENTRIC EXERCISES

1. Identify a contract program within your organization, and research how they operationally delineate MISSION SYSTEM and SUPPORT SYSTEM operations for their deliverable system, product, or service. Report your findings, observations, and results.

Chapter 21

System Operational Capability Derivation and Allocation

21.1 INTRODUCTION

When the system phases and modes of operation are defined, SEs must answer several key questions:

1. *What operational capabilities are required for each phase and mode of operation that the MISSION SYSTEM and SUPPORT SYSTEM must provide and perform?*
2. *How are those capabilities allocated and flowed down to system entities—namely the PRODUCT and SUBSYSTEM—at various levels of abstraction?*

Our discussion in this section introduces and explores an analytical approach for deriving, allocating, and flowing down system capabilities. The approach is presented from an *instructional perspective* to explain the basic concept. *Actual implementation* can be achieved with automated tools or other methods tailored to your application.

What You Should Learn from This Chapter

- What is system *operational capability*?
- What are *required operational capabilities (ROCs)*?
- What is an *Operational Capability Matrix*?
- What is the *relationship* between modes of operation and system capabilities?

Definition of Key Terms

- **Design and Construction Constraints** User or Acquirer constraints such as size, weight, color, safety, material properties, training, and security imposed on all or specific elements of the deliverable system or product.
- **System Capabilities Matrix** A matrix method that enables the identification of required operational capabilities by analytical investigation of operations and interactions for each mode of operation and MISSION SYSTEM and SUPPORT SYSTEM elements, OPERATING ENVIRONMENT elements, and design and construction constraints. The matrix also identifies the unique architectural configuration required to support the mode of operation.
- **System Operational Capability** A *use case* based operation or task that performs an action to produce a specific performance-based outcome. Note that a system *capability* represents

218 Chapter 21 System Operational Capability Derivation and Allocation

the *potential* to perform an action. In contrast, an *operational capability* may integrate several physical system capabilities to produce a specific outcome to achieve a mission objective.

21.2 CONCEPT OVERVIEW

To better understand how system operational capability derivation and allocation are performed, let's establish the relationships between the organizational mission and the system. Figure 21.1 provides a graphical framework.

Organizational Roles and Mission Objectives (1) are accomplished by MISSION SYSTEM Operations (3) synchronized with a Mission Event Timeline (MET) (2). MISSION SYSTEM operations (3) interact with the OPERATING ENVIRONMENT *system elements* (4). During these interactions MISSION SYSTEM Operations (3) consist of Phases of Operation (5), each of which may consist of several Modes of Operation (6).

For the MISSION SYSTEM to perform its operations (3), it must provide System Operational Capabilities (7). Each of these capabilities is translated into System Performance Requirements (8) that are allocated to applicable MISSION SYSTEM elements (9) (PERSONNEL, EQUIPMENT, etc.). The system's stakeholders may also identify Design and Construction Constraints (10) that are levied on one or more System Performance Requirements (8).

System Performance Requirements (8), allocated to the MISSION SYSTEM elements (9), are used to formulate and select the MISSION SYSTEM and the System Architectural Configuration(s) (11) of the system element level. System Performance Requirements (8) are then allocated to each selected System Architectural Configuration (11). Each System Architectural Configuration (11) is verified against the System Performance Requirements (8) to formally demonstrate that the requirements have been satisfied.

Based on this framework, let's begin with our first discussion topic, *identifying and deriving system operational capabilities*.

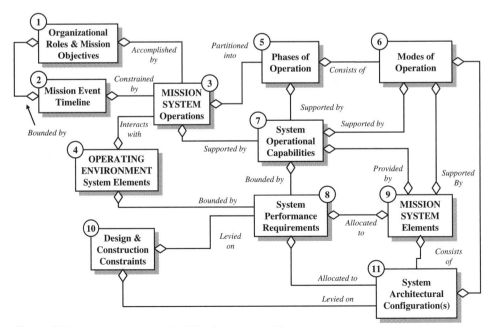

Figure 21.1 System Operational Capability Derivation & Allocation

21.3 REQUIRED OPERATIONAL CAPABILITIES

Every *use case* represents an application of the system, product, or service employed by the End User to conduct organizational missions, operations, or tasks. Therefore, each use case represents a *required operational capability* (ROC).

Use cases often express very broad, general statements of User expectations. These expectations require clarification via analytical WHAT IF scenarios. Consider the following example:

EXAMPLE 21.1

Suppose a use case characterizes a required operational capability for a DVD player to respond to user commands when a DVD is inserted. WHAT IF the user inserts a single-sided DVD upside down?

As illustrated earlier in Figure 17.1, use case scenarios represent HOW the User might *apply*, *use*, *misuse*, or *abuse* the system, product, or service. As such, a use case scenario may motivate System Developers to specify lower level required operational capabilities mandating:

1. A *warning* or *caution* to the User about an unacceptable, dangerous, or hazardous action, safety violation, or condition and request a decision.
2. Automatical correction of the problem, if practical.

Therefore, the system design must be sufficiently robust to respond to the action or condition without *failure* or degraded performance. This is why we organize this information and relationships as illustrated in Figure 21.2. Note how each cell representing a logical physical relationship has its own unique identifier. Ultimately, the required operational capabilities will be translated into singular *SPS or item development specification* (IDS) requirements statements.

Where: (#) = Reference to description ▨ = Not Applicable	System Phases of Operation					
	Pre-Mission		Mission		Post Mission	
MISSION SYSTEM Operations	MISSION SYSTEM Elements	SUPPORT SYSTEM Elements	MISSION SYSTEM Elements	SUPPORT SYSTEM Elements	MISSION SYSTEM Elements	SUPPORT SYSTEM Elements
3.0 Deploy System	(1)	(2)	(3)	(4)	(5)	(6)
4.0 Conduct Training Decision	(7)	(8)	(9)	(10)	(11)	(12)
5.0 Mission Notification Decision	(13)	(14)	(15)	(16)	(17)	(18)
6.0 Configure System for Mission	(19)	(20)	(21)	(22)	(23)	(24)
7.0 Assess Mission Readiness	(25)	(26)	(27)	(28)	(29)	(30)
9.0 Mission Go-Ahead Decision	(31)	(32)	(33)	(34)	(35)	(36)
10.0 Conduct System Mission	(37)	(38)	(39)	(40)	(41)	(42)
13.0 Assess Mission & System Performance	(43)	(44)	(45)	(46)	(47)	(48)
14.0 Deactivate System Decision	(49)	(50)	(51)	(52)	(53)	(54)
16.0 Replenish System Decision	(55)	(56)	(57)	(58)	(59)	(60)
17.0 Conduct System Training	(61)	(62)	(63)	(64)	(65)	(66)
18.0 Redeploy System Decision	(67)	(68)	(69)	(70)	(71)	(72)
20.0 Perform System Maintenance	(73)	(74)	(75)	(76)	(77)	(78)
21.0 Deactivate / Phase-Out System	(79)	(80)	(81)	(82)	(83)	(84)

Figure 21.2 Operational Tasks to Phases Mapping Representing Required Operational Capabilities

21.4 IDENTIFYING AND DERIVING SYSTEM OPERATIONAL CAPABILITIES

Expanding on our discussion in Chapter 19, *System Phases, Modes, and States of Operation*, we need to answer a key question: *WHAT are the required operational capabilities for each phase and mode of operation that the MISSION SYSTEM* and the *SUPPORT SYSTEM must provide to support?* We can answer this question by creating a System Capability Matrix such as the one shown in Figure 21.2. Observe that the operations names shown in the left column are based on those defined earlier in our System Operations Model discussion of Figure 18.1.

The System Capability Matrix *maps* system operations to phases of operation. The phases of operation are subdivided into *pre-mission, mission*, and *post-mission* phases for the SYSTEM OF INTEREST (SOI). Each SOI phase is further subdivided into MISSION SYSTEM elements and SUPPORT SYSTEM elements.

Each intersecting cell in the matrix contains a unique reference identifier (ID). This identifier links the matrix cell to an *explicit* description that *scopes* and *describes* what capabilities are provided. Thus, for Operation 3.0 Deploy System, (1) represents the pre-mission capabilities required of the MISSION SYSTEM and (2) represents the capabilities required of the SUPPORT SYSTEM.

When matrix mapping is employed, we may not know explicitly WHAT each system element's *contribution* and *relationship* are to the overall performance of the operational task. However, by deductive reasoning and experience it is known that an entity or associative relationship exists. The intent, as indicated by the reference IDs, is to simply *acknowledge* that each element has some level of contribution and relationship of accountability.

To facilitate the definition of these *acknowledgments*, ID numbers serve as symbolic notations of the reference descriptions. These descriptions may begin *informally* and evolve to a formalization. Remember, the ID entries describe *system operations* that require operational capabilities and are translated into one or more *specification requirements*. The matrix shown in Figure 21.2 serves as a valuable system analysis tool to indicate HOW Operations 3.0 through 21.0—namely Figure 18.1 of the *System Operations Model*—are supported by capabilities and performance requirements associated with each MISSION SYSTEM or SUPPORT SYSTEM during each phase of operation.

Author's Note 21.1 *When you create a matrix such as Figure 21.2, the contents can become very busy, making it difficult to discern which tasks are applicable. For illustrative purposes, let's assume we have a system where some operational tasks are not applicable to specific system elements.*

We could arbitrarily leave the cell empty, as do most people. The problem is that reviewers may not know if the SE or analyst considered the applicability and intentionally left the cell blank, ignored it, or simply overlooked it. Therefore, AVOID deleting the ID. Rather, note in the ID's description:

1. *If the operation is applicable to the MISSION SYSTEM or SUPPORT SYSTEM.*
2. *The rationale as to why the operation is not applicable. Later, if it is determined that the operation is applicable, the "placeholder" allows you to add the capability and avoid having to renumber the document test.*

Disciplined SE practices require that you at least *acknowledge* the consideration. One method to accomplish this is to *avoid* deleting the ID and shade the background in dark gray as shown in Figure 21.2. This way tasks that are *not applicable* (N/A) are explicitly identified.

Additionally, if the reviewer wants to know WHY the task is not applicable, they can use the reference ID to link to the rationale for the decision. There will often be instances where either requirements change or document reviewers determine that this decision did not include a key consideration of the other factors. Thus, the act of leaving the ID intact in a description enables the document to be easily updated, as required, and the change incorporated into the linked documentation without the IDs having to be renumbered.

System Operations Dictionary

Mapping matrices such as Figure 21.2 should be supported by a *System Operations Dictionary* that provides a brief description of *how* each task is accomplished by the MISSION SYSTEM and/or SUPPORT SYSTEM. Dictionary descriptions of the matrix serve to bound the derivation of the MISSION SYSTEM and SUPPORT SYSTEM capabilities and performance requirements.

Concluding Point

The casual reader may remark that it is impractical to document all 84 IDs shown in Figure 21.2. Yet, in completing a system development effort, you will *unknowingly* spend numerous hours on each of these 84 items. Those who view this matrix as *impractical* are the same people who often have to *explain* to program management during the System Integration, Test, and Evaluation (SITE) Phase WHY certain items have been *inadvertently overlooked* or *ignored* during the design phase. For those programs, SITE requires two to three times the nominally required time. Perform your analyses "up front" and make sure the 84 items in the figure are documented, scoped, and well communicated. The matrix approach simply provides an analytical structure to improve your chances of success.

21.5 ALLOCATING AND FLOWING DOWN SYSTEM REQUIREMENTS

The preceding discussion provides a method for identifying system operational capabilities required for each SYSTEM level *mode of operation*. The capabilities represent WHAT is required to accomplish specific objectives associated with each *mode of operation*. Therefore, the MISSION SYSTEM and SUPPORT SYSTEM must be capable of being configured with capabilities and levels of performance to support each of these *modes of operation*. To better understand HOW these capabilities are defined, refer to Figure 21.3.

MISSION SYSTEM Modal Operations and Interactions

You should recognize the key elements of Figure 21.3 from an earlier description of Figure 21.2. In the earlier figure we mapped *modes of operation* to the MISSION SYSTEM and SUPPORT SYSTEM for each of the three phases of operation—pre-mission, mission, and postmission. Recall that the IDs enable us to *link* system capabilities to descriptions defined in a *System Operations Dictionary*.

Figure 21.3, as an expansion of Figure 21.2, enables us to an answer to some basic questions. *For each of a system's modes of operations (left column):*

1. WHICH phase(s) of operation does the mode support?
2. WHAT operations and interactions occur with each of the MISSION SYSTEM and OPERATING ENVIRONMENT system elements?
3. WHAT are the system design and construction constraints that limit these operations and interactions?
4. WHAT architectural configuration satisfies items 1, 2, and 3 above?

222 Chapter 21 System Operational Capability Derivation and Allocation

Where:
- (X) = Reference to Required Operational Capabilities
- ▓ = Scenario-Based Operational Capabilities

Operational Mode	PHASE			SYSTEM ELEMENTS						OP. ENVIRON.				Design & Construct. Constraints	Architectural Configuration
	Pre-Mission	Mission	Post-Mission	Mission Resources	Procedural Data	Equipment	Personnel	Facilities	System Responses	Friendly I/Fs	Man-Made Systems: System Threats	Induced	Natural		
POWER-OFF	(1)	(2)	(3)	(4)	(5)	(6)	(7)	(8)	(9)	(10)	(11)	(12)	(13)	(14)	(15)
POWER-UP INITIAL.	(16)	(17)	(18)	(19)	(20)	(21)	(22)	(23)	(24)	(25)	(26)	(27)	(28)	(29)	(30)
CONFIGURE	(31)	(32)	(33)	(34)	(35)	(36)	(37)	(38)	(39)	(40)	(41)	(42)	(43)	(44)	(45)
CALIBRATE/ALIGN	(46)	(47)	(48)	(49)	(50)	(51)	(52)	(53)	(54)	(55)	(56)	(57)	(58)	(59)	(60)
TRAINING	(61)	(62)	(63)	(64)	(65)	(66)	(67)	(68)	(69)	(70)	(71)	(72)	(73)	(74)	(75)
NORMAL OPS	(76)	(77)	(78)	(79)	(80)	(81)	(82)	(83)	(84)	(85)	(86)	(87)	(88)	(89)	(90)
ABNORMAL OPS	(91)	(92)	(93)	(94)	(95)	(96)	(97)	(98)	(99)	(100)	(101)	(102)	(103)	(104)	(105)
SAFING	(106)	(107)	(108)	(109)	(110)	(111)	(112)	(113)	(114)	(115)	(116)	(117)	(118)	(119)	(120)
ANALYSIS	(121)	(122)	(123)	(124)	(125)	(126)	(127)	(128)	(129)	(130)	(131)	(132)	(133)	(134)	(135)
MAINTENANCE	(136)	(137)	(138)	(139)	(140)	(141)	(142)	(143)	(144)	(145)	(146)	(147)	(148)	(149)	(150)
POWER-DOWN	(151)	(152)	(153)	(154)	(155)	(156)	(157)	(158)	(159)	(160)	(161)	(162)	(163)	(164)	(165)

Figure 21.3 System Operational Capability Identification

From Figure 21.3, similar charts are developed for the PRODUCT level of abstraction. The process continues to lower levels until all aspects of the system design solution are defined. For example, the PRODUCT's unique EQUIPMENT element *operations* and *interactions* ultimately are allocated to its HARDWARE and SOFTWARE elements. Consider the following example:

EXAMPLE 21.2

For the NORMAL OPERATIONS mode, Capability (81) describes operational capabilities required for the EQUIPMENT element. Capability (86) identifies operational and system capabilities to be established for the MISSION SYSTEM to *avoid, camouflage, conceal,* or *defend* itself from external system threats. Capability (87) identifies any system level operations required for the system to perform in the prescribed INDUCED ENVIRONMENT.

Consider the next example:

EXAMPLE 21.3

The Space Shuttle conducts space-based missions in an INDUCED ENVIRONMENT that includes space debris. So, if you wanted to read about system capabilities required to cope with and operate in the INDUCED ENVIRONMENT, Capability (87) should provide the answer. If not, it needs to be defined.

One of the unique attributes of Figure 21.3 is that the IDs in any column can be "collected" for a given MISSION SYSTEM element or OPERATING ENVIRONMENT. This collection of *operations* and *interactions* can then be translated into requirements documented in specific sections of the *System Performance Specification (SPS)*.

The Importance of the Operational Capability Matrix

On the surface, Figure 21.3 appears to be just another matrix for mapping modes of operation to the *system elements*, OPERATING ENVIRONMENT, and Design and Construction Constraints to satisfy analytical curiosity. The mapping serves three very important purposes:

1. It provides a *methodical* means for *organizing* and *covering* all relevant aspects of a system design solution.
2. It provides a *mechanism* for SEs to "collect" system operational capability needs on an attribute basis by column for all phases and modes of operation.
3. Once requirements are documented, the "collection" process can be reversed to trace back to the source of a requirement.

Recall from our earlier discussion that the system *levels of abstraction*—such as SYSTEM, PRODUCT, SUBSYSTEM, ASSEMBLY, and SUBASSEMBLY—consist of one or more *physical* instances of *items* referred to as configuration items (CIs), commercial-off-the-shelf (COTS), and nondevelopmental items (NDIs). As we derive requirements from the operational capabilities identified in Figure 21.3, these requirements are *allocated* and *flowed down* to subsequent levels by using similar mapping techniques at each level.

EXAMPLE 21.4

EQUIPMENT element capabilities and performance identified at the SYSTEM level are *derived*, *allocated*, and *flowed* down to subsequent PRODUCT, SUBSYSTEM, ASSEMBLY, SUBASSEMBLY, and PART levels. Each physical component of the system, regardless of *level of abstraction*, must support one or more modes of operation discussed here.

EXAMPLE 21.5

A potentiometer—which is an adjustable resistance device—installed on a printed circuit board may only support the CALIBRATE mode.

Final Thoughts

With time, a system analyst becomes proficient in being able to mentally *assimilate* the information and *synthesize* the Operational Capability Matrix. Although SEs may not realize it, we subconsciously *assimilate* and *synthesize* this information all the time.

For those SEs who can assimilate this information, you have a *challenge*. Although you may be recognized by management and your peers for this ability, the *challenge* is *communicating* and *explaining*, via matrices such as the one illustrated, WHAT YOU KNOW to others on the development team so that the team benefits.

Author's Note 21.2 *When working in an Integrated Product Team (IPT) and other environments, prepare yourself to be challenged by other disciplines concerning the utility of this approach. Many view the matrix as analytical bureaucracy. This may require a training session for the team.*

Organizational management has a tendency to recognize "hero" personalities who have this level of ability. However, program and system development success requires more than philosophical wizards walking about with a lot of information in their heads; the information must be *documented*, *communicated*, and *reduced to practice* for the personnel.

224 Chapter 21 System Operational Capability Derivation and Allocation

Organizations and programs that are built around "heroes" often get into trouble during the System Integration, Test, and Evaluation (SITE) Phase and focus blame on the other personnel. The blaming rationale is that the others should have known WHAT the so-called hero knew but failed to properly communicate or explain. From the hero's perspective, the idea was *so intuitively obvious that no explanation was required*.

EXAMPLE 21.6

During a dry run for a Critical Design Review (CDR), which is the final technical design review, one of the presenters describes HOW they anticipated testing a particular operational requirement related to software. Immediately one of the software SEs asks, "Where did that requirement come from? It isn't documented in the specification." The presenter replies, "I thought it was so obvious it didn't have to be documented." No one, should be immune from employing sound engineering practices including documentation coordination, and communication.

21.6 COLLECTING SYSTEM ELEMENT REQUIREMENTS

Recall the mapping of system operational *modes to: phases of operation*, *system elements*, the OPERATING ENVIRONMENT, and Design Construction Constraints in Figure 21.3. We are now ready to begin summarizing system capability and performance requirements. To facilitate our discussion, these modes of operation are diagramed in Figure 21.4. In this figure the Operational Capability Matrix is represented as a small graphic in the upper left-hand corner. Cells within the matrix link to a description of capabilities and performance required to support each *mode of operation*.

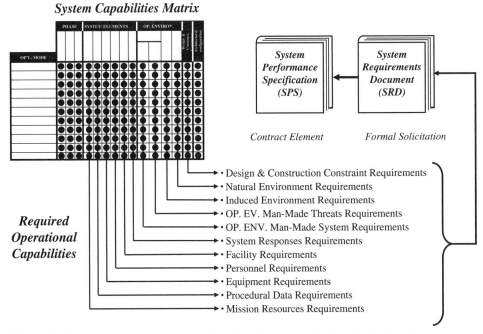

Figure 21.4 Collecting System Operational Capability Requirements for Specification Development

- Vertical columns *represent* the collection of required operational capabilities for the MISSION SYSTEM elements, OPERATING ENVIRONMENT elements, Design and Construction Constraints, etc.
- The operational capabilities of the vertical columns, denoted by the black-filled circles, are "collected" and translated into specific types of requirements as shown in the lower center of the graphic.

Acquirer's *translate*, *organize*, and *incorporate* these required operational capabilities into a *System Requirements Document (SRD)* used for formal solicitation of proposals. Subsequently, at contract award, the *System Performance Specification (SPS)* becomes the basis for System Developers to reverse the process by extracting the various requirements shown in the bottom center of the graphic.

System Element Requirements Allocations and Flow Down

As the System Developer analyses each SPS requirement, the question is: *WHAT system elements—such as EQUIPMENT or PERSONNEL—are accountable for providing WHAT required operational capabilities and levels of performance to satisfy the SPS requirement?* We answer this question via a process referred to as requirements *allocation* and *flow down*.

When we allocate the Level 0 system requirements identified in Figure 21.4 to the Level 1 MISSION SYSTEM or SUPPORT SYSTEM, we do so as illustrated in Figure 21.5. Then, MISSION SYSTEM or SUPPORT SYSTEM requirements are *allocated* and *flowed down* to the respective *system elements*.

EQUIPMENT System Element Allocations

For those requirements allocated to the EQUIPMENT element and documented in the SPS, the question becomes: *WHAT system entities—such as PRODUCT, SUBSYSTEM, ASSEMBLY, SUB-*

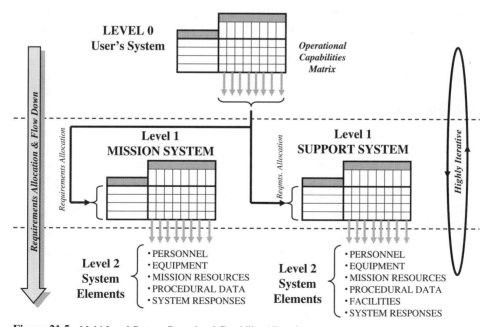

Figure 21.5 Multi-Level System Operational Capability Allocations

226 Chapter 21 System Operational Capability Derivation and Allocation

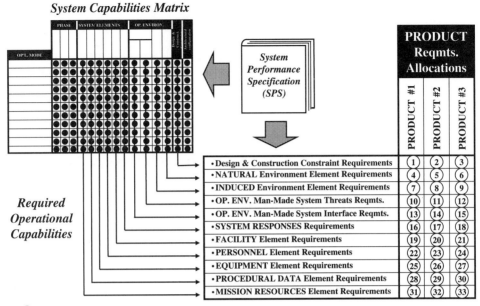

Figure 21.6 System Requirements Allocations to Product Level

ASSEMBLY, and PART levels—are accountable for providing the EQUIPMENT level operational capabilities and levels of performance to satisfy each SPS requirement? We answer this question by expanding each set of requirements allocations to lower levels. Figure 21.6 illustrates our discussion.

Each requirement allocated to the EQUIPMENT element is allocated to PRODUCT 1, PRODUCT 2, and PRODUCT 3 as shown at the right side of the graphic. Each capability number *represents* the various types of SPS requirements allocated to PRODUCT 1, PRODUCT 2, and PRODUCT 3. Once the allocations are made, the set of requirements allocated to each PRODUCT are documented in the topical section of the respective *PRODUCT Item Development Specification (IDS)*, as shown in Figure 21.7.

Top-down/Bottom-up Iterations

The preceding discussion focused on the top-down *allocation* and *flow down* of requirements to various *system levels of abstraction*. However, despite the nomenclature this is not a linearly sequential process. It is a *highly iterative* process whereby requirements allocations between levels must be reconciled within the architecture for that particular *item* and traded off within the system capability and performance constraints. The result is a set of *highly iterative* top-down/bottom-up/left-right/right-left set of decision-making activities. Therefore, the oval icon in the upper left center portion of Figure 21.7 symbolizes this *top-down/bottom-up* process for all levels of system decomposition.

21.7 SUMMARY

In summary, this chapter introduced the concept of deriving and allocating required operational capabilities to multi-level system entities. We began with the system's modes of operation and identified the various operational capabilities and interactions required for each mode. This included the *system elements* for the

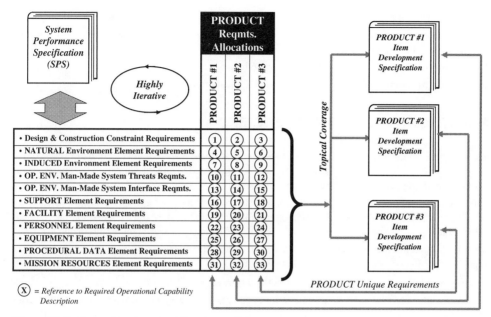

Figure 21.7 System Requirements Allocations to Subsystems

MISSION SYSTEM and SUPPORT SYSTEM, OPERATING ENVIRONMENT, and Design and Construction Constraints.

Next we described how the required operational capabilities were aggregated and documented in the *System Performance Specification (SPS)*. We also learned that SPS requirements are allocated and flowed down to system entities at various system *levels of abstraction* and documented in their respective *Item Development Specification (IDS)*.

We also described *how* the set of required operational capabilities for each mode of operation must be provided by a unique configuration of *system elements* referred to as its *architecture*. The convolution of mode-based architectural configurations is referred to as the system architecture.

Does an SE have to perform all of the analytical work? You and your organization have to answer this question. With time and experience, you will learn to mentally assimilate and process this concept. Remember, the methods described in this chapter aim to *explain a concept for instructional purposes*. The challenge questions SEs must answer are:

1. *How do we assure a program's management that all aspects of a system's specification or design solution are sufficiently addressed to ensure program success.*
2. *What tools, methods, and techniques do we use to provide that assurance and level of confidence without getting "analysis paralysis."*

For many applications, automated tools based on relational databases are available to support the concept described here.

GENERAL EXERCISES

1. Answer each of the *What You Should Learn from This Chapter* questions identified in the *Introduction*.
2. Refer to the list of systems identified in Chapter 2. Based on a selection from the preceding chapter's General Exercises or a new system selection, apply your knowledge derived from this chapter's topical discussions.

(a) Create a System Capabilities Matrix for the selected item that identifies the item's modes of operation.
(b) For each mode of operation, pick one or more *system elements* and identify what operational capabilities are required to support the mode.
(c) Repeat item (b) for design and construction constraints.
(d) Repeat item (b) for OPERATING ENVIRONMENT interactions.

ORGANIZATIONAL CENTRIC EXERCISES

1. Contact a program within your organization and research *how* they analyzed, allocated, and flowed down their SPS requirements to the various system levels of abstraction. Document your findings and observations. Avoid specifying programs and individuals by name.
   (a) What methods, such as the System Capabilities Matrix, did the program use to *ensure* that: 1) all required operational capabilities were addressed in the SPS and 2) they fully understood the implications of satisfying the requirements as evidenced by derivation of lower level requirements.
   (b) Were the SPS requirements identified via an organized analytical process or random accumulations of requirements? Report your findings and observations.
2. For the program in item 1 above. Based on the method(s) selected:
   (a) What types of tools were used (e.g., internally developed database, spreadsheet, requirements management tool)?
   (b) Is this practice common across all programs?
   (c) How many people were involved in the exercise? Was any training provided?
   (d) To what extent was the tool employed?
   (e) How did the amount of time available factor into the usage of the tool?
   (f) What lessons learned did the program collect from the exercise?

Chapter 22

The Anatomy of a System Capability

22.1 INTRODUCTION

Every man-made system provides operational capabilities to support accomplishment of the User's organizational or personal missions. Your mission, as an SE, is to ensure that system capabilities support this mission as discussed in Chapter 13 *Organizational Roles, Missions, and System Applications*. Operational capabilities, as system assets, *characterize* the mechanical, electrical, optical, chemical, or processing features that enable a system to function, process MISSION RESOURCES, make decisions, and achieve a required level of success based on performance.

If you ask most SEs what a *capability* is, the typical response includes the usual *function* with one or more *performance bounding* elements. Certainly a capability includes these elements; however, those elements *characterize* the *outcome* of a capability. From an engineering perspective, a *capability* is broader in scope than simply a functional element, especially in large, complex systems. It represents a physical potential—strength, capacity, endurance—to perform an outcome-based action for a given duration under a specified set of OPERATING ENVIRONMENT conditions.

This section explores the anatomy of a *system operational capability*. We begin our discussion with the introduction of the System Capability Construct, a graphical template that enables system analysts and SEs to model a system capability's behavior. As you will discover, a system *capability* is a "system" in its own *context* and has its own set of pre-mission, mission, and postmission phases. We investigate each phase of a capability and the tasks performed.

What You Should Learn from This Chapter

1. What are the *phases of operation* of a capability?
2. Differentiate automated from semi-automated capabilities.
3. What is the *System Capability Construct*?
4. Why is the System Capability Construct important?
5. Recreate the System Capability Construct, and describe each of its task flows and interdependencies.
6. What is meant by capability *exception handling*?
7. Why is *exception handling* required?

System Analysis, Design, and Development, by Charles S. Wasson
Copyright © 2006 by John Wiley & Sons, Inc.

Definition of Key Terms

- **Automated or Semi-automated Capability** A capability that has been mechanized or pre-programmed to execute a series of tasks on receipt of external data or triggering event such as an interrupt.

22.2 BOUNDING AN OPERATIONAL CAPABILITY

If you ask a sampling of engineers *how* they *specify a system's capabilities*, most will respond, "We write requirements." Although this response has a degree of correctness, requirements are a merely *mechanisms* for documenting and communicating WHAT the User requires in terms of acceptance at delivery. The response you should hear from SEs is: "We specify system capabilities and levels of performance to achieve specific outcomes required by the User."

When engineers focus on writing *requirements statements* (i.e., requirements centric approach) rather than specifying capabilities (i.e., capability centric approach), requirements statements are what you get. When you attempt to analyze specifications written with a *requirements centric* approach, you will encounter a *wish list* domain that is characterized by *random, semi-organized thoughts*; *overlapping, conflicting, replicated,* and *missing* requirements; ambiguous statements subject to interpretation; compound requirements; contract statement of work (CSOW) tasks; mixtures of goals and requirements, and so on.

In contrast, specifications written by SEs who *employ model-based* requirements derived from coherent structures of system capabilities generally produce documents that *eliminate* or *reduce* the number and types of *deficiencies* noted above. These specifications also tend to require less maintenance and facilitate verification during the system integration and test phase of the program.

If you analyze and characterize *work products of* the *requirements centric approach*, several points emerge:

1. Lack of understanding of the end product (e.g., bounding system capabilities) and its interactions with its OPERATING ENVIRONMENT.
2. Lack of training and experience in how to identify and derive capability-based requirements.
3. Poor understanding of the key elements of a requirements statement.

We will address the last two points in later chapters. Our discussion in this chapter focuses on the first item; understanding system capabilities.

Referral More information about writing requirements will be presented in Chapter 33 *Requirements Statement Development* practices.

Implicit and Explicit Meanings of a Capability

The term "capability" for most systems has both *implicit* and *explicit* meanings. *Implicitly*, the term infers the *unrealized potential* to perform an action. *Explicitly*, a capability is *realized* when the system performs its planned action(s) in its prescribed OPERATING ENVIRONMENT as intended by its developers.

> **EXAMPLE 22.1**

A simple device such as a log *implicitly* has the *unrealized potential* or capability to displace heavy objects. *Explicitly*, the log has a *realized capability* to physically displace heavy objects WHEN controlled by a human with a fulcrum and manipulated by human actions. To successfully complete the desired human outcome—to displace a heavy object—the log must possess a bounded and integrated set of physical characteristics and properties—capability—to perform the action.

Automated or Semi-automated Capabilities

In the preceding example the set of *actors*—human–log–fulcrum—were *integrated* and *mechanized* into a set of semi-automated actions intended to achieve a specified result, which was to displace a heavy object. If the fulcrum were implemented as a programmed, repeatable assembly line device, we could refer to it as *automated*. Thus the *integration*, *mechanization*, and *automation* of *compatible* and *interoperable* system element actions into a "working system" capability provide the basis for our next discussion.

22.3 THE SYSTEM CAPABILITY CONSTRUCT

A system's operational capability can be modeled using a basic construct. The *construct* consists of sequential and concurrent control flow operations required to: 1) initialize, 2) implement, and 3) complete the capability or action. This is particularly useful when modeling *automated or semi-automated* sequences of complex system operations that are common to many human-made systems.

If you observe and analyze *how* an automated or semi-automated, *cyclical* capability is implemented, you will discover that the *construct* has its own life cycle with pre-mission, mission, and post-mission phases. In general, an automated or semi-automated capability is *characterized* by:

1. An initial set of preparatory operations, conditions, or configurations.
2. A set of primary operations that are executed to accomplish the capability.
3. A set of closure or termination operations.

To illustrate *how* these three sets of operational phases and operations are implemented, refer to Figure 22.1.

Exploring the System Capability Construct

On inspection of Figure 22.1, observe that the model flows from the INITIAL STATE to the FINAL STATE via a series of sequential and concurrent operations. Thus a system can initiate a capability to perform a single action or loop continuously until directed to stop by the HIGHER ORDER SYSTEM capability—the DO UNTIL (condition). To better understand how the capability is implemented, let's investigate each of its operations.

Operation 1.0: Activate Capability Decision. The basic System Capability Construct begins with a simple "activate capability" decision control point that determines whether the capability is to be *activated* or *enabled*. If the capability is not required and is presently *deactivated* or *disabled*, control flow transfers to the FINAL STATE. If a capability is to be *activated*, control flow transfers to Operation 2.0, Initialize the Decision and begin activation.

Operation 2.0: Initialize Decision. From a *pre-mission* perspective, the first operation of a capability should be to establish any *initial conditions* that may be required to accomplish the capability including Initialization. The initial conditions may include data retrieval, initialization of variables, calibration, and alignment. As a result an initialize decision is made.

232 Chapter 22 The Anatomy of a System Capability

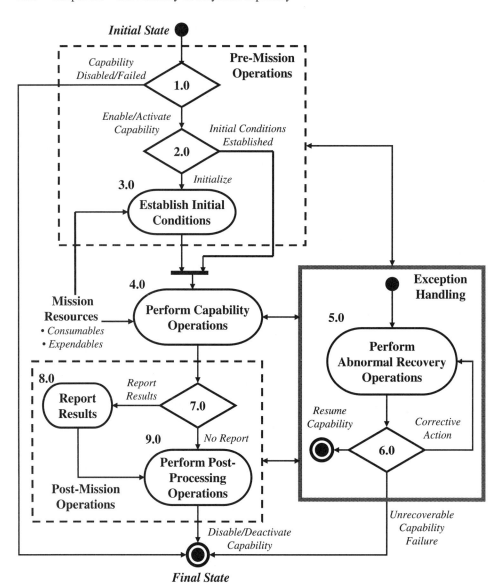

Figure 22.1 Basic System Entity Capability Construct

EXAMPLE 22.2

A process control computer system may require initialization of specific parameters, variables, or positioning of a mechanical arm on some device prior to executing the capability action—namely moving the mechanical arm.

If the Initialize decision is *Yes* or TRUE, *control flow* transfers to Operation 3.0, which is to establish initial conditions. If the Initialize decision is *No* or FALSE, *control flow* progresses to Operation 4.0, Perform Capability Operations.

Author's Note 22.1 *The FALSE decision result is intended to accommodate cyclical capability actions—namely DO UNTIL (condition)—whereby the capability has already been initialized in a previous cycle of the construct.*

Operation 3.0: Establish Initial Conditions. Operation 3.0 to establish the initial conditions depends on the type of capability application.

EXAMPLE 22.3

Computational solutions may require initialization of variables to default values or settings. Mechanical devices may have a mechanical arm that requires pre-positioning (i.e., alignment) to a calibrated reference point, step, or stop. Devices with heating or cooling elements may require activation and stabilization over a prescribed period of time.

Some applications may also require that some form of initial or readiness health status check be conducted before, during, or following performance of the capability.

EXAMPLE 22.4

A daily operational readiness test (DORT) may be required to be performed at the beginning of each workday.

Finally, some systems may also include a requirement to *report* the results of the Establish Initial Conditions (Operation 3.0) as a health check status back to some central collection point such as a computer, file, or system operator I/F. When the initial conditions established by Operation 3.0 have satisfied a pre-determined set of *exit criteria* or encountered a *triggering event* or an interrupt by an external stimulus, such as a system operator input, *control flow* transfers to Operation 4.0, Perform Capability Operations.

Operation 4.0: Perform Capability Operations

Operation 4.0 Perform Capability Operations represents the core execution of the capability. The operation illustrated here is a high-level *abstraction* that represents the multiple levels of subordinate operations required to perform the primary mission. As the system's capability is being performed, replenishment of the MISSION RESOURCES element assets such as consumables and expendables are required to support and sustain system operations.

Error conditions and latent defects such as design *deficiencies*, *flaws*, or *malfunctions* can deter mechanical capability and force the system to shift to an ABNORMAL operating condition or state. Abnormal conditions include:

1. Nonfatal, degraded performance operations.
2. Emergency operations.
3. Catastrophic operations that require employment of special actions.

If an abnormal condition triggers a disruption of mechanical capability, control flow transfers to Operation 5.0, Perform Abnormal Recovery Operations.

When the mechanical operations are successfully completed and determined to satisfy a set of exit criteria, *control flow* transfers to Operation 7.0, Report Results decision.

Operation 5.0: Perform Abnormal Recovery Operations

When an *error*, *latent defect*, *exception*, or *malfunction triggers* an abnormal condition, the system may immediately shift to some form of recovery operation. In general, these conditions are called *exception handling* or error correction.

EXAMPLE 22.5

Software logs an error condition and attempts to resume normal processing or restarts. Data communicaitons links detect errors and request retransmission or attempt to correct data bits. Watchdog timers are employed by hardware designers as a periodic heartbeat that restarts the system if not reset within a specified timeperiod such as 1 second.

If an *ABNORMAL condition* occurs, the system should be expected to log or record the event as well as the conditions under which the event occurs. In this manner, an historical record documents objective evidence to support reconstruction of the chain of events to better understand *how* and *why* the system encountered the exception. Analysis of results and sequences of actions or conditions leading up to the event may indicate that an *operational scenario* of a specific system mode of operation had not been anticipated by the SE.

EXAMPLE 22.6

A flight data recorder, referred to as the "black box," records vital aircraft systems information such as aircraft configuration, sequences of events, and flight conditions.

When recovery operations satisfy a set of *exit criteria*, control flow transfers to Operation 6.0, Resume Capability *decision control point*.

Operation 6.0: Resume Capability Decision

Operation 6.0 Resume Capability decision makes the determination whether normal capability operations can be resumed. In general, control flow can transfer to any one of three options: 1) resume capability, 2) perform corrective action, or 3) exit due to failure. If the recovery has been successful and satisfies pre-determined *exit criteria*, control flow *resumes* to Operation 4.0, Perform Capability Operations.

If recovery is *impractical* or *impossible* under the current conditions, the capability is *disabled or deactivated* for the remainder of the mission or until the condition that created it changes. If a capability is *disabled* or *deactivated*, control flow transfers to FINAL STATE. When this occurs, some systems may require software flags or indicators to be set to indicate the condition and disable the capability for the remainder of the mission. For hardware systems, power cables may be disconnected, maintenance tags may be attached to the equipment, and/or locks may be installed to prevent usage until the *fault* is corrected.

If the recovery has not been completed and the condition is *recoverable*, control flow transfers back to Operation 5.0, Perform Abnormal Recovery Operations.

Operation 7.0: Report Results Decision

When Operation 4.0 Perform Capability Operations is complete, control flow progresses to Operation 7.0, Report Results *decision control point*. If the decision is *Yes* or TRUE, control flow transfers to Operation 8.0, Report Results. If the decision is *No* or FALSE, control flow transfers to Operation 9.0, Perform Postprocessing Operations.

Author's Note 22.2 *In the context of systems, a system's mechanical capability may not have a reporting requirement—as in Operation 8.0. However, a PERSONNEL system element capability, operating concurrently with a mechanical capability and with each part performing its own capability construct, may have a requirement to report results of mechanical capability. For example, astronauts operating the Space Shuttle's mechanical robotic arm may observe and verbally report status, results, and completion of mission operations tasks.*

Operation 8.0: Report Results

When required, Operation 8.0 Report Results *communicates* the outcome of operations performance or status and health checks. The "report" in this context is a generic term that refers to audible, visual, mechanical, or electrical methods for reporting results. Remember, a report gives *informational*, *cautionary*, or *warning* indications, depending on the operating condition or situation.

EXAMPLE 22.7

Audible results include PERSONNEL reporting the results of an operation. A gunshot indicates the shooting of a projectile, the slam of a door indicates the closure of a mechanism, and so on. *Visual* examples include a burst of light from an explosion, a static or flashing signal light from an aircraft, and so on. *Mechanical* examples include a movement of a mechanical arm during a hard copy printout, and *electrical* examples include a spark or smoke.

When the reporting operations are complete, satisfying pre-determined exit criteria, control flow transfers to Operation 9.0, Perform Postprocessing Operations.

Operation 9.0: Perform Postprocessing Operations

When completed, the final operations of the System Capability Construct is to:

1. Place the capability in a safe and secure *standby*, *stowed*, or *dormant* physical state or condition to protect the capability from intrusion or interference by external capabilities or threats.
2. *Posture* or *prepare* the capability for its next *iterative* cycle.

EXAMPLE 22.8

A mechanical arm must be stowed to a specific condition, clearing out a software buffer, resetting or reconfiguring a computer I/O port, and so on. When the postprocessing operations are complete, control flow progresses to FINAL STATE, at least for this cyclical iteration.

EXAMPLE 22.9

The Space Shuttle's robotic arm is *retracted* and *stowed* in its Cargo Bay at the completion of a mission for the return flight back to Earth.

Concluding Point

A system is actually an integrated set of multi-level capabilities to achieve a higher level capability. To illustrate this point, consider the Car-Driver system shown in Figure 22.2. Here, the Driver has a capability construct that interacts with the overall Car capability construct.

236 Chapter 22 The Anatomy of a System Capability

To further understand how the capabilities interact, refer to Figure 22.3. The figure illustrates the car's various capability options that can be exercised at the driver's discretion. Each of the capabilities (i.e., engine, lights, etc.) has its own high-level capability construct (Figure 22.1). From an SE perspective, the key is to: 1) identify and define each capability and 2) *synchronize* the capabilities into a coherent system operation. This includes the *allowable actions* and *prohibited actions* illustrated in Figure 19.6.

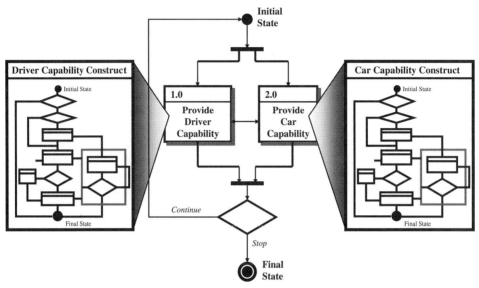

Figure 22.2 Car-Driver System Operations

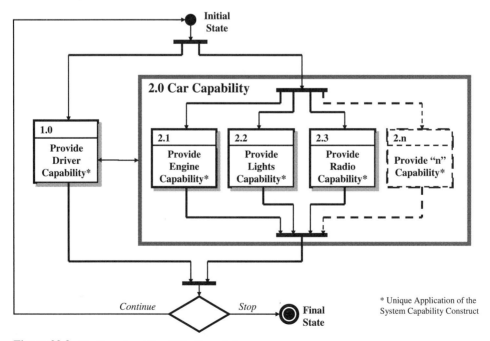

Figure 22.3 Car Operational Capability Expanded

22.4 SYSTEM OPERATIONAL CAPABILITY ANALYSIS RULES

Our discussion of the *automated* or *semi-automated* system capability construct and its application as a generic template enables us to establish the System Operational Capability analysis rules identified in Table 22.1.

Table 22.1 System operational capability analysis guidelines

ID	Topic	System Analysis and Design Rules
22.1	Multi-phase Operations	Every automated or semi-automated system capability consists of at least three types of operations: pre-mission, mission, and postmission.
22.2	Capability activated decision	Each capability must be enabled or activated to perform its intended mission. If not, the capability remains DEACTIVATED or DISABLED until the decision is revisited in the next cycle.
22.3	Capability initialized decision	Each automated or semi-automated system capability, when ACTIVATED or ENABLED, may require initialization to establish a set of initial conditions that enable the capability to perform its mission. If initialization is not required, workflow progresses to the next cycle, perform capability operations.
22.4	Perform capability operations	Each automated or semi-automated system capability must perform a primary mission that focuses on accomplishing a specific system level mission objective (i.e., the SPS requirement).
22.5	Perform exception handling	Each automated or semi-automated system capability should provide a mechanism for recognizing, recording, and processing exceptions and errors.
22.6	Perform abnormal recovery operations	Every automated or semi-automated system capability should provide a mechanism for the capability to safely recover from processing exceptions and errors without destructive consequences to the system, life, property, or the environment.
22.7	Resume capability decision	As each recovery operation is attempted, the system must decide whether additional attempts are justified.
22.8	Report results decision	When a capability completes its operations, other capabilities external to the capability may require automatic or manual reporting.
22.9	Report results decision	Every automated or semi-automated system capability may require a mechanism for reporting accomplishment of the capability's mission and outcome.
22.10	Perform Postprocessing operations	Every automated or semi-automated system capability may require a set of operations or actions to safely and securely store or stow the capability to protect it from external threats or itself within resource constraints.
22.11	Cycle back to INITIAL STATE	Once a capability has cycled through its planned operations, it returns to its INITIAL STATE or repeats, if so commanded.

The Importance of the System Capability Construct

At the start of this chapter we contrasted traditional engineering approaches that focus on *writing requirements* with modern-day engineering approaches that focus on *specifying capabilities*. Our discussion of the System Capability Construct serves as compelling *objective evidence* as to WHY it is important to structure and specify a capability and then translate it into a requirements statement. Operations within the structure serve as a *graphical checklist* for specification requirements—namely WHAT is to be accomplished and HOW WELL (performance) without stating HOW it is to be implemented.

Author's Note 22.3 *For those who think that we have violated a cardinal rule of system specifications by specifying HOW to implement a solution, our answer is no. We have not specified HOW. Instead, we have translated System Capability Construct operations or tasks into text-based requirements statements that identify WHAT must be accomplished. The HOW occurs when you specify:*

1. *The physical configuration implementation.*
2. *The logical sequences of operations depicted in the construct.*

Referral More information on *specification development methods* as provided in Chapter 32 on *System Specification* practices.

As our discussion illustrated, a capability consists of a series of operations—functional tasks, decisions, inputs, and outcomes. Each of these operations is *translated* into a specific requirement statement and is *integrated* into a set of requirements that bound the total set of capabilities. Without the *capability centric* focus, the end product—e.g. specification—is nothing more than a set of random, loosely coupled text statements with missing requirements representing overlooked operations within the capability construct.

Does this mean every operation within the construct must have a requirements statement? No, you have to apply good judgment and identify which operations require special consideration by designers during the capability's implementation.

Remember the old adage: *If you do not tell someone WHAT you want, you cannot complain about WHAT gets delivered*. If you *forget* to specify a specific operational aspect of a capability, the System Developer will be pleased to accommodate that requirement for a price—and in some cases, a very *large* price. Therefore, do your homework and make sure all capability requirements are complete. The capability construct provides one approach for doing this, but the approach is *only* as *good* as YOU *define* it.

Our discussion introduced the concept of *automated* or *semi-automated system* capabilities. We described *how* most human-made system capabilities can be modeled using the System Capability Construct as a template. The *construct* provides an *initial* framework for describing requirements that *specify* and *bound* the capability.

Resource Utilization Considerations

Now consider situations whereby physical device resources may be limited due to expense, weight, and size. The solution may require round-the-clock operations in the form of shifts. Examples include situations where several people utilize single pieces of equipment or where space restrictions limit how many people can work in close proximity without interference as in the Space Shuttle or International Space Station. As a result some workers may perform the primary mission capabilities while others are in a "sleep cycle"—all concurrent operations.

22.5 GUIDING PRINCIPLES

In summary, the preceding discussions provide the basis with which to establish guiding principles that govern the implementation of a system capability.

Principle 22.1 Every operational capability has a System Capability Construct that models the capability's action-based operations, tasks, and external interactions.

Principle 22.2 Every operational capability requires an external trigger—a cue or a stimulus—to initiate its outcome-based processing.

Principle 22.3 Every operational capability, as an integrated system, consists of three sequential phase-based actions:

1. Pre-mission phase—initialization.
2. Mission phase—application-based performance.
3. Post-mission phase—analysis and deactivation.

Principle 22.4 Every capability requires consideration of HOW exceptions to NORMAL and ABNORMAL operations will be handled.

Principle 22.5 On completion of tasking, every capability should report notification of successful completion.

22.6 SUMMARY

This chapter introduced the System Capability Construct and discussed its application to systems. Our discussion covered the construct's operations and control flows, and equated its structure to real world examples. We also emphasized the importance of the construct as a *graphical checklist* for specifying capability requirements statements in terms of WHAT must be accomplished and HOW WELL without specifying HOW the requirement was to be implemented.

GENERAL EXERCISES

1. Answer each of the *What You Should Learn from This Chapter* questions identified in the *Introduction*.
2. Refer to the list of systems identified in Chapter 2. Based on a selection from the preceding chapter's *General Exercises* or a new system selection, apply your knowledge derived from this chapter's topical discussions. If you were the designer of a specific capability, using Figure 22.1 as a guide,
   (a) Describe the set of tasks the capability should perform.
   (b) How would you handle exceptions?
   (c) How are outcome based results of the capability reported? In what format and media?
   (d) Translate each of the capability tasks into a set of operational capability requirements.

ORGANIZATIONAL CENTRIC EXERCISES

1. Contact a program within your organization. Interview SEs concerning how they define and implement a capability.
   (a) How do their designers accommodate the various operations or tasks of the System Capability Construct?
   (b) Without making them aware of this chapter's discussions or Figure 22.1, have they synthesized this concept on their own or just unconsciously do things ad hoc based on lessons learned from experience?
   (c) Present your findings and observations.

Chapter 23

System Analysis Synthesis

Throughout Part I we sequenced through topical series of chapters that provide an analytical perspective into HOW to THINK about, *organize*, and *characterize* systems. These discussions provide the foundation for Part II System Design and Development Practices, which enable us to translate an abstract *System Performance Specification (SPS)* into a physical system that can be *verified* and *validated* as meeting the User's needs. So, HOW did Part I System Analysis Concepts provide this foundation?

23.1 SYNTHESIZING PART I ON SYSTEM ANALYSIS CONCEPTS

Part I concepts were embodied in several key themes that systems analysts and SEs need to understand when developing a new *system*, *product*, or *service*.

1. WHAT are the *boundary conditions* and *constraints* imposed by the User on a system, product, or service in terms of missions within a prescribed OPERATING ENVIRONMENT?
2. Given the set of *boundary conditions* and *constraints*, HOW does the User envision deploying, operating, and supporting the system, product, or service to perform its missions within specific time limitations, if applicable?
3. Given the *deployment, operation, support,* and *time constraints* planned for the system, product, or service, WHAT is the set of outcome-based behaviors and responses required of the system to accomplish its missions?
4. Given the set of *outcome-based behaviors* and *responses* required of the system to accomplish its mission, HOW is the deliverable system, product, or service to be physically implemented to perform those missions and demonstrate?

To better understand HOW Part I's topical series and chapters supported these themes, let's briefly explore each one.

Theme 1: The User's Mission

Boundary conditions and constraints for most systems are established by the organization that *owns* or *acquires* the system, product, or service to accomplish missions with one or more outcome-based performance objectives. The following chapters provide a topical foundation for understanding organizational boundary conditions and constraints.

System Analysis, Design, and Development, by Charles S. Wasson
Copyright © 2006 by John Wiley & Sons, Inc.

Chapter 13: *Organizational Roles, Missions, and System Applications*
Chapter 14: *Understanding the System's Problem, Opportunity, and Solution Spaces*
Chapter 15: *System Interactions with Its OPERATING ENVIRONMENT*
Chapter 16: *System Mission Analysis*

Theme 2: Deployment, Operations, and Support of the System

Once the organization's vision, boundary conditions, and constraints are understood, we addressed HOW the User envisions deploying, operating, and supporting the system to perform its missions. The following chapters provide a topical foundation for understanding HOW systems, products, or services are deployed, operated, and supported.

Chapter 17: *System Use Cases and Scenarios*
Chapter 18: *System Operations Model*
Chapter 19: *System Phases, Modes, and States of Operation*

Theme 3: System Behavior in Its OPERATING ENVIRONMENT

Given the *deployment, operation, support,* and *time constraints* planned for the system, product, or service, we need to identify the set of outcome-based *behaviors* and *responses* required of the system to accomplish its missions. The following chapters provide a topical foundation for understanding HOW *systems, products,* or *services* are expected to behave and interact with their OPERATING ENVIRONMENT.

Chapter 20: *Modeling System and Support Operations*
Chapter 21: *System Operational Capability Derivation and Allocation*
Chapter 22: *The Anatomy of a System Capability*

Theme 4: Physical Implementation of the System

Based on an understanding of *outcome-based behaviors* and *responses* required of the system to accomplish its mission, the question is: *HOW do we physically implement a system, product, or service to perform those missions?* The following chapters provide a topical foundation for understanding HOW systems, products, or services are physically implemented.

Chapter 8: *The Architecture of Systems*
Chapter 9: *System Levels of Abstraction and Semantics*
Chapter 10: *System of Interest (SOI) Architecture*
Chapter 11: *Operating Environment Architecture*
Chapter 12: *System Interfaces*

By inspection, these themes range from the *abstract* concepts to the *physical* implementation; this is not coincidence. This progression is intended to show HOW SEs evolve a system design solution from *abstract vision* to *physical realization*.

After examining this list, you may ask: *Why did we choose to talk about system architectures early in an order that puts it last in this list?* For instructional purposes, system architectures represent the physical world most people can relate to. As such, architectures provide the *frame of reference* for semantics that are key to understanding Chapters 13–22.

23.2 INTRODUCING THE FOUR DOMAINS OF SOLUTION DEVELOPMENT

If we *simplify* and *reduce* these thematic groupings, we find that they represent four *classes* or *domains* of solutions that characterize HOW a system, product, or service is designed and developed, the subject of Part II. Table 23.1 illustrates the mapping between the *Part I's systems analysis concepts* themes and the *four domain solutions*.

There are several key points to be made about the mapping. *First*, observe that objectives 1 and 2 employ the User as the "operative" term; Objectives 3 and 4 do not. *Does this mean the User is "out of the loop"*? Absolutely not! Table 23.1 communicates that the User, Acquirer, and System Developer have *rationalized* and *expressed* WHAT is required. Given that direction, the system development contract imposes *boundary conditions* and *constraints* on developing the system. This communicates to the System Developer *"Go THINK about this problem and TELL us about your proposed solution in terms of its operations, behaviors, and cost-effective implementation."* Since Table 23.1 represents how a system evolves, User involvement occurs *explicitly* and *implicitly* throughout all of the themes. Remember, if the User had the capabilities and resources available, such as expertise, tools, and facilities to satisfy Objectives 3–4, they would have already independently developed the system.

Second, if: 1) a SYSTEM has four domains of solutions and 2) the SYSTEM, by definition, is composed of integrated sets of components working synergistically to achieve an objective greater that the individual component objectives, *deductive reasoning* leads to a statement that each of the components ALSO has four domains of solutions, all LINKED, both *vertically* and *horizontally*.

The four themes provide a framework for "bridging the gap" between a User's abstract *vision* and the *physical realization* of the system, product, or service. Thus, each theme builds on decisions established by its *predecessor* and *expands* the level of detail of the evolving system design solution as illustrated at the left side of in Figure 23.1. This allows us to make several observations:

Table 23.1 Linking Part I *System Analysis Concepts* themes into Part II *System Design* and *Development Practices* semantics

ID	Part I Thematic Objectives	Domain Solutions
23.1	Objective 1—WHAT are the *boundary conditions* and *constraints* imposed by the User on a system, product, or service in terms of missions within a prescribed OPERATING ENVIRONMENT?	Requirements Domain Solution
23.2	Objective 2—Given the set of *boundary conditions* and *constraints*, HOW does the User envision *deploying*, *operating*, and *supporting* the system, product, or service to perform its missions within specific time limitations, if applicable?	Operations Domain Solution
23.3	Objective 3—Given the *deployment*, *operation*, *support*, and *time constraints* planned for the system, product, or service, WHAT is the set of outcome-based *behaviors* and *responses* required of the system to accomplish its missions?	Behavioral Domain Solution
23.4	Objective 4—Given the set of outcome-based *behaviors* and *responses* required of the system to accomplish its mission, HOW is the deliverable system, product, or service to be physically implemented to perform those missions and demonstrate?	Physical Domain Solution

244 Chapter 23 System Analysis Synthesis

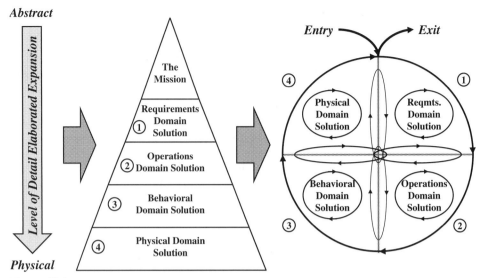

Figure 23.1 Development and Evolution of a SYSTEM's/Entity's Solution Domains

- The mission (i.e., the *opportunity/problem space*) forms the basis for the User to establish the Requirements Domain Solution (i.e., the *solution space*).
- The Requirements Domain Solution forms the basis for *developing* and *maturing* the Operations Domain Solution.
- The *evolving* Operations Domain Solution forms the basis for *developing* and *maturing* the Behavioral Domain Solution.
- The *evolving* Behavioral Domain Solution forms the basis for *developing* and *maturing* the Physical Domain Solution based on physical components and technologies available.

From a *workflow perspective*, the design and development of the system solution *evolves* and *matures* from the *abstract* to the *physical* over time. However, the *workflow progression* consists of numerous feedback loops to preceding solutions as System Analysts and SEs mature the solutions and reconcile *critical operational and technical issues (COIs/CTIs)*. As a result, we *symbolize* the system solution domains as shown at the right side of Figure 23.1.

23.3 SYSTEM DOMAIN SOLUTION SEQUENCING

Figure 23.2 provides a way to better understand how the system domain solutions evolve over time. As shown, the Requirements Domain Solution is initiated first, either in the form of a contract *System Performance Specification (SPS)* or a System Developer's *item development specification*. Here is how the sequencing occurs:

- When the Requirements Domain Solution is understood and reaches a level of maturity sufficient to develop concepts of operation, initiate the Operations Domain Solution.
- When the Operations Domain Solution reaches a *level of maturity* sufficient to define *relationships* and *interactions* among system capabilities, initiate the Behavioral Domain Solution.

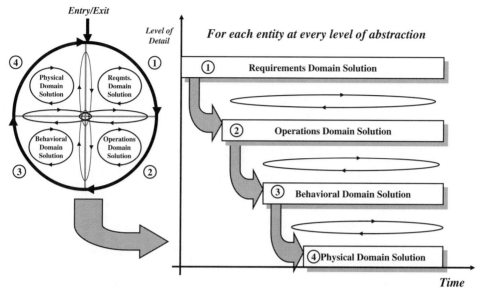

Figure 23.2 System Solution Domain Time-Based Implementation

- When the Behavioral Domain Solution reaches a *level of maturity* sufficient to allocate the behavioral capabilities to physical components, initiate the Physical Domain Solution.
- Once initiated, the Requirements, Operations, Behavioral, and Physical Domain Solutions evolve *concurrently*, *mature*, and *stabilize*.

23.4 SUMMARY

In this chapter we synthesized our discussions in *Part I on* system analysis concepts and established the foundation for *Part II on* system design and development. The introduction of the Requirements, Operations, Behavioral, and Physical Solution Domains, coupled with chapter references in each domain, encapsulate the key system analysis concepts that enable us to THINK about, communicate, analyze, and organize *systems*, *products*, and *services* for design and development. With this foundation in place, we are now ready to proceed to *Part II System Design and Development Practices*.

Part II

Systems Design and Development Practices

EXECUTIVE SUMMARY

Part II, *System Design and Development Practices*, builds on the foundation established in Part I *System Analysis Concepts* and consists of 34 chapters organized into six series of practices. The six series consist of:

- System Development Strategies Practices
- System Specification Practices
- System Design and Development Practices
- Decision Support Practices
- System Verification and Validation Practices
- System Deployment, Operations, and Support Practices

As an introductory overview, let's explore a brief synopsis of each of these practices.

System Development Strategy Practices

Successful system development requires establishing an insightful strategy and supporting workflow that employs proven practices to enable a program to efficiently progress from contract award to system delivery and acceptance. The *System Development Strategy Practices*, which consists of Chapter 24–27, provide insights for establishing a program strategy.

Our discussions describe how a program employs verification and validation concepts introduced in Part I to create a workflow that translates multi-level specifications into a physical design solution that leads to delivery of systems, products, or services. We explore various development methods such as the waterfall approach, incremental development, evolutionary development, and spiral development. We also dispel a myth that V & V are only performed after a system has been integrated and tested; V & V are performed continuously from contract award through system delivery and acceptance.

Given an understanding of *System Development Strategy Practices*, we introduce the cornerstone for system design and development via the *System Specification Practices*.

System Specification Practices

System design and development begins with the derivation and development of system specifications and requirements that bound the User's solution space subject to technology, cost, schedule,

---

System Analysis, Design, and Development, by Charles S. Wasson
Copyright © 2006 by John Wiley & Sons, Inc.

248 Part II Systems Design and Development Practices

support, and risk constraints. The *System Specification Series*, which consist of Chapters 28–33, explore what a specification is; types of specifications; how specifications are analyzed and developed; and how specification requirements are analyzed, derived, developed, and reviewed.

The *System Specification Practices* provide the cornerstone for our next topical discussion, *System Design and Development Practices*.

System Design and Development Practices

The design and development of a system requires that the developers establish an in-depth understanding of WHAT the user is attempting to accomplish and select a solution from a set of viable candidates based on decision factors such as technical, technology, support, cost, schedule, and risk.

The *System Design and Development Practices* series consists of Chapters 34–46 and cover a diverse range of system design and development practices. Our discussions include: understanding the operational utility, suitability, effectiveness, and availability requirements; formulation of domain solutions; selection of a system architecture; configuration identification; system interface design; standards and conventions; and design and development documentation.

The *System Design and Development Practices* require timely data to support informed decision making that the RIGHT system solution is selected. This brings us to our next topic, *Decision Support Practices*, which provide the data.

Decision Support Practices

The design and development of integrated sets of system elements requires analytical support to provide data and ensure that the system design balances technical, technology, support, cost, schedule, and risk considerations. The *Decision Support Practices* series, which consist of Chapters 47–52, provide mechanisms that range from analyzes to prototypes and demonstrations to provide timely data and recommendations.

Our discussions address analyses; statistical variation influences on system design; system performance budgets and margins; system reliability, availability, and maintainability; system modeling and simulation; and trade study analysis of alternatives.

System design and development requires on-going integrity assessments to ensure that the system is being designed correctly and will satisfy the user's operational need(s). This brings us to our next topic, *Verification and Validation Practices*, which enable us to assess the integrity of the evolving system design solution.

Verification and Validation Practices

System design and development requires answering two key questions: 1) Is the system being designed and developed RIGHT—in accordance with the contract requirements and 2) Does the system satisfy the user's operational needs? The *Verification and Validation Practices* series, which consist of Chapters 53 through 55, enable the system users, acquirer, and developers to answer these questions from contract award through system delivery and acceptance.

Our discussions explore what verification and validation are; describe the importance of technical reviews to verify and validate the evolving and maturing system design solution; and address how system integration, test, and evaluation plays a key role in performing V & V. We introduce verification methods such as inspection/examination, analysis, test, and demonstration that are available for verifying compliance to specification requirements.

Once a system is verified, validated, and delivered for final acceptance, the user is ready to employ the system to perform organizational missions. This brings us to our next topic, *System Deployment, Operations, and Support Practices*.

System Deployment, Operations, and Support Practices

People often believe that SE and analysis end with system delivery and acceptance by the user; SE continues throughout the operational life of the system, product, or service. The *System Deployment, Operations, and Support Practices* series, which consist of Chapters 56 and 57, provide key insights into system and mission applications and performance that require system analyst and SE assessments, not only for corrective actions to the current system but requirements for future systems and capabilities.

Our discussions explore how a system is deployed including site selection, development, and activation; describe key considerations required for system integration at a site into a higher level system; address how system deficiencies are investigated and form the basis for acquisition requirements for new systems, products, or services; and investigate key engineering considerations that must be translated into specification requirements for new systems.

Chapter 24

System Development Workflow Strategy

24.1 INTRODUCTION

The award of a system development contract to a System Developer or Services Provider signifies the beginning of the System Development Phase. This phase covers all activities required to meet the provisions of the contract, produce the end item deliverable(s), and deploy or distribute the deliverables to the designated contract delivery site.

On contract award, the Offeror transforms itself from a proposal organization to a System Developer or Service Provider organization. This requires the organization to demonstrate that they can competently deliver the proposed system on time and within budget in accordance with the provisions of the contract. This transformation is best captured in a business jest: *"The good news is we won the contract! The bad news is WHAT have we done to ourselves?"*

Our discussion of this phase focuses on *how* a proposed system is developed and delivered to the User. We explore *how* the System Developer or Service Provider evolves the visionary and abstract set of User requirements through the various *stages* of system development to ultimately produce a physical system. The "system" may be country, a space shuttle, a mass mailing service, a trucking company, a hospital, or a symposium. The important point to keep in mind is that the duration of the System Development Phase may last from a few weeks or months to several years.

Author's Note 24.1 *The System Development Phase described here, in conjunction with the System Procurement Phase, may be repeated several times before a final system is fielded. For example, in some domains, the selection of a System Developer may require several sequences of System Development Phase contracts to evolve the system requirements and down select from a field of qualified contractors to one or two contractors. Such is the case with spiral development.*

For a System Service Provider contract, the System Development Phase may be a preparatory time to develop or adapt reusable system operations, processes, and procedures to support the contract's mission, support services for the System Operations Phase. For example, a healthcare insurance provider may win a contract to deliver "outsourced" support services for a corporation's insurance program. The delivered services may be a "tailored" version similar to programs the contractor has administered for other organizations.

Once you have mastered the concepts discussed in this section, you should have a firm understanding of *how* the SE process should be implemented and *how* to manage its implementation.

System Analysis, Design, and Development, by Charles S. Wasson
Copyright © 2006 by John Wiley & Sons, Inc.

252 Chapter 24 System Development Workflow Strategy

What You Should Learn from This Chapter

1. What are the *workflow* steps in system development?
2. What is the *verification and validation* (V&V) strategy for system development?
3. How does the V&V strategy relate to the *system development workflow*?
4. Why is the V and V strategy important?
5. What is the *Developmental Configuration*?
6. When does the *Developmental Configuration* start and end?
7. What is a *first article* system?
8. What is *developmental test and evaluation (DT&E)*?
9. How is DT&E performed?
10. When is DT&E performed during the System Development Phase?
11. Who is responsible for performing DT & E?
12. What is *operational test & evaluation (OT&E)*?
13. When is OT&E performed during the System Development Phase?
14. What is the *objective* of OT&E?
15. Who is responsible for performing OT&E?
16. What is the System Developer's role in OT&E?

Definitions of Key Terms

- **Developmental Test and Evaluation (DT&E)** "Test and evaluation performed to:
  1. Identify potential operational and technological limitations of the alternative concepts and design options being pursued.
  2. Support the identification of cost-performance trade-offs.
  3. Support the identification and description of design risks.
  4. Substantiate that contract technical performance and manufacturing process requirements have been achieved.
  5. Support the decision to certify the system ready for operational test and evaluation." (Source: MIL-HDBK-1908, Section 3.0, Definitions, p. 12)

- **Developmental Configuration** "The contractor's design and associated technical documentation that defines the evolving configuration of a configuration item during development. It is under the developing contractor's configuration control and describes the design definition and implementation. The developmental configuration for a configuration item consists of the contractor's released hardware and software designs and associated technical documentation until establishment of the formal product baseline." (Source: MIL-STD-973 (Canceled), *Configuration Management*, para. 3.30)

- **First Article** "[I]ncludes preproduction models, initial production samples, test samples, first lots, pilot models, and pilot lots; and approval involves testing and evaluating the first article for conformance with specified contract requirements before or in the initial stage of production under a contract." (Source: DSMC—*Test & Evaluation Management Guide*, Appendix B, *Glossary of Test Terminology*)

- **Functional Configuration Audit (FCA)** "An audit conducted to verify that the development of a configuration item has been completed satisfactorily, that the item has achieved the performance and functional characteristics specified in the functional or allocated configuration

identification, and that its operational and support documents are complete and satisfactory." (Source: IEEE 610.12-1990 *Standard Glossary of Software Engineering Terminology*)
- **Independent Test Agency (ITA)** An independent organization employed by the Acquirer to represent the User's interests and evaluate how well the *verified* system satisfies the User's *validated* operational needs under field operating conditions in areas such as *operational utility*, *suitability*, and *effectiveness*.
- **Operational Test and Evaluation (OT&E)** Field test and evaluation activities performed by the User or an Independent Test Agency (ITA) under actual OPERATING ENVIRONMENT conditions to assess the operational utility, suitability, and effectiveness of a system based on validated User operational needs. The activities may include considerations such as training effectiveness, logistics supportability, reliability and maintainability demonstrations, and efficiency.
- **Physical Configuration Audit (PCA)** "An audit conducted to verify that a configuration item, as built, conforms to the technical documentation that defines it." (Source: IEEE 610.12-1990 *Standard Glossary of Software Engineering Terminology*)
- **Quality Record (QR)** A document such as a memo, e-mail, report, analysis, meeting minutes, and action items that serves as *objective evidence* to commemorate a task-based action or event performed.

Phase Objective(s)

The *primary objective* of the System Development Phase is to translate the contract and *System Performance Specification (SPS)* requirements into a *physical*, deliverable system that has been:

1. *Verified* against those requirements.
2. *Validated* by the User, if required.
3. Formally accepted by the User or the Acquirer, as the User's contract and technical representative.

24.2 SYSTEM DEVELOPMENT VERIFICATION AND VALIDATION STRATEGY

During our discussion of *system entity concepts* in Figure 6.2 we explored the basic concept of system *verification* and *validation*. Verification and validation (V&V) provides the basis for a conceptual strategy to ensure the *integrity* of an evolving SE design solution. Let's expand on Figure 6.2 to establish the technical and programmatic foundation for our discussion in this chapter. Figure 24.1 serves as a navigation aid for our discussion for a *closed loop V&V system*.

System Definition and System Procurement Phases V&V

When the User identifies an Operational Need (1), the User may employ the services of an Acquirer to serve as a *contract* and *technical representative* for the procurement action. The operational needs, which may already be documented in a *Mission Needs Statement (MNS)*, are translated by the Acquirer into an *Operational Requirements Document (ORD)* (2) and validated in collaboration with the User. The ORD becomes the basis for the Acquirer to develop a *System Requirements Document (SRD)* (3) or *Statement of Objectives (SOO)*. The SRD/SOO specifies the technical requirements for the formal solicitation—namely the Request for Proposal (RFP)—in an OPEN competition to qualified Offerors.

254 Chapter 24 System Development Workflow Strategy

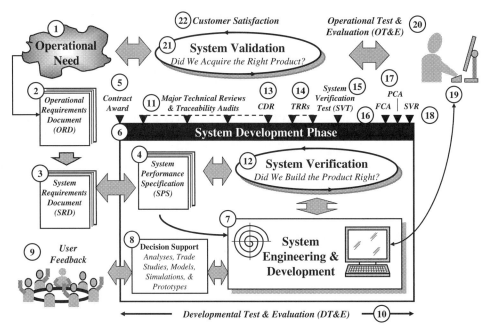

Figure 24.1 System V & V—Programmatic Perspective

Offerors analyze the SRD/SOO, derive and develop a *System Performance Specification (SPS)* (4) from the SRD/SOO (2), and submit the SPS as part of their proposal. When the Acquirer makes a final source selection decision, a System Development Agreement (6) is formally established at the time of contract award (5).

System Development Phase V&V

The SPS (4) provides the technical basis for developing the deliverable system or product via System Engineering and Development (6) activities. Depending on the maturity of the requirements, the System Developer may employ *spiral development* and other strategies to develop the system design solution. In support of the system engineering and development (6) activity, Decision Support (8) performs analyses and trade studies, among other such activities, with inputs and preliminary assessments provided from User Feedback (9), such as *validation* on the implementation of requirements.

As the SE design evolves, System Verification (12) methods are continually applied to assess the requirements *allocation*, *flow down*, and *designs* at all *levels of abstraction*—at the PRODUCT, SUBSYSTEM, ASSEMBLY, SUBASSEMBLY, and PART levels. *Design verification* activities include Developmental Test and Evaluation (DT&E) (10) and Major Technical Reviews and Traceability Audits (11). The purpose of these verification activities is to assess and monitor the *progress*, *maturity*, *integrity* and *risk* of the SE design solution. Baselines are established at critical *staging* or *control points*—using technical reviews—to capture formal baselines of the evolving *Developmental Configuration* to facilitate decision making.

Author's Note 24.2 *Although it isn't explicitly shown in Figure 24.1, validation activities continually occur within the System Developer's program organization. Owners VALIDATE lower level design solution implementations in terms of documented use case-based requirements.*

Design requirements established at the Critical Design Review (CDR) (13) provide the basis for procuring and developing components. As each component is completed, the item is *verified* for compliance to its current *design requirements* baseline.

Successive levels of components progress through levels of System Integration, Test, and Evaluation (SITE) and *verified* against their respective *item development specifications (IDSs)*. Test Readiness Reviews (TRRs) (14) are conducted at various levels of integration to verify that all aspects of a configuration and test environment are ready to commence testing with low risk.

When the SYSTEM level of integration is ready to be *verified* against the SPS (4), a System Verification Test (SVT) (15) is conducted. The SVT must answer the question *"Did we build the system or product RIGHT?"*—in accordance with the SPS (4) requirements.

Following the SVT, a Functional Configuration Audit (FCA) (16) is conducted to *authenticate* the SVT results, via *quality records* (QRs), as compliant with the SPS (4) requirements. The FCA may be followed by a Physical Configuration Audit (PCA) (17) to *authenticate* by physical measurement *compliance* of items against their respective *design requirements*. On completion of the FCA (16) and PCA (17), a System Verification Review (SVR) (18) is conducted to *certify* the results of the FCA and PCA.

Depending on the *terms* and *conditions* (T&Cs) of the System Development Agreement (6), completion of the SVR (18) serves as prerequisite for final system or product *delivery* and *acceptance* (19) by the Acquirer for the User. For some agreements an Operational Test and Evaluation (OT&E) (20) may be required. In preparation for the OT&E (20), the User or an Independent Test Agency (ITA) *representing* the User's interests may be employed to conduct scenario-based field exercises using the system or product under actual or similar OPERATING ENVIRONMENT conditions.

During OT&E (20), Acquirer System Validation (21) activities are conducted to answer the question *"Did we acquire the RIGHT system or product?"*—as documented in the ORD or the SOO, whichever is applicable. Depending on the scope of the contract (5), *corrective actions* may be required during OT&E (20) for any *design flaws*, *latent defects*, *deficiencies*, and the like. Following OT&E (20), the ITA prepares an assessment and recommendations.

Author's Note 24.3 *Although the System Development contract (6) may be complete, the User performs system verification and validation activities continuously throughout the System Development and system Operations and Support (O&S) phases of the system product life cycle. V&V activities expand to encompass organizational and system missions. As competitive and adversarial threats in the OPERATING ENVIRONMENT evolve and maintenance costs increase, "gaps" emerge in achieving organizational and system missions with existing capabilities. These degree of urgency to close these gaps subsequently leads to the next system or product development or upgrade to existing capabilities.*

Now that we have established the V&V strategy of system development, the question is: HOW do we implement it? This brings us to our next topic, *Implementing the system development phase*.

24.3 IMPLEMENTING THE SYSTEM DEVELOPMENT PHASE

The *workflow* during the system development phase consists of five sequential workflow processes as illustrated in Figure 24.2. The processes consist of:

1. System Design Process
2. Component Procurement and Development Process

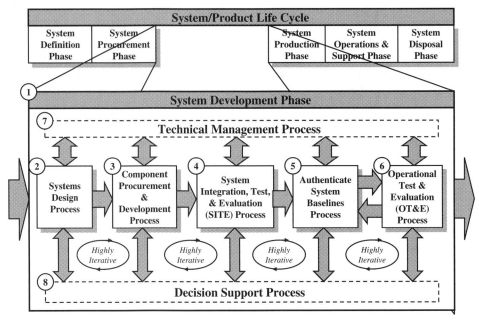

Figure 24.2 The System Development Process Work flow

3. System Integration, Test, and Evaluation (SITE) Process
4. Authenticate System Baseline Process
5. Operational Test, and Evaluation (OT&E) Process

While the general workflow appears to be sequential, there are highly iterative feedback loops that connect to predecessor segments.

The System Development Phase begins at contract award and continues through deliverable system acceptance by the Acquirer and User. During this phase the approach that enabled the System Developer or Service Provider to convince the Acquirer that the organization can perform on the contract is implemented. Remember those brochureware phrases: *well-organized*, *seamless organization*, *highly efficient*, *highly trained* and *performing teams*; no problem, and so on.

The System Development Phase includes those technical activities required to translate the contract performance specifications into a physical system solution. We refer to the initial system(s) as the *first article* of the *Developmental Configuration*. Throughout the phase, the *highly iterative* system design solution evolves through a progression of maturity stages. Each stage of maturity typically consists of a series of design reviews with *entry* and *exit criteria* supported by analyses, prototypes, and technology demonstrations. The reviews culminate in design baselines that capture snapshots of the evolving *Developmental Configuration*. When the system design solution is formally approved at a Critical Design Review (CDR), the *Developmental Configuration* provides the basis for component procurement and/or development.

Procured and developed components are inspected, integrated, and *verified* against their respective *design requirements-drawings-* and *performance specifications* at various levels of integration. The intent of verification is to answer the question: *"Did we develop the system RIGHT?"*—according to the specification requirements. The integration culminates with a System Verification Test (SVT) against the *System Performance Specification (SPS)*. Since the System Development Phase focuses on development of the system, product, or service, testing throughout SVT is referred to as *Developmental Test and Evaluation (DT&E)*.

When the *first article* system(s) of the *Developmental Configuration* has been *verified* as meeting its SPS requirements, one of two options may occur, depending on contract requirements. The system may be deployed to either of the following:

1. Another location for *validation testing* by the User or an Independent Test Agency (ITA) representing the User's interests.
2. The User's designated field site for installation, checkout, integration into the user's Level 0 system, and final acceptance.

Validation testing, which is referred to as *Operational Test and Evaluation (OT&E)*, enables the User to make a determination if they *specified* and *procured* the RIGHT SYSTEM to meet their validated operational needs. Any deficiencies are documented as *discrepancy reports (DRs)* and resolved in accordance with the *terms and conditions (Ts&Cs)* of the contract.

After an initial period of system operational use in the field to correct *latent defects* such as *design flaws*, *errors* and *deficiencies* and collect field data to *validate* system operations, a decision is made to begin the System Production Phase, if applicable. If the User does not intend to place the system or product in production, the Acquirer formally accepts system delivery, thereby initiating the System Operations and Support (O&S) Phase.

Technical Management Process

The technical orchestration of the System Development Phase resides with the Technical Management Process. The *objective* of this process is to plan, staff, direct, and control product-based team activities focused on delivering their assigned items within technical, technology, cost, schedule, and risk constraints.

Decision Support Process

The Decision Support Process supports all aspects of the System Development Phase process decision-making activities. This includes conducting analyses, trade studies, modeling, simulation, testing, and technology demonstrations to collect data to validate models and provide prioritized recommendations to support informed decision making within each of the workflow processes.

Exit to System Production Phase or System Deployment Phase

When the *System Development Phase* is completed, the workflow progresses to the System Production Phase or System Operations and Support (O&S) Phase, whichever is applicable.

Guidepost 24.1 *Based on a description of the System Development Phase workflow processes, let's investigate HOW the V&V strategy is integrated with the workflow progression.*

24.4 APPLYING V&V TO THE SYSTEM DEVELOPMENT WORKFLOW

So far we have introduced the System Development Phase processes and described the sequential workflow. Each of these processes enables the System Developer to accomplish specific objectives such as:

1. Select and mature a design from a set of viable candidate solutions based on an analysis of alternatives.

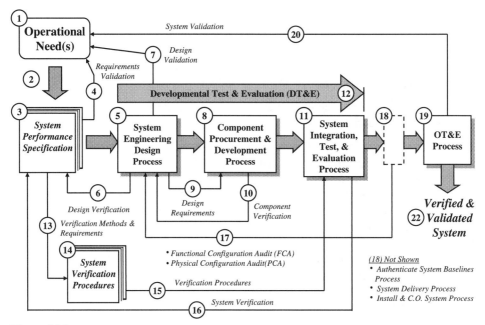

Figure 24.3 Development Process Context System Verification and Validation Concept

2. Procure, fabricate, code, and test PART level components.
3. Perform multi-level system integration, test, and evaluation.
4. Verify items at each level of integration satisfy specification requirements.
5. Validate, if applicable, the integrated SYSTEM as meeting User operational needs.

Until the system is delivered and accepted by the Acquirer, the Developmental Configuration, which captures the various design baselines, is always in a state of evolution. It may require redesign or rework to correct *latent defects*, *deficiencies*, *errors*, and the like. So how do we minimize the impacts of these effects? This brings us to the need for an integrated verification and validation (V&V) strategy. To facilitate our discussion, Figure 24.3 provides a framework.

System Performance Specification (SPS) V&V Strategy

During the System Procurement Phase, *Operational Needs* (1) identified by the User and Acquirer are documented (2) in the *System Performance Specification (SPS)* (3). This is a critical step. The reason is that by this point the Acquirer, in collaboration with the User, has partitioned the organizational *problem space* into one or more *solution spaces*.

Each *solution space* is bounded by requirements specified in its SPS. If there are any errors in tactical or engineering judgment, they *manifest* themselves in the requirements documented in the SPS. Therefore the challenge question for the Acquirer, User, and ultimately the System Developer is: *Have we specified the RIGHT system—solution space—to satisfy one or more operational needs—problem space? How do we answer this question?*

SPS requirements should be subjected to *Requirements Validation* (4) against the Operational Need (1) to validate that the *right solution space* description has been *accurately* and *precisely* bounded by the SPS (3).

Author's Note 24.4 *A word of caution: any discussions with the User and Procurement Team regarding the System Performance Specification (SPS) requirements validation requires tactful professionalism and sensitivity. In effect, you are validating that the Acquirer performed their job correctly. On the one hand they may be grateful for you identified any potential deficiencies in their assessment. Conversely, you may highly offend! Approach any discussions in a tactful, well-conceived, professional manner.*

SE Design V&V Strategy

When the SPS requirements have been *validated*, the SPS (3) serves as *originating* or *source* requirements inputs to system design. During the system design, *Interim Design Verification* (6) is performed on the evolving system design solution by tracing allocated requirements back to the SPS and prototyping design areas for RISK mitigation and *critical operational* or *technical issue* (COI/CTI) resolution. *Design Validation* (7) activities should also be performed to confirm that the User and Acquirer, as the User's technical representative, agree that the evolving System Design Solution satisfies their needs.

Author's Note 24.5 *The Interim Design Verification (6) and Interim Design Validation (7), or design "verification and validation," are considered complete when the system has been verified, validated, and legally accepted by the User via the Acquirer in accordance with the terms of the contract. Therefore the term "interim" is applied.*

Design verification and validation occurs throughout the SE Design Process. *Validation* is accomplished via: 1) technical reviews (e.g., SDR, SSR, PDR, and CDR) and 2) technical demonstrations. Communications media such as conceptual views, sketches, drawings, presentations, technical demonstrations, and/or prototypes are used to obtain Acquirer and User *validation acceptance* and *approval*, as appropriate. On completion of a system level CDR, workflow progresses to Component Procurement and Development (8).

Component Procurement and Development V&V Strategy

During Component Procurement and Development (8), *design requirements* from the System Design (5) serve as the basis for procuring, fabricating, coding, and assembling system components. Each component undergoes *component verification* (10) against its *Design Requirements (5)*. As components are *verified*, workflow progresses to System Integration, Test, and Evaluation (SITE) (11).

System Integration, Test, and Evaluation (SITE) V&V Strategy

When components complete verification, they enter System Integration, Test, and Evaluation (11). Activities performed during this process are often referred to as *developmental test and evaluation* (DT&E) (12). DT&E occurs throughout the entire System Development Phase, from System Design (5) through SITE (11). The purpose of DT&E in this context is to *verify* that the system and its embedded subsystems, HWCIs/CSCIs, assemblies, and parts are *compatible* and *interoperable* with themselves and the system's external interfaces.

To accomplish the SITE Process, *Verification Methods and Requirements (13)* defined in the SPS (3) and multi-level *item development specifications* (IDSs) *are* used to develop *Verification Procedures* (14). *Verification methods*—consisting of *inspection, analysis, test, demonstration, and similarity*—are defined by the SPS for each requirement and used as the basis for *verifying* compliance. One or more *detailed test procedures* (14) that prescribe the test environment configuration—such as the OPERATING ENVIRONMENT (initial and dynamic), data inputs, and expected test results—support each *verification method*.

During SITE (11), the System Developer formally tests the SYSTEM with representatives of the Acquirer and User as witnesses. Multi-level system *verification* activities, at appropriate *integration points* (IPs) review (15) test data and results against the *verification procedures* (14) and expected results specified in the appropriate *development specifications*. When the system completes SITE (11), a formal System Verification Test (SVT) corroborates the system's capabilities and performance against the SPS.

Authenticate System Baselines V&V Strategy—First Pass

When the SVT is completed, workflow progresses to Authenticate System Baselines (18). The process contains two authentication processes that may be performed at different times, depending on contract requirements. The first *authentication* consists of a Functional Configuration Audit (FCA) (17). Using the SPS and other development specification requirements as a basis, the FCA reviews the results of the *As-Designed*, *Built*, and *Verified* system that has just completed the SVT to verify that it fully complies with the SPS functional and performance requirements. The FCA may be conducted at various levels of IPs during the SITE Process.

On successful completion of the FCA, the system may be deployed to a User's test range or site to undergo Operational Test and Evaluation (OT&E) (19). On completion of OT&E, the system may reenter the Authenticate System Baselines process for the second pass.

Validate System Process V&V Strategy

Up to this point, the system is *verified* by SVT and audited by FCA to meet the SPS requirements. The next step is to *validate* that the *As-Designed*, *Built*, and *Verified* system *satisfies* the User's Operational Needs (1) as part of the Operational Test and Evaluation (OT&E) (20) activities.

System validation activities (20) demonstrate *how well* the fielded system performs missions in its *intended* operational environment as originally envisioned by the User. Any system *latent* defects and deficiencies discovered during system *validation* are recorded as problem reports and submitted to the appropriate decision authority for disposition and corrective action, if required.

Author's Note 24.6 *During system validation, a determination is made that an identified deficiency is within the scope of the original contract's SPS. This is a critical issue. For example, did the User or Acquirer overlook a specific capability as an operational need and failed to document it in the SPS? This point reinforces the need to perform a credible Requirements Validation (4) activity prior to or immediately after Contract Award to AVOID surprises during system acceptance. If the deficiency is not within the scope of the contract, the Acquirer may be confronted with modifying the contract and funding additional design implementation and efforts to incorporate changes to correct the deficiency.*

Authenticate System Baselines V&V Strategy—Second Pass

During the *second pass* through Authenticate System Baselines (18), the *As Designed, Built, Verified, and Validated* (20) configuration system is subjected to a Physical Configuration Audit (PCA) (17). PCA *audits* the *As-Designed, Built, and Verified* physical system to determine if it fully complies with its design requirements such as drawings and parts lists. On successful completion of the PCA (17), a System Verification Review (SVR) is conducted to:

1. Certify the results of the FCA and PCA.
2. Resolve any outstanding FCA/PCA issues related to those results.
3. Assess readiness-to-ship decision.

On successful completion of the OT&E Process (19), a *Verified* and *Validated* System (22) should be ready for delivery to the User via formal *acceptance* by the Acquirer.

Guidepost 24.2 *Integrating the V&V strategy into the System Development Phase workflow describes the mechanics of ensuring the evolving Development Configuration is progressing to a plan. However, performing to a plan does not guarantee that the system can be completed successfully on schedule and within budget. The development, especially for large, complex systems, must resolve critical operational and technical issues (COIs/CTIs), each with one or more risks. So how do we mitigate these risks? This brings us to our next topical discussion, the roles of the Developmental Test and Evaluation (DT&E) and the Operational Test and Evaluation (OT&E).*

24.5 RISK MITIGATION WITH DT&E AND OT&E

Satisfactory completion of system development requires that a robust strategy be established up front. We noted earlier that although the workflow appears to be sequential, each process consists of *highly iterative* feedback loops with each other as illustrated by Figure 24.4. To accomplish this, two types of testing occur during the System Development Phase: 1) Developmental Test & Evaluation (DT & E) and 2) Operational Test & Evaluation (OT & E).

Developmental Test and Evaluation (DT&E)

Developmental testing (DT) serves as a risk mitigation approach to ensure that the evolving system design solution, including its components, complies with the *System Performance Specification (SPS)* requirements. DT focuses on two themes:

1. *Are we building the system or product right—meaning using best practices in compliance with the SPS.*
2. *Do we have a design solution that represents the best, acceptable risk solution for a given set of technical, cost, technology, and schedule constraints?*

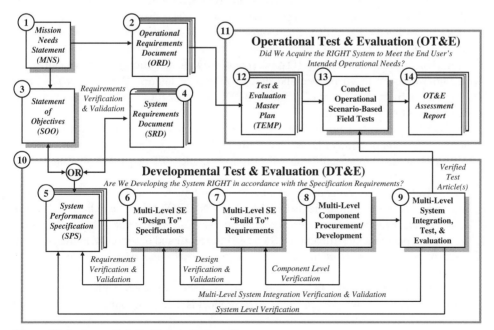

Figure 24.4 System V & V—Technical Perspective

The DSMC *T&E Management Guide* states that the objectives of DT&E are to:

1. Identify potential operational and technological capabilities and limitations of the alternative concepts and design options being pursued;
2. Support the identification of cost-performance tradeoffs by providing analyses of the capabilities and limitations of alternatives;
3. Support the identification and description of design technical risks;
4. Assess progress toward meeting *critical operational issues (COIs)*, mitigation of acquisition technical risk, achievement of manufacturing process requirements and system maturity;
5. Assess validity of assumptions and conclusions from the analysis of alternatives (AOA);
6. Provide data and analysis in support of the decision to certify the system ready for operational test and evaluation (OT&E);
7. In the case of automated information systems, support an information systems security certification prior to processing classified or sensitive data and ensure a standards conformance certification.

(Source: Adapted from *DSMC Test & Evaluation Management Guide*, App. B, p. B-6)

DT&E is performed throughout System Design Process, Component and Procurement Process, and the SITE Process. Each process task verifies that the *evolving* and *maturing* system or product design solution—the *Developmental Configuration*—fully complies with the SPS requirements. This is accomplished via reviews, *proof of principle* and *proof of concept* demonstrations, technology demonstrations, engineering models, simulations, brass boards, and prototypes.

On completion of *verification*, the physical system or product enters OT&E, whereby it is validated against the User's documented operational need.

Operational Test and Evaluation (OT&E)

Operational test and evaluation (OT&E) activities are typically conducted on large, complex systems such as aircraft and military acquirer activity systems. The theme of OT&E is: *Did we acquire the RIGHT system or product to satisfy our operational need(s)*? OT&E consists of subjecting the *test articles* to actual field environmental conditions with operators from the User's organization. An Independent Test Agency (ITA) designated by the Acquirer or User typically conducts this testing. To ensure independence and *avoid* conflicts of interest, the contract *precludes* the System Developer from *direct participation* in OT&E; the System Developer may, however, provide maintenance support, if required.

Since the OT&E is dependent on *how well* the system's Users perform with the new system or product, the ITA or System Developer train the User's personnel to safely *operate* the system. This may occur prior to system deployment following the SVT or on arrival at the OT&E site.

During the OT&E, the ITA trains the User's personnel in how to conduct various operational *use cases* and *scenarios* under actual field OPERATING ENVIRONMENT conditions. The use cases and scenarios are structured to evaluate system *operational utility, suitability, availability,* and *effectiveness*. ITA personnel instrument the SYSTEM to record and observe the human–system interactions and responses. Results of the interactions are *scored, summarized,* and *presented* as recommendations.

On successful completion of the DT&E and OT&E and the follow-on Authenticate System Baselines Process, the *verified* and *validated* system or product is delivered to the Acquirer or User for final acceptance.

24.6 GUIDING PRINCIPLES

In summary, the preceding discussions provide the basis with which to establish the guiding principles that govern system development workflow strategy practices.

Principle 24.1 A system development strategy must have three elements:

1. A strategy-based roadmap to get from Contract Award to system delivery and acceptance supported by incremental verification and validation.
2. A plan of action for implementing the strategy.
3. Documented objective evidence that you performed to the plan via work product quality records.

Principle 24.2 System verification and validation applies to every stage of product development workflow beginning at Contract Award and continuing until system delivery and acceptance.

Principle 24.3 Developmental test and evaluation (DT&E) is performed by the System Developer to mitigate Developmental Configuration risks; Users employ the operational test and evaluation (OT&E) to determine if they acquired the right system.

24.7 SUMMARY

During our discussion of the system development workflow strategy we introduced the system development phase processes. The System Development Phase processes include:

1. System Design
2. Component Procurement and Development
3. System Integration, Test, and Evaluation (SITE)
4. Authenticate System Baseline
5. Operational Test, and Evaluation (OT&E)

Based on the System Development Phase processes, we described an overall workflow strategy for verification and validation. This strategy provides the high-level framework for transforming a User's *validated* operational need into a deliverable system, product, or service.

We introduced the concepts of developmental test and evaluation (DT&E) and operational test and evaluation (OT&E). Our discussion covered how DT&E and OT&E serve as key verification and validation activities and their relationship to the system development workflow strategy.

GENERAL EXERCISES

1. Answer each of the *What You Should Learn from This Chapter* questions identified in the Introduction.
2. Using a system listed in Table 2.1, develop a description of the activities for each System Development Phase process to be employed and integrated into an overall V & V strategy.

ORGANIZATIONAL CENTRIC EXERCISES

1. Research your organization's command media for guidance and direction in implementing the *System Development Phase* from an SE perspective.
   (a) What requirements are levied on SE contributions?

(b) What overall process is required and how do SEs contribute?

(c) What SE work products and quality records are required?

(d) What verification and validation activities are required?

2. Contact a small, medium, and a large contract program within your organization. Interview the Technical Director or Project Engineer to identify the following information:

   (a) Request the individual to graphically depict their development strategy?

   (b) What factors drove them to choose the implementation strategy?

   (c) What were some of the lessons learned from developing and implementing the strategy that would influence their approach next time?

   (d) How was the V & V strategy implemented?

REFERENCES

Defense Systems Management College (DSMC). 1998. DSMC *Test and Evaluation Management Guide*, 3rd ed. Defense Acquisition Press Ft. Belvoir, VA.

IEEE Std 610.12-1990. 1990. *IEEE Standard Glossary of Modeling and Simulation Terminology*. Institute of Electrical and Electronic Engineers (IEEE) New York, NY.

MIL-STD-973. 1992. Military Standard: *Configuration Management*. Washington, DC: Department of Defense (DoD).

MIL-HDBK-1908B. 1999. *DoD Definitions of Human Factors Terms*. Washington, DC: Department of Defense (DOD).

ADDITIONAL READING

Defense Systems Management College (DSMC). 2001. *Glossary: Defense Acquisition Acronyms and Terms, 10th ed.* Defense Acquisition University Press Ft. Belvoir, VA.

MIL-HDBK-470A. 1997. *Designing and Developing Maintainable Products and Systems.* Washington, DC: Department of Defense (DoD).

MIL-STD-1521B (canceled). 1985. Military Standard: *Technical Reviews and Audits for Systems, Equipments, and Computer Software.* Washington, DC: Department of Defense (DoD).

ASD-100 Architecture and System Engineering. 2003. *National Air Space System—Systems Engineering Manual.* Washington, DC: Federal Aviation Administration (FAA).

Defense Systems Management College (DSMC). 2001. *Systems Engineering Fundamentals.* Defense Acquisition University Press Ft. Belvoir, VA.

Chapter 25

System Design, Integration, and Verification Strategy

25.1 INTRODUCTION

Our discussion of the *system development workflow strategy* established a sequence of highly interdependent processes that depict the workflow for *translating* the *System Performance Specification (SPS)* into a system design solution. The strategy provides a *frame of reference* to:

1. *Verify* compliance with the SPS requirements.
2. *Validate* that the deliverable system satisfies the User's validated operational needs.

The workflow strategy identified two key processes that form the basis for designing and developing a system: the System Design Process and the System Integration, Test, and Evaluation (SITE) Process.

This section focuses on the strategies for implementing the System Design Process and the SITE Process. Each strategy expands each process via lower level processes. Finally, we integrate the two strategies into an overall strategy referred to as the V-Model for system deployment.

What You Should Learn from This Chapter

1. What is the basic *strategy* for implementing the System Design Process of the System Development Phase?
2. What is the basic *strategy* for implementing the SITE Process of the System Development Phase?
3. What is the *V-Model* of system design and development?
4. What is an *integration point*?
5. How do the system design process strategy and the SITE Process integrate?

Definitions of Key Terms

- **Corrective Action** The set of tasks required to correct or rescope specification contents, errors, or omissions; designs flaws or errors; rework components due to poor workmanship, defective materials or parts; or correct errors or omissions in test procedures.
- **Discrepancy Report (DR)** A report that identifies a condition in which a document or test results indicate a noncompliance with a capability and performance requirement specified in a *performance* or *item development specification*.

System Analysis, Design, and Development, by Charles S. Wasson
Copyright © 2006 by John Wiley & Sons, Inc.

- **Integration Point** Any one of a number of confluence points during the System Integration, Test, and Evaluation (SITE) Process where two or more entities are integrated.
- **V-Model** A graphical model that illustrates the time-based, *multi-level* strategy for (1) decomposing specification requirements, (2) procuring and developing physical components, and (3) integrating, testing, evaluating, and verifying each set of integrated components.

25.2 SYSTEM DESIGN PROCESS STRATEGY

The System Design Process of the System Development Phase employs a *highly iterative*, top-down/bottom-up/lateral process. Since the process requires *analysis* and *decomposition/expansion* of *abstract*, *high-level* SPS requirements into lower levels of detail for managing complexity, each lower level design activity subsequently occurs in later time increments. Figure 25.1 illustrates this sequencing. Each horizontal bar in the figure:

1. Represents design activities that may range from very small to very large level of efforts (LOEs) over time.
2. Includes preliminary activities that ramp up with time.

This graphic is representative of most development programs. Some programs may be *unprecedented* and require more maturity at higher levels until lower level design activities are initiated as evidenced by the white, horizontal bars. Other systems may be *precedented* and *reuse* some or most of an existing design solution. Thus preliminary design activities at all levels may be initiated with a small effort at Contract Award and ramp up over time. Given this backdrop, let's describe the strategy for creating the multi-level design solution.

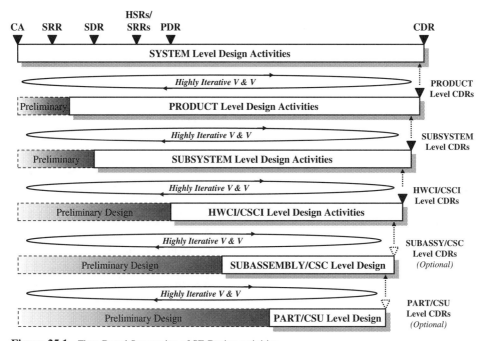

Figure 25.1 Time-Based Sequencing of SE Design activities

Guidepost 25.1 *At this point we have established the theoretical approach for performing system design. This brings us to the next topic, implementing the System Design Process.*

25.3 IMPLEMENTATION OF THE SYSTEM DESIGN PROCESS

The initial implementation of the System Design Process occurs during the System Procurement Phase of the system/product life cycle. During the System Procurement Phase, the System Developer proposes *solution* responses to the Acquirer's formal *Request for Proposal (RFP)* solicitation.

When the System Development Phase begins at Contact Award (CA), the System Design Process is repeated to develop refinements in the PROPOSED system design solution. These refinements are implemented to accommodate requirements changes that may have occurred as part of the final contract negotiation and maturation of the proposed design solution.

The actual implementation involves several design approaches. Among these approaches are Waterfall, Incremental Development, and Spiral Development which are introduced in a later chapter. The approach selected depends on criteria such as:

1. Level of understanding of the *problem* and *solution spaces*.
2. The *maturity* of the SPS requirements.
3. Level of risk.
4. *Critical operational* or *technical issues* (COIs/CTIs).

Some people may categorize Figure 25.1 as the Waterfall Model for system development. While it may appear to resemble a waterfall, it is not a true waterfall in design. The Waterfall Model *presumes* each level of design to be completed just before the next level is initiated (as illustrated in Figure 27.1). The *fallacy* with the Waterfall approach is:

- You must perform lower level analysis and preliminary design to be able to understand the requirements decisions at a higher level and their lower level ramifications.
- As a typical entry criterion for the Critical Design Review (CDR), the total system design is expected to have a reasonable level of maturity sufficient to commit component procurement and development resources with *acceptable risk*. Remember, at CDR most *system designs solutions* are UNPROVEN—that is, as the fully integrated system. Therefore, the *system level design solution* is NOT considered *officially* complete UNTIL the system has been formally accepted in accordance with the *terms and conditions* (T&Cs) of the contract.

As Figure 25.1 illustrates, each level of design occurs *concurrently* at various levels of effort throughout the System Design Process. Since every design level incrementally evolves to maturity over time toward CDR, the level *of activity* of each design activity bar diminishes toward CDR. By CDR, the quantity of requirements and performance allocation changes should have *diminished* and *stabilized* as *objective evidence* of the *maturity*.

Concurrent, Multi-level Design Activities

Each activity (bar) in Figure 25.1 consists of a shaded Preliminary Design segment followed by a Design Activity segment.

Preliminary Design segments are shaded from left to right, with darker shading to represent *increasing* level of effort (LOE) work activity. These preliminary design activities may involve analyses, modeling, and simulation to investigate lower level design issues to support higher level decision making by one or more individuals on a part-time or full-time basis. Consider the following example:

EXAMPLE 25.1

An Integrated Product Team (IPT) may be tasked or subcontractor contracted to perform preliminary analysis and design with the *understanding* that a task or contract *option* may be to shift work activities at any level from feasibility studies to actual design activities.

The distinction between Preliminary Design (shaded) and Design Activities is graphically separated for discussion purposes. Within System Developer organizations, there is no break between segments—just an expansion in LOE from one or two people to five or ten.

Design Maturation Reviews

Throughout the System Design Process, program *formal* and *informal* reviews are conducted to assess the *maturity*, *completeness*, *consistency*, and *risk* of the *evolving* system design solution.

Referral A detailed description of these reviews appears in Chapter 54 *Technical Review Practices* used in the *verification and validation* phase of the design process.

Guidepost 25.2 Given an overview of the *system design strategy*, let's shift our focus to the *system integration, test, and evaluation (SITE) strategy*.

25.4 IMPLEMENTING THE SYSTEM DESIGN PROCESS STRATEGY

To illustrate HOW the system design activities are implemented, consider the example illustrated in Figure 25.2. In the figure the SPS (1) is analyzed, a SYSTEM level Engineering Design (3) is selected, and requirements are *allocated* (2) to PRODUCTS A and B. SPS requirements allocated to PRODUCT B are *flowed down* (5) and captured via the *PRODUCT B Development Specification* (7). *PRODUCT B Development Specification* requirements are then traced (6) back to the *source* or *originating* requirements of the SPS.

To fully understand the implications of requirements allocated to PRODUCT B, a design team initiates PRODUCT B's Engineering Design (9). PRODUCT B's Engineering Design (9) is selected from a set of viable candidates. *PRODUCT B's Development Specification* requirements are then *allocated* (8) to SUBSYSTEMS B1 and B2.

At each level, *formal* and *informal* technical review(s) are conducted to verify, (4), (10), (16), and (22) that the *evolving* designs *comply* with the requirements allocated from their respective specifications. The process *repeats* until all levels of abstraction have been *expanded* into detailed designs for review and approval at the SYSTEM level CDR.

Guidepost 25.3 *At this point we have investigated the strategy that depicts how the SPS is translated into a detailed design for approval at the CDR. Next we explore the SITE strategy that demonstrates that the various levels of integrated, physical components satisfy their specification requirements.*

25.5 SYSTEM INTEGRATION, TEST, AND EVALUATION (SITE) STRATEGY

The *system design strategy* is based on a hierarchical decomposition framework that partitions a complex system *solution space* into lower level *solution spaces* until the PART level is reached.

25.5 System Integration, Test, and Evaluation (SITE) Strategy

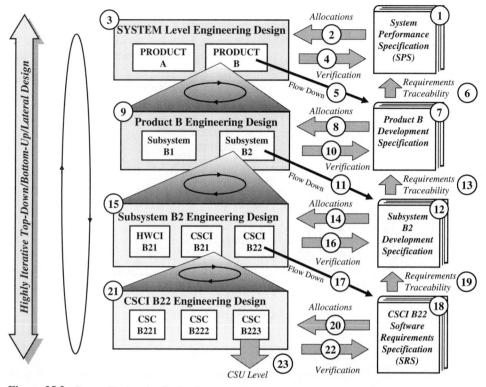

Figure 25.2 System Engineering Design Strategy

For system integration we reverse this strategy by integrating the PART level solutions into higher levels of complexity. Thus we establish the fundamental strategy for SITE.

The SITE Process is implemented by integrating into successively higher levels physical components that have been *verified* at a given level of abstraction. We refer to each integration *node* as an Integration Point (IP). Figure 25.3 provides an illustrative example of the time-based graphical sequence of our discussion.

Multi-level Integration and Verification Activities

Suppose that we have a system that consists of multiple levels of abstraction. Physical hardware PARTS and/or computer software units (CSUs) that have been *verified* are integrated into higher level hardware SUBASSEMBLIES and/or computer software components (CSCs). Each SUBASSEMBLY/CSC is then formally *verified* for compliance to its respective *specification* or *design* requirements. The process continues until SYSTEM level integration and verification is *completed* via a formal System Verification Test (SVT).

Applying Verification Methods to SITE

Each integration step employs INSPECTION, ANALYSIS, DEMONSTRATION, or TEST *verification methods* prescribed by the respective *development specification*. Verification methods are implemented via *acceptance test procedures (ATPs)* formally approved by the program and Acquirer, if applicable, and maintained under program CM control. Representatives from the System Developer, the Acquirer, to the User organizations, as appropriate, participate in the formal event and *witness* verification of each requirement.

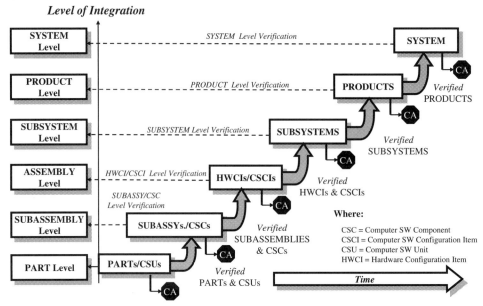

Figure 25.3 System Integration, Test, and Evaluation (SITE) Sequencing

Author's Note 25.1 *The level of formality required for witnessing verification events varies by contract and organization. For some systems, certified System Developer testers at lower levels may be permitted to verify some components without a quality assurance (QA) witness. At higher levels the Acquirer may elect to participate and invite the User. Consult your contract and organization's policies and protocols for specific guidance. For some systems, critical technologies may require involvement of all parties in formal verification events at lower levels.*

Correcting Design Flaws, Errors, Defects, and Deficiencies

If *discrepancies* between *actual* results and *expected* results occur, *corrective actions* are initiated. Consider the following example:

EXAMPLE 25.2

Corrective actions include:

1. Modification or updating of the specification requirements.
2. Redesign of components.
3. Design changes or error corrections.
4. Replacement of defective parts or materials, etc.
5. Workmanship corrections.
6. Retraining of certified testers, etc.

When the verification results have been approved and all *critical Discrepancy Reports (DRs)* are closed, the item is then integrated at the next *higher* level. In our example, the next level consists of hardware configuration items (HWCIs) and computer software configuration items (CSCIs).

25.6 IMPLEMENTING THE SITE PROCESS STRATEGY

Our previous discussions presented the top-down, multi-level, specification driven, SE Design Process. In this chapter we present a bottom-up, "mirror image" discussion regarding HOW the multi-level SITE Process and *system verification* are accomplished.

Our discussion of the system integration, test, and verification (SITE) strategy uses Figure 25.4 as a reference. Let's begin by highlighting key areas of the figure.

- The left side of the chart depicts the multi-level specifications used by the System Engineering (SE) Design Process to create the system design solution.
- The right side of the chart depicts a hierarchical structure that represents *how* system components at various levels begin as verified hardware PARTS or computer software units (CSUs) are integrated to form the SYSTEM at the top of the structure.
- The horizontal gray arrows between these two columns represent the linking of verification methods, acceptance test procedures (ATPs), and verification results at each level.

Author's Note 25.2   *As previously discussed in the SE Design Process, the contents of this chart serve as an illustrative example for discussion purposes. Attempting to illustrate all the combina-*

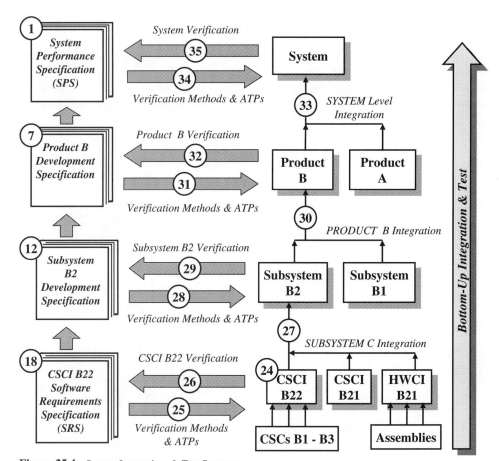

Figure 25.4   System Integration, & Test Strategy

272 Chapter 25 System Design, Integration, and Verification Strategy

tions and permutations for every conceivable system in a single chart can be confusing and impractical. You, as a practicing Systems Engineer, need to employ your own mental skills to apply this concept to your own business domain, systems, and applications.

Guidepost 25.4 *The preceding discussions focused on the individual system design and SITE strategies. Next we integrate these strategies into a model.*

25.7 INTEGRATING SYSTEM DESIGN AND DEVELOPMENT STRATEGIES

Although the system design and SITE strategies represent the overall system development workflow, the progression has numerous *feedback loops* to perform corrective actions for design flaws, errors, and deficiencies. As such, the two strategies need to be integrated to form an overall strategy that enables us to address the feedback loops. If we integrate Figures 25.1 and 25.3, Figure 25.5 emerges and forms what is referred to as the V-Model of system development.

The V-Model is a *pseudo time-based* model. In general, workflow progresses from *left* to *right* over time. However, the *highly iterative* characteristic of the System Design Process strategy and *verification corrective action* aspects of SITE may require returning to a preceding step. Recall from above that the *corrective actions* might involve a re-working of lower level specifications, designs, and components. So, as the corrective actions are implemented over time, workflow does progress from *left* to *right* to delivery and acceptance of the system.

Author's Note 25.3 *This point illustrates WHERE and HOW system development programs become "bottlenecked," consuming resources without making earned value work progress because of re-work. It also reinforces the importance of investing in up-front SE as a means of minimizing and mitigating re-work risks! Despite all of the rhetoric by local heroes that SE is a non–value-*

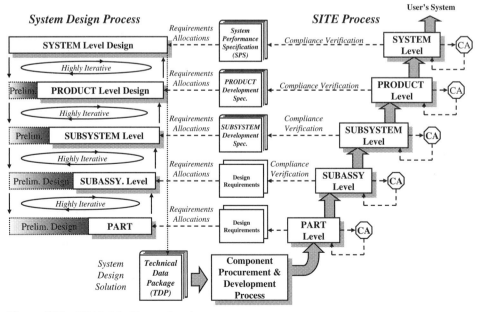

Figure 25.5 "V" Model of System Development

added activity, SITE exemplifies WHY homegrown ad hoc *engineering efforts falter and program cost and schedule performance reflects it.*

Final Thought

Although we have not covered it in this chapter, some programs begin work from a very abstract Statement of Objectives (SOO) rather than an SPS. Where this is the case, *spiral development* is employed to reiterate the V-Model for incremental builds intended to mature knowledge about the SYSTEM requirements. We will discuss this topic in Chapter 27 on *system development* Models.

25.8 GUIDING PRINCIPLES

In summary, the preceding discussions provide the basis with which to establish the guiding principles that govern system design, integration, and verification strategy practices.

Principle 25.1 System design is a highly iterative, collaborative, and multi-level process with each level dependent on maturation of higher level specification and design decisions.

Principle 25.2 A system design solution is not contractually complete until it is verified as compliant with its Acquirer approved System Performance Specification (SPS). Technically it is not complete until all latent defects are removed, but most systems exist between these two extremes.

Principle 25.3 The number of latent defects in the fielded system is a function of the thoroughness of the effort—time and resources—spent on system integration, test, and evaluation (SITE).

25.9 SUMMARY

During our discussion of the *system design, integration, and verification strategy*, we described the SE design strategy that analyzes, allocates, and flows down *System Performance Specification (SPS)* requirements through multiple levels of abstraction to various *item development specifications* and item architectural designs. Next we described a strategy for integrating each of the procured or developed items into successively higher levels of integration. At each level each *item's* capabilities and levels of performance are to be *verified* against their respective specifications.

1. The SE process strategy provides a multi-level model for *allocating* and *flowing down* SPS requirements to lower levels of abstraction.
2. Unlike the Waterfall Model, the SE Process strategy *accommodates* simultaneous, multi-level design activities including Preliminary Design activities.
3. Whereas a design at any level may be formally *baselined* to promote stability for lower level decision making, a design at any level is still subject to formal change management modification through the end of formal acceptance for delivery.
4. The SE design strategy includes multiple *control points* to verify and validate decisions prior to commitment to the next level of design activities.
5. The SITE Process implements a strategy that enables us to integrate and verify lower level components into successively higher levels until the system is fully assembled and verified.
6. The SITE Process strategy includes breakout points to implement corrective actions that often lead back to the SE Design Process.
7. *Corrective actions* may require revision of lower level specifications, redesign, rework of components, or retraining of test operators to correct for design flaws and errors, deficiencies, discrepancies, etc.

274 Chapter 25 System Design, Integration, and Verification Strategy

8. The integration of the System Design Process and SITE Process strategies produces what we refer to as the V-Model of system development.
9. The V-Model, which represents a common model used on many programs, may be performed numerous times, especially in situations such as spiral development to evolve requirements to maturity.

GENERAL EXERCISES

1. Answer each of the *What You Should Learn from This Chapter* questions identified in the *Introduction*.

ORGANIZATIONAL CENTRIC EXERCISES

1. Research your organization's command media for direction and guidance in developing SE design and system integration, test, and evaluation (SITE) strategies. Report your findings.
2. Contact a small, medium, and a large program within your organization. Interview the Lead SE or Technical Director to understand what strategies the program used for SE design and system integration, test, and evaluation (SITE) and sketch a graphic of the strategy. For each type of program:
   (a) How did the strategy prove to be the right decision.
   (b) How would they tailor the strategy next time to improve its performance?

Chapter 26

The SE Process Model

26.1 INTRODUCTION

If you were to survey organizations to learn about what methods they employ to develop systems, products, or services, the responses would range from *ad hoc* methods to *logic-based* methodologies. Humans, by nature, generally deplore structured methods and will naively go to great lengths to avoid them without understanding: 1) WHY they exist and 2) HOW they benefit from them. While ad hoc methods may be proved successful on simple, small systems and products, the *scalability* of these methods to large, complex programs employing dozens or hundreds of people results in *chaos* and *disorder*. So, the question is: *Does a simple methodology exist that is scalable and can be applied for all size programs*?

At the beginning of Part II, we introduced the basic workflow that System Developers employ to transform the abstract *System Performance Specification (SPS)* into a deliverable, physical system undergo. Although we described that workflow in terms of its six processes, the workflow does not capture HOW TO create the system, product, or service. Only how it evolves like a production line from *conceptualization* to *delivery* over time.

Our introduction of the *system solution domains* at the conclusion of Part I presented the Requirements, Operations, Behavioral, and Physical Domain Solutions, their sequential development, and interrelationships. The system solution domains enable us to describe the key elements of a system, product, or service solution. So the challenge question is: *HOW do we create a logical method that enables us to*:

1. *Develop a system, product, or service?*
2. *Apply it across all system development phase workflow processes?*

This chapter introduces the SE Process Model, its underlying methodology, and application to developing an SE design for a system or one of its components. Our discussion begins with a graphical depiction and accompanying descriptions of the SE Process Model and its methodology. Since the model is described by two characteristics: *highly iterative* and *recursive*, we illustrate how the model's internal elements iterate and show how the model applies to multiple levels of abstraction within the system design process. We provide an example of HOW the model applies to entities at various levels of abstraction.

Finally, we illustrate HOW the application of the model produces an integrated framework that represents the multi-level system design workflow progression via the Requirements, Operations, Behavioral, and Physical Domain Solutions. The last point illustrates, by definition, a system composed of *integrated* elements *synergistically* working to achieve a *purpose* greater than their individual purpose-focused capabilities.

System Analysis, Design, and Development, by Charles S. Wasson
Copyright © 2006 by John Wiley & Sons, Inc.

What You Should Learn from This Chapter

1. What is the *SE Process Model*?
2. What are the *key elements* of the SE Process Model?
3. How do the elements of the SE Process Model interrelate?
4. What are the steps of the underlying *SE Process Model methodology*?
5. What is meant by the Process Model's *highly iterative* characteristic?
6. What is meant by the Process Model's *recursive* characteristic?

Definitions of Key Terms

- **Behavioral Domain Solution** A technical design that:
  1. Represents the proposed logical/functional solution to a specification.
  2. Describes the entity relationships between an entity's logical/functional capabilities including external and internal interface definitions.
  3. Provides traceability between specification requirements and logical/functional capabilities.
  4. Is traceable to the multi-level Physical Domain Solution via physical configuration items (PCIs).
- **Iterative Characteristic** A characterization of the interactions between each of the SE Process Model's elements as an entity's design solution evolves to maturity.
- **Operations Domain Solution** A unique view of a system that expresses HOW the System Developer, in collaboration with the User and Acquirer, envision deploying, operating, supporting, and disposing of the system to satisfy a *solution space* mission objectives and, if applicable, safely return the system to a home base or port.
- **Physical Domain Solution** A technical design that: 1) represents the proposed physical solution to a specification, 2) describes the entity relationships between hierarchical, *physical configuration items* (PCIs)—namely the physical components of a system—including external and internal interface definitions, 3) provides traceability between specification requirements and PCIs, and 4) is traceable to the multi-level Behavioral design solution via *functional configuration items* (FCIs).
- **Recursive Characteristic** An attribute of the SE Process Model that enables it to be applied to any system or entity within a system regardless of level of abstraction.
- **Requirements Domain Solution** A unique view of a system that expresses: 1) the hierarchical framework of specifications—namely a specification tree—and requirements, 2) requirements to capability linkages, and 3) *vertical* and *horizontal* traceability linkages to User *originating* or *source requirements* and between specifications and their respective requirements.
- **SE Process Model** A construct derived from a *highly iterative*, problem solving-solution development methodology that can be applied *recursively* to multiple levels of system design.
- **Solution Domain** A requirements, operations, behavioral, or physical viewpoint of the development of a multi-level entity or configuration item (CI) required to translate and elaborate a set of User requirements into a deliverable system, product, or service.

26.2 SE PROCESS MODEL OBJECTIVE

The objective of the SE Process Model is enable SEs to *transform* and *evolve* a User's abstract operational need(s) into a physical system design solution that *represents* the *optimal* balance of technical, technology, cost, schedule, and support solutions and risks.

Brief Background on SE Processes

Since World War II several types of SE processes have evolved. Organizations such as the US Department of Defense (DoD), the Institute of Electrical and Electronic Engineers (IEEE), Electronics Industries Alliance (EIA), and the International Council of System Engineers (INCOSE) have documented a series of SE process methodologies. The more recent publications include US Army FM 770-78, Mil-Std-499, commercial standards IEEE 1220-1998, EIA-632 and ISO/IEC 15288. Each of these SE process methodologies highlights the aspects its developers considered fundamental to engineering practice.

Although the SE processes noted above advanced the state of the practice in SE, in the author's opinion, no single SE process captures the actual steps performed in engineering a system, product, or service. As is the case with a recipe, SEs and organizations often formulate their own variations of how they view the SE process based on what works for them. This chapter introduces an SE Process Model validated through the experiences of the author and others. Note the two terms: *process* and *model*.

In this chapter's Introduction, we considered a general workflow progression that translates a User's abstract operational needs into a physical system, product, or service solution. We can state this progression to be a *process*. However, the workflow process steps required to engineer systems must involve *highly iterative feedback loops* to preceding steps for reconciliation actions. The SE Process is more than simply a sequential end-to-end process. The SE Process is an embedded element of a *problem-solving/solution-development model* that transforms a set of inputs and operating constraints into a deliverable system, product, or service. Therefore, we apply the label SE Process Model.

Entry Criteria

Entry criteria for the SE process are established by the system/product life cycle phase that implements the SE Process. In the case of the System Development Phase, the SE Process Model is applied with the initiation of each *entity* or configuration item (CI's) SE design. This includes the SYSTEM, PRODUCT, SUBSYSTEM, ASSEMBLY, SUBASSEMBLY, and PART levels.

26.3 SE PROCESS MODEL METHODOLOGY

We concluded Part I with an introduction to the *system solution domains*, consisting of Requirements, Operations, Behavioral, and Physical Domain Solutions. Although the domain solutions provide a useful means to characterize a system or one of its entities, individually, they do not help us create the total system solution. So, *how do we do this*?

We can solve this challenge by creating a system development model that enables us to translate the User's vision into a preferred solution. However, the domain solutions are missing two key elements:

1. Understanding the *opportunity/problem space* and the relationship of the *solution space*.
2. Optimizing the domain solutions to achieve mission objectives.

278 Chapter 26 The SE Process Model

If we integrate these missing elements with the sequencing of the system solution domains, we can create a methodology that enables us to apply it to any entity, regardless of level of abstraction. The steps of the methodology are:

Step 1: Understand the entity's opportunity/problem and solution spaces.
Step 2: Develop the entity's Requirements Domain Solution.
Step 3: Develop the entity's Operations Domain Solution.
Step 4: Develop the entity's Behavioral Domain Solution.
Step 5: Develop the entity's Physical Domain Solution.
Step 6: Evaluate and optimize the entity's total design solution.

When we depict these steps, their initial sequencing, and interrelationships graphically, Figure 26.1 emerges. We will refer to this as the SE Process Model.

Before we proceed with describing the SE Process Model, let's preface our discussion with several key points:

1. The description uses the term, *entity*, to denote a *logical/functional* capability or physical item such as PRODUCT, SUBSYSTEM, ASSEMBLY, SUBASSEMBLY, and PART. You could apply the term, *component*. However, there may be some *unprecedented* systems in which physical components may not emerge until later in a design process. Therefore, we use the term *entity*.

2. The act of partitioning a *problem space* into lower level *solution spaces* is traditionally referred to in SE as *decomposition*. Since decomposition connotes various meanings, some people prefer to use the term *expansion*.

3. Since the model *applies* to any *level of abstraction*, role-based terms such as Acquirer, User, and System Developer are contextual. For example, the Acquirer (role) of a system con-

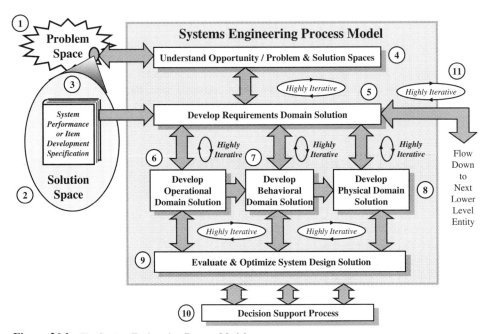

Figure 26.1 The System Engineering Process Model

tracts with a System Developer (role). Within the contractor's (System Developer role) program organization, a PRODUCT level team (Acquirer role) allocates requirements to Development Team (System Developer role) to develop and deliver a SUBSYSTEM. The SUBSYSTEM development team, acting as an Acquirer (role), may procure some of the SUBSYSTEM's components from various vendors (System Developer roles).

4. For simplicity, the description addresses a one-time procurement for a system, product, or service. Some complex system development efforts may consist of a series *spiral development* deliverables or contracts. In these cases, the requirements may be initially *immature* thereby necessitating *stages* of development to *mature* the requirements to a level *necessary* and *sufficient* for final system development. Thus the SE Process Model is reapplied to all levels of abstraction for each iteration of the spiral. These approaches serve to *reduce* system development *risk*.

Step 1: Understand the Entity's Opportunity/Problem and Solution Spaces

The first step of the SE Process Model is to simply understand the *entity's* opportunity/problem and solution spaces. System analysts and SEs need to understand and validate: *HOW the User intends to use the system as well as WHAT expectations are levied on the entity to achieve higher-level mission objectives*. This requires understanding:

1. The *entity's* contextual role in the next higher level *solution space*—namely the User's level 0 system or the deliverable SYSTEM, PRODUCT, SUBSYSTEM, SUBASSEMBLY, and other levels.
2. HOW the User plans to *deploy*, *operate*, *support*, and *dispose* of the system—namely *use cases* and *scenarios*.
3. The system's interfaces with external systems in its OPERATING ENVIRONMENT such as HUMAN-MADE, NATURAL, and INDUCED ENVIRONMENTS.
4. The system's mission event timeline (MET) or allocations.
5. The expected *outcomes* from system interactions with its OPERATING ENVIRONMENT.
6. Products, by-products, and services comprising the system outputs required to accomplish those outcomes.

Step 2: Develop the Entity's Requirements Domain Solution

As the understanding of the entity's opportunity/problem and solution spaces *evolves* and *matures*, the next step is to *Develop the Requirements Domain Solution*. The requirements, which *specify* and *bound* the entity's *solution space*, document the Acquirer's (role) required system capabilities via *Statement of Objectives (SOO)* or *System Performance Specification (SPS)*. Within the System Developer's program, lower level PRODUCT, SUBSYSTEM, ASSEMBLY, or SUBASSEMBLY *item development specification (IDS)*, as applicable, capture the entity's requirements.

As illustrated by the SE Process Model in Figure 26.1, the entity's *Requirements Domain Solution*:

1. Serves as the frame of reference for deriving the *Operations, Behavioral, and Physical Domain Solutions*.
2. Iterates with *understand the opportunity/problem and solution spaces*.
3. Iterates with the *Operations, Behavioral*, and *Physical Domain Solutions*.

4. Integrates with a higher level User, SYSTEM, PRODUCT, ASSEMBLY, etc., level *Requirements Domain Solution* (Figure 26.6).
5. May be expanded or decomposed into at least two or more lower level entity *Requirements Domain Solutions* and documented in *development specifications*.
6. Provides the decision criteria used by the *Evaluate and Optimize the System Design Solution* step to assess design compliance (e.g., verification, consistency, and traceability).

The *Requirements Domain Solution* consists of a hierarchical set of requirements derived from a User's *source* or *originating* requirements, typically in a contract *System Performance Specification (SPS)*. Any entity requirement that is:

1. Too abstract and broad.
2. Complicates *implementation* and *verification*, is simplified by *decomposing* it into two or more lower level SIBLING or DERIVED requirements.

Derived requirements more explicitly define WHAT is required and HOW WELL. Thus they become more manageable and eliminate the possibility of multiple interpretations. As a result each *derived* requirement *simplifies* and *clarifies* WHAT must be accomplished to satisfy a portion of the higher *parent* requirements.

Since requirements express User (role) expectations for acceptance of each entity, at a *minimum*, each requirement is assigned a *verification method* such as inspection, analysis, demonstration, or test. Additional verification criteria such as the level where a specific requirement will be verified and verification conditions may also be added. The set of verification *methods*, *levels*, and *criteria* that serve as the basis for delivery acceptance are documented as a key section of the entity's specification.

Step 3: Develop the Entity's Operations Domain Solution

As each entity's *Requirements Domain Solution* evolves to maturity, System Developers formulate and mature the *Operations Domain Solution*. Operational concepts are synthesized and documented in the entity's *Concept of Operations (ConOps)* document or *Theory of Operations*. The ConOps:

1. Identifies the Level 0 User's operational architecture that includes the MISSION SYSTEM(s) and SUPPORT SYSTEM(s).
2. Identifies *friendly*, *benign*, and *hostile* systems and threats in the OPERATING ENVIRONMENT that interact with the system.
3. Identifies system operations and tasks required to accomplish the mission.
4. Synchronizes the tasks with the *mission event timeline (MET)* or allocations.
5. Identifies *products*, *by-products*, or *services* required to achieve mission outcomes.

Referring to Figure 26.1, an entity's *Operations Domain Solution*:

1. Implements requirements allocated and derived from the entity's *Requirements Domain Solution*.
2. Documents work products that serve as inputs for deriving the entity's *Behavioral Domain Solution*.
3. Integrates with a higher level User, SYSTEM, PRODUCT, ASSEMBLY, and SUB-ASSEMBLY *Operations Domain Solution* (Figure 26.6).
4. May be *expanded* or *decomposed* into at least two or more lower level entity *Operations Domain Solutions* (Figure 26.6).

5. Iterates with the *Requirements and Behavioral Domain Solutions* for completeness, consistency, and traceability.
6. Is assessed for consistency, compliance, and performance by the *Evaluate and Optimize the System Design Solution*.

Step 4: Develop the Entity's Behavioral Domain Solution

As each entity's *Operations Domain Solution* evolves to maturity, System Developers formulate and mature the *Behavioral Domain Solution*. The *Behavioral Domain Solution* describes WHAT is to be accomplished in terms of *logical* interactions and sequences of tasks or processing required to produce the desired outcomes. This includes:

1. Identification of the required *capabilities*—namely *functions* and *performance*.
2. Constructing system interaction and sequence diagrams that depict HOW the SYSTEM is envisioned to *react* and *respond* to various external stimuli and cues from its OPERATING ENVIRONMENT.
3. Synchronizing those interactions, products, by-products, and services with the MET.
4. Analyzing, modeling, and simulating capability sequences and performance to ensure compliance with requirements and balance performance allocations.

Referring to Figure 26.1, an entity's *Behavioral Domain Solution*:

1. Is allocated requirements from and must be traceable to the entity's *Requirements Domain Solution*.
2. Documents *work products* that serve as the inputs for the developing *Physical Domain Solution*.
3. Integrates with a higher level User, SYSTEM, PRODUCT, ASSEMBLY, and SUBASSEMBLY *Behavioral Domain Solution* (Figure 26.6).
4. May be expanded or decomposed into at least two or more lower level entity *Behavioral Domain Solutions* (Figure 26.6).
5. Iterates with the *Operations* and *Physical Domain Solutions*.
6. Is assessed for consistency, compliance, and performance by the *Evaluate and Optimize the System Design Solution*.

Step 5: Develop the Physical Domain Solution

As the entity's *Behavioral Domain Solution* evolves to maturity, System Developers formulate and mature the *Physical Domain Solution*. This solution describes HOW the *Behavioral Domain Solution* is implemented via multi-level physical components. This requires:

1. Formulating several viable candidate architectures for the entity.
2. Conducting analyses and trade studies to evaluate and score the merits of each architecture relative to a set of pre-defined decision criteria.
3. Selecting a preferred architecture from a viable set of alternatives.
4. Establishing performance budgets and safety margins.
5. Finalizing selection of components to satisfy the entities identified within the architecture.
6. Translating the physical architecture into a detailed design—such as assembly drawings, schematics, wiring diagrams, and software design.

282 Chapter 26 The SE Process Model

7. Assessing compatibility and interoperability with external systems in the entity's OPERATING ENVIRONMENT.

Referring to Figure 26.1, each entity's *Physical Domain Solution*:

1. Is *allocated* requirements from and must be *traceable* to the entity's *Requirements Domain Solution*.
2. Integrates with a higher level User, SYSTEM, PRODUCT, ASSEMBLY, and SUB-ASSEMBLY *Physical Domain Solution* (Figure 26.6).
3. May be expanded or decomposed at least two or more lower level *entity Physical Domain Solutions* (Figure 26.6).
4. Is assessed for consistency, compliance, and performance by the *Evaluate and Optimize the System Design Solution*.

Step 6: Evaluate and Optimize the Entity's Total Design Solution

As the *Physical Domain Solution* evolves and matures, the next step is to *Evaluate and Optimize the System Design Solution*. The purpose of this step is to *verify* and *validate* that the entity's *Requirements, Operations, Behavioral*, and *Physical Design Solutions*:

1. Are consistent with each other.
2. Fully comply with and are traceable to its *Requirements Domain Solution*.

These objectives are accomplished via the following:

1. Technical reviews.
2. Technical audits.
3. Prototype development.
4. Modeling and simulation.
5. Proof of concept or technology demonstrations.

You will encounter people who contend that you *cannot optimize* a system for a diverse set of operating scenarios and conditions, it can only be *optimal*: let's differentiate the two viewpoints.

For a prescribed set of operating conditions and priorities, you can theoretically *optimize* a system. The challenge is that *these conditions are often independent and statistically random occurrences in the OPERATING ENVIRONMENT*. As a result a system, product, or service performance may not be *optimized* for all sets of random variable conditions. So people characterize the SYSTEM's performance as optimal in dealing with these random variables.

There is an unwritten rule that says that most human attempts fall short of their goals. Assuming you have realistic goals, WHAT you accomplish depends on WHAT you strive to achieve. So in Step 6, *Evaluate and Optimize System Design Solution*, we use the term *optimize* to communicate WHAT we strive to achieve. Given human performance history in goal achievement, striving to simply be *optimal* will probably produce a result that is *less than* optimal, an even less desirable outcome.

26.4 DECISION SUPPORT TO THE SE PROCESS MODEL

Decision support practices such as analyses, trade studies, prototypes, demonstrations, models, and simulations, etc. are employed to provide recommendations for technical decisions that *bound* the system's *solution space*—such as the Requirements Domain Solution compliance.

26.5 EXIT CRITERIA

Since the SE Process Model is *highly iterative* and subject to development time constraints of the *entity*, *exit criteria* are determined by:

1. *Level of maturity* required of the *entity* being designed and its required work products—specifications, designs, verification procedures, etc.
2. *Level of criticality* and practicality in correcting *discrepancies* between specification requirements and verification data subject to cost and schedule constraints.

26.6 WORK PRODUCTS AND QUALITY RECORDS

The SE Process Model supports the development of numerous system/product life cycle phase *work products* and *quality records*. When applied to a *phase specific* process, the SE Process produces four categories of *work products* for each entity: 1) the Requirements Domain Solution, 2) the Operations Domain Solution, 3) the Behavioral Domain Solution, and 4) the Physical Domain Solution.

General examples of *work products* and *quality records* include specifications, specification trees, architectures, analyses, trade studies, drawings, technical reports, verification records, and meeting minutes. Refer to Chapters 37–40 for details of each solution description that follows for specific *work products* and *quality records*.

Author's Note 26.1 *People often confuse the purpose of any SE Process. They believe that the SE Process is established to create documentation; this is erroneous! The purpose of the SE Process Model is to establish a methodology to solve problems and produce a preferred solution that satisfies contract requirements subject to technical, cost, schedule, technology, and risk constraints. Work product and quality record documentation are simply enablers and artifacts of the process, a means to an end, not the primary focus.*

Author's Note 26.2 *Remember, if you focus on producing documentation, you get documentation and a design solution that may or may not meet requirements. If you focus on producing a design solution via the SE Process Model, you should arrive at a design solution that satisfies requirements supported by documentation artifacts that prove the integrity and validity of the solution.*

26.7 SE PROCESS MODEL CHARACTERISTICS

The SE Process Model is characterized as being *highly iterative* and *recursive*. To better understand these characteristics, let's explore each.

Highly Iterative Characteristic

The SE Process Model, when applied to a specific entity within a system *level of abstraction*, is characterized as *highly iterative* as illustrated in Figure 26.2. Although each of the multi-level steps of the methodology has a sequential workflow progression, each step has *feedback* loops that allow a return to preceding steps and reassess decisions when issues occur later along the workflow. As a result the feedback loops establish the *highly iterative* characteristic as illustrated in Figure 26.3.

284 Chapter 26 The SE Process Model

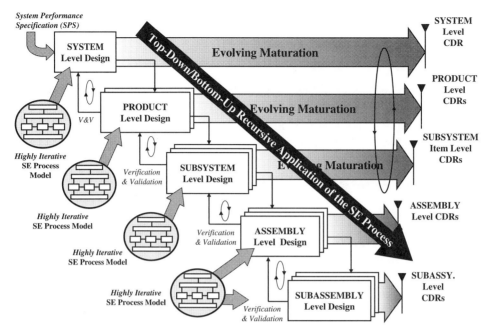

Figure 26.2 Multi-Level System Engineering Design

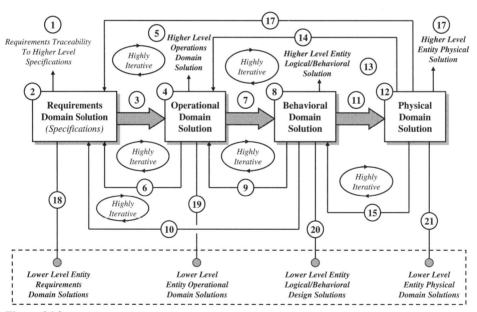

Figure 26.3 Entity/Item SE Design Flow

Recursive Characteristic

Referring to Figure 26.2, observe that the SE Process Model is applied to every level of abstraction. We call this its *recursive* characteristic. Thus the same model applies to the SYSTEM level as its does at the SUBASSEMBLY level. Let's explore this point further.

We can simplify Figure 26.3 as shown in Figure 26.4. Note that each system development program begins with the *System Performance Specification (SPS)* and evolves through all four solution domains. This graphic with its alternating gray quadrants and R (requirements), O (operations), B (behavioral), and P (physical) symbols—in short, ROBP—will be used as an *icon* to symbolize the *four solution domains* in our later discussions on *system design practices*.

Applying the SE Process Model to System Development

Now let's suppose that we have the system shown in Figure 26.5. The SYSTEM consists of PRODUCTS 1 and 2. PRODUCT 1, a large, complex design, consists of SUBSYSTEMs 1 and 2. PRODUCT 2, a simpler design, consists of hardware configuration item (HWCI) 21 and computer software configuration item (CSCI) 21. We apply the SE Process Model to each entity as illustrated by the boxes. Application of the SE Process Model to each entity continues until the SE design is mature and ready to commit to implementation. Figure 26.6 illustrates the state of the system design solution at completion.

Here we see a multi-level framework that depicts the horizontal workflow progression over time. Vertically, the Requirements, Operations, Behavioral, and Physical Domain Solutions are decomposed into various levels of abstraction. Collectively the framework graphically illustrates a system, which by definition, is the integration of multiple levels of capabilities to a higher level purpose that is greater than their individual capabilities.

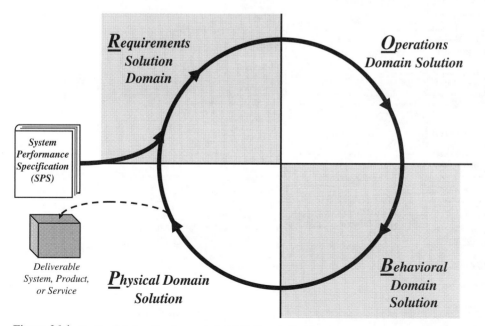

Figure 26.4 System Solution Development via Multi-Domain Iterations

286 Chapter 26 The SE Process Model

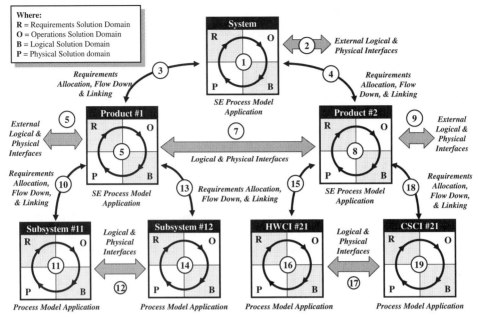

Figure 26.5 Multi-Domain SE Process Iterations

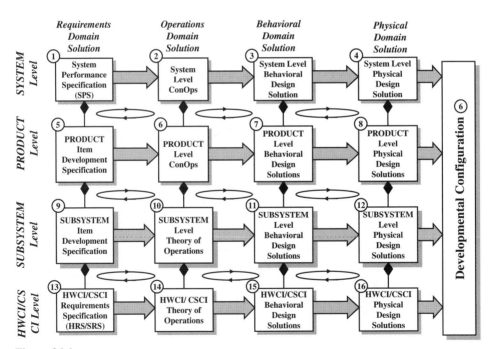

Figure 26.6 System Design Framework

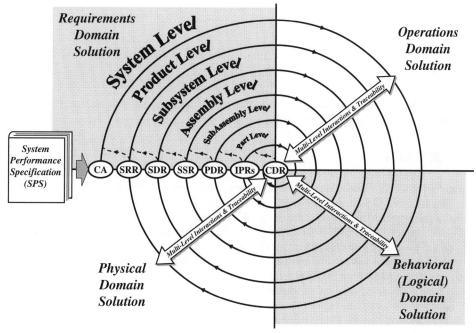

Figure 26.7 The System Development Spiral

26.8 EVOLVING AND REVIEWING THE SYSTEM SOLUTIONS

Based on the *highly iterative* and *recursive* characteristics, System Developers evolve the system design solution over time from the SPS into a series of workflow progressions through each level of abstraction until the system design solution is *initially* complete. Figure 26.7 illustrates how the total system design solution *evolves* through the domain solutions at each level of abstraction and culminates with the Critical Design Review (CDR).

Symbolically, the inner loops of the spiral represent *increasing* levels of detail until; the CDR is conducted. Each loop of the spiral *culminates* in a technical review that serves as a *critical staging* or *control point* for *commitment* to the next level of detail. Each loop includes a breakout point to permit *reconciling* changes with previous levels and to continue to evolve and mature the higher level solutions until CDR.

Author's Note 26.3   *The context of Figure 26.7 is for the period between Contract Award and CDR when the total design solution is approved and released for component procurement and development. However, the system design solution IS NOT finalized until the first article system or product has been integrated, tested, verified, validated (optional), and accepted by the Acquirer or User.*

26.9 GUIDING PRINCIPLES

In summary, the preceding discussions provide the basis with which to establish the guiding principles that govern the implementation of the SE Process Model.

Principle 26.1 Problem solving and solution development lead to an optimal design solution; simply creating a point design solution does not always indicate problem solving.

Principle 26.2 An entity's design solution is composed of four domain solutions: Requirements, Operations, Behavioral, and Physical.

Principle 26.3 As a workflow, system design is the *highly iterative*, multi-level, transformation of an entity's requirements into operations, behavior, and physical implementation.

26.10 SUMMARY

This section introduced the SE Process Model and its structure. In our discussion we described how the model integrates the Requirements, Operations, Behavioral, and Physical Domain Solutions into a *highly iterative* framework that can be applied to the design of entities at all levels of abstraction. We noted that the SE Process Model's multi-level application attribute is referred to as its *recursive* characteristic.

The first step of the model is to *Understand* the *Opportunity/Problem* and *Solution Spaces* to appreciate the context of the requirements allocated to each entity. As a *highly iterative* model, we described how the model incorporates the workflow from the Requirements Domain to the Operations Domain to the Behavioral Domain to the Physical Domain. Since design solutions must be traceable to their requirements allocations as documented in the entity's specification, we illustrated how the Requirements Domain Solution links to the Operations, Behavioral, and Physical Domain Solutions. Since every design solution must be evaluated and optimized, we illustrated how *the Evaluate* and *Optimize* the *Entity's* Design Solution activity supports the Operations, Behavioral, and Physical Domain Solutions.

GENERAL EXERCISES

1. Answer each of the *What You Should Learn from This Chapter* questions identified in the *Introduction*.
2. Refer to the list of systems identified in Chapter 2. Based on a selection from the preceding chapter's General *Exercise* or a new system, selection, apply your knowledge derived from this chapter's topical discussions. Describe how the SE Process Model is applied to identify the following:
   (a) The system's *opportunity/problem space* and *solution space(s)*.
   (b) The Requirements Domain Solution.
   (c) The Operations Domain Solution.
   (d) The Behavioral Domain Solution.
   (e) The Physical Domain Solution.

ORGANIZATIONAL CENTRIC EXERCISES

1. Research your local command media for SE process requirements.
   (a) Does your organization have a standard SE Process?
   (b) How are SEs within the organization trained to apply the SE Process?
   (c) Compare and contrast the organization's SE process with the one described here.
   (d) How are multidisciplined SEs trained to apply the process?
2. Research the following SE processes created over several decades. Develop a paper that describes each SE process, compare and contrast the differences; note evolutionary changes over time, and contrast with your own experiences.
   (a) US Army Field Manual FM-770-78
   (b) MIL-STD-499

(c) IEEE 1220–1998
(d) International Council on Systems Engineering (INCOSE)
(e) ANSI/EIA 632

3. Contact technical programs within your organization and interview personnel concerning what SE process or methodology they employed to develop their systems or products. Report your findings and observations.

ADDITIONAL READING

ANSI/EIA 632-1999. 1999. *Processes for Engineering a System*. Electronic Industries Alliance (EIA). Arlington, VA.

IEEE 1220-1998. 1998. *IEEE Standard for the Application and Management of the Systems Engineering Process*. Institute of Electrical and Electronic Engineers (IEEE). New York, NY.

International Council on System Engineering (INCOSE). 2000. INCOSE *System Engineering Handbook*, Version 2.0. Seattle, WA.

ISO/IEC 15288. *System Engineering—System Life Cycle Processes*. International Organization for Standardization (ISO). Geneva, Switzerland.

FM-770-78. 1979. Field Manual: *System Engineering*. Washington, DC: US Army.

MIL-STD-499B (cancelled draft). 1994. *Systems Engineering*. Washington, DC: Department of Defense (DoD).

Defense Systems Management College (DSMC). 2001. *Systems Engineering Fundamentals*. Defense Acquisition University Press Ft. Belvoir, VA.

Sheard, Sarah A. *The Frameworks Quagmire, A Brief Look*, 1997, Software Productivity Consortium, Herndon, VA.

Chapter 27

System Development Models

27.1 INTRODUCTION

Our discussions up to this point have viewed the *System Performance Specification (SPS)* as an ideal requirements document with mature, well-defined requirements. In reality, however, requirements range from those that are well defined to those that are very *immature*, or *ambiguous*. So, *how does an SE deal with this wide variety of requirements maturity?* Actually there are several *system development models* that provide a basis for dealing with the requirements *maturity* problems.

This chapter introduces and investigates various system development models that represent various methodologies for developing systems. The models include:

1. The Waterfall Development Model
2. The Evolutionary Development Model
3. The Incremental Development Model
4. The Spiral Development Model

Our discussions describe each model, identify how each evolved, highlight flaws, and provide illustrative, real world examples.

You may ask why topics such these are worthy of discussion in an SE book. There are two good reasons:

- *First*, you need a toolkit of system development approaches that enables you to address a variety of requirements definition and maturity challenges.
- *Second*, you need to fully understand each of these models, their origin, attributes and flaws to make sure you have the RIGHT approach to specific types of system development.

As an SE, you need to fully understand HOW the models are applied to various system development scenarios. You will encounter people throughout industry that refer to these models via *buzzwords* but often lack an understanding of each type of model's background and application.

What You Should Learn from This Chapter

1. What is a *system development model*?
2. What are the *four primary system development models*?
3. What is the *Waterfall Development Model*, its characteristics, and shortcomings?
4. What is the *Evolutionary Development Model*, its characteristics, and shortcomings?

System Analysis, Design, and Development, by Charles S. Wasson
Copyright © 2006 by John Wiley & Sons, Inc.

5. What is the *Incremental Development Model*, its characteristics, and shortcomings?
6. What is the *Spiral Development Model*, its characteristics, and shortcomings?
7. How is the *SE Process Model* applied to these development models?

Definitions of Key Terms

- **Evolutionary Development Strategy** A development strategy used to develop "a system in builds, but differs from the Incremental Strategy in acknowledging that the user need is not fully understood and all requirements cannot be defined up front. In this strategy, user needs and system requirements are partially defined up front, then are refined in each succeeding build." (Source: Former MIL-STD-498, para. G.3)

- **Grand Design Development Strategy** A development strategy that is "essentially a 'once-through, do-each-step-once' strategy. Simplistically: determine user needs, define requirements, design the system, implement the system, test, fix, and deliver." (Source: Former MIL-STD-498, para. G.3)

- **Incremental Development** "A software development technique in which requirements definition, design, implementation, and testing occur in an overlapping, iterative (rather than sequential) manner, resulting in incremental completion of the overall software product." (Source: IEEE 610.12-1990)

- **Incremental Development Strategy** A development strategy that "determines user needs and defines the overall architecture, but then delivers the system in a series of increments ("software builds"). The first build incorporates a part of the total planned capabilities, the next build adds more capabilities, and so on, until the entire system is complete." (Source: *Glossary: Defense Acquisition Acronyms and Terms*)

- **Spiral Development Strategy** A development strategy that "develops and delivers a system in builds, but differs from the incremental approach by acknowledging that the user need is not fully formed at the beginning of development, so that all requirements are not initially defined. The initial build delivers a system based on the requirements as they are known at the time development is initiated, and then succeeding builds are delivered that meet additional requirements as they become known." (Source: *Glossary: Defense Acquisition Acronyms and Terms*)

- **Spiral Model** "A model of the software development process in which the constituent activities, typically requirements analysis, preliminary and detailed design, coding, integration, and testing, are performed iteratively until the software is complete." (Source: IEEE 610.12-1990)

- **Waterfall Model** "A model of the software development process in which the constituent activities, typically a concept phase, requirements phase, design phase, implementation phase, test phase, and installation and checkout phase, are performed in that order, possibly with overlap but with little or no iteration." (Source: IEEE 610.12-1990)

- **Waterfall Development Strategy** A strategy in which "development activities are performed in order, with possibly minor overlap, but with little or no iteration between activities. User needs are determined, requirements are defined, and the full system is designed, built, and tested for ultimate delivery at one point in time." (Source: *Glossary: Defense Acquisition Acronyms and Terms*)

27.2 SYSTEM DEVELOPMENT MODELS

The trends of increasing technical complexity of the systems, coupled with the need for *repeatable* and *predictable* process methodologies, have driven System Developers to establish development models.

Background Leading to System Development Models

Systems and products prior to the 1950s were hardware intensive systems. Processing was accomplished via electromechanical devices that implemented logical and mathematical computational processes. By the 1960s, analog and digital small-, medium-, and large-scale integrated circuits, coupled with modular design methods, enabled developers to improve the reliability and accuracy of the computations with some limited software involvement. During these years when *design errors* or *changes* occurred, the cost of making corrections in mechanical hardware and electronic circuitry was becoming *increasingly expensive* and *time-consuming*.

With the introduction of microprocessor technologies in the early 1970's, software became a viable alternative to system development. Conceptually, capabilities that had to be implemented with hardware could now be implemented more easily and quickly in software. As a result system design evolved toward flexible, reconfigurable systems that enable System Developers and maintainers to target unique field applications simply by tailoring the software.

The shift toward software intensive systems advanced faster than the methods used to produce the software. As the software applications became larger and more complex, so did the risks and problem areas. System Developer organizations became more sensitive to the *challenges* and *risks* associated with meeting contract cost, schedule, and technical performance criteria and the risks of failure to meet those criteria. Additionally the costs of *rework* and *poor quality* became central *issues*, especially in areas such as reliability, safety, and health. The need for *repeatability* and *predictability* in development process was also a *critical issue*.

To meet these challenges, system development models began evolving in the software domain and have since expanded to the broader *engineering of systems* domain as overall system development approaches.

System Development Models

System and product development approaches generally follow *four* commonly used models or hybrids. The most common models include the: 1) Waterfall Development Model, 2) Evolutionary Development Model, 3) Incremental Development Model, and 4) Spiral Development Model. Dr. Barry Boehm, who has written numerous technical books and articles about software development, provides some key insights regarding the evolution, strengths, and weaknesses of the models. To facilitate our discussion, we will reference some of Dr. Boehm's observations.

Author's Note 27.1 *Historically, you will find many instances of problem solving–solution-development methodologies that emerge within a discipline and are equally applicable to other disciplines. In the discussions that follow, you will discover how software development, as an emerging discipline, tackled a problem space and produced an evolving set of methodologies that are equally applicable to SE and hardware engineering.*

27.3 THE WATERFALL DEVELOPMENT MODEL

The *Waterfall Development Model* represents one of the initial attempts to characterize software development in terms of a model. Today, the Waterfall Model *exemplifies* how many organizations develop systems and products. Figure 27.1 provides an illustration of this model.

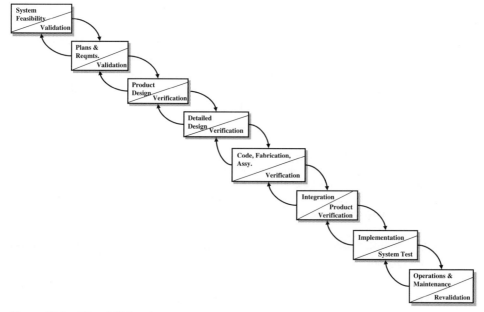

Figure 27.1 "Waterfall" Development Model
Source: Figure 1, "A Spiral Model of Software Development and Enhancement" by Dr. Barry boehm, IEEE Computer May 1988, p. 62.

Author's Note 27.2 *The term "waterfall" has always been a misnomer and tends to confuse many people. The term reflects the graphical top-down, diagonal representation rather than the actual implementation. As we will see in a later chapter, the earlier stages do represent expansion of levels of design detail over time. The latter stages, beginning with "Integration," represent the "upside" of the V-Model discussed in Figure 25.5. Unlike the Waterfall Model, the V-Model is implemented with highly iterative and recursive feedback loops within and between levels of abstraction.*

In the Waterfall approach, "development activities are performed in sequential order, with possibly minor overlap, and minimal or no iteration between activities. User needs are determined, requirements are defined, and the full system is designed, built, and tested for ultimate delivery at one point in time. Some people refer to this as a stage-wise model." (Source: *Glossary: Defense Acquisition Acronyms and Terms*)

27.4 THE EVOLUTIONARY DEVELOPMENT MODEL

In general, the *Evolutionary Development Model* is based on the premise that "stages consist of expanding increments of an operational software product, with the directions of evolution being determined by operational experience." (Source: Boehm, p. 63) This conception is based on an evolutionary strategy of a system or product development through a series of *pre-planned product improvement* (P3I) releases.

Evolutionary development provides a potential solution for Acquirers, Users, and System Developers. As discussed in an earlier section, some systems/products are *single-use items*; others are longer term, *multi-application* items. For some mission and system applications, you generally know at system acquisition what the requirements are. In other applications, you may be able to define a few up-front objectives and capabilities. Over time, the fielded system/product requires new capabilities as problem/opportunity spaces evolve.

Some systems, such as computers, become *obsolete* in a very short period of time and are discarded. From a business perspective, the cost to upgrade and maintain the devices is *prohibitive* relative to purchasing a new computer. In contrast, some Users, driven by *decreasing* budgets and *slow* changes in the external environments, may use systems and products far beyond their original intended service lives. Consider the following example:

EXAMPLE 27.1

The US Air Force B-52 aircraft is postured to achieve the longest service life of any aircraft. During its lifetime, the aircraft's systems and missions have evolved from their *initial operational capability (IOC)* when the aircraft was first introduced to the present via *capability upgrades*. The projected service lifespan far exceeds what the aircraft's innovators envisioned.

The Evolutionary Development Model is based on the premise that a system's capabilities evolve over time via a series upgrades.

Fallacies of the Evolutionary Development Model

Conceptually, the Evolutionary Development Model may be suited for some applications; however, it, too, has its fallacies. Dr. Boehm (p. 63) notes the following points:

Fallacy 1: The Evolutionary Development Model is difficult to delineate from the *build*, *test*, and *fix* approaches "whose spaghetti code and lack of planning were the initial motivation for the "waterfall model."

Fallacy 2: The Evolutionary Development Model stemmed from the "often unrealistic" assumption that the system would always provide the flexibility to accommodate *unplanned evolution paths*.

Regarding fallacy 2, Dr. Boehm (p. 63) states "This assumption is unjustified in three primary circumstances:

1. Circumstance in which several independently evolved applications must subsequently be closely integrated.
2. Information sclerosis" cases, in which temporary work-around for software deficiencies increasingly solidify into unchangeable constraints on evolution, ...
3. Bridging situations, in which the new software is incrementally replacing a large existing system. If the system is poorly modularized, it is difficult to provide a good sequence of 'bridges' between old software and the expanding increments of new software."

27.5 THE SPIRAL DEVELOPMENT MODEL

Because of inherent flaws in the *Evolutionary Development Model*, coupled with a lack of understanding, maturity, and risk in system requirements up front, Dr. Boehm introduced the *Spiral Development Model* illustrated in Figure 27.2.

Spiral development employs a series of *highly iterative development* activities whereby the deliverable end product of each activity may not be the deliverable system. Instead, the evolving set of knowledge and subsequent system requirements that lead to development of a deliverable

27.5 The Spiral Development Model

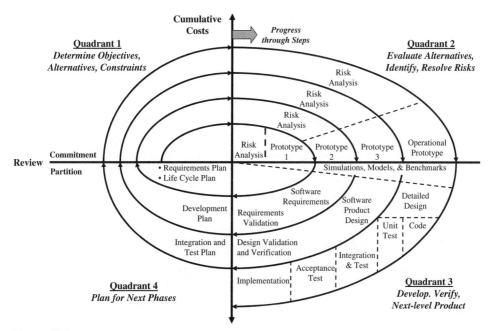

Figure 27.2 Boehm's Spiral Development Model
Source: Figure 2, "A Spiral Model of Software Development and Enhancement" by Dr. Barry Boehm, IEEE Computer May 1988, p. 64. Reprinted with permission.

system contribute to the maturing design solution. The knowledgebase evolves via *technology* or *proof of concept demonstrations* to a level of maturity worthy of: 1) introduction to the marketplace and 2) production investments from an acceptable risk perspective.

DSMC describes the model in this manner. "The spiral approach also develops and delivers a system in builds, but differs from the incremental approach by acknowledging that the user need is not fully formed at the beginning of development, so that all requirements are not initially defined. The initial build delivers a system based on the requirements, as they are known at the time development is initiated, and then succeeding builds are delivered that meet additional requirements as they become known. (Additional needs are usually identified and requirements defined as a result of user experience with the initial build)." (Source: *Glossary: Defense Acquisition Acronyms and Terms*)

Spiral Model Description

The development *spiral* consists of four quadrants as shown in Figure 27.2:

Quadrant 1: Determine objectives, alternatives, and constraints.
Quadrant 2: Evaluate alternatives, identify, resolve risks.
Quadrant 3: Develop, verify, next-level product.
Quadrant 4: Plan next phases.

Although the spiral, as depicted, is oriented toward software development, the concept is equally applicable to systems, hardware, and training, for example. To better understand the scope of each spiral development quadrant, let's briefly address each one.

Quadrant 1: Determine Objectives, Alternatives, and Constraints

Activities performed in this quadrant include:

1. Establish an understanding of the system or product objectives—namely *performance, functionality*, and *ability to accommodate change*. (Boehm, p. 65)
2. Investigate implementation alternatives—namely design, reuse, procure, and procure/modify.
3. Investigate constraints imposed on the alternatives—namely technology, cost, schedule, support, and risk.

Once the system or product's objectives, alternatives, and constraints are understood, Quadrant 2 (Evaluate alternatives, identify, and resolve risks) is performed.

Quadrant 2: Evaluate Alternatives, Identify, Resolve Risks

Engineering activities performed in this quadrant select an alternative approach that best satisfies technical, technology, cost, schedule, support, and risk constraints. The focus here is on *risk mitigation*. Each alternative is *investigated* and *prototyped* to reduce the risk associated with the development decisions. Boehm (p. 65) describes these activities as follows:

> ... This may involve prototyping, simulation, benchmarking, reference checking, administering user questionnaires, analytic modeling, or combinations of these and other risk resolution techniques.

The outcome of the evaluation determines the next course of action. If *critical operational and/or technical issues* (COIs/CTIs) such as *performance* and *interoperability* (i.e., external and internal) risks remain, more detailed prototyping may need to be added before progressing to the next quadrant. Dr. Boehm (p. 65) notes that if the alternative chosen is "operationally useful and robust enough to serve as a low-risk base for future product evolution, the subsequent risk-driven steps would be the evolving series of evolutionary prototypes going toward the right (hand side of the graphic) ... the option of writing specifications would be addressed but not exercised." This brings us to Quadrant 3.

Quadrant 3: Develop, Verify, Next-Level Product

If a determination is made that the previous prototyping efforts have resolved the COIs/CTIs, activities to develop, verify, next-level product are performed. As a result, the basic "waterfall" approach may be employed—meaning concept of operations, design, development, integration, and test of the next system or product iteration. If appropriate, *incremental development* approaches may also be applicable.

Quadrant 4: Plan Next Phases

The *spiral development model* has one characteristic that is common to all models—the need for advanced *technical planning* and *multidisciplinary* reviews at *critical staging* or *control points*. Each cycle of the model culminates with a technical review that assesses the status, progress, maturity, merits, risk, of development efforts to date; resolves *critical operational and/or technical issues* (COIs/CTIs); and reviews plans and identifies COIs/CTIs to be resolved for the next iteration of the spiral.

Subsequent implementations of the spiral may involve lower level spirals that follow the same quadrant paths and decision considerations.

27.6 THE INCREMENTAL DEVELOPMENT MODEL

Sometimes the development of systems and products is constrained by a variety of reasons. Reasons include:

1. Availability of resources (expertise, etc.)
2. Lack of availability of interfacing systems
3. Lack of funding resources
4. Technology risk
5. Logistical support
6. System/product availability

When confronted by these constraints, the Users, the Acquirer, and the System Developer may be confronted with formulating an *incremental development* strategy. The strategy may require establishing an *Initial Operational Capability (IOC)* followed by a series of *incremental development* "builds" that enhance and refine the system or product's capabilities to achieve a *Full Operational Capability (FOC)* by some future date. Figure 27.3 illustrates how *incremental development* is phased.

The *DoD Glossary of Terms* characterizes incremental development as follows: "The incremental approach determines user needs and defines the overall architecture, but then delivers the system in a series of increments—i.e., *software builds*. The first build incorporates a part of the total planned capabilities, the next build adds more capabilities, and so on, until the entire system is complete." (Source: *Glossary: Defense Acquisition Acronyms and Terms*)

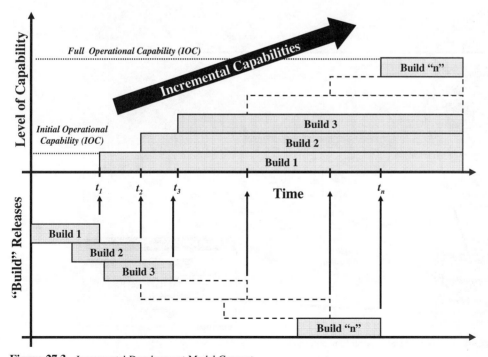

Figure 27.3 Incremental Development Model Concept

298 Chapter 27 System Development Models

Implementation

Implementation of the *Incremental Development Model* requires that a sound "build" strategy be established "up front." As each *build* cycle is initiated, development teams establish unique system requirements for each *build*—either by separate specification or by delineated portions of the *System Performance Specification (SPS)*. Each *build* is designed, developed, integrated, and tested via a series of overlapping V-Models, as illustrated in Figure 27.4.

The Challenge for Systems Engineering

When implementing the *Incremental Development Model* approach, SEs, in collaboration with other disciplines, must:

1. Thoroughly analyze and partition each "build's" system requirements based on a core set of requirements.
2. *Flow down* and *allocate* the requirements to PRODUCTS, SUBSYSTEMS, or ASSEMBLIES.
3. Schedule the "builds" over time to reflect User priorities (capability "gaps, resources, schedules, etc.).

Incremental "builds" may include integrating new components into the system and upgrading existing components. The challenge for SEs is to determine *how* to establish and partition the initial set of functional capabilities and integrate other capabilities over time without disrupting existing system operations. Intensive interface analysis is required to ensure the "build" integration occurs in the proper sequence and the support tools are available.

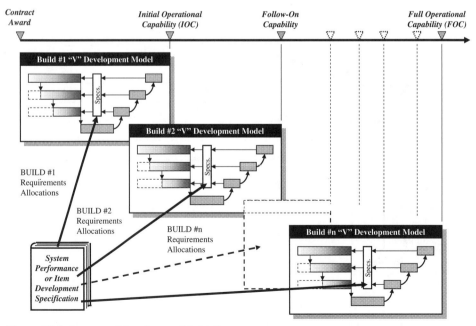

Figure 27.4 Incremental Development Model Concept

27.7 SYSTEM VERSUS COMPONENT DEVELOPMENT STRATEGIES

Recognize that any system development programs may employ several different development strategies, depending on the SYSTEM or entity. You may find instances where the SYSTEM is developed using *incremental development* of which one or more of its "builds" may employ another development strategy such as *spiral development*.

If you analyze most systems, you will find SUBSYSTEMS, ASSEMBLIES, and SUB-ASSEMBLIES that:

1. Have well-defined requirements.
2. Employ mature, off-the-shelf technologies and design methods that have been around for years.
3. Are developed by highly experienced developers.

Other SUBSYSTEMS and ASSEMBLIES may have the opposite situation. They may have:

1. Ill-defined or immature requirements.
2. Employ immature technologies and design methods.
3. Inexperienced developers.
4. Or, all of these.

Where this is the case, you may have to select a specific development strategy that enables you to reduce the risk of development. Consider the following example:

EXAMPLE 27.2

Over the years automobile technology has evolved and increased in complexity. Today, we enjoy the benefits of new technologies such as fuel injection systems, anti-lock braking systems (ABS), front wheel drive, air bag safety restraint systems (SRS), crumple zones, GPS mapping, and so on. All facets of automobile design have changed over the years. However, for illustration purposes, imagine for a moment that the fundamental automobile at higher levels of abstraction did not change drastically. It still has four doors, a passenger compartment, trunk, windshield, and steering. However, the maturation of the major technologies noted earlier required strategies such as *spiral development* that *enabled* them to *mature* and *productize* technologies such as ABS and SRS for application and integration into *Evolutionary Development Model* of the basic automobile.

27.8 GUIDING PRINCIPLES

In summary, the preceding discussions provide the basis with which to establish the guiding principles that govern system development strategy practices.

Principle 27.1 Select a development strategy based on the maturity of the SYSTEM/entity requirements, risk, budgeted schedule and costs, and User delivery needs.

Principle 27.2 A SYSTEM level development strategy may be different from an entity's development strategy, depending on the maturity of the requirements, technologies, processes, and experience and skill levels of the developers.

Principle 27.3 Each development strategy selection should include rationale for: 1) its selection and 2) for not selecting other strategies.

Principle 27.4 A system development program may consist of a mixture of development models, each chosen to uniquely satisfy SYSTEM or component development requirements and development constraints.

27.9 SUMMARY

Our discussion in this chapter provided an orientation on the various *System Development Strategy practices*. System and product development approaches require implementing a smart strategy that enables you to meet technical, cost, and schedule requirements with acceptable risk as well as User operational needs. Selecting the *RIGHT* system or product development strategy is a key competitive step. From an SE perspective, you should be familiar with the basic attributes of each type of model and understand how to apply it to meet your specific application's needs.

GENERAL EXERCISES

1. Answer each of the *What You Should Learn from This Chapter* questions identified in the *Introduction*.
2. Refer to the list of systems identified in Chapter 2. Based on a selection from the preceding chapter's General *Exercise* or a new system selection, apply your knowledge derived from this chapter's topical discussions.
   (a) What type of system development model would you recommend for developing the system?
   (b) For each of the other models not selected, provide supporting rationale on their degree of relevance and why they should not be used.
3. Identify three types of systems or system upgrades that may be ideal candidates for a *Spiral Development Model* strategy.
4. Identify three types of systems or system upgrades that may be ideal candidates for a *Waterfall Development Model* strategy.
5. Identify three types of systems or system upgrades that may be ideal candidates for *Incremental Development Model*.
6. In your own words and experiences, pick a system and describe how any of the following approaches might apply for someone who is unfamiliar with the methods?
   (a) Waterfall Approach
   (b) Evolutionary Approach
   (c) Incremental Approach
   (d) Spiral Approach

ORGANIZATIONAL CENTRIC EXERCISES

1. Contact two or three contract development programs within your organization.
   (a) Interview key personnel to discuss the type(s) of development strategies that were employed and the rationale for each.
   (b) Document and report your results.

REFERENCES

BOEHM, BARRY W. 1998. *A Spiral Model of Software Development and Enhancement*. IEEE Computer Society Press.

Defense Systems Management College (DSMC). 2001. *Glossary: Defense Acquisition Acronyms and Terms, 10th ed.* Defense Acquisition University Press. Ft. Belvoir, VA.

IEEE Std 610.12-1990. 1990. *IEEE Standard Glossary of Modeling and Simulation Terminology*. Institute of Electrical and Electronic Engineers (IEEE). New York, NY.

MIL-STD-498 (cancelled). 1994. Military Standard: *Software Development and Documentation*. Washington, DC: Department of Defense (DoD).

ADDITIONAL READING

BOEHM, BARRY W. 1982. *Software Engineering Economics*. Englewood Cliffs, NJ: Prentice-Hall.

Chapter 28

System Specification Practices

28.1 INTRODUCTION

The formal mechanism for specifying WHAT capabilities a system is required to provide and HOW WELL the capabilities are to be performed is the *System Performance Specification (SPS)*. The SPS establishes the formal technical requirements of the contract between the Acquirer, as the User's *contract* and *technical representative*, and the System Developer.

Many people *erroneously* believe specifications are documents used on the "front end" of a program to design the system; this is only partially true. Specifications serve as the basis for decision making throughout the System Development Phase. They:

1. Represent human attempts to *translate, bound, and communicate* the User's prescribed *solution space* into a language of text and graphics for capability (i.e., functional and performance) requirements to produce a physical *system*, *product*, or *service* that satisfies the intended operational need.

2. Serve as a *frame of reference* for decision making by establishing the thresholds for evaluating and verifying technical compliance as a precursor for final system acceptance and delivery.

Specification development requires support from a multi-level system analysis process. The analysis decomposes bounded *solution space* capabilities into manageable, lower level specifications for the SUBSYSTEMS, PRODUCTS, and ASSEMBLIES that ultimately form the totality of the system.

This section introduces *specification practices* that establish the multi-level, integrated *framework* of specifications required to develop a system, product(s), or services. Our discussions introduce the various types of specifications, what they contain, and how they relate to one another. The discussion includes a *general* specification outline that can be used as reference model.

What You Should Learn from This Chapter

1. What is a *specification*?
2. What makes a *"good" specification*?
3. Describe the evolution of specifications from initial system concept to *System Performance Specification (SPS)*.
4. What are the *basic types of specifications*?
5. How does each type of specification apply to system development?
6. What is a *specification tree*?

System Analysis, Design, and Development, by Charles S. Wasson
Copyright © 2006 by John Wiley & Sons, Inc.

7. How is the *specification tree* structured?
8. What is the *standard format* for most specifications?

Definitions of Key Terms

- **Detail specification** "A specification that specifies design requirements, such as materials to be used, how a requirement is to be achieved, or how an item is to be fabricated or constructed. A specification that contains both performance and detail requirements is still considered a detail specification." (Source: MIL-STD-961D, para. 3.9, p. 5)
- **Development Specification** "A document applicable to an item below the system level which states specification performance, interface, and other technical requirements in sufficient detail to permit design, engineering for service use, and evaluation." (Source: Kossiakoff and Sweet, *System Engineering*, p. 447)
- **Deviation** "A written authorization, granted prior to the manufacture of an item, to depart from a particular performance or design requirement of a specification, drawing or other document for a specific number of units or a specified period of time." (Source: DSMC, *Defense Acquisition Acronyms and Terms*, Appendix B *Glossary of Terms*)
- **Interface Specification** "A specification, derived from the interface requirements, that details the required mechanical properties and/or logical connection between system elements, including the exact format and structure of the data and/or electrical signal communicated across the interface." (Source: Kossiakoff and Sweet, *System Engineering*, p. 449)
- **Performance Specification** "A specification that states requirements in terms of the required results with criteria for verifying compliance, but without stating the methods for achieving the required results. A performance specification defines the functional requirements for the item, the environment in which it must operate, and interface and interchangeability characteristics." (Source: MIL-STD-961D, para. 3.29, p. 7)
- **Source or Originating Requirements** The set of requirements that serve as the publicly released requirements used as the basis to acquire a system, product, or service. In general, a formal *Request for Proposal (RFP)* solicitation's *Statement of Objectives (SOO)* or *System Requirements Document (SRD)* are viewed as *source* or *originating requirements*.
- **Specification** A document that describes the essential requirements for items, materials, processes, or services of a prescribed *solution space*, data required to implement the requirements, and methods of verification to satisfy specific criteria for formal acceptance.
- **Specification Tree** "The hierarchical depiction of all the specifications needed to control the development, manufacture, and integration of items in the transition from customer needs to the complete set of system solutions that satisfy those needs." (Source: Former MIL-STD-499B Draft (cancelled), Appendix A, *Glossary*, p. 39)
- **Tailoring** "The process by which individual requirements (sections, paragraphs, or sentences) of the selected specifications, standards, and related documents are evaluated to determine the extent to which they are most suitable for a specific system and equipment acquisition and the modification of these requirements to ensure that each achieves an optimal balance between operational needs and cost." (Source: MIL-STD-961D, para. 3.40, p. 8)
- **Waiver** "A written authorization to accept a configuration item (CI) or other designated item, which, during production, or after having been submitted for inspection, is found to depart from specified requirements, but nevertheless is considered suitable "as is" or after rework by an approved method." (Source: DSMC, *Defense Acquisition Acronyms and Terms*, Appendix B, *Glossary of Terms*)

28.2 WHAT IS A SPECIFICATION?

Development of any type of requirements requires that you establish a firm understanding of:

1. WHAT is a specification?
2. WHAT is the purpose of a specification?
3. HOW does a specification accomplish a specific objective?

If you analyze the definition of a *specification* provided in this chapter's *Definition of Key Terms*, there are three key parts of this definition. Let's briefly examine each part.

- First—"... essential *requirements for items, materials, processes, or services* ..." Specifications are written not only for physical deliverable *items* but also for *services* and multi-level components, *materials* that compose those components, and procedural *processes* required to convert those materials into a usable component.
- Second—"... data *required to implement the requirements* ..." System development is often constrained by the need to adhere to and comply with other statement of objectives (SOO), design criteria, specifications, and standards.
- Third—"... *methods of verification to satisfy specific criteria for formal acceptance.*" Specifications establish the formal technical agreement between the Acquirer and System Developer regarding HOW each requirement will be formally *verified* to demonstrate the physical SYSTEM/entity has fully achieved the specified capability and associated level of performance. Note that verifying the achievement of a requirement may only satisfy incremental criteria required by contract for formal Acquirer acceptance. The contract may require other criteria such as installation and checkout in the field; field trails and demonstrations; operational test and evaluation (OT&E); resolution of outstanding *defects*, *errors*, and *deficiencies*. Remember, *System Performance Specifications (SPS)* are *subordinate* elements of the contract and represent only a portion of criteria for deliverable system acceptance and subsequent contract completion.

Given these comments concerning the definition of a specification, let's establish the objective of a *specification*.

Objective of Specifications

The *objective* of a specification is to document and communicate:

1. WHAT essential operational capabilities are required of an item (SYSTEM, PRODUCT, SUBSYSTEM, etc.)?
2. HOW WELL the capabilities must be performed?
3. WHEN the capabilities are to be performed?
4. By WHOM?
5. Under WHAT prescribed OPERATING ENVIRONMENT (i.e., bounded ranges of ACCEPTABLE inputs and environmental conditions)?
6. WHAT *desirable* or *acceptable* results or outcomes (e.g., behavior, products, by-products, and services) are expected?
7. WHAT *undesirable* or *unacceptable* results or outcomes are to be *minimized* or *avoided*?

Delineating Specifications and Contract Statements of Work (CSOWs)

People often have problems delineating a specification from a *Contract Statement of Work (CSOW)*. As evidence of this confusion, you will typically see CSOW language such as activities, tasks, and work products written into specifications. So, *what is the difference between the two types of documents*?

The CSOW is an Acquirer's contract document that specifies the *work activities to be accomplished and work products* to be *delivered* by the System Developer, subcontractor, or vendor to *fulfill* the *terms and conditions* (Ts&Cs) of the contract. In contrast, the *specification* specifies and bounds the capabilities, characteristics, and their associated levels of performance required of the deliverable *system*, *product*, or *service*.

Requirements versus Specification Requirements

Some people often use the term "requirements" while others refer to "specification requirements." *What is the difference?* The answer depends on the context. The contract, as the overarching document, specifies *requirements* that encompass, among other things specification requirements, schedule requirements, compliance requirements, and cost requirements. For brevity, people often *avoid* saying specification requirements and shorten the form to simply "requirements." As we will see, for systems with multiple levels of specifications and specifications within levels, general usage of the term "requirements" must be cast in terms of context—*of which specification is the frame of reference*.

28.3 WHY DO WE NEED SPECIFICATIONS?

A common question is: *WHY do we need specifications*. The best response is to begin with an old adage: *If you do not tell people WHAT you WANT, you can't complain about WHAT they deliver.* "But that's not WHAT we asked for" is a frequent response, which may bring a retort *"we delivered what you put on a piece of paper . . . [nothing!]."* There is also another adage, which raises grammarians' eyebrows, that states: *If it isn't written down, it never happened!*

In any case, if the technical wrangling continues, legal remedies may be pursued. The legal community seeks to unravel and enlighten:

1. WHAT did you specify via the contract, meeting minutes, official correspondence, conversations, etc.?
2. WHO did you talk to?
3. WHERE did you discuss the matter?
4. WHEN did you discuss it?
5. HOW did you document WHAT you agreed to?

The best way to avoid these conflicts is to establish a mechanism "up front" prior to Contract Award (CA). The mechanism should capture the technical agreements between the Acquirer and the System Developer to the mutual satisfaction and understanding of both parties. From a technical perspective, that *mechanism* is THE *System Specification Performance (SPS)*.

The conflicts discussed above, which characterize the Acquirer and System Developer/Services Provider—Subcontractor—Vendor interfaces, are not limited to organizations. In fact, the same issues occur *vertically* within the System Developer's program organization between the SYSTEM, PRODUCT, SUBSYSTEM, ASSEMBLY, SUBASSEMBLY, and PART levels of abstraction. As a result, the allocation and flow down of requirements from the *System Performance Specification (SPS)* to lower levels requires similar technical agreements between SE design teams. This occurs as each allocated and assigned *problem space* is decomposed into lower level *solution spaces* bounded by specifications.

Therefore, WHY do we need specifications? To *explicitly articulate and communicate* in a language that:

1. Employs terms that are *simple* and *easy* for the Acquirer, User, and System Developer *stakeholders to understand*.
2. Expresses essential features and characteristics of the deliverable product.
3. Avoids the need for "open" interpretation or further clarification that may lead to potential conflict at final system, product, or service acceptance.

28.4 WHAT MAKES A GOOD SPECIFICATION?

People commonly ask *What constitutes a GOOD specification*? "Good" is a relative term. What is GOOD to one person may be judged as POOR to someone else. Perhaps the more appropriate question is *"What makes a well-defined specification 'good'?"*

SEs often respond to this question with comments such as *"It's easy to work with," "we didn't have to make too many changes to get it RIGHT," "we didn't have any problems with the Acquirer at acceptance,"* and so on. However, some *explicit* characteristics make *well-defined* specifications standout as *models*. So let's examine two perspectives of well-written specifications.

General Attributes of Specifications

Well-written specifications share several general attributes that contribute to the specification's success.

Standard Outline. *Well-written specifications* are based on *standard* outline topics that are recognized by industry as best practices. The outline should cover the spectrum of *stakeholder* engineering topics to ensure all aspects of technical performance are addressed. We will address this topic in more detail later in this chapter.

Ownership Accountability. *Well-written specifications* are *assigned to* and *owned* by an individual or a team that is *accountable* for the implementation of the specification requirements, maintenance, verification, and final acceptance of the deliverable system, product, or *service* of the specification.

Baseline Control and Maintenance. Specifications are baselined at a strategic *staging* or *control points* when all *stakeholders* are in mutual agreement with its contents.

Referral For more information about *event-based control points* for specification review, approval, and release, refer to Chapter 46 *System Design and Development Documentation Practices* (refer to Figure 46.2) and Chapter 54 on *Technical Review Practices*.

Once the baseline is established, specifications are *updated* and *verified* through a *stakeholder* review agreement process (i.e., configuration control) that ensures that the document properly reflects the current consensus of stakeholder requirements.

Traceable to User Needs. *Well-written specifications* should be *traceable* to a User's prescribed *solution space* and fulfill *validated operational needs* derived from a *problem space* relative to the organization's mission and objectives.

Written in a Language and Terms That Are Simple and Easy to Understand. *Well-written specifications* are written in a language that uses terms that are *well defined* and are *easy to understand*. In general, many qualified people should independently read any requirement statement and emerge with the same *interpretation* and *understanding* of technical performance requirements.

Feasible to Implement. A *well-defined* specification must be feasible to implement within *realistically achievable* technologies, skills, processes, tools, and resources with *acceptable* technical, cost, schedule, and support risk to the Acquirer and the System Developer or Services Provider.

28.5 UNDERSTANDING TYPES OF SPECIFICATIONS

When SEs *specify* the items, materials, and processes required to support *system*, *product*, or *service* development, *how is this accomplished*? These work products are specified via a hierarchical set of specifications that focus on:

1. Defining the requirements for multi-level system items.
2. Supporting specifications for *materials* and *processes* to support development of those items.

The hierarchical set of specifications is documented via a framework referred to as the *specification tree*. We will discuss the *specification tree* later in this section. To understand the hierarchical structure within the *specification tree*, we need to first establish the types of specifications that may appear in the framework.

The classes of specifications include the following:

- General or performance specifications
- Detail or item development specifications
- Design or fabrication specifications
- Material specifications
- Process specifications
- Product specifications
- Procurement specifications
- Inventory item specifications
- Facility interface specifications

Figure 28.1 illustrates the primary types of specifications.

28.6 PUTTING SPECIFICATIONS INTO PERSPECTIVE

To place all of these various types of specifications into perspective, let's use the example illustrated in Figure 28.2. Requirements for a SYSTEM, as documented in the *System Performance Specification (SPS)*, are *allocated* and *flowed down* one or more levels to one or more items—such

308 Chapter 28 System Specification Practices

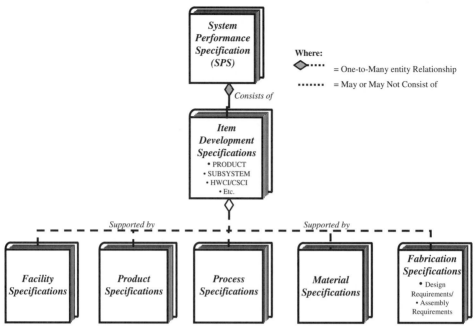

Figure 28.1 Entity Relationships of for Various types of Specifications

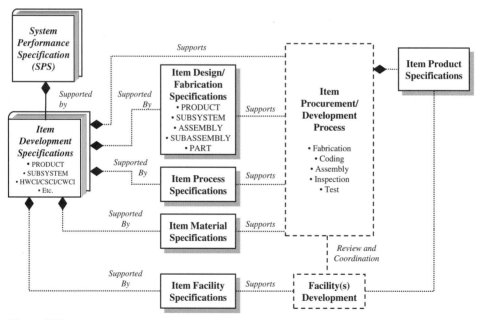

Figure 28.2 Application of Various Types of Specifications to Item Development

as PRODUCTS, SUBSYSTEMS, ASSEMBLIES. Requirements for these items are captured in their respective *development* or procurement *specifications* that document the "*As Specified*" configuration.

As the highly iterative, multi-level, and recursive design effort evolves, SEs develop one or more *design or fabrication specifications* to capture the attributes and characteristics of the physical PARTS to be developed. In the design effort one or more *process specifications* and *material specifications* are developed to aid in the procurement, fabrication, coding, assembly, inspection, and test of the item. The collective set of baselined specifications represent the "As Designed" Developmental Configuration that documents HOW to develop or procure the item.

The set of specifications generated for each item or configuration item (CI) serve as the basis for its development, as well as facilities, either internally or via external subcontractors or vendors. When the CI completes the System Integration, Test, and Evaluation (SITE) phase that includes *system verification*, the "As Verified" configuration is documented as the product baseline and captured as the CI's *product specification*.

28.7 THE SPECIFICATION TREE

The multi-level *allocation* and *flow down* of requirements employs a hierarchical framework that *logically* links SYSTEM entities *vertically* into a structure referred to as the *specification tree*. The right side of Figure 28.3 provides an example of a specification tree.

Specification Tree Ownership and Control

The *specification tree* is typically *owned* and *controlled* by a program's Technical Director or a System Engineering and Integration Team (SEIT). The SEIT also functions as a Configuration Control Board (CCB) to manage changes to the *current* baseline of a specification.

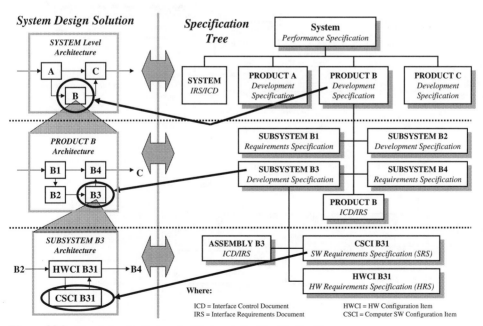

Figure 28.3 Correlating the System Architecture to the Specification Tree

310 Chapter 28 System Specification Practices

Linking the Specification Tree to the CWBS and System Architecture

People often mistakenly develop the *specification tree* as an *independent* activity unrelated to the system architecture and *contract work breakdown structure (CWBS)*. This is a serious error! The *specification tree* and the CWBS should reflect the primary structure of the system architecture and be linked accordingly. To illustrate this point, consider the graphic shown in Figure 28.3.

Guidepost 28.1 *Now that we have established the specification tree as the framework for linking SYSTEM entity specifications, let's shift our focus to understanding the structure of most specifications.*

28.8 UNDERSTANDING THE BASIS FOR SPECIFICATION CONTENT

The learning process for most engineers begins with a specification outline. However, most engineers lack exposure and understanding as to HOW the specification outline was derived. To eliminate the gap in this learning process, let's back away from the details of outlines and address WHAT specifications should specify. Let's begin by graphically depicting a system entity and the key factors that drive its *form, fit,* and *function*. Figure 28.4 serves this purpose.

Author's Note 28.1 *Remember, the term SYSTEM or entity is used here in a generic sense. By definition, SEGMENTs, PRODUCTs, SUBSYSTEMs, ASSEMBLIES, HWCIs, and CSCIs are SYSTEMs. Thus, the discussion here applies to any level of abstraction.*

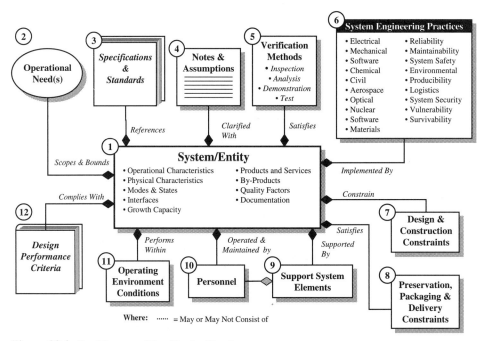

Figure 28.4 Key Elements of Specification Development

Specifying the System of Interest (SOI)

The *SYSTEM entity*, for which the specification is to be written, is shown in the central part of the figure. If we analyze systems, we will find there are several key attributes that characterize a system entity or item. These attributes include: 1) operational characteristics, 2) physical characteristics, 3) modes and states, 4) interfaces, 5) growth capacity, 6) products and services, 7) by-products, 8) mass properties, 9) quality factors, and 10) documentation. The intent here is to characterize the entity's *behavioral* and *physical* characteristics and properties.

The challenge question is, however, *HOW do we arrive at this set of attributes*? The answer resides in a variety of external factors that *influence* and *constrain* the entity. Let's investigate these external factors.

External Factors That Drive SYSTEM Requirements

The process of specifying a SYSTEM entity's requirements is driven by external factors that bound and constrain the entity's solution.

Operational Need (2). The purpose of a SYSTEM entity is to satisfy a User(s) operational needs. Thus, the specification must specify requirements for capabilities and level of performance that have been derived from and are traceable to the User's *operational needs*.

Specifications, Standards, and Statutory Constraints (3). The design of any SYSTEM entity often requires strict compliance with existing specifications, standards, and statutory constraints that may include interfacing systems, workmanship, and materials.

Notes and Assumptions (4). Since requirements are specified with text and graphics, they often call for contextual *clarifications*. Where some requirements are *unknown*, assumptions may be used. Additionally graphical *conventions* may have to be clarified. Therefore, specifications should include Notes and Assumptions to provide a definition, context and usage information, and terms and *conventions*.

Verification Methods (5). Requirements that specify key attributes of a SYSTEM entity are often written in language and terms the system designers understand. The challenge is, however, *HOW do you verify and validate that the physical entity complies with the requirements*? To solve this dilemma, specifications include a section that specifies requirements for entity *verification* and *validation (User option)*.

System Engineering Practices (6). Successful system development requires that *best practices* derived from *lessons learned* and engineering discipline be *consistently* applied to *minimize* risk. Therefore, specifications invoke system engineering practices to ensure that the deliverable product will achieve SYSTEM requirements.

Design and Construction Constraints (7). Specifications do more than communicate WHAT and HOW WELL a SYSTEM/entity must be accomplished. They communicate the *constraints* that are levied on the SYSTEM and development decisions related to SYSTEM operations and capabilities. We refer to these as *design and construction constraints*. In general, *design and construction constraints* consist of *nonfunctional* requirements such as size, weight, color, mass properties, maintenance, safety restrictions, human factors, and workmanship.

Preservation, Packaging, and Delivery (8). When the SYSTEM is to be delivered, care must be taken to ensure that it arrives fully capable and available to support operational missions. Preservation, packaging, and delivery requirements specify how the deliverable SYSTEM/entity is to be prepared, shipped, and delivered.

SUPPORT SYSTEM Element Requirements (9). MISSION SYSTEMS require sustainable pre-mission, mission, and postmission *support test equipment* (STE) at critical *staging events* and *areas*. This may require the use of *existing* facilities and support equipment such as *common support Equipment* (CSE) or *peculiar support Equipment* (PSE) or the need to develop those items. Therefore, specifications include SUPPORT SYSTEM element requirements.

PERSONNEL Element Requirements (10). Many SYSTEMs typically require the PERSONNEL Element for "hands-on" operation and control of the SYSTEM during pre-mission, mission, and postmission operations. Additionally, human-machine trade-offs must be made to *optimize* system performance. This requires delineating and specifying WHAT humans do best versus WHAT the EQUIPMENT element does best. Therefore, specifications identify the skill and training requirements to be levied on the PERSONNEL Element to ensure human-system integration success.

Operating Environment Conditions (11). Every entity must be capable of performing missions in a prescribed OPERATING ENVIRONMENT at a level of performance that will ensure mission success. Therefore, specifications must define the OPERATING ENVIRONMENT conditions that drive and bound entity capabilities and levels of performance.

Design Performance Criteria (12). SYSTEM entities are often required to operate within performance envelopes, especially when simulating the performance of or interfacing with the physical SYSTEMs. When this occurs, the specification must invoke external performance requirements that are characterized as *design criteria*. In these cases a *Design Criteria List (DCL)* for the interfacing systems or system to be simulated is produced to serve as a reference.

28.9 UNDERSTANDING THE GENERALIZED SPECIFICATION STRUCTURE

We discussed earlier in this practice the need to employ a *standard* outline to prepare a specification. Standard specifications typically employ a seven-section outline. The basic structure includes the following elements:

Front Matter

Section 1.0: Introduction
Section 2.0: Referenced Documents
Section 3.0: Requirements
Section 4.0: Qualification Provisions
Section 5.0: Packaging, Handling, Storage, and Delivery
Section 6.0: Requirements Traceability
Section 7.0: Notes

We will address these requirements in more detail in Chapter 32 *Specification Development Practices*.

Author's Note 28.2 *As is ALWAYS the case, consult your contract for guidance on performance specification formats—and your Contract Data Requirements List (CDRL) item formats. If guidance is not provided, confer with the Program's Technical Director or organization's Contracting Officer (ACO).*

28.10 SPECIFICATION DEVELOPMENT EVOLUTION AND SEQUENCING

To better understand the development and sequencing of specifications, we use the structure of the System Development Process *workflow* as illustrated in Figure 24.2. If we generalize the System Development Process in time, Figure 28.5 illustrates HOW the family of specifications evolves over time from Contract Award through System Verification Test (SVT).

28.11 GUIDING PRINCIPLES

In summary, the preceding discussions provide the basis with which to establish the guiding principles that govern specification approaches.

Principle 28.1 Specifications state WHAT the SYSTEM/entity is to accomplish and HOW WELL; CSOWs specify tasks that the System Developer is to perform and the work products to be delivered. Preserve and protect the scope of each document.

28.12 SUMMARY

In our discussion of the *specification practices* we identified key challenges, issues, and methods related to developing system specifications. As the final part of the concept, we introduced the structure for a general specification. Given a fundamental understanding of specifications, we are now ready to explore types of requirements documented in a specification via the next topic on *understanding system requirements*.

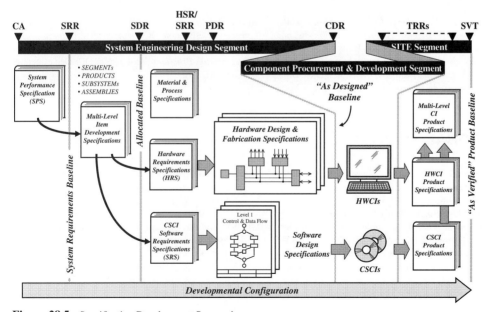

Figure 28.5 Specification Development Sequencing

314 Chapter 28 System Specification Practices

GENERAL EXERCISES

1. Answer each of the *What You Should Learn from This Chapter* questions identified in the *Introduction*.
2. Refer to the list of systems identified in Chapter 2. Based on a selection from the preceding chapter's General *Exercise* or a new system, selection, apply your knowledge derived from this chapter's topical discussions. If you were a consultant to an Acquirer that had ample resources:
   (a) What types of specifications would you recommend as Contract Data Requirements List (CDRL) items. Provide rationale for your decisions.
   (b) Annotate each of the specification outline topics with a brief synopsis of what you would recommend the Acquirer's SEs address in the *System Performance Specification (SPS)*.

ORGANIZATION CENTRIC EXERCISES

1. Research your organization's command media for guidance concerning development of specifications.
   (a) What requirements do the command media impose on contract programs?
   (b) What does the media specify about timing for developing and approving specifications?
2. Contact a contract program within your organization.
   (a) What specifications are required to be delivered as *Contract Data Requirements List (CDRL)* items?
   (b) What delivery requirements are levied on the program specification CDRL deliveries during the System Development Phase?
   (c) Does the program have a specification tree?
   (d) Contrast the specification tree structure with the multi-level system architecture. Do they identically match? How do they differ? What is the program's rationale for the difference?
   (e) What specifications are required that are viewed as unnecessary?
   (f) What specification did the program develop that were not required by contract?

REFERENCES

Defense Systems Management College (DSMC). 2001. DSMC *Glossary: Defense Acquisition Acronyms and Terms*, 10th ed. Defense Acquisition University Press. Ft. Belvoir, VA.

Kossiakoff, Alexander, and Sweet, N. William 2003. *Systems Engineering Principles and Practice.* New York: Wiley-InterScience.

MIL-STD-499B (cancelled dRAFT). 1994. *Systems Engineering.* Washington, DC: Department of Defense (DoD).

MIL-STD-961D. 1995. *DOD Standard Practice for Specification Practices.* Washington, DC: Department of Defense (DoD).

ADDITIONAL READING

Defense Systems Management College (DSMC). 1998. DSMC *Test and Evaluation Management Guide*, 3rd ed. Defense Acquisition Press. Ft. Belvoir, VA.

IEEE Std 610.12-1990. 1990. *IEEE Standard Glossary of Modeling and Simulation Terminology.* Institute of Electrical and Electronic Engineers (IEEE). New York, NY.

MIL-HDBK-245D. 1996. *Handbook for Preparation of the Statement of Work.* Washington, DC: Department of Defense (DoD).

MIL-STD-480B (canceled). 1988. *Military Standard: Configuration Control: Engineering Changes, Deviations, and Waivers.* Washington, DC: Department of Defense (DoD).

MIL-STD-490A (canceled). 1985. *Military Standard: Specification Practices.* Washington, DC: Department of Defense (DoD).

SD-15. 1995. *Performance Specification Guide.* Washington, DC: Department of Defense (DoD).

Chapter 29

Understanding Specification Requirements

29.1 INTRODUCTION

Specifications establish the agreement of the technical capabilities and levels of performance required for a system to achieve its mission and objectives within a prescribed *solution space*. As such, specifications *represent* human attempts to *specify* and *bound* the prescribed *solution space* that will enable the User to accomplish their organizational mission and objectives.

This chapter provides the foundation for specifying *system, product*, or *service* requirements. We explore the various categories and types of stakeholder and specification requirements—such as operational, capability, nonfunctional, interface, verification, and validation requirements. We expand the discussion of operational requirements and link them to the fours modes of operation—such as NORMAL, ABNORMAL, EMERENCY, and CATASTROPHIC.

Specification requirements that bound a *solution space* are hierarchical and interrelated. We discuss the hierarchical structure and relationships among the various types of requirements. We illustrate why specifications prepared in an ad hoc manner are prone to problems of *missing*, *misplaced*, *conflicting*, and *duplicated* requirements. These problems represent risk areas that SEs need to understand and recognize.

What You Should Learn from This Chapter

1. What is a *requirement*?
2. What is a *source* or *originating* requirement?
3. What is a *stakeholder* requirement?
4. What is an *objective* requirement?
5. What is a *threshold* requirement?
6. What are the categories of specification requirements?
7. What are *operational* requirements?
8. What are *capability* requirements?
9. What are *nonfunctional* requirements?
10. What are *interface* requirements?
11. What are *verification* requirements?
12. What are *validation* requirements?

System Analysis, Design, and Development, by Charles S. Wasson
Copyright © 2006 by John Wiley & Sons, Inc.

13. What are requirement priorities?
14. What are the four types of *operational* requirements?
15. What are four common problems with requirements?

Definitions of Key Terms

- **Nonfunctional Requirement** A statement of a static, nonbehavioral requirement that characterizes an attribute—such as appearance, and weight, dimensions—of an entity.

- **Objective Requirement** The objective value is that value desired by the user, which the PM is contracting for or otherwise attempting to obtain. The objective value could represent an operationally meaningful, time critical and cost-effective increment above the threshold for each program parameter. (Adapted from the DoD Acquisition Program Preparation Instruction; Source: DoD 5000.2R Draft, Jan. 1, 2001, Section 1.2)

- **Operational Requirement** A statement of a required capability the represents an integrated set of operations needed to satisfy a specific mission objective and outcome within a specified time frame and prescribed OPERATING ENVIRONMENT.

- **Performance Requirement** "A criterion that identifies and quantifies the degree to which a particular attribute of a function must be accomplished, and the conditions under which this capability is to be achieved." (Source: Kossiakoff and Sweet, *System Engineering*, p. 451)

- **Threshold Requirements** Requirements stated with a *minimum* acceptable value that, in the user's judgment, is necessary to satisfy the need. If threshold values are not achieved, program performance is seriously degraded, the program may be too costly, or the program may no longer be timely. (Adapted from the DoD Acquisition Program Preparation Instruction; Source: DoD 5000.2R Draft, Jan. 1, 2001, Section 1.2)

- **Requirement Elicitation** The act of identifying and collecting stakeholder requirements through understanding of the *problem* and *solution spaces* such as via personal interviews and observation.

- **Requirement Stakeholder** Anyone who has a stake or vested interest in *identifying, defining, specifying, prioritizing, verifying*, and *validating* system capability and performance requirements. Requirements stakeholders include all personnel responsible for system definition, procurement, development, production, operations, support, and retirement of a system, product or service over its life cycle.

- **Requirement** "Any condition, characteristic, or capability that must be achieved and is essential to the end item's ability to perform its mission in the environment in which it must operate is a requirement. Requirements must be verifiable." (Source: SD-15 *Performance Specification Guide*, p. 9)

- **Verification Requirement** A statement of a method and conditions for demonstrating successful achievement of a system, product, or service capability and its *minimum/maximum* level of performance.

- **Validation Requirement** A statement of an approach and/or method to be employed to demonstrate and confirm that a *system*, *product*, or *service* satisfies a User's documented operational need.

29.2 WHAT IS A REQUIREMENT?

The heart of a specification resides in the requirements. Each requirements statement serves to *specify* and *bound* the deliverable system, product, or service to be developed or modified. The prerequisite to developing requirements is to first identity WHAT capability is required. Therefore, you need to understand HOW requirements are *categorized*.

Understanding Types of Requirements

People often think of requirements in a generic sense and fail to recognize that requirements also have specific *missions* and *objectives*. If you analyze the requirements stated in the specifications, you will discover that requirements can be grouped into various *categories*. This discussion identifies the types of requirements and delineates usage of terms that sometimes complicate the application.

Source or Originating Requirements

When an Acquirer formally releases a requirements document that *specifies* and *bounds* the *system* or entity for procurement, the specifications are often referred to as the SOURCE or ORIGINATING requirements. These requirements can encompass multiple categories of requirements in a single document or several documents, such as a *System Requirements Document (SRD)* and a *Statement of Objectives (SOO)*.

One of the problems in applying the term *source* or *originating requirements* is that it is *relative*. Relative to WHOM? From an Acquirer's perspective, the *source* or *originating* requirements should be *traceable* to the User's *validated* operational need. These needs may evolve through a chain of decision documents that culminate in a procurement specification as illustrated earlier in Figure 24.5.

From a System Developer's perspective, *source* or *originating requirements* are applied to the Acquirer's procurement specification such as the *Request for Proposal (RFP)*, SRD, or SOO used to acquire the system, product, or service.

Stakeholder Requirements

A specification captures all *essential* and *prioritized* stakeholder requirements that FIT within User's budget and development schedule *constraints*.

Requirements Stakeholder Objectives. The primary objectives of the *requirements stakeholders* are to ensure that:

1. All requirements essential to their task domain are identified, analyzed, and documented in a *System Performance Specification (SPS)*.
2. Specific requirements are *accurately* and *precisely* specified and given an equitable priority relative to other stakeholder requirements.
3. Domain/disciplinary requirements are *clear, concise, easy to understand, unambiguous, testable, measurable, and verifiable*.

Requirements Stakeholder Elicitation and Documentation. Since *stakeholder* requirements reside on both sides of the contract interface boundary, each organizational entity—User, Acquirer, and System Developer—is accountable for stakeholder requirements identification. So, *HOW does this occur?*

318 Chapter 29 Understanding Specification Requirements

The Acquirer is typically *accountable* for establishing a *consensus* of agreement, preferably before the contract is awarded, of all User operational stakeholders. Specifically, *stakeholders* with *accountability* for the system *outcomes* and *performance* during the System Production, System Deployment, System Operations and Support (O&S), and System Disposal Phases of the system/product life cycle.

Author's Note 29.1 *The System Developer also has contract stakeholder performance and financial interests in clarifying any internal stakeholder requirements that relate to the System Development and System Production, if applicable, Phases of the contract. Thus, the System Developer stakeholder requirements must be addressed prior to submitting the SPS as part of a proposal as well as during contract negotiations.*

Warning! Read, fully understand, and comprehend WHAT you are signing the contract to perform.

29.3 SPECIFICATION REQUIREMENTS CATEGORIES

Requirements can be organized into various categories to best capture their intended use and objectives. Typical categories include: 1) *operational requirements*, 2) *capability and performance requirements*, 3) *nonfunctional requirements*, 4) *interface requirements*, 5) *design and construction requirements*, 6) *verification requirements*, and 7) *validation requirements*. Let's briefly describe each of these categories of requirements.

Operational Requirements

Operational requirements consist of high-level requirements related to system mission objectives and behavioral interactions and responses within a prescribed OPERATING ENVIRONMENT and conditions. In general, WHAT is the entity expected to accomplish?

Capability Requirements

Capability requirements specify and bound a *solution space* with functional/logical and performance actions each system entity or item must be capable of producing such as outcome(s), products, by-products, or services. Traditionally these requirements were often referred to as *functional requirements* and focused on the *function* to be performed. *Capability requirements*, however, encompass both the *functional/logical* substance and *level of performance* associated with HOW WELL the function must be performed.

Nonfunctional Requirements

Nonfunctional requirements relate to physical system attributes and characteristics of a system or entity. *Nonfunctional requirements* DO NOT perform an action but may influence a specific operational *outcome* or *effect*. For example, does bright yellow paint (nonfunctional requirement) improve safety?

Interface Requirements

Interface requirements consist of those statements that *specify* and *bound* a system's direct or indirect connectivity or *logical* relationships with external system entities *beyond* its own physical boundary.

Verification Requirements

Verification requirements consist of requirements statements and methods to be employed to assess system or entity compliance with a *capability*, *performance parameter*, or *nonfunctional requirement*. *Verification requirements* are typically stated in terms of *verification methods* such as *inspection*, *analysis*, *demonstration*, and *test*.

Validation Requirements

Validation requirements consist of mission-oriented, use case scenario statements intended to describe WHAT must be performed to clearly demonstrate that the RIGHT *system* has been built to satisfy the User's intended operational needs. *Validation requirements* are typically documented in an *Operational Test and Evaluation (OT&E) Plan* prepared by the User or an Independent Test Agency (ITA) representing the User. Since *use cases* represent HOW the User envisions using the SYSTEM, a use cases document would serve a comparable purpose. In general, *validation* should demonstrate that *critical operational* and *technical issues* (COIs/CTIs) have been resolved or minimized.

29.4 REQUIREMENTS PRIORITIES

When requirements of *elicited*, each stakeholder places a value on the importance of the requirement to enable them to achieve their organizational missions. We refer to the value as a *priority*. The realities of system development are that *EVERY requirement has a cost to implement and deliver*. Given limited resources and stakeholder values, bounding the *solution space* requires reconciling the cost of the desired requirements with the available resources. As a result, requirements must be *prioritized* for implementation.

Threshold and Objective Requirements

One method of expressing and delineating requirements priorities is to establish *threshold* and *objective* requirements.

- *Threshold requirements* express a *minimum acceptable* capability and level of performance.
- *Objective requirements* express goals *above* and *beyond* the *threshold* requirements level that the Acquirer expects the System Developer to *strive* to achieve.

EXAMPLE 29.1

A User may require a *minimum* level of acceptable SYSTEM performance—threshold requirement—with an *expressed* desire to achieve a specified higher level. Achievement of a higher level objective requirement may be dependent on the maturity of the technology and a vendor's ability to *consistently* and *reliably* produce the technology.

Threshold and *objective* requirements must be consistent with the *Operational Requirements Document (ORD)* and the *Test and Evaluation Master Plan (TEMP)*.

When you specify *threshold* and *objective* requirements, you should *explicitly* identify as such. There are several ways of doing this.

Option 1: Explicitly label each *threshold* or *objective* requirement using parenthesis marks such as XYZ Capability (Threshold) and ABC Capability (Objective).

Option 2: Specify "up front" in the specification that unless specified otherwise, all requirements are "threshold requirements." Later in the document when *objective requirements* are stated, the following label might be used ". . . ABC Capability (Objective). . . ."

In any case, you should ALWAYS define the terms (i.e., *threshold* and *objective*) in the appropriate section of the specification. A *third option* might involve a summary matrix of *threshold* and *objective* requirements. A *shortcoming* of the matrix approach is that it is physically located away from the stated requirement. This approach, which may be confusing to the reader, creates extra work flipping back and forth between text and the matrix. KEEP IT SIMPLE and tag the statement (*threshold* or *objective*) with the appropriate label.

29.5 TYPES OF OPERATIONAL REQUIREMENTS

When specification developers derive and develop requirements, human tendencies naturally focus on the SYSTEM operating *ideally*—meaning normally and successfully. In the *ideal* world this may be true, but in the real world systems, products, and services have finite reliabilities and life cycles. As such, they may not always function normally. As a result, specifications must specify *use case* requirements that cover phases and modes for a prescribed set of OPERATING ENVIRONMENT scenarios and conditions.

Phases of Operation Requirements

Systems, products, and services perform in operational phases. As such, SEs must ensure that the requirements that *express* stakeholder expectations for each phase are adequately addressed. At a *minimum*, this includes pre-mission phase, mission phase, and postmission phase operations.

Modes of Operation Requirements

Each *phase of operation* includes at least one or more modes of operation that require a specified set of capabilities to support achievement of mission phases of operation objectives. If you investigate and analyze most MISSION SYSTEM operations, you will discover that the SYSTEM must be prepared to cope with four types of OPERATING ENVIRONMENT scenarios and conditions: (1) *NORMAL*, (2) *ABNORMAL*, (3) *EMERGENCY*, and (4) *CATASTROPHIC*. Let's define the context of each of these conditions:

NORMAL Operations

NORMAL operations consist of the set of MISSION SYSTEM activities and tasks that apply to SYSTEM capabilities and performance operating within their specified performance limits and resources.

ABNORMAL Operations. *ABNORMAL operations* consist of a set of SYSTEM operations and tasks that focus on *identifying*, *troubleshooting*, *isolating*, and *correcting* a functional or physical capability condition. These conditions represent levels of performance that are outside the tolerance band for NOMINAL mission performance but may not be *critical* to the safety of humans, property, or the environment.

EMERGENCY Operations. *EMERGENCY operations* consist of a set of SYSTEM operations and tasks that focus exclusively on *correcting* and *eliminating* a *threatening, hazardous to safety*, physical capability condition that has the potential to pose a major health, safety, financial, or security risk to humans, organizations, property, or the environment.

CATASTROPHIC Operations. *CATASTROPHIC operations* consist of a set of operations or tasks performed following a *major SYSTEM malfunction event* that may have *adversely* affected the health, safety, financial, and security of humans, organizations, property, and the environment in the immediate area.

Relationships between Operating Condition Categories

Requirements for SYSTEM operating conditions must not only *scope* and *bound* the *use case* scenario or the condition but also the transitory *modes* and *states* that contributed to the conditions. To better understand this statement, refer to Figure 29.1.

As indicated in the figure, a SYSTEM operates under NORMAL operations. The challenge for SEs is to derive reliability, availability, maintainability, and spares requirements to support onboard *preventive* and *corrective* maintenance operations to sustain NORMAL operations. If a *condition* or *event* occurs prior to, during, or after a mission, the SYSTEM may be forced to transition (1) to ABNORMAL operations. NORMAL operations can conceivably encounter an emergency and transition (3) immediately to EMERGENCY operations. Analytically, Figure 29.1 assumes some ABNORMAL conditions precede the EMERGENCY.

During ABNORMAL operations, a SYSTEM or its operators institute recovery operations, correct or eliminate the condition, and hopefully return to NORMAL operations (2). During ABNORMAL operations, an *emergency condition* or *event* may occur forcing the SYSTEM into a state of EMERGENCY mode operations (3). Depending on the system and its post emergency health condition, the SYSTEM may be able to shift back to ABNORMAL operations (4).

If EMERGENCY operations are *unsuccessful*, the SYSTEM may encounter a *catastrophic event*, thereby requiring CATASTROPHIC mode operations (6).

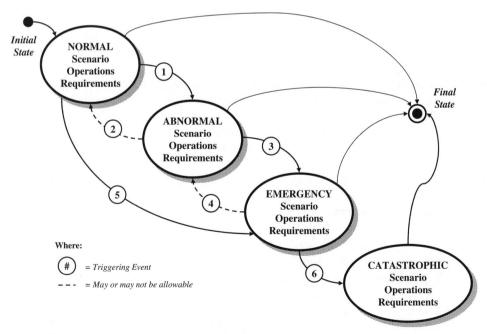

Figure 29.1 System Operations Requirements Coverage

How Do These Categories Relate to SYSTEM Requirements?

Depending on the SYSTEM and its mission applications, the *System Performance Specification (SPS)* must specify capability and performance requirements that cover these conditions to ensure the safety of humans, property, and the environment. *Will you find a section in the SPS titled NORMAL, ABNORMAL, EMERGENCY, or CATASTROPHIC operations*? Generally, no. Instead, requirements for these types of *conditions* are distributed throughout the SPS. As a matter of good practice, however, it is recommended that Acquirer and System Developers maintain some type of documentation that links SPS requirements to these conditions to ensure proper coverage and consideration.

Failure to Define Requirements for these Categories

History is filled with events that exemplify a SYSTEM's *inability* to COPE with physical conditions in its OPERATING ENVIRONMENT simply because these conditional and scenario requirements were *ignored, overlooked*, or *assumed* to have only a remote probability of occurrence.

Allocation of Conditional sRequirements to System Elements

When you develop the SPS, keep in mind that the SYSTEM must be capable of accommodating all types of NORMAL, ABNORMAL, EMERGENCY, or CATASTROPHIC conditions within the practicality of budgets. Note the operative word "SYSTEM" here. Remember, a SYSTEM encompasses ALL system elements. Ultimately, when EQUIPMENT is specified, SEs also allocate responsibility and accountability for some conditional requirements to the PROCEDURAL DATA, PERSONNEL, MISSION RESOURCES, or SUPPORT SYSTEM elements that satisfy safety requirements and reduce development costs.

29.6 UNDERSTANDING THE REQUIREMENTS HIERARCHY

System Performance Specification (SPS) requirements generally begin with high-level, mission-oriented requirements statements. These high-level statements are then elaborated into successively lower level requirements that *explicitly clarify*, *specify*, and *bound* the intent of the higher level requirements. In effect, a hierarchical structure of requirements emerges.

Contrary to what many people believe, requirements should not be stated as random *thoughts* or wish lists. *Objective evidence* of this mindset is illustrated in the way specifications are often written. We will discuss this point later in *Specification Development Practices*. Instead, specifications should be *structured* and *organized* via a framework of hierarchical links. Thus, the framework provides a referential mechanism to trace requirements back to one or more *source* or *originating requirements*.

We can depict the hierarchical structure of these requirements graphically as shown in Figure 29.2. Note that each of the requirements is labeled using a convention that depicts its lineage and traceability to higher level requirements.

EXAMPLE 29.2

Parent capability requirement, R1, has three SIBLING capability requirements, R11, R12, and R13. The labeling convention simply adds a right-most digit for each level and begins the numerical sequence with "_1." Thus, you can follow the "root lineage" for each requirement by decoding the numerical sequence of digits. The lineage of requirement, R31223, is based on the following decompositional "chain": R3 $\Rightarrow$ R31 $\Rightarrow$ R312 $\Rightarrow$ R3122 $\Rightarrow$ R31223.

29.7 Common Problems with Specification Requirements

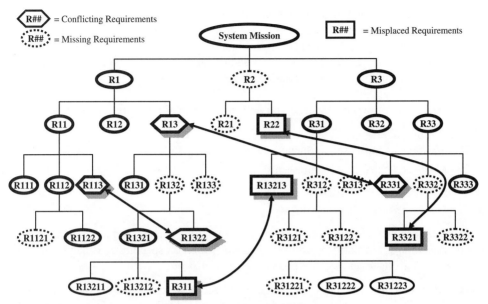

Figure 29.2 Requirements Hierarchy Tree Illustrating Conflicting Requirements

Author's Note 29.2 *The hierarchy shown is best described as ideal—meaning all requirements are properly aligned to higher level requirements. For discussion purposes, we will restrict the simplicity of this diagram to generic requirements. Requirements are derived from and must be traceable to measures of effectiveness (MOEs), measures of suitability (MOSs), and measures of performance (MOPs) at the system and mission levels.*

Referral For more information about MOEs, MOS, and MOPs, refer to the *Operational Utility, Suitability,* and *Effectiveness Practices* discussion in Chapter 34.

The SYSTEM's mission represents the highest level requirement in Figure 29.2. The mission is *scoped, bounded,* and *described* by three high-level capability requirements, R1, R2, and R3. When a single requirement, such as R12, R32, and R33, effectively ends the chain, the term "leaf" requirement is sometimes used.

Guidepost 24.1 *From this basic understanding of a theoretical capability requirement hierarchy we are now ready to investigate common problems with specification requirements.*

29.7 COMMON PROBLEMS WITH SPECIFICATION REQUIREMENTS

Specifications often have *imperfections* and are *unintentionally* released with a number of *errors, defects,* and *deficiencies.* To illustrate these conditions, let's identify a series of *undesirable* conditions that commonly plague specifications.

We noted earlier that some people develop specifications as *random* sets of thoughts or *wish lists* organized to a *standard* outline structure. Specifications of this type typically *evolve* from informal to formal brainstorming sessions. Requirements captured during the session are published for review. During the session reviewers *lobby* other reviewers to support additions of their respective desires to the structure.

324 Chapter 29 Understanding Specification Requirements

Author's Note 29.3 Requirements should NOT be added to a specification *unless* budgetary resources are provided. Remember, each requirement costs money to *implement*—be it hardware or software. At the SPS level, requirements changes should be managed as contract modifications. Within the System Developer's organization, any additional requirements should include commensurate cost and schedule modification considerations.

If you analyze specification *work products*, you will discover they typically *exhibit* at least four major types of problems:

Problem 1: *Missing* capability requirements
Problem 2: *Misplaced* capability requirements
Problem 3: *Conflicting* specification requirements
Problem 4: *Duplicated* requirements

Figure 29.2 illustrates these conditions that often result in *errors*, *defects*, and *deficiencies*.

These four *conditions* exemplify a few of the *common problems* specification analysts and SEs encounter. This further illustrates two key points:

- WHY it is important to formally train SEs, specification requirements analysts, and *stakeholders* in HOW to analyze, prepare, and review the specifications before you contractually commit to their implementation.
- WHY it is important to conduct specification reviews with all *stakeholders*. People who casually *read* specifications in a linear "front to back" or sectional manner for proper grammar and text usage typically *overlook* or *fail* to assimilate and recognize these conditions.

29.8 GUIDING PRINCIPLES

In summary, our discussions in this section provide the basis with which to establish the several guiding principles that govern specification requirements practices.

Principle 29.1 Every specification must specify requirements that address four types of system conditions:

1. Normal operations
2. External system failures
3. Degraded operations
4. Internal system failures

29.9 SUMMARY

Our discussion of specification requirements provided an introduction into the types of requirements contained within specifications statements. Later, chapters will use this foundation to determine the requirements for internally developed configuration items (CIs) or the procurement of nondevelopmental items (NDIs) from external vendors.

GENERAL EXERCISES

1. Answer each of the *What You Should Learn from This Chapter* questions identified in the *Introduction*.
2. Refer to the list of systems identified in Chapter 2. Based on a selection from the preceding chapter's General *Exercises* or a new system selection, apply your knowledge derived from this chapter's topical discussions. Identify and describe the four types operations—NORMAL, ABNORMAL, EMERGENCY, and CATASTROPHIC, as applicable—that apply to the selected system.

ORGANIZATION CENTRIC EXERCISES

1. Contact several contract programs within your organization. Request an opportunity to analyze the *System Performance Specification (SPS)* for each program and answer the following questions:
   (a) Identify five examples of *operational* requirements.
   (b) Identify five examples of *capability* requirements.
   (c) Identify five examples of *nonfunctional* requirements
   (d) Identify five different examples of *verification* requirements.
   (e) Identify five examples of *design and construction* constraints.
   (f) Are *threshold* and *objective* requirements identified?
2. Interview program technical management and SEs. How were requirements in the *System Performance Specification (SPS)* elicited and collected from stakeholders? Document your *findings* and *observations*?
3. What lessons learned did program personnel learn in the following areas and how did they resolve the issue?
   (a) Missing requirements
   (b) Misplaced requirements
   (c) Conflicting requirements
   (d) Duplicated requirements
   (e) Nonfunctional requirements
   (f) Verification requirements
4. What types of metrics are used to track specification *defects* and *deficiencies*?
5. Select a specification on a contract program for analysis. Using the concepts discussed in this chapter, as a consultant to the specification developer, identify defects and deficiencies in the specification and suggest recommendations for improvement.
6. Research contract specifications and identify those that specify the precedence of requirements for decision making.
7. For a contract program, identify who the SPS stakeholders are based on the specified requirements. Based on those stakeholders, prioritize them in terms of an estimate based on requirements count.

REFERENCES

DoD 5000.2R (Interim Regulation). 2000. *Mandatory Procedures for Major Defense Acquisition Programs (MDAPs) and Major Automated Information System (MAIS) Acquisition Programs.* Washington, DC: Department of Defense (DoD).

KOSSIAKOFF, ALEXANDER, and SWEET, WILLIAM N. 2003. *Systems Engineering Principles and Practice.* New York: Wiley-InterScience.

SD-15. 1995. *Performance Specification Guide.* Washington, DC: Department of Defense (DoD).

ADDITIONAL READING

IEEE Std 610.12-1990. 1990. *IEEE Standard Glossary of Modeling and Simulation Terminology*. Institute of Electrical and Electronic Engineers (IEEE). New York, NY.

Defense Systems Management College (DSMC). 2001. *Glossary: Defense Acquisition Acronyms and Terms, 10th ed.* Defense Acquisition University Press. FT, BELVOIR, VA

Chapter 30

Specification Analysis

30.1 INTRODUCTION

When systems, products, and services are acquired, the Acquirer typically provides a *System Requirements Document (SRD)* or *Statement of Objectives (SOO)* with the formal *Request for Proposal (RFP)* solicitation. The SRD/SOO *expresses* the required set of capabilities and performance Offerors use as the basis to submit solution-based proposals. The challenge for Acquirers and System Developers is to formulate, derive, and negotiate a *System Performance Specification (SPS)* that:

1. *Concisely* and *completely* bounds the solution space.
2. Is *well understood* by all stakeholders.
3. Establishes the basis for deliverable system, product, or service technical acceptance.

Our discussion in this chapter describes various *Specification Analysis Practices*. We explore various methods and techniques used to analyze specification requirements for completeness from several perspectives. We introduce common specification practice *deficiencies* and investigate *methods* for identifying, tracking, and resolving these deficiencies. We also consider semantic ambiguities such as *comply* versus *conform* versus meet that Acquirers and System Developers employ that do have significance in interpretation.

What You Should Learn from This Chapter

1. How do you methodically *analyze* a specification?
2. What are some common types of *specification requirements deficiencies*?
3. When requirements deficiencies are identified, how should you resolve them?
4. What does it mean to *comply* with a requirement?
5. What does it mean to *conform* to a requirement?
6. What does it mean to *meet* a requirement?

30.2 ANALYZING EXISTING SPECIFICATIONS

Our previous discussion focused on HOW an Acquirer might develop their procurement *System Requirements Document (SRD)* or the System Developer might developer lower level *item development specifications*. However, *WHAT if a specification already exists*? HOW does the Acquirer or a System Developer candidate analyze the specification for *completeness*, *reasonableness*, and feasibility?

There are two key contexts regarding analysis of specifications:

1. Acquirer verification PRIOR TO formal solicitation.
2. System Developer analysis during the proposal process and following Contract Award (CA).

Let's investigate these contexts further.

Acquirer Perspective

Acquirers start by reducing the risk of the procurement action. *How can this be accomplished?* It is by releasing a high-quality draft specification that *accurately*, *precisely*, and *completely* specifies and bounds the *solution space* system, product, or service. Before release, the Acquirer must thoroughly review the specification for *accuracy*, *completeness*, *specificity*, and *legal* purposes. Key review questions to ask might include:

1. *Have all User System Deployment Phase, System Operations and Support (O&S) Phase, and System Disposal Phase stakeholder requirements been adequately identified, prioritized, scoped, and specified?*
2. *Have we bounded the CORRECT solution space within the problem space?*
3. *Have we identified the RIGHT system to fill the prescribed solution space and cope with OPERATING ENVIRONMENT?*
4. *Does this specification adequately, accurately, and precisely specify the selected solution space?*
5. *If we procure a system based on these requirements, will the deliverable product satisfy the User's intended operational needs?*
6. *Can the system specified be developed within the total life cycle cost—such as acquisition, operations and support (O&S), and retirement costs—budgets that are available?*

What happens if you inadequately address these and other questions? Later, if it is determined that the requirements have latent *defects* such as *errors*, *deficiencies*, or *omissions*, the cost to modify the contract can be very expensive. To minimize specification risk, Acquirers often release a pre-solicitation draft specification for qualified candidate Offeror comments.

System Developer Perspective

In contrast, the System Developer must reduce contract cost, schedule, and technical risk. To do this, specification analysis must answer many questions. Key review questions might include:

1. *Do we fully understand the scope of the effort we are signing up to perform?*
2. *Do the system requirements, as stated, specify a system that satisfies the User's operational needs? If not, what approach must we use to inform them?*
3. *Have we thoroughly investigated and talked with a representative cross section of the User community to validate their requirements and needs?*
4. *Do we understand the problem the User is attempting to solve by procuring this system? Does the specification bound the problem or a sympton of the problem?*
5. *Can the requirements, as stated, be verified within reasonable expectations, cost, schedule, and risk?*
6. *Do these requirements mandate technologies that pose unacceptable risks?*

30.3 SPECIFICATION ANALYSIS

Given the Acquirer and System Developer perspectives, *HOW do they approach analysis of the specification*? The answer encompasses the methods and techniques identified in earlier questions in Chapters 3 to 29. Due to the broad scope of this answer, we will briefly address some high level approaches you can apply to specification analysis.

Visually "Inspect" the Specification Outline

Examine the outline STRUCTURE for *missing* sections and topics that are *crucial* to developing a system of the type specified.

Perform System Requirements Analysis (SRA)

Perform a System Requirements Analysis (SRA) to understand WHAT the system is expected to do. Ask key questions such as:

1. *Do the list of requirements appear to be generated as a feature-based "wish list" or reflect structured analysis?*
2. *Do the requirements follow standard guidelines discussed later in Requirements Statement Development Practices?*
3. *Do the requirements appear to have been written by a seasoned subject matter expert (SME) or semi-knowledgeable person?*
4. *Do the requirements adequately capture User operational needs? Are they necessary and sufficient?*
5. *Do the requirements unnecessarily CONSTRAIN the range of viable solutions?*
6. *Are all system interface requirements identified?*
7. *Are there any TBDs remaining in the specification?*
8. *Are there any critical operational or technical issues (COIs/CTIs) that require resolution or clarification?*

Perform Engineering Graphical Analysis

1. Based on the requirements, as stated, can we draw a simple graphic of the system and its interactions with its OPERATING ENVIRONMENT?
2. Are there any obvious "holes" in the graphic that are not specified as requirements in the specification?

Hierarchical Analysis

1. Are there any misplaced, overlapping, duplicated, or conflicting requirements?
2. Are the requirements positioned and scoped at the right levels?
3. Are there any "holes" in the set of requirements?

Technology Analysis

Do the specification requirements indicate a willingness or unwillingness by the Acquirer to consider and accept new technologies or solutions?

Competitive Analysis

Do the specification requirements favor or target a competitor's *products*, *services*, or organizational capabilities?

Modeling and Simulation Analysis

If appropriate, is it worthwhile to develop models and simulations as decision aids to analyze system performance issues?

Verifying Specification Requirements

1. Are there any requirements that are unreasonable, unverifiable, or cost prohibitive using the verification methods specified?
2. Does verification require any special test facilities, tools, equipment, or training?

Validating Specification Requirements

When SEs analyze specifications, especially those prepared by external organizations, most engineers presume the specification has been prepared by someone who:

1. *Understands* the User's *problem space* and *solution space(s)*.
2. Accurately *analyzes*, *translates*, and *articulates* the solution space into requirements that can be implemented economically with acceptable risk, and so forth.

Exercise CAUTION with this mindset! AVOID assuming anything UNTIL you have VALIDATED the specification requirements.

30.4 DEALING WITH CONTRACT SPECIFICATION DEFICIENCIES

Human systems, even with the best of intentions, are not perfect. Inevitably, every contract *System Performance Specification (SPS)* has *blemishes*, *degrees of goodness*, *strong* and *weak* points. Although the *degree of goodness* has an academic connotation, "goodness" resides in the minds and perceptions of the Acquirer and System Developer. Remember, *one person's work of art may be viewed by another person as unorganized rambling*.

Discussions by both parties reach a point whereby *willingness* to entertain contract modification to *eliminate* specification *blemishes* or *deficiencies* are *rejected*. The Acquirer may want changes but is reluctant to request changes due to the System Developer taking advantage of the situation via cost changes.

Conversely, the System Developer may WANT changes but the Acquirer is *unwilling* to allow any changes for FEAR of the *unknown* that may result from the changes. Even when both parties agree, there may be latent SPS deficiencies that lie *dormant* and go *undiscovered* until late in the System Development Phase of the contract. The best that can occur is for both parties to *accommodate* each other's wishes at *no cost*, assuming that is the *appropriate* and *reasonable* solution.

Regardless of the scenario you may have a situation where the *System Performance Specification (SPS)* contains *defects, deficiencies, or errors* and the Acquirer refuses to modify the contract. *What do you do?*

One solution is to create an electronic *System Design Notebook (SDN)*; some people refer to this as a *Design Rationale Document (DRD)*. Why do you need an SDN or DRD? You need a *mechanism* to record design assumptions and rationale for requirements allocations and design criteria.

Under the *Terms and Conditions (Ts&Cs)* of the contract, the System Developer must perform to the SPS requirements that are flowed down to lower level *development specifications*. Therefore, *lacking* a definitive set of SPS requirements, you may want to consider an AT RISK solution that *expresses* your organization's *interpretation* of the *ambiguous* requirements. A copy of the document should be provided through contracting protocol to the Acquirer's Contracting Officer (ACO).

Author's Note 30.1 *The point above highlights the need to THINK SMARTLY "up front" about requirements and AVOID this situation. You will find executives and impatient people who insist that you move on and not worry about interpretations. "Besides, it's perfectly clear to me!" BEWARE! Any time investment and energy spent up front clarifying SPS requirements BEFORE they become contract obligations will be significantly less costly than after Contract Award.*

When you conduct the System Requirements Review (SRR), SPS *defects* and *deficiencies* should be addressed with the Acquirer. The requirements *defects* and *deficiencies* discussion and decisions should be recorded in the SRR meeting minutes. If the Acquirer *refuses* to allow corrections via contract modification, they may acknowledge via the Acquirer Contracting Officer (ACO) your chosen approach. Therefore, document your *design assumptions* and include *open* distribution or accessibility of that portion of the SDN to the Acquirer.

Author's Note 30.2 *Every contract and situation is different and requires decision making on its own merits. Professionally and technically speaking, the Acquirer and System Developer should emerge from the SRR with no outstanding issues. The reality is:*

1. *The Acquirer may have to settle for a negotiated acceptance of the system AS IS with known deficiencies at system delivery.*
2. *The System Developer may be UNABLE to perform to the terms and conditions (Ts&Cs) of the contract.*

All stakeholders need to emerge from the contract as winners! Therefore, AVOID this problem and resolve it up front during the proposal phase or not later than the SRR and throughout the System Development Phase, as appropriate.

30.5 COMMON SPECIFICATION DEFICIENCIES

If you analyze specification requirements practices in many organizations, there are a number of deficiencies that occur frequently. These include:

Deficiency 1: Failure to follow a standard outline
Deficiency 2: Specification reuse risks
Deficiency 3: Lack of specification and requirements ownership
Deficiency 4: Specifying broad references
Deficiency 5: Reference versus applicable documents
Deficiency 6: Usage of ambiguous terms
Deficiency 7: Missing normal, degraded, and emergency scenario requirements
Deficiency 8: Specification change management
Deficiency 9: Over-underspecification of requirements
Deficiency 10: References to unapproved specifications

Deficiency 11: Failure to track specification changes and updates
Deficiency 12: Failure to appropriate time for specification analysis
Deficiency 13: Requirements applicability–configuration effectivity
Deficiency 14: Dominating personality specification writers

Deficiency 1: Failure to Follow a Standard Specification Outline

Many specification issues are traceable to a lack of commitment to establish and employ standard specification development outlines and guidelines. Standard outlines represent organized lessons learned that reflect *problem are*as or *issues* that someone else has encountered and must be corrected in future efforts. Over time they incorporate a broad spectrum of topics that may or may not be applicable to all programs. The natural tendency of SEs is to *delete* nonapplicable sections of a standard specification outline. Additionally, management often dictates that specific topics are to be deleted because "We don't want to bring it to someone's attention that we are not going to perform (topic)."

The reality is standard outlines include topics that are intended to *keep you out of trouble*! A *cardinal rule* of system specification practices requires you to provide rationale as to WHY a standard outline topic is not applicable to your program. The rationale communicates to the reader that you:

1. Considered the subject matter.
2. Determined the topic is not relevant to your system development effort for the stated rationale.

Thus, if someone more knowledgeable determines later that "Yes, it is relevant," you can correct the applicability statement. *This is true for plans, specifications, and other types of technical decision-making documents.*

Problems arise when SEs purposefully delete sections from a standard outline. Once deleted, the section is *"out of sight, out of mind."* Since contract success is dependent on delivering a properly designed and developed SYSTEM on schedule and within budget, you are better off identifying a topical section as "Not Applicable." Then, if others determine that it is applicable, at least you have some lead time to take corrective action BEFORE it is too late.

If you follow this guideline and go into a System Requirements Review (SRR), any non-applicability issues can be addressed at that time. All parties emerge with a record of agreement via the *conference minutes* concerning the applicability issue.

Author's Note 30.3 Remember, *the cost to correct specification errors and omissions requirements increases almost exponentially with time after Contract Award.*

Deficiency 2: Specification Reuse Risks

Some engineers pride themselves in being able to quickly "assemble a specification" by duplicating legacy system specifications without due operational, mission, and system analysis. Guess what? Management takes great pride in the practice, as well. AHEAD of schedule, LIFE is ROSY! Later SEs discover that key requirements were *overlooked* or *ignored*, were not estimated, and have significant cost to implement. Guess what? Management is very unhappy!

Where *precedented* systems exist and are used as the basis to create new specifications with only minor modifications, the practice of plagiarizing existing specifications may be acceptable. However, be *cautious of the practice*. Learn WHEN and HOW TO apply specification REUSE effectively.

Deficiency 3: Lack of Specification and Requirements Ownership

Specification requirements are often ignored due to a lack of ownership. Two SEs argue that each thought the other was responsible for implementing the requirement. *Every* requirement in *every* specification should have an OWNER who either generated the requirement or is *accountable* for its implementation and verification.

Deficiency 4: Specifying Broad References

Specification writers spend the bulk of their time *wordsmithing* and *correcting* documents and very limited time, if any time, on systems analysis—even less on specification references. References are often inserted at the last minute. Why?

Typically, those same SEs do not have the time to thoroughly research the references. They make broad references such as "in accordance with MIL-STD-1472 *Human Engineering*" because:

1. Management is demanding completion.
2. We'll "clean it up" later.
3. We don't understand WHAT the reference means, but it sounds GOOD; "*saw it in another spec one time so let's use it,*" etc.

Since this is a specification and the System Developer is required by contract to implement the provisions of the SPS, *are you prepared to PAY the bill to incorporate ALL provisions of Mil-Std-1472, for example?* Absolutely not! With luck, the problem may work itself out during the proposal phase via the Offeror formal question and answer process.

This problem is a challenge for the Acquirer and System Developers.

1. The referenced document may be *outdated* or *obsolete*. Professionally speaking, this can be embarrassing for the organization and specification developer.
2. Unskilled specification writers—despite 30 years of experience in other nontopic areas—may inadvertently make technical decisions that may have legal, safety, and risk ramifications.
3. System Developers often fail to properly research Request for Proposal (RFP) references, thereby costing significant amounts of money to implement the reference as stated in the *System Performance Specification (SPS)*.

Do yourself and your organization a favor. *Thoroughly research RFP references*, REQUEST Acquirer (role) *clarifications* or *confirmations*, and then document in brief form WHAT the reference EXPLICITLY requires. Remember, *UNDOCUMENTED Acquirer comments are magically FORGOTTEN when things go WRONG!*

Deficiency 5: Reference versus Applicable Documents

Referenced documents refer to those documents—namely specifications and standards—*explicitly* specified in the Section 3.0 requirements and listed in Section 2.0 *Referenced* or *Applicable Documents* of the specification. In contrast, *Applicable Documents* are those containing relevant information to the topic but are not REQUIRED or REFERENCED by the specification. AVOID listing these documents in a specification's Section 2.0 *Referenced Documents*.

Deficiency 6: Usage of Ambiguous Terms

Specification writers are notorious for writing requirements statements that employ *ambiguous* words that are *subject* or *open* to *interpretation*. In accordance with specification practices that promote *explicitness, ambiguous words* and *terms* should be avoided or defined. Consider the following example:

EXAMPLE 30.1

Simulation community specifications often include terms such as *realistic representation* and *effective training*.

The challenge for the Acquirer and System Developer is: *How do you know WHEN a "realistic representation" or "effective training" has been successfully achieved*? The answer is to explicitly *define* the terms. WHAT *subjective criteria* will the User, subconsciously in their minds and perceptions, use to determine successful achievement of the requirements?

Deficiency 7: Missing Scenario Requirements

Most SEs prepare specifications for the "ideal" world. The reality is systems and interfaces fail, sometimes with CATASTROPHIC consequences. When specifications are written, make sure the requirements are specified to address NORMAL, DEGRADED, EMERGENCY, and, if appropriate, CATASTROPHIC operations.

Author's Note 30.4 *Remember, system safety design issues demand proper consideration to minimize risk to the SYSTEM and its personnel (people). This includes direct or indirect impacts by the system and its operation or lack thereof on the public, physical property, and the environment.*

Deficiency 8: Specification Change Management

Once a specification is approved, baselined, and released, changes must be formally *approved* and communicated to *stakeholders*—among them management and SE designers. Because of a lack of communications, two problems can occur.

First, program and technical management often lack an appreciation of the need to *communicate* specification changes to program personnel. Ironically, the technical integrity of the program is at risk, which is the very topic management seems to be averse to. *Second*, when management does communicate changes, Program personnel often IGNORE announcements about specification changes and continue to work with a previous version. *Programs need to learn how to better communicate*! Verify that program documentation communications are *clearly* communicated and understood!

Deficiency 9: Over-Underspecification of Requirements

Specification developers are often confronted with uncertainty regarding the adequacy of the requirements. Specification requirements can be *overspecified* or *underspecified*.

The way to AVOID *over-underspecification* involves three criteria:

1. Focus on specifying the system item's *performance envelope*—meaning boundaries for capabilities and performance.
2. AVOID details that specify HOW the capability-performance envelope is to be implemented.
3. AVOID specifying capabilities more than one level below the entity's level of abstraction.

Deficiency 10: References to Unapproved Specifications

When the System Developer's organization plans to develop several multi-level *item development specifications*, the efforts must be synchronized. One of the ironies of specification writing is a *perception* that multi-level specifications can be written *simultaneously* to cut development time. This is an *erroneous perception* that ultimately leads to technical CHAOS—conflicts and inconsistencies! The sequencing and approval of specification development at lower levels is dependent on *maturation* and approval of higher level specifications. There are ways of accomplishing this, assuming teams at multiple levels communicate well and mature decisions quickly at higher levels.

Deficiency 11: Failure to Track Specification Changes and Updates

System integrity *demands* change management process discipline: *track* approved specification updates, incorporate changes *immediately*, and *notify* all stakeholders of the latest changes. This includes proper *versioning* to enable stakeholders to determine currency.

Deficiency 12: Failure to Appropriate Time for Specification Analysis

One of the ironies of system development is *failure to allocate* the proper amount of time to analyze or develop specifications.

Wasson's Law. *The time allocated by management for most program and technical decision making tasks is inversely proportional to the significance of the tasks or decision to the deliverable product, User, or organization.*

Three of the most crucial *specification analysis* tasks during a proposal effort are *understanding*:

1. WHAT problem the User is trying to solve.
2. WHAT system, product, or service does the Acquirer's formal solicitation's SRD/SOO specify.
3. WHAT you have committed yourself and your organization to via the draft *System Performance Specification (SPS)* submitted as the proposal response.

Despite their significance, these three tasks often fall back in the priority list behind multi-level managerial briefings, which are important, and other "administrative" tasks.

Deficiency 13: Requirements Applicability—Configuration Effectivity

Some specification requirements may only be applicable to specific configurations and units—that is, configuration *effectivity*. Where this is the case, label the requirement as only applicable to the configurations A, B, C, etc., or serial numbers XXXX to YYYY. If this occurs, include a statement to the reader on the cover and in the Section 1.0 *Introduction* that this specification *applies* to configurations A, B, and C and serial number effectivity XXXX through YYYY. Then, list the serial numbers.

Deficiency 14: Dominating Personality Specification Writers

As much as we aspire to objectively neutralize the personality factor in specification writing, personalities sometimes have a dominating influence over specification development. *There are ego-*

tists whom programs defer to because of their personalities rather than their technical competence. If you allow this practice to occur on your program, you may be *jeopardizing* your own success as well as that of the program.

There are times when dominating personalities will "bully" requirements through that do not make sense—*"Makes sense to me . . . why can't you see the same?"* As a result there may be highly competent reviewers who have a "gut instinct" that something is *not* right yet are unable to *articulate* or *identify* the problem for corrective action. If this scenario occurs, view it as a *risk indicator*, take notice, and get the situation corrected technically. *Level* the playing field. *Keep* the team focused on *technical solutions* and resolving issues!

30.6 CLARIFYING AND RESOLVING SPECIFICATION ISSUES AND CONCERNS

Specifications often contain requirements that require *clarification* to ensure the proper *interpretation* and *understanding* of the requirement. Some requirements, however, raise *critical operational issues (COIs)* or *critical technical issues (CTIs)*, that present major challenges to implementation. The issues may reflect technical, technology, cost, schedule, or support risks as well as TBDs, and so on.

Requirements issues and the need for *clarifications* occur BEFORE and AFTER Contract Award. Let's briefly explore the handling of issues during these two time frames.

Issue Resolution Prior to Contract Award

Acquirers often release the draft version of a formal solicitation—such as a *Request for Proposal (RFP)*—for review and comment. The draft SRD may contain various specification requirements issues and the need for clarifications.

The first step of any specification analysis should be to identify and tag all requirements requiring clarification—technical, cost, schedule, technology, and support risks, and so on. Remember, *publicly surfacing issues and clarifications during the RFP process may potentially tip competitors regarding your proposal strategy.* Therefore, the proposal team must make a decision regarding which issues to submit for clarification and which to give additional internal analysis and/or follow-up.

The key point is that *Offerors (System Developers) must thoroughly analyze, scrutinize, and resolve all requirements issues and clarifications BEFORE they submit their specification as part of the proposal and sign a contract.*

Issue Resolution after Contract Award (ACA)

Requirements issue resolution after Contract Award (ACA) varies by contract and Acquirer. The WILLINGNESS to modify specification requirements language may depend on how recently the Contract Award has occurred.

Prior to the System Requirements Review (SRR)

If the System Requirements Review (SRR) has not been conducted, there may be an opportunity to address the need for *clarifications* or to *correct deficiencies*. Even then, some Acquirers may be *reluctant* to agree to specification changes via contract modifications.

From the Acquirer's perspective, the need to change always goes back to the proposal response prior to Contract Award ". . . *WHY wasn't the matter addressed at that time or during contract negotiations?"* The problem is exacerbated if the Offeror (System Developer) voluntarily proposed specification requirements language without adequate analysis or consideration. The situation potentially has degrees of organizational, professional, and technical embarrassment.

Post–SRR Requirements Issues

There are occasions when *latent* specification requirements *defects* go *undiscovered* until after SRR, generally due to poor analysis. Acquirers may reluctantly consider and accept engineering change proposals (ECPs) for requirements changes. Depending on the situation, the only workable solution may be to request a *deviation* to the specification requirements.

Final Points

When specification requirements issues are discovered, surfacing the issue to the surprise of the Acquirer program manager in a major technical review may have consequences. *People, in general, do not like surprises, especially in public forums.* Although the handling of an issue depends on the personnel and organizations involved, the best advice may be for the System Developer's program or technical director to *informally* introduce the issue *off-line*—meaning in private conversation—with the Acquirer program manager prior to a review. Depending on the outcome and response, the System Developer may choose to address the issue *formally* through normal procurement channels.

Finally, the *reluctance* and *willingness* of the Acquirer to *entertain* the *idea* of specification requirements changes may be driven by having to implement a contract modification that requires numerous approvals and justifications, a laborious and career risk process. It also challenges the initiator(s) to *explain* to your management WHY you *failed* to *recognize* this situation and *rectify* it during contract negotiations.

Tracking Requirements Issues and Clarifications

When specification requirements *issues* and *clarifications* are identified, it is critical to bring all of them to closure quickly. One mechanism for tracking closure status is to establish a metric that *represents* the number of outstanding COIs, CTIs, TBDs, TBSs, and clarifications.

30.7 REQUIREMENTS COMPLIANCE AND CONFORMANCE

As a System Developer, one of the most common questions posed by management to SEs is: *Can we meet the specification's requirements?* The response typically involves words such as *comply*, *conform*, and *meet*. Since SEs often lack training on the proper usage of the terms, you will find these terms are used interchangeably. So, *what do the terms mean?*

If you delve into the definitions of *comply*, *conform*, or *meet* in most dictionaries, you may emerge in a *state of confusion*. The reason is that most dictionary definitions of these terms employ either of the other two terms as part of the definition, resulting in circular references. So, to bring some clarity to this confusion, consider the following explanations.

Compliance

The term, *compliance, is often* used in references to *requirements compliance, process compliance*, and *regulatory compliance*. In general, compliance infers STRICT *adherence* or *obedience* to the "letter" of a requirement with no exceptions. Despite the term's intent you will often find degrees of compliance. Within the system development contracting domain, there is an *expectation* that any organization that refuses to or is unable to comply with specification requirements formally notifies the Acquirer and documents the *exception*, its supporting rationale, and proposed remedy. Failure to achieve full compliance often results in contract performance penalties or a negotiated reduction in contract payments.

EXAMPLE 30.2

People are expected to *comply* with the letter of the law or requirement regarding statutes and regulations established by international, federal, state, and local governmental authority.

Conformance

We often hear the phrase "We will conform to your requirements." The response communicates, "We have standard organizational practices—such as processes, methods, and behavioral patterns—that we use on a regular basis. However, to promote harmony among team members and a positive relationship, we will ADAPT or ADOPT—that is, to conform to—a set of processes, methods, and behavioral patterns to be mutually acceptable to the other party."

EXAMPLE 30.3

You visit a foreign country and discover their eating habits are different from yours. You have a choice: 1) *conform* to their practices, 2) go without food supply, or 3) bring your own chef and food.

"Meet" Requirements

You often hear someone say they or their organization can meet the requirements. *What does this mean?* In general, the term *meet* carries a *minimum threshold* connotation. In effect they are stating, *"We will meet your MINIMUM requirements."*

To summarize our discussion in an SE context, a subcontractor might tell a System Developer, "For this subcontract, we will *conform* to your documentation system's review and approval process. When we submit our documents for review and approval, we will *comply* with the subcontract's instructions for document format and submittal via the Contracting Officer."

30.8 GUIDING PRINCIPLES

In summary, the preceding discussions provide the basis with which to establish the guiding principles that govern specification analysis practices.

Principle 30.1 Undocumented verbal agreements vaporize when either party in an agreement gets into trouble.

30.9 SUMMARY

We began our discussion of *specification analysis* by addressing Acquirer and System Developer perspectives for a specification. We introduced various methods that System Developers employ to provide a high-level assessment of the "goodness" of the specification.

Next we introduced common deficiencies such as *missing*, *misplaced*, *overlapping*, or *duplicated* requirements that plague many specifications. We described how specification issues and concerns should be resolved between the System Developer and Acquirer prior to and after Contract Award.

Last, we delineated three terms—*comply*, conform, and *meet*—that contractors express interchangeably but have different connotations.

GENERAL EXERCISES

1. Answer each of the *What You Should Learn from This Chapter* questions identified in the *Introduction*.

ORGANIZATIONAL CENTRIC EXERCISES

1. Research your organization's command media. What guidance or requirements are levied on contract programs concerning specification analysis? Document your findings.
2. Contact several contract programs within your organization. Interview the Technical Director, Project Engineer, or Lead SEs to answer the following questions:
    (a) How were system requirements analyzed during the proposal effort that culminated in the development of the program's *System Performance Specification (SPS)*?
    (b) What deficiencies—missing, misplaced, conflicting, or duplicated requirements—did personnel find in the Acquirer's *System Requirements Document (SRD)* provided with the solicitation? How were the deficiencies resolved?
    (c) Were there any requirements issues and or requirements that required clarification? Did the System Developer or the Acquirer write these requirements? How were these resolved? Did the Acquirer refuse to make changes to the SPS to resolve or clarify issues? How did the program deal with the refusal to modify SPS requirements?

Chapter 31

Specification Development

31.1 INTRODUCTION

Performance and *development specifications* serve as the formal mechanism for an Acquirer (role) to:

1. S*pecify* WHAT capabilities a system or entity is required to provide.
2. *Bound* HOW WELL the capabilities are to be performed.
3. *Identify* external interfaces the SYSTEM/entity must accommodate.
4. *Levy* constraints on the solution set.
5. *Establish* criteria concerning HOW the System Developer is to demonstrate compliance as a precursor for delivery and acceptance.

At the highest level, the *System Performance Specification (SPS)* or *Statement of Objectives (SOO)* establish a contract's *technical agreement* between the Acquirer, as the User's technical representative, and the System Developer. This section introduces *Specification Development Practices* and expands on the standard outline discussion in Chapter 28 *System Specification*.

When tasked to write a specification, most engineers have a tendency to approach specification development as if it were a test.

1. Identify the SYSTEM OF INTEREST (SOI).
2. What is the system's mission?
3. What are its modes and states of operation?
4. What functions should the SYSTEM/entity perform?
5. Identify the system's interfaces?
6. Any weight restrictions?
7. Any special considerations about safety?
8. Any other constraints concerning the way the SYSTEM/entity is to be designed and constructed?
9. What about training?

This chapter introduces and explores some of the more common approaches to specification development. Our discussion introduces a logical strategy that provides the basis for structuring a specification. Using this strategy, you can create an outline that best fits your organization.

Based on an understanding of methods for creating a specification and its topical outline, we shift our focus to investigating an example approach for creating a meaningful specification. We

---

System Analysis, Design, and Development, by Charles S. Wasson
Copyright © 2006 by John Wiley & Sons, Inc.

leverage our discussions in Chapter 18 *System Operations Model* and use it as an analytical tool to organize requirements within the specification outline structure via *model-based structured analysis*. We conclude with a suggested a checklist for developing specifications and conducting specification reviews.

What You Should Learn from This Chapter

1. What are some common approaches to developing specifications?
2. What is the *architecture-based specification development approach*?
3. What is the *performance-based specification development approach*?
4. What is the *feature-based specification development approach*?
5. What is the *reuse specification development approach*?
6. What is the *model-based development* approach to specification development?
7. How are specification reviews performed?

Definitions of Key Terms

- **Architecture-Based Approach** A structured analysis approach that employs: 1) conceptual system phases and modes of operation and 2) a multi-level logical architecture (i.e., entities and interfaces) as the framework for specifying system capabilities and performance requirements.
- **Performance-Based Approach** A specification development approach that analytically treats a system or entity as a box and specifies condition-based inputs, outputs, behavioral capabilities and responses, and constraints for developing specification requirements.
- **Specification Review** A technical review by stakeholders, peers, and subject matter experts (SMEs) to assess the *completeness*, *accuracy*, *validity*, *verifiability*, *producibility*, and *risk* of a specification.
- **Reuse-Based Approach** An approach that *exploits* or *plagiarizes* an existing specification that may or may to be applicable to a specific application and uses it as the basis for creating a new specification. This approach, which is typically prone to *errors* and *omissions*, is highly dependent on the specification writer's and reviewer's knowledge and expertise to identify and correct *errors* and *omissions*.
- **Feature-Based Approach** An ad hoc brainstormed approach to specification requirements development. This approach, by virtue of its *feature-based* nature, is subject to *omissions* in the hierarchy of requirements and is highly dependent on reviewer assimilation and recognition of the *omissions* to correct them.

31.2 UNDERSTANDING WHAT A SPECIFICATION COMMUNICATES

Specifications are not intended to be Pulitzer Prize winning novels on the bestseller list. However, like novels, they do require some insightful *forethought* to bring a degree of *continuity* and *coherency* to a set of mundane, disjointed outline topics.

There are numerous ways of structuring a specification outline. Rather than elaborate on commonly available specification outlines, let's take a different approach and THINK about WHAT we need to *communicate* in the specification. Then, based on that discussion, *translate* the information into a meaningful outline structure that best fits your organization or application.

Standard Outline Templates

As with any commonly used documentation, start with a standard outline, tailor, and apply it across all specifications for a contract program. If you do not have one, consider establishing one in your organizational command media.

If you analyze and implement most specification outlines, you will discover two things:

1. People who are not practitioners with in-depth experience often create specification outlines used in many organizations. Outline topics are *mismatched* and *disorganized*. While the outline satisfies an organizational ego, it may be *useless* to *stakeholders*—namely the practitioners and readers.

2. Specification outline creators tend to structure outlines from an academic perspective that makes the outline *unwieldy* for practitioners. You can organize a specification with many levels of hollow academic structure that addresses the first requirement at the fourth, fifth, or sixth level. Imagine having such as high-level statement as "*The system shall perform the following capabilities;*" appear in Section 3.X.X.X.X. Since specifications for large, complex systems often require four to eight or ten levels of detail, placing the first requirements at the fifth level makes the document IMPRACTICAL to read and reference.

The point is to *organize* the outline structure in a *meaningful* way that posts major requirements at least by the third level.

Specification Outline

To facilitate our understanding as to WHAT a specification communicates, let's use the following structure to guide our discussion. You may decide to restructure these topics to better suit your needs:

1.0 INTRODUCTION
2.0 REFERENCED DOCUMENTS
3.0 REQUIREMENTS
    3.1 Operational Performance Characteristics
        3.1.1 System (Level 1 System) Entity Definition
        3.1.2 System Mission(s)
        3.1.3 Phases of Operation
        3.1.4 Mission Reliability
        3.1.5 System Maintainability
        3.1.6 System Availability
        3.1.n (Other Mission Related Topics)
    3.2 SYSTEM/Entity Capabilities (Level 1)
        3.2.1 System Capability Architecture
        3.2.2 Phase-Based Capability Application Matrix
        3.2.3 through **3.2.n** (Individual Capabilities)
    3.3 System Interfaces
        3.3.1 External Interfaces (Level 1)
        3.3.2 Internal Interfaces (Level 2)

3.4 Design and Construction Constraints

3.5 Personnel and Training

3.6 Operations and Support (O&S)

3.7 Transportability

3.8 Technical Documentation

3.9 Precedence and Criticality of Requirements

4.0 QUALIFICATION PROVISIONS

4.1 Responsibility for Verification

4.2 Verification Methods

4.3 Quality Conformance Inspections

4.4 Qualification Tests

5.0 PACKAGING

6.0 REQUIREMENTS TRACEABILITY

7.0 NOTES

7.1 Acronyms and Abbreviations

7.2 Definitions

7.3 Assumptions

Guidepost 32.1 *As a general rule, most organizations employ some form of the outline structure above. Beyond this point, specification development depends on the organization or developer's preferred approach.*

We now shift our focus to understanding HOW many organizations and individuals develop specifications.

31.3 SPECIFICATION DEVELOPMENT APPROACHES

Organizations and SEs employ a number of approaches to development of specifications. Typical approaches include:

1. *Feature-based approach*
2. *Reuse-based approach*
3. *Performance-based approach*
4. *Model-based approach*

Let's explore a brief description each type beginning with the most informal, the *feature-based approach*.

The Feature-Based Approach

Feature-based specifications are essentially ad hoc brainstormed lists of requirements. People who LACK formal training in specification development commonly use a *feature-based* approach. Specifications developed in this manner are often just *formalized* wish lists. Although *feature-based specifications* may use standard specification outlines, they are often *poorly organized* and *prone* to *missing*, *misplaced, conflicting or contradictory*, and *duplicated* requirements.

Advantages of the Feature-Based Approach. The *feature-based* approach enables developers to quickly *elicit* and *collect* requirements inputs with a *minimal* effort. Specification developers spend very little time *analyzing* and *understanding* potential implications and impacts of the requirements. When time is limited or a new system is being developed from minor modifications of an existing system, this method may be *minimally acceptable*. Every system is different and should be evaluated on a case-by-case basis. *Remember, it is easy to rationalize erroneous logic that ignores common sense to meet schedule and budget constraints.*

Disadvantages of the Feature-Based Approach. The disadvantages of the *feature-based approach* are that the *random method* produces a smattering of hierarchical requirements with:

1. Compound requirements written in paragrah style prose.
2. Conflicts with other requirements.
3. Multiple instances of duplication.
4. Vague, ambiguous requirements statements open to interpretation.

Figure 29.2 illustrates how this approach might affect the quality of the specification and result in DEFICIENCIES such as *missing*, *misplaced*, *conflicting*, or *duplicated* requirements.

The Reuse-Based Approach

The *reuse-based approach* simply *exploits* or *plagiarizes* an existing specification or integrates "snippets" from several specifications. The underlying assumption is that existing specifications are reference models. This can potentially be a big *mistake*! THINK about it! The *source* specification you plan to use as a starting point may be of *poor quality*!

The *reuse-based approach* is often used under the guise of *economy*—meaning saving time and money. The developers *fail* to *recognize* that corrections to the product design due to specification *deficiencies*, such as *overlooked* and *incorrect* requirements, often cost more than employing *model-based structured analysis* discussed later in this chatper.

Specification reuse often occurs within a product line, which is fine. You are working from a known entity accepted and refined by the organization. However, specification reuse occurs *across* product domains and organizations that may be *unrelated*, which may cause *significant* risk.

Author's Note 31.1 *The reuse-based approach is generally the standard repertoire for most new engineers, especially if they have not been given formal training in the proper ways to develop specifications. A manager tasks an engineer to develop a specification. Confronted with meeting schedule commitments, the engineer decides to contact someone who might have an existing specification and simply ADAPT it to meet the needs of the task. Constructively speaking, this approach becomes the DEFAULT survival mechanism for the engineer without ever learning the proper way. Some engineers spend their entire careers without ever learning methods that are more appropriate. Unfortunately, most organizations and managers contribute to the problem. They lack knowledge themselves and fail to recognize the need to provide the proper training.*

The risk of using the *reuse-based approach* is the *source* or "model" specification may be intended for a totally different system, system application, or mission. The only commonalities between applications may be the high-level topics of the outline. As a result, the specification could potentially contain two types of *flaws*:

1. References to *inappropriate* or *obsolete* external specifications and standards.
2. Retention of requirements that are not relevant or applicable.

Even if the requirements are topically relevant, they may *miss* or *over-/underspecify* the capabilities and levels of performance required for the new system's field application.

Guidepost 31.2 *Our discussion of the performance-based and model–based approaches provide a better method of developing a specification. As we will see, the model-based approach presumes some knowledge of the SYSTEM/entity architecture and employs the model-based approach to specify entities with the architecture.*

The Performance-Based Approach

The *performance-based approach* specifies SYSTEM/entity capability requirements in terms of *performance boundary conditions* and *transactions*. The SYSTEM/entity is automatically treated as a simple *box* with inputs and outputs, as illustrated in Figure 31.1.

The specification's Section 3.0 *Requirements* identifies the SYSTEM/entity:

1. Relationship to User Level 0/Tier 0 systems—or external interfaces
2. System missions, phases, modes/use cases
3. ACCEPTABLE and UNACCEPTABLE inputs
4. Capabilities required to transform the inputs into *performance-based outcomes*
5. ACCEPTABLE and UNACCEPTABLE outputs—behavior, products, by-products, and services
6. Design and construction constraints

Performance specifications represent the preferred *approach* to specification development for many applications, particularly *unprecedented* systems. By AVOIDING design specific requirements, the Acquirer provides the System Developer with the flexibility to *innovate and create* any number of architectural solutions within contract cost, schedule, and risk constraints. Depending on the Acquirer's intent, *performance-based specifications* require extensive System Developer/Subcontractor's *structured analysis* and *derivation* of requirements to select a preferred system architecture.

Most Acquirers developing an *unprecedented system—unproved*—tend to favor the *performance-based specification* approach as the initial step of a multi-phase acquisition strategy where

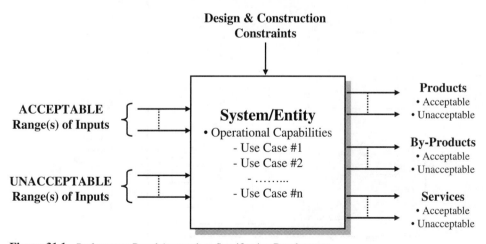

Figure 31.1 Performance-Based Approach to Specification Development

346 Chapter 31 Specification Development

requirements may be *unknown* or *immature*. The strategy may employ a series of spiral development contracts to evolve and mature the system requirements. Consider the following example:

EXAMPLE 31.1

An Acquirer plans to develop an *unprecedented* system. After due consideration, the Acquirer decides to establish a multi-phase acquisition strategy. Phase 1 of the acquisition strategy results in the award of a *performance-based specification* contract to develop initial prototypes for testing, collecting, and analyzing performance data; selecting a system architecture; and producing a set of requirements as the *work product* for a Phase 2 follow-on prototype or system. There may even be several other spiral development phases, all focused on derisking the final system development.

As the Phase 1 requirements and system architecture mature over one or more contracts, the Acquirer may shift to our next topic, the *model-based structured analysis approach*.

The Model-Based Structured Analysis Approach

The *model-based structured analysis approach* focuses on *specifying* and *bounding* capabilities and performance for elements within a defined SYSTEM/entity model-based architectural framework, as illustrated in Figure 31.2. This approach is often referred to as *model-based structured analysis*. The SYSTEM level is still treated as a "box." However, the SYSTEM is analytically decomposed into an architecture of interrelated entities or capabilities. Each SYSTEM architectural entity—SUBSYSTEM—or capability can be treated two ways:

1. As a performance-based entity using the performance-based approach.
2. Or, decomposed to lower level architectures of entities or capabilities.

To illustrate this approach, consider the following example:

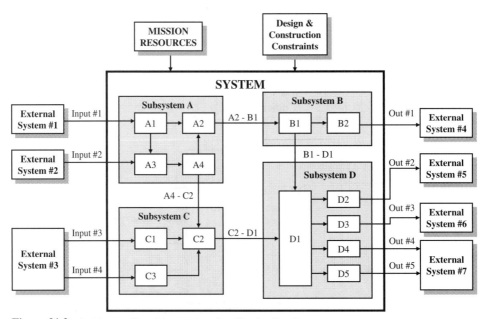

Figure 31.2 Architecture-Based Approach to Specification Development

EXAMPLE 31.2

A *System Performance Specification (SPS)* is to be developed for a vehicle. Section 3.1 addresses operational performance and characteristics assembled in the same manner as the *performance-based approach*. Next, specification developers structure Section 3.2 *Capabilities* based on the architectural elements:

3.2 Capabilities
- **3.2.1** Vehicle Frame System
- **3.2.2** Body System
- **3.2.3** Propulsion System
- **3.2.4** Fuel System
- **3.2.5** Electrical System
- **3.2.6** Cooling System
- **3.2.7** Steering System
- **3.2.8** Entertainment System
- **3.2.9** Storage System, etc.

Implementing the Model–Based Analysis Approach. Suppose that a specification development team or developer creates an analytical architecture for a SYSTEM that includes SUBSYSTEM's A through D, as illustrated in Figure 31.2. SUBSYSTEM A consists of capabilities A1 through A4, SUBSYSTEM B consists of capabilities B1 and B2, and so forth. The team constructs an architectural model that expresses the relationships between:

- The SYSTEM and External Systems 1 through 5.
- SUBSYSTEMs A through D.
- Capabilities A1–A4, B1–B2 and so forth within each SUBSYSTEM.

Author's Note 31.2 *Note the use of the term "architectural model." The specification development team creates a model of the system that is informally or formally controlled by the team for their exclusive use.* Based on the analysis, the specification developer(s) structure specification Section 3.2 *Capabilities* as follows:

3.2 System Capabilities
- **3.2.1** Capability A
    - **3.2.1.1** Capability A1
    - **3.2.1.2** Capability A2
    - **3.2.1.3** Capability A3
    - **3.2.1.4** Capability A4
- **3.2.2** Capability B
    - **3.2.2.1** Capability B1
    - **3.2.2.2** Capability B2
- **3.2.3** Capability C
    - **3.2.3.1** Capability C1
    - **3.2.3.2** Capability C2
    - **3.2.3.3** Capability C3

3.2.4 Capability D
    3.2.4.1 Capability D1
    3.2.4.2 Capability D2
    3.2.4.3 Capability D3
    3.2.4.4 Capability D4

The model represents the configuration of capabilities that accommodates all instances of SYSTEM/entity use cases and scenarios. This means that for any use case or scenario, the analysts can trace the "thread" from input through each capability to produce a performance-based outcome.

Using this method, the specification developer(s) translate WHAT capabilities the SYSTEM is required to provide via *text* statements positioned in the specification outline. For analysis purposes, you may decide to insert specification paragraph references in the appropriate graphical model element *without* dictating a specific configuration solution. Each capability is then decomposed into *multi-level* subcapabilities that form the basis for *outcome-based performance* requirement statements.

For specifications developed for implementation within the System Developer's organization such as configuration items (CIs), you may decide to include Figure 31.2 in the document. However, if you intend to procure the SYSTEM or one or more SUBSYSTEMs from external vendors, you may be *dictating* a specific solution.

Later, if the you or the vendor determines that the graphic *overlooked* a key capability, you may be confronted with *modifying* the contract and *paying* additional money to incorporate the *missed* capability. The downside here is that it gives the vendor the opportunity to recover costs of other capabilities they overlooked in their *original* estimate at your *expense*.

Additionally, if the vendor develops the system you mandated graphically and it *fails* to satisfy your needs, the vendor's response will be *"We built the system YOU contracted us to develop."* The bottom line is: Exercise CAUTION when inserting architectural graphics into specifications, especially SYSTEM/entity architecture graphics.

By creating an analytical architecture such as Figure 31.2 to support specification development, you improve your chances of expressing the capabilities you desire *without* mandating a specific solution. IF you perform the analysis well and translate it into capability-based requirements, the reader with some insight should be able to "reverse engineer" the system graphic. The difference is: *You haven't told the System Developer HOW TO design the system.*

Applying the Model-Based Analysis Approach. *Model-Based specifications* are well-suited for *precedented* system applications where there is an established architecture—such as airframe, propulsion system, and cockpit. However, when the specification developers specify the primary architectural components, they may *limit* the potential for *new* and *innovative* architectures that may be able to *exploit* new technologies and methods such as combining two traditional components into one.

Additionally, there is ALWAYS the risk that requirements specified for a SUBSYSTEM may *unintentionally constrain* the design of an interfacing SUBSYSTEM. As a result, cost and schedule impacts may be incurred. Simply state the architectural component capabilities as *performance-based* entities. Consider the following example:

EXAMPLE 31.3

Traditionally, a commercial aircraft crew consisted of a pilot, a copilot, and a navigator. As a paradigm, most aircraft had three crewmembers. However, technology advances, in combination with the need to reduce operating costs, contributed to the *elimination* of the navigator position. Thus, the paradigm shifted to integrating the NAVIGATOR function into the pilot and copilot roles and designing navigation systems to support those two roles.

Using the example above, mentally contrast a *model-based specification* approach that required three crewmember positions versus a *performance-based specification* approach that simply identifies three functional roles and thereby leaving the physical implementation to the System Developer.

Guidepost 32.3 *At this point we have introduced some of the more common approaches to specification development. Our next discussion delves into HOW we create specifications.*

31.4 UNDERSTANDING THE SPECIFICATION DEVELOPMENT PARADIGM

Most people have the perception that specification developers *write* a specification as if they were writing a novel as THE only work product of the exercise. They provide no supporting analyses or rationale as to HOW they arrived at the capabilities and levels of performance specified in the document. This is a *paradigm* that exemplifies WHY many contracts and system development efforts "get off to the wrong start," beginning at Contract Award.

Specification development is a *multi-level, concurrent, system analysis*, conceptual design effort that requires documented rationale of decisions. Some organizations and SEs say, *"We do this. If Joe is writing a specification and has a TBD that needs to be replaced with a numeric value, we perform a 'back of the envelope' analysis or go into the lab and run a simulation, and give him a number. So the problem is solved!"* That's NOT what we are referring to here.

Our point is that when you develop a specification, we want to know WHAT architectural and behavioral analyses you performed and the source criteria that *compelled* you to make *informed* decisions concerning the capability-based requirement. How did you:

1. AVOID *missing*, *misplaced*, *conflicting*, or *duplicated* specification requirements.
2. Support requirement allocations to lower level specifications.

The answer resides in implementation of the SE Process Model, as illustrated in Figure 31.3. Let's briefly describe the process.

SE Process Model Implementation

From an implementation perspective, the System Engineering and Integration Team (SEIT):

1. Implements the SYSTEM level SE Process Model and analyzes the SPS.
2. Creates the SE Process Model's Operational, Behavioral, and Physical Domain Solutions.

During the creation of these domain solutions, the SEIT may request Decision Support assistance to obtain data to support requirements allocations to PRODUCT level development specifications. This cycle repeats at lower levels as the respective development or integrated product teams (IPTs) implement the SE Process Model, as illustrated in Figure 31.3.

Now, let's investigate HOW the SE Process Model relates to specification development at each level.

350 Chapter 31 Specification Development

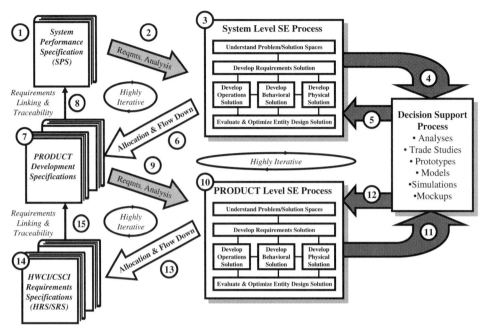

Figure 31.3 Multi-Level SE Process Model Requirements Analysis, Allocation, and Flow Down Process

Applying the SE Process Model to Specification Development

We can apply the preceding strategy to our earlier discussion of the model-based approach. Figure 31.4 illustrates this application of the specification development strategy for the system illustrated in Figure 31.2.

For any specification, the SEIT, IPT, and other teams analyze multi-level User requirements such as a *Statement of Objectives (SOOs), System Requirements Document (SRD), SPS*, and *development specifications*. Based on the analysis, which includes understanding the *problem* and *solution space*, use cases, scenarios, and the like, the team identifies the SYSTEM/entity I/O Model capabilities (2). Given those capabilities, they formulate a set of viable candidate architectures and select the most suitable one (4) based on pre-defined selection criteria.

At this point, the accountable team has to answer the question: *HOW does the selected architecture relate to higher level source requirements or specification capabilities*? They answer the question by creating a simple matrix (7) that *allocates* or *links* architectural elements to derived capabilities. Based on those *allocations*, requirements are *flowed down* to lower level specifications. It is important to note here that the requirements *allocated* and *flowed down* are high-level translations of WHAT the User wants the entity to accomplish and HOW WELL.

To illustrate HOW *model-based* specification development occurs from a multi-level perspective, let's return to the model-based approach. Let's investigate HOW lower level specifications can be developed using SUBSYSTEM A of Figure 31.2 as an example. Figure 31.5 provides a graphical illustration.

- The SEIT analyzes the SPS, employs the SE Process Model to create an I/O Model of multi-level capabilities, selects a SYSTEM architecture, allocates capabilities to architectural elements, and flows requirements allocations down to the *PRODUCT A3 Development Specification*.

31.4 Understanding the Specification Development Paradigm   351

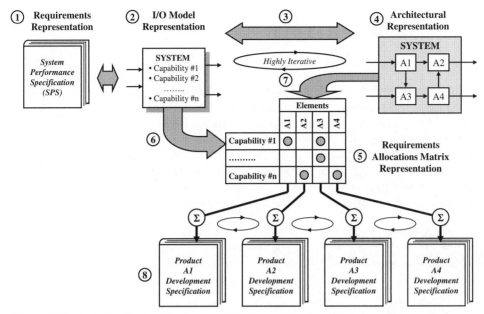

Figure 31.4 Basic Specification Requirements Allocation and Flowdown Method

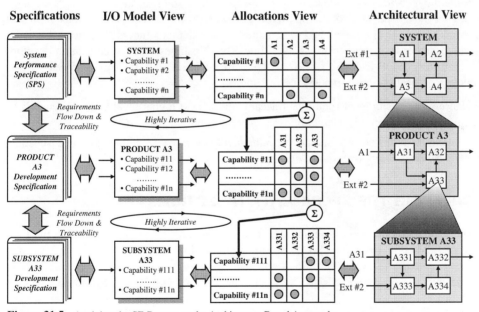

Figure 31.5 Applying the SE Process to the Architecture-Based Approach

- The PRODUCT A3 IPT analyzes its requirements allocations, employs the SE Process Model to create an I/O Model of multi-level capabilities, selects the PRODUCT A3 architecture, allocates capabilities to architectural elements, and flows requirements allocations down to the *SUBSYSTEM A33 Development Specification*.

If we employed the traditional Waterfall Model discussed earlier in Chapter 27 *System Development Models*, these steps would be performed sequentially—starting at a higher-level specification.

Its design would be complete before progressing to the next level. As the *highly iterative* ovals indicate, specification development is a multi-level effort.

Realities

As the preceding discussion illustrates, specification development is a *highly iterative, multi-level process*. Obviously to more *efficiently* and *effectively* utilize personnel resources, requirements allocations at higher levels have to reach a *level of maturity* before initiating lower level specification development efforts. So, there may be some offsets in the start of specifications at multiple levels.

In any case, CONCURRENT specification development drives the need to investigate lower level *critical operational and technical issues* (COIs/CTIs) to *understand* and *reconcile* WHAT needs to be specified at higher levels. You may ask: *WHY is this?*

At each level of SYSTEM decomposition and specification development, each requirement has:

1. A priority to the User.
2. A cost to implement.
3. A level of risk.
4. A schedule time element to implement.

So, you need to develop multi-level specifications for an internal-to-the-program "*go investigate and determine the feasibility of implementing these capabilities and provide us feedback*" activity. Whereas higher level specifications tend to be *abstract*, lower level specifications represent WHERE items are physically developed or procured from external vendors. You need this feedback to mature higher level specifications over time. This illustrates WHY specification development is a multi-level TOP-DOWN, BOTTOM-UP, LEFT-RIGHT, RIGHT-LEFT process. When you complete specification development:

1. Requirements at any level should be traceable to the next higher level and subsequently to the Acquirer's *source* or *originating* procurement requirements—*SOO* or *SRD*.
2. Specification requirements should be *realistically achievable* within cost and schedule budgetary *constraints* with *acceptable* risk.

31.5 CREATING THE SPECIFICATION SECTIONS

When we develop a specification, we employ an outline structure to communicate HOW the User wants the SYSTEM/entity to operate. So, the specification Section 3.0 consists of Section 3.1 which addresses *operational characteristics*. For *performance-based specifications*, you may choose to bypass Section 3.2 on *System Capabilities*, shift the outline, and have a Section 3.2 *System Interfaces*.

The process of developing specification Section 3.1 *Characteristics* is illustrated in Figure 31.6. The figure shows HOW the *operational entity relationships* identified in Figure 31.2 enable us to structure and translate operational capabilities into multi-level capability-based requirements statements. Since most specifications are developed for *precedented* systems, they employ Section 3.1, as stated above, and elaborate specific architectural element requirements in Section 3.2 on *System Entity Capabilities*. This allows the Acquirer to express requirements that relate to specific architectural capabilities. The specification developer(s) create an analytical architecture such as Figure 31.2. Based on the architecture, they organize specification Section 3.2 into a hierarchical structure and translate architectural capabilities into requirements, as illustrated in Figure 31.7.

31.5 Creating the Specification Sections   353

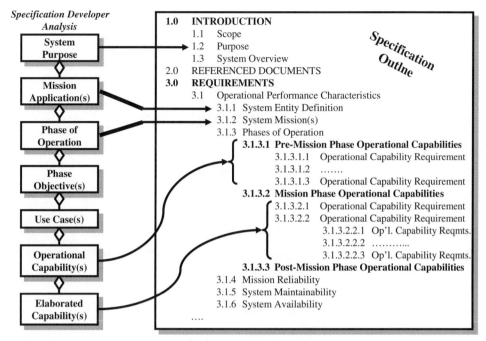

Figure 31.6 Defining a Specification's Section 3.1 Operational Characteristics (Level 1 system)

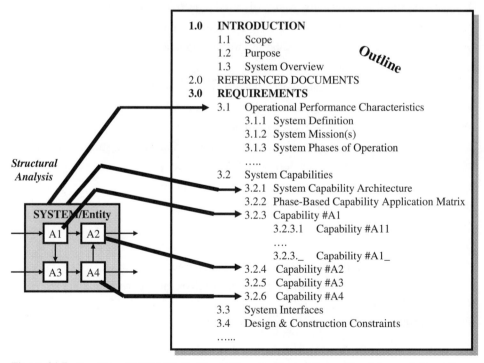

Figure 31.7 Specifying SYSTEM/Entity Capabilities using the Architecture-Based Approach (Level 2 System/Entity)

Final Thoughts

Specification development approaches ARE NOT foolproof. Approaches are ONLY as GOOD as the analysts who *identify* and *specify* the requirements. However, the *model-based structured analysis approach* is superior to the *feature-based* method for ensuring more complete coverage of SYSTEM capability requirements during all phases and modes of operation.

31.6 SPECIFICATION DEVELOPMENT CHECKLIST

SYSTEM/entity specifications offer a number of challenges to their developers. Based on lessons learned from developing specifications, here are some suggestions:

1. AVOID stating CSOW tasks as specification requirements.
2. Understand the User's *problem* and *solution spaces*.
3. Validate operational requirements with the User.
4. State requirements using terminology familiar to the stakeholders; create lower level specifications using terminology familiar to the developers.
5. Specify the RIGHT *solution space* and system.
6. Adequately bound the *solution space* and OPERATING ENVIRONMENT.
7. Identify system threats and priorities.
8. AVOID dictating a *system design solution*.
9. Specify requirements that state WHAT rather than HOW.
10. Delineate threshold *requirements* from goal-based objective requirements.
11. Explicitly identify and bound external reference requirements.
12. Select the minimum verification method to prove compliance at the least cost.
13. AVOID writing incomplete requirements.
14. AVOID writing subjective requirements that are open to interpretation.
15. AVOID duplicating, conflicting, and missing requirements.
16. AVOID requiring unnecessary precision and accuracy.
17. Ensure requirements *consistency* and *completeness* within and across specifications.
18. Balance technical, technology, support, cost, and schedule risk.
19. Assign specification development and ownership accountability.
20. Identify and scope verification methods.
21. Establish and maintain *requirements traceability*.
22. Provide definitions of key terms, notes, and assumptions to clarify meanings.
23. Prioritize requirements for implementation and cost purposes.
24. Include system block diagrams (SBDs) if they CLARIFY and DO NOT conflict with text-based requirements.
25. Elicit stakeholder requirements and review comments.

31.7 SPECIFICATION REVIEWS

When a specification reaches a level of maturity, conduct a *specification review* with stakeholders. There are several ways of conducting these reviews.

Line-by-Line Reviews

Some individuals and organizations conduct specification reviews on a line-by-line basis that consumes hours.

Reviewer Comments–Issues Approach

An alternative approach is to distribute the document electronically, request that comments be inserted as changes in a different color font, and returned. Review the comments and incorporate those that make sense; follow up with the stakeholder for comments that require clarification.

Based on the collection of review comments, identify any major issues and conduct a review with the stakeholders. The purpose of this review is to simply resolve *major issues*. Based on the results of the review, update the document for the next review or approval for baselining and release. This approach avoids consuming hours of stakeholder time changing "happy" to "glad" on a line-by-line basis.

Review Records

Whichever approach is best for the review, document the list of stakeholder attendees, approved changes, decisions, and action items in conference minutes.

Specification Baselines

You may ask: *When should certain types of specifications be baselined*? First, ALWAYS consult your contract for specific requirements. As a general rule, specifications are baselined and released, as listed in Table 31.1. Figure 46.2 provides a graphical view of the sequences.

31.8 GUIDING PRINCIPLES

In summary, our discussions in this chapter provide the basis with which to establish the several key principles that govern specification development practices.

Table 31.1 Example specification release time frames

Technical Review	Specifications Approved
System Requirements Review (SRR)	System Performance Specification (SPS)
System Design Review (SDR)	PRODUCT/SUBSYSTEM development specifications
Hardware Specification Review (HSR)	Hardware configuration item (HWCI) requirements specifications
Software Specification Review (SSR)	Computer software configuration item (CSCI) requirements specifications
Preliminary Design Review (PDR)	• Lower level development specifications, as applicable • Facility interface specifications (FIS)

Principle 31.1 Anyone can WRITE to a specification outline; however, DEVELOPING a specification requires premeditated forethought supported by informed analysis.

Principle 31.2 Every specification must have an assigned owner accountable for its development, implementation, and maintenance.

Principle 31.3 Every specification expresses two aspects of requirements:

1. WHAT the SYSTEM/entity is to accomplish and HOW WELL.
2. Qualification provisions for verifying requirements compliance.

Principle 31.4 Use system specification terminology that is familiar to the Acquirer and User. Apply terminology familiar in developers in lower level specifications.

Principle 31.5 SHALL requirement statements express mandatory actions required to achieve compliance; GOALS express nonmandatory or voluntary desires to strive for.

31.9 SUMMARY

Our discussion of *specification development practices* identified the key challenges, issues, and methods related to developing system specifications. As the final part of the concept, we introduced a basic methodology for preparing specifications. The methodology provides an operational approach to specification development based on HOW the User intends to use the system. We are now ready to focus on developing the system solution as a response to the *System Performance Specification (SPS)* requirements.

GENERAL EXERCISES

1. Answer each of the *What You Should Learn from This Chapter* questions identified in the *Introduction*.
2. Refer to the list of systems identified in Chapter 2. Based on a selection from the preceding chapter's *General Exercises* or a new system selection, apply your knowledge derived from this chapter's topical discussions. Specifically identify the following:
   (a) Annotate HOW you believe the *System Performance Specification (SPS)* Section 3.0 outline should be structured.
   (b) Pick a component of the system. Annotate the Section 3.0 outline for the components's development specification.

ORGANIZATIONAL CENTRIC EXERCISES

1. Contact several contract programs within your organization. Interview the Technical Director, Project/Chief Engineer, or lead SEs and answer the following questions:
   (a) What type of specifications are required by the contract?
   (b) How was each specification developed—by feature-based, reuse, or structured analysis?
   (c) Based on the interviewee(s) opinion, if they were to redevelop the specification, what would they do differently?
2. Research your organization's command media. What guidance or direction is provided concerning development of specifications and approaches?

ADDITIONAL READING

IEEE Std. 1220–1998. 1998. *IEEE Standard for Application and Management of the Systems Engineering Process*. Institute of Electrical and Electronic Engineers (IEEE). New York, NY.

International Council on System Engineering (INCOSE). 2000. INCOSE *System Engineering Handbook*, Version 2.0. Seattle, WA.

MIL-STD-490A (canceled). 1985. *Military Standard: Specification Practices*. Washington, DC: Department of Defense (DoD).

MIL-STD-961D. 1995. *DoD Standard Practice for Defense Specifications*. Washington, DC: Department of Defense (DoD).

Chapter 32

Requirements Derivation, Flow Down, Allocation, and Traceability

32.1 INTRODUCTION

Development of the *System Performance Specification (SPS)* typically represents only a small portion of a large, complex system's hierarchy of requirements. The challenge for SEs is: *How do we establish the lower level specifications that enable system developers to ultimately create PART level designs?*

This chapter introduces the *requirements derivation, allocation, flow down, and traceability practices* that enable us to create the specification-based hierarchy of system requirements. Our discussions delineate the terms requirements derivation, allocation, and flow down. We explore how requirements are derived, identify a methodology for deriving requirements, and how to apply the methodology.

What You Should Learn from This Chapter

1. What is *requirements derivation*?
2. How do you *derive* requirements?
3. What is requirements *allocation*?
4. How do you *allocate* requirements?
5. How do you *flow down* requirements?
6. How do you trace requirements? To where?

Definitions of Key Terms

- **Leaf Requirement** The lowest level derived requirement for a specified capability.
- **Requirements Allocation** The act of assigning accountability for implementation of a requirement to at least one or more lower level contributing *system elements*—such as EQUIPMENT, PERSONNEL, and FACILITIES—or embedded items within those elements.
- **Requirements Derivation** The act of decomposing an abstract *parent* requirement into lower level objective, performance-based *sibling* actions. Collective accomplishment of the set of derived "sibling" actions constitutes satisfactory accomplishment of the "parent" requirement.

System Analysis, Design, and Development, by Charles S. Wasson
Copyright © 2006 by John Wiley & Sons, Inc.

- **Requirements Flow Down** The act of transferring accountability for implementation of a requirement or portion thereof to a lower system level of abstraction—such as SYSTEM to PRODUCT or PRODUCT to SUBSYSTEM.
- **Requirements Testing** The act of evaluating the lineage, content, and quality of a requirement statement relative to a pre-defined set of requirements development criteria. The purpose is to determine if a requirement is *ambiguous*, *testable*, *measurable*, *verifiable*, and *traceable*.
- **Requirements Traceability** The establishment of bottom-up, multi-level linkages between lower level specification requirements back to one or more *source* or *originating requirements*.
- **Requirements Verification** The act of proving by verification methods such as inspection, analysis, demonstration, or test that a logical or physical *item* clearly complies with allocated capability and performance requirements documented in its *performance* or *item development specification*, as applicable. *Requirements verification* occurs throughout the System Procurement Phase and System Development Phase of the system/product life cycle.
- **Requirements Validation** An activity performed by the User or Acquirer to ensure that the stated requirements *completely*, *accurately*, and *precisely* address the RIGHT *problem space*, specify and bound the User's intended operational needs as delineated by the *solution space*.

32.2 UNDERSTANDING HOW REQUIREMENTS ARE DERIVED

The identification and derivation of requirements for various multi-level system items requires more than simply documenting text statements. The process is a *highly iterative and immersive* environment that must *consider*, *reconcile*, and *harmonize* a large number of constraints. Perhaps the best way to describe the process is by depicting the environment using the graphic shown in Figure 32.1. The following discussion serves as an overview for the Requirements Derivation Process, our next topic, symbolized at the top center of the chart.

Requirements Derivation via the System Engineering Process

The focal point for requirements identification and derivation is the SE Process Model as symbolized at the center of the chart. As the SE Process Model is applied to each level of abstraction and entity, requirements are *allocated* and *flowed down* from higher level specifications to lower level item development specifications (IDS). As the multi-level SYSTEM solution evolves, requirements are allocated and flowed down via lower level *item development specifications*.

You should also recall that the *Decision Support Process* supports the SE Process Model, as illustrated in Figure 26.1. Technical decisions related to requirements derivation may require in-depth analyses and trade studies to be performed. In turn, completion of the decision support analyses and trade studies may require development of prototypes, models, and simulations to provide data to support informed decision making.

Relationships and Interfaces with Other Processes

The *Requirements Derivation Process*, symbolized at the top center of Figure 32.1, is actually part of the *Develop Requirements Domain Solution* within the SE Process Model. The *Requirements Derivation Process*, which is more than simply a set of sequential decision steps, as noted by the "clouds", is characterized by *highly iterative* interfaces with the following SE Process activities:

360 Chapter 32 Requirements Derivation, Flow Down, Allocation, and Traceability

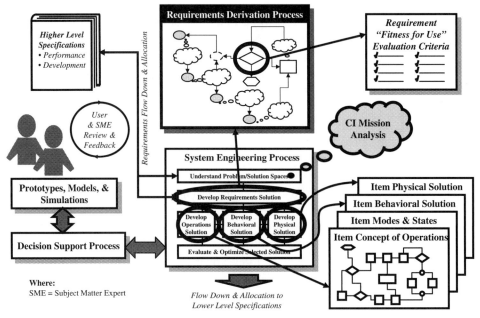

Figure 32.1 Requirements Derivation Process Environment

- Understand the Entity's Problem and Solution Spaces.
- Develop the Entity's Operational Domain Solution.
- Develop the Entity's Behavioral Domain Solution.
- Develop the Entity's Domain Solution.

The Understand Item Problem and Solution Spaces activity *demands* that requirements developers thoroughly *understand* and *appreciate*:

1. HOW the SYSTEM or *item*, for which requirements are being developed, is intended to fill a given *solution space*.
2. HOW that *solution space* relates to a higher level *problem space*.

This brings us to *HOW* the Requirements Derivation Process "fits" within this environment. Thus, SEs must employ mission analysis methods and techniques to understand the context, sequencing, and dependencies of the *solution space* within the *problem space*.

Reconciling WHAT the Customer WANTS, NEEDS, Can AFFORD, and WILLING to PAY

The flow down of requirements from higher-level specifications expresses WHAT the Acquirer or higher level System Developer teams WANT. The Requirements Derivation Process must also *reconcile* WHAT the User (role) WANTs versus WHAT they NEED versus WHAT they can AFFORD versus WHAT they are WILLING to PAY as illustrated in Figure 32.2.

Finally, within this same context, understand the system entity's *solution space* in terms of WHAT is required.

The User's Dilemma ......

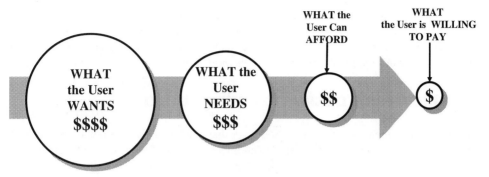

...... *the System Developer's Challenge.*

Figure 32.2 Qualifying Customer Needs

- *How does the User intend to use the system entity?*
- *How do we bound the system entity's operational effectiveness, utility, and suitability relative to User's WANTS, NEEDS, can AFFORD, and WILLINGNESS to PAY?*

Let's explore this further.

Understanding Each Requirement's Intent and Meaning

As higher level specification requirements are *allocated* and *flowed down* to the system entities, SEs must *analyze* each requirement to understand its *relationship* and *contribution* to bounding the system entity's *solution space*. Each requirement, by definition, *identifies*, *specifies*, and *bounds* a system *item* capability and its level of performance. The analysis must answer questions such as:

1. WHAT outcomes must be achieved to satisfy this requirement?
2. WHEN is the capability to be provided to achieve the desired outcome?
3. HOW WELL must each required capability be performed?
4. Under WHAT scenarios and conditions should each capability be performed?
5. WHAT is the relationship of this capability to others?

The answers to these *thought provoking* questions must evolve *harmoniously* through preliminary graphical and text descriptions such as the item *concept of operations (ConOps)*, item modes and states, item behavioral/logical solution, and item physical solution. These descriptions enable SEs deriving the requirements to gain multi-perspective views that scope of the domain and context of the requirement.

Concluding Point

As we will see, deriving requirements statements is just one portion of the process. The other portion is developing requirements statements for capabilities and levels of performance that can be:

1. Easily communicated and understood by the implementers.
2. Traced back to higher level source requirements.
3. Physically verified.

Thus, pre-defined *fitness-for-use* requirement evaluation criteria must be established to ensure a level of quality as well as maintain integrity of the requirements traceability. Given that requirements derivation is a *chain* of decisions, each dependent on the *integrity* and *quality* of its parent requirement's derivation, requirements *traceability* is paramount for ensuring system *integrity*.

Based on this fundamental understanding of the requirements derivation environment, let's begin our discussion of the *requirements derivation methodology*.

32.3 REQUIREMENTS DERIVATION METHODOLOGY

The derivation of requirements requires a methodology that can be easily communicated and produces *repeatable*, *predictable* results. To facilitate an understanding of the need for the methodology, let's first understand the mindsets and environment surrounding the process.

Derivation Paradigms

Requirements derivation, as an abstract label, falls into the category of topics of every day vocabularies. Yet, few engineers actually understand WHAT is required to derive the *requirements*. *Objective evidence* of this point is illustrated in specifications based on multi-level Wiley & CW feature-based specifications.

The problem is that *this label DOES NOT educate newcomers without some form of understanding its origin*. Ironically, many people learn to flaunt the buzzword to impress their peers with their SE vocabulary without ever understanding WHAT the label means.

In SE, *requirements derivation* refers to *elaborating* an abstract parent requirement statement into a set of lower level, objective, performance-oriented *sibling* requirements in which collective accomplishment constitutes satisfaction of the *parent* requirement.

The Requirements Derivation Methodology

One approach to requirement derivation focuses on a line of *investigative* questions tempered by seasoned experience and observations. The purpose of these questions is to identify key elements that will be used as a *methodology* for deriving requirements. Read, understand, tailor, and apply this methodology accordingly to your specific business needs and applications. The steps of this methodology include:

Step 1: Analyze a requirement (1).
Step 2: Identify outcomes (2)—behaviors, products, by-products, and services.
Step 3: Derive capabilities, behavioral responses, and performance from outcomes (3).
Step 4: Identify and bound capability operating scenarios and conditions (4).
Step 5: Bound capability levels of performance (5).
Step 6: State the requirement (6).
Step 7: Identify the verification method(s) for the requirement (7).
Step 8: Validate the requirement (8).
Step 9: Proceed to the next requirement (10).

Figure 32.3 provides a graphical illustration of these steps.

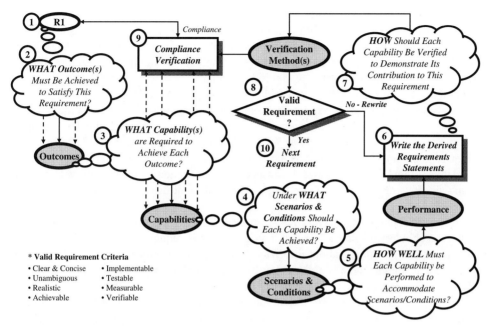

Figure 32.3 Performance Requirements ID & Derivation

32.4 APPLYING THE REQUIREMENTS DERIVATION METHODOLOGY

After reading through the methodology, you might ask: *How can we apply the requirements derivation methodology to develop a specification that may contain hundreds of requirements?* The answer resides in seasoned SE *experience*.

Most seasoned SEs imprint a methodology such as this in their minds. Then, when called upon to develop a specification, their subconscious *thought processes* automatically cycle through this methodology. The more *experience* you have, the more *proficient* you become in applying the methodology.

User Interface Requirements Derivation

The *requirements derivation methodology* applies to most requirements situations. There are some instances, however, that pose challenging questions when deriving User interface requirements. When addressing the SYSTEM level mission requirements, we characteristically use requirements phrases such as "... *user friendly* ... *intuitively obvious* ... *effective training* ... *realistically representative of the real world* ... *easy to maintain* ... *aesthetically pleasing*".

As humans, engineers have both a case of the *smarts* and the *uncertainties* regarding these terms. We generally have a "gut instinct," through our own personal experiences and frustrations, as to WHAT the specification writer is attempting to communicate. Conversely, we also have uncertainty as to WHAT this term or phrases requires. HOW do we verify a SYSTEM's *user friendliness*, *intuitively obvious* interactions, *effective training*, and the like? There are ways of dealing with these problems.

Generally, engineers ask two questions regarding *developing subjective* requirements:

1. *Is it okay to include subjective requirements statements in the specification?*
2. *If we leave the statement in, how do we verify SYSTEM compliance?*

The answer to the first question, depending on the situation, is generally a *qualified* yes. The answer to the second question requires statement of subordinate requirements that explicitly state WHAT *user friendliness*, *intuitively obvious*, and *effective training*, for example, mean to the User. Then, structure the *parent* requirement statement to indicate compliance with all of the *sibling* requirements constitutes accomplishment of the *parent* requirement.

This may seem a bit surprising but the world is far from perfect. Although we like to espouse beliefs that we *only* write *objective* requirements, there are times when *subjective* requirements, when properly stated and clarified, are *effective* in communicating WHAT the User desires. Remember, one of the key objectives of the *System Performance Specification (SPS)* is to communicate requirements in a *language* that is easily understood by all *stakeholders*—the User, the Acquirer, and the System Developer. Lower level specifications written for an internal system development team—the Integrated Product Team (IPT)—are a different matter and have a different set of stakeholders.

Ironically, *explicit* statements of *objective requirements* in language that *is* unfamiliar *or* misunderstood by stakeholders may be worse that having *subjective* requirements. If the Acquirer and Users have "*warm, fuzzy feelings*" about *user friendly* phrases in specifications, for example, this may be acceptable as long as there is a *consensus* by ALL stakeholders that *compliance* with supporting *sibling* requirements constitutes the basis for compliance with the *subjective requirement*. There must be *qualifying* and *clarifying* requirements that *explicitly* state HOW the stakeholders will know WHEN *user friendliness* has been achieved to the satisfaction of the Acquirer and User.

Visual Configuration Requirements

When developing specifications, there may be some requirements that are best depicted graphically rather than by using text. Consider the following examples:

EXAMPLE 32.1

Visual display layouts, panel layouts, safety icons, graphics, and graphical plots overcome the difficulty of attempting to describe a graphic image in words that are open to creating false impressions and are open to interpretation.

The reality is that attempting to describe displays and panel layouts using text language is often *impractical* and produces *ambiguous* results. A solution may be to simply specify that the display, panel layouts, and other layouts match or use Figure XX as a guide. Then, insert the graphic as Figure XX in the specification.

The preferred approach to this type of problem is through a *performance specification* that uniquely bounds human engineering requirements for display, panel, safety, and other layouts. This approach leaves many decisions "open" to collaborative interpretation by the System Developer. The fact is that Users either have some idea concerning WHAT they WANT or expect the System Developer to possess the expertise in ergonomic design to offer options from which the user can decide.

The best way to bring a *convergence* and *consensus* of the minds is to simply develop "rapid prototypes"—such as actual displays, viewgraphs and so forth—using conventional presentation software tools for the User to evaluate, validate, and offer recommendations. This is a classic case of *spiral development* whereby the requirements may be undefined, and numerous iterations are required to "get the requirements right." The product of the exercise is a set of graphics that capture the *consensus* of the User desires within the specification.

Referral For more information about *spiral development*, refer to Chapter 27 on *System Development Models* practices.

Performance Characteristics Requirements

The development of systems often requires extensive engineering analysis data and derived from reference models, experimentation, and data collection.

EXAMPLE 32.2

Examples include performance characteristics of propulsion systems and behavior of materials.

The problem is *HOW* do you communicate performance characteristics in a language-based specification? One solution resides in *developing* language-based specification requirements that reference graphical plots and tables that characterize data or a performance envelope.

A Word of Caution! *If you specify a capability that has a behavioral response represented by a linear or nonlinear graphical plot over a range of values, you may be expected to verify the system response at each point along the line for the respective inputs. Therefore, when you include such as plot in a specification, provide only the specific data points along the curve that are to be formally verified, such as end and middle points.*

Final Thoughts

Requirements derivation focuses on decomposing a higher level objective or capability into lower level supporting capabilities that can be translated into design requirements. On the surface this appears to be a mundane process in which each requirement leads to the next. As you derive lower level requirements closer to the ASSEMBLY/CSC level, there is another method you can employ to ensure that all aspects of a capability are derived.

You should recall our discussion of Chapter 22 on the *Anatomy of a System Capability*. Figure 22.1 illustrated the basic construct for a generic capability. The point is that you should *consider using this figure to facilitate some of the deliberations of requirements derivation, especially at lower levels such as development specification levels for SUBSYSTEMS or lower, as appropriate.*

32.5 REQUIREMENTS ALLOCATION, FLOW DOWN, AND TRACEABILITY

The cornerstone for developing specification requirements resides in the integrity of the requirements *allocation*, *flow down*, and *traceability*.

Requirement Allocation and Flow Down

Requirements allocation, flow down, and traceability can be illustrated by a simple construct such as the one shown in Figure 32.4. For this illustration, let's assume that the *SUBSYSTEM A Development Specification* includes use case based capability requirement A_11. The requirement *specifies* and *bounds* a capability to input, process, and display data. SEs perform a *use case analysis* and construct a *use case thread* that characterizes HOW this capability can be accomplished. This way they manage to obtain the required *performance-based* outcome. Based on the analysis, they translate each of the *use case* steps into Capability Requirements, A_111, A_112, and A_113.

This illustration provides insights for deriving capability requirements within a specification. Now, suppose that we have a SYSTEM level requirement that must be implemented ACROSS several SUBSYSTEMs, each with its own *development specification*.

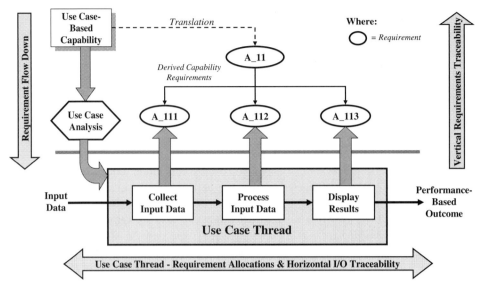

Figure 32.4 Intra-specification Requirements Traceability

Author's Note 32.1 *As you allocate and flow down requirements observe a basic WHY–HOW construct. From any requirement position within the hierarchy between the top and bottom, moving upward within the hierarchy represents WHY the requirement exists. Moving downward elaborates HOW the requirement is accomplished.*

Requirements Derivation, Allocations, and Flow Down Across Entity Boundaries

When you *allocate* and *flow down* requirements across entity specification boundaries. The process becomes more challenging, as illustrated in Figure 32.5. SEs, as facilitators of the requirements allocation and flow down process, must ensure that the *use case thread* is traceable across ALL applicable specifications via their respective requirements.

Suppose that we have a SYSTEM level capability requirement, SYS_11. The SEIT analyzes the requirement, performs a use case analysis, and derives sibling requirements A_11 and B_11.

Author's Note 32.2 *At this point, the SEIT has not allocated requirements A_11 and B_11 to SUBSYSTEMs A and B. We designate each for now as A_11 and B_11 to delineate the uniqueness of each requirement.*

Use case analysis reveals that Requirements A_11 and B_11 require further decomposition. So, without becoming immersed into whether requirement A_11 should be allocated to SUBSYSTEM A or B, SEs construct a *use case thread*. Based on that analysis, SEs derive:

- Requirements A_111 and A_112 from A_11.
- Requirements B_111, B_112, and B_113 from B_11.

Subsequently these requirements are *allocated* to SUBSYSTEMs A and B and *flowed down* to their respective *development specifications*.

So, in terms of what the SYSTEM is expected to accomplish, *what does this mean*? It means:

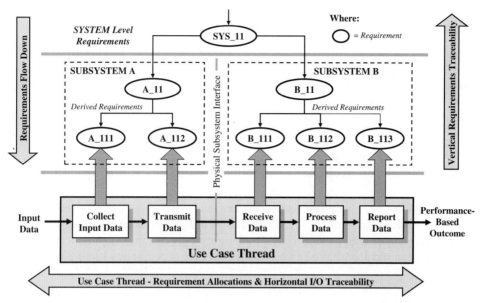

Figure 32.5 Inter-Specification Requirements Traceability

- If you want to accomplish the performance-based outcome specified by SYSTEM Requirement SYS_11, SUBSYSTEM A must accomplish Requirement A_11 and provide the result (outcome) to SUBSYSTEM B, which must accomplish requirement B_11 and produce the final result (outcome).

How do SUBSYSTEMs A and B accomplish requirements A_11 and B_11 respectively?

- SUBSYSTEM A accomplishes requirement A_11 by performing requirements A_111 and A_112.
- SUBSYSTEM B accomplishes requirement B_11 by performing requirements B_111, B_112, and B_113.

Consider the following example:

EXAMPLE 32.3

Suppose that we have a simple system that has a remote sensor (SUBSYSTEM A). The sensor collects input data and transmits it to a central site as illustrated in Figure 32.5.

- The central site (SUBSYSTEM B) receives and processes data to produce a report. So, the *Remote Sensor Development Specification* must state that it: 1) collects data (A_111) and 2) transmits data to the central site (A_112).
- The *Central Site Development Specification* includes the remainder of the requirements use case thread to: 1) receive data (B_111), 2) process data (B_112), and 3) report data (B_113).

32.6 REQUIREMENTS TRACEABILITY

When requirements are derived, the *allocation* and *flow down* process should be *verified* via bottom-up requirements traceability linkages to higher level requirements and ultimately to the *source* or *originating* requirements.

Individual, Contract, and Organizational Requirements Traceability

Anecdotal evidence suggests that engineers and organizations tend to learn and mature in requirements traceability via a three-stage process. The degree of *evolution* depends on the *size*, *complexity*, and *risk* of SYSTEMs being developed and an organizational and individual desire and *willingness* to *improve* performance.

Stage 1: There is general recognition of the need to *organize* and *structure* requirements using a standard specification outline. Over a period of time, the individual and organizational lessons learned, such as *missing*, *misplaced*, and *conflicting requirements*, indicate that a new level of requirements implementation capability is required. This leads to Stage 2.

Stage 2: The individual or organization recognizes that multi-level system requirements involve vertical *requirements traceability* via lineage with high-level requirements. Again, over time and many painful experiences in implementing specification requirements, there is a recognition and appreciation that vertical requirements traceability is one-dimensional. This leads to Stage 3.

Stage 3: The individual and organization recognize the need for two-dimensional requirements traceability, VERTICALLY through the requirements hierarchy to the source or originating requirements and HORIZONTALLY via the use case thread traceability.

What Happens If You Fail to Do This?

You may ask, "*Why bother to do all of this consistency thread checking?*" If we limit the requirements *derivation*, *allocation*, and *flow down* to only vertical linkages from Capability Requirement B_11 to Capability Requirement B_111, B_112, and B_113:

1. What will the text of language of each of these requirements state?
2. Will there be any objective evidence of WHAT outcome is expected when capability requirement B_111 is implemented, and WHAT is the relationship of this outcome to capability requirements A_112 and B_112?
3. Will capability requirement A_112 indicate WHAT it expects from capability requirement A_111, and WHAT transformational processing is required to produce an output acceptable for requirement B_111?

Generally, no. Limiting requirements *allocation* and *flow down* to strictly VERTICAL entity relationships DOES NOT guarantee that teams implementing requirements B_111 through B_113 will:

1. *Communicate with each other.*
2. *If they do communicate, incorporate the results into design documentation.*

Unfortunately, many SEs do not discover these *discontinuities* in system *use case* threads until the system is in integration and test when the cost to correct is prohibitively expensive.

32.7 GUIDING PRINCIPLES

In summary, our discussions in this chapter provide the basis to establish the several key principles that govern practices requirement derivation, flow down, allocation, and traceability.

Principle 32.1 When deriving SYSTEM/entity requirements, understand:
1. What the User NEEDs.
2. What the User WANTs.
3. What the User can AFFORD.
4. What the User is WILLING to pay.

Principle 32.2 Specification requirements allocations ensure vertical requirements traceability; *use case* threads integrate those allocations into horizontal workflows that produce SYSTEM level performance-based outcomes.

32.8 SUMMARY

In our discussion of the *requirements derivation, allocation, flow down, and traceability practices*, we described a methodology for deriving multi-level specification requirements. We considered the importance of *vertical requirements traceability* based on requirements allocations and flow down. We showed also the importance of *horizontal requirements traceability* based on use case continuity thread checks.

GENERAL EXERCISES

1. Answer each of the *What You Should Learn from This Chapter* questions identified in the *Introduction*.
2. Refer to the list of systems identified in Chapter 2. Based on a selection from the preceding chapter General *Exercises* or a new system, apply your knowledge derived from this chapter's topical discussions.
   (a) Using the system architecture previously created, identify PRODUCTs or SUBSYSTEMs that will require development specifications.
   (b) Create a SYSTEM level, use case based requirement concerning system operation.
   (c) Derive *sibling* requirements.
   (d) *Allocate* and *flow down* the requirements to PRODUCTS or SUBSYSTEMs.
   (e) Perform a use case continuity I/O thread check between the respective PRODUCT or SUBSYSTEM requirements.

ORGANIZATIONAL CENTRIC EXERCISES

1. Research your organization's command media. What guidance and direction is provided for deriving, allocating, flowing down, and tracing requirements?
2. Contact a contract program within your organization and interview the Technical Director, Project Engineer, and Lead SEs to answer the following questions:
   (a) How were requirements derived?
   (b) Did the contract mandate traceability of requirements?
   (c) Did the program use a requirements management tool to enable traceability between specifications? If not, how were requirements traced?
   (d) What analysis tools were used to reconcile and balance performance allocations?
   (e) What did personnel learn from the exercise?

Chapter 33

Requirements Statement Development

33.1 INTRODUCTION

Once a *solution space* is identified, the challenge for SEs is being able to *accurately* and *precisely* bound and specify it via the *System Performance Specification* (SPS) or entity *development specification*. Our discussion on *specification development practices* provided an approach for identifying types of specification requirements.

This chapter focuses on the formulation and development of those requirements statements. Our discussions introduce and explore various methods for preparing requirements statements. We identify a syntactic structure to facilitate definition of a requirement. The discussions emphasize the need to define the substantive content of a requirement before focusing on the grammar. To facilitate preparation of requirements, we provide requirements development guidelines and discuss how to "test" a requirement.

We conclude the chapter with a discussion of one of the challenges of specification requirements development: knowing WHEN a set of requirements is necessary and sufficient. We explore the need to *minimize* the quantity of requirements and avoid *overspecification* and *underspecification*.

What You Should Learn from This Chapter

1. What are the attributes of a "good" requirement?
2. When is a "requirement" officially recognized as a requirement?
3. What are some *suggestions* for preparing "good" requirements?
4. What is a *common pitfall* in preparing requirements statements?
5. What *methodology* should one use to create a requirements statement?
6. What *criteria* do you use to "test" a requirement?
7. How do you identify a *requirement's verification method(s)*?
8. What is a *Requirements Verification Matrix (RVM)*?
9. How do you develop an RVM?
10. What is *requirements minimization*?
11. How do you *minimize requirements*?

System Analysis, Design, and Development, by Charles S. Wasson
Copyright © 2006 by John Wiley & Sons, Inc.

Definition of Key Term

- **Specification Language** "A language, often a machine-processable combination of natural and formal language, used to express the requirements, design, behavior, or other characteristics of a system or component. For example, a design language or requirements specification language." (Source: IEEE 610.12-1990)

33.2 UNDERSTANDING THE STRUCTURE OF GOOD REQUIREMENTS

People often say that a specification has a *good* set of requirements. The *subjective* usage of the term "good" makes SEs cringe. "Good" for WHOM relative to WHAT "fitness-for-use" standard(s)? The User? The System Developer? If good requirements are relative to the User, WHAT makes a "*bad* requirement bad?" Does "bad" mean *poorly stated, open to interpretation, unachievable, immeasurable, unverifiable, and untestable*?

What these people are attempting to convey is: GOOD requirement statements are *well defined, explicit*, and can be implemented with low risk of *misinterpretation*. They are *well defined* and *explicitly* stated in a *clear* and *concise* manner that results in the same interpretation by two or more independent readers with comparable disciplinary skills.

Understanding the Contents of a Capability Requirement

A *well-defined requirement* should consist of several key characteristics, depending on application. It must explicitly *express* the following, as applicable:

1. The *use case* based operational capability to the performed.
2. The *source* of the action.
3. WHAT scenario or conditions for performing the action.
4. WHAT is expected as a performance-based outcome (i.e., WHAT products or services are to be produced).
5. WHAT by-products are to be AVOIDED.
6. WHEN the capability or action is to be initiated.
7. WHO is the expected recipient(s) of the action.
8. WHAT are the expected *result(s)* and outcome(s).
9. HOW is the outcome to be measured and verified.

Let's examine each of these characteristics further.

Use Case Based Operational Capability. Requirements statements should originate with *use case based* operational capabilities. Lower level requirements should elaborate specific action-based outcomes that contribute to achievement of the use case as well as any *probable* scenarios anticipated.

Source of the Action. The *source* of the *action* for a capability requirement is the noun-based object—person, place, or thing—that triggered or initiated the capability.

Operating Environment Conditions and Scenarios. Each capability requirement must be bounded by prescribed OPERATING ENVIRONMENT scenarios that *represent* the *threshold* or triggering *events* or *cues* that *initiate* or *stimulate* a capability's behavioral response.

Outcome Produced by the Capability Action. When a capability is initiated, an *action* must be performed. The *outcome* attribute of the requirement identifies the behavioral actions—meaning the products or services that result from the capability.

Product and/or Service Outcome. The subject of a requirement is a capability that represents an *expected* response action to *transform* and/or *process* information, energy, or data, into (produce) an outcome, product/service, or *behavioral* response.

Action Response Time. Time constraints between the *triggering* stimulus—cue or event—and the required capability response from the system. The question is: *WHAT period of performance is permissible between the triggering event and the output response*?

Recipient(s) of the Action. The intended recipient(s) of the capability's outcome(s) must be identified, recommended and bounded including their capability to receive the action.

Format the Action Response. The format of the action response to *cooperative, benign, or hostile* interactions must be compatible, interoperable, or commensurate with the *rules of engagement*. Consider the following example:

EXAMPLE 33.1

When two computers communicate with another, both must implement an interface protocol that includes messages formatted in accordance with an *interface control document (ICD), interface requirements specification*, or industry standard.

EXAMPLE 33.2

If an adversary commits a *hostile* action, the defender may *respond* with an action-based response commensurate with *rules of engagement* (i.e., allowable actions and measured responses).

Expected Outcome and Results of the Action. Producing a capability response is merely a means to an end. Therefore, the requirement must also define WHAT the *expected outcome* or *action* is to be commensurate with the behavioral response—a measured level of performance such as magnitude or priority.

Requirement Verification. Every specified requirement must be *linked* to one or more Section 4.0 verification requirements—*inspection, analysis, demonstration,* and *test.*

33.3 WHEN DOES A REQUIREMENT BECOME OFFICIAL?

Our previous discussions focused on the content of well-defined requirements. A key question commonly asked is: *WHEN is a requirement recognized as a requirement*? There are two stages of answers to the question: 1) *unofficial* and 2) *official.*

Simply identifying this information in a statement is only a prerequisite to official stakeholder *recognition* and *acceptance*. This first stage merely establishes the statement as RELEVANT and WORTHY of consideration as a requirement. That takes us to the next stage, official *approval* and subsequent *release.*

Many SEs erroneously believe that simply preparing a requirement statement makes the requirement *complete*. Preparing a requirement statement DOES NOT mean that the statement will *pass* the requirement *completeness criteria* discussed later in this chapter. WHY is it necessary to test completeness? There are two key factors to consider.

First, identifying the verification method(s) forces the requirement developer to THINK about HOW the requirement can be *verified*. If you can write a one- or two-sentence verification test plan—for example, compare A to B—you MAY have a *legitimate* basis for the requirement. Conversely, if you have difficulty identifying a test plan, maybe you should consider *rewriting* and *rescoping* the requirement statement.

Author's Note 33.1 *Engineers tend to defer identification of the verification method(s) up to the start of the system integration, test, and evaluation (SITE) phase. This is not adequate! You must identify the verification methods early as part of the requirements preparation process, not only for cost estimates but also to validate the reasonableness of the requirement.*

Second, identifying the verification method *prematurely* may force an early revision to the baselined specification. WHY? Since most systems have multiple levels of specifications, analysis must be performed one or more levels below the current level to better understand HOW the higher level requirement can be verified.

Regarding the second point, an approach may be to identify a *preliminary* set of verification methods when the requirements are derived. This will provide an initial completeness *check* on the requirement. Then, continue the analysis to lower levels. Prior to the need to baseline the higher level requirements, review and reconcile the preliminary verification methods at the higher level.

While the potential ramifications for *premature* baselining are known—such as increased costs due to stability of decisions—*human procrastination* can pose an even greater risk. Whether intentional or not, program schedules often become a convenient excuse for SEs to shift identification of verification methods to the back of the priority list.

People, in general, tend to AVOID spending time on an effort unless they recognize its importance to technical management or are directed to do so. Then, the verification methods are haphazardly thrown together just prior to the System Integration, Test, and Evaluation (SITE) phase. The bottom line is: *If System Developers do not identify specification verification methods up front, you will suffer the consequences during SITE.*

33.4 PREPARING THE REQUIREMENT STATEMENT

Specification developers often become consumed with attempting to state too many requirements at once. When you develop requirements, there is actually a three-step process that you should follow:

Step 1: Focus on the capability that is to be communicated in the requirement and its supporting attributes.

Step 2: Focus on the *syntax* of the primary contents of the requirement.

Step 3: Focus on the *grammar*.

Let's elaborate on each of these points.

Step 1: Identify the Key Elements of the Requirement

Earlier in this chapter we identified key attributes that describe the content of most requirements (source of the action, etc.). For each of these attributes, identify in short notation (verb–noun form) WHAT the requirement attributes are to communicate. Consider the example shown in Table 33.1. Here we describe the fundamentals of a SYSTEM level requirement related to a car design.

Notice that Table 33.1 focuses on WHAT must be accomplished; *grammar* comes later as illustrated in the example that follows.

EXAMPLE 33.3

The vehicle shall have the capability to safely and comfortably transport a driver and four passengers on-demand to a destination 200 miles away during all weather conditions.

Step 2: Develop the Draft Requirements Statement

Humans unintentionally *overcomplicate* the performance and development specification requirements process. The formulation and development of a requirements statements involves three aspects: 1) content, 2) syntax, and 3) grammar. One of the most common problems in requirements statement development is that people *attempt* to perform all three of these aspects *simultaneously*. As a result, when you focus on *syntax* and *grammar*, you may lose focus on the *content*. One solution to this problem is to develop primitive *requirements* statements.

Create Primitive Requirements Statements. *Primitive requirements statements* focus on the content of the requirement. The best way to develop *primitives* is to use a tabular approach such as the one illustrated in Table 33.2. Here, we assign a unique requirement ID—SPS-136—to

Table 33.1 Example requirements attributes definition approach for a car-driver system

Requirement Statement Attribute	Requirement Contents
Required use case capability	Transportation
Source of the action	Vehicle
Expected outcome/results of the action	Safe and comfortable travel to/from a planned destination
Operating environment constraints	All weather conditions, −30°F to +130°F
Action response time	On-demand; immediately available
Recipient(s) of the action	Driver and four passengers
Format of the action	Seated

Table 33.2 Table for identifying primitive requirements

Requirement ID	Capability to Be Provided	Relational Operators	Boundary Constraints, Tolerances, or Conditions	Notes
SPS-136	Capability 24	• Not to exceed • Less than or equal to • Greater than or equal to • In accordance with (IAW) • Not exceed • And so forth	• 50 pounds • 68° Fahrenheit • 25° Celsius • 6 g's • 6 nautical miles (NM) • 25 Hertz (Hz) • 10.000 volts dc +/− 0.010 volts dc • Sea state 3 • 12 megabytes (Mb)	

Table 33.3 Examples of primitive requirements using the tabular approach

Requirement ID	Capability to be Provided	Relational Operators	Boundary Constraints, Thresholds, Tolerances, or Conditions	Notes
SYS_1	The SYSTEM **shall** operate from an external power source	Rated at	• 110 vac +/− 10%	Assuming a TBD load
SYS_123	The gross weight of the SYSTEM **shall** ...	Not exceed	• 50 pounds	
SYS_341	The SYSTEM **shall** operate ...	Throughout a range of	• −10°F to +90°F • +20% to +100% humidity	
SYS_426	On receipt of a command from the operator, the SYSTEM **shall** produce reports formatted	In accordance with	• Regulation 126 Section 3.2.1	
SYS_525	The SYSTEM **shall** sense overload conditions that ...	Exceed	• +10 amps dc	

Capability 24, which represents a specific use case. The remainder of the columns focus attention on substantive content such as types of relational operators; boundary constraints, tolerances, or conditions; and notes.

Notice how Table 33.2 focuses on the *primary* content of the requirement and *AVOIDS* becoming mired in grammar. Once the elements of the requirements are established, the next step is to translate the contents into a syntax statement.

Develop the Syntactical Statement Structure. Step 2 translates the *primitive* requirement into a text phrase as illustrated in Table 33.3.

Step 3: Refine the Requirements Statement Grammar

The third step requires the syntactical statements in Table 33.3 to be transformed into a *clear* and *concise* grammatical statements that are easily *read* and *understood*.

33.5 IDENTIFYING THE REQUIREMENT VERIFICATION METHOD

One of the key elements of specification development is establishing *agreement* between the Acquirer and the System Developer as to HOW to *prove* that the specification Section 3.*X Requirements* have been satisfied. One of the greatest potential risks in system development is for the System Developer to prepare or conduct acceptance tests for the Acquirer and both parties DISAGREE on the method of verification. Disagreements at acceptance testing result in significant impacts on System Developer contract costs, schedules, and Acquirer fielding and support costs schedules.

To preclude this scenario, specifications include a Section 4.0 *Qualification Provisions* section that *explicitly* captures the Acquirer and System Developer agreement on verification methods. The agreement is documented as a *Requirements Verification Matrix (RVM)*.

376   Chapter 33   Requirements Statement Development

Author's Note 33.2   *Remember, the terms Acquirer and System Developer are used in a "role-based" context. These terms, as used above, identify the contract level roles and relationships. They also apply in a similar context between SYSTEM and PRODUCT, PRODUCT and SUBSYSTEM development teams; System Developer and Subcontractors, and so on. Why? As each team flows down requirements to lower level item development specifications, the implementer of those requirements must demonstrate to their Acquirer (role)—the higher level team—that the requirements have been satisfied.*

Before we address the Requirements Verification Matrix (RVM), let's address HOW verification methods are selected.

Referral   For more information about verification methods, refer to Chapter 53 System Verification and Validation practices.

Verification Method Selection Process

In general, SEs have four basic verification methods available to demonstrate that a system, product, or service meets a specific requirement. These methods include INSPECTION or EXAMINATION, ANALYSIS, DEMONSTRATION, and TEST. A fifth method, SIMILARITY, is permitted by some organizations.

Verification method selection requires insightful forethought. *Why? First*, verification decisions range from simple visual inspections to highly complex tests that include analysis. This can be very costly and consumes valuable schedule time. *Second*, common sense asks WHY you would want to go to the trouble and expense of performing a TEST when a simple visual INSPECTION is sufficient.

So, *how do SEs select a method*? A time-cost driven methodology provides a basis for the verification method of decision making as illustrated in Figure 33.1.

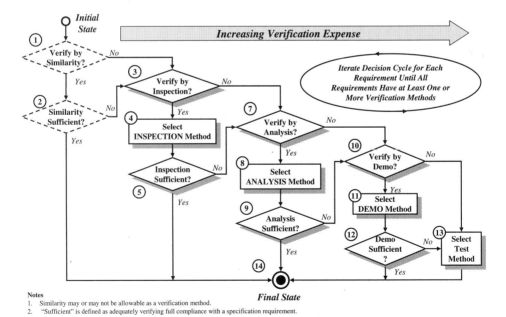

Figure 33.1   Verification Method Selection Process

Verification by SIMILARITY. Some contracts or organizations permit *verification* by SIMILARITY. In this case a previously *verified* design is permitted to be *reused* to develop a new system or product, provided that the design's performance data and *quality records* (CM, QA, etc.) remain valid. This means:

1. No changes have been made to the design since its official verification.
2. No safety critical problems have been encountered in fielded units.
3. Design requirements of the new system or product do not exceed the reusable component's "As Verified" requirements.

One of the reasons many organizations *discourage* verification by SIMILARITY is that it requires INSPECTION or EXAMINATION of records, which may no longer be available. The best approach may be to verify the requirement by INSPECTION/EXAMINATION and show that the reused design has not changed.

Verification by INSPECTION or EXAMINATION Decision. Verification by INSPECTION requires a questioning exercise that has two potential *outcomes*:

- Is INSPECTION/EXAMINATION *necessary* to prove that a physical *entity* complies with a stated specification requirement? If the answer is YES, select INSPECTION/EXAMINATION as a verification method.
- Is the INSPECTION/EXAMINATION *sufficient* to prove that the requirement and its associate level of performance have been successfully accomplished? If the answer is YES, proceed to identification of verification methods for the next requirement. If not, proceed to the verification by ANALYSIS decision.

Verification by ANALYSIS Decision. Verification by ANALYSIS requires a questioning exercise that has two potential *outcomes*:

- Is ANALYSIS *necessary* to prove that a physical *entity* or data collected from its formal testing prove that it complies with a stated specification requirement? If the answer is YES, select ANALYSIS as a verification method.
- Is ANALYSIS *sufficient* to prove that the requirement and its associate level of performance have been successfully accomplished? If the answer is YES, proceed to identification of verification methods for the next requirement. If not, proceed to the verification by DEMONSTRATION decision.

Verification by DEMONSTRATION Decision. Verification by DEMONSTRATION requires a questioning exercise that has two potential *outcomes*:

- Is DEMONSTRATION *necessary* to prove by formal observation that a physical *entity* produces *repeatable* and *predictable* outcome(s) without having to record formal measurements to prove that it complies with a stated specification requirement? If the answer is YES, select DEMONSTRATION as a verification method.
- Is DEMONSTRATION *sufficient* to prove that the requirement and its associate level of performance have been successfully accomplished? If the answer is YES, proceed to identification of verification methods for the next requirement. If not, proceed to the verification by TEST decision.

Verification by TEST Decision. If any of the preceding questions indicate that a formal test is required to collect data to verify that the physical entity produces *repeatable* and *predictable* outcome(s) to prove that it complies with a stated specification requirement, then TEST must be selected as a verification method for a specific requirement.

Given this understanding of HOW verification methods are selected, let's proceed with HOW the methods are documented via the RVM.

33.6 WILEY WHAT IS A REQUIREMENTS VERIFICATION MATRIX (RVM)?

A Requirements Verification Matrix (RVM) is a tabular listing that maps each Section 3.X requirement to specific Section 4.X *Verification Methods*—namely inspection, analysis, demonstration, and test.

RVM Example. To illustrate the structure of a RVM, let's use a hypothetical example. Assume that requirement 3.1.1 for Capability A states: *The system shall perform Capability A, which consists of the capabilities specified in the subparagraphs below*:

- 3.1.1.1 Capability A1 ...
- 3.1.1.2 Capability A2 ...
- 3.1.1.3 Capability A3 ...
- 3.1.1.4 Capability A4 ...

Table 33.4 gives an example of an RTM entry for this requirement.

As illustrated in the table, Capability A is *verified* by INSPECTION that three *sibling* capability requirements—3.1.1.1, 3.1.1.2, 3.1.1.3, and 3.1.1.4 have been verified. Each of these *sibling* capability requirements is *verified* as follows:

- 3.1.1.1 Capability A1 is *verified* using the INSPECTION method.
- 3.1.1.2 Capability A2 is *verified* using the INSPECTION and ANALYSIS methods.
- 3.1.1.3 Capability A3 is *verified* using the DEMONSTRATION method.
- 3.1.1.4 Capability A4 is *verified* using the INSPECTION and TEST methods.

Since Capability A's requirement is satisfied by *verification* of its *sibling* requirements, Capability A's verification methods are INSPECTION that the sibling requirements have been verified.

Most SEs develop the RVM structure and progress down the list one item at a time by checking the verification methods boxes. *You simply cannot do this for every system application.* The identification of requirements *verification methods* is highly iterative and must be reconciled between multiple levels or chains of requirements.

Table 33.4 Specification Section 4.0 Requirements Traceability Matrix (RTM) example

Specification Requirement Paragraph	Verification Methods			
	Inspection	Analysis	Demonstration	Test
3.1.1 Capability A	X			
3.1.1.1 Capability A1	X			
3.1.1.2 Capability A2	X	X		
3.1.1.3 Capability A3			X	
3.1.1.4 Capability A4	X			X

Requirements Management Tools. Traditional specifications include a separate table for Section 4.0 Qualification Provisions. The difficulty with this approach is that the specification developer and readers must flip back and forth between Section 3.0 requirements and Section 4.0 verification methods. From an SE perspective, this method is *very inefficient*.

The best approach is to use a requirements management tool, preferably one that is based on an object-oriented, relational database. The tool enables the verification methods to be stated in a spreadsheet display as additional columns of the requirement. Table 33.5 illustrates this approach. Some tools simplify this approach and use "pick lists" for each cell in one column instead of five columns.

In the example above, note the following:

- Each requirement has its own unique identifier such as SYS_136 or 4692xxx.
- The full text of each requirement is stated.
- Each requirement can be easily related to its verification method(s).

Table 33.5 represents a *view* that can be obtained from an object-oriented, relational database tool. If you have a small technical program and do not have a tool of this type, you could use a spreadsheet to accomplish similar results.

A Word about Requirements Management Tools. Traditionally, Acquirers prefer word processor-based documents. Why? They are easier to control as part of the contract and require skills and software applications most *stakeholders* possess. The SE problem is, however, *HOW do you link requirements in a word processor document to another specification, engineering drawing, graphic, and so on*?

With today's Web-based technologies, you will say this can be done easily with document links. However, conventional document links do not allow you to display or printout the "thread" of text statements that are traceable across documents. Requirements management tools provide this unique capability. You can live without a requirements management tool. However, the tool saves hours when verifying requirements traceability. This is why RT tools provide an advantage

Table 33.5 Requirements management tool RVM report

Requirements ID	Requirements Statement	Method of Verification				
		N/A	Inspect	Analysis	Demo	Test
SYS_136	**3.1.1 Capability A** The system shall perform (Capability A), which consists of the capabilities specified in the subparagraphs below:		X			
SYS_137	**3.1.1.1 Capability A1** The system shall (Capability A1) . . .		X	X		
SYS_138	**3.1.1.2 Capability A2** The system shall (Capability A2) . . .			X		
SYS_139	**3.1.1.3 Capability A3** The system shall (Capability A3) . . .				X	
SYS_140	**3.1.1.4 Capability A4** The system shall (Capability A4) . . .					X

380 Chapter 33 Requirements Statement Development

and *leverage* your own ability and time. You can *convince* yourself to *believe* you cannot *afford* the tool. However, the capabilities and productivity of your personnel and organization will *pay* dividends, especially for large, complex system development efforts.

Verification Level. We like to think that simply identifying verification methods for a requirement is sufficient. However, some requirements may not be *verifiable* at every entity or CI level. Consider the following example:

EXAMPLE 33.4

You are assigned a task to create *SUBSYSTEM A Development Specification*. Ideally, you would like to verify SUBSYSTEM A as a *work product*. There may be cases whereby it is *impractical* to *complete* verification of SUBSYSTEM A as a work product until it is integrated into the higher level SYSTEM. For example, it may be impractical to *stimulate* one of SUBSYSTEM A's interfaces during verification tests at the SUBSYSTEM level. As a result, you may want to indicate in the RVM, for internal record keeping, the *level of abstraction* at which *verification* of a specific requirement will be accomplished. Table 33.6 provides a tabular illustration.

As illustrated in Table 33.6, requirement 3.1.1.1 will be verified at the SUBSYSTEM level. Requirement 3.1.1.4, which requires SUBSYSTEM A to be integrated with SUBSYSTEM B, will be verified at the SYSTEM level. Thus, the verification of requirement 3.1.1 cannot be completed until the 3.1.1.4 TEST method is performed at SYSTEM level integration.

Table 33.6 RVM Example illustrating Verification Level for Each Requirement

Requirements ID	Requirement Statement	Verification Level	Verification Methods				
			N/A	Inspect	Analysis	Demo	Test
SYS_136	**3.1.1 Capability A** The system shall perform (Capability A) Capability A, which consists of the capabilities specified in the subparagraphs below:	See subparagraphs below					
SYS_137	**3.1.1.1 Capability A1** The system shall provide (Capability A1) . . .	SUBSYSTEM		X	X		
⋮	⋮	⋮					
SYS_140	**3.1.1.4 Capability A4** The system shall provide (Capability A4) . . .	SYSTEM					X

33.7 REQUIREMENTS DEVELOPMENT GUIDELINES

Requirements can be characterized a range of *attributes*. These *attributes* include legal, technical, cost, priority, and schedule considerations. Let's examine each of these briefly.

Suggestion 1: Title Each Requirement Statement

Requirements tend to lose their identities as statements within a larger document. This is particularly troublesome when the need arises to rapidly search for an instance of a requirement. Make *it* easy for reviewers and users of specifications to *easily locate* requirements. Give each requirement a "bumper sticker" title. This provides a mechanism for capture in the specification's table of contents and makes identification easier.

Suggestion 2: Use the Word "Shall"

Requirements, when issued as part of a contract, are considered *legally binding* and *sufficient* for procurement action when *expressly stated* with the word "shall." Engineers have a REPUTATION for expressing requirements with the words "will," "should," and "must." Those terms are *expressions of intent*, not a *mandatory* or *required action*. To facilitate specification readability, consider **boldfacing** each instance of the word "*shall*" (e.g., **shall**).

Author's Note 33.3 *There is increasing evidence of less discipline in specification wording by using words such as "will." The general viewpoint is that the "context" of the statement of the statement is what is important. For example, "... the system will perform the tasks specified below. ..." If you subscribe to this notion, it is critically important that you establish "up front" in the specification HOW the term "will" is to be interpreted relative to being "mandatory." Whatever convention you and your organization adopt, you must be consistent throughout the specification. The best practice is to use the word SHALL. In any case, ALWAYS seek the advice of your organization's legal counselors before committing to a specification that contains ambiguous wording that may be subject to legal interpretation.*

Suggestion 3: References to Other Sections, Specifications, and Standards

Specification sections often reference other sections within the document. Typically, the specification will state, "Refer to Section 3.4.2.6." If you do this, consider including the title of the section with the sectional reference. Why? The structures of specifications often change as topics are added or deleted. When this occurs, the sections are renumbered. Thus, the original "*Refer to Section 3.4.2.6.*" may now reference an unrelated requirement topic. By noting the sectional title to the reference, the reviewer can easily recognize a conflict between the referencing and target sections. The same applies to references to external standards and specifications.

Suggestion 4: References to External Specifications and Standards

When you reference external specifications and standards, include the title and document number. Normally the title and document identification should be stated in Section 2.0 Referenced Documents. However, there may be instances of documents with similar titles that may be confusing. If this occurs, references as follows: "*XYZ System Performance Specification (SPS)* (Document No. 123456)."

Suggestion 6: Operational and Technical Capabilities and Performance Characteristics

Operational and *technical capabilities and characteristics* requirements should *explicitly* describe the required system capabilities and levels of performance. The *DSMC Test and Evaluation Management Guide* defines these terms as follows:

- **Required Operational Characteristics** "System parameters that are primary indicators of the system's capability to be employed to perform the required mission functions, and to be supported." (Source: DSMC T&E Mgt. Guide, Appendix B, *DoD Glossary of Test Terminology*, p. B-17)
- **Required Technical Characteristics** "System parameters selected as primary indicators of achievement of engineering goals. These need not be direct measures of, but should always relate to the system's capability to perform the required mission functions, and to be supported." (Source: DSMC T&E Mgt. Guide, Appendix B, *DoD Glossary of Test Terminology*, p. B-17)

Remember, OPERATIONAL and TECHNICAL characteristics should be *measurable*, *achievable*, *realistic*, *testable*, and *verifiable*.

Suggestion 7: Avoid Paragraph-Based Text Requirements

Engineers default to traditional writing methods when preparing requirements statements. They write requirements statements in paragraphs with theme sentences, compound statements, and lists. When specifying systems, *isolate* requirements as *singular testable requirements*.

Paragraph-based declarative statements in specifications are acceptable for Section 1.0 *Introduction*. The problem is System Developers must spend valuable time *separating* or *parsing* compound requirements into SINGULAR requirements in Section 3.x. Do yourself and the System Developer (contractor, subcontractor, etc.) a favor and generate system specifications with *singular requirements* statements. This saves valuable time that can be better spent on higher priority tasks and creates a consistent format that promotes *readability*, *understandability* and *verifiability*.

Suggestion 8: Eliminate Compound Requirements Statements

When requirements statements specify compound requirements, allocation of the requirement becomes confusing and verification becomes difficult. *Why?* If multiple requirements are specified in a single requirements statement and they are allocated to different system entities, the challenge is: *HOW do you link which portion of a requirement statement to a specific target entity?*

When the system must be verified, HOW can you check off a requirement as complete (i.e., verified) when portions remain to be verified due to missing equipment, and the like? The only alternative is to uniquely number each requirement within the statement. When it comes time to verify the requirement statement, complete verification of all requirements embedded within the statement is required to satisfy verification of the whole statement.

Suggestion 9: Eliminate the Word "ALL"

Specification writers are often consumed with the word "all." Eliminate uses of the word "all" in a specification! Appreciate the legal significance of stating that ". . . *ALL instances of an XYZ will be verified*. . . ." Contemplate HOW large in scope the ALL universe is? *Remember*, ALL is one of those limitless words, at least for practical purposes. As a specification writer, your mission is to *specify* and *bound* the *solution space* as simply and as practical, not open it up for verifying every conceivable scenario or instance of an entity. ALL, in a *limitless* context, is boundless and is subject to interpretation. Do yourself a favor and AVOID usage of ALL.

Suggestion 10: Eliminate the Term and Abbreviation "Etc."

Et cetera is another word that we use constantly. Most engineers discover early in their careers there will always be someone who "nit picks" about miniscule *instances* of science or a natural occurrence and chastise the engineer for missing it. Thus, engineers learn to "cover all cases" by using the term "etc." Again, when you develop specifications, your mission is to *specify* and *bound* the solution space. When you state, *"As a minimum, the system shall consist of a, b, c, etc.,"* the Acquirer could say later, *"Well, we also want a "d, e, and f."* You may reply we did not bid the cost of including a "d, e, and f," which may then bring a response *"You agreed to the requirements and "etc." means we can request anything we want to."* Do yourself a favor and *eliminate* all instances of etc. in specifications.

Suggestion 11: Avoid the Term "And/Or"

Specification writers often specify requirements via enumerated lists that include the term "and/or."

EXAMPLE 33.5

For example, *". . . the system shall consist of capabilities: A, B, C, and/or D."*

Be explicit. Either the system consists of A, B, C, or D or it does not. If not, so state and *bound* exactly WHAT the system is to contain. Remember, specifications must state *exactly* WHAT capabilities are required at SYSTEM delivery and acceptance.

33.8 "TESTING" DERIVED REQUIREMENTS STATEMENTS

Many people are surprised to find that you can "test" requirements. Requirement *testing* takes the form of *technical compliance* audits with organizational standards and conventions, such as coding standards and graphical conventions. Modeling and simulation provides another method for testing requirements. Execution of those models and simulations provides insights into the reasonableness of a requirement, performance allocation, potential conflicts, and difficulty in verification.

Requirement testing also includes *inspection* and *evaluation* of each requirements statement in accordance with pre-defined *criteria*.

Requirement Validation Criteria

When testing the *validity* of a requirement by *inspection* or *evaluation*, there are a number of criteria that can be applied to determine the *adequacy* and *sufficiency* of the requirement. Table 33.7 provides an example listing of key criteria.

The criteria stated in Table 33.7, though examples, are reasonably comprehensive. You are probably thinking *how* can you evaluate a specification with potentially hundreds of requirements using these criteria.

First, seasoned SEs imprint these criteria subconsciously in their minds. With experience you will learn to test specification requirements rapidly. *Second*, for large specifications, obviously groups of SEs must be participants in the testing and analysis exercise. This further demonstrates the *criticality of training* all SEs in the proper methods of requirements writing and review to ensure a *level of confidence* and *continuity* in the results.

Table 33.7 Requirement validation criteria

ID	Criteria	Criteria Question
1.	Unique identity	Does the requirement have its own unique identity such as a title that reflects the subject capability and unique tracking identifier number?
2.	Singleness of purpose	Does the requirement specify and bound one and only one capability or are there compound requirements that should be broken out separately?
3.	Legitimacy and traceability	Is this a legitimate requirement that reflects a capability traceable to the User's intended operational needs or simply a random thought that may or may not be applicable?
4.	SOW language avoidance	Does this requirement include language that logically belongs within the *Contract Statement of Work (CSOW)*?
5.	Appropriateness	Does this requirement fall within the scope of this specification or does it belong in another specification?
6.	Hierarchical level	If this requirement is within the scope of this specification, is the requirement positioned at the right level within the requirements hierarchy?
7.	User priority	What is the User's priority level for this requirement—can live without, nice to have, desirable, or mandatory?
8.	Realism	Is the requirement realistic?
9.	Achievability	If the requirement is realistic, can it be achieved economically and at acceptable risk with available technology?
10.	Feasibility	Can the requirement be implemented within reasonable need priorities and budgetary cost without limiting the minimum required set of requirements?
11.	Explicitness	Is the requirement stated in unambiguous language—in text and graphics?
12.	Readability and understandability	Is the requirement simply stated in language that is easily understood by the document's *stakeholders* and ensures that any combination of *stakeholders* emerge with the same interpretation and understanding of WHAT is required after reading the requirement?
13.	Completeness	Does the requirement adequately satisfy the structural syntax criteria identified in an earlier section?
14.	Consistency	Is the requirement consistent with system terminology used: **1.** Throughout the specification? **2.** Among specifications within the specification tree? **3.** By the Acquirer and User?
15.	Terminology	Does the requirement contain any terms that require scoping definitions?
16.	Assumptions	Does the requirement make assumptions that should be documented in the *Notes and Assumptions* section of the specification?
17.	Conflicts	Does this requirement conflict with another requirement?
18.	Objectivity	Is the requirement stated *objectively* and avoids *subjective* compliance assessments?
19.	Accuracy	Is the requirement stated in language that accurately bounds the subject capability and level of required performance?

(continued)

Table 33.7 *continued*

ID	Criteria	Criteria Question
20.	Precision	Is the precision of the required level of performance adequate?
21.	Testability	If required, can a test or series of tests be devised and instrumented economically with available resources— knowledge, skills, and equipment?
22.	Measurability	If testable, can the results be measured directly or indirectly derived by analysis of test results?
23.	Verifiability	If the results of the test can be analyzed either directly or indirectly, can the results be verified by *inspection*, *analysis*, *demonstration*, or *test* to prove full compliance with the requirement?
24.	Level of risk	Does this requirement pose any significant technical, technology, cost, schedule, or support risks?

33.9 REQUIREMENTS MINIMIZATION

SEs often struggle with the rhetorical question how many requirements do you need to specify a system. *There are no specific guidelines or rules; only* disciplined and seasoned *experience*. Contrary to many things in life, specification *quality* is not measured by the QUANTITY of refinements and features.

Frightening? Yes! But think about it! Every layer of requirements adds *restrictions, complexity, cost, and schedule* that limit the System Developer's flexibility and options to *innovate* and achieve lower costs, schedule implementation, and risk.

How many requirements should an ideal specification have? Hypothetically, the answer could be ONE requirement; however, a one-requirement specification has limited utility. We can nevertheless state that a properly prepared specification is one that has the *minimal* number of requirements but yet enables the User/Acquirer to procure a system that can be *verified* and *validated* to meet their operational capability and performance needs.

How do we emerge with this idyllic specification? The answer may be found in *performance specifications*. *Performance specifications* enable SEs to treat a SYSTEM like a box with inputs and outputs (Figure 31.1). The intent is to *scope* and *bound* the behavior of a SYSTEM relative to scenarios and conditions in its prescribed OPERATING ENVIRONMENT. This is accomplished via *measures of effectiveness (MOEs) and measures of suitability (MOSs)* with supporting *measures of performance (MOPs)*. By treating the system characteristics as a "performance envelope," we have avoided specifying *how* the system is to be designed.

Referral For more information regarding *measures of effectiveness (MOEs)*, *measures of suitability (MOSs)*, and *measures of performance (MOPs)*, refer to Chapter 34 *Operational Utility, Suitability*, and *Effectiveness* practices.

There is one problem, however, with bounding the SYSTEM at the "performance envelope" level. If you cast the *performance specification* in a manner that allows too much *flexibility*, the results may *unacceptable*, especially to the Acquirer. This leaves the System Developer with the challenge or *interpretation* of determining WHAT specific capabilities the SYSTEM is to provide and HOW WELL each capability is to be provided. This can be *good* or *bad*, depending on your role as an Acquirer or System Developer/Subcontractor.

Since all capabilities may not be considered to be of *equal* importance by the User, especially in terms of constrained budgets, the User's priority *list* may be different from the System Developer's PRIORITY *list*.

Author's Note 33.4 *Remember, the User is interested in optimizing capabilities and performance while minimizing technical, cost, and schedule risk. The System Developer, depending on type of contract, is also interested in minimizing technical, cost, and schedule risk while optimizing profitability. Thus, unless both parties are willing to work toward an optimal solution that represents a win-win for both parties, conflicts can arise regarding these priority viewpoints and organizational objectives can clash.*

How do we solve this dilemma? The Acquirer may be confronted with having to specify additional requirements that more *explicitly* identify the key capabilities, levels of performance, as well as priorities. The challenge then becomes: *At what level does the Acquirer stop specifying requirements?* Ultimately, the Acquirer could potentially OVERSPECIFY *or* UNDERSPECIFY the system, depending on budget factors. Additionally, there may be other options for the Acquirer to work around the problem via "decision control or approval" tasks incorporated into the *Contract Statement of Work (SOW)*.

So, *what is the optimal number of requirements?* There are no easy answers. Philosophically, however, we may be able to describe the answer in our next topic, the optimal system requirements concept.

Optimal System Requirements Concept

Anecdotal data suggest that a notional profile can be constructed to express the ideal quantity of specification requirements. To illustrate this notional concept, consider the graphic shown in Figure 33.2.

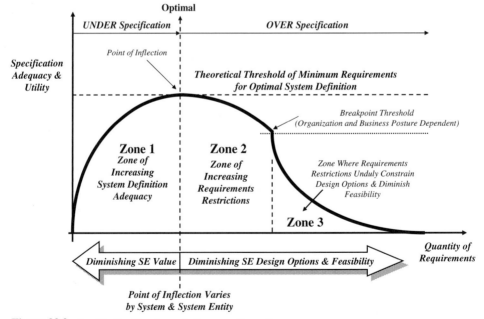

Figure 33.2 Specification Requirements Coverage Concept

The figure consists of a graphical profile with three curved line segments. For discussion purposes, we will identify the area under each segment as a *zone*. Beginning at the intersection of the *X*- and *Y*-axes, every legitimate requirement that is identified at the top hierarchical levels increases the adequacy of system definition toward a theoretical *optimal level*. We characterize this theoretical point as the *point of inflection* in the curve's slope.

At the *point of inflection*, we should have an *optimal* number of requirements. Hypothetically, the quantity of requirements should be *technically necessary* and *sufficient* to specify a SYSTEM with the desired capabilities and levels of performance. At this level the requirements are *minimally sufficient* to *specify* and *bound* the User's intended operational needs.

Beyond the *point of inflection* lies the *Zone of Increasing Requirements Restrictions*. As the slope of the curve indicates, requirements can be added but at the expense of *specification adequacy* and *utility—over specification*. Each additional requirement restricts the SE design options and may increase technical, cost, schedule, technology, and support risk.

As the quantity of requirements continues to increase to the right, you finally reach a *breakpoint* for Zone 3. Zone 3 represents the region where the requirements become *too restrictive*. Thus, the requirements *unduly restrict* SE design options and *severely limit feasibility* of the system. Generally, when this occurs, the Acquirer discovers this problem during the draft proposal stage. To alleviate the risk, the Acquirer may remove some requirements due to prohibitive technical, cost, schedule, technology, and support costs and risks.

The problem described here is not unique to the Acquirer. The same problem challenges the System Developer program organization, not just at the SYSTEM level but at multiple levels of specifications at lower levels. Every requirement at every level has a *cost to implement* and a *cost to verify* besides their schedule and risk implications. This impacts SE Design Process, the Component Procurement and Development Process and the System Integration, Test, and Evaluation (SITE) Process of the System Development Phase of the system/product life cycle.

Does this concept answer the question: *What is the optimal number of specification requirements*? No. However, it illustrates some hypothetical conditions—the *points of inflection* and *breakpoints*—that should be within your mindset. The bottom line is: writing specifications requires more thought than simply writing "random thoughts" within schedule constraints. You need to THINK about WHAT you require and the potential ramifications on the System Developer.

Author's Note 33.5 *As a discipline, industry is consumed with metrics and comparing them with everyone else. While there may be some rough order of magnitude (ROM) relevance, do not lose sight that the "end game" is for the Acquirer, System Developer, and Subcontractors to respectively come to a mutual understanding and agreement on WHAT is to be delivered and emerge with a POSITIVE experience. Every system is different and should be evaluated on its own merits. If you make system-to-system comparison, make sure the two systems are comparable in form, fit, and function.*

Final Thoughts

The intent of the preceding discussion has been to heighten your awareness of a *key theme*: when specifying requirements, strive for the optimal number of requirements that is *sufficient in quantity to adequately to cover the subject but few in quantity to avoid increasing costs, risks, and development time and reducing the flexibility of equally qualified design options.*

33.10 GUIDING PRINCIPLES

In summary, the preceding discussions provide the basis to establish guiding principles that govern requirements statement development practices.

Principle 33.1 Every system requirement must be unique within the system, singular in purpose, consistent with other requirements, and nonconflicting.

Principle 33.2 Every requirement must be explicitly understandable, realistic, achievable, consistent, testable, measurable, and verifiable.

Principle 33.3 Every requirement must have an assigned owner who is accountable for its implementation and maintenance.

Principle 33.4 A requirement is NOT a requirement until it:

1. Is traceable to source or originating requirements via an RVM.
2. Satisfies requirements validation criteria (Table 33.7).
3. Is assigned one or more verification methods.
4. Is accepted by a consensus of its stakeholders.
5. Is approved and released for implementation.

Principle 33.5 Every requirement:

1. Has a value and priority to the User.
2. Constrains design solution options.
3. Increases risk.
4. Has a cost to implement and maintain.

33.11 SUMMARY

Our discussion of the *requirements statement development practices*:

- Described the key attributes that characterize a good requirement.
- Introduced a basic methodology for preparing requirements statements via a three-step approach that avoids common problems of developing requirement content and grammar simultaneously.
- Provided a list of suggestions for developing good requirements.
- Highlighted the need to focus on singular requirements statements that meet specific criteria.
- Described how to develop a *Requirements Verification Matrix (RVM)*.
- Discussed when a requirement should be officially recognized as a requirement.
- Described the process for selecting requirement verification methods.

GENERAL EXERCISES

1. Answer each of the *What You Should Learn from This Chapter* questions identified in the *Introduction*.
2. Refer to the list of systems identified in Chapter 2. Based on a selection from the preceding chapter's General *Exercises* or a new system selection, apply your knowledge derived from this chapter's topical discussions. Specifically identify the following:
   (a) Based on four to six use case based capabilities identified for the SYSTEM, write requirements statements for each capability.
   (b) For each capability requirement, derive the next level of sibling requirements that constitute what must be completed to achieve the higher level capability.

(c) Create a Requirements Verification Matrix (RVM), and identify the verification methods required to prove compliance with each requirement.

(d) Pick two requirements. Using Table 33.1 as a reference, create your own table for each requirement and append a separate column that rationalizes how the respective requirement satisfies the criteria.

ORGANIZATIONAL CENTRIC EXERCISES

1. Select a contract specification or one developed by your organization and test it using the criteria described in this section.
2. Research your organization's command media for guidance or direction in preparing requirements statements. Document your findings.
3. Contact a contract program within your organization.
   (a) What types of specification review methods were used to review specification requirements?
   (b) What guidance did the program provide to specification developers regarding requirements development guidelines or criteria?

REFERENCES

Defense Systems Management College (DSMC). 1998. *Test and Evaluation Management Guide*, 3rd ed. Defense Acquisition Press, Ft. Belvoir, VA.

IEEE Std 610.12-1990, 1990. *IEEE Standard Glossary of Modeling and Simulation Terminology*. Institute of Electrical and Electronic Engineers (IEEE). New York, NY

ADDITIONAL READING

KAR, PHILLIP, and BAILEY, MICHELLE. 1996. *Characteristics of Good Requirements*. International Council on Systems Engineering (INCOSE) Requirements Working Group. 1996 INCOSE Symposium.

Chapter 34

Operational Utility, Suitability, and Effectiveness

34.1 INTRODUCTION

When you develop systems, there are four basic questions the User, Acquirer, and System Developer need to answer:

1. *If we invest in the development of this system, product or service, will it have UTILITY to the User in accomplishing their organizational missions?*
2. *If the system has UTILITY, will it be SUITABLE for the User's mission application(s) and integrate easily into their business model?*
3. *If the system has UTILITY and is SUITABLE for the application, will it be operationally AVAILABLE to perform the mission when tasked?*
4. *If the system has UTILITY to the organization, is SUITABLE for the application, and is AVAILABLE to perform its mission when tasked, will it be EFFECTIVE in performing mission and accomplishing mission objectives with a required level of success?*

The ultimate test of any system, product, or service is its *mission* and *system effectiveness* in performing User missions and accomplishing mission objectives. Failure to perform within prescribed OPERATING ENVIRONMENT *conditions* and *constraints* places operational, financial, and survival risks on the Users, their organizations, and the public. One of System Engineering's greatest challenges for SEs is being able to translate *operational* and *system effectiveness* objectives into meaningful capability and performance requirements that developers understand and can implement. The *challenge* is exacerbated by a lack of formal training on this topic to help SEs understand HOW to perform the practice.

This section introduces one of the most challenging areas to SE. In fact, it serves as the core of requirements identification, analysis, allocation and flow down. As a general practice, this area is filled with buzzwords that verbose SEs pontificate with limited knowledge, Why? Typically, the recipients of the verbose pontification know less about the topic and do not bother to seek *clarification* and *understanding*.

Author's Note 34.1 *Operational AVAILABILITY is part of this discussion thread. However, since AVAILABILITY is a mathematical function of reliability and maintainability, we will defer the AVAILABILITY discussion. The topic is a key element of the Reliability, Availability, and Maintainability (RAM) Practices discussion in Chapter 50 of the Decision Support Series.*

System Analysis, Design, and Development, by Charles S. Wasson
Copyright © 2006 by John Wiley & Sons, Inc.

What You Should Learn from This Chapter

1. What is a *measure of effectiveness (MOE)*?
2. What is a *measure of suitability (MOS)*?
3. What is a *technical performance measure (TPM)*?
4. What is a *technical performance parameter (TPP)*?
5. What is the *relationship* between MOEs, MOSs, MOPs, and TPMs?
6. Who is *responsible* for MOEs, MOSs, MOPs, and TPMs?
7. How do you *track and control* MOEs, MOSs, MOPs, and TPMs?
8. What is the *relationship* between TPMs and risk items?
9. When do TPMs *trigger* risk items?

Definitions of Key Terms

- **Critical Issues** "Those aspects of a system's capability, either operational, technical, or other, that must be questioned before a system's overall suitability can be known. Critical issues are of primary importance to the decision authority in reaching a decision to allow the system to advance into the next phase of development." (Source: DSMC, *Glossary: Defense Acquisition Acronyms and Terms*)
- **Critical Operational Issue (COI)** "Operational effectiveness and operational suitability issues (not parameters, objectives, or thresholds) that must be examined in operational test and evaluation (OT&E) to determine the system's capability to perform its mission. A COI is normally phrased as a question that must be answered in order to properly evaluate operational effectiveness (e.g., 'Will the system detect the threat in a combat environment at adequate range to allow successful engagement?') or operational suitability (e.g., 'Will the system be safe to operate in a combat environment?')." (Source: DSMC, *Glossary: Defense Acquisition Acronyms and Terms*)
- **Figure of Merit (FOM)** "The numerical value assigned to a measure of effectiveness, parameter, or other figure, as a result of an analysis, synthesis, or estimating technique." (Source: DSMC, *Glossary: Defense Acquisition Acronyms and Terms*)
- **Key Performance Parameters (KPPs)** "Those capabilities or characteristics so significant that failure to meet the threshold can be cause for the concept or system selected to be reevaluated or the program to be reassessed or terminated." (Source: DSMC, *Test and Evaluation Management Guide*, Appendix B, Glossary of Test Terminology, p. B-10)
- **Technical (Performance) Parameters (TPPs)** "A selected subset of the system's technical metrics tracked in Technical Performance Measurement (TPM). Critical technical parameters relate to critical system characteristics and are identified from risk analyses and contract specifications. Technical parameters examples (include):
  1. Specification requirements.
  2. Metrics associated with technical objectives and other key decision metrics used to guide and control progressive development.
  3. Design-to-cost targets.
  4. Parameters identified in the acquisition program baseline or user requirements documentation."
  (Source: Former MIL-STD-499B DRAFT, Appendix A, Glossary, p. 41)

- **Technical Performance Measurement (TPM)** "The continuing verification of the degree of anticipated and actual achievement of technical parameters. TPM is used to identify and flag the importance of a design deficiency that might jeopardize meeting a system level requirement that has been determined to be critical. Measured values that fall outside an established tolerance band require proper corrective actions to be taken by management.
    a. **Achievement to Date** Measured progress or estimate of progress plotted and compared with the planned progress at designated milestone dates.
    b. **Current Estimate** The value of a technical parameter that is predicted to be achieved with existing resources by the end of contract (EOC).
    c. **Technical Milestone** Time period when a TPM evaluation is accomplished. Evaluations are made to support technical reviews, significant test events and cost reporting intervals.
    d. **Planned Value** Predicted value of the technical parameter for the time of measurement based on e planned profile.
    e. **Planned Value Profile** Profile representing the projected time-phased demonstration of a technical parameter requirement.
    f. **Tolerance Band** Management alert limits placed each side of the planned profile to indicate the envelope or degree of variation allowed. The tolerance band represents the projected level of estimating error.
    g. **Threshold** The limiting acceptable value of a technical parameter; usually a contractual performance requirement.
    h. **Variation** Difference between the planned value of the technical parameter and the achievement-to-date value derived from analysis, test, or demonstration."
    (Source: INCOSE *Handbook*, Section 12, Appendix, Glossary, p. 46–47)

Scope of Discussion

Our discussion in this chapter employs hierarchical entity relationships to address the *objective* aspects—*operational and system effectiveness*—of system development. Though somewhat *subjective*, similar approaches and methods can be applied to *operational utility and suitability*.

Based on this introduction, let's begin our discussion with an overview of *MOEs, MOSs, MOPs, and TPM relationships*.

34.2 OVERVIEW OF MOE, MOS, MOP, AND TPM RELATIONSHIPS

In most engineering environments, SEs are tasked to identify *Technical Performance Measures (TPMs)* for their SYSTEM, PRODUCT, and SUBSYSTEM. As the SEs struggle, program management laments as to WHY such as simple task can be so difficult. There are two reasons.

First, most engineers are not trained to understand WHAT a TPM is, HOW it originates, and WHY TPMs are important—an organizational management and training system failure. *Second*, the Lead Systems Engineer (LSE) and the System Engineering and Integration Team (SEIT) that oversee the technical program have not performed their job of linking lower level TPMs to critical MOEs and MOSs.

So, to most engineers, TPMs are bureaucratic exercises of *randomly* selecting requirements that may or may not have relevance to overall system performance from their assigned development specifications.

The terms *measures of effectiveness (MOEs)*, *measures of suitability (MOSs)*, *measures of performance (MOPs)*, and *technical performance measures (TPMs)* should be integrally linked. Unfortunately, they typically exist as unrelated *information fragments* in separate documents and may only be linked mentally by a few reviewers. Given this situation, the potential for risk abounds in the form of *overlooked* or *missing* requirements dropped along the path of getting to the *System Performance Specification (SPS)*. Let's explore this point further.

The relationships between MOEs, MOSs, MOPs, and TPMs can be very confusing. Part of the difficulty arises from a lack of CLEAR understanding of *how* MOEs, MOSs, MOPs, and TPMs originate. MOEs, MOSs, MOPS, and TPMs serve as critical benchmark metrics for deriving operational capabilities and requirements. The parameters provide *traceability* links with the intent of delivering a physical system that fully *complies* with the Acquirer's SPS requirements and satisfies the User's validated operational needs.

Linking MOEs, MOSs, MOPs, and TPMs

Figure 34.1 depicts the entity relationships between MOEs, MOPs, MOSs, and TPMs. Observe the partitioning of the figure into two basic parts as denoted by the vertical dashed line: the User/Acquirer domain on the LEFT side and the System Developer domain shown on the RIGHT side.

Author's Note 34.2 *Some Acquirers may derive the SOO from the ORD in lieu of the SRD. In this context the SOO serves as the source or originating requirements for the SPS.*

When a new system is conceived, the User formulates a high-level *Statement of Objectives (SOO)* (1). Users that lack in-depth knowledge of SE and acquisition expertise may employ an Acquirer to represent their system development technical and contract interests. The Acquirer collaborates with the Users to understand their *problem space* and bound a *solution space*. As part of this process, the User requirements are elaborated in an *Operational Requirements Document (ORD)* (2). The ORD, which typically focuses on application of the proposed system, describes HOW the User envisions employing the system to perform their organizational mission(s) and achieve the mission objectives.

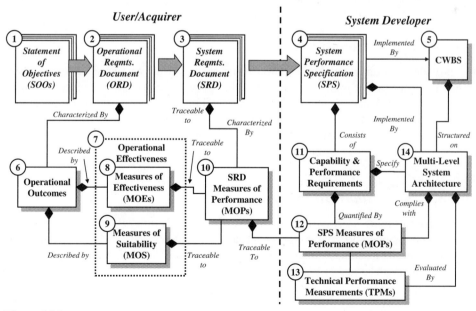

Figure 34.1 MOE, MOS, MOP, & TPM Relationships

The ORD (2) is oriented around a set of Operational Outcomes (6) required to achieve the organizational mission success. Each operational outcome may consist of least one more *measures of effectiveness (MOEs)* and/or at least one or more *measures of suitability (MOSs)*. Collectively, the MOEs and MOSs constitute what is referred to as *Operational Effectiveness* (7). Remember, the focus up to this point has been on HOW *the User intends to employ the system, product, or service* with MOE and MOSs as indicators of *operational effectiveness* and *success*. The challenge for the Acquirer SEs is to translate the MOEs and MOSs into specification requirements legally sufficient for procurement and development.

The translation of ORD MOEs (8) and MOSs (9) into performance requirements is documented in a *System Requirements Document (SRD)* (3). More specifically, each ORD MOE (8) and MOS (9) is translated into a set of requirements, each of which is bounded by SRD *measures of performance (MOPs)* (10).

During the System Procurement Phase of the system/product life cycle, System Developer Offerors analyze the SRD requirements. Typically, the Offerors are required to propose a *System Performance Specification (SPS)* (4) that will serve as the technical basis for the system development contract. Each Offeror reviews the SRD and its MOPs and formulates a proposed solution consisting of the *Contract Work Breakdown Structure (CWBS)* (5), SPS Capability and Performance Requirements (11), and a Multi-level System Architecture (14). SPS MOPs (12) bound each SPS capability and are traceable to the SRD MOPs (10). The Offeror may be asked to select a list of critical *technical performance measures (TPMs)* (13) to track from the list of SPS MOPs (12).

When the contract begins, the SPS requirements are *analyzed*, *allocated*, and *flowed down* to multi-level *development specifications* that specify at least one or more the system architectural elements (14) and their associated *items* or configuration items (CIs). Each *development specification* consists of at least one more MOPs that are traceable back to the SPS MOPs.

During the System Development Phase of the system/project life cycle, developers plot analytically-based TPM predictions. The TPMs provide a basis for determine HOW WELL the system architecture element designs will be able to achieve the *development specification* and SPS requirements. TPM status is typically reviewed at each of the major technical reviews—such as SRR and SDR. When *actual* TPM performance data become available, either via prototypes or demonstrations during integration and testing, that data will be plotted.

MOE and MOP Relationships

Figure 34.2 provides an example of MOE to MOP relationships. Here an airline's Mission Phase is partitioned into subphases of flight. Each subphase has primary and supporting objectives to be accomplished based on *use cases*. Each *use case* is translated into an operational capability requirement and referred to as a MOE. Each MOE is documented in the SPS and decomposed into MOPs that are allocated to PRODUCTs, SUBSYSTEMs, and so forth, and then allocated and flowed down their respective *item development specifications* (IDSs).

Specifying MOEs in the System Performance Specification (SPS)

Given the example in Figure 34.2, *how do we translate the MOPs into SPS requirements*? Each phase and subphase of operation has a specific measure of effectiveness (MOE) that must be achieved. Thus we specify a summary requirement for each MOE in the SPS paragraph 3.1, System Phases of Operation, as illustrated in Figure 34.3.

34.2 Overview of MOE, MOS, MOP, and TPM Relationships

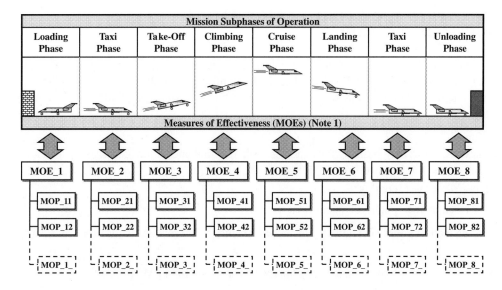

Note 1: To keep the graphic simple, only one MOE is illustrated for each subphase. Subphases may have several objectives and supporting use cases, each of which is assessed by an MOE.

Figure 34.2 Aircraft Example

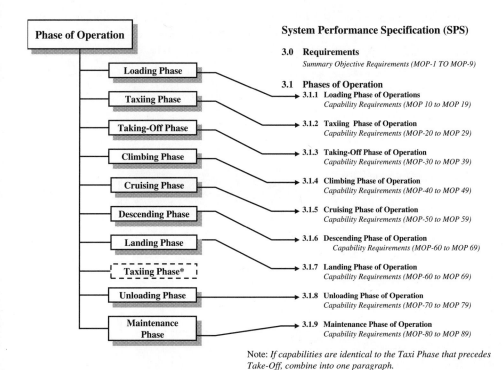

Figure 34.3 Contrasting Views of TPMs Mapping System Effectiveness to Operational Phase MOEs

34.3 PLOTTING TECHNICAL PERFORMANCE MEASUREMENTS (TPMS)

Once the *measures of performance (MOPs)* for each *item* and configuration item (CI) are established, the next step is to track MOP values. This is accomplished a number of ways such as numerical value reporting. The best practice is to plot the values over time graphically as shown in Figure 34.4.

TPMs should be:

1. Tracked on a weekly basis by those accountable for implementation—such as by development teams or Integrated Product Teams (IPTs).
2. Reported at least monthly.
3. Reviewed at each major technical review.

Regarding the last point, note how the development team or IPT at each major technical review reported:

1. Here's the level of performance we *projected* by analysis for TPM XYZ.
2. Here's the level of performance we have *achieved* to date.
3. Here's the level of performance we *expect* to achieve by the *next* review.
4. Here's the corrective action plan for how we expect to *align* today's level of performance with projected performance and acceptable control limits.

Note also how the TPM values consist of *analytical* projections through CDR with a level of risk associated with achievement. Between CDR and TRR the physical components become available for system integration and test. As such, *actual* values are *measured* and become the basis for final TPM tracking.

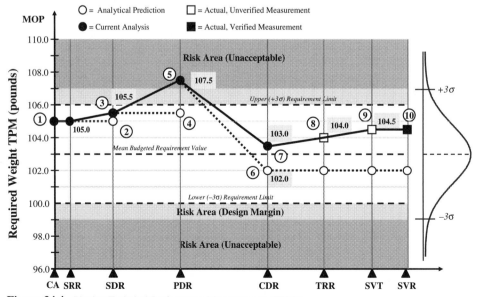

Figure 34.4 Plotting Technical Performance Measurements (TPMs)

Recognize that system component performance varies due to mass properties, manufacturing tolerances, and so forth. This is why the nominal TPM value represents the *mean* of a Normal Distribution. The challenge for SEs is, for a given TPM value: *What are the allowable upper and lower control limits for a given entity—PRODUCT or SUBSYSTEM—that do not diminish overall system performance?* Based on the results of this decision, $+3\sigma$ and -3σ or other applicable thresholds are established for triggering multi-level risk items and mitigation plans. Within each $+3\sigma$ and -3σ threshold, design safety margins may be established. To facilitate viewing, the design margins zones should be YELLOW; anything outside the $+3\sigma$ and -3σ thresholds should be RED.

Author's Note 34.3 *Some requirements may only have single "... shall not exceed..." or "... shall not be less than..." thresholds. As such, the TPM plot should so reflect these UPPER or LOWER control limits.*

TPMs along the development path pose potential risks, especially if the development team or IPT is unable to achieve the analytical projections. When this occurs, each TPM should be required to have risk thresholds that trigger risk items for tracking. If the risk becomes significant, risk item mitigation plans may be required to provide a risk profile for reducing the risk over time and bring it in line with specification requirements. The Program's *risk management plan* should definitize this process and threshold criteria for triggering risk item for tracking and mitigation plans.

Selecting TPMs

TPMs can easily become a very time-consuming activity, especially for reporting purposes. Obviously, every MOP in a specification CANNOT be tracked. So you need to select those of critical importance. *How is this accomplished?*

The Lead SE and SEIT should determine those critical SYSTEM level MOEs and MOPs that have a major impact on achieving mission and system objectives. Then, collaborate with the respective teams to select four to six TPMs from each development specification that have traceability to SPS level MOEs.

34.4 TPM CHALLENGES

TPM tracking involves several challenges. Let's explore a few.

Challenge 1: Bureaucratic Metrics Tracking

TPMs have two *levels of criticality*. *First*, TPMs serve as visual indicators to alert Integrated Products Teams (IPTs) or developers to potential technical trouble. *IPT Leads need to clearly understand this*. Otherwise, the effort is perceived as nothing but bureaucratic metrics tracking performed to impress the Acquirer. *Second*, as a Lead SE, Project Engineer, or Technical Director, you need *early indicators* that to provide a level of confidence that the system is going to perform as specified and designed. If not, you need to know sufficiently in advance to take corrective action.

Challenge 2: Select TPMs Wisely

IPTs often randomly select a few TPMs to satisfy the metrics-tracking task. Selecting the "easiest to achieve" TPMs causes you to believe your own success rhetoric. Select the most difficult, potential SHOWSTOPPER TPMs to make sure the most critical risk areas stay fully addressed through proactive tracking.

Challenge 3: Withholding Actual TPM Data

If a TPM becomes a risk item, IPT Leads often continue to plot *analytical* predictions to *avoid* the realities of *actual data* that pose political and technical risks. *Don't play games*! Your obligation as an SE is to report *factual, existing* data. If you have political problems, deal with the matter in other ways. Remember, if you are not meeting performance levels now and you choose to *ignore* potentially *major* problems, wait until you attempt to *explain* those problems to technical, program, and executive management when the *cost to correct* is very expensive.

Conversely, program management should recognize objective reporting as constructive behavior. Avoid "punishing the messenger." Focus on constructive techical solutions that lead to success. Remember-designer/developer success leads to program management success.

Challenge 4: TPM "Shelf Life"

TPMs, especially lower level ones, have a *shelf life*. During design and early in the system integration, test, and evaluation phase (SITE), low-level TPMs are critical indicators of performance that may affect overall system performance and effectiveness. Once a TPM requirement has been *verified*, the necessity to TRACK the TPM may be pointless, *unless* something fails within the system. Remember, lower level MOPs were derived from higher level MOPs. Once a lower level MOP has been verified and its configuration item (CI) or *items* have been integrated into the next higher level, which has its own TPM, you should not continue to track the MOP. There may be *exceptions*; use informed judgment.

Challenge 5: TPM Reporting

Your contract should specify WHAT TPMs the Acquirer requires you to report. There may be other TPMs that you need to track internally. Depending on the relationship and maturity of the Acquirer organization, some System Developers open all TPMs for review. *Exercise caution* when doing this. Some Acquirer organizations are more *mature* than others in treating the OPENNESS with admiration.

A Word of Caution Make sure your *openness* about TPMs doesn't become a basis for those with *political agendas* to make major *issues* out of minor TPM excursions. Recognize political potholes!

34.5 Guiding Principles

In summary, the preceding discussions provide the basis with which to establish the guiding principles that govern operational utility, suitability, and effectiveness practices.

Principle 34.1 Quantify each measure of effectiveness or suitability (MOE/MOS) with at least one or more measures of performance (MOPs).

Pricipal 34.2 Select TPMs for a specific item that are key contributors and performance affecters for the next higher level system—SYSTEM, PRODUCT, SUBSYSTEM, and so forth.

34.6 SUMMARY

As an SE you need to understand how system effectiveness is determined and decomposed into *measures of effectiveness (MOEs)* and *measures of suitability (MOSs)*, each with *measures of performance (MOPs)* that are documented in various requirements documents. We addressed how TPMs are used to track *planned* versus *actual* MOP TPM values and the importance of thresholds for triggering risk items and associated *risk mitigation plan (RMP)*.

In closing, people will often state that TPMs are too laborious to perform and track. If this is the case, then how can you and your organization deliver a system, product, or service that meets the User's operational needs without performing the practice?

GENERAL EXERCISES

1. Answer each of the *What You Should Learn from This Chapter* questions identified in the *Introduction*.
2. Refer to the list of systems identified in Chapter 2. Based on a selection from the preceding chapter's General *Exercises* or a new system selection, apply your knowledge derived from this chapter's topical discussions. Specifically identify the following:
   (a) Based on the system's use cases, identify the system's MOEs.
   (b) Based on the system's use cases, identify the system's MOSs.
   (c) For each MOE and MOS, identify derived requirements.
   (d) For each requirement, identify MOPs.
   (e) For TPM tracking purposes, which MOPs would you track and provide supporting rationale?

ORGANIZATIONAL CENTRIC EXERCISES

1. Research your organization's command media.
   (a) What minimum requirements does the organization impose for TPM tracking?
   (b) What tools and methods are recommended for tracking TPMs?
   (c) What TPM training does the organization provide and who is accountable?
2. Identify a small, a medium, and a large contract program within your organization. Interview program personnel concerning TPMs.
   (a) Does the *Contract Statement of Work (CSOW)* require TPM tracking?
   (b) What requirements does the CSOW impose on TPM reporting?
   (c) How did the program select TPMs?
   (d) How were TPMs linked to risk tracking?
   (e) What lessons did the program learn from TPMs?

REFERENCES

DoD 5000.59-M. 1998. *DoD Modeling and Simulation (M&S) Glossary*. Washington, DC: Department of Defense (DoD).

Defense Systems Management College (DSMC). 1998. *DSMC Test and Evaluation. Management Guide*, 3rd ed. Defense Acquisition Press. Fr. Belvoir, VA.

Defense Systems Management College. 2001. *Glossary: Defense Acquisition Acronyms and Terms*, 10th ed. Defense Acquisition University Press. Ft. Belvoir, VA.

International Council on System Engineering (INCOSE). 2000. INCOSE *System Engineering Handbook*, Version 2.0. Seattle, WA.

MIL-STD-499B (canceled DRAFT). *Systems Engineering*. Washington, DC: Department of Defense (DoD).

ADDITIONAL READING

Defense Systems Management College (DSMC). 2001. *Systems Engineering Fundamentals*. Defense Acquisition University Press Fr. Belvoir, VA.

IEEE Std. 1220-1998. 1998. *IEEE Standard for Application and Management of the Systems Engineering Process*. Institute of Electrical and Electronic Engineers (IEEE). New York, NY.

Chapter 35

System Design Objectives

35.1 INTRODUCTION

When SEs engineer systems, the general mindset is to *propose* and *develop* solutions that solve User *solution spaces* within *problem spaces*. The problem with this mindset is that it lacks a central focal point that captures WHAT the User needs or seeks. For example: *Is the User concerned about Growth? Reliability? Manueverability?* And so forth? If you do not understand: 1) WHAT the User needs and 2) WHAT priorities they place on those needs, you are just going through a "check-the-box" exercise. So, *how do SEs AVOID this mindset*?

This section introduces design *objectives* that are often key drivers in system development. These objectives form the basis for proposal and system development activities responding to Acquirer formal solicitations and contracts. The broad, far-reaching ramifications of technical decisions made in support of these objectives clearly focuses on the need to integrate subject matter experts (SMEs), as key members of Integrated Product Teams (IPTs), to support system development.

Our discussions in this section identify and summarize each of these objectives. From a User's perspective, the system design solution exemplifies one or more of these objectives. As such, these objectives should be highlighted as key themes in a formal request for proposal solicitations. In response, Offerors should *highlight* their key features and capabilities of their proposed system solution coupled with their experience in developing comparable solutions for other Users.

What You Should Learn from This Chapter

What are the design ramifications of the following objectives?

1. Design to value (DTV)
2. Design to cost (DTC)
3. Design for usability
4. Design for single-use/multi-use applications
5. Design for comfort
6. Design for interoperability
7. Design for transportability
8. Design for mobility
9. Design for maneuverability
10. Design for portability

System Analysis, Design, and Development, by Charles S. Wasson
Copyright © 2006 by John Wiley & Sons, Inc.

11. Design for growth and expansion
12. Design for reliability
13. Design for availability
14. Design for producibility
15. Design for mission support
16. Design for deployment
17. Design for training
18. Design for vulnerability
19. Design for lethality
20. Design for survivability
21. Design for efficiency
22. Design for effectiveness
23. Design for reconfigurability
24. Design for integration, test, and evaluation
25. Design for verification
26. Design for maintainability
27. Design for disposal
28. Design for security and protection
29. Design for safety

Definitions of Key Terms

- **Availability** A measure of the degree to which an item is in an operable and committable state at the start of a mission when the mission is called for at an unknown (random) time. (The item's state at start of a mission includes the combined effects of the readiness-related system R&M (Reliability & Maintainability) parameters, but excludes mission time.) (Source: MIL-HDBK-470A, Appendix G, *Glossary*, p. G-2)
- **Efficiency** "The degree to which a system or component performs its designated functions with minimum consumption of resources." (Source: IEEE 610.12-1990)
- **Maintainability** "The ability of an item to be retained in, or restored to, a specified condition when maintenance is performed by personnel having specified skill levels, using prescribed procedures and resources, at each prescribed level of maintenance and repair." (Source: DoD *Glossary: Defense Acquisition Acronyms and Terms*)
- **Portability** (1) Hardware: The relative ease of moving a piece of EQUIPMENT under specific conditions from one location to another. (2) Software: The relative ease of moving

a computer software configuration item (CSCI) from one type of computer platform to another.

- **Producibility** "The relative ease of manufacturing an item or system. This relative ease is governed by the characteristics and features of a design that enables economical fabrication, assembly, inspection, and testing using available manufacturing techniques." (Source: DoD *Glossary: Defense Acquisition Acronyms and Terms*)
- **Reconfigurability** The ability of a system, product, or service configuration to be modified manually or automatically to support mission objectives.
- **Reliability** "The ability of a system and its parts to perform its mission of a specified duration under specific operating conditions without failure, degradation, or demand on the support system." (Source: Adapted from DoD *Glossary: Defense Acquisition Acronyms and Terms*)
- **Serviceability** "A measure of the degree to which servicing of an item will be accomplished within a given time under specified conditions." (Source: DoD Glossary: Defense Acquisition Acronyms and Terms)
- **Simplicity** "The degree to which a system or component has a design and implementation that is straightforward and easy to understand." (Source: IEEE 610.12-1990)
- **Supportability** "The degree of ease to which system design characteristics and planned logistic resources, including the logistic support (LS) elements, allow for the meeting of system availability and wartime utilization requirements." (Source: DoD *Glossary: Defense Acquisition Acronyms and Terms*)
- **Survivability** "The capability of a system and its crew, if applicable, to avoid or withstand a hostile Man-Made, Natural, and Induced OPERATING ENVIRONMENT without suffering an abortive impairment of its ability to accomplish its designated mission." (Adapted from DoD *Glossary: Defense Acquisition Acronyms and Terms*)
- **Susceptibility** "The degree to which a device, equipment, or weapon system is open to effective attack due to one or more inherent weaknesses. Susceptibility is a function of operational tactics, countermeasures, probability of enemy fielding a threat, etc. Susceptibility is considered a subset of survivability." (Source: DoD *Glossary: Defense Acquisition Acronyms and Terms*)
- **Sustainability** "Sustainment includes, but is not limited to, plans and activities related to supply, maintenance, transportation, sustaining engineering, data management, configuration management, manpower, training, safety, and health. This work effort overlaps the Full Rate Production and Deployment work effort of the Production and Deployment phase. See also Logistics Support and Logistics Support Elements." (Source: DoD *Glossary: Defense Acquisition Acronyms and Terms*)
- **System Safety** "The application of engineering and management principles, criteria, and techniques to optimize safety within the constraints of operational effectiveness, time, and cost throughout all phases of the system life cycle." (Source: MIL-STD-882D *System Safety*, para. 3.13, p. 2)
- **System Security** The level of protection that characterizes a system, product, or service's ability to reject intrusion and access by external threats or unauthorized systems.
- **Testability** "The degree to which a requirement is stated in terms that permits establishment of test criteria and performance of tests to determine whether those criteria have been met." (Source: IEEE 610.12-1990)

- **Transportability** "The capability of materiel to be moved by towing, self-propulsion, or carrier through any means, such as railways, highways, waterways, pipelines, oceans, and airways. (Full consideration of available and projected transportation assets, mobility plans and schedules, and the impact of system equipment and support items on the strategic mobility of operating military forces is required to achieve this capability.)" (Source: *DSMC Definition of Terms*)
- **Usability** "The ease with which a user can learn to operate, prepare inputs for, and interpret outputs of a system or component." (Source: IEEE Std. 610.12-1990)
- **Vulnerability** "The characteristics of a system that cause it to suffer a definite degradation (loss or reduction of capability to perform the designated mission) as a result of having been subjected to a certain (defined) level of effects in an unnatural [human-made] hostile environment. Vulnerability is considered a subset of survivability." (Source: DoD *Glossary: Defense Acquisition Acronyms and Terms*)

35.2 COMMONLY APPLIED SYSTEM DESIGN OBJECTIVES

The Introduction identified a list of objectives that influence system design decisions. The discussions that follow scope the context of each objective.

Author's Note 35.1 *Two key points: First, specifications presumably document all of the Acquirer's key detail requirements. You can reach a point where you "can't see the forest for the trees," as such immersion in the details tend to obscure WHAT is important to the User. So, once you read a specification, talk with the User via Acquirer contracting protocol to determine the key objectives that matter most and affect System Developer decision making. System design objectives provide those proverbial "forest" level insights. Second, COLLABORATE with the User to prioritize the objectives.*

Design-for-Value (DTV) Objective

Users and Acquirers with limited budgets are often challenged to obtain the most system capabilities and performance for their budget. The *Design-for-Value (DTV)* objective is a *critical issue* for these customers. *How do you define DTV?*

1. Determine WHAT the User needs.
2. Prioritize those needs in terms of relative importance.
3. Map those needs to the *System Performance Specification (SPS)* requirements.

You can identify a single solution or a variety of solutions and offer them as the basis of Cost as an Independent Variable (CAIV). Methods such as Quality Function Deployment (QFD) can be also used to support value-oriented decisions.

Design-to-Cost (DTC) Objective

Users expect some systems to have budgetary *"per unit cost"* limit especially mass-produced systems and products. Where this is the case, the *Design-to-Cost (DTC)* objective becomes a critical driver in the *engineering of systems*. Once the DTC is established at the SYSTEM Level, design cost constraints are *allocated* and *flowed* down through all system levels of abstraction to the PART level. This is a *highly iterative* top-down/bottom-up/left-right/right-left process with refinements of solutions and cost allocations until all entity technical design requirements, cost constraints, and risks are "in balance."

Design-for-Usability Objective

Durable systems and products require a high *usability*. *Usability*, as viewed by the User, is often vague and ambiguous. Classic descriptions include *user friendly interfaces* . . . , *easy to use* . . . , *easy to understand* . . . , *easy to control* . . . , *easy to drive over rough terrain* . . . , *easy to lift* . . . , *easy to carry* . . . , and so forth. From an SE perspective, it is critical that the *Design-for-Usability* objective be *accurately* and *precisely* bounded in specification requirements.

Rapid prototypes may need to be developed to present to Users for *usability* evaluation and constructive feedback. User decisions must then be captured as graphical displays or text requirements. Application examples include:

1. Special considerations for the physically challenged.
2. Special equipment for *ingress* and *egress* from a vehicle.

Design-for-Single-use/Multi-use Applications Objective

Users acquire most systems and products for *multi-use applications*.

Multi-use systems and products must be designed for durability and maintenance over a specified number of usage cycles. The *Design-for-Multi-use Applications* objective requires particular attention in areas such as modularity and interchangeability. In contrast, *single-use applications* may require a focus on design and construction.

Design-for-Comfort Objective

Vehicular systems developed for Users such as operators and passengers for travel purposes over long periods of time require a level of comfort to minimize fatigue, boredom, and so on. As in the *Design-for-Usability* objective, requirements must be explicitly identified, scoped, and bounded. The *Design-for-Comfort* objective is a key driver in the development of systems such as homes, vehicles, spacecraft, and offices.

Design-for-Interoperability Objective

Systems and products that must be *interoperable* with other systems and products must have a *Design-for-Interoperability* objective. If the Design-for-Interoperability objective is selected, strict interface *design standards* and *protocols* must be established and controlled.

Design-for-Transportability Objective

Vehicle systems, military troops, et al require design considerations based on a *Design-for-Transportability* objective. This requires understanding WHO/WHAT is to be transported, WHAT space and carrying capacity is required, HOW the cargo is to be secured, WHO requires WHAT access to the cargo and WHEN, WHAT protection mechanisms are required from environmental and human-made system threats.

Design-for-Mobility Objective

Vehicle systems and military troops are Users that require design considerations based on a *Design-for-Mobility* objective. The design considerations must include HOW the system will be secured for mobility, physically moved, HOW often, and HOW the system is to be secured when it reaches its destination.

Design-for-Maneuverability Objective

Systems often require the capability to navigate through their OPERATING ENVIRONMENT. This requires steering mechanisms that enable the operators to *physically* or *remotely* move the vehicle from one location, change orientation relative to a frame of reference, or make directional vector headed changes.

Design-for-Portability Objective

Some systems and products must be developed with a *Design-for-Portability* objective. *Portability* concerns two key *contexts*: 1) physical properties and 2) software. *Portability* in one context refers to physical properties of a system or product—such as the acceptable size and weight that a human can lift, move, or transport. *Portability* in a software context refers to the ability to integrate and execute computer software configuration items (CSCIs) on various types of computer systems with minimal adaption.

Design-for-Growth and Expansion Objective

Systems, products, and services often require the capability to be expanded to accommodate future upgrades. This may require additional processing or propulsion power, flexibility to increase storage capacity of *expendables* or *consumables*, increased data communications ports or bandwidth, or organizational or geographical expansion.

Design-for-Reliability Objective

Every system, product, and service has an *intrinsic value* to Users and stakeholders in being able to support organizational missions and objectives. Depending on the stakes and OPERATING ENVIRONMENT, mission and system *reliability* relative to achieving mission success may be a critical issue.

Design-for-Availability Objective

The first criterion for systems, products, and services success is *system readiness* to perform the mission when called upon. Some systems are required to be available *on demand*.

Establish a *Design-for-Availability* objective commensurate with the available budget and mission. This includes implementation of daily operational readiness tests (DORTs), built-in tests (BITs), built-in test equipment (BITE), indicators, and display meters. These *activities* and *design considerations* can provide early *indications* of potential problem areas and thus enable corrective action to be taken in advance of mission needs.

Design-for-Producibility Objective

Systems and products planned for production must be producible in manner that:

1. Has *acceptable risk*.
2. *Repeatable* and *predictable* processes and methods.
3. Can be produced within budgets and schedules at a reasonable profit.

Engineering development *first articles* are often *functional demonstrators* that may include lesser quality materials, improvised components, and "add-on" instrumentation to support performance verification. Production items may not require these additional items or weight that limit performance and increase cost. A *Design-for-Producibility* objective often drives a search for alternative *materials* and *processes* to *reduce* cost and risk and *improve* or *maintain* system performance.

Design-for-Mission Support Objective

Multi-use applications require continuing support throughout their planned lifecycle. A *Design-for-Mission Support* objective ensures that appropriate design considerations are given to replenishment of *expendables* and *consumables* and to *corrective* and *preventive* maintenance.

Thus, *system support concepts* are critical for establishing *Design-for-Mission Support* requirements.

Design-for-Deployment Objective

Many systems, products, and services require a *Design-for-Deployment objective* to support deployment or shipment to duty station mission areas. Design considerations include: tie-down chains, anchors, accelerometer sensor instrumentation packages, controlled environmental atmospheres, and shock and vibration proofing. Additionally, enroute deployment constraints such as bridge heights and maximum weights, road grades, and hazardous waste routes constraints must be factored into the decisions.

Design-for-Training Objective

Most systems and products require some form of *Design-for-Training* objective. Where this is the case, TRAINING is a *critical operational issue (COI)*, not only for the students but also for the instructors, general public, and environment. Additionally, scoring and debriefs of training sessions are important. Therefore, the *Training Concept* becomes a key input into system and product requirements. In some cases systems and products are modified to include instructor controls.

Design-for-Vulnerability Objective

Systems that operate in *hostile threat environments* or can be *misused* or *abused* by the operator(s) should have a *Design-for-Vulnerability* objective. This applies to buildings, safes, vehicles, computers, and electrical circuits.

When a system, product, or service is anticipated to be *vulnerable* to OPERATING ENVIRONMENT threats, mission analysis and use case analysis should identify the threats, threat scenarios, and prioritize design capabilities to *resist* or *minimize* the effects of those threats. This area exemplifies our discussion of *generalized* and *specialized solutions* covered in Part I, *System Interfaces*. The *generalized solution* acknowledges the interaction; the *specialized solution* incorporates key capabilities or features to protect the system and its operators from external threats.

Design-for-Lethality Objective

Some systems such as munitions and missiles are intended to penetrate vulnerable areas and DESTROY mechanisms that enable the targeted system to survive. The *Design-for-Lethality* objective focuses on the system design and material characteristics that enable a system to achieve this objective.

Design-for-Survivability Objective

Some systems and products are required to operate in *harsh*, *hostile* environments. They must be capable of surviving to complete the mission and as applicable, return safely. Examples include: thermal insulation, layers of armor, elimination of single points of fallure (SPF), and so forth. Systems such as these require a *Design-for-Survivability* objective.

Design-for-Efficiency

Some systems and products require focused consideration on efficient utilization of resources. In these cases a *Design-for-Efficiency* objective must be established.

Design-for-Effectiveness

One of the key contributors to User satisfaction is a system, product, or service's degree of *effectiveness*. Did the missile hit and destroy the target? *Does the flight simulator improve pilot effectiveness*? Establish *Design-for-Effectiveness* objectives where the *effectiveness* is the key determinate for mission success.

Design-for-Reconfigurability

Some systems and products are designed to accommodate a variety of missions as well as a quick turnaround between missions. Therefore, some may have to be reconfigurable within a specified time frame.

Design-for-Integration, Test, and Evaluation Objective

Modular system and products that undergo multiple levels of integration and test require a *Design-for-Integration, Test, and Evaluation* objective. Ideally, you would isolate each configuration item (CI) and test it with *actual*, *simulated*, *stimulated*, or *emulated* interfaces. If the system or product is to be integrated at various facilities, special interface considerations should be given to physical constraints and equipment, and the tools available should be investigated.

Finally, EQUIPMENT-based systems and products must be designed to facilitate multi-level system verification and validation. This requires the incorporation of temporary or permanent test points and access ports for calibration and alignments among other things.

Design-for-Verification Objective

Once the system or product is integrated, tested, and ready to be verified, designers must consider *HOW* the item will be verified. Some requirements can be physically verified; other requirements may not. Establish a *Design-for-Verification* objective to ensure all data required for verification are accessible and can be easily measured.

Design-for-Maintainability Objective

Single-use, *multi-use*, and *multi-purpose systems* require some form of *Design-for-Maintainability* objective. This may include *preventive maintenance*, *corrective maintenance*, calibration, upgrades, and refurbishment. Key design considerations include maintenance operator access and clearances for hands, arms, tools, and equipment. Additional considerations include the availability of electrical power, need for batteries, or electrical generators, external air sources for aircraft while on the ground, and so on. These considerations emphasize the need for a *Maintenance Concept* to provide a framework for deriving *Design-for-Maintainability* requirements.

Design-for-Disposal Objective

Systems and products that employ nuclear, biological, or chemical (NBC) materials and ultimately wear out, become exhausted, damaged, or destroyed intentionally or by accident. In any case, the system or product requires a *Design-for-Disposal* objective. This includes mechanisms for monitoring and removal of hazardous materials such as NBC substances or traces. For items that can be *reclaimed* and *recycled*, special tools and equipment—categorized as peculiar support equipment (PSE)—may be required.

Design-for-Security-and-Protection Objective

Various systems and products require a *Design-for-Security-and-Protection* objective that limits SYSTEM or product access to only authorized individuals or organizations. Design considerations include layers of armor, Internet firewalls, and authorized accounts and passwords.

Design-for-Safety Objective

The application of SE requires strict adherence to laws, regulations, and engineering principles and practices that promote the safety of system and product stakeholders—the operators, maintainers, general public, personal property, and environment. The *Design-for-Safety* objective focuses on ensuring that systems, products, and services are safe to deploy, operate, maintain, and dispose of. This includes not only the physical product but also establishing training and instructional procedures, cautionary warnings and notices, and potential consequences for violation.

Quality Function Deployment (QFD)

One method for sorting out customer needs and priority values is Quality Function Deployment (QFD). Where time permits, you are encouraged to consider and investigate QFD as part of your analysis.

35.3 GUIDING PRINCIPLES

In summary, the preceding discussions provide the basis with which to establish the guiding principles that govern system design objectives practices.

Principle 35.1 Every User has key Operations and Support (O&S) Phase objectives that a SYSTEM/entity design must satisfy; collaborate with Users to understand their needs and prioritize them.

35.4 SUMMARY

Our discussions in this chapter highlighted the need to establish *system design objectives practices* to ensure that User operational needs are met. These objectives form the basis for developing *Mission Needs Statements (MNSs)*, *Statements of Objectives (SOOs)*, *Operational Requirements Document (ORD)*, *System Requirements Documents (SRDs)*, and *System Performance Specifications (SPS)*.

GENERAL EXERCISES

1. Answer each of the *What You Should Learn from This Chapter* questions identified in the *Introduction*.
2. Refer to the list of systems identified in Chapter 2. Based on a selection from the preceding chapter's *General Exercises* or a new system selection, apply your knowledge derived from this chapter's topical discussions. A User has employed your services to recommend the driving design "to/for" objectives for the selected system or product.
   (a) What are the top three or five objectives you would recommend?
   (b) How would you prioritize each objective?

(c) Prepare a statement for a formal Request for Proposal (RFP) that expresses each objective and its relative priority.

ORGANIZATIONAL CENTRIC EXERCISES

1. Contact a contract program in your organization.
    (a) What are the User's objectives for the system, product, or service?
    (b) How were the objectives expressed? RFP? Contract?
    (c) What are the objectives?
    (d) How is the program implementing the objectives?
    (e) What lessons has the program learned from this exercise?

REFERENCES

IEEE Std 610.12-1990. 1990. *IEEE Standard Glossary of Software Engineering Terminology*. Institute of Electrical and Electronic Engineers (IEEE). New York, NY.

Defense Systems Management College (DSMC) Ft. Belvoir. VA. 2001. *Glossary: Defense Acquisition Acronyms and Terms*, 10th ed. Defense Acquisition University Press.

MIL-STD-882D. 2000. *Standard Practice for System Safety*. Washington, DC: Department of Defense (DoD).

Chapter 36

System Architecture Development

36.1 INTRODUCTION

All human-made and living systems, by definition, are composed of interacting elements. Each element has its own unique identity, capabilities, and characteristics, integrated into a purposeful framework specifically arranged to accomplish a function or mission. The integrated, multi-level framework of elements and combinations of elements represent the SYSTEM's *architectural configuration* or simply *architecture*.

This chapter explores the development of system architectures. As discussed in *Part I on System Analysis Concepts*, the system architecture is an aggregate abstraction consisting of four classes of architectures: 1) a requirements architecture, 2) an operations architecture, 3) a behavioral architecture, and 4) a physical architecture.

Our discussions in this chapter establish the foundation for developing a system architecture. Since the fundamental concepts of the architectural frameworks were covered in *Part I on System Architecture Concepts*, this chapter focuses attention on physical configuration unique topics. Topics include *centralized*, *decentralized*, and *distributed* processing; fault tolerant architectural design; environmental, safety, and health (ES&H) considerations, fire detection and suppression; and security and protection.

What You Should Learn from This Chapter

1. What is a *system architecture*?
2. What are the key attributes of an *architecture*?
3. What are the primary *architectural views* of a system?
4. Define the *semantics* of architectures.
5. What is *centralized control* processing architecture?
6. What is *decentralized control* processing architecture?
7. What is *distributed* processing architecture?
8. What is a *fault tolerant architectural design*?
9. What are some architectural power system considerations?
10. What are some architectural *environmental, safety, and health (ES&H) considerations*?
11. What are some *fire detection and suppression architectural configuration considerations*?
12. What are some *security and protection architectural considerations*?

System Analysis, Design, and Development, by Charles S. Wasson
Copyright © 2006 by John Wiley & Sons, Inc.

Definitions of Key Terms

- **Architect** (System) The person, team, or organization responsible for *innovating* and *creating* a system configuration that provides the best solution to User expectations and a set of requirements within technical, cost, schedule, technology, and support constraints.
- **Architecting** "The activities of defining, documenting, maintaining, improving, and certifying proper implementation of an architecture." (Source: IEEE Std. 1471-2000, para. 3.3, p. 3)
- **Architectural Description (AD)** "A collection of products to document an architecture." (Source: IEEE Std. 1471-2000, para. 3.4, p. 3)
- **Architecture** A graphical *model* or *representation*, such as an interpretative artistic rendering, a technical drawing, or a sketch, of a specific view of a system that communicates the *form*, *fit*, or *function* of a SYSTEM, its operational elements, and interfaces as envisioned by its developer.
- **Centralized Architecture** An architecture that "uses a central location for the execution of the transformation and control functions of a system." (Source: Buede 2000, p. 231)
- **Concerns** "Those interests which pertain to the system's development, its operation or any other aspects that are critical or otherwise important to one or more stakeholders. Concerns include system considerations such as performance, reliability, security, distribution, and evolvability." (Source: IEEE Std. 1471-2000, para. 4.1, p. 4)
- **Decentralized Architecture** An architecture characterized by "multiple, specific locations at which the same or similar transformational or control functions are performed." (Source: Buede, 2000, p. 231)
- **Open System Architecture** "A logical, physical structure implemented via well defined, widely used, publicly-maintained, non-proprietary specifications for interfaces, services, and supporting formats to accomplish system functionality, thereby enabling the use of properly engineered components across a wide range of systems with minimal changes." (Source: Former MIL-STD-499, Appendix A, *Glossary*, p. 37)
- **Open Systems Environment (OSE)** "A comprehensive set of interfaces, services and supporting formats, plus aspects of interoperability of application, as specified by information technology standards and profiles. An OSE enables information systems to be developed, operated and maintained independent of application specific technical solutions or vendor products." (Source: Adapted from DSMC, *Glossary: Defense Acquisition Acronyms and Terms*)
- **View** "A representation of a whole system from the perspective of a related set of concerns." (Source: IEEE Std. 1471-2000, para. 3.9, p. 3)
- **Viewpoint** "A specification of the conventions for constructing and using a view. A pattern or template from which to develop individual views by establishing the purposes and audience for a view and the techniques for its creation and analysis." (Source: IEEE Std. 1471-2000, para. 3.10, p. 3)

36.2 WHAT IS AN ARCHITECTURE?

IEEE Std. 1471-2000 (para. 3.5, p. 3) defines an *architecture* as "*The fundamental organization of a system embodied in its components, their relationships to each other, and to the environment, and the principles guiding its design and evolution.*"

Through our educational systems, most people associate architecture with beautiful classical buildings with ornamented façades that are tracable back to the Greek and Roman antiquity. What is missing from the educational paradigm is a universal definition that encompasses building architectures, systems, products, and services. Using the IEEE definition as a backdrop, if you analyze the educational paradigm of *architecture*, you soon discover that architecture represents the *totality* of three elements common to systems, products, and services. These elements are *form*, *fit*, and *function*.

An architecture *exposes* key features of a system, product, or service and *expressively* communicates via interpretative artistic renderings or graphics HOW those features interrelate within the overall framework and its OPERATING ENVIRONMENT. Note the term "exposes." However, just because an architectural element or object is visually *exposed* DOES NOT infer frequency of usage.

System, product, or service architectures depict the summation of a system's entities and capabilities levels of abstraction that support all phases of *deployment*, *operations*, and *support*. Some entities may be an integral part of a system's phases of operation 100% of the time; other entities may be only used 1% of the time. Depending on the mission or system application, the system, product, or service architecture can be abstracted to expose only those capabilities unique to the mission.

System Architects

In most professional domains, the *system architect* is expected to possess licensed credentials, preferably by some form of certification accorded by a state of residency. Part of this process is to demonstrate to *decision authorities* that the system architect has the experience, knowledge, and understanding of artistic, mathematical, and scientific principles to translate a User's vision into a system, product, or service within the constraints of performance standards, laws, and regulations established by society.

As in the case of the educational architecture paradigm above, there is a paradigm for system architects. Whereas most people think of architects as being individuals, teams and organizations may be referred to as architects.

Formulating the System Architecture

System architecting requires years of experience in application-dependent knowledge and technology. Due to the diversity of systems, product, and services, there are no specific ways to formulate an architecture. There are guidelines that, in combination with experience, enable us to formulate system, product, or service architectures.

In recent years the Institute of Electrical and Electronic Engineers (IEEE) issued IEEE Standard-1471-2000, *IEEE Recommended Practice for Architectural Description of Software-Intensive Systems*. IEEE-Std-1471-2000 established as a conceptual framework for developing architectural descriptions of *software intensive* systems.

IEEE 1471-2000 (para. 1.1, p. 1) defines software intensive systems as those "... *where software contributes essential influences to the design, construction, deployment, and evolution of the system as a whole*." Although IEEE 1471 is a software standard, the conceptual framework presented is equally applicable to all types of systems—electrical, electronic, mechanical, optical, and so forth. Figure 36.1 illustrates the standard's framework.

IEEE 1471 provides a key construct that *exposes* several key terms that serve as the framework for formulating a system, product, or service architecture. Specifically the terms are: *architectural description*, *viewpoints*, *views*, and *concerns*.

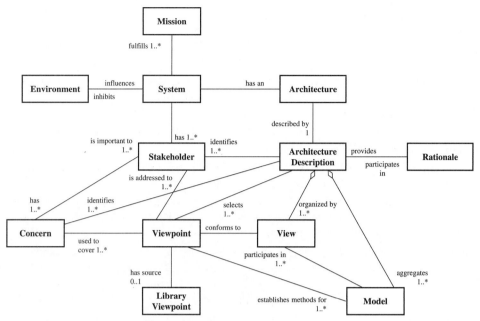

Figure 36.1 IEEE-Std-1471-2000 Architectural Description
Source: IEEE-Std-1471-2000 Figure 1: "Conceptual Model of Architectural Description," p. 5. Reprinted with permission of IEEE.

- **Architectural Description** "An architectural description selects one or more viewpoints for use. The selection of viewpoints typically will be based on consideration of the stakeholders to whom the AD is addressed and their concerns." (Source: IEEE Std. 1471-2000, para. 4.1, p. 4)
- **Viewpoint** "A viewpoint establishes the conventions by which a view is created, depicted and analyzed. In this way, a view conforms to a viewpoint. The viewpoint determines the languages (including notations, model, or product types) to be used to describe the view, and any associated modeling methods or analysis techniques to be applied to these representations of the view. These languages and techniques are used to yield results relevant to the concerns addressed by the viewpoint." (Source: IEEE Std. 1471-2000, para. 4.1, p. 4)
- **View** "A view may consist of one or more architectural models. Each such architectural model is developed using the methods established by its associated architectural viewpoint. An architectural model may participate in more than one view." (Source: IEEE Std. 1471-2000, para. 4.1, p. 4)
- **Concerns** "Those interests which pertain to the system's development, its operation or any other aspects that are critical or otherwise important to one or more stakeholders. Concerns include system considerations such as performance, reliability, security, distribution, and evolvability." (Source: IEEE Std. 1471-2000, para. 4.1, p. 4)

Attributes of an Architectural Description

An Architectural Description *exposes* and *expresses* the architecture of a system, product, or service via standard *attributes* and *conventions*. These include:

- **Architectural Entities** An architecture *exposes* the operational elements such as *objects* or *actors*—which can be persons, places, things, roles, or capabilities—that comprise a SYSTEM or entity, regardless of the level of abstraction, and interacts *synergistically* to perform the entity's mission.

- **Hierarchical Level of Abstraction** An architecture expresses an entity's operational, behavioral, and physical context within the User's *system of systems* (SoS) for a given *level of abstraction* (Level 0, Level 1, Level 2, etc.).

- **Unique Identity** Architecture expresses HOW a SYSTEM's *object* or *actor* capabilities are uniquely identified via reference designators.

- **Interactions with its OPERATING ENVIRONMENT** An architecture *expresses* how the SYSTEM entity *interacts* and *interoperates* with external systems in its OPERATING ENVIRONMENT based on external inputs, stimuli, or cues—for example, inputs such as event-based interrupts, raw materials, and information—as well as its SYSTEM RESPONSES—its behavioral patterns, products, by-products, or services.

- **Completeness** An architecture *expresses* the integrated set of entities, capabilities, and interactions for a specific entity required to satisfy a prescribed set of User mission *use cases* and *operational scenarios*. This is accomplished in terms of *incremental* or *phased* capabilities evolving from: 1) an initial operational capability (IOC) through a series of "builds" to a full operational capability (FOC) or 2) a single grand design.

- **Architectural Views** An entity's architecture is characterized by four types of views: 1) a requirements view, 2) an operations view, 3) a behavioral view, and 4) a physical view.

Now that we have established WHAT a *system architecture* is and its attributes, let's explore the HOW the *architectural description* is represented.

Architectural Description Representation Methods

For most applications, *system architectures* are communicated via three mechanisms: 1) three-dimensional and two-dimensional artistic renderings of buildings, 2) block diagrams, and 3) hierarchy trees. Most SE applications employ block diagrams such as system block diagrams (SBDs), architecture block diagrams (ABDs), and functional block diagrams (FBDs) as the primary mechanism for communicating a SYSTEM or entity's architecture.

Block diagrams depict *horizontal* peer level and external relationships within a given system level of abstraction. Vertical linkages to higher level parent or lower level siblings are referenced by symbology but not shown. For example, a system entity, as a level of abstraction, may include a symbol on or next to its box to denote lower levels exist.

In contrast, hierarchy *trees* enable us to depict vertical relationships that infer *levels of abstraction*; they do not communicate, however, *direct relationships* and *interactions* among peers. Figure 36.2 contrats the two approaches.

Architectural Description Views and Concerns

When tasked to develop the *architectural description*, engineers naturally gravitate to debates over WHAT *tools* and *methods* to use. Traditionally, SEs prefer to use functional methods; software engineers focus on *object-oriented methods*, and so forth. Instead, the focus should be on WHAT is to be described in terms of *views*. Once these views are identified, then you can decide whether a *viewpoint* is best described using *functional*, *object-oriented*, or some other methods. The selection depends on WHAT architectural *view* best *communicates* the System Architect's solution to the User's vision in a manner that is *accepted* by the User.

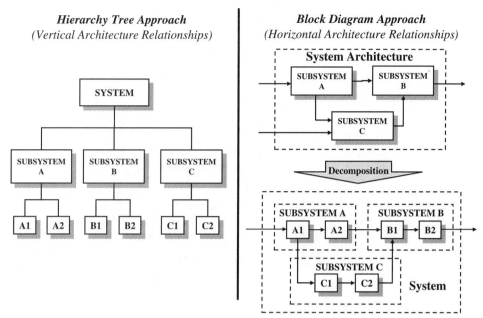

Figure 36.2 System Architecture Presentation Methods

36.3 ARCHITECTURAL VIEWS OF A SYSTEM

There are numerous engineering perspectives of a system's architecture. In general, the perspectives are reflective of the *views* and *concerns* of *key stakeholders* of the system, product, or service as illustrated in Figure 36.3. Your job as a system analyst or SE is to integrate these views.

Earlier we introduced the concept of the four solution domains: Requirements, Operations, Behavioral, and Physical. Each solution domain represents a unique view of a system representing a consensus of its stakeholders. Let's define each view:

- **Requirements Architecture View** A *representation* of the hierarchy and traceability of an entity's specification requirements that bound its phases and modes of operation, capabilities, characteristics, design and construction constraints, and verification methods.

- **Operations Architecture View** A *representation* of HOW the MISSION SYSTEM and SUPPORT SYSTEM *operational assets* are employed—meaning deployed, operated, and supported—by the User in their Level 0 HIGHER ORDER SYSTEM. The *operational architecture* may include lower level training, maintenance, and support architectures that graphically represent their respective concepts.

- **Logical/Functional Architecture View** A *representation* of the logical/behavioral entity *relationships* and *interactions*—meaning behavior, products, by-products, and services—that express HOW the MISSION SYSTEM capabilities are envisioned to respond to external *stimuli* and *cues* for *hypothesized* scenarios within its prescribed OPERATING ENVIRONMENT.

- **Physical Architecture View** A *representation* of HOW an entity is physically composed, constructed, configured, and interfaced to respond to external stimuli and cues to achieve the desired outcome-based responses.

416 Chapter 36 System Architecture Development

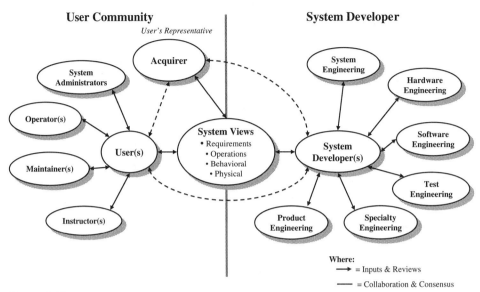

Figure 36.3 Stakeholder Views of a System

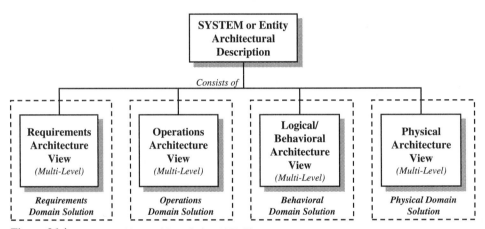

Figure 36.4 System Architectural Description (AD) Elements

Figure 36.4 illustrates these relationships as elements of an overall system *architecture description*. The SYSTEM architecture description (AD) consists of a *multi-level* requirements architecture, a *multi-level* operations architecture, a *multi-level* logical/functional architecture, and a *multi-level* physical architecture.

Architectural Responses to the Solution Space

If we translate the IEEE construct shown in Figures 36.1 with the views of a system illustrated in Figure 36.4, Figure 36.5 results. Here, we see the architectural views as satisfying the *solution space(s)* based on stakeholders, views, and concerns.

Guidepost 36.1 *At this point we have established the general relationships between the four solution domain architectures with the solution space. Now let's explore the interdependencies of these views further.*

36.3 Architectural Views of a System

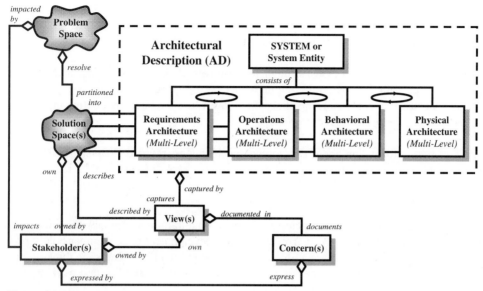

Figure 36.5 System Solution Viewpoints and Views

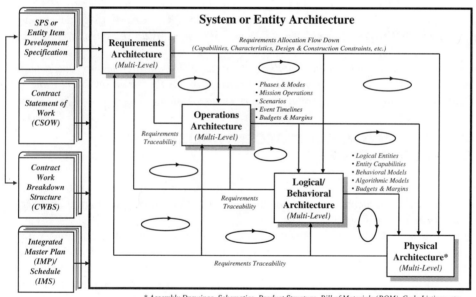

Figure 36.6 System Architecture Element Relationships

Architectural View Interdependencies

The requirements, operations, behavioral, and physical architectures are *highly integrated* and *interdependent* shown in Figure 36.6. Notice that the *highly iterative* interdependencies are depicted via an N^2 ($N \times N$ matrix) diagram. When you develop the architectures, it is very important for you to maintain traceability with the *Contract Statement of Work (CSOW)*, *Contract Work Breakdown Structure (CWBS)*, *Integrated Master Plan (IMP)*, and *Integrated Master Schedule (IMS)* or their contract documents.

Author's Note 36.1 *The point here is that architecture development encompasses more than simply creating graphical views of the system. The architectural description serves as the cornerstone for the CWBS, IMP, and IMS. It must also be consistent with the CSOW.*

As system development progresses through lower levels of design over time, architectural attributes such as capabilities or operations, requirements, performance budgets and safety margins, and design and construction constraints are allocated to lower level architecture entities and flowed down via entity *item development specifications*.

Referral For more information about developing specifications, refer to Chapter 31 *Specification Development Practices*.

Closing Points

One of the ambiguities of SE and deficiencies of organizational training or the lack thereof occurs when system architectures are developed. You may hear a development team member boldly proclaim they are going to *"develop a system architecture."* The problem is listeners are thinking one type of architecture and the doer has a "pet" architecture. As a result, the architectural work product may or may not suit the development team's needs.

One way to *avoid* this situation is to express the four domain solutions in terms of *views*. As each *solution* is formulated as illustrated in Figure 36.6, the team has a good idea of WHAT the deliverable architecture will depict. So, when someone boldly PROCLAIMs to be developing a system architecture, ASK: *WHICH architectural views and viewpoints do you intend to EXPRESS*.

Guidepost 36.2 *Given a fundamental understanding in WHAT an architecture and architectural description are, we now shift our focus to key considerations that drive selection of the type of architecture suitable for a system, product, or service application.*

36.4 CENTRALIZED VERSUS DECENTRALIZED CONTROL ARCHITECTURES

Once a system's interfaces are identified, most system architecture development activities begin with HOW the system is structured for *communications* and *decision-making*. Figure 36.7 serves as a reference for our discussion.

Chapters 8 through 12 *System Architecture Concepts* discussed system interactions with its OPERATING ENVIRONMENT. The discussion highlighted various types of *command* and *control* (C^2) interactions that included open loop and closed loop systems. For the C^2 mechanism, a key question is: *HOW do we efficiently and effectively implement C^2?* This requires a determination of the need for a *centralized* versus *decentralized* or *distributed* control architectures. So, what are these?

Centralized Control Architectures

Centralized control architectures, as illustrated on the LEFT side of Figure 36.7, consist of a single processing mechanism. For most applications the mechanism interfaces to remote access ports or sensors via mechanical, electronic, or optical types of devices. Consider the following examples:

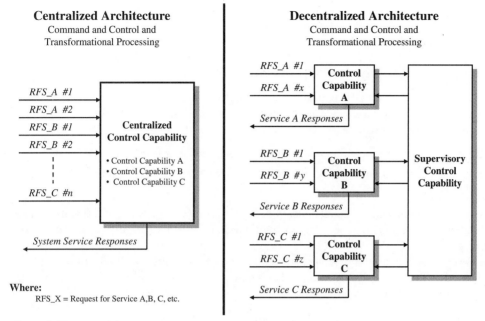

Figure 36.7 Centralized and Decentralized Architectures

EXAMPLE 36.1

Video surveillance systems route multi-channel, real-time camera video back to a multi-screen command center staffed with security personnel.

EXAMPLE 36.2

A power control circuit breaker panel distributes and controls power from an external service line to various circuits throughout an office building or home.

Limitations of Centralized Control. *Centralized control architectures* are fine for many applications such as the examples cited above. However, they do have limitations. They are a potential *single point of failure* (SPF).

As a SPF, some applications may require vast lengths of wiring to remote sensors that increase weight. For applications such as aircraft where an SPF may be a critical risk, additional weight, assuming it can be accommodated, *translates* into *increased* fuel consumption, *increased* fuel tank capacity, and *reduces* payload weight.

There are several approaches to solving this *problem space*. Example solutions include:

1. AVOID performance degradation and provide for expansion and growth by *decentralization* of processing functions.
2. REDUCE weight by deploying decision-making mechanisms at key locations and interconnecting the mechanisms via network-based configuration nodes.
3. AVOID risk due to potential SPFs by implementing control mechanism redundancies.

Author's Note 36.2 *Based on the SPF discussion, we should note that the processing mechanism, as illustrated at the right side of Figure 36.7, might consist of redundant processors as a means of improving the mechanism's reliability.*

Decentralized Control Architectures

Decentralized control architectures partition key control functions and deploy them via remote processing mechanisms that service input/output (I/O) requests as illustrated at the right side of Figure 36.7. The deployment may require:

1. Remote, dedicated processing to support a specific sensor or suite.
2. Retaining a central supervisory function to oversee each of the decentralized computing functions.

Depending on the mission and system application, *decentralized* functions might reside in the same rack at a site, be physically stationed throughout a building, or geographically separated across a country or around the world.

The *decentralized functions* may be allocated to hardware or software entities or dynamically assigned based on processing loads. As a result, overall system performance is improved but at the expense and risk of adding more processors. Consider the following example:

EXAMPLE 36.3

An organization operates technical support centers to assist customers in implementing the organization's products. In one approach, geographic sites are dedicated to addressing specific product questions. Since calls for a specific product: 1) may not occur in a uniform distribution and 2) may have surge periods, it may be *inefficient* and *unprofitable* to employ large staffs at a single site. So, a central phone system places customer calls in a *first-come–first-served* queue and assigns each call to *"the next available representative or technician"* that may be available in any one of several different product support sites.

For some applications, I/O processors may be identified in an architecture to off-load *mundane* I/O processing tasks such as data communications, printing, or interrupts from the main processor. This allows the supervisory processor to concentrate on higher level performance intensive tasks.

Client-Server Architectures

For system applications that require desktop or Web-based access to a central repository of information, *client-server architectures* are employed. In this case a processor is dedicated to processing client requests for data, retrieving the data from a central repository, and disseminating the data to the client. Applications such as this, which include internal organizational intranets and Web-based sites, are helpful for contract programs that need to provide access to program and contractor data to authorized Acquirer/Users, System Developers, subcontractors, and vendors.

Network Architectures

Organizations and systems often have need for personnel/entities to share and access common repositories of information as well as e-mail. Because of the cost and risk of having to connect dedicated wires from a central C^2 system throughout the facility, high-speed serial data communications networks are employed. Local Area Networks (LANs) are used to service clients within a facility, LANs for geographically separated facilities may be connected into wide area networks (WANs). Consider the following examples:

EXAMPLE 36.4
Commercial network systems employ Broadband, DSL, Ethernet, among other data communications. Military aircraft on-board networks employ MIL-STD-1553/1773 data communications networks; 1553 for wire-based applications, 1773 for fiber-optic applications.

36.5 FAULT TOLERANT ARCHITECTURES

The challenge in developing any type of system is creating a system architecture that is sufficiently ROBUST to *tolerate* and *cope* with various types of internal and external vulnerabilities. Then, when confronted with these conditions, the processing must be able to continue without significant *performance degradation* or *catastrophic failure*.

Depending on system design objectives, there are numerous ways of developing fault-tolerant architectures. Typically, a failure modes and effects analysis (FMEA) addresses potential failure modes, system effects, single points of failure (SPF), and so on. A more comprehensive expansion of a FMEA includes a *criticality analysis* (FMECA) to identify specific components that require close attention. The FMEA/FMECA assess and recommend *compensating provisions*—namely design modifications and operating procedures to mitigate failure conditions.

Referral A summary of FMEA/FMECA is provided in Chapter 50 *System Reliability, Availability*, and *Maintainability* (RAM) practices.

The FMEA/FMECA, in conjunction with the system's mission reliability requirement and resource budgets, influence the system architecture approach. Specification Developers are notorious for specifying redundancy requirements without thorough analysis of the current system architecture to determine if redundancy is a *necessary* and *sufficient* condition to satisfy mission reliability requirements. Avoid mandating design methods in specifications that increase cost and risk without having *compelling, fact-based evidence* that motivates such actions.

This brings us to a key point: *key areas for developing fault tolerant systems*.

Key Areas for Developing Fault Tolerant Systems

In general, design flaws and internal component *faults* or *malfunctions* can cause or lead to system *failures*. Examples include:

1. Inadequate system architecture selection.
2. Lack of system stability during various OPERATING ENVIRONMENT conditions.
3. Internal component failures due to OPERATING ENVIRONMENT conditions, surges, or long-term effects.
4. *Latent defects* due to improper or inadequate testing.
5. Faulty control logic.
6. Unknown modes and states.
7. Preoccupation with trivial, molecular level computation processing.
8. Poor work practices.
9. Improper operation results from abuse, misuse, or misapplication of the system or product.
10. Physical breaks in resource or data communications interfaces or supplies.
11. Lack of *preventive* and *corrective* maintenance.

12. Physical intrusion such as hacking, spamming, malicious mischief, and sabotage by unauthorized users and threats.

For systems in which a failure may be catastrophic, the system architecture decision making should include design OPTION considerations such as *redundancies*. Again, *compensating actions* should only be implemented after a determination that design reliability with a SPF is inadequate.

Guidepost 36.3 *At this point we have established the commonly used types of system architectures. We now shift our focus to understanding how to improve the reliability of an architectural implementation.*

36.6 TYPES OF REDUNDANCIES

For most applications we can classify redundancies in terms of architectural *configuration* and component *implementation*.

Architectural Configuration Redundancy

Architectural configuration redundancy consists of configuring redundant components either as fully operational "on-line" or "off-line" devices during all or portions of mission phases. There a three primary types of architectural redundancy: 1) operational, 2) cold or standy, and 3) k out of n systems.

Operational Redundancy. *Operational redundancy* configurations employ backup items that are *activated* or *energized* throughout the operating cycle of the system or product. All primary and redundant elements operate simultaneously for a total of *n* elements. Some people refer to this as *hot* or *active redundancy*. Consider the following example:

EXAMPLE 36.5

An aircraft may require a minimum number of engines to be operating within specified performance limits for takeoff, cruising, or landing; independent inertial navigation systems required to continue a mission; or tandem train locomotives for specific geographic and loading conditions.

During SYSTEM operations, redundant items can also be configured to operate concurrently and even share the loads. If one of the redundant items fails, the other item(s) assumes the load performed by the failed item and continues to perform the required capability(s). In most cases, the failed component is left in place or, assuming it does not interfere or create a safety hazard until *corrective maintenance* is available.

Cold or Standby Redundancy. *Cold or standby redundancy* consists of components that are not *energized*, *activated*, or *configured* into the system until the primary item fails. If the primary item fails, the standby item is connected automatically or manually through direct or remote operator intervention. Consider the following example:

EXAMPLE 36.6

Cold or standby redundancies include an emergency brake on an automobile, adding mass transportation vehicles (trains, buses, etc.) to support surges in consumer demand, emergency backup lighting switched on during a power failure, and backup power generation equipment.

k-*out-of*-n Systems Redundancy. This is a hybrid system whereby there are a total of n elements in the system but only k elements (k out of n) are required to operate during specific phases of a mission. For example, a car needs four of its five tires, including spare, to be in safe operating condition to complete a trip.

You may ask: *What is the difference between operational redundancy and* k-*out-of*-n *redundancy*. *Operational redundancy* assumes all components are configured into the system, can only be turned on/off, but not easily removed during a mission such as an engine on an aircraft or gyro. K-*out-of*-n *redundancy* factors in spares as replacements such as the car tire example.

Redundancy Implementation Approaches

Once a configuration redundancy concept is selected, the next step is to determine HOW to physically implement the concept, either by *identical* components or *unlike* components that are similar in function.

Like Redundancy Configuration. *Like redundancy* is implemented with identical components—vendor product model part numbers—that are employed in an *operational* or *standby redundancy* configuration.

Unlike Redundancy Configuration. *Unlike redundancy* consists of components that are *not physically identical*—different vendors—but provide the functionality and performance required to perform the capability.

Both implementation approaches offer *advantages* and *disadvantages*. If the components used for *like redundancy* are *sensitive* to certain OPERATING ENVIRONMENT characteristics, having identical components may not be a solution. If you purposely choose *unlike redundancy* and the same situation occurs, redundancy may only exist over a limited operating range if one component has higher reliability. If components identical only in function and performance are qualified over the full operating range, *unlike redundancy* may offer advantages.

Reducing Component SPF Risk

One method of reducing the component SPF risk requires operating *identical* or *redundant* components in several types of configurations. Redundancy type examples for electronic systems include:

- Processing redundancy
- Voted k out of n redundancy
- Data link redundancy
- Service request redundancy

Figure 36.8 provides a graphical view to support the discussions that follow.

Processing Redundancy. Returning to an earlier discussion of *decentralized processing* via *distributed components*, detection of a processor failure and dynamic reallocation of processing tasks to an available processor enhances fault tolerance. So SUBSYSTEMs A and B both include redundant processing components, A' and B'.

Voted "k *out of* n" Component Redundancy. Some systems have redundant, peer-level processors that employ an *operational hot* or *active redundancy* configuration. Individual processor results Subsystems A and B are routed through a central decision-making mechanism that determines if k out of n results agree. If k out of n results agree, transmit the results to a specific destination.

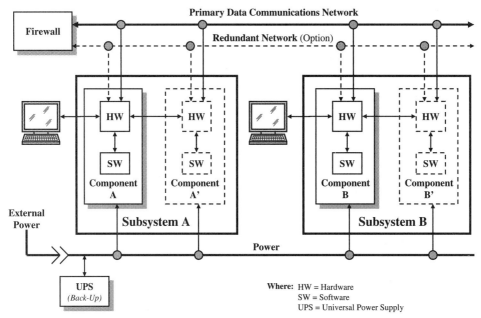

Figure 36.8 System Architecture Element Relationships

Data Link Redundancy. There is an old adage: "*Systems break at their interfaces.*" From an interface reliability perspective, this adage holds true. System developers often take great pride in creating ELEGANT system designs that employ redundant processing components. Then, they connect the redundant components to an external interface that is a *single point of failure* (SPF).

One way to AVOID this problem is to employ redundant networks as shown in Figure 36.8. Obviously, if interconnecting components such as cables are placed in a stable/static position, not subjected to stressful OPERATING ENVIRONMENT conditions, and interfaced properly, there is a good chance that additional *independent* connections are *unnecessary* and can be avoided.

For applications that employ satellite links or transmission lines that may be switched periodically, it may be necessary to employ backup links, such as land lines, as a contingency.

Service Request Redundancy. Some systems may be designed to automatically transmit messages one or more times. Using Figure 36.8 as an example, SUBSYSTEMs A and B automatically retransmit messages to each other. Others may issue service requests to repeat messages, acknowledgments, or data responses. As illustrated in Figures 15.3 and 15.4, this example is equally applicable to external systems.

36.7 ERROR CORRECTION CONSIDERATIONS

Some systems employ data communication protocols that request retransmission IF errors are detected. In general, *error correction* may or may not be considered an *explicit* redundancy method. However, when it is employed, it provides comparable benefits.

Guidepost 36.4 *The development of a system architecture requires more than simply innovation and creation, it also requires other architectual considerations:*

1. *Compliance with local, state, federal, and international statutes and regulations.*
2. *Sustainment of resources.*
3. *Recognition of the cause and effect the architecture has on the public and the environment.*

We now shift the discussion to these considerations.

36.8 OTHER ARCHITECTURAL SELECTION CONSIDERATIONS

Developing a system architecture to provide capabilities to support all phases of the mission is only part of what is required. The architecture must also include other key considerations. These include:

- Power source architectural considerations.
- Environmental, safety, and health (ES&H) architectural considerations.
- Fire detection and suppression architectural considerations.
- System security architectural considerations.

Power System Architectural Considerations

The preceding discussions highlight HOW to enhance the fault tolerance of the systems and products we build. If they are electrically powered, no matter how elegant the redundancy solution, it only works WHEN power is applied. The loss of power involves several issues:

1. Safe storage of critical mission and system data immediately following the event to prevent data loss.
2. Safe evacuation of personnel from facilities to prevent injury or loss of life.
3. Sustainment power to critical processes that must process to completion and place the system in a SAFE mode.

Sustaining Operations and EQUIPMENT Resulting from Loss of Power. When a power loss event occurs, systems require a finite amount of time to store mission and system data. To ensure a continuation of power for a specified time, rechargeable batteries or an uninterruptible power supply (UPS), offer potential solutions. Depending on mission and system application, alternative power solutions include external fuel-based generators, solar panels, fuel cells, and other technologies.

Power Quality Considerations. Another factor that requires architectural consideration is power quality. Power surges, brownouts, overvoltage, noise, and stability conditions wreak havoc with some systems that require power conditioning. So, make sure these considerations are fully addressed by the architecture within resource constraints.

Environmental, Safety, and Health (ES&H) Architectural Considerations

System architecting, in general, tends to focus on EQUIPMENT architectures rather than the *effects* of the EQUIPMENT on the Users, public, and NATURAL ENVIRONMENT. Therefore, when evaluating a system architecture, the system architect and others should factor in considerations for environmental, safety, and health (ES&H).

EXAMPLE 36.7

At a *minimum*, considerations of the effects of EQUIPMENT by-products on the User(s), public, and environment should include the following considerations:

- Moisture
- Condensation
- Water
- Shock and vibration
- Atmospheric pressure
- High pressures
- Noise
- Refuse
- Spills
- Leaks
- Laser radiation
- Nuclear waste
- Ergonomics

EXAMPLE 36.8

At a *minimum*, safety examples include the following considerations:

- Walk space
- Hazardous conditions
- Ingress and egress
- Emergency exits
- Warning notices and cautions
- Visual and audio alarms
- Electrical shock protection
- Perimeter fencing
- Lockout tags on damaged or out-of-calibration equipment
- Doors and stairwells
- Video surveillance
- Grounding schemes

EXAMPLE 36.9

At a *minimum*, health examples include the following considerations:

- Toxic chemicals and fumes
- Air quality
- Ergonomics
- Noise

Fire Detection and Suppression Architectural Considerations

Another key architectural consideration is *fire detection* and *suppression systems*. Since personnel, equipment, and facility safety are paramount, the system architecture should include features that enable a rapid response when fires are detected including suppression systems that eliminate the source of the fire following personnel evacuation.

System Security Architectural Considerations

For those systems that involve sensitive or classified data, system security should be a key architectural consideration. This includes *reasonable measures* for physical security, operational security, communications security, and data security.

Final Thought

All considered effects should include both the short-term effects and the long-term effects. *Compensating actions* for any effects may require accomplishment via one or more of the SYSTEM's architectural system elements—EQUIPMENT, PERSONNEL, PROCEDURAL DATA, and so forth.

36.9 GUIDING PRINCIPLES

In summary, the preceding discussions provide the basis with which to establish the guiding principles that govern system architecture development practices.

Principle 36.1 An *architecture* expresses stakeholder *concerns* via *views* that comply with *viewpoint* conventions for constructing the view.

Principle 36.2 Every SYSTEM/entity has four architectural views: requirements, operations, behavior, and physical, each consistent with and traceable to the other.

Principle 36.3 System architectural redundancy is a design method for achieving a challenging reliability requirement, and not a specification requirement.

Principle 36.4 An architecture represents the totality of the integrated configuration of system capabilities required to perform its use cases and cope with use case scenarios without regard to element frequency of usage.

Principle 36.5 Every system architecture must be compliant with its specification requirements and applicable laws, regulations, and cultural values.

Principle 36.6 Every system architecture must factor in considerations for the environment, safety, and health.

36.10 SUMMARY

The discussion of *system architecture development practices* of this chapter serve as a precursor for the Requirements, Operations, Behavioral, and Physical Domain Solutions that follow. System architectures provide the means to communicate the key stakeholder concerns and engineering viewpoints required to develop each SYSTEM or entity's four *solution domains*—consisting of Requirements, Operations, Behavioral, and Physical Domain Solutions.

During our discussions we introduced and defined the origin of the term architecture as well as other architecture-related terms. We discussed the need to delineate the type of system architecture—operational, behavioral, and physical—when someone boldly proclaims an intent to develop the "system's" architecture.

Specific architecture view implementations are described in the respective *Requirements Domain Solution*, *Operations Domain Solution*, *Behavioral Domain Solution*, and *Physical Domain Solution practices* that follow.

GENERAL EXERCISES

1. Answer each of the *What You Should Learn from This Chapter* questions identified in the *Introduction*.
2. Refer to the list of systems identified in Chapter 2. Based on a selection from the preceding chapter's *General Exercises* or a new system selection, apply your knowledge derived from this chapter's topical discussions. Specifically identify the following:
   (a) Who are the system's stakeholders?
   (b) What are their views and concerns of the system's architecture?
   (c) How would these views and viewpoint concerns be captured in the Requirements, Operations, Behavioral, and Physical Domain architectures?
   (d) What are potential trade-off areas require consideration in formulating the architecture?
3. Identify an example of each of the types of system architectures and design options listed below:
   (a) Centralized architecture
   (b) Decentralized architecture
   (c) Hot redundancy
   (d) Standby redundancy
   (e) k out of n redundancy
   (f) Like redundancy
   (g) Unlike redundancy

ORGANIZATION CENTRIC EXERCISES

1. Research your organizations command media.
   (a) What requirements are levied on programs for development of system architectures?
   (b) How are system architectures to be evaluated?
2. Contact a small, a medium, and a large contract program in your organization.
   (a) What types of system architectures are used?
   (b) Does the architecture include *redundant* elements? Were they required by contract?
   (c) Are there redundant elements; what is the rationale?
   (d) What compelling evidence—such as meeting reliability, availability, and maintainability (RAM) requirements—drove them to redundancy?
   (e) Was redundancy a requirement in the SPS or development specification? Why?

REFERENCES

BUEDE, DENNIS M. 2000. *The Engineering Design of Systems*. New York: Wiley.

Defense Systems Management College (DSMC). 2001. *Glossary: Defense Acquisition Acronyms and Terms*, 10th ed. Defense Acquisition University Press, Ft. Belvoir, VA.

IEEE 1471-2000. 2000. *IEEE Recommended Practice for Architectural Description of Software-Intensive Systems*. Institute of Electrical and Electronic Engineers (IEEE) New York, NY.

"Systems always break at their interfaces." (Anonymous)

ADDITIONAL READING

LEITCH, R.D. 1988. *BASIC Reliability Engineering Analysis*. Stoneham, MA: Butterworth.

PIDD, MICHAEL. 1998. *Computer Simulation in Management Science*, 4th ed. Chichester: Wiley.

RECHTIN, EBERHARDT. 1991. *Systems Architecting: Creating and Building Complex Systems*. Englewood Cliffs, NJ: Prentice-Hall.

RECHTIN, EBERHARDT, and MAIER, MARK W. 2000. *Systems Architecting: Creating and Building Complex System*, 2nd ed. Boca Raton, FL: CRC Press.

Chapter 37

Developing an Entity's Requirements Domain Solution

37.1 INTRODUCTION

The Requirements Domain Solution *bounds* an entity's—SYSTEM, PRODUCT, SUBSYSTEM, etc.—*solution space* within technical, technology, support, cost, and schedule constraints and risks. The Requirements Domain Solution specifies:

1. WHAT capabilities and performance characteristics are required from the *system*, *product*, or *service*.
2. WHAT levels of performance are expected—and HOW WELL.
3. *System element* accountability for accomplishing capability-based requirements.
4. WHEN the capability is required.
5. Under WHAT OPERATING ENVIRONMENT conditions and interactions.
6. WHAT outcomes or results are expected to satisfy the User's operational needs and successfully achieve the system and mission objectives.

This chapter expands on the Requirements Domain Solution description introduced in our discussion of the *SE Process Model* in Chapter 26. Our discussion addresses the Requirements Domain Solution: objectives, key elements, sequence in the SE Process Model work flow, development responsibility, dependencies, development methodology, challenges, and work products.

What You Should Learn from This Chapter

1. What is the *objective* of the Requirements Domain Solution?
2. What are the *key elements* of the Requirements Domain Solution?
3. What is the relationship of the Requirements Domain Solution to the SE Process Model?
4. Explain the *relationships* of the Requirements Domain Solution to the Operations, Behavioral, and Physical Domain Solutions?
5. What is the *methodology* used to develop the Requirements Domain Solution?
6. What are the steps of the methodology employed to derive the Requirements Domain Solution?
7. How do you develop the Requirements Domain Solution architecture?

System Analysis, Design, and Development, by Charles S. Wasson
Copyright © 2006 by John Wiley & Sons, Inc.

8. What are the *exit criteria* of the Requirements Domain Solution?
9. What are some of the *challenges* in developing the Requirements Domain Solution?
10. What *work products* do the Requirements Domain Solution activities produce?
11. How is the Requirements Domain Solution is *verified* and *validated*?

Requirements Domain Solution Development Activity Objective(s)

The objectives of the Requirements Domain Solution development activity are to:

1. *Accurately*, *precisely*, *consistently*, and *completely* bound the *solution space* and identify the required *capabilities*—the functions and performance, and *characteristics* required to satisfy the User's (contextual role) *validated* operational need(s).
2. Provide *objective evidence* as a work product to support entity verification and formal acceptance.

Author's Note 37.1 *The usage of the term "contextual role" above refers to multi-level development. For example, at the SYSTEM Level, the Acquirer and System Developer organizations establish a contractual agreement for requirements via the System Performance Specification (SPS). Within the System Developer's program organization, the System Engineering and Integration Team (SEIT) establishes item development specifications (IDS) with PRODUCT level integrated product teams (IPTs), other development teams, or issues subcontracts to develop PRODUCTS. In this latter context, the SEIT serves as an Acquirer (role) and the IPT, development teams, or subcontractors serve as System Developers (role) for their respective PRODUCTS.*

Key Elements of the Requirements Domain Solution

The key elements of the Requirements Domain Solution and their interrelationships include the following entities: problem space, solution space, operating constraints, capabilities, *Mission Event Timeline (MET)*, specifications, verification methods, functions, measures of performance (MOP), critical operational/technical (COIs/CTIs) issues, and clarifications.

Requirements Domain Solution Dependencies

As discussed in earlier sections, the Requirements Domain Solution development is a multi-level, *highly iterative* process that requires close *collaboration* and *coordination* with the evolving Operations, Behavioral, and Physical Domain Solutions as shown in Figure 26.3. *Iterative analysis and technical reviews* are required, preferably by independent reviewers, to ensure that the SYSTEM OF INTEREST (SOI) is properly balanced—meaning *optimal*—from an overall development and life cycle perspective in terms of technical, cost, schedule, support, and risks.

Requirements Domain Solution Development Sequencing

As the initial stage of SYSTEM and entity development, the Requirements Domain Solution development occurs ahead of the Operations, Behavioral, and Physical Domain Solutions as shown in Figure 23.2. As the technical specification of a contract or task agreement, this solution ultimately establishes the legal basis for determining formal contract-based acceptance of the physical *system*, *product*, or *service* by the Acquirer and User.

Requirements Domain Solution Responsibility

Responsibility for the Requirements Domain Solution resides with the program's Technical Director or Project Engineer and is usually delegated to a Lead SE for the program. The Lead SE facilitates the development of the Requirements Domain Solution by ensuring that all *operational* and *disciplinary* stakeholder interests are *represented* in the development and review of the *System Performance Specification (SPS)* and each entity's *item development specifications*.

Once the specification requirements baselines are *approved* and *released*, Configuration Management (CM) administers formal change management in accordance with direction from a Configuration Control Board (CCB) or Software Configuration Control Board (CCB) of SYSTEM or entity *stakeholders*.

37.2 DEVELOPING THE REQUIREMENTS SOLUTION ARCHITECTURE

The core infrastructure for design, development, and acceptance of any *system*, *product*, or *service* resides in the requirements. The requirements *scope* and *bound* each entity's *operational*, *behavioral*, and *physical* capabilities, performance characteristics, and properties. From the first abstract requirement describing an operational *objective* or *need* through the lowest level of system design (nuts and bolts, code, etc.), the requirements must be linked to allow traceability back to their *source* or *originating requirements*. This point establishes a fundamental rule of *traceability* that *governs* the basis for formulation of a system design solution:

- *Each requirement has a cost to implement within contract or task cost and schedule performance measurement baseline constraints. If any requirement is not traceable back to a source or originating requirement, either eliminate the requirement or renegotiate cost and schedule constraints and replan the effort.*

We refer to the *hierarchical* and *horizontal* requirements infrastructure as the Requirements Architecture. In practice, the *Specification Tree* provides the framework for linking multi-level specification requirements.

Referral For more information about specifications, refer to Chapters 28 to 33 *System Specification Practices*.

37.3 REQUIREMENTS SOLUTION DEVELOPMENT METHODOLOGY

The Requirement Domain Solution is developed as a key element of the SE Process Model as shown in Figure 26.1. The *highly collaborative* and *iterative* integration of the Operations Domain Solution with the Operations, Behavioral, and Physical Domain Solutions can be *chaotic* and *confusing*.

We can minimize the *chaos* and *confusion* by applying an *iterative* methodology that enables us to create the Requirements Domain Solution. The *methodology represents* one of many approaches to developing the Requirements Domain Solution. View these steps as an example strategy and tailor them to suit your own business domain and systems application.

Step 1: Understand the problem and solution space(s).

Step 2: Capture and bound entity requirements.

Step 3: Analyze and reconcile entity specification requirements.

Step 4: Derive and assimilate entity capabilities.
Step 5: Resolve requirements issues and clarifications.
Step 6: Verify and validate the Requirements Solution.
Step 7: Establish and maintain a Requirements Solution baseline.

Step 1: Understand the Problem and Solution Space(s)

The first step in developing the Requirements Domain Solution is to *understand*:

1. WHAT *problem* or *issue* the User is attempting to solve.
2. WHAT specification requirements are required to bound the *solution space* within technology, cost, schedule and risk constraints.

Conduct a mission and operations analysis, field interviews with the User, analyze existing system trouble reports, and understand the OPERATING ENVIRONMENT and its players and threats.

Step 2: Capture and Bound Entity Requirements

If an entity's *item development specification* requirements are undefined, you need to *capture*, *allocate*, and *flow down* higher level requirements to the entity. *Accurately*, *concisely*, *consistently*, and *completely* bound the entity's required capabilities and their associated levels of performance as well as document *operational*, *physical*, *design*, and *construction* constraints. Work with the requirements *stakeholders* to prioritize the specification requirements.

As entity requirements are captured, *bound* each one within the framework of the entity's *required operational capabilities* and characteristics. *Does each requirement specify*:

1. WHAT action is to be performed?
2. WHO/WHAT mechanism is accountable for performing the action?
3. WHEN the action is to be performed?
4. HOW WELL the action is to be performed?
5. Under WHAT operating scenarios and conditions?
6. WHAT outcome or result is expected from the action?

Finally, perform a reasonableness check on each requirement.

How do we do this? Formulate a mini-verification plan. Write a one- or two-sentence verification strategy to *validate* that the requirement can be *verified*—that it is measurable, testable, and realistically achievable—within allocated resource constraints—budgetary, schedule, facilities, and test environment constraints.

Step 3: Analyze and Reconcile Entity Specification Requirements

As the entity's requirements are established, analyze the requirements to ensure that you understand *WHAT*:

1. *Capabilities* are required of the deliverable *system*, *product*, or *service*.
2. *Levels of performance* bound the capabilities.
3. *Physical characteristics*—such as size, weight, and reliability.
4. *Design and construction constraints* are levied on those capabilities.
5. *Verification methods* are required to verify achievement of each capability.

434 Chapter 37 Developing an Entity's Requirements Domain Solution

Since specifications sometimes contain *missing*, *misplaced*, *duplicated*, or *conflicting* requirements, the requirements analysis should strive to *discover*, *clarify*, or *resolve* any deficiencies, issues, and concerns. The *work product* of the exercise should be a set of specification requirements that are *necessary* and *sufficient* to support development of the entity's Operations, Behavioral, and Physical Domain Solutions.

Step 4: Derive and Assimilate Entity Capabilities

Ideally, the initial *System Requirements Document (SRD)* provided in the formal *Request for Proposal (RFP)* solicitation, the *System Performance Specification (SPS)*, and entity *item development specifications* explicitly *identify* and *bound* the required capabilities of the proposed solution via requirements statements. In general, there should be a *one-to-one correlation* between the requirements and the required capabilities. Best practices recommend that specification requirements state *one and only one* requirement for a capability as illustrated by Figure 37.1.

Referral For more information regarding development of requirements statements, refer to Chapter 33 *Requirements Statement Development Practices*.

The *realities* of system development, however, indicate this is not always the case. When specification developers write requirements statements as paragraphs containing *multiple sentences* with *compound requirements*, SEs must delineate and assimilate the required capabilities. Sometimes this is easy; other times not.

If you are confronted with a *feature-based* specification written as a RANDOM set of thoughts, you may have to derive a set of capabilities using the method shown in Figure 37.1 and *link* the capabilities back to specification requirements. The physical linking can be accomplished with a spreadsheet or preferably with an object-oriented (OO) requirements traceability tool based on a relational database.

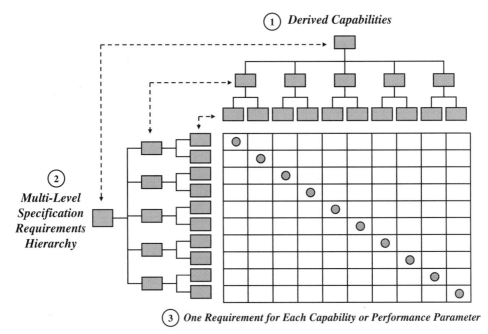

Figure 37.1 Requirements Driven Capabilities

Referral For more information regarding feature-based specifications, refer to Chapter 31 *Specification Development Practices*.

To illustrate *how* this occurs, consider the table shown in Figure 37.2. The left side provides a tabular listing of specification requirements that include a reference identifier (REQ_ID), specification paragraph number, and requirements statement. *Ideally* the specification requirements are written as *singular* requirements statements that specify *one and only one operational* or *functional* capability.

We can extract individual capabilities from requirements statements containing *singular* and *compound* requirements as shown on the right side of Figure 37.2. The capabilities are then assimilated into a hierarchy of capabilities, each *linked* to applicable specification requirements.

Step 5: Resolve Requirements Issues and Clarifications

Multi-level decision making requires that issues at higher levels be resolved quickly.

1. Investigate the *accuracy*, *consistency*, and *completeness* of the set of requirements and interfaces to other sets of requirements.
2. Resolve any *critical operational issues (COIs)* or *critical technical issues (CTIs)*.
3. Resolve any *missing*, *misplaced*, *conflicting*, *duplicated*, or *compound* requirements.
4. Verify that each requirement is applicable to the entity. If not, place it in the appropriate entity's *item development specification*.
5. Eliminate any *irrelevant* or *non-valued-added* requirements; *clarify* any ambiguous requirements.
6. Verify that each requirement is *realistic*, *achievable*, *measurable*, *testable*, and *verifiable*.

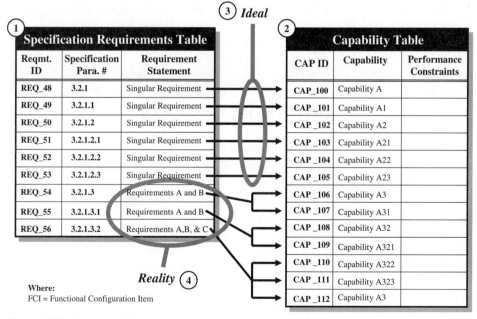

Figure 37.2 Extracting FCIs from Specification Requirements

Step 6: Verify and Validate Requirements Solution

Throughout the development of the Requirements Domain Solution, you should *continuously verify*:

1. The integrity of SPS or *item developmental specification* requirements to User source and contract documentation.
2. Relevance of each requirement to the entity as bounded by the contract or task scope.
3. Achievement of each requirement within technology, cost, schedule and risk constraints.
4. Whether all requirements issues and concerns have been resolved to the satisfaction of all *stakeholders*.
5. Whether the evolving Operations, Behavioral, and Physical Domain Solutions are traceable to and fully *compliant* with the Requirements Domain Solution.

The Requirements Domain Solution should also be *validated* with Users and other key *stakeholders* via Technical Interchange Meetings (TIMS), interviews, technology demonstrations, rapid prototypes, and models and simulations.

When the Requirements Domain Solution reaches maturity, technical review events such as the System Requirements Review (SRR), Hardware Specification Review (HSR), Software Specification Review (HSR), and In-Process Reviews (IPRs) should be conducted. The events should *communicate* and obtain *consensus* buy-in from key *stakeholders*, as appropriate, of the Requirements Domain Solution.

Referral For more information about *technical reviews*, refer to Chapter 54 *Technical Review Practices*.

Step 7: Establish and Maintain Requirements Solution Baseline

Once the Requirements Domain solution is approved, establish an entity Requirements Baseline—such as a System Requirements Baseline—for requirements allocations to lower levels, future decision making, and change control of the requirements. Incorporate each entity's Requirements Baseline into the evolving *Developmental Configuration*.

Using this methodology, let's identify some of the *Requirements Domain Solution challenges*.

37.4 REQUIREMENTS DOMAIN SOLUTION CHALLENGES

When the Requirements Domain Solution is developed, there are several challenges that the User, Acquirer, and System Developer(s) need to address. Consider the following challenges:

Challenge 1: Understanding of the Problem Space

Do we thoroughly understand and have we articulated the problem or issue—the *problem space*—the User is attempting to solve?

Challenge 2: Development of the "Right" System

Do the requirements specify *necessary* and *sufficient* capabilities, performance, and characteristics for a deliverable *system*, *product*, or *service* that fulfills the *solution space(s)*?

Challenge 3: Stakeholder Operational Needs

If we develop the system according to these requirements, will the deliverable *system*, *product*, or *service* fulfill the User's intended operational needs—*validation* of achievement?

Challenge 4: Stakeholder Interest Representation

Do the requirements *represent* the consensus and prioritized interests of key *stakeholders* and life cycle phases within resource constraints?

Challenge 5: Requirements Verifiability

Are the requirements *realistic*, *achievable*, *testable*, *measurable*, and *verifiable*?

Challenge 6: Fitness-for-Use Criteria

Do the requirements, as stated, satisfy "fitness-for-use" criteria required by the Operations, Behavioral, and Physical Domain Solutions?

Challenge 7: Development Risks

Do the requirements, as stated, pose any *unacceptable* technical, technology, operational, support, cost, and schedule risks?

Challenge 8: Contract Risks

Can the requirements, as stated, be implemented by a competent organization with a given level of capability within the allowable cost and schedule constraints?

37.5 REQUIREMENTS DOMAIN SOLUTION WORK PRODUCTS

The Requirements Domain Solution:

1. Is characterized via a hierarchical *specification tree* of multi-level entity *development, process, and material specifications*, System Performance Specification (SPS) as well as User operational needs.
2. Consists of a set of derived capabilities with linkages to specification requirements.
3. Includes a *Requirements Traceability Matrix (RTM)* that documents requirements *allocation* to lower levels and vertical *traceability* to the SPS and User operational needs, and *horizontal traceability* between requirements.
4. Is supported by working papers and tools such as analyses, models, and simulations that document the technical and scientific basis for informed decision making.

37.6 GUIDING PRINCIPLES

In summary, the preceding discussions provide the basis with which to establish the guiding principles that govern developing an entity's Requirements Domain Solution practices.

Principle 37.1 The first step in developing any SYSTEM/entity solution begins with creating, bounding, and specifying the requirements domain solution.

Principle 37.2 Every requirement must be traceable to a *source* or *originating* requirement and a cost to implement; untraceable requirements should be eliminated or linked to a source requirement.

37.7 SUMMARY

Our discussion of the *Requirements Domain Solution* highlighted the key elements and *work products* of the solution space. Remember the Requirements Domain Solution is *highly collaborative* and *iterative* with the Operations, Behavioral, and Physical Domain Solutions. Requirements established for this solution bound the solution space from which the Operations, Behavioral, and Physical Solutions are developed.

GENERAL EXERCISES

1. Answer each of the *What You Should Learn from This Chapter* questions identified in the *Introduction*.
2. Refer to the list of systems identified in Chapter 2. Based on a selection from the preceding chapter's *General Exercises* or a new system selection, apply your knowledge derived from this chapter's topical discussions. Specifically identify the following:
   (a) Create a multi level system architecture hierarchy diagram.
   (b) Derive the system's specification tree from the system architecture
   (c) Describe and bound what development specifications are required for the system and provide rationale.

ORGANIZATIONAL CENTRIC EXERCISES

1. Research your organization's command media. What guidance is provided regarding developing of the *Requirements Domain Solution*? Document and report your results.
2. Contact a contract program within your organization. Interview the lead SEs and research HOW the program: 1) formulated its *Requirements Domain Solution*, 2) manages the requirements baselines, and 3) links the *Requirements Domain Solution* to the Operations Domain Solution.
3. Select one of your organization's system specifications and develop an architectural framework (hierarchy) for the *Requirements Domain Solution*.
   (a) How well do the requirements and capabilities map?
   (b) Are there missing, misplaced, conflicting, or duplicated requirements?
   (c) Test the syntax of each requirement. Does the set of requirements comply with criteria discussed later in Chapter 33 on requirements statement development?
   (d) Identify any technical issues in the set of requirements.

Chapter 38

Developing an Entity's Operations Domain Solution

38.1 INTRODUCTION

At this point, you should have an understanding of the Requirements Domain Solution. The next step is to derive the entity's Operations Domain Solution. The Operations Domain Solution focuses on HOW the User intends to *deploy*, *operate*, and *support* the SYSTEM OF INTEREST (SOI)—namely the MISSION SYSTEM and SUPPORT SYSTEM—to perform organizational missions and achieve mission objectives.

This chapter addresses the *Operations Domain Solution* identified in the SE Process Model. Our discussions address objectives, key elements, and sequence in the SE Process Model workflow, development responsibility, dependencies, development methodology, challenges, and work products.

What You Should Learn from This Chapter

1. What is the *objective* of the Operations Domain Solution?
2. What are the *key elements* of the Operations Domain Solution?
3. What is the *relationship* of the Operations Domain Solution to the SE Process Model?
4. What is the *relationship* of the Operations Domain Solution with the Requirements, Behavioral, and Physical Domain Solutions?
5. What is the *relationship* between required operational capabilities and mission operations?
6. What *methodology* is employed to develop the Operations Domain Solution?
7. What are the *work products* that represent the Operations Domain Solution?
8. How do you *verify* and *validate* the Operations Domain Solution?

Definitions of Key Terms

- **Operational Architecture** A model-based *representation* depicting HOW role-based system elements (actors) are integrated to deliver *products*, *by-products*, or *services* to achieve performance-based organizational mission objectives outcomes.
- **Operational Asset** A physical system, product, or service that an organization employs to perform or support missions and achieve organizational objectives.

440 Chapter 38 Developing an Entity's Operations Domain Solution

- **Operations Function** "Tasks, actions, and activities performed with available resources to accomplish defined mission objectives and tasks environments planned or expected." (Source: Adapted from former MIL-STD-499 DRAFT, *Appendix A, Glossary*, A.3 Definitions, p. 37)

Operations Domain Solution Development Objective

The objective of the Operations Domain Solution activity is to *identify*, *formulate*, and *document* HOW the User envisions deploying, operating, supporting, and disposing of the deliverable *system*, *product*, or *service* to accomplish organizational missions and successfully achieve the mission objectives.

Key Elements of the Operations Domain Solution

The *Operations Domain Solution* consists of a number of key elements that require consideration during the development of the solution. Operational elements include missions, mission objectives, Mission Event Timeline (MET), operational assets, OPERATING ENVIRONMENT entities, use cases, and use case scenarios.

Operations Domain Solution Dependencies

The multi-level Operations Domain Solution is *highly iterative* and requires close *collaboration* and *coordination* with the development of the Requirements, Behavioral, and Physical Domain Solutions as shown in Figure 26.3. *Iterations* are also required to ensure that the SYSTEM OF INTEREST (SOI) is properly balanced—that is, *optimal*—from an overall development and life cycle perspective—in terms of technical, cost, schedule, support, and risks.

Operations Domain Solution Development Sequencing

The *Operations Domain Solution* evolves slightly behind but *concurrently* with the Requirements Domain Solution and ahead of the Behavioral and Physical Domain Solutions as shown in Figure 23.2.

Operations Domain Solution Development Responsibility

Responsibility for the Operations Domain Solution resides with the program's Technical Director or Project Engineer and is usually delegated to a Lead SE for the program. The Lead SE facilitates the development of the Operations Domain Solution by ensuring that all *operational* and *disciplinary* stakeholder interests are *represented* in the *operational architecture* and System Concept of Operations (ConOps).

As key elements of the Operations Domain Solution are *approved* and *released*, Configuration Management (CM) administers formal change management in accordance with direction from a Configuration Control Board (CCB) or Software Configuration Control Board (SCCB) of SYSTEM or entity *stakeholders*.

38.2 DEVELOPING THE OPERATIONAL ARCHITECTURE

The Operations Domain Solution is structured around a framework referred to as the *operational architecture*. The operational architecture identifies the entities that interact during all phases of the mission. These entities include persons, places, roles, or things—that is, objects. The Unified Modeling Language (UML™) refers to these as *actors*. Whereas Figure 17.2 symbolizes *actors* as analytical "stick figures," the operational architecture typically employs "clip art" objects such as

38.3 Operations Domain Solution Development Methodology

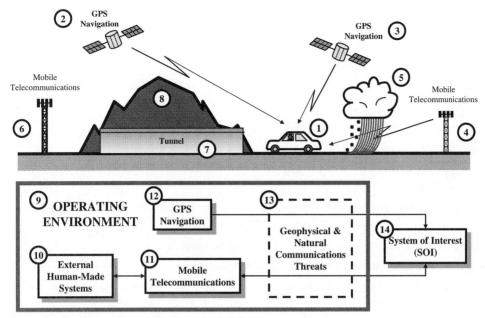

Figure 38.1 Mobile Transportation System Operational Architecture Example

buildings, trucks, and computers that more closely resemble the real world. Figure 38.1 provides an illustration of a Car-Driver System and how it communicates with other systems.

While operational architectures can be developed for any entity within a system, most are developed for the upper system levels of abstraction—SYSTEM, PRODUCT, or SUBSYSTEM—from a User's perspective. Consider the following example:

EXAMPLE 38.1

An operational architecture of a printed circuit board in a desktop computer system may be of limited interest to most computer Users. In contrast, NASA's JPL Mars Exploration Program users are very interested in various aspects of the operational architectures for landing systems and deployment of their rovers including HOW events during those phases may impact the designs of their EQUIPMENT.

The key elements of an *operational architecture* originate and are derived using the System Element Architecture template shown in Figure 10.1. In general, the intent of the *operational architecture* is to *EXPOSE* the actors, *relationships*, and *interactions* of the MISSION SYSTEM, SUPPORT SYSTEM, and their OPERATING ENVIRONMENT. Typically, two or more interoperable, organizational assets—such as the MISSION SYSTEM and SUPPORT SYSTEM—are operationally configured and integrated to conduct missions to achieve outcome-based results.

38.3 OPERATIONS DOMAIN SOLUTION DEVELOPMENT METHODOLOGY

The Operations Domain Solution is developed as a key element of the SE Process Model as shown in Figure 26.1. The *highly collaborative* and *iterative* integration of the Operations Domain Solution with the Requirements, Behavioral, and Physical Domain Solutions is often very *chaotic* and *confusing*.

We can minimize the *chaos* and *confusion* by applying an iterative methodology that enables us to create the Operations Domain Solution. The *methodology represents* one of many approaches for developing the Operations Domain Solution. View these steps as an example and tailor them to suit your business domain and systems applications.

Step 1: Conduct a mission analysis.
Step 2: Identify system elements and actors.
Step 3: Develop actor-based operational architecture.
Step 4: Develop system operations workflow sequences.
Step 5: Allocate mission operations to phases of operation.
Step 6: Establish the Mission Event Timeline (MET).
Step 7: Translate mission operations into system use cases and scenarios.
Step 8: Identify the system modes and states of operation.
Step 9: Derive system capabilities from use cases and scenarios.
Step 10: Develop the system Concept of Operations (ConOps).
Step 11: Resolve critical operational and technical issues (COIs/CTIs) and risks.
Step 12: Verify and validate operational solution.
Step 13: Establish and maintain the Baseline Concept Description (BCD).

Referral For additional information about the SE Process Model, refer to Chapter 26.

Step 1: Conduct a Mission Analysis

The System Architect, in collaboration with the System Engineering and Integration Team (SEIT), begins by developing a full understanding of the *System Performance Specification (SPS)* operational requirements. During the formulation of the concept, the System Architect or LSE *collaborates* with various system *stakeholders* to fully understand and *validate* their operational needs and vision for the new system within the scope established by the contract and SPS.

Stakeholder analysis is accomplished in conjunction with a *mission analysis* of the Level 0 SYSTEM—the User's SYSTEM. The purpose of the analysis is to understand HOW the User envisions operating and maintaining the MISSION SYSTEM to achieve organizational objectives within the prescribed OPERATING ENVIRONMENT. The analysis should:

1. Identify types of missions and system use cases for each type of mission.
2. Address all aspects of MISSION SYSTEM operations including *pre-mission*, *mission*, and *post-mission* operations
3. Identify OPERATING ENVIRONMENT threats and *most likely* scenarios within each phase *and mode* of operation.

Mission analysis should be conducted in conjunction with *Understand and Bound the Problem and Solutions Space* step of the SE Process Model and Requirements Domain Solution methodology. The analysis should identify any *gaps* between the SPS or entity's *item development specification* requirements and WHAT *capabilities* and *characteristics* are required to successfully accomplish the mission and objectives.

If "gaps" exist between MISSION operational needs and documented requirements, confer with the entity's owner or program's Technical Director. As appropriate, the program has a professional obligation to inform the Acquirer and User regarding the gaps and provide a recommended *course of action* to resolve the "gaps" to support an informed decision by the Acquirer based on collaborative consultation with the User.

The *mission analysis* should also produce a *Mission Event Timeline (MET)* that establishes the time-based performance benchmark for allocating and assessing mission operations. The MET and operations should be *synchronized*, *modeled*, and *simulated* for resolution of conflicts. Below the SYSTEM level, internal timelines should be established and allocated as entity capability *performance budgets* and *margins*.

Most likely and *worst-case scenarios* anticipated to be encountered during the *pre-mission*, *mission*, and *post-mission* operations should also be *identified* and prioritized in terms of *likelihood* or *probability* of *occurrence*. These include:

1. MISSION SYSTEM interruptions and malfunctions.
2. SUPPORT SYSTEM maintenance and supply interruptions.
3. Hostile, worst-case NATURAL and INDUCED ENVIRONMENTS.

Step 2: Identify System Element Actors

Based on the *mission analysis*, assimilate and identify what types of *use case* operational *actors*—such as SUPPORT SYSTEM facilities, supply, training, and data—are required to support application of the MISSION SYSTEM to achieve organizational missions and objectives during all phases of operation—*pre-mission*, *mission*, and *post-mission*.

Step 3: Develop Actor-Based Operational Architecture

Using the actors identified above, FORMULATE an *operational architecture* that depicts the User's Level 0 SYSTEM implementation of the MISSION SYSTEM within the prescribed OPERATING ENVIRONMENT. Tools such as N × N (Refer to Figure 39.2) matrices and N^2 diagrams provide an excellent means of graphically depicting the interactions among operational assets.

Step 4: Develop System Operations Workflow Sequences

Given the operational architecture, develop an *operational* workflow that depicts the integrated MISSION SYSTEM and SUPPORT SYSTEM elements sequences required to:

1. *Prepare for, configure, train, and deploy* for a mission.
2. *Conduct* the mission.
3. *Recover* and *maintain* the MISSION SYSTEM following the mission.

For this activity, functional flow block diagrams (FFBDs), entity relationship diagrams (ERDs), and UML™ sequence diagrams and interaction diagrams are among the methods used to describe the overall mission workflow and operations. The Operations Domain Solution is captured as a series of concurrent, multi-level, integrated System Element operations—such as EQUIPMENT, PERSONNEL, and SUPPORT—that interact as a system to perform the system's mission and accomplish its objectives. Figures 18.1 and 18.2 serve as examples.

Author's Note 38.1 *Note that the description above represents an initial starting point. As the Behavioral and Physical Domain Solutions evolve and mature, downstream design decisions may force some rethinking and revision of the operations workflow. In some cases this may continue after system deployment when Users have had an opportunity to "find a better way or approach" to more efficiently and effectively perform mission operations. Obviously, we strive to "find a better way or approach" up front before we design the system, product, or service. Stakeholder* *assessments of models, simulations, prototypes, and demonstrations are excellent tools for ferreting out some of these operational issues during early system development.*

Step 5: Allocate Mission Operations to Phases of Operation

Based on the operational workflow sequences documented in Step 4, allocate each *operation* or *task* to the pre-mission, mission, and post-mission phases of operation. For each phase of operation:

1. Identify specific mission-related objectives to be accomplished.
2. Mark the beginning of each phase of operation with a key event.
3. Identify key milestones that represent critical *control, staging, or waypoints* to assess progress in accomplishing phase objectives.

Since *operations* and *tasks* may be *vague*, create an *Operations Dictionary* to document, scope, and identify the objectives, inputs/outputs, outcomes, and excepted results of each operation.

Step 6: Establish the Mission Event Timeline (MET)

The identification of critical *control, staging, or waypoints* in Step 5 leads to a key question. *For a given system phase and mode of operation, what are the SYSTEM performance time constraints that must be met to achieve the mission, phase, and mode objectives?* The answer to this question requires that we establish operational *performance budgets* and *design safety margins* derived from the *Mission Event Timeline (MET)* for each entity. The MET establishes the critical *staging points* where the MISSION SYSTEM and SUPPORT SYSTEM must be synchronized to:

1. Be configured, prepared, and in a *state of mission readiness*—or system availability—to conduct the mission—pre-mission.
2. Conduct the mission—mission.
3. Extract mission data and prepare and deploy the system for the next mission—post-mission.

For each of the phases of operation identified in Step 6, establish the *Mission Event Timeline (METs)* that depicts the sequences and timing of the phases of operation and their key milestones. Figures 16.1 and 49.2 serve as examples.

Referral For more information about *performance budgets and safety margins*, refer to Chapter 49 *System Performance Analysis, Budgets, and Safety Margins Practices*.

Step 7: Translate Mission Operations into System Use Cases and Scenarios

For each mission operation, identify system *use cases* that depict HOW the User envisions performing the operation. For each *use case*, identify the *most likely* and *worst-case scenarios* that are anticipated to occur. For each *use case*:

1. IDENTIFY the attributes discussed in Chapter 17 *System Use Cases and Scenarios*.
2. QUANTIFY the risk—*probability of* success times the *consequences of failure*.
3. PRIORITIZE for resource allocation.

Realistically, the expanse of potential or desired *use cases* is much greater than the available organizational development resources or contract value. Therefore, once the *uses cases* are identified, *prioritize* and *negotiate* to "fit" within your resource constraints.

Author's Note 38.2 *The previous point MUST be accomplished during the proposal phase of a system. Once the contract is awarded and your organization failed to reconcile the "cost to win" with "price to win" during the proposal effort, and signed up to everything, negotiation favors the Acquirer. Do yourself and your organization a favor, investigate these priorities during the proposal phase; negotiate PRIOR TO Contract Award.*

In general, it is *impractical* for the User, Acquirer, and System Developer to address every conceivable operational scenario that could occur. However, to ignore *most likely* scenarios leaves the SOI *vulnerable*, not only from a mission perspective but also the cost of the EQUIPMENT and PERSONNEL safety and lives. You can, however, plan and budget for the *most probable or likely scenarios*. The challenge is: *How do you identify these most probable or likely scenarios*?

The prioritized results should be discussed with User *stakeholders* and reconciled with the available budget. *The most probable or likely scenarios* that are determined to be within the scope of the contract should be documented. Once the *most probable or likely scenarios* are identified for each mode, link them to the system phases, modes, and use cases of operation.

Unanticipated or Unscheduled MISSION SYSTEM Events and Conditions. Depending on a SYSTEM's mission and application *use cases*, systems often encounter *unanticipated or unscheduled events and conditions* (scenarios, malfunctions, etc.) in their OPERATING ENVIRONMENT. Scenarios originate with: 1) the SOI, 2) NATURAL ENVIRONMENT, 3) INDUCED NATURAL ENVIRONMENT, or 4) external HUMAN-MADE SYSTEMS. From a *system safety*, *reliability*, *vulnerability*, and *survivability* perspective, the system should be designed with a *degree of robustness* that enables it to tolerate, respond to, and survive these situations. You may be asking the question: *WHY should we be concerned with use case based scenarios*? The answer lies in the response: *WHAT are the necessary and sufficient level of capabilities and robustness that will enable the system solution to respond and react to the most likely scenarios and achieve its required mission reliability*?

Author's Note 38.3 *Note the operative term "robustness." You will often hear System Developers refer to their "robust" design. But, what does it mean? In general, the term refers to the system's ability to cope with and tolerate internal and external failures as well as OPERATING ENVIRONMENT threats during the conduct of its mission and still accomplish most or all mission objectives.*

If your analysis reveals that certain *scenarios* are not within the scope of your contract, the System Developer has a *professional* and *ethical obligation* to inform and discuss the matter with the Acquirer via contracting protocol. A remedy should be negotiated, as appropriate.

Referral For additional information about system phases and modes of operation refer to Chapter 19 *System Phases*, *Modes*, and *States of Operation*.

Step 8: Identify the System Modes and States of Operation

Abstract *mission operations* and their *use cases* into modes of operation are illustrated in Figure 19.2. For each *mode of operation*, identify the triggering events (scenarios, interrupts, malfunctions, time-based events, etc.) that force a transition to the other modes of operation.

Author's Note 38.4 *It is very important to delineate the context of usage of modes of operation. For the User's Level 0 SYSTEM, phases and modes represent the operational aspects of employing the SYSTEM OF INTEREST as an element of the Level 0 system operational architecture.*

446 Chapter 38 Developing an Entity's Operations Domain Solution

As requirements are flowed down to the System Performance Specification (SPS), the System Developer may create a set of modes of operation unique to each of the deliverable system, product, or service physical configurations. The names of those modes may or may not be identically matched to the specification.

Finally, for each mode of operation, identify the system's operational states and conditions that are allowable. Examples include ON/OFF, loading, accelerating, braking, and inspecting.

Step 9: Derive System Capabilities from Use Cases and Scenarios

For each *use case* and its *most likely*, *probable*, and *worst case scenarios* within each mode of operation, derive the hierarchical set of system capabilities. Figure 38.2 provides an example. A spreadsheet matrix or a relational database should be used. Remember, since operational capabilities are *linked* to the *Requirements Domain Solution*, the linking of system capabilities to modes establishes traceability from SPS or entity *item development specification* (IDS) requirements to the modes of operation. We refer to these as *modal capabilities*.

Step 10: Develop the System Concept of Operations (ConOps)

Using the information developed in the steps above, develop a SYSTEM *Level Concept of Operations (ConOps)* to document HOW the MISSION SYSTEM and SUPPORT SYSTEM are envisioned to perform organizational missions. This includes developing the sequential and concurrent mission operations workflows. Some organizations refer to this effort as the *operational concept description (OCD), baseline concept description (BCD),* or simply ConOps.

At a *minimum*, the ConOps should address the following:

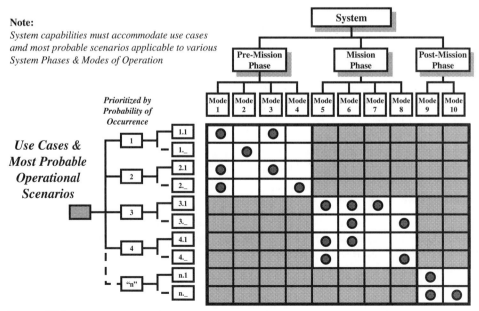

Figure 38.2 Modal Scenarios Influencing System Design Solution

1. System operational architecture.
2. MISSION SYSTEM actor roles, responsibilities, and authorities.
3. MISSION SYSTEM modes of operation, use cases, and use case scenarios.
4. SUPPORT SYSTEM actor roles, responsibilities, and authorities.
5. *Pre-mission*, *mission*, and *post-mission* operations workflows.
6. *Operations Dictionary* that identifies, scopes, and bounds each operation.
7. Mission Event Timeline (MET).

Referral For additional information about the ConOps, refer to Chapter 18 *System Operations Model*.

Step 11: Resolve Critical Operational and Technical Issues and Risks

As the Operations Domain Solution evolves, the System Developer inevitably encounters various *critical operational* and *technical issues* (COIs/CTIs), each with a level of *criticality* and *urgency*. Each *issue* and *risk* should be resolved quickly to enable lower level decision making to proceed. This includes:

1. Capturing in a database.
2. Linking to the respective *contract work breakdown structure (CWBS)* element.
3. Resolving, as appropriate, with the Acquirer and User via contracting protocol.

Step 12: Verify and Validate Operational Solution

Throughout the development of the Operations Domain Solution, the System Developer should continuously *verify* and *validate* the integrity of the evolving solution. This is accomplished via traceability to the Requirements Domain Solution and User source requirements. *Coordinate* and *collaborate* with the Acquirer and User—the stakeholders—as applicable, via technical interchange meetings (TIMS), technical reviews and the like.

As the Operations Domain Solution matures, conduct internal process reviews (IPRs) and technical reviews with stakeholders, prototypes, and demonstrations. Technical reviews such as the System Design Review (SDR), Hardware and Software Specification Reviews (HSRs/SSRs), Preliminary Design Reviews (PDRs), and Critical Design Reviews (CDRs) provide a forum for reviewing the Operations Domain Solution at various stages of development maturity.

Referral For additional information about the technical reviews, refer to Chapter 54 *Technical Review Practices*.

Step 13: Establish and Maintain the Baseline Concept Description (BCD)

Once the system ConOps is approved, the document and relevant supporting documentation—the operational architecture, modes and states of operations, and mission event timeline (MET)—are baselined by configuration management and incorporated into the evolving *developmental configuration*. The collection of information constitutes the Baseline Concept Description (BCD).

Author's Note 38.5 *Although the discussion above addressed the methodology in a linear step-by-step basis, these steps are highly iterative and have implicit feedback loops to preceding steps.*

Similar iterations occur between the Operations Domain Solution and the Requirements, Behavioral, and Physical Domain Solutions until the total system design matures.

38.4 OPERATIONS DOMAIN SOLUTION DEVELOPMENT CHALLENGES

When the Operations Domain Solution is developed, there are several challenges that the User, Acquirer, and System Developer(s) need to address. Examples include:

Challenge 1: ConOps Acceptance

Do the stakeholders agree with the *system Concept of Operations* (ConOps) document?

Challenge 2: Use Case Identification and Priorities

Do the stakeholders agree with: 1) the use cases and 2) use case priorities?

Challenge 3: Most Likely or Worst-Case Scenarios and Conditions

Do the system phases and modes of operation reflect the anticipated OPERATING ENVIRONMENT application and *most likely* or *worst-case* scenarios and conditions?

Challenge 4: "Gaps" in Operational Capabilities

Based on a review of the system phases, modes, and states of operation, are there any missing capabilities that have not been specified or included in the cost estimate?

Challenge 5: "Fitness-for-Use" or Acceptance Criteria

Are the capabilities allocated to system phases and modes of operation *necessary* and *sufficient* to achieve the mission, phase, and mode of operation objectives?

Challenge 6: Performance Timeline(s)

Does the Mission Event Timeline (MET) allocated to system phases and modes of operation pass the reasonableness test?

38.5 OPERATIONS DOMAIN SOLUTION WORK PRODUCTS

The Operations Domain Solution is documented via a series of *work products* that *represent* HOW the User intends to deploy and employ the MISSION SYSTEM to accomplish various organizational missions and scenarios. The overall Operations Domain Solution is documented initially in the *System Concept of Operations (ConOps)* document.

Author's Note 38.6 *Most people tend to limit their thinking of the Operations Domain Solution to the SYSTEM level. As requirements are allocated to lower levels by the respective entity/item development specification owners, EACH entity will have its own unique operational concept description (OCD) of how it will operate and support to fulfill the higher level entity's operations. For example, a PC board or movable part should have its own unique OCD or Theory of Operation that serves as the foundation for deriving the Behavioral and Physical Domain Solutions.*

38.6 GUIDING PRINCIPLES

In summary, the preceding discussions provide the basis to establish guiding principles that govern development of an entity's Operations Domain Solution.

Principle 38.1 The Operations Domain Solution is a derivative of the Requirements Domain Solution and serves as the foundation for developing the Behavioral Domain Solution.

Principle 38.2 The Operations Domain Solution must be *consistent with* and *traceable to* the Requirements, Behavioral, and Physical Domain Solutions.

38.7 SUMMARY

The development of the Operations Domain Solution is *highly iterative* with the Requirements, Behavioral, and Physical Domain Solutions as they evolve. As the Operations Domain Solution matures, it serves as the high-level abstraction from which the Behavioral Domain Solution is derived.

The Operations Domain Solution is:

1. Derived from the system capability and performance requirements established by the Requirements Domain Solution.
2. Establishes the operational architecture that expresses how system elements interact.
3. Establishes the system phases and modes of operation and associated *use cases*.
4. Creates the Mission Event Timeline (MET).
5. Maps required operational capabilities to the system phases and modes of operation.
6. Defines the system capability operations and tasks.
7. Allocates MET performance to system capability operations and tasks.

Based on an understanding of the Operations Domain Solution practices, we are now ready to discuss the development of the Behavioral Domain Solution practices.

NOTE

The *Unified Modeling Language* (UML®) is a registered trademark of the Object Management Group (OMG).

GENERAL EXERCISES

1. Answer each of the *What You Should Learn from This* Chapter questions identified in the *Introduction*.
2. Refer to the list of systems identified in Chapter 2. Based on a selection from the preceding chapter's General *Exercises* or a new system selection, apply your knowledge derived from this chapter's topical discussions. You have been hired as a consultant by an Acquirer to make recommendations for the system selected.
    (a) Develop a system operations model that captures HOW the system is envisioned to be operated based on interviews with the User.
    (b) Develop an outline for a ConOps to be used to describe the system operations model.
    (c) Describe how the *Operations Domain Solution* links to its *Requirements Domain Solution*.

ORGANIZATIONAL CENTRIC EXERCISES

1. Research your organization's command media. What guidance is provided regarding development of the *Operations Domain Solution*? Document and report your results.
2. Contact a contract program within your organization. Interview the Lead SEs and research HOW the program:
   (a) Formulated its *Operations Domain Solution*.
   (b) Manages the requirements baselines.
   (c) Links the *Requirements Domain Solution* to the *Operations Domain Solution*.
3. Select one of your organization's system specifications and develop a high-level *Operations Domain Solution*.

REFERENCE

MIL-STD-499B (cancelled draft). 1994. *Systems Engineering*. Washington, DC: Department of Defense (DoD).

ADDITIONAL READING

DI-IPSC-81430. 1994. *Operational Concept Description*. DoD Data Item Description (DID). Washington, DC: Department of Defense (DoD).

Chapter 39

Developing an Entity's Behavioral Domain Solution

39.1 INTRODUCTION

As the *Operational Domain Solution* evolves and matures, the next stage is to establish the entity's *Behavioral Domain Solution*. During this stage of SE design we elaborate HOW the User might envision the system sensing, reacting to, processing inputs, and responding to stimuli and cues internal and external to the system. We do this by identifying WHAT behavioral interactions occur between *Operational Domain Solution* elements to accomplish mission objectives. This stage also represents what may be the:

1. Most critical and neglected stage of SE design.
2. Source of many problems that do not surface until System Integration, Test, and Evaluation (SITE).

When new system development is started, engineers often begin at the *wrong end* of the system design solution—which is the *Physical Domain Solution*. They tend to prematurely focus on hardware and software designs, graphical user interfaces (GUIs), data communications rates, and hardware and software selection. They do this without stopping to understand:

1. HOW the system is expected to *react* and *respond* to operator and external stimuli.
2. HOW WELL the *responses are to be performed*.

As a design paradigm, this is *a model for disaster*. The design paradigm is exacerbated by *inexperienced, egotistical* decision makers who charge ahead and mandate that if you are not *"cutting metal, coding software, and soldering PC board components"* the FIRST week after Contract Award, the program is behind schedule.

Author's Note 39.1 *The design paradigm noted above can be characterized as a "hobby shop" approach. There are well-founded instances where the paradigm may be valid—as minor modifications to existing equipment. The context of this observation relates to complex, medium- to large-scale system development programs. Well-disciplined and experienced SEs understand the fallacies and pitfalls of this approach and understand how to tailor the methodology to fit within program constraints without diluting the integrity of the methodology.*

The message here is do the job RIGHT "up front" for $X+ or pay $2X+ and consume 2X+ schedule attempting to *correct* these *premature* and *immature* approaches to system develop-

---

System Analysis, Design, and Development, by Charles S. Wasson
Copyright © 2006 by John Wiley & Sons, Inc.

ment. The decision ultimately comes down to *"Are you running a hobby shop for amateurs or a highly efficient and effective performing engineering organization?"*

This chapter expands on the *Behavioral Domain Solution* description provided in our discussion of the *System Solution Domains*. Our discussion addresses the components of the *Behavioral Domain Solution*: objectives, key elements, sequence in the SE Process Model workflow, development responsibility, dependencies, development methodology, challenges, and work products.

What You Should Learn from This Chapter

1. What is the *objective* of the Behavioral Domain Solution?
2. What are the *key elements* of the Behavioral Domain Solution?
3. What is the *relationship* of the Behavioral Domain Solution to the SE Process Model?
4. Explain the *relationship* of the Behavioral Domain Solution to the Requirements, Operations, and Physical Domain Solutions?
5. What is the *relationship* between specification requirements and capabilities?
6. What is the *methodology* used to develop the Behavioral Domain Solution?
7. What are the *work products* that represent the Behavioral Domain Solution?
8. How do you *verify* and *validate* the Behavioral Domain Solution?

Definitions of Key Terms

- **Transaction** "In software engineering, a data element, control element, signal, event, or change of state that causes, triggers, or initiates an action or sequence of actions." (Source: IEEE 610.12-1990)
- **Transaction Analysis** "A software development technique in which the structure of a system is derived from analyzing the transactions that the system is required to process." (Source: IEEE 610.12-1990)

Behavioral Domain Solution Objective

The objective of the *Behavioral Domain Solution* development activity is to *conceptualize*, *formulate*, and *translate* the *Operations Domain Solution* use cases, scenarios, and capabilities into a set of behavioral capabilities, interactions, and responses that support all *phases of operation*—pre-mission, mission, and post-mission.

39.2 KEY ELEMENTS OF THE BEHAVIORAL DOMAIN SOLUTION

The Behavioral Domain Solution is characterized by several elements that provide the basis for formulating system behavioral *reactions* and *responses*. These include scenarios, stimuli, capabilities, resources, constraints, and responses. Capabilities are implemented by one or more system operations that may consist of at least two or more system tasks. Each system operation and task:

1. Communicates via one or more interactions.
2. Is bounded by at least one or more performance budgets and safety margins.

Referral For more information about performance budgets and safety margins, refer to Chapter 49 *System Analysis, Performance Budgets, and Safety Margins Practices*.

The Need for the Behavioral Domain Solution

You may ask: *why do we need a Behavioral Domain Solution?* The answer lies in understanding HOW System Developers, as humans, approach design problems.

As humans, we have differing levels of training, knowledge, and experience. Within a design team, some members focus on WHAT the system is required to accomplish—its *behavior*—while others are focused on HOW a circuit will be designed, WHAT data exchange protocol to use, or WHAT software operating system language is to be used—its *implementation*.

As an SE, you have to exhibit leadership and establish control of these wanderings and diversions; otherwise, frustration, chaos, and havoc can prevail. So get the group focused on one task at a time. First, understand WHAT the system is required to accomplish via the *Behavioral Domain Solution*. Then, as the *Behavioral Domain Solution* gains some maturity, shift to the *second task*—concurrent development of the *Physical Domain Solution* that addresses HOW to implement the Behavioral Domain Solution via physical components.

Behavioral Domain Solution Dependencies

The multi-level Behavioral Domain Solution is *highly iterative* and requires close *coordination* and *collaboration* with the Requirements, Operations, and Physical Domain Solutions as they as shown in Figure 26.3. *Iterations* are also required to ensure that the SYSTEM OF INTEREST (SOI) is optimally balanced from an overall development and life cycle perspective of technical, cost, schedule, support, and risks.

Behavioral Domain Solution Development Sequencing

The Behavioral Domain Solution is a multi-disciplined activity lead by the entity's Lead SE. The Behavioral Domain Solution evolves slightly behind but concurrently with the Requirements and Operations Domain Solutions and ahead of the Physical Domain Solution, as shown in Figure 23.2.

Behavioral Domain Solution Development Responsibility

Responsibility for the Behavioral Domain Solution resides with the program's Technical Director or Project Engineer and is usually delegated to a Lead SE for the program. The Lead SE facilitates the development of the Behavioral Domain Solution by ensuring that all *operational* and *disciplinary* stakeholder interests are *represented* in the *logical/functional architecture* and *its description*.

As key elements of the Behavioral Domain Solution are *approved* and *released*, Configuration Management (CM) administers formal change management in accordance with direction from a Configuration Control Board (CCB) or Software Configuration Control Board (CCB) of SYSTEM or entity *stakeholders*.

39.3 DEVELOPING THE BEHAVIORAL ARCHITECTURE

As the SYSTEM or entity's operational architecture *evolves* and *matures*, the next step is to understand HOW the SYSTEM and its *entities*—the UML® actors—*interact* and *respond* to various operator and external system *stimuli* and *cues*. The foundation for this understanding begins with:

1. The entities identified in the operational architecture.
2. Capabilities specified in its specification—such as *System Performance Specification (SPS)* or *item development specification*.

At this stage of the SYSTEM or entity's development, the entity is simply an *abstraction*. As an analytical abstraction, we refer to it as having *logical* or *associative* relationships. The challenge here is creating a logical/functional framework that captures these relationships. We refer to it as the *logical* or *functional architecture*.

Foundation for the Logical Architecture

The *logical* or *functional architecture* exposes the key, multi-level logical or functional capabilities and interrelationships between the *Operations Domain Solution* elements. The logical/functional architecture provides a framework to depict the behavioral *interactions* of HOW the SYSTEM or entity responds to its OPERATING ENVIRONMENT for various missions and scenarios.

As the SYSTEM or entity processes inputs, it exhibits behavioral patterns and characteristics. To accomplish the internal processing, a SYSTEM or entity provides one or more capabilities to transform the set of inputs into *behavior, products, by-products,* and *services*. If the system is required to interact with *friendly, benign,* or *hostile* external entities in its OPERATING ENVIRONMENT, additional capabilities ensure a level of *interoperability* with or *survivability* during interactions with those entities.

Referral For more information about system interactions, refer to Chapter 15 System Interactions with Its Operating Environment and Chapter 43 *System Interface Analysis, Design,* and *Control Practices.*

As we analyze this environment, we discover there are *virtual* and *physical* system entities or *actors* (UML®)—namely *people, capabilities, operations, functions, events, roles, constraints,* and *controls*—that *interact* with each other or represent the *behavior, products, by-products,* or *services* produced by the system. We generically classify *people, capabilities, operations, functions, events, roles, constraints,* and *controls* as *objects, entities, or actors* within the SYSTEM. In general, the actors, objects, or entities are identified as *nouns*—the same as persons, places, or things.

The logical/functional architecture portrays WHAT *relationships* and *interactions* are required to contribute capabilities to enable the SYSTEM to perform its missions. Note that this characterization DOES NOT address HOW these *objects* and *interactions* will be physically implemented; the Physical Domain Solution addresses HOW via its physical architecture. As the cliché goes, *function* always precedes *form* and *fit*.

Logical/Functional Architecture Development Methodology

We have established that the logical/functional architecture exposes the objects, entities, or actors that exist within a given system *level of abstraction* and class entity—such as PRODUCT SUBSYSTEM, and ASSEMBLY and their interdependency relationships. The question is: *HOW do we develop the architecture?*

One approach is to employ a simple methodology that consists of the steps listed below:

Step 1: Identify logical objects or entities.
Step 2: Identify each entity's capabilities.
Step 3: Create a logical interactions matrix.
Step 4: Create the logical/functional architecture.

Let's elaborate further on each of these steps.

Step 1: Identify Logical Objects or Entities. The first step in developing a *logical or functional architecture* is to simply identify the SYSTEM or entity's required capabilities; create a list.

39.3 Developing the Behavioral Architecture

Then, translate the list into a hierarchy tree similar to the one shown in Figure 39.1, which illustrates a Car-Driver SYSTEM.

Step 2: Identify Each Entity's Capabilities. Once you identify the SYSTEM or entity's key capabilities, the next step is to establish their interrelationships. The method used depends on the size and complexity of the SYSTEM or entity. *If you attempt to start with system block diagram SBD, the leap may be too challenging.* You may discover that you are not only contending with the relationships but also the graphics. Interconnecting lines may be criss-crossing everywhere, making the diagram unnecessarily complex and confusing.

Step 3: Create a Logical Interactions Matrix. So, *how do we model system logical/functional and interactions*? First, we need to understand which capabilities have *logical associations*. We can do this by creating a matrix as shown in the upper left section of Figure 39.2. The $N \times N$ or N^2 matrix allows us to link one capability to another as well as system stimuli—the inputs—and responses—the outputs. Capability A has a *logical entity relationship* to Capability B, B to C, and so forth. For large, complex systems, a summary matrix is invaluable as a foundation for the next step.

Author's Note 39.2 *Although it appears to create an extra step, the matrix serves as an important analytical pairwise comparison method. Under the pairwise method we focus on one element and then determine if it has entity relationships with each of the remaining elements across a row of the matrix. The next step involves creating a graphical rendering of the logical architecture for presentation purposes. If we had attempted to create the next step first, we would have lacked a clean method of ensuring every interaction is considered, which is the strength of the matrix and N^2 approaches.*

Step 4: Create the Logical/Functional Architecture. Once the logical interaction matrix is established, we translate the interactions into a graphic referred to as an N^2 diagram as shown in the lower right portion of Figure 39.2. Each of the capabilities represents a multi-level set of

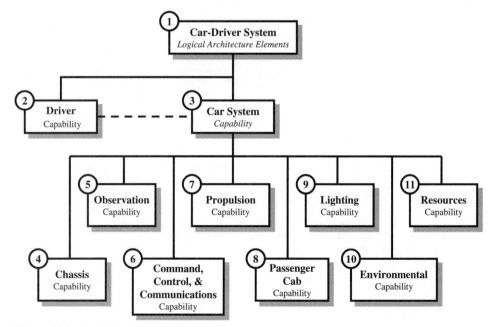

Figure 39.1 Car-Driver System Logical Elements

456 Chapter 39 Developing an Entity's Behavioral Domain Solution

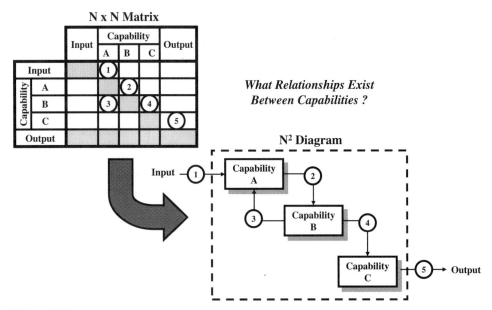

Figure 39.2 Behavioral Solution Capability Interactions

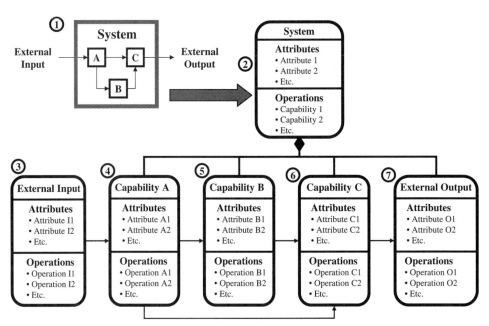

Figure 39.3 Object-Oriented Representation of Logical Entities

operations or *tasks* that perform the actions required to produce the *expected outcome* and *levels of performance*. Both the N × N and N^2 methods enable us to focus on *interactions* without being distracted with "packaging" the graphic.

An alternative to the N^2 method is the object-oriented (OO) approach shown in Figure 39.3. This method illustrates each capability's *name*, *attributes*, and *operations*. Similar annotations could be added to the N^2 diagram shown in Figure 39.2.

Guidepost 39.1 *At this point we have introduced HOW to construct the basic logical/functional architecture. The key here is identifying the primary capabilities of the SYSTEM or entity. HOW we identify these elements depends on whether a system is precedented or unprecedented—thereby avoiding unnecessarily analyses if the basic architectural elements are known.*

Precedented System Logical/Functional Architecture Development

As discussed in Part I, most human-made systems are *precedented* systems. Their proven architectures and technologies improve on existing designs. Consider the following example:

EXAMPLE 39.1

As a *precedented* system, a car has a chassis, engine, and so forth. Although you need to avoid *paradigms* that limit creative design solutions and help us to "think outside the box," we do not have to functionally re-derive a car's fundamental design—create the chassis, engine, and body from a blank piece of paper. We simply defer to a conventional vehicle architecture as an *initial starting point*.

Hypothetically, if you had two control groups and tasked each to create the *logical architecture* for the same system, the graphics may and probably will be different. *Is one architecture RIGHT and the other WRONG*? Generally, no. ALWAYS have multiple approaches to solving the same class of engineering problem. We see this in commercial products such as automobiles, computers, and TVs. In general, they fill similar *market solution space* abstractions—in this example, transportation, steering, and the like.

Unprecedented System Logical/Functional Architecture Development

In contrast, *unprecedented systems* may require new "*thinking outside the box*" approaches to a SYSTEM's logical/functional architecture. Consider the following example:

EXAMPLE 39.2

Although the NASA had previously landed unmanned probes on the Moon, the Apollo Lunar Lander was a *first* in the design of a manned lunar landing and launch craft.

EXAMPLE 39.3

NASA's Skylab program represented a first for human habitation in a space-based laboratory environment, particularly in performing human tasks in a weightless environment for an extended period of time.

Where *unprecedented systems* are the case, the development team derives the logical/functional architecture from *abstract* mission or scientific objectives—namely "*Do great and wonderful science.*" In these cases SEs, in collaboration with principal investigators, *derive* and *extract* the SYSTEM's capabilities. Some Acquirers may initiate study contracts or *proof of concept* demonstration contracts, deriving SYSTEM capabilities via a series of "down select" contracting stages.

Modeling the Logical/Functional Architecture

Logical/functional architectures employ *model-based representations* via system block diagrams (SBDs), object diagrams, among other graphical methods, to expose SYSTEM or entity embedded elements and their *relationships*. However, SBDs DO NOT *expose* the *logical* sequences of *data exchanges* or *behavioral interactions* that occur between *actors* or *objects*. Therefore, behavioral modeling tools are employed, in combination with timelines and performance budget and safety margin allocations, to evaluate the overall *effectiveness* of each candidate architecture solution. Behavioral modeling tools include N^2 diagrams, UML® interaction diagrams, functional flow block diagrams (FFBDs), and *control* and *data flows*.

You may ask: *WHY do we need to create a model of a logical/functional architecture*? Logical/functional architecture modeling and its internal capability processing are crucial for:

1. Driving out or exposing undiscovered requirements.
2. Establishing SYSTEM or entity *performance budgets and margins*.
3. "Load" balancing to optimize system performance.
4. *Allocating* and *flowing down* requirements to lower levels via *item development specifications*.

Referral   For more information about *performance budgets and safety margins*, refer to Chapter 49 *System Analysis*, *Performance Budgets*, and *Design Safety Margins*.

Referral   For more information about *modeling architectures*, refer to Chapter 20 *Modeling System* and *Support Operations Practices* and Chapter 51 *System Modeling* and *Simulation*.

Final Thoughts

Logical entity relationships enable us to communicate the *associations* or *connectivity* required between two or more logical entities based on a *validated* operational *need*. Since these are *need-based interactions*, the emphasis is on:

1. WHO *interacts* with WHOM.
2. WHAT is to be *interacted* or *communicated*.
3. WHEN the behavioral interactions occur.
4. Under WHAT conditions.

Based on this discussion, let's describe the basic methodology used to develop the Behavioral Domain Solution. Tailor the methodology to fit your specific business needs.

Guidepost 39.2   *At this juncture, we have described HOW to create the logical/functional architecture that serves as the framework for the Behavioral Domain Solution. We now shift our focus to the methodology for creating the overall Behavioral Domain Solution for a SYSTEM or entity.*

39.4 BEHAVIORAL DOMAIN SOLUTION DEVELOPMENT METHODOLOGY

The Behavioral Domain Solution is developed as a key element of the SE Process Model. The *highly collaborative* and *iterative* convolution of the Behavioral Domain Solution with the Requirements, Operations, and Physical Domain Solutions is often very *chaotic* and *confusing*.

39.4 Behavioral Domain Solution Development Methodology

We can *minimize* the *chaos* and *confusion* by applying an *iterative* methodology that enables us to create the Behavioral Domain Solution. View these steps as an example strategy and tailor them to suit your business domain and systems application.

Referral   For additional information about the SE Process Model, refer to Chapter 26 *SE Process Model*.

Author's Note 39.3   *The Behavioral Domain Solution is one area where there are multiple approaches and methods for defining the solution and its work products. There seems to be no consensus regarding a RIGHT approach or method; we all have our own approach or recipe.*

Our discussions here are intended to focus on the strategy or methodology of Behavioral Solution Development. The takeaway from this discussion should be key insights as to HOW you might approach developing a Behavioral Domain Solution. Then, formulate an approach that works for you, your team, and system development application, and apply the appropriate tools to support your methodology.

The methodology consists of the following steps:

Step 1:   Establish the multi-mode logical architecture.
Step 2:   Model mode-based system interactions.
Step 3:   Allocate entity performance budgets and design safety margins.
Step 4:   Analyze system failure modes and effects.
Step 5:   Evaluate and optimize system behavioral performance.
Step 6:   Resolve critical operational and technical issues (COIs/CTIs).
Step 7:   Verify and validate behavioral domain solution.
Step 8:   Establish and maintain the entity behavioral solution baseline.

Let's elaborate further on each of these steps.

Step 1: Establish the Multi-Mode Logical Architecture

The first step in developing the *Behavioral Domain Solution* is to understand the *entity relationships* between system capabilities. The relationships and their associated interactions provide the basis to construct a framework referred to as the logical/functional architecture. Based on the MISSION SYSTEM, SUPPORT SYSTEM, and OPERATING ENVIRONMENT interactions, *formulate* a logical architecture that *integrates* mode-based capabilities, behavior, and interactions. If appropriate, identify alternative logical architectures. *Model*, *simulate*, *evaluate*, and *conduct trade-offs* of the set of architectures. Select the logical architecture that best *represents* the proper balance of system performance for all modes of operation. Figure 31.2 provides a high level example of a logical architecture that captures associative relationships between system capabilities.

Step 2: Model Mode-Based System Interactions

Using the high-level *interactions* as a foundation, capture the MISSION SYSTEM's interactions as a function of phase and mode of operation. Based on each mode's use cases, model the behavioral interactions "thread" from input to output through the system's capabilities as shown in Figure 39.4.

460 Chapter 39 Developing an Entity's Behavioral Domain Solution

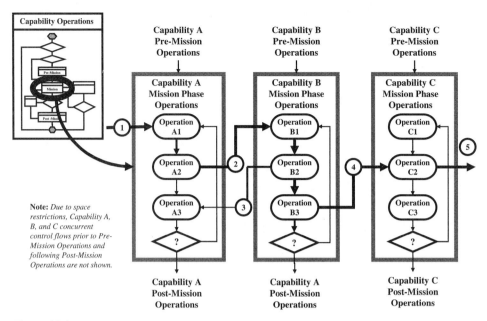

Figure 39.4 Capability Interactions

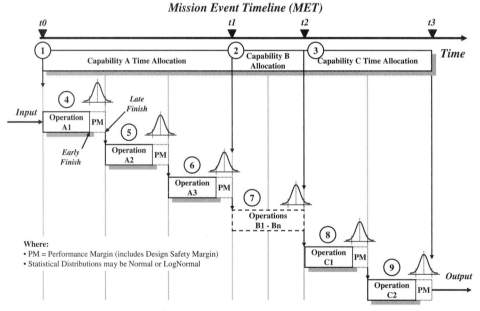

Figure 39.5 Capability "Thread" Time Performance Allocations

Step 3: Allocate Entity Performance Budgets and Safety Margins

For each *phase*, *mode*, *use case*, and *scenario* of operation, link capability interactions and performance to time-constrained performance—such as the *Mission Event Timeline (MET)*. Allocate performance BUDGETS and design safety MARGINS to each logical entity. Include timing performance allocations derived from the *Mission Event Timeline (MET)*, as shown in Figure 39.5.

Step 4: Analyze System Failure Modes and Effects

For those systems that are *mission critical* or may have the potential to damage EQUIPMENT, PERSONNEL, the public, or the environment, conduct a Failure Modes and Effects Analysis (FMEA). For *mission critical* elements that may involve hazards that are high risk, expand the FMEA into a Failure Modes, Effect, and Criticality Analysis (FMECA). Former Mil-Std-1629A *Military Standard Procedures for Performing a Failure Modes, Effects, and Criticality Analysis* provides guidance.

Step 5: Evaluate and Optimize System Behavioral Performance

Evaluate and *optimize* Behavioral Domain Solution performance relative to the SPS or each entity's *item development specification* requirements. For each phase and mode of operation, *validate* that each mode's capabilities and performance interactions will withstand and survive the *most likely* and *worst case scenarios* subject to technical, technology, cost, and schedule constraints.

Step 6: Resolve Critical Operational and Technical Risks and Issues

As the Behavioral Domain Solution evolves, *identify*, *clarify*, and *resolve critical operational* and *technical issues* (COIs/CTIs) and risks. Link the issues and risks to the appropriate CWBS elements and specification requirements. When and where appropriate, collaborate with the Acquirer and User to resolve the COIs/CTIs via contracting protocol.

Step 7: Verify and Validate Behavioral Domain Solution

Throughout the development of the Behavioral Domain Solution, *continuously verify* and *validate* the *completeness* and *integrity* of the evolving solution via document reviews, technical reviews, prototypes, technology demonstrations, and modeling and simulation. Verify that all aspects of the Behavioral Domain Solution are *traceable* back to the *source* or *originating* requirements as documented in the *System Requirements Document (SRD)*—by way of a formal *Request for Proposal (RFP)* solicitation—or in the contract, as applicable via the system's hierarchical specification tree.

Step 8: Establish and Maintain the Entity Behavioral Solution Baseline

As the SYSTEM and each entity's Behavioral Domain Solution are approved, establish a formal baseline to serve as the basis for requirements *allocation* and *flow down*, future technical decision making, and change control. Incorporate the Behavioral Domain Solution and its updates into the evolving *Developmental Configuration*.

39.5 BEHAVIORAL DOMAIN SOLUTION DEVELOPMENT CHALLENGES

When the Operations Domain Solution is developed, there are several challenges that the User, Acquirer, and System Developer(s) need to address. Examples include:

Challenge 1: Requirements traceability
Challenge 2: Stakeholder collaboration
Challenge 3: Stakeholder reviews

Challenge 4: Critical issue risk mitigation
Challenge 5: Baseline management
Challenge 6: Realism
Challenge 7: Behavioral solution system description
Challenge 8: CWBS traceability

Let's explore some of the key questions that represent challenge to developing the Behavioral Domain Solution.

Challenge 1: Requirements Traceability

Is the Behavioral Domain Solution traceable to the following?

1. *Requirements Domain Solution System* documented in *Performance Specification (SPS)* and development specifications.
2. *Operations Domain Solution* documented in System *Concept of Operations (ConOps)*, *Mission Event Timelines (METs)*, system phases and modes of operation, etc.
3. *Physical Domain Solution* documented in Physical system architecture, Bills of Materials (BOMs) etc.

Challenge 2: Stakeholder Collaboration

Did you consult and collaborate with key stakeholders in developing the Behavioral Domain Solution?

Challenge 3: Stakeholder Reviews

Have the key stakeholders reviewed and approved, as appropriate, portions or all of the Behavioral Solution?

Challenge 4: Critical Issue Risk Mitigation

Have all critical operations, technical, technology, support, cost, and schedule risks been identified and mitigated?

Challenge 5: Baseline Management

Have the Behavioral Domain Solution work products been incorporated into the evolving Allocated Baseline of the Developmental Configuration?

Challenge 6: Realism

Can the Behavioral Solution implementation be *realistically achieved* with physical components and technologies with an *acceptable* level of risk within the available cost and schedule constraints?

Challenge 7: Behavioral Solution System Description

Has the Behavioral Domain Solution been *adequately* documented in the *System/Subsystem Design Description (SSDD)* or *Interface Design Description (IDD)* to a level of detail that permits procurement or development and maintenance of the system or entity?

Challenge 8: CWBS Traceability

Are the elements of the Behavioral Domain Solution traceable to the *Contract Work Breakdown Structure (CWBS)* elements?

Behavioral Domain Solution Work Products

Key *work products* of the *Behavioral Domain Solution* consist of:

1. Interaction diagrams for multi-level system capabilities.
2. UML® sequence diagrams representing processing capabilities.
3. A logical or functional architecture—such as FFBDs and ERDs.
4. Capability performance timelines.
5. Allocations of time-based performance to operations and tasks required to implement the capability.

39.6 GUIDING PRINCIPLES

In summary, the preceding discussions provide the basis with which to establish the guiding principles that govern development of an entity's Behavioral Domain Solution practices.

Principle 39.1 The Behavioral Domain Solution is a derivative of the Operations Domain Solution and provides the foundation for developing the Physical Domain Solution.

Principle 39.2 The Behavioral Domain Solution must be *consistent with* and *traceable* to the Requirements, Operations, and Physical Domain Solutions.

39.7 SUMMARY

Our discussions of the *Behavioral Domain Solution* highlighted the need for and the evolution of the Behavioral Domain Solution and its logical architecture. We noted the importance of the following:

1. Identify the logical entities that provide system capabilities.
2. Understand WHAT capabilities the logical entities are intended to provide.
3. Express HOW the logical entities *relate* and *interact*.
4. Link logical performance to system performance metrics.

This solution, as typical of the others, progresses to maturity over time. This is accomplished by a lot of *collaborative iterations*, *interactions*, and *negotiations* with the Requirements, Operations, and Physical Domain Solutions.

Based on this discussion and understanding, we are now ready to address the *Physical Domain Solution*.

NOTE

The *Unified Modeling Language* (UML®) is a registered trademark of the Object Management Group (OMG).

GENERAL EXERCISES

1. Answer each of the *What You Should Learn from This Chapter* questions identified in the *Introduction*.
2. Refer to the list of systems identified in Chapter 2. Based on a selection from the preceding chapter's General *Exercises* or a new system selection, apply your knowledge derived from this chapter's topical discussions. You have been hired as a consultant by an Acquirer to make recommendations for the system selected.
   (a) Identify the actors of the system's Behavioral Domain Solution.
   (b) Develop the system's logical or functional architecture.
   (c) Create a matrix to captures the logical entity relationships and interactions.
   (d) Develop high-level *interaction diagrams* and *sequence diagrams* that depict actor interactions.

ORGANIZATIONAL CENTRIC EXERCISES

1. Research your organization's command media. What guidance is provided regarding developing of the *Behavioral Domain Solution*? Document and report your results.
2. Contact a contract program within your organization. Interview the lead SEs and research HOW the program:
   (a) Formulated its *Behavioral Domain Solution*.
   (b) Documented the *Behavioral Domain Solution*.
   (c) Links the *Behavioral Domain Solution* to the *Requirements, Operations, and Physical Domain Solutions*.
3. Select one of your organization's system specifications and create a high-level *Behavioral Domain Solution*.

REFERENCES

IEEE Std 610.12-1990. 1990. *IEEE Standard Glossary of Modeling and Simulation Terminology.* Institute of Electrical and Electronic Engineers (IEEE). New York, NY.

MIL-STD-1629A (Cancelled) 1984. *Military Standard Procedures for Performing a Failure Mode, Effects, and Criticality Analysis.* Washington, DC: Department of Defense (DoD).

ADDITIONAL READING

BUEDE, DENNIS M. 2000. *The Engineering Design of Systems: Models and Methods* New York: Wiley.

LANO, R.J. 1977. *The N^2 Chart. System and Software Requirements Engineering.* IEEE Computer Society Press.

Chapter 40

Developing an Entity's Physical Domain Solution

40.1 INTRODUCTION

Based on the preceding discussions of the Requirements Domain Solution, Operations Domain Solution, and Behavioral Domain Solution, we have:

1. Established a SYSTEM/entity's required operational capabilities.
2. Developed a *system Concept of Operations (ConOps)* describing its system phases and use cases or modes of operation based on its System Operations Model.
3. Developed behavioral capability interactions and defined their relationships and time/performance based interactions.

The question now is: *HOW do we implement the* Behavioral Domain Solution *via physical components* such as PRODUCTS, SUBSYSTEMS, and ASSEMBLIES? The answer resides in the *Physical Domain Solution.*

The *Physical Domain Solution* represents:

1. A *preferred solution* selected from a set of viable candidate solutions.
2. The *physical implementation* of the Behavioral Domain Solution in terms of PERSONNEL, FACILTIES, EQUIPMENT, and PROCEDURAL DATA.

When operating, the Physical Domain Solution *validates* the *correctness* of engineering *form, fit,* and *function assumptions* and *decisions*. Though analytical up to this point, SEs for the User, Acquirer, and System Developer must ask the following types of system delivery questions:

1. VERIFICATION *Did the System Developer inplement the SYSTEM/entity RIGHT (i.e., correctly) in compliance with its specifications and design requirements?*
2. VALIDATION *Did we, the Acquirer, specify and bound the RIGHT product for the User's operational need?*

This chapter addresses the Physical Domain Solution: objectives, key elements, sequence in the SE Process Model workflow, development responsibility, dependencies, development methodology, challenges, and work products.

System Analysis, Design, and Development, by Charles S. Wasson
Copyright © 2006 by John Wiley & Sons, Inc.

What You Should Learn from This Chapter

1. What is the *objective* of the Physical Domain Solution?
2. What are the *key elements* of the Physical Domain Solution?
3. What is the *relationship* of the Physical Domain Solution to the SE Process Model?
4. What is the *relationship* of the Physical Domain Solution with the Requirements, Operations, and Physical Domain Solutions?
5. What is the *relationship* between specification requirements and capabilities?
6. What is the *methodology* used to develop the Physical Domain Solution?
7. What are the *work products* that represent the Physical Domain Solution?
8. How is the Physical Domain Solution is *verified* and *validated*?

Definitions of Key Terms

- **Component** A physical entity that *implements* one or more *logical entities* and its *capabilities* to produce SYSTEM RESPONSES.
- **Physical Architecture** A multi-level configuration used that identifies the physical components of the system, how they interconnect, and contribute to produce *behavior*, *products*, *by-products*, and *services* to fulfill the system's mission and accomplish its objectives.

As a generic entity, an *item* applies to: 1) externally procured products, such as commercial off-the-shelf (COTS) items and nondevelopmental items (NDI), and 2) *internally* developed systems or products, such as configuration items (CIs) or modified COTs/NDIs.

Referral For more information about items, refer to Chapter 42 *System Configuration Identification Practices*.

Author's Note 40.1 *Since most people are comfortable with the term, "component," we use it here as a generic representation of any physical item such as a PRODUCT, SUBSYSTEM, or ASSEMBLY. Later, in our system configuration identification discussion, we use the term "item" recognized by Configuration Management when referring to generic physical entities. For now, component should be adequate.*

40.2 KEY ELEMENTS OF THE PHYSICAL DOMAIN SOLUTION

The Physical Domain Solution has several key elements that provide the basis for selecting physical components. The physical system architecture is characterized by *one-to-many* relationships with *Standard Operating Practices and Procedures (SOPPs)*, physical elements, constraints, inputs, and behavioral responses and resources.

Physical Domain Solution Development Activity Objectives

The objectives of the Physical Domain Solution development activity are to:

1. *Select* a physical architecture from a set of viable alterative solutions.
2. S*elect* physical components to satisfy architecture element requirements.
3. *Translate* component requirements into detailed design requirements that can be used to *procure*, *develop*, or *modify* the components.

4. *Optimize* the performance of the preceding items to achieve an optimal balance of technical, cost, and schedule performance with acceptable risk to the stakeholders.

Based on this introduction, let's begin our discussion with the key elements of the Physical Design Solution.

Physical Domain Solution Dependencies

The multi-level Physical Domain Solution is *highly iterative*. It requires close *coordination* and *collaboration* with the Requirements, Operations, and Behavioral Domain Solutions as they as shown in Figure 26.3. *Iterations* are also required to ensure that the SYSTEM OF INTEREST (SOI) is properly balanced (i.e., *optimal*) from an overall development and life cycle perspective in terms of technical cost, schedule, support, and risks.

Physical Domain Solution Development

The Physical Domain Solution evolves slightly behind but *synchronized* with the Requirements, Operations, and Behavioral Domain Solutions as shown in Figure 23.2. This solution ultimately defines the *physical* system, product, or service that is to be *verified* and delivered in accordance with the *Terms and Conditions (Ts&Cs)* of the contract and its *System Performance Specification (SPS)* or an entity's *item development specification (IDS)*.

Physical Domain Solution Development Responsibility

The System Architect or Lead SE *collaborates* with the System Engineering and Integration (SEIT) members to formulate and select the *Physical Domain Solution* from a set of viable candidate alternatives. The System Architect, via the SEIT, oversees the development of the *Physical Domain Solution* and its integration with the Requirements, Operations, and the Behavioral Domain Solutions.

40.3 DEVELOPING THE PHYSICAL ARCHITECTURE

As the Behavioral Domain Solution and its logical/functional architecture evolve and mature, SEs direct their attention to formulating and selecting a *physical architecture*. Physical architectures *represent* the transition *point* from the abstract analytical world to the physical world. *Whereas the Behavioral Domain Solution focuses on WHAT is the SYSTEM or entity required to accomplish, the Physical Domain Solution focuses on HOW to identify the physical configuration and entities to implement WHAT is to be accomplished.*

Formulating and Selecting Candidate Physical Architectures

In general, as discussed in Chapter 39 about the logical/functional architecture, the physical architecture is selected from a set of *viable* candidate architectures based on a pre-defined set of *decision criteria* and *weights*. The previous statement sounds simple; however, it may be *misleading* in terms of HOW the physical architecture is selected. *How so?*

The reality is the physical architecture should *represent* the best value, acceptable risk, *minimal* life cycle support costs approach to fulfilling the *System Performance Specification (SPS)* or entity's *item development specification* (IDS) requirements.

Developing the Product Structure Tree. The first step in developing the physical architecture is to identify generic physical items that comprise the SYSTEM or entity. Create a Product

468 Chapter 40 Developing an Entity's Physical Domain Solution

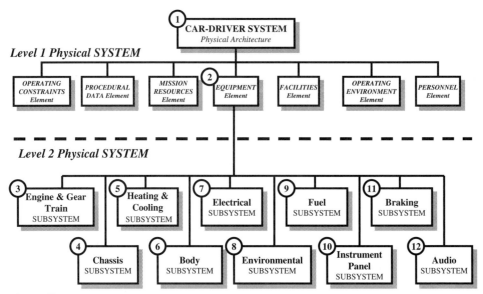

Figure 40.1 System Element-Based Physical Architecture

Structure hierarchy tree such as the one shown in Figure 40.1 for a Car-Driver System. Although SEs often generically refer to this as a hierarchy tree, the proper designation is the SYSTEM'S *Product Structure*. Ultimately, the *Product Structure* captures the hierarchical *pieces and parts* of the physical item.

When you develop the *Product Structure*, a good *rule of thumb* is to avoid having more than six to eight components at any level of abstraction. If you have more than this number, you may have inadvertently mixed components from different levels of abstraction. Tailor appropriately to your own system application.

Author's Note 40.2 *Several things should be noted here.*

- *Figure 40.1 represents the "first pass" at a physical architecture structure, an initial starting point.*
- *Once the make/buy/modify decisions are made, physical items from the initial hierarchical structure may be combined or separated as peer level items on subsequent passes and shown appropriately in the physical architecture diagram.*

Developing the Physical Architecture SBD. As the *Product Structure* evolves, create the physical architecture's system block diagram (SBD). Physical items at the SYSTEM level or entities within a given level of the Product Structure are linked via physical entity interfaces to their OPERATING ENVIRONMENT.

Linking the CWBS to the Physical Architecture

As the physical architecture evolves, develop the *Contract Work Breakdown Structure (CWBS)* Mission Equipment Element to reflect the physical architecture's *levels of abstraction* and physical configuration items (PCIs). Figure 40.1 illustrates a *decomposition* or *expansion* of the EQUIPMENT Element in terms of the physical architecture. Note that it *links* to and provides the PRIMARY hierarchical structure for the CWBS Mission Equipment Element.

Author's Note 40.3 *Observe that we noted "PRIMARY hierarchical structure." The reason is the CWBS also includes the (entity) Integration, Test, and Checkout Element at each Integration Point (IP) within in the Mission Equipment hierarchy.*

COTS/NDI Physical Item Realities

COTS/NDI suppliers typically develop hardware and software for marketplace needs; not your unique program needs unless your organization is a strategically *valued customer*. Therefore, you will have to investigate and evaluate potential COTS/NDI candidate hardware and software products and determine which one best fits your needs, assuming they do.

Since the COTS/NDI items *may or may not* provide the required level of *functionality* you require, you may have to *supplement* the *lack of functionality* with internal development. Organizations often refer to *adapting* COTS/NDIs to meet interface requirements as "wrappers."

So, HOW does this affect the architecture decisions? It means that the initial physical architecture:

1. *Product structure* may have to be revised to reflect the new mix of internal development and COTS/NDI items for a specific entity—such as PRODUCT, SUBSYSTEM, or ASSEMBLY.
2. The entity's architecture (SBD, ABD, FBD, object, etc.) may have to be modified to reflect the item selections and their interfaces.

Physical Architecture Attributes

Physical architectures are more than simply solution configurations; most have to support a variety of User needs. If we analyze the spectrum of User requirements, we identify several attributes that warrant consideration when developing specifications. Buede (2000, p. 233), for example, identifies several key attributes that include:

- **Flexibility** The capability of an architecture to be reconfigured to support a variety of mission applications.
- **Scalability** The capability to expand processing capabilities and performance to accommodate a range of mission applications.
- **Fault Tolerance** "The capability to adjust operations when one of the hardware or software elements fails." (Source: Buede, 2000, p. 233)
- **Open Architecture** The ability to accept architectural elements based on interface standards of *modularity* and *interchangeability*.
- **Transparency** An operational characteristic that enables the User to interact with the architecture without requiring knowledge of its true physical architecture.

Human–System Interface (HSI) Decision Making

Most human-made systems require humans to perform some or all of the *command* and *control* (C^2) functions. From an SE perspective this requires *human-system interfaces* (HSIs).

Philosophically, the intent is to optimize system performance by *trading-off* and *allocating* required SYSTEM capabilities between the PERSONNEL and EQUIPMENT elements. In effect, capability allocation decisions focus on optimizing overall system performance based on WHAT HUMANS DO BEST versus WHAT EQUIPMENT-BASED SYSTEMS DO BEST. Operationally, human operators must have the capability to *manually* override and control the system subject to various *safety* and *security* constraints.

40.4 PHYSICAL DOMAIN SOLUTION DEVELOPMENT METHODOLOGY

The Physical Domain Solution is developed as a key element of the SE Process. The *highly collaborative* and *iterative* convolution of the Physical Domain Solution with the Requirements, Operations, and Behavioral Domain Solutions is often very *chaotic* and *confusing*.

We can *minimize* the *chaos* and *confusion* by applying an *iterative* methodology that enables us to create the Physical Domain Solution. The *methodology represents* one of many approaches to developing the Physical Domain Solution. View these steps as an example and tailor them to suit your business domain and systems application.

The methodology consists of the following steps:

- **Step 1:** Formulate candidate physical architectures.
- **Step 2:** Allocate capabilities and requirements to physical architecture items.
- **Step 3:** Evaluate and select physical architecture.
- **Step 4:** Perform make versus buy decisions.
- **Step 5:** Identify physical architecture configuration states.
- **Step 6:** Establish physical item performance budgets and design safety margins.
- **Step 7:** Link physical configuration states to phases and modes of operation.
- **Step 8:** Allocate specialty engineering requirements.
- **Step 9:** Conduct a FMEA or a FMECA.
- **Step 10:** Assess environmental, safety, health, and security compliance.
- **Step 11:** Professional engineering design certification.
- **Step 12:** Evaluate and optimize physical system performance.
- **Step 13:** Resolve critical operational and technical risks and issues.
- **Step 14:** Verify and validate physical solution performance.
- **Step 15:** Establish and maintain physical solution baseline.

Step 1: Formulate Candidate Physical Architectures

When we develop systems, the intent is to minimize development and life cycle costs and meet or exceed *System Performance Specification (SPS)* or *item development specification* requirements. From an accounting perspective, these costs are of two types: *nonrecurring* (i.e., developmental) or *recurring* (i.e., operational). In either case, both contribute to *total ownership cost (TOC)*. The User's objective is to *minimize* both the *recurring* and the *non-recurring* costs. HOW the system is physically implemented, its operational performance, and its ability to be maintained are *crucial drivers* in the cost equation.

The strategy to *minimize* the *total ownership cost (TOC)* drives the need to find *alternative* solutions or combinations of solutions. As a result, SEs should begin development of the Physical Domain Solution by *formulating* a number of *viable* candidate architectures. Analyze the Behavioral Domain Solution and formulate viable physical architectures that may be candidates for selection as the physical system architecture. *Remember*, the intent is to identify the Physical Domain Solution that provides the *best balance* among technical, technology, development and life cycle cost, schedule, and support requirements and risks.

Although every system and application is unique, identify at least *two to four* viable candidates. Formulate, analyze and construct each candidate architecture using methods such as the matrix, N^2 diagram, system block diagram (SBD), and Unified Modeling Language (UML®).

Author's Note 40.4 *Use common sense when identifying candidate physical system architectures. A good rule of thumb is to identify two to four candidate architectures or variations. Some architectures may be top contenders; others may not. Since trade studies consume valuable resources, perform a high-level trade study selection and identify the top two to four candidates. Then, apply resources for an in-depth trade study on the two MOST PROMISING candidates.*

Step 2: Allocate Capabilities and Requirements to Physical Architecture Items

Based on the behavioral interactions between SYSTEM or entity capabilities developed as part of the Behavioral Domain Solution, *allocate* and *link* the capabilities to physical architecture items and configuration items (CIs). Since *capabilities* link to *requirements*, linking requirements to physical *items* also provides a linkage to capabilities, especially with automated tools such as requirements management tools.

As you formulate each candidate architecture, make sure each one incorporates the *logical entity capabilities* identified in the Behavioral Domain Solution. We do this by allocating the capabilities to the appropriate physical architecture elements via a matrix mapping method as shown in Figure 40.2.

Author's Note 40.5 *As a reminder, DO NOT allocate highly interdependent capabilities across physical CI boundaries. (For more information on this topic, refer to the discussion of System Interface Analysis, Design, and Control Practices in Chapter 43 and Figure 40.4).*

Logically group and allocate *closely coupled* interactions to a single item such as a PRODUCT, a SUBSYSTEM, or an ASSEMBLY. This avoids *degrading* system performance and *increasing* the likelihood of technical risks due to *disconnected* or *disrupted* interfaces. Figure 40.3 illustrates how

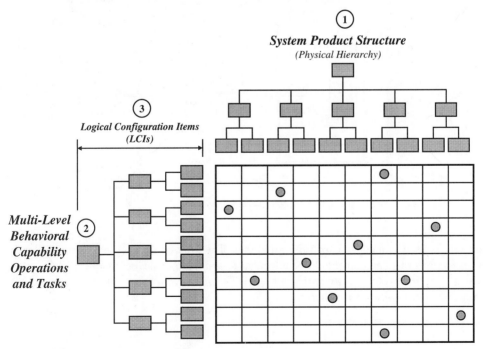

Figure 40.2 Behavioral Allocations to Product Structure Items

472 Chapter 40 Developing an Entity's Physical Domain Solution

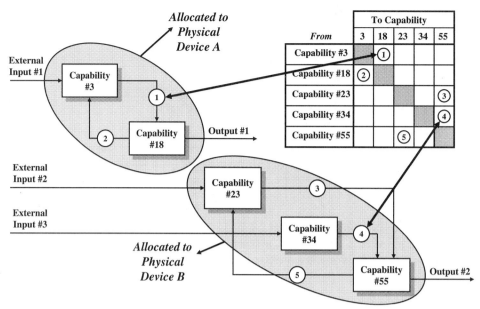

Figure 40.3 Example—N2 Chart Logical Grouping

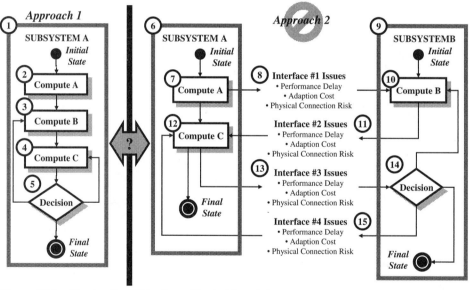

Figure 40.4 Allocaing Capabilities Across Physical Boundaries

an N^2 diagram is used to analyze capability groupings. The gray ovals represent areas of closely coupled interactions that can and should be allocated to a single physical component. Remember, a key objective of this exercise is to *minimize* the *level of interaction* across physical item boundaries as illustrated in Figure 40.4. Otherwise, we:

1. Degrade system performance.
2. Increase risk.

3. Increase costs due to increased complexity.
4. Reduce reliability.

Make sure that closely coupled capabilities such as CSCIs are allocated to a single physical item, not across multiple items. Figure 40.4 illustrates HOW splitting closely coupled interactions *across* physical interface boundaries creates potential performance problems.

Investigate and establish standards to promote *modularity*, *interchangeability*, and *interoperability*. Determine the type of architecture to be implemented to meet requirements—*centralized* versus *decentralized/distributed*, client-server, and so forth.

Step 3: Evaluate and Select the Physical Architecture

Evaluate each candidate SYSTEM or entity's physical system architecture by conducting *trade studies* using:

1. Pre-defined architecture selection criteria—such as technology, performance, cost, schedule, support and risk.
2. Criteria weights.
3. Models and simulations provided by *decision support* (refer to Chapter 51).

Prioritize results of the architecture evaluations for recommendations. Some level of OPTIMIZATION may be required as part of the evaluation. Based on recommendations from the trade study, select a preferred *physical architecture*.

Finally, periodically stop what you are doing. Back away from being immersed in the details—to avoid *analysis paralysis*—and perform a *reality* and *reasonableness check*. Ask yourself: *Is the configuration as simple as it can be? Is there any aspect we may have ignored, overlooked or diminished in level of significance? What are we not asking ourselves that later on we may wish we had? Does the Physical Domain Solution meet the minimum requirements?*

Step 4: Perform Make versus Buy Decisions

For each item within the architecture, determine the best approach for implementation. Chapter 42 on *System Configuration Identification Practices* presents various decision making options such as reuse of existing components, procurement of COTS/NDI, and internal development.

Step 5: Identify Physical Architecture Configuration States

As the Physical Domain Solution *matures*, we may discover that the physical system architecture implementation represents the convolution of several configurations. These configurations may range from physical add-on payload components—such as the satellites in the Space Shuttle's Cargo Bay—to physical *reconfiguration* of the system or product's structure—such as wings flaps and landing gear on aircraft—during various *phases* and *modes of operation*.

In effect, the total physical system design represents the integration of these various configurations into a single architecture. Thus, the architecture integrates the physical implementation of the capabilities required to support a specific phase or mode(s) of operations. Once completed, the physical configuration *states* constitute the final part of the system phases, modes, and states of operation.

Author's Note 40.6 *Engineers who lack formal SE training often pick up the jargon of "phases, modes, and states" without understanding the appropriate time to identify system operational and physical states during the evolution of the system solution. As a result, you will hear people erro-*

neously refer to identifying the system's physical (instead of operational) states of operation as one of the initial steps of the SE design process. Although there may be instances where this can occur, a behavioral FUNCTION drives the system's physical domain's FORM and FIT configuration.

Step 6: Establish Physical Item Performance Budgets and Design Safety Margins

For each capability allocated to and implemented by a physical item, *allocate* and *assign performance budgets* and *design safety margins* to behavioral interactions—inputs, outputs, interfaces, and so forth.

Referral For more information about *performance budgets and margins*, refer to Chapter 49 on *System Performance Analysis*, *Budgets*, and *Safety Margins Practices*.

Step 7: Link Physical Configuration States to Phases and Modes of Operation

For each mode of operation, identify and link each physical system architecture configuration *state* to one or more modes of operation. Thus, for a given mode of operation, configure the physical system architecture into one or more *physical states* to provide the *necessary* and *sufficient* capabilities to support the *use cases* associated with that mode.

Referral For more information about *phases and modes of operation, refer to Chapter 19 on System Phases, Modes, and States of Operation*.

Step 8: Allocate Specialty Engineering Requirements

The selection of the physical architecture and its elements simply provide the framework for developing the Physical Domain Solution. Specialty engineering requirements—such as reliability, availability, and maintainability (RAM); producibility; environmental, safety, and health (ESH); human engineering; vulnerability and survivability; and lethality—must be *analyzed*, *allocated*, and *flowed down* to SYSTEM components at all levels of abstraction.

Referral For more information about *RAM, refer to Chapters 47 through 52 on Decision Support Practices*.

Step 9: Conduct FMEA or FMECA

For system architectures that have *mission critical* components or may have the potential to damage EQUIPMENT, PERSONNEL, the general public, or the environment, review and update the Failure Modes and Effects Analysis (FMEA) initiated as part of the Behavioral Domain Solution. If necessary, expand the FMEA into a Failure Modes, Effects, and Criticality Analysis (FMEA). Former MIL-STD-1629A *Military Standard Procedures for Performing a Failure Modes, Effects, and Criticality Analysis* provides guidance. Ask key questions:

1. What can go *wrong* with the EQUIPMENT or one of its components or interfaces?
2. How can the EQUIPMENT be *misapplied*, *misused*, or *abused*?
3. Is the system *stable* in all phases, modes, and states of operation?

Step 10: Assess Environmental, Safety, Health, and Security Compliance

Review all elements of the Physical Domain Solution relative to environmental, safety, and health (ES&H) as well as security compliance. Ensure that:

1. All EQUIPMENT components are properly marked with appropriate *safety* and *security* notices (warnings, cautions, etc.).
2. Supporting PROCEDURAL DATA (i.e., operators and user's manuals) are properly documented with this information.

Avoid just "reading" the documentation; investigate HOW the guidance can be *misinterpreted*.

Step 11: Professional Engineering Design Certification

For those design applications such as buildings, bridges, and structures that require certification by a licensed and registered, practicing, professional engineer, employ the services of these professionals. Make sure that all aspects of the Physical Domain Solution, as well as other domain solutions, are fully addressed as part of the certification review, analysis, and approval process.

Step 12: Evaluate and Optimize Physical System Performance

EVALUATE and OPTIMIZE the physical performance of the SYSTEM or entity based on criteria established by the SPS or entity's *item development specification*. As appropriate, solicit User inputs and participation in SE design activities that provide *subjective look* and *feel* feedback related to system *optimization* efforts of operator-based EQUIPMENT element. Record and *assess* any *discrepancies* relative to *criticality* and scope of the contract or task activities and resources.

Step 13: Resolve Critical Operational and Technical Issues (COIs/CTIs) and Risks

As the Physical Domain Solution evolves, identify and capture *critical operational and technical issues* (COIs/CTIs) and risks. Where appropriate, resolve the issues with the Acquirer and User via contracting protocol. As appropriate, develop prototypes, mockups, and demonstrations that solicit constructive feedback from the Users.

Step 14: Verify and Validate Physical Solution Performance

Throughout the development of the Physical Domain Solution, continually *verify* and *validate* the *completeness* and *integrity* of the evolving solution via document reviews, technical reviews, prototypes, technology demonstrations, and modeling and simulation.

Conduct a SYSTEM level or Entity—PRODUCT, SUBSYSTEM, and ASSEMBLY—Critical Design Review (CDR) to assess the *status*, *progress*, *maturity*, and *risks* with the *Physical Domain Solution*. Initiate corrective action as necessary.

Referral For more information about *technical reviews*, refer to Chapter 54 Technical Reviews.

Step 15: Establish and Maintain Physical Solution Baseline

As the SYSTEM and each entity's *Physical Domain Solution* are reviewed and approved, if applicable, by the Acquirer, the acceptance of CDR documentation, presentations, meeting minutes, and closure of action items enable establishment of a formal baseline. The baseline serves as the basis for requirements *flow down* and *allocation*, future technical decision making, and change control. Incorporate the Physical Domain Solution into the *Developmental Configuration*.

40.5 LINKING THE BEHAVIORAL AND PHYSICAL ARCHITECTURES

As the overall system solution matures, the behavioral (i.e., logical) and physical system architectures are linked via requirements traceability. To see these links, let's focus on the Requirements, Behavioral, and Physical Domain Solutions. Figure 40.5 provides a graphical view using an entity relationship diagram (ERD).

40.6 PHYSICAL DOMAIN SOLUTION CHALLENGES

When the Physical Domain Solution is developed, there are several challenges that the User, Acquirer, and System Developer(s) need to answer. The challenges manifest themselves as questions SEs need to be able to answer.

Challenge 1: Solution Space Validation

In its physical manifestation, will the Physical Domain Solution satisfy the User's *validated* operational need?

Challenge 2: Technical Design Integrity

Are all elements of the Physical Domain Solution fully traceable back to the source *System Performance Specification (SPS)* requirements?

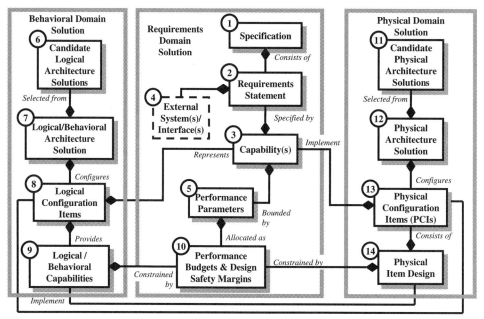

Figure 40.5 Requirements, Logical, & Physical Domain Solution Relationships

Challenge 3: Multi-domain Solution Agreement

Is the Physical System Solution *consistent with* and *traceable to* the Requirements, Operations, and Behavioral Domain Solutions?

Challenge 4: Risk Identification and Mitigation

Does the Physical Domain Solution impose any *unacceptable* technical, technology, support, operations, cost, or schedule risks?

Challenge 5: Environment, Safety, and Health (ESH)

Does the Physical Domain Solution pose any *noncompliant* or *unacceptable* environment, safety, and health issues and concerns to the operators or the general public?

Challenge 6: System Solution Stability

Is the Physical Domain Solution *stable* and *controllable* in all system phases, modes, and states of operation?

Challenge 7: System Support

Is the system, product, or service *supportable*—in terms of maintenance, training, and logistics—within budgetary cost, time, and resource constraints?

Challenge 8: Interfaces

For each interface, have the appropriate allocation decisions been made to ensure that interface performance has been *optimized* based on the inherent capabilities of the interfacing components. Are humans doing what they do best and machines doing what they do best? Do the interfaces meet *security* and *survivability criteria*?

Challenge 9: System Optimization

Has physical system performance for all or specific modes of operation been *optimized* based on a set of pre-defined set of weighted criteria supported by stakeholders?

Challenge 10: Phases and Modes of Operation

Does the Physical Domain Solution provide all of the capabilities required to support MISSION SYSTEM operations and *most likely* and *worst case scenarios* in all planned phases and modes of operation?

40.7 PHYSICAL DOMAIN SOLUTION WORK PRODUCTS

The Physical Domain Solution defines and documents the *physical characteristics* as *design requirements* that will be used to *procure, fabricate, assemble, code, inspect,* and *test* the components that will be integrated to form the system to be *verified* and *validated*.

Work products that capture the *Physical Domain Solution* consist of the following:

- Item development specifications (IDS)
- Performance budgets and design safety margins

- Physical system architecture
- Interface control documents (ICDs)
- Interface design descriptions (IDDs)
- Assembly drawings
- Wiring diagrams
- Schematics
- Bill of materials (BOM)
- Program design language (PDL)
- Source code
- Software design descriptions (SDDs)
- System descriptions
- Analyses
- Trade studies
- Technical reports
- Requirements traceability matrices (RTMs)
- Prototypes
- Technical demonstrations
- Models and simulations
- Test data and results
- Timing diagrams
- Engineering change proposals (ECPs)
- Change requests (CRs)
- Test plans and procedures
- Briefings
- Conference minutes and action items

40.8 GUIDING PRINCIPLES

In summary, the preceding discussions provide the basis with which to establish the guiding principles that govern development of an entity's Physical Domain Solution practices.

Principle 40.1 The Physical Domain Solution represents a physical configuration of items that implement one of more elements of the Behavioral Domain Solution allocated to the items.

Principle 40.2 The Physical Domain Solution must be *consistent with* and *traceable to* the Requirements, Operations, and Physical Domain Solutions.

Principle 40.3 Investigate ALL Physical Design Solution documentation, especially User guides, for ways the guides can be *misinterpreted.*

Principle 40.4 Increases in physical item part counts decrease system reliability; investigate ways to reduce parts counts to improve system reliability.

40.9 SUMMARY

We described the system solution development methodology steps used to derive the physical system architecture. The discussion described how the Behavioral Domain Solution provides the basis for:

1. Formulating candidate physical system architectures.
2. Evaluating, recommending, and selecting the physical system architectures.
3. Linking the physical architecture to the requirements, operational, and behavioral architectures.
4. Optimizing the physical system architectures.
5. Linking the physical system architectures to system phases and modes of operation.

GENERAL EXERCISES

1. Answer each of the *What You Should Learn from This Chapter* questions identified in the *Introduction*.
2. Refer to the list of systems identified in Chapter 2. Based on a selection from the preceding chapter's General *Exercises* or a new system selection, apply your knowledge derived from this chapter's topical discussions. Specifically identify the following:
   (a) The entity's physical architecture solution.
   (b) Describe how it traces to its Requirements, Operations, and Behavioral Domain Solutions.
   (c) Describe the type of architecture (central, distributed, client server, etc.) best suited for this application?

ORGANIZATIONAL CENTRIC EXERCISES

1. Research your organization's command media. What guidance is provided regarding development of the *Physical Domain Solution*? Document and report your results.
2. Contact a contract program within your organization. Interview the lead SEs and research HOW the program (a) formulated its *Behavioral Domain Solution*, (b) manages the requirements baselines, and (c) links to the *Requirements Domain Solution*, *Operations Domain Solution*, and *Behavioral Domain Solution*.
3. Select one of your organization's system specifications and develop a high-level *Physical Domain Solution*.

REFERENCES

MIL-STD-1629A (cancelled). 1984. *Military Standard Procedures for Performing a Failure Modes, Effects, and Criticality Analysis.* Washington, DC: Department of Defense (DoD).

BUEDE, DENNIS M. 2000. *The Engineering Design of Systems: Models and Methods.* New York: Wiley.

ADDITIONAL READING

LANO, R.J. 1977. *The N^2 Chart*. System and Software Requirements Engineering. IEEE Computer Society Press.

Chapter 41

Component Selection and Development

41.1 INTRODUCTION

The allocation of *System Performance Specification (SPS)* and *item development specification* requirements to multi-level items is a *highly iterative* process driven by component selection decisions. In general, System Developers have to answer the question: WHAT *is the best value, lowest cost, acceptable risk approach to selecting components to meet contract requirements?*

1. *Are there in-house, reusable components already available?*
2. *Do we procure commercially available components from external vendors?*
3. *Are there commercially available components that require only minor modifications to meet our requirements?*
4. *Do we procure commercially available components from external vendors and modify them in-house or have the vendor modify them?*
5. *Do we obtain components from the User as Acquirer furnished property (AFP)?*
6. *Do we create the component in-house as new development?*

Depending on the outcome of these questions and the item's required capabilities, the initial requirements may have to be reallocated to supplement capabilities of commercial components.

Our discussion in this section focuses on *component selection and development practices* that drive SE decision making. After a brief discussion of options available for system development, we introduce the concepts of commercial off-the-shelf (COTS) and nondevelopmental items (NDIs). Next, we define a methodology that describes a decision-making method for selecting the strategy for component development. We conclude with a summary of driving issues that influence COTS/NDI selection.

Author's Note 41.1 *Typically the next chapter on* system configuration identification *should precede component selection and development. However, configuration identification employs terms such as COTS and NDI, which are defined in this chapter. Therefore, we consider this chapter's material first.*

What You Should Learn from This Chapter

1. What are the six primary approaches for developing components?
2. What is a *commercial off-the-shelf (COTS) item*?
3. What is a *nondevelopmental item (NDI)*?
4. What is *Acquirer furnished property (AFP)*?
5. What is the *source* of AFP?
6. How are COTS items procured?
7. How are NDIs procured?
8. What are the steps of the *component selection methodology*?
9. What are the *classes* of criteria to be considered when selecting components from external vendors?

Definitions of Key Terms

- **Acquirer Furnished Property (AFP)** Physical assets such as equipment, data, software, and facilities provided by the User or other organizations via the Acquirer by contract to the System Developer for modification and/or integration into a deliverable system, product, or service.
- **Commercial Off-the-Shelf (COTS) Product** "A standard product line item that is offered for general sale to the public by a vendor and requires no unique modifications or maintenance over the life cycle of the product to meet the needs of the procuring agency (Acquirer)." (Source: Adapted from DoD/FAR and DSMC, *Glossary of Terms*)
- **Legacy System** An existing system that may or may not be operational.
- **Make–Buy–Modify Decisions** Technical decisions that determine whether to develop an item or lower level item internally, procure as a COTS/NDI or subcontract item from an external vendor, or procure an item from an external vendor and modify internally to meet item development specification requirements.
- **Nondevelopmental Item (NDI)** A COTS item that has been modified or adapted (i.e., customized or tailored) to meet procurement specification requirements for application in a specific OPERATING ENVIRONMENT.
- **Out-of-the-Box Functionality** Specified capabilities and levels of performance inherent to a COTS product or NDI without additional development.
- **Outsourcing** A business decision to procure systems, products, or services from external organizations based on cost avoidance resource availability, or other factors.

41.2 REDUCING SYSTEM COSTS AND RISK

One of the key objectives of SE is to minimize *development* and *life cycle costs* as well as risk. Achievement of these objectives requires insightful strategies that include selection of components.

Most engineers enter the workforce with noble aspirations to innovate and create *elegant* designs. Although this is partially true, it is also a reflection of misunderstood priorities. New design should be a LAST resort after attempts to find existing components that meet specification requirements have been exhausted. So, *what should the priorities be*? This brings us to *item design implementation options and priorities*.

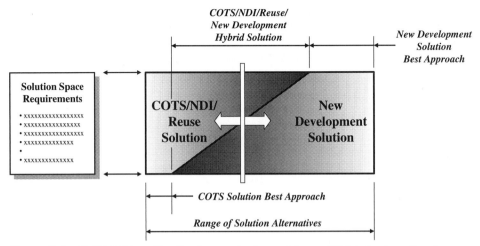

Figure 41.1 COTS/NDI/Reuse/New Development Application Mixes to Fill Solution Space Requirements

Item Design Implementation Options and Priorities

The underlying philosophy of system design and development begins with a simple question: *Are there available commercial or other products or services that are more cost effective, less risky, that will satisfy requirements allocated to a specific entity of item?* Answering this question requires exploration of at least six sequential design options that require consideration:

1. Employ Acquirer furnished property (AFP).
2. Reuse existing in-house component designs.
3. Procure commercially available vendor products.
4. Procure commercially available vendor products that can be easily modified by the vendor to meet our requirements.
5. Procure commercially available products and modify the products in-house.
6. Develop new designs.

Author's Note 41.2 *Be advised that reuse of existing designs may present legal and contractual issues regarding the type of funding used to develop the product, data rights, and so forth. ALWAYS consult with your program, contracts, legal, and export control organizations for guidance in these areas.*

Based on these design options, the system design solution for any item in the system, regardless of *level of abstraction*, may consist of one or a combination of these implementations. As a result the COTS/NDI/new development composition of any design implementation depends on the application as illustrated in Figure 41.1. Note that design implementation ranges from all commercial to all new development. So, the challenge for the System Engineering and Integration Team (SEIT) and component selection decision makers is to determine the RIGHT combination of item implementation options that REDUCE overall SYSTEM development and life cycle *costs* and *risk*.

41.3 COMMERCIAL PRODUCT TYPES

The preceding discussion addressed commercially available products as design implementation options. In general, there are two basic classes of commercial products: *commercial off-the-shelf (COTS) items* and *nondevelopmental items (NDIs)*. Let's scope each of these classes.

Commercial Off-the-Shelf (COTS) Items

COTS products represent a class of products that can be procured by the public from a vendor's catalog by part number. Procurement of COTS products is accomplished via a purchase order or other procurement mechanism. Generally, vendors provide a *Certificate of Compliance (CofC)* verifying that the product meets its published specification requirements.

Nondevelopmental Items (NDIs)

NDIs represent COTS products that have been *modified* or *customized* to meet a set of *application requirements*. Procurement of NDIs is accomplished via a purchase order that references a procurement specification that *specifies* and *bounds* the capabilities and performance of the modified COTS item. Prior to delivery, the Acquirer verifies the NDI, in the presence of Acquirer (role) witnesses, typically for compliance to its procurement specification.

41.4 COMPONENT SELECTION METHODOLOGY

The selection of components to fulfill *item development specification (IDS)* requirements requires a methodology that enables the item's development team to *minimize* technical, cost, schedule, and risk. In general, the selection process involves answering the questions posed during the introduction to this practice. So, *HOW do we answer these questions*?

One solution is to establish a basic methodology that enables selection from a range of alternatives. There are a number of ways the methodology can be created. The methodology described below is one example.

COTS Selection Methodology

The methodology used to select *item* components can be described in an six-step, *highly iterative* process as illustrated in Figure 41.2.

- **Step 1:** Identify candidate components.
    - **Step 1.1.** Identify potential in-house reusable solutions.
    - **Step 1.2.** Assess in-house solution(s) feasibility, capabilities, and performance.
    - **Step 1.3.** Identify potential COTS/NDI product solution(s).
    - **Step 1.4.** Assess COTS/NDI solution(s) feasibility, capabilities, and performance.
    - **Step 1.5.** Investigate the feasibility of modifying COTS products in-house.
- **Step 2:** Evaluate system impact of component approaches.
- **Step 3:** Validate component selection approaches (repeat Step 1).
- **Step 4:** Solicit and evaluate vendor COTS/NDI proposals.
- **Step 5:** Select component development approach.
- **Step 6:** Implement component selection decision.

Component Selection Summary

When the component selection and development decision-making process are complete, each item within the multi-level hierarchy of the system architecture and the CWBS may have a combination of various types of components such as in-house, COTS, and NDI that satisfy specification requirements allocated to the item.

Now that we have established a basic methodology, let's examine some of the *driving issues that influence COTS/NDI selection*.

484 Chapter 41 Component Selection and Development

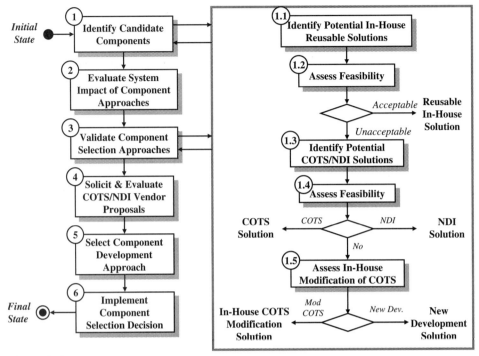

Figure 41.2 Example Component Selection Methodology

41.5 DRIVING ISSUES THAT INFLUENCE COTS/NDI SELECTION

COTS/NDI products *may* or *may not* be applicable to your contract application. Only you, your organization, and the Acquirer/User, if appropriate, can make that determination. Let's look at examples of some of the types of questions, for your tactful edification, that you should consider asking when selecting COTS/NDI products.

Caution Every set of Acquirer requirements, applications, COTS/NDI products, and perspectives is unique. Consult with SMEs in your organization or employ the services of a respected, credible consultant to assist you in formulating the list of questions and thoroughly investigate potential COTS products solutions before you commit to a decision.

COTS Product Line Example Questions

1. What is the *heritage* and *maturity* of the COTS product line and family?
2. What is the size of the COTS product user base?
3. What organizations or industries are the primary users of the COTS product?
4. What are the current technology trends relative to the product line directions?
5. At what stage of maturity is the COTS product in its maturity cycle as well as the marketplace?

6. How long has this version of the COTS product been produced?
7. Is the COT product in *alpha* and *beta* testing? If not how long ago did this occur?

Customer Satisfaction Example Question

8. Is the vendor *willing* to provide a list of customer references to discuss their experiences with the product?
9. What is the *degree of customer satisfaction* with current COTS product and prior versions?
10. Do customer testimonials, complaints, and applications of usage correlate?
11. Is the customer satisfaction based on using the product in the operating environment prescribed by the vendor?

Corporate Commitment to and Stability of the COTS Product Example Questions

12. How long will the vendor commit *on paper* to: 1) producing and 2) supporting the version the COTS product under consideration?
13. How financially stable is the vendor and parent company?
14. What is the vendor and parent company's commitment to the COTS product line and family?
15. How STABLE is the vendor's workforce (i.e., turnover) that develops and supports the COTS product.
16. For future reference, who are the vendor's SMEs and how long they have been with the organization and COTS product?
17. What are their roles relative to the COTS product?

COTS Product Design Example Questions

18. Assuming you have some level of access to the vendor's documentation, what is the quality and depth of COTS product documentation?
19. What *degree of verification* and *validation* has been performed on the COTS product?
20. Was the *verification* and *validation* performed internally, or did the vendor depend on the user community to "find the defects"?
21. Does the COTS product have documented and accessible test points, test hooks, entry/exit points in the design? Are these available to the developers and maintainers?
22. Does the COTS product use an industry standard interface that *fully complies* with the standard or just a *subset* of the standard? What are the compliance exceptions.
23. What *liberties*—meaning *interpretations* and *assumptions*—has the vendor taken with implementing the standard?
24. Is the vendor willing to *modify* the COTS product interface to meet your system or product's interface? To what degree?
25. What systems or products does the vendor certify that the COTS product is compatible and interoperable with?
26. What are the current *known defects* in the existing COTS product? Are there plans and priorities to correct them?
27. How many known and *latent defects* remain in the current "new and improved" product version?

28. How many *undocumented* and *untested* "features"—*latent defects* that can't be or will not be corrected—remain in the COTS product?
29. What detailed design information and "on call" support is available to support System Developer integration of the COTS product into their system or product? Are there fees required for this support?
30. Will future versions of the COTS product be forward and backward compatible with the current version being considered?

COTS Product Production Example Questions

31. What is the quality and discipline of the COTS vendor's quality assurance (QA) organization as well as its configuration management system and version control, assuming it exists?
32. Are the COTS products *serialized* and *tracked* for *upgrades* and *recalls*?

COTS Product Support Example Questions

33. What degree of 24/7 support (i.e., 24 hours per day/ 7 days per week) is the vendor willing to support the COTS product? From what country, time zone, and hours of availability; Internet on-line support and documentation; "live" support via 800 numbers or "read what is *on-line*"?
34. What level of *responsiveness* to technical support is the vendor willing to provide (4 hours, one-week, etc.)?
35. What level of accessibility to COTS product SMEs is the vendor willing to provide?
36. Does the COTS product have operations and support (O&S) and technical manuals?
37. Are O&M and technical manuals delivered with the COTS product, available freely on-line, or do you have to purchase them? If purchase, how readily available are they?
38. Are *alignment* and *calibration* procedures and data documented and available to System Developers?
39. Does the vendor provide field service support? How responsive? What constraints (days of the week, holidays, hours, etc.)?
40. If you purchase the COTS/NDI product, are you required to contact the vendor to perform field service on site to "*remove* and *replace*, *align*, and *calibrate*," or can the System Developer perform this?
41. Who pays for field service expenses?

COTS Product Warranty Example Questions

42. Is each COTS product covered by an *expressed* or *implied* warranty? Will the vendor provide a copy?
43. What actions or physical modifications by the Acquirer may *invalidate* or *void* the warranty?
44. What are *acceptable* System Developer modifications to the COTS product that do not *void* the *warranty*?
45. Is the vendor willing to modify the COTS product or do they recommend third parties?
46. What impact do third-party modifications have on the on warranty?

COTS Product Procurement Example Questions

47. Does the COTS product require usage licenses—software, hardware, and export control?
48. Are licenses based on a per platform basis or is a *site license* available subject to a maximum number of "floating" users?
49. If a site license is available, how is the number of users *restricted* (simultaneous, total population, number of available "keys," etc.)?
50. Are there "other bundled" products included in the COTS product or must be procured—for a *small* additional cost? Do they require licenses?
51. Is there a minimum *buy quantity* requirement for the COTS product?
52. If there is a *minimum buy*, is this on a per purchase basis, cumulative over the year, or cumulative over several years?
53. What quantities are the price break thresholds?
54. What product quality assurance processes are in place to ensure the quality of the COTS product?
55. What *certifications* are the vendor willing to provide regarding the integrity of the product and its materials?
56. Are COTS product specifications, processes, and test procedures available for review and inspection?

Bottom Line

57. Always ask *what you need to know* about the product that *you did not ask*.
58. Finally, *caveat emptor*—let the buyer beware!

41.6 GUIDING PRINCIPLES

In summary, the preceding discussions provide the basis with which to establish the guiding principles that govern practices.

Principle 41.1 System design should be a last resort option AFTER all practical efforts have been made to identify reusable, COTS, NDI, or AFE components to meet entity requirements.

Principle 41.2 When procuring COTS/NDI, thoroughly INVESTIGATE and UNDERSTAND the development and life cycle costs and support; otherwise, CAVEAT EMPTOR!

41.7 SUMMARY

In general, COTS/NDI solutions may be RIGHT for you and your application; in other cases, they MAY NOT be. As an SE leading the selection process, you must make *informed decisions* based on:

1. The *facts*.
2. Trade-offs to meet requirements, minimize development and life cycle costs, and reduce risk to an acceptable level.
3. Other User's application experiences.

COTS products can be very powerful tool to reducing development and life cycle costs; they can become a problem as well. Perhaps the best way to THINK of COTS is to use a *mirage* analogy: *make sure that WHAT you FIND when you implement the product matches the virtual image you perceived.* Whatever decision you make, you will have to live with your action(s) and consequences. Research component selections and vendors carefully and thoroughly, incorporate flexibility for contingency planning, and make wise choices. The bottom line is: *caveat emptor*—let the buyer beware!

GENERAL EXERCISES

1. Answer each of the *What You Should Learn from This Chapter* questions identified in the *Introduction*.

ORGANIZATIONAL CENTRIC EXERCISES

1. Research you organizational command media guidance for guidance about component selection and development.
  (a) What guidance is stated regarding make, procure, procure, and modify decisions?
  (b) What criteria are to be applied to these decisions?
  (c) What guidance is provided and approval process is required for procuring COTS, NDI, and AFP products?
2. Contact several contract programs or product lines within your organization. Investigate the following questions and document your findings and observations.
  (a) Does the program use internal and externally supplied components?
  (b) If COTS/NDI products were used, how were they selected?
  (c) How were the selection decisions documented?
  (d) How were COTS/NDI products procured?
  (e) What lessons learned has the program learned from COTS/NDI procurement?
  (f) What changes would they make next time?
3. Contact an organization that provides COTS products to the marketplace. Identify issues the COTS vendor must resolve in selling their products to System Developers.

ADDITIONAL READING

Federal Acquisition Regulation (FAR). Parts 6, 10, 11, 12, and 14. January 1995.

Office of Federal Procurement Policy. 1995. *Guide To Best Practices For Past Performance*, interim ed. Washington, DC: Government Printing Office.

DSMC. 1995. *Integrated Logistics Support Guide*. Fort Belvoir, VA: Defense Systems Management College (DSMC) Press.

SD1. Standardization Directory. Published 4 times per year. Defense Standardization Program, Washington, DC: Department of Defense.

SD-2. 1996. *Buying Commercial and Nondevelopmental Items: A Handbook*. Washington, DC: Department of Defense, Office of the Undersecretary of Defense for Acquisition and Technology.

SD-5. 1997. *Market Research: Gathering Information about Commercial Products and Services*, Washington, DC: Department of Defense, Office of the Undersecretary of Defense (Industrial Affairs and Installtions).

SD-15. 1995. *Performance Specification Guide*, Defense Standardization Program, Washington, DC: Department of Defense, Office of the Assistant Secretary of Defense for Economic Security.

Army Materiel Command Pamphlet 715-3. 1994. *The Best Value Approach to Selecting a Contract Source*, Vol. 5. Washington, DC: Government Printing Office.

Chapter 42

System Configuration Identification

42.1 INTRODUCTION

Systems, as multi-level abstractions of integrated capabilities, require a structural framework and building blocks to integrate those capabilities. We refer to the *structural framework* as the *system's architecture* and the *building blocks* as *items*. The challenge for SEs is to determine HOW to:

1. *Conceptualize*, *formulate*, and *select* the RIGHT architectural framework, especially for large highly complex systems.
2. Partition the architecture into the respective levels of interconnected *items*.

SEs approach this challenge by partitioning large, complex problems into smaller, multi-level problems that can be easily *managed* and *solved*. We refer to the multi-level partitioning as *hierarchical* decomposition *or* expansion. To accomplish a decomposition of the architectural framework, we apply *requirements analysis*, *functional analysis*, and *object analysis* methods and techniques to decompose the multi-level specifications into a set of hierarchical capabilities. Each capability is subsequently *allocated to* and *performed by* physical components we categorize as the system elements—EQUIPMENT, PERSONNEL, and so on.

During the hierarchical decomposition, several types of make versus buy versus buy-modify decisions are made. The challenge question is: *How do we translate System Performance Specification (SPS) capabilities of the Behavioral Domain Solution into manageable EQUIPMENT items that can be designed, developed, procured, and the Behavioral Domain Solution and modified, and integrated to fulfill the system's overall capability?*

This chapter describes HOW elements of a system's architectural configuration are identified and designated for configuration tracking. Our discussions begin with establishing the configuration management semantics and explain why these terms are often confusing. We explore HOW architectural items are selected from external vendor *commercial off-the-shelf* (*COTS*) *items* or *non-developmental items* (*NDIs*), or *Acquirer furnished property* (*AFP*), or they can be developed in-house from legacy designs or new development. We provide illustrations of how *items* are assigned to development teams. We conclude with a discussion of configuration baselines—when they are established—and contrast SE and configuration management (CM) viewpoints of each one.

What You Should Learn from This Chapter

1. What is a system or product *configuration*?
2. What is the *Developmental Configuration*?
3. What is an *item*?

4. What is a *configuration item* (*CI*)?
5. What is the relationship between *items* and *CIs*?
6. Who is responsible for *identifying* CIs?
7. How are CIs *selected*?
8. How do CIs relate to the *specification tree*?
9. How do COTS items and NDIs relate to items and CIs?
10. What is a *baseline*?
11. What is meant by *configuration effectivity*?
12. What is the relationship between *baselines* and the *Developmental Configuration*?
13. Describe the *evolution* of the Developmental Configuration and its baselines.
14. What is a *line replaceable unit* (*LRU*)?
15. What are *mission-specific* and *infrastructure* CIs?
16. What is the "As Specified" Configuration, and when is it established?
17. What is the "As Designed" Configuration, and when is it established?
18. What is the "As Built" Configuration and when is it established?
19. What is the "As Verified" Configuration, and when is it established?
20. What is the "As Validated" Configuration, and when is it established?
21. What is the "As Maintained" Configuration, and when is it established?
22. What is the "As Produced" Configuration, and when is it established?

Definitions of Key Terms

- **Allocated Baseline** "The initially approved documentation describing a configuration item's (CI) functional and interface characteristics that are allocated from those of a higher level CI; interface requirements with other CIs; design restraints; and verification required to demonstrate the achievement of specified functional and interface characteristics. Allocated baseline consists of the development specifications that define functional requirements for each CI." (Source: DSMC, *Glossary: Defense Acquisition Acronyms and Terms*)

- **Baseline** A collection of configuration-controlled, specification and design requirements documentation for a configuration item (CI) or set of CIs that represents the current, approved, and released design.

- **Baseline Management** "In configuration management, the application of technical and administrative direction to designate the documents and changes to those documents that formally identify and establish baselines at specific times during the life cycle of a configuration item." (Source: IEEE 610.12-1990)

- **Computer Software Configuration Item (CSCI)** "An aggregation of software that satisfies an end use function and is designated for separate configuration management by the acquirer. CSCIs are selected based on trade-offs among software function, size, host or target computers, developer, support concept, plans for reuse, criticality, interface considerations, need to be separately documented and controlled, and other factors." (Source: Former MIL-STD-498, p. 5)

- **Configuration** "(1) The performance, functional, and physical attributes of an existing or planned product, or a combination of products. (2) One of a series of sequentially created variations of a product." (Source: ANSI/EIA 649-19116, Sec. 3, p. 3)

- **Configuration Item (CI)** "An aggregation of hardware, firmware, computer software, or any of their discrete portions, which satisfies an end use function and is designated by the (Acquirer) for separate configuration management. Configuration items may vary widely in complexity, size, and type. Any item required for logistic support and designated for separate procurement is a CI." (Source: Adapted from *DSMC Glossary: Defense Acquisition Acronyms and Terms*)
- **Configuration Management** "A management process for establishing and maintaining consistency of a product's performance, functional, and physical attributes with its requirements, design and operational information throughout its life." (Source: ANSI-EIA-649-1998, p. 4)
- **Effectivity** "A designation defining the product range (e.g., serial, lot numbers, model, dates) or event at which a change to a specific product is to be (or has been) effected or to which a variance applies." (Source: ANSI-EIA-649-1998, p. 4)
- **Hardware Configuration Item (HWCI)** "An aggregation of hardware that satisfies an end use function and is designated for separate configuration management by the Acquirer." (Source: Adapted from former MIL-STD-498, p. 5)
- **Item** Any physical component of a system such as a PRODUCT, SUBSYSTEM, ASSEMBLY, SUBASSEMBLY, or PART.
- **Line Replaceable Unit (LRU)** "A unit designed to be removed upon failure from a larger entity (product or item) in the operational environment, normally at the organizational level." (Source: MIL-HDBK-470A, Appendix G, *Glossary*, p. G-8)
- **Logical Configuration Item (LCI)** An optional designation assigned to a specific capability.
- **Physical Configuration Item (PCI)** An item that represents the physical instance of a configuration item (CI). Each item is assigned a part number and may be serialized.
- **Product Structure** Refer to definition provided in Chapter 9 *System Levels of Abstraction and Semantics*.
- **Requirements Baseline** "Documentation describing a system's/segments functional characteristics and the verification required to demonstrate the achievement of those specified functional characteristics. The system or segment specification establishes the functional baseline" (Source: DSMC *Glossary: Defense Acquisition Acronyms and Terms*)
- **Variance, Deviation, Waiver, Departure** "A specific written authorization to depart from a particular requirement(s) of a product's current approved configuration documentation for a specific number of units or a specified time period. (A variance differs from an engineering change in that an approved engineering change requires corresponding revision of the product's current approved configuration documentation, whereas a variance does not.)" (Source: ANSI/EIA-649-1998, Section 3.0, *Definitions*, p. 6)

Items—Building Blocks of Systems

Large, complex systems are developed by groups of people working as Integrated Product Teams (IPTs) or development teams with assigned roles and responsibilities for producing various system products. Depending on the *size* and *complexity* of the system or *item* being developed, teams are assigned tasks such as specifying, designing, developing, integrating, and verifying various components within the system. This presents some significant challenges:

1. *How do you partition the architecture of the EQUIPMENT element into multiple levels and instances of manageable physical items?*

492 Chapter 42 System Configuration Identification

2. *How can people establish semantics that enable them to communicate with others about the types of items they are developing or procuring from external vendors?*

3. *How can people communicate about the evolution of system components that go through various stages from abstract system specification entities into physical components used to build the system?*

The solution resides in integrated *building blocks* referred to as *items*, *configuration items* (*CIs*), *hardware configuration items* (*HWCIs*), and *software configuration items* (*CSCIs*). Each of these *building blocks* represents semantics used to identify abstract entities within a system and their evolution from SPS to deliverables.

42.2 UNDERSTANDING CONFIGURATION IDENTIFICATION SEMANTICS

Configuration identification knowledge for most SEs typically comes from informal exposure via verbal discussions in meetings, on-the-job-training (OJT), and observation over many years. Most engineers have little or no training in the principles of configuration management; most are self-taught through observation and knowledge of general CM standards. By these means they feel able to proclaim themselves to be configuration experts.

These rudimentary skills provide basic insights about system architectural configuration decisions. However, *meanings* and *applications* of the terms are often convoluted. Instead of seeking insightful guidance from *competent* CM personnel, technical leads exercise their authority, make decisions, and blunder down the *road of regrets*, while CM and others expend endless energy contending with the consequences of these decisions. So, to *minimize* the confusion, let's begin by informally introducing key terms. Figure 42.1 depicts entity relationships to support our discussions.

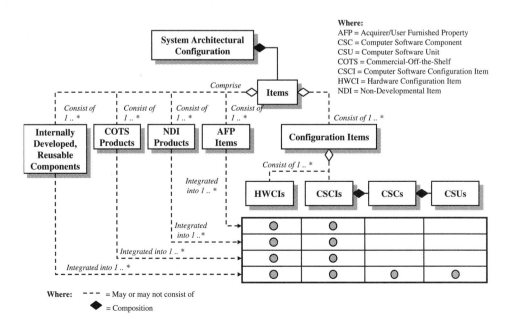

Figure 42.1 System Configuration Identification Elements

Component Origins

Every entity within a SYSTEM, regardless of *level of abstraction* is referred to as an ITEM. If you analyze most systems, you will discover that items originate from at least six different methods:

1. Procured from a vendor's catalog.
2. Procured from a vendor's catalog and customized/adapted in-house.
3. Customized or modified versions of items found in a vendor's catalog.
4. Developed in-house by human intellect or procured as components or raw materials from suppliers.
5. Developed in-house from customizations of existing, legacy designs.
6. Provided by the Acquirer in accordance with the *terms* and *conditions* (Ts&Cs) of the system development contract and integrated into the SYSTEM design.

Observe two types of themes above. First, *items* are procured externally, and second, *items* are developed internally.

Externally Acquired or Procured Components

As introduced in Chapter 41, items procured from a vendor's catalog are referred to as *commercial off-the-shelf* (*COTS*) items. *COTS* items *customized* or *modified* for a specific application are referred to as *nondevelopmental items* (*NDIs*). Acquirer provided items are referred to as *Acquirer furnished property* (*AFP*).

Author's Note 42.1 When the System Developer receives contract-based AFP from the User via the Acquirer, each item must be recorded, tracked, and controlled in accordance with the terms and conditions (Ts&Cs) of the contract. Planned modifications to AFP typically require authorization by the Acquirer's Contracting Officer (ACO). Make sure the Ts&Cs clearly delineate WHO is accountable for:

1. Providing AFP documentation.
2. AFP failures while in the possession of the System Developer.

Configuration Items (CIs)

When a System Developer decides to develop a MAJOR item such as a PRODUCT, SUBSYSTEM, or ASSEMBLY in-house, the program designates the item as a *configuration item* (*CI*). CIs require a specification that *specifies* and *bounds* the CI's capabilities and performance.

A CI, as a major item, integrates lower level components that may consist of: AFP, COTS items, NDIs, and two other types developed by the System Developer in-house:

- Hardware configuration items (HWCIs)
- Computer software configuration items (CSCIs)

HWCIs and CSCIs

HWCIs are major *hardware* items and CSCIs are major *software applications* designated for configuration control.

- As CIs, HWCIs and CSCIs may include COTS items, NDIs, internally developed or legacy items, or combinations of these.

- Each HWCI is specified via requirements documented in an *HWCI Requirements Specification* (*HRS*).
- Each CSCI is specified via requirements docuemented in a *CSCI Software Requirements Specification* (*SRS*).

Firmware

Some processor-based applications such as single board computers (SBC) employ software encoded into an integrated circuit (IC) referred to as *firmware*. Firmware ICs, which are nonvolatile memory device types, may be implemented with *single-use*, read only, and reprogrammable devices. *Firmware* represents a hybrid instance of an *item* that evolves from a CSCI into an HWCI, as illustrated in Figure 42.2.

Initially, the software program executed by the SBC is developed as a CSCI software application and debugged on laboratory prototype SBC hardware, using emulators and other devices. When the software application reaches *maturity* and is ready for final integration, the CSCI's code is electronically programmed into the firmware IC device. Once programmed, the firmware IC is:

1. Designated as an HWCI.
2. Assigned a part number, serial number, and version.

Both the CSCI and HWCI are controlled in accordance with the CM procedures.

Guidepost 42.1 *The preceding discussions introduced the semantics of configuration identification. Some of these semantics apply to ANY level of abstraction. You should recall from our discussion in Chapter 9 System Levels of Abstraction and Semantics that one organization's SUBSYSTEM might be another organization's SYSTEM. Our follow-on discussions illustrate WHY the referential nature of configuration identification semantics when applied to levels of abstraction result in confusion.*

Configuration Semantics Synthesis

To understand how configuration identification semantics relate to multi-level system architectures, Figures 42.1 and 42.3 depict the entity relationships. Table 42.1 provides a listing of entity relationship rules that govern the implementation of this graphic.

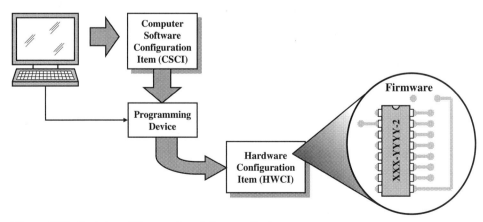

Figure 42.2 Evolution of Firmware from Software to Hardware

42.2 Understanding Configuration Identification Semantics

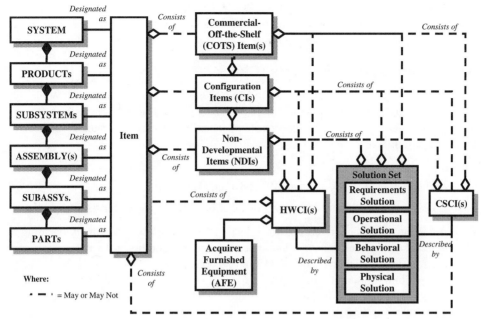

Figure 42.3 Item/Configuration Items (CIs) Compositional Entity Relationships

Table 42.1 Configuration semantics rules

Rule	Title	Configuration Identification and Development Rule
42.1	Items	Every entity within a system, regardless of level of abstraction, is referred to as an *item*.
42.2	Configuration items	Designate major *items* selected for internal development and configuration control.
42.3	Configuration items	*Items* may originate from several types of sources: 1. Replicated from existing, internally developed component designs. 2. Acquired as COTS, NDI, or AFP. 3. Acquired as COTS, NDI, or AFP and modified in-house. 4. Developed as new designs such as an HWCI(s) or CSCI(s).
42.4	CI composition	A CI's composition may consist of one or more COTS products, NDIs, HWCIs, CSCIs, AFPs, or combinations thereof.
42.5	CI specifications	Develop a *performance* or *development specification* for each *item* designated as a CI.
42.6	HWCIs and CSCIs	Develop an *HWCI Requirements Specification (HRS)* for each HWCI; Develop a *CSCI Software Requirements Specification (SRS)* for each CSCI.
42.7	CI solution set	Develop a Requirements Domain, Operations Domain, Behavioral Domain, and Physical Domain Solutions for each CI, HWCI, and CSCI.
42.8	CSCIs	The product structure of each CSCI consists of at least two or more CSCs that consist of at least two or more CSUs.
42.9	CI ownership	Assign *accountability* for the design, development, and integration and verification for each CI, HWCI, and CSCI to an individual or IPT.

Guidepost 42.2 *At this juncture we have established the context and composition of CIs. The question is: HOW do we determine which items should be designated as CIs. This brings us to our next topic on the selection of configuration items.*

Selection of Configuration Items (CIs)

The preceding discussions established that CIs are MAJOR *items* developed in-house. While this is an important criterion, CIs often require additional considerations. The best approach for selecting CIs is to simply establish a set of selection criteria. Then, perform a reasonableness check to ensure that the selection:

1. Is logical.
2. Provides the proper visibility for technical, cost, and schedule tracking.
3. Exposes development activities at a level that can be used for assessing risk.

Some organizations establish specific criteria for selecting CIs that go beyond simply deciding to develop an *item* internally. These decisions should be made in collaboration with a program's configuration manager, a subject matter expert (SME) in this domain.

The selection of configuration items often varies from one organization or business domain to another. To standardize thinking about *selecting CIs*, MIL-STD-483A (cancelled) offers the following guidance in selecting CIs:

1. *Is it a critical high risk, and/or a safety item?*
2. *Is it readily identifiable with respect to size, shape, and weight (hardware)?*
3. *Is it newly developed?*
4. *Does it incorporate new technologies?*
5. *Does it have an interface with hardware or software developed under another contract?*
6. *With respect to form, fit, or function, does it interface with other configuration items whose configuration is controlled by other entities?*
7. *Is there a requirement to know the exact configuration and status of changes to it during its life cycle?*

(Source: Former MIL-STD-483A, para. 170.9, pp. 118–119)

Configuration Item (CI) Boundaries

Contrary to some beliefs, *items* and CIs SHOULD NOT be *partitioned* across physical device boundaries; this is a violation of good design practice as noted in our earlier discussion of Figure 40.4. In general, CIs:

1. Are *bounded* by a development specification, *HWCI Requirements Specification* (*HRS*) or *CSCI Requirements Specification* (*SRS*).
2. Reside within the boundaries of a physical item—such as SYSTEM, SUBSYSTEM, ASSEMBLY, or SUBASSEMBLY as a computer system, printed circuit board, software application, and so forth.
3. Must be verified against their respective *HRS or SRS*.

To illustrate this point, consider the hypothetical example below:

EXAMPLE 42.1

As an example of an INCORRECT approach, an organization has a contract to develop a word processor software application. Using the INCORRECT approach, the software CSCI would have CSCs implemented across physical items:

1. A desktop computer (HWCI)
2. The printer (HWCI)
3. Other (HWCIs) on a network

So, if CIs, HWCIs, and CSCIs require stand-alone *performance* or *development specifications* to specify and bound an integrated item that is SELF-CONTAINED within a physical item, *How do we* test *the software, as an item, when it* resides *across several HWCIs (boundaries)*?

Remember, CIs are tested and verified as *standalone test articles*—as a black box with inputs and outputs—with all the necessary functionality self-contained. Adhering to the boundary constraints specified above, there is an expectation that the CSCI will be tested on a single HWCI device such as a desktop computer. Obviously, this is not achievable since the CSCI's CSCs are implemented in several HWCI devices (desktop computer, printer, network, etc.) across the system. The CORRECT approach requires a CSCI on each device.

Configuration Identification Responsibility

Configuration identification, as an informed, multi-discipline, decision-making process, requires collaboration with those stakeholders. Contrary to what many people believe, it IS NOT a decision by ONE individual exercising their discretionary authority in a vacuum without inputs from the key decision stakeholders. As the multi-level entity architectures evolve, the Configuration Manager (CM) / Software Configuration Manager (SCM), Lead SE, and other SEs *collaborate* to select the CORRECT approach for identifying *items* and CIs that will endure the *test of time* and AVOID the junk heap of POOR decisions.

Guidepost 42.3 *At this point we have established the basic set of configuration identification semantics and how they are applied to multi-level system architectures. These discussions highlighted the need during internal development to prepare a development specification for each CI, HWCI, and CSCI. For the first article on Developmental Configurations, this is a straightforward process. Two key questions:*

1. *HOW are these specifications maintained for production systems that evolve over time as new capabilities and refinements are added to established designs?*
2. *How do these impacts affect systems or products that are already fielded that may require RETROFITTING?*

This brings us to our next topic, configuration effectivity.

Configuration Effectivity

Production systems may *evolve* over several years as newer technologies, capabilities, and improvements are incorporated into the evolutionary system design solution. As such, the physical configuration changes. So the question becomes: *HOW do we delineate the changes in configuration to a given CI, HWCI, or CSCI?* Configuration management addresses these configurations via a concept referred to as *configuration effectivity*.

Every CI, HWCI, and CSCI is labeled with a unique identifier that delineates it from all others. This occurs at two levels: 1) model number and 2) serial numbers. So, when physical configurations change over the years, some organizations simply reference the effectivity beginning with Serial Number (S/N) XXX; others append a "dash number" to the model number—such as Model 123456-1—to indicate a specific version. Most organizations today affix barcode labels to CIs, HWCIs, and CSCIs to facilitate automated version or configuration tracking.

Versioning provides the System Developer a couple of options. It allows *evolutionary* tracking of a product line over its life span, and it provides a means to account for special *customizations* delivered to an Acquirer. In lieu of model numbers and versioning, some vendor's employ contract numbers and serial numbers to designate items.

You may ask: *Why is this important to SE?*

Effectivity-Based Specifications

During the development of a system or product, multi-discipline SEs prepare *item development specifications* (*IDSs*) for PRODUCTS, SUBSYSTEMS, HWCIs, and CSCIs that form the Developmental Configuration. Although cost is a key constraint, most *first article* systems or products MAY NOT represent the *most cost-effective solution* due to schedule and other constraints. *First articles* are simply a solution that meets specification requirements. Typically, each deliverable is assigned contract/model numbers and serial numbers.

If a system is planned for production, Product Engineering initiates efforts to reduce the *recurring* per unit cost via design improvements, component and material selection and procurement, manufacturing methods, and so forth. Ultimately, the improvements culminate in a *revised item development specification* with effectivity beginning with S/N XXX forward.

Once production starts, CIs, HWCIs, and CSCIs evolve over time. Whereas during the original system development, revisions to the Developmental Configuration specifications were issued when changes occurred. So, when production item changes occur, you have to *revise* the specification level and the *effectivity*.

When this occurs, the program has two choices: create a new specification unique to a configuration, or designate model and serial number *effectivity* on the cover page and *annotate* specification requirements unique to the effectivity within the document.

42.3 ALIGNING ITEMS AND CIs WITH THE SPECIFICATION TREE

Once CIs are identified, the key question is: *HOW do you communicate their location within the system's product structure?* The answer is: CIs should be explicitly identified in the *specification tree* based on derivation from the system architecture as shown in Figure 42.4.

Here, the SEIT analyzes the *System Performance Specification* (*SPS*) requirements to create the SYSTEM level architecture, as shown in the lower left corner of the graphic. The architecture consists of PRODUCTs A and B. PRODUCT B, which consists of ITEMs B_1 through B_3, will be developed internally, and is designated as a CI.

As the system architecture evolves, the *specification tree* evolves, as shown on the right side of the graphic. Based on the designation of CIs and *items, item development specifications* are identified.

- PRODUCT A has its own *PRODUCT A development specification.*
- PRODUCT B, as a CI, has its own unique *PRODUCT B development specification* that includes requirements for ITEMs B_1 through B_3.

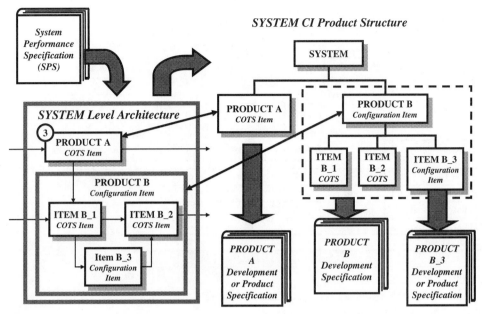

Figure 42.4 System Configuration Documentation

- ITEM B_3 is designated as a CI for development and, as such, is required to have its own unique *ITEM B_3 development specification*.

42.4 ASSIGNING OWNERSHIP OF ITEMS AND CIs

As the *specification tree* evolves, CI development ACCOUNTABILITY should be assigned to owners such as development teams or Integrated Product Teams (IPTs). Figure 42.5 provides an illustrative example. Note how the system architecture decomposes along *product structure* lines. This is a key point, especially the operative word "product."

For programs that establish Integrated Product Teams (IPTs), each IPT focuses on "product" development and collaborates with interfacing IPTs developing items that interface to their assigned product. For example, IPT 1 collaborates with IPT 2 on mutual interface design issues.

Accountability for developing ONE product is assigned to *ONE and only one* IPT. Depending on the *size*, *complexity*, and *risk* of the multi-level items, an IPT may be assigned *accountability* for one or more products as illustrated in Figure 42.5. Accountability for developing PRODUCTs A and B, which have a moderate degree of *complexity* and *risk*, is assigned to IPT 1. In contrast, accountability for PRODUCT C is assigned to IPT 2 due to its *complexity* and *risk*. This brings us to a final point.

Programs often get into trouble because SEs develop the IPT organizational structure first and then leave the IPTs to identify the architectural configuration with limited, if any, oversight by the SEIT. In this example, PRODUCTs A and B, by virtue of accountability by IPT 1 would be bundled together, regardless of the lack of physical interfaces and be idenitifed as a PRODUCT or SUB-SYSTEM. Then, the IPT will attempt to develop a *development specification* for the conglomeration.

In many cases PRODUCTs A and B are unrelated, thereby indicating NO interfaces. Yet, the IPT will be required to verify both PRODUCTs together as a "black box." *Avoid* and *correct* these system configuration identification decisions. Often these decisions are made by local "heroes" with

500 Chapter 42 System Configuration Identification

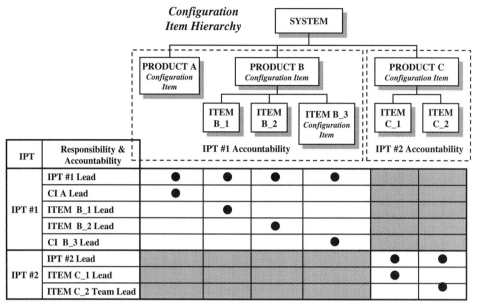

Figure 42.5 CI Assigned Responsibility & Accountability

an ounce of knowledge about system architecture development and IPT implementation yet wield authority and create programmatic situations that severely impact contract performance.

42.5 RECOGNIZING TYPES OF ARCHITECTURAL ITEM BOUNDARIES

The Industrial Revolution introduced new concepts for standardizing and reproducing *modular* and *interchangeable* components via predictable methods in order *to* leverage the benefits of *economies of scale*. Our discussions on SYSTEM, PRODUCT, SUBSYSTEM, ASSEMBLY, SUBASSEMBLY, and PART levels of abstraction expound on these themes.

The concept of modularity can easily lead to the SE mindset that all *items* and CIs are constructed as modular *plug and play* boxes. The System of Systems (SOS) approach further reinforces the mindset of "box" CIs integrated into a higher level system. However, there are systems whereby INTEGRATION occurs ACROSS the traditional "box" boundaries.

In general, systems often consist of two classes of PRODUCTs/SUBSYSTEMs:

1. *Mission specific* PRODUCTs/SUBSYSTEMS.
2. *Infrastructure* PRODUCTs/SUBSYSTEMS that *transcend* the mission-specific SUBSYSTEM boundaries.

Figure 42.6 illustrates this type of architecture. Consider the following example:

EXAMPLE 42.2

Office building systems, as MISSION SYSTEMs, consist of well-defined architectural "box" boundaries. Hierarchically, we could refer to the individual floors of office buildings as SUBSYSTEM level CIs. However, *what about the plumbing and electrical, heating, ventilation, and air conditioning (HVAC) system CIs that*

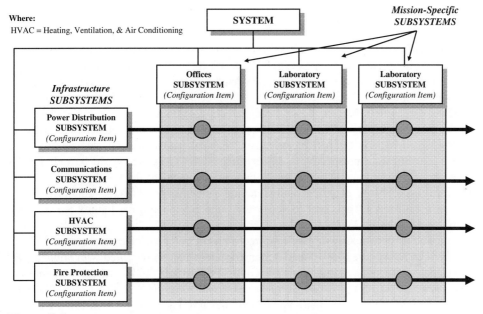

Figure 42.6 System Configuration Identification with Cross-Cutting CIs

TRANSCEND all of the floors and office boundaries? We can declare the floors and offices as *mission-based* SUBSYSTEMS. *Infrastructure* SUBSYSTEMS such as HVAC systems, would plumbing, electrical, and other such systems, *transcend* each floor.

Systems such as aircraft and automobiles, exhibit similar configurations. Fuel systems and electrical systems, for example, *traverse* the entire structure.

Relevance to SE

Now you are probably asking: *WHY IS this relevant to SE*? After all, this is simply a means of creating a SYSTEM architecture. The importance of this point is estimating the cost of a system during the System Procurement Phase. SEs must:

1. *Recognize* the existence of *mission-specific* and cross-cutting *infrastructure* systems.

2. *Appreciate* the need to establish a consensus on system configuration boundaries and delineate WHAT IS/IS NOT included within each *mission-specific* and *infrastructure* boundary via a *CWBS Dictionary* that scopes the contents.

The focal point for *organizing* cost estimating efforts, establishing work performance benchmarks, and measuring *planned* versus *actual* work progress is the *Contract Work Breakdown Structure* (*CWBS*). As we have reiterated numerous times, the CWBS, especially its Mission Equipment Element, should be *derived* from the SYSTEM's architecture. Unfortunately, most organizations do just the *reverse*. They haphazardly create the CWBS and then try to contort the system architecture to match the flawed CWBS!

Returning to our office building example, HOW do you estimate costs on the basis of:

1. Hierarchical categories of rooms with embedded plumbing, electrical, HVAC, and structural components?

2. Hierarchical *mission-specific* rooms and separate *infrastructure* structural, electrical, plumbing, and HVAC systems?

For systems such as buildings, homes, and aircraft, the *CWBS Dictionary* scopes *infrastructure systems* such as structural, electrical, plumbing, HVAC, and communications as separately cost accountable CWBS elements. This is driven by that fact that for electrical, plumbing, HVAC, network, and audio-visual communications elements, there are contractors who specialize in these areas.

Contemplate what would happen if we structured the CWBS to merge home construction *mission-specific* and *infrastructure* SUBSYSTEMs. The foyer, living room, kitchen, and bedroom SUBSYSTEMs would include internal electrical, mechanical, and plumbing elements. By this approach, from a verification perspective, each of these SUBSYSTEMS would be individually *inspected* and *verified*. Now imagine what would happen if you called the building inspectors to "inspect and verify" the HVAC, electrical, and structural elements of the Dining Room SUBSYSTEM and called them back a week later to verify the same elements of the Family Room SUBSYSTEM. The approach is FLAWED!

The bottom line is: decompose your system into a hierarchy of logical/behavioral and physical elements that can be specified, developed, procured, integrated, and verified with *minimal sets* of interdependencies. Recognize that *some* PRODUCTs/SUBSYSTEMs are mission-specific; others are *infrastructure* SUBSYSTEMs that transcend *mission-specific* SUBSYSTEMS. Structure the CWBS to accommodate not only the SYSTEM architecture but also cost estimating and contracting efforts.

Additionally, this approach enables separate specifications for infrastructure CIs that can be easily subcontracted out to vendors who specialize in those areas. If the infrastructure entities were embedded within a *mission-specific* CI, SEs would have to partition out requirements for infrastructure entities and create separate procurement specifications. The *extra work is simply unnecessary*.

42.6 MULTIPLE INSTANCES OF CI IMPLEMENTATION

Although we tend to think that every *item* in a system is unique, systems and products often have multiple instances of a single CI throughout the system. One of the roles of SE and the SEIT is to reduce development cost, schedule, and risk. You do this by investigating the evolving system design solution and searching for opportunities to STANDARDIZE components and interfaces. The bottom line is: AVOID REINVENTING THE WHEEL by creating specialized CI designs that can be satisfied by one common CI design.

42.7 CONFIGURATION BASELINES

System development, as evidenced by previous discussions, requires translation of abstract *System Performance Specification* (*SPS*) requirements into a physical solution. The translation *decomposes* or *expands* abstract complexity into more manageable lower levels of detail. Ultimately, a detailed design matures to a point whereby *design requirements* such as drawings and software design are sufficient in detail to support procurement, fabrication, and coding of all items in the SYSTEM.

Lower levels of design decisions, as refinements of higher level decisions, are totally dependent on maturing *stability* of the higher level design decisions. Otherwise, the entire solution evolves into multi-level *chaos*. So, as higher level decisions stabilize, it is important to capture and control the *state* of the evolving solution referred to as the *Developmental Configuration*.

The Developmental Configuration

The *Developmental Configuration* is characterized by a series of configuration "snapshots" that capture the evolving system design's solution at various stages of maturity. Each configuration enables SEs to maintain intellectual control of the *evolving* and *maturing* system design's solution by controlling changes to the configuration. Therefore, the scope of the Development Configuration spans the time period from Contract Award until a system is committed to production.

From an SE perspective, there are six types of configuration classifications as listed below that represent a level of maturity at various stages of development.

Stage 1: "As Specified" Configuration
Stage 2: "As Allocated" Configuration
Stage 3: "As Designed" Configuration
Stage 4: "As Built" Configuration
Stage 5: "As Verified" Configuration
Stage 6: "As Validated" Configuration
Stage 7: "As Maintained" Configuration
Stage 8: "As Produced" Configuration

Let's briefly synopsize each of these SE configurations.

- **"As Specified" Configuration** Represents the state of the *SPS* and lower level specifications as *captured* and *maintained* via the System Requirements Baseline.
- **"As Allocated" Configuration** Represents the *allocation* of requirements from the System Requirements Baseline specifications to items.
- **"As Built" Configuration** Represents the state of the Developmental Configuration for *first articles* at any level of abstraction prior to formal verification. Generally, the "As Built" Configuration employs serial number effectivity configuration control methods to match the physical system to its "As Designed" Configuration.
- **"As Designed" Configuration** Represents the current, approved Developmental Configuration baseline derived from the "*As Specified*" Configuration or System Requirements Baseline. The "*As Designed*" Configuration is initially captured at the System Design Review (SDR) and maintained to incorporate any change modifications to the baseline.
- **"As Verified" Configuration** Represents the state of the physical SYSTEM at completion of its formal System Verification Review (SVR) in which the "*As Specified*," "*As Designed*," and "*As Built*" Configurations mutually agree with each other and comply with specification requirements.
- **"As Validated" Configuration** Represents the state of the physical system as VALIDATED by the User, or an Independent Test Agency (ITA) representing the user, during the Operational Test and Evaluation (OT&E) under prescribed field operating environments and conditions.
- **"As Maintained" Configuration** Represents the state of the physical system configuration that is currently operated and maintained by the User in the field. From an SE and a configuration management perspective, the FAILURE to keep the "*As Maintained*" Configuration documentation current is a MAJOR RISK area, especially for developmental systems planned for production.

- **"As Produced" Configuration** Represents the state of the Production Baseline used to manufacture the system or product in production quantities.

Author's Note 42.2 *The "As Maintained" Configuration is a crucial point for SE. It documents the configuration of the fielded operational system(s). It is not uncommon for the User to become LAX due to the lack of initiative or budgets in maintaining the "As Maintained" Configuration documentation. The result is a physical system or product that DOES NOT identically match its configuration documentation, which is a MAJOR risk factor for the System Developer.*

Developmental Configuration Staging or Control Points

As the *Developmental Configuration* evolves through a series of design and development stages, it is important to establish *agreement* among the Acquirer, User, and System Developer about the *evolving* and *maturing* system design solution. Depending on the type of contract and WHAT provisions each party has in the decision-making process, we do this via *staging* or *control points*.

Staging or *control points*, which consist of major technical review events, are intended to represent *stages of maturity* of the evolving system design solution as it advances into lower levels of abstraction or detail over time. We do this via a series of *Configuration Baselines*.

From a configuration management perspective, there are four configuration baselines:

- System Requirements or *Functional* Baseline
- Allocated Baseline
- Product Baseline
- Production Baseline

Note that our discussions identified two perspectives of the *evolving* and *maturing* Developmental Configuration: 1) an SE configuration perspective and 2) a configuration management perspective. Despite the semantics both sets correlate as illustrated in Table 42.2.

42.8 GUIDING PRINCIPLES

In summary, the preceding discussions provide the basis with which to establish the guiding principles that govern system configuration identification practices.

Principle 42.1 Every component in a SYSTEM is an item. Some are designated as configuration items (CIs) for internal development; others as commercial off-the-shelf (COTS) items or nondevelopmental items (NDIs) for external procurement.

Principle 42.2 Development of a configuration item (CI) should only occur when all other COTS/NDI alternatives have proved to be impractical, not cost effective, or noncompliant with technical, cost and schedule requirements.

Principle 42.3 Every SYSTEM/Entity has seven SE design configurations—As Specified, As Designed, As Built, As Verified, As Validated, As Maintained, and As Produced—that must be current and consistent.

Principle 42.4 Every SYSTEM has three primary Developmental Configuration Baselines: 1) a System Requirements or Functional Baseline, 2) a Allocated Baseline, and 3) a Product Baseline. Systems in production have a Production Baseline.

Table 42.2 Correlating SE configuration and CM baseline perspectives

Staging or Control Point	SE Configuration Perspective	CM Baselines
System Requirements Review (SRR)	As Specified (system)	System Entity Requirements or Functional Baseline
System Design Review (SDR)	As Allocated As Designed (system/subsystem)	Allocated Baseline
Hardware Specification Review (HSR)	As Specified (HWCIs)	
Software Specification Review (SSR)	As Specified (CSCIs)	
Preliminary Design Review (PDR)	As Designed (HWCIs/CSCIs)	
Critical Design Review (PDR)	As Designed	
Completion of Component Procurement and Development	As Built	
Test Readiness Reviews (TRRs)	As Verified (components)	
System Verification Review (SVR)	As Verified (system)	Product Baseline
Formal Qualification Review (FQR)	As Validated (system)	
Pre-Production Readiness Review (PPRR)	As Maintained	
Production Readiness Review (PRR)	As Produced	Production Baseline

Principle 42.5 As a contract element, the Acquirer owns and controls the *System Performance Specification* (SPS); the System Developer maintains the SPS in accordance with the contract direction authorized and provided by the Acquirer Contracting Officer (ACO).

42.9 SUMMARY

During *system configuration identification discussions*, we addressed how various elements of the EQUIPMENT system element are designated via a set of semantic terms. These are the terms that SEs use as a common language to communicate. The intent is to enable the recipient to understand the context of each application.

GENERAL EXERCISES

1. Answer each of the *What You Should Learn from This Chapter* questions identified in the *Introduction*.
2. Refer to the list of systems identified in Chapter 2. Based on a selection from the preceding chapter's *General Exercises* or a new system selection, apply your knowledge derived from this chapter's topical discussions. Specifically identify the following:
   (a) Based on an analysis of the multi-level system architecture, which items would you designate as *configuration items*?
   (b) Which items might be procured as COTS or NDIs?

ORGANIZATIONAL CENTRIC EXERCISES

1. Research your organization's command media. Identify what requirements are levied by the organization on programs regarding CIs, HWCIs, CSCIs, COTS, and NDIs?
2. Contact an internal contract program. Inquire as to how items, CIs, HWCIs, CSCIs, COTS, and NDIs are implemented in the program.
   (a) Does the contract specify criteria for selecting CIs?
   (b) Correlate CIs, HWCIs, and CSCIs with the specification tree.
   (c) Report your findings and observations.
3. For the contract program noted above, investigate how *firmware* is developed, managed, and controlled.
4. Research Internet sites for instances of how other organizations specify and control CIs, HWCIs, CSCIs, COTS, NDIs, and firmware.

REFERENCES

ANSI/EIA-649-1998. 1998. *National Consensus Standard for Configuration Management*, Electronic Industries Alliance (EIA). Arlington, VA

IEEE Std 610.12-1990. 1990. *IEEE Standard Glossary of Modeling and Simulation Terminology.* Institute of Electrical and Electronic Engineers (IEEE). New York, NY.

Defense Systems Management College (DSMC). 2001. *Glossary: Defense Acquisition Acronyms and Terms*, 10th ed. Defense Acquisition University Press Ft. Belvoir, VA.

MIL-STD-483A (canceled). 1985. *Configuration Management Practices for Systems, Munitions, and Computer Programs.* Washington, DC: Department of Defense.

MIL-STD-498 (canceled). 1994. *Software Development and Documentation.* Washington, DC: Department of Defense (DoD).

MIL-STD-973 (canceled). 1992. *Configuration Management.* Washington, DC: Department of Defense (DoD).

ADDITIONAL READING

DI-IPSC-8 1448A, 1999. DoD Data Item Description (DID). *Firmware Support Manual (FSM).* Washington, DC: Department of Defense (DoD).

MIL-HDBK-61. 1997. *Configuration Management Guidance.* Washington, DC: Department of Defense (DoD).

Federal Aviation Administration (FAA), ASD-100 Architecture and System Engineering. *National Air Space System—Systems Engineering Manual.* Washington, DC: FAA.

Chapter 43

System Interface Analysis, Design, and Control

43.1 INTRODUCTION

The identification, design, and control of system interfaces are key activities for system architecture development. The capability of the SYSTEM and *item* interfaces to cooperatively or defensively *interact* and *interoperate* with external systems within the context of their OPERATING ENVIRONMENT often determines mission survival and success.

Our discussion expands that of Chapter 12, *System Interfaces*, into HOW SEs translate abstract interface requirements into specification requirements. Based on those requirements, analytical, scientific, engineering, and management principles enable SEs to design the interface. The foundation for interface design builds on our discussions of Chapter 8 *The Architecture of Systems* and Chapter 21 *System Operational Capability Derivation and Allocation*.

We continue our discussion of *generalized* and *specialized* interfaces, define an interface design methodology, investigate system ownership and control, address the need for standardization, and define how interfaces are documented. We conclude with a discussion of the challenges and issues in defining the system interfaces.

What You Should Learn from This Chapter

1. How are interfaces identified for and within a SYSTEM?
2. How do you analyze interface interactions?
3. How is the *System Capability Construct* applied to interface design?
4. What is the *methodology* for designing interfaces?
5. Who *owns* and *controls* SYSTEM and item interfaces at various levels of abstraction?
6. What is an *Interface Requirements Specification (IRS)*?
7. What is an *Interface Control Document (ICD)*?
8. What is an *Interface Design Description (IDD)*?
9. How do you decide to develop and IRS, ICD, and/or IDD?
10. What is an *Interface Control Working Group (ICWG)*?
11. Who chairs the ICWG?
12. What are some rules for *analyzing*, *designing*, and *controlling* SYSTEM or entity interfaces?

System Analysis, Design, and Development, by Charles S. Wasson
Copyright © 2006 by John Wiley & Sons, Inc.

Definitions of Key Terms

- **Compatibility** Refer to the definition provided in Chapter 15 *System Interactions with Its Operating Environment.*
- **Coupling** "The manner and degree of interdependence between software modules. Types include common-environment coupling, content coupling, control coupling, data coupling, hybrid coupling, and pathological coupling." (Source: IEEE 610.12-1990)
- **Interface Control** Refer to the definition provided in Chapter 12 *System Interfaces.*
- **Interface Ownership** The assignment of accountability to an individual, team, or organization regarding the definition, specification, development, control, operation, and support of an interface.
- **Line Replaceable Unit (LRU)** Refer to definition in Chapter 42 on *System Configuration Identification Practices.*

43.2 IDENTIFYING AND ANALYZING INTERFACE INTERACTIONS

Once *logical entity relationships* between the SYSTEM and external systems or between internal items are identified, the next step is to *analyze* and *bound* the interactions. In Chapter 12, *System Interfaces,* we noted that physical interfaces can be characterized as *mechanical*, *electrical*, *optical*, *acoustical*, *environmental*, *chemical*, *biological*, and *nuclear*. The question is: *How do we specify and bound the operational and physical characteristics of the interface.* Let's answer each part of this question separately.

Specifying and Bounding Interface Operational Characteristics

Interface operational characteristics are derived using UML® *sequence diagrams* as illustrated in Figure 17.3. These diagrams, coupled with a *Mission Event Timeline (MET)*, provide analytical insights into HOW the interfacing entities *interact* and *interoperate*.

Specifying and Bounding Interface Physical Characteristics

Analysis of *interacting* systems requires investigation of a variety of classes of interactions. For most systems the classes of interfaces include:

- Electrical
- Mechanical
- Optical
- Chemical
- Biological
- Acoustical
- Human
- Mass Properties

One of the challenges of physical interface analysis is that SEs and analysts become *intrigued* and *immersed* by a specific class of interaction and tend to overlook or ignore other classes that may become SHOWSTOPPERS. One method of analyzing interfaces employs a matrix approach as illustrated in Figure 43.1.

Author's Note 43.1 *The matrix provides a framework for illustrating the thought processes required to understand all of the performance effecters that influence design considerations. Based on these thought processes, your job as a system analyst or SE is to determine which one(s) of the effecters warrants consideration for specific SYSTEM applications.*

The matrix maps interactions between a MISSION SYSTEM interface classes (rows) and the OPERATING ENVIRONMENT interface classes (columns). Since both domains of system elements have comparable classes of interfaces, SEs divide each domain into the various categories. Note also that the OPERATING ENVIRONMENT includes the NATURAL, INDUCED and HUMAN-MADE SYSTEMS elements. To facilitate the analysis, we assign a unique identifier to each *interaction* (row-column intersection).

Thus, for each interaction, at least one or more specification requirements are written to *specify* and *bound* the *interaction* and the *expected outcome* and performance of each interaction. To illustrate this point, consider the following example:

EXAMPLE 43.1

In an environmentally controlled laboratory environment, electrical (class) interactions—such as electromagnetic radiation (EMI) and noise—are *likely* to occur and may have an effect on the test articles or instrumentation. In contrast, chemical (class) interactions—such as salt spray—do not occur naturally in a laboratory.

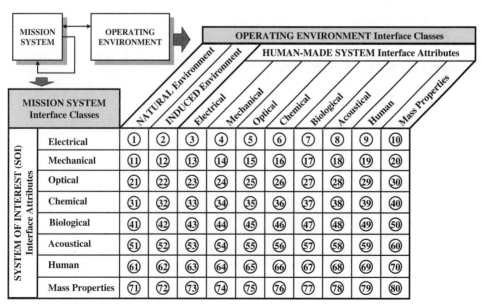

Figure 43.1 SYSTEM OF INTEREST (SOI) Interface Interactions Analysis Matrix

Practical Realities

In *theory*, this approach seems logical; however, *is it practical to develop an analysis such as this within contract or task resource and time constraints*? The answer depends on your situation. In general, most seasoned SEs *subconsciously imprint* this analytical method into memory based on personal experience. The challenge is assimilating all *relevant interactions* from memory without overlooking any condition.

If your contract or task is resource and time limited, you might consider using a template such as this as a quick checklist to identify the *most likely* or *probable* interactions. In sharp contrast, some SYSTEMS may have inherent safety risks with potential consequences for human health and safety, property, the environment, or survival of the enterprise. *You and your organization must weight the cost to perform and merits of this analytical task versus the legal, financial*, and other *consequences of IGNORING ALL likely interface interactions in practical terms.*

Guidepost 43.1 *Based on the preceding discussion, we have identified and characterized the attributes of interface interactions. The question is: WHAT inherent SYSTEM or entity interface capabilities and levels of performance are required to successfully:*

1. *Be compatible or interoperable with the external SYSTEMS or entities.*
2. *Avoid threat vulnerabilities related to these interactions?*

This brings us to our next topic, *understanding system interface design solutions*.

43.3 GENERALIZED VERSUS SPECIALIZED INTERFACE DESIGN SOLUTIONS

Physical interface solutions occur at two levels: 1) as *generalized solutions* and 2) as *specialized solutions*. *Generalized* solutions represent a "first-pass" physical implementation that appears "on the surface" to be adequate for most applications. There may be special circumstances that require further interface design *refinements* or *robustness*; we refer to these *specialized* solutions. Let's explore each of these further.

Generalized Interface Solutions

Generalized interface solutions represent the analytical world whereby that analyst identifies a *logical entity relationship*, association, or potential relationship between two entities such as a MISSION SYSTEM and its OPERATING ENVIRONMENT. Consider the following example:

EXAMPLE 43.2

Analytically, a car has a potential *logical entity relationship* with other vehicles, trees, and people in its OPERATING ENVIRONMENT. The recognition of this association is an acknowledgement that:

1. A physical relationship exists between the entities.
2. Further interface analysis must determine the *level of significance* and *outcomes* of the *interactions* to the MISSION SYSTEM and, if necessary, on the interfacing entity—its *vulnerability* and *survivability*.

Specialized Interface Solutions

Once a determination is made that a *logical entity relationship* exists between entities, the associations are analyzed to determine their effects on the interface and subsequently the MISSION SYSTEM. Let's return to the preceding car example.

EXAMPLE 43.3

Engineers determined many years ago that vehicles require front and rear bumpers. After many years of accidents and federal mandates, automobile manufacturers *upgraded* bumper designs to include shock absorbers that reduce the effects of impact with other vehicles and objects. Although the new bumpers solved one problem, vehicle occupants continued to have the risk of injuries due to head-on crashes. Subsequently, seat belts were added to vehicles. Although seat belts reduced injuries and saved lives, vehicle occupants continued to suffer life-threatening injuries. So, car designs incorporated air bag systems into the driver's side of the vehicle. Later air bags were added to protect the front seat passenger. Today additional specialized solutions are being added to incorporate side impact air bags. New evidence indicates emergency medical services (EMS) personnel have the potential risk of injury from air bag deployment while attending to accident victims trapped in the wreck. So, specialized solutions to operational needs continue to evolve.

This example illustrates HOW a *generalized solution* representing a *logical entity relationship* *evolves* into *specialized solutions* via physical design implementations.

Guidepost 43.2 *Based on an understanding of generalized and specialized interfaces, we shift our attention to HOW an interface is implemented.*

43.4 APPLYING THE SYSTEM CAPABILITY CONSTRUCT TO INTERFACES

As a prerequisite to our discussion here, revisit the key points noted in Chapter 22 *The Anatomy of System Capability*. Figure 22.1 serves as a guide for our discussion.

The *System Capability Construct* provides a framework to describe HOW a system capability such as an interface is:

1. Initiated.
2. Performs its mission—resulting in a product, service, or action.
3. Configures itself for *repetitive cycle* or *dormancy* until initiated again.

In performing these actions, an interface capability assesses health and readiness status, makes a determination of degraded performance or failure conditions, reports those conditions, and attempts to recover from those conditions.

43.5 INTERFACE DESIGN METHODOLOGY

Key elements of interface design practices can be summarized as a five-step methodology that fits most applications. The steps are:

Step 1: Identify the SOI–OPERATING ENVIRONMENT relationships.
Step 2: Develop the SYSTEM or item architecture.
Step 3: Characterize the logical entity relationships of the architecture.

Step 4: Characterize the operational interface use cases.

Step 5: Characterize the physical interface characteristics.

Let's address each of the steps of the methodology.

Step 1: Identify the SOI–OPERATING ENVIRONMENT Relationships

The initial step of the methodology is to identify those entities within the User's OPERATING ENVIRONMENT that *represent* the interfaces relevant to the SYSTEM OF INTEREST (SOI).

Step 2: Develop the SYSTEM or Item Architecture

Construct the SYSTEM or entity architecture to depict its *logical* (associative) and/or *physical* relationships with external systems.

Step 3: Characterize the Logical Entity Relationships of the Architecture

Based on the SYSTEM or entity architecture or validated User needs analysis, characterize the *logical entity relationships* between internal and external entities—such as HUMAN-MADE SYSTEMS and the OPERATING ENVIRONMENT. In general, this step formally describes the interface.

Step 4: Characterize the Operational Interface Use Cases

For each SYSTEM level or entity interface, identify the key operational characteristics of the interface.

1. WHO are the interface stakeholders?
2. WHAT is to be exchanged across the interface—content, forces, and directional flow?
3. WHEN and HOW frequently is the interface to be employed—continuous or intermittent connectivity?
4. WHERE is the interface to be employed?
5. HOW will the interface be controlled, security and privacy methods?
6. Under WHAT conditions is the interface to be employed?

Step 5: Characterize the Physical Interface Characteristics

Using the operational interface attributes as the basis, *identify* the key *operational* and *physical characteristics* of the interface. For data communications interfaces, interface attributes include connectors, pin-outs, wiring diagrams, grounding and shielding, protocol, timing and synchronization, data formats, handshakes, addressing, encryption, and standards as noted in Table 43.1.

Step 6: Develop and Document the Interface Design Solution

As the *operational*, *logical*, and *physical* attributes of the interface are identified and characterized, capture the results in an *Interface Control Document* (*ICD*) or *Interface Design Description* (*IDD*). Document the design rationale, trade-offs, and so forth.

Table 43.1 Interface attributes and descriptions

ID	Attribute	Attribute Description
43.1	Physical Interface Type	Characterize the interface in terms of its electrical, mechanical, optical, chemical, or environmental attributes.
43.2	Operational States and Modes	Identify the interface *modes and states of operation* such as STANDBY, POWER OFF, stowed, and disconnected.
43.3	Directionality	Determine the directionality of the interface for various modes of operation. Some interfaces are unidirectional, bi-directional, etc.
43.4	Interface Protocol	Investigate any protocol requirements to implement a specific protocol for sending and receiving messages such as Identification Friend or Foe (IFF) and standards.
43.5	Frequency of Usage	Determine the frequency of usage such as statistical distributions of usage during hours of operation, peak periods; *asynchronous* and *synchronous* communications, etc.
43.6	Reserve Capacity	Depending on the frequency of usage, some interfaces may require memory storage to accommodate message buffering prior to transmission or after receipt of messages. Electrical power and mechanical systems require similar analogies to serve as design safety margins to sustain performance.
43.7	Strength	Some data system interfaces separated by long distances impose *minimum* electrical, mechanical, or optical strength requirements such as amplitude, power, and pressure that require drivers to sustain signal strength.
43.8	Specialty Requirements	Depending on technical risk, investigate interface specialty-engineering requirements that include reliability, availability, maintainability, vulnerability, survivability, etc.
43.9	Technology	For some systems, characterize the interface technology such as laser, sensor, data communications, chemical, and radiation.
43.10	Training	Due to the nature of some interfaces, identify special training for operators and maintainers.
43.11	Cost	Assess WHAT level of capability you can provide within budgetary cost and schedule constraints.
43.12	Risk	Identify the level of risk associated with operating the interface including probability of occurrence and consequences of failure.

43.6 SPECIFYING AND BOUNDING SYSTEM INTERFACE REQUIREMENTS

Before we address *specifying* and *bounding* interface requirements, let's establish WHERE and HOW interface requirements are documented. Interface requirements are typically specified in the *System Interfaces* section of a *System Performance Specification* (*SPS*) or *item development specification*; typically, Section 3.X. In some cases, software interface requirements may be specified in an *interface requirements specification (IRS)*.

The Interface Specification

What is an IRS? DoD Data Item Description (DID) DI-IPSC-81434 states *"The Interface Requirements Specification (IRS) specifies the requirements imposed on one or more systems, subsystems, Hardware Configuration Items (HWCIs), Computer Software Configuration Items (CSCIs), manual operations, or other system components to achieve one or more interfaces among these entities. An IRS can cover any number of interfaces.... The IRS can be used to supplement the System/Subsystem Specification (SSS) . . . and Software Requirements Specification (SRS) . . . as the basis for design and qualification testing of systems and CSCIs."* [Source: DoD Data Item Description (DID) DI-IPSC-81434, p. 1]

People ask: *If the interface requirements are stated in the SPS or item development specification, WHY do you need a separate interface specification?* The answer has two contexts:

1. Interfaces *external* to the SYSTEM.
2. Interfaces *internal* to a SYSTEM.

Ideally, interface requirements for a SYSTEM should be documented in a single requirements specification, preferably the SPS, or a lower level *item development specification (IDS)* for a configuration item (CI). The answer to the question depends on:

1. WHAT your contract requires.
2. Other factors related to protecting System Developer sensitive data such as contract specifications, intellectual property, proprietary data, and security.

Contract Requirements for an Interface Specification

ALWAYS investigate and READ the Request for Proposal (RFP) and its *Contract Data Requirements List (CDRL)* regarding interface specification requirements, if any. If there are no IRS CDRL items, incorporate interface requirements into the SPS or *item development specification*, depending on application. This avoids:

1. The *necessity* and *expense* of maintaining and verifying two separate documents—SPS/IDS and an interface specification—with common boilerplate.
2. Coordinating separate document reviews and approvals.

From a system perspective, you want to VERIFY the system as an entity using *one* specification with:

- One set of *acceptance test procedures (ATPs)*.
- One *Requirements Verification Matrix (RVM)*.
- One set of ATP data.

Simultaneously verifying requirements stated in TWO separate specifications compounds the tasks, paperwork, and coordination. You may respond that you could have two specifications and a single set of ATPs and ATP results. This is true, but you still have to account for HOW you can TRACK these separate/combined results. This leads to WHY not specify all SYSTEM or entity interface requirements in one document *unless* there is a *compelling* need to isolate the requirements in a separate document. *Keep it simple*!

Is an IRS Necessary for Internal Development? No. In fact a *common problem* of many programs is having ill-informed decision makers declaring in their proposal they will develop *"an IRS for EVERY external and internal interface."* This is naive, expensive, and unnecessary.

There are, however, reasons—such as volume of requirements, data security classification, and procurement reasons—for a separate IRS. By default, define the interface requirements in the SPS or *item development specification*. Then, if there are compelling reasons to isolate the total set of interface requirements, then and only then consider creating a separate IRS.

Is an IRS Necessary for External Development? An IRS can be helpful in contracting relationships with User organizations and subcontractors. In this case the IRS *expresses* a willingness to abide by a set of ground rule requirements that represent a consensus of the interfacing stakeholders. Ultimately, the IRS requirements evolve into interface solutions documented in hardware ICDs and/or software IDDs.

Standard IRS Templates. The DoD employs data item descriptions (DIDs) to communicate deliverable data requirements. DID DI-IPSC-81434 addresses hardware configuration items (HWCIs) and computer software configuration items (CSCIs) for software intensive systems.

Guidepost 43.3   *Given an understanding of HOW we specify and bound interface requirements, we shift our focus to Interface Ownership and Control.*

43.7 INTERFACE OWNERSHIP AND CONTROL

Unlike system elements or entities that are assigned to individuals or development teams—such as Integrated Product Teams (IPTs)—HOW do you *partition* or *share ownership* of the interface? There are a couple of ways to assign ownership:

1. *Ad hoc* approach.
2. *Structured analysis* approach.

Ad Hoc Approach to Interface Ownership

Technical directors and project engineers often approach interface ownership by *tasking* the two interfacing parties to *"work it out on their own."* Depending on the situation and personalities, this approach works in some cases; in other cases, it is very *ineffective*.

Several potential PROBLEMS can arise with this approach.

1. Conflicts occur when the interfacing parties are *unable* or *unwilling to agree* on HOW to implement the interface.
2. One personality *dominates* the other, thereby creating a technical, cost, or schedule *suboptimization* that favors of the dominating party.

If this chaos or dominance continues, the dominating party may suboptimize the SYSTEM. There is, however, a better approach to interface ownership and control that resolves the problems created by the ad hoc ownership approach.

Structured Interface Ownership and Control

To avoid the problems of the ad hoc approach, assign interface ownership and control to the individual or development team accountable for the *System Performance Specification (SPS)* or *item development specification* and the associated architecture. The interfacing parties are elements WITHIN the architecture.

Some systems have external interfaces that require both *technical* and *programmatic* solutions between organizations. This brings us to our next topic, *Interface Control Working Groups (ICWGs)*.

Interface Control Working Groups (ICWGs)

Once the *stakeholders* agree on the requirements and implementation of an interface, *HOW do they exercise control over changes to an interface?* Some organizations establish an Interface Control Working Group (ICWG) that functions as a Configuration Control Board (CCB) or makes recommendations to a CCB. Generally, ICWGs are typically established for external system interfaces that involve the User community and external systems. Depending on the size of the contract, the ICWG and CCB may consist of the same *stakeholders*. In other cases the ICWG may consist of technical representatives from User and/or contractor organizations.

As an organizational element, the ICWG defines the *operational*, *behavioral/logical*, and *physical* characteristics of the interface and documents them in a hardware ICD or a software IDD. Interface stakeholders review the ICD or IDD and make recommendations to submit to the next higher level team as the interface owner for approval and release.

When approved, the ICWG chairperson requests the program's Configuration Manager to baseline interface documentation such as an IRS and formally release it for usage. After baselining, if subsequent changes are required, the ICWG reviews the changes and recommends approval to the program's Configuration Control Board (CCB).

43.8 INTERFACE DESIGN DOCUMENTATION

Interface designs are documented via a number of methods such as system block diagrams (SBDs); entity relationship diagrams (ERDs); schematic, wiring, and timing diagrams; tables; and descriptions. For large systems, this may consist of hundreds of pages of documents that must be individually controlled. *HOW can this be accomplished reasonably?*

SEs use several types of documents to capture and control interface technical descriptions. These documents include:

1. *Interface Control Documents* (*ICDs*) for HARDWARE interfaces.
2. *Interface Design Descriptions* (*IDDs*) for SOFTWARE interfaces.

For small programs the *System/Segment Design Description* (*SSDD*) or *Software Design Description* (*SDD*) may suffice to document interface definitions rather than having separate ICDs/IDDs. These documents serve as a single location for finding interface design details. Use the appropriate document that enables you to reduce cost and help the developers by MINIMIZING the total number of documents requiring configuration control.

Interface Design Descriptions (IDDs)

For software applications document interfaces in an *Interface Design Description (IDD)*. This approach is often used for software CSCIs where there is a need to isolate CSCI specific interface design descriptions as a separate document.

Interface Control Documents (ICDs)

The most common approach for documenting interfaces is via an *Interface Control Document* (*ICD*). As a general rule, an ICD documents hardware interfaces. An ICD:

1. Ranges in length from a single- or multi-page document, such as a control drawing or a computer listing, to a volume of details.
2. Serves as a detailed design solution *response* to SPS, item *development specification*, or IRS requirements.

3. Documents the electrical, mechanical, or optical characteristics of the interface.
4. Includes operational descriptions, schematic wiring diagrams, assembly drawings, and connector detail drawing including pin outs.
5. Employs standard graphical symbols that may be dictated by contract or established as industry standards.

When all interface *stakeholders* agree with the contents of the ICDs, the documents are approved, baselined, placed under configuration control, and released for formal decision making. ICDs are controlled by a Configuration Control Board (CCB) or delegated to an ICWG.

Interface Control Document (ICD) Outline

ICDs often developed in *free form* unless the contract or organization specifies a specific format. There are numerous ways to structure an ICD, depending on the application. Perhaps the most important aspects of the outline can be derived from the traditional specification outline such as:

- Section 1.0 *Introduction*
- Section 2.0 *Referenced Documents*
- Section 3.0 *Requirements*

Beyond this basic structure, attention should be focused on Section 3.0 *Requirements*, which addresses the physical characteristics of interfaces such as mechanical, electrical, and optical.

How Many ICDs/IDDs?

One of the early challenges in identifying interfaces is determining how many ICDs/IDDs are required. For each item, regardless of *level of abstraction*, consider the following options illustrated in Figure 43.2:

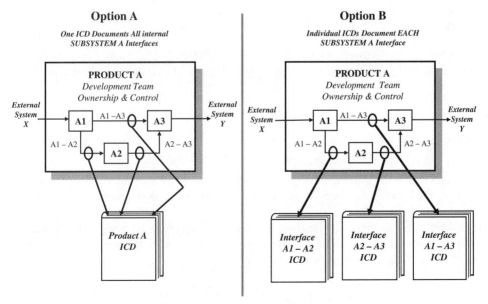

Figure 43.2 SYSTEM/PRODUCT Level Interface Definition Options

- **Option A** *Do we create a single ICD that defines all of PRODUCT A interfaces?*
- **Option B** *Do we create individual ICDs for each PRODUCT A internal interface—such as A1–A2 ICD, A2–A3 ICD, or A1–A3 ICD?*

There are several ways to answer these questions.

First, each document you create *increases* costs to *maintain*. As a general rule, AVOID creating separate ICDs/IDDs unless:

- The amount of information becomes *unwieldy*—because of large page counts.
- There is a need to *isolate* details due to intellectual property, proprietary data, and security reasons—to limit *authorized* access to a specific interface based on a NEED TO KNOW justification.

At the PRODUCT level and below, start with a single ICD for all interfaces internal to the item. You may ask: *Why not external interfaces?* There are two reasons: *First*, interfaces external to an item are *owned* and *controlled* by the next higher-level team as part of their architecture for which the item of interest is an element. The SEIT owns and controls PRODUCT A's external interfaces. *Second*, publishing an item's interfaces in a single document is convenient for the reader, assuming that the document is not large.

Why Stand-alone ICDs and IDDs?

The preceding ICD discussion opens up a new question: *WHY do you need separate ICDs and IDDs for each interface?*

First, there are no rules that say you cannot have a single interface document that covers BOTH hardware ICD and software IDD details. Potential readers can rationalize the need to have all details about a single interface in one document without having to *research* or *retrieve* several documents.

In general, you would expect the COST of producing and maintaining one document should be *less* than maintaining two documents two documents. However, distributing changes to all stakeholder approvers, regardless of whether they are affected by the changes, can be *unwieldy* and *costly*.

What do the ICD/IDD users say?

- Software engineers and programmers are not interested in reading through pages of electrical schematics, wiring diagrams, and mechanical drawings.
- Hardware engineers are not interested in reading through tabular listings of software data.

So, in answer to the question:

1. Identify all of an item's internal interfaces.
2. Collaborate with the stakeholders, preferably as an Integrated Product Team (IPT).
3. Address the utility of standalone versus integrated ICDs and IDDs.
4. If a decision is made that the item requires hardware and software interfaces, decide if all hardware or software interfaces should be documented in a one or several ICDs/IDDs.

Guidepost 43.4 *Our focus up to this point has been on individual interfaces. This leaves the potential for every interface to be unique which increases cost and risk. We now shift our focus to minimizing these two factors via Interface Standardization.*

43.9 INTERFACE STANDARDIZATION

Interface design, as is the case with any form of SYSTEM design, focuses on meeting the specified requirements while minimizing cost, schedule, technical, technology, and support risks. Every time you embark on designing a new interface solution, you must be prepared to mitigate the risks of an *unproven* interface.

One way of reducing the impacts of these risks is to employ design solutions that are already *proven*. Additionally, any technology solution you choose may be subject to becoming obsolete in a short period of time. In sharp contrast, consumer products in the marketplace, especially computers, demand that SYSTEMs be designed to *accept* technology *upgrades* to maintain system capabilities and performance without requiring completely new systems.

One of the ways industry addresses this marketplace need is with standard interfaces that promote line replaceable unit (LRU) *modularity*, *interchangeability*, *flexibility*, and *maintainability*. What does this mean?

A computer contains printed circuit boards (LRUs) that interface via connectors with a standard bus structure implemented on a motherboard (LRU). As new processor technologies or other capabilities are introduced, the User replaces a PC board (LRU) with a newer one with identical or greater capability. Thus, interface standards provide an opportunity to *leverage* new technologies and capabilities while keeping costs and risk low.

Guidepost 43.5 *Based on the preceding discussions, we are ready to investigate some of the challenges and issues in analyzing, designing, and controlling interfaces.*

43.10 INTERFACE DEFINITION AND CONTROL CHALLENGES

Interface definition, design, development, operations, and support activities often face challenges that are common across many systems. Let's identify and discuss some of these key challenges.

Challenge 1: Lack of external interface commitments
Challenge 2: Lack of interface ownership and control
Challenge 3: Identification and vulnerability of threat interfaces
Challenge 4: Human and environmental safety and health risks
Challenge 5: Lack of compatibility and interoperability
Challenge 6: Lack of interface availability
Challenge 7: Lack of interface reliability
Challenge 8: Lack of interface maintainability
Challenge 9: Interface vulnerability to external threats
Challenge 10: Mitigating interface integrity compromises and failures
Challenge 11: External electrical power—availability, quality, and backup
Challenge 12: Analog and digital signal grounding and shielding
Challenge 13: Interface electromagnetic emissions

Challenge 1: Lack of External Interface Commitments

Contracts are awarded every day whereby the Acquirer states in the *System Performance Specification* (*SPS*) "the system shall provide the capability to interface with External System XYZ." On investigation, the Acquirer may not have agreement or commitment from the owner(s) of external system XYZ to allow the SYSTEM OF INTEREST (SOI) to connect. This is normally the Acquirer's responsibility and should have been worked out prior to release of the formal solicitation. Ironically, the Acquirer levies responsibility for "working the commitment" on the System Developer, who signed up to the *terms and conditions* (*Ts&Cs*) of the contract.

In some cases this approach may be is *preferable*, especially if the System Developers has:

1. The expertise, capabilities, and resources.
2. Established relationships with the interfacing parties.

Therefore, it may be *acceptable* to contract with the System Developer to perform this task. The *critical issue* occurs when external system XYZ's owner organization is also part of the Acquirer's organization. Thoroughly investigate WHAT *agreements* and *commitments* have been made by the Acquirer or User with external system owners to integrate the system at delivery BEFORE the contract is signed.

Challenge 2: Interface Ownership and Control

Each system interface must be assigned an owner(s)—be it an individual, organization, or Interface Control Working Group (ICWG). Those accountable MUST control interface definition, design; development; system integration, test, and evaluation (SITE); system operation and maintenance; or disposal. As the *accountable* owner, the individual or organization is responsible for *reviewing* and *approving* changes to the interface design baseline as well as provide oversight of SYSTEM operations, maintenance, and training.

Challenge 3: Identification and Vulnerability of Threat Interfaces

System interface design is based on a pre-conceived set of known interfaces. The reality is some operational system interfaces—such as military systems or networks—are *vulnerable* to external threats and attacks. These systems must contend with the *unknowns* and the *unknown—unknowns*. Acquirers and System Developers must work with the Users and their supporting organizations to thoroughly:

1. Understand and anticipate a system's threats.
2. Define HOW the SYSTEM interfaces will *cope* with those *threats*.

Challenge 4: Human and Environmental Safety and Health Risks

Any type of operational system may pose potential threats to the safety, health, and welfare of humans, property or the environment. When designing system interfaces, thoroughly analyze potential scenarios—such as exhaust emissions, toxic chemicals, and leaking fluids—that may lead to health and safety risks and require remediation to an *acceptable* level.

Challenge 5: Lack of Compatibility and Interoperability

Assuming each interface is *known* to exist and has an accountable owner, the SYSTEM or entity interface must be *compatible* and *interoperable* with its OPERATING ENVIRONMENT system elements *operationally*, *behaviorally*, and *physically*.

Challenge 6: Lack of Interface Availability

Interface availability is a critical issue in two key contexts: *internally* and *externally*. *Internally* each interface must be *available*—the state of readiness—when activated to perform its intended mission. If external systems or their interfaces fail, you must consider HOW your SYSTEM will *respond* or *adapt* to the failure and derive MISSION RESOURCES Element data from contingency sources.

Referral For more information about *availability*, refer to Chapter 50 on *Reliability, Availability*, and *Maintainability Practices*.

Challenge 7: Lack of Interface Reliability

If an interface capability is available when required, the question is: *Can the interface reliably perform its intended mission to the level of performance required by the SYSTEM or entity(s) specification?* Each interface must be designed with a level of reliability that will ensure accomplishment of mission tasks and sustainment of that capability throughout the mission. Supporting topics include interface *fault tolerance* and *redundancy*.

Referral For more information about *reliability*, refer to Chapter 50.

Challenge 8: Lack of Interface Maintainability

To *minimize downtime* between SYSTEM missions you must ensure that the interfaces are maintainable with a specified level of skills and tools commensurate with the mission phase of operation—pre-mission, mission, and post-mission.

Referral For more information about *maintainability*, refer to Chapter 50.

Challenge 9: Interface Vulnerability to External Threats

After a SYSTEM is deployed, the owner(s) must continuously *monitor* the performance of the mechanisms and processes used to operate the interface. Additionally, the owners must also assess the *susceptibility* and *vulnerability* of the interface to evolving or potential threats in the SYSTEM's OPERATING ENVIRONMENT. Where appropriate, *specialized solutions* may be implemented.

Challenge 10: Mitigating Interface Integrity Compromises and Failures

The integrity of an interface is dependent on HOW WELL its design *mitigates* compromises and failures via *specialized* interfaces. Interface security mechanisms include examples such as: lockable access plates, safety chains, retractable wheels, bulkheads, filters, shields, deflectors, cable hooks on aircraft, safety nets, safety rails, drogue chutes, parachutes, pressure relief valves, emergency cutoff valves, and emergency power down.

Challenge 11: External Electrical Power—Availability, Quality, and Backup

Engineers often focus on the "fun stuff" of internal design of their SYSTEM of INTEREST—the SYSTEM, PRODUCT, SUBSYSTEM, and so forth. They procrastinate on researching external interfaces such as electrical power sources, their attributes, and quality. In general, they assume that 110 vac or +28 vdc electrical power will *always be available and all we have to do is plug our device in.*" Given these examples of power types, engineers often *overlook* the *subtleties* such as 50 Hz versus 60 Hz versus 400 Hz as well as tolerances placed on the magnitudes and frequencies. Power factors are also a consideration. Finally, a key question is: *Will the power source be continuous or have periodic operational hours, blackouts, etc., during peak hours of operation?*

DO NOT ASSUME that the external system *will* ALWAYS *be available* WHEN you need it until you have thoroughly investigated and analyzed the interface. Additionally, establish a documented agreement that *represents* a commitment by the power source's owner to allocate power budget resources and allow your system to be integrated with the power source(s). The same is true for power quality and filtration.

Finally, assess the need for:

1. Backup power to *avoid* data loss during a mission.
2. Alert *notification* that electrical power has been lost to initiate power down actions to *minimize* data loss or damage to equipment.

Challenge 12: Analog and Digital Signal Grounding and Shielding

Analog and digital grounding and shielding interfaces have similarities to the electrical power issue noted above—such as procrastination in performing the analysis, depth of analysis, and resource commitments. Typically, only a single power ground is available from the external source. This is particularly *problematic* when your SYSTEM must collect measurement data or transmit data relative to power ground that is corrupted with switching noise or ground loops. Thoroughly investigate:

1. WHAT types of external grounding systems are available.
2. HOW the external system implements power and signal ground.
3. WHAT other systems experienced and discovered when interfacing to this source.

Challenge 13: Interface Electromagnetic Emissions

Electronic power and signal interfaces often emit electromagnetic signals that may couple to other data sensitive devices or are tracked by external surveillance systems. Most engineers tend to think of signal shielding and grounding from a cabling perspective. However, signal shielding and grounding also applies to signal sources contained within mechanical enclosures as well as facilities.

43.11 GUIDING PRINCIPLES

In summary, the preceding discussions provide the basis with which to establish the guiding principles that govern system interface analysis, design, and control practices.

Principle 43.1 Specifications for and interfaces between systems and items are *owned* and *controlled* by the organization accountable for the architecture that depicts the items.

Principle 43.2 Document the operational, physical, and data requirements of every interface within an entity's specification; create a separate software IRS only if there are compelling reasons.

Principle 43.3 ICDs document hardware interfaces; IDDs document software interfaces.

43.12 SUMMARY

During our discussion of *system interface analysis*, *design*, *and control* we:

1. Described HOW interfaces are identified and documented in an IRS, ICD, and IDD.
2. Established a methodology for identifying and defining system interfaces.
3. Discussed HOW ICWGs control system interfaces.
4. Provided common examples of interface standards.
5. Identified common challenges and issues in interface definition and control.

GENERAL EXERCISES

1. Answer each of the *What You Should Learn from This Chapter* questions identified in the *Introduction*.
2. Refer to the list of systems identified in Chapter 2. Based on a selection from the preceding chapter's General *Exercises* or a new system selection, apply your knowledge derived from this section's topical discussions. Using the system architecture previously developed in Part I, assume you are required to prepare a *Technical Management Plan* (*TMP*) that describes how you intend to manage interfaces:
   (a) Prepare a description of guidance you want to convey in the TMP that describes HOW interfaces will be identified, owned, and controlled.
   (b) Identify how interfaces are to be documented in terms of IRSs, ICDs, and IDDs. Describe what will be documented in each document.

ORGANIZATIONAL CENTRIC EXERCISES

1. Research your organization's command media.
   (a) What minimum requirements does the media impose on interface identification, ownership, and control.
   (b) Are programs required to use a standard IRS, ICD, or IDD outline?
   (c) What is the structure of the organization's IRS, ICD, or IDD?
   (d) Who controls the IRS, ICD, or IDD outlines on each contract program?
2. Contact a small, a medium, and a large program in your organization.
   (a) What IRS, ICD, or IDD requirements did their Contract Data Requirements List (CDRL) impose on the program?
   (b) Did the contract specify an outline format?
   (c) If not, what type of outline was used on each type of document?
   (d) How is interface ownership and control established?
   (e) What new lessons learned did program personnel discover that they would avoid on the next program?

REFERENCES

DI-IPSC-81434A, 1999. *Interface Requirements Specification* (*IRS*). Data Item Description (DID). Washington, DC: Department of Defense (DoD).

DI-IPSC-81436. *Interface Design Description* (*IDD*), Data Item Description (DID). Washington, DC: Department of Defense (DoD).

IEEE Std 610.12-1990. 1990. *IEEE Standard Glossary of Modeling and Simulation Terminology*. Institute of Electrical and Electronic Engineers (IEEE). New York, NY.

MIL-STD-480B (cancelled). 1988. *Military Standard Configuration Control: Engineering Changes, Deviations, and Waivers*. Washington, DC: Department of Defense (DoD).

ASD-100 Architecture and System Engineering. 2003. *National Air Space System—Systems Engineering Manual*, Federal Aviation Administration. Washington, DC: FAA.

Chapter 44

Human–System Integration

44.1 INTRODUCTION

Human-made systems, products, and services inevitably require some form of human *interaction* and *control* throughout all phases of operation. As technologies advance and deployment, operations, and support costs increase, we continually strive to automate systems to *minimize* the number of human interactions to improve productivity, efficiency, effectiveness, and reduce costs.

For most organizations, the *focus* of contracts is typically on producing EQUIPMENT Element systems and products. Given that *focus*, the PERSONNEL Element, which deploys, operates, and supports the EQUIPMENT Element, is often given token "*lip service*" when requirements are allocated from the *System Performance Specification (SPS)*. Yet, the EQUIPMENT Element, as a nonliving object, is totally dependent on the PERSONNEL Element to: 1) prepare the system for a mission, 2) conduct the mission, and 3) perform postmission actions. When the PERSONNEL Element is addressed, there is often an *imbalance* between PERSONNEL performance and EQUIPMENT performance.

This chapter explores a system's PERSONNEL–EQUIPMENT interactions that influence development of a *system*, *product*, or *service*. As an overview, our discussions begin with the System Operations Model introduced in Chapter 18. The model's *operations* and *tasks* provide a framework for addressing a simple question: *WHICH tasks are best performed by the EQUIPMENT, PERSONNEL, or combinations of the two?* Since PERSONNEL requirements are *allocated* to operators and maintainers on a TASK basis, we introduce attributes of tasks that must be investigated and specified. Typically, this requires analyses and trade studies of Human-In-the-Loop (HITL) interactions to understand human performance relative to overall system or product performance.

Notice *The information provided herein is intended to heighten a basic SE AWARENESS of factors that influence human–system interface design. ALWAYS employ the services of a competent, highly qualified Human Factors Engineer to perform human–system interface design.*

What You Should Learn from This Chapter

1. What is *anthropometry*?
2. What are *haptics*?
3. What are *ergonomics*?
4. What are the classes of *human–system interfaces*?
5. What types of input/output (I/O) devices are available for human–system interfaces?
6. What are the key attributes of human tasks?

System Analysis, Design, and Development, by Charles S. Wasson
Copyright © 2006 by John Wiley & Sons, Inc.

7. What is Human Factors Engineering (HFE)?
8. What are the seven *elements of Human–System Integration (HSI)*?
9. What are *areas of concern* related to each HSI element?
10. What are the five *types of human factors*?
11. What are the common types of *human characteristics* associated with human factors?
12. What are the key analytical techniques for analyzing human–system interfaces?
13. What are the key areas of interest related to HFE and their areas of concern?

Definitions of Key Terms

- **Anthropometry** "The scientific measurement and collection of data about human physical characteristics and the application (engineering anthropometry) of these data to the design and evaluation of systems, equipment, and facilities." (Source: MIL-HDBK-1908B, *Definitions*, para. 3.0, p. 6)
- **Anthropometrics** "Quantitative descriptions and measurements of the physical body variations in people. These are useful in human factors design." (Source: MIL-HDBK-470A, Appendix G: *Glossary*, p. G-2)
- **Ergonomics** The multi-disciplinary science concerned with the study of work and how to apply knowledge of human capabilities, performance, and limitations to workplace design via human-system interfaces and interactions.
- **Haptic** "Refers to all the physical sensors that provide a sense of touch at the skin level and force feedback information from muscles and joints." (Source: DoD 5000.59-M *Modeling and Simulation [M&S] Glossary*, Part II, Glossary-A, Item 241, p. 117)
- **Haptics** "The design of clothing or exoskeletons that not only sense motions of body parts (e.g., fingers) but also provide tactile and force feedback for haptic perception of a virtual world." (Source: DoD 5000.59-M *Modeling and Simulation [M&S] Glossary*, Part II, Glossary-A, Item 242, p. 117)
- **Human Factors** "A body of scientific facts about human characteristics. The term covers all biomedical and psychosocial considerations; it includes, but is not limited to, principles and applications in the areas of human engineering, personnel selection, training, life support, job performance aids, and human performance evaluation." (Source: MIL-HDBK-1908B, *Definitions*, para. 3.0, p. 17)
- **Human Performance** "A measure of human functions and action in a specified environment, reflecting the ability of actual users and maintainers to meet the system's performance standards, including reliability and maintainability, under the conditions in which the system will be employed." (Source: MIL-HDBK-1908B, *Definitions*, para. 3.0, p. 18)
- **Man-Man Interface (MMI)** "The actions, reactions, and interactions between humans and other system components. This also applies to a multistation, multiperson configuration or system. Term also defines the properties of the hardware, software or equipment which constitute conditions for interactions." (Source: MIL-HDBK-1908B, *Definitions*, para. 3.0, p. 21)
- **Safety Critical** "A term applied to any condition, event, operation, process, or item whose proper recognition, control, performance, or tolerance is essential to safe system operation and support (e.g., safety critical function, safety critical path, or safety critical component)." (Source: MIL-STD-882D, para. A.3.2.10, p. 9)

526 Chapter 44 Human–System Integration

- **Task Analysis** "A systematic method used to develop a time-oriented description of personnel-equipment/software interactions brought about by an operator, controller or maintainer in accomplishing a unit of work with a system or item of equipment. It shows the sequential and simultaneous manual and intellectual activities of personnel operating, maintaining or controlling equipment, in addition to sequential operation of the equipment. It is a part of system engineering analysis where system engineering is required." (Source: MIL-HDBK-1908B, *Definitions*, para. 3.0, pp. 31–32)

- **User-Computer Interface (UCI)** "The modes by which the human user and the computer communicate information and by which control is commanded, including areas such as: information presentation, displays, displayed information, formats and data elements; command modes and languages; input devices and techniques; dialog, interaction and transaction modes; timing and pacing of operations; feedback, error diagnosis, prompting, queuing and job performance aiding; and decision aiding." (Source: MIL-HDBK-1908B, *Definitions*, para. 3.0, p. 34)

44.2 APPROACH TO HSI

Our approach to this section is to provide an *awareness* of the types of technical decisions SEs need to understand concerning the development of systems that require *human-in-the-loop* (HITL) interations with a system. Since Human Factors Engineering (HFE) is a specialty engineering discipline, we approach HSI from an SE perspective as illustrated in Figure 44.1. Specifically, our topics of discussion include:

- Human interface classes
- Human factors
- Human System Integration (HSI) Elements
- HSI issue areas
- Task attributes

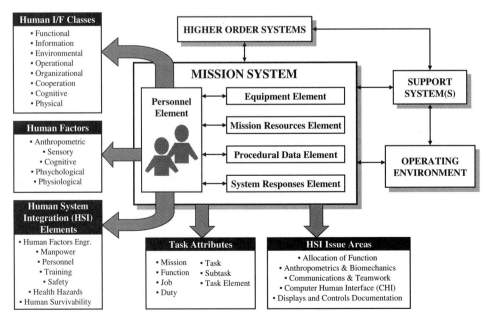

Figure 44.1 SE Human Factors Considerations

44.3 HUMAN–SYSTEM INTERFACES

Chapters 18 through 20 *System Operations Concepts* introduced the System Operations Model. The model provides an initial framework for defining HOW a system's pre-mission, mission, and post-mission phases of operation might be constructed. The framework consists of *serial* and *concurrent operations* and *tasks* to be performed to accomplish OBJECTIVES established for each phase of operation.

The System Operations Model framework presents a challenge: *WHICH individual or combinations of System Elements—such as EQUIPMENT or PERSONNEL—should be allocated requirements for performing the OPERATIONS and TASKS*? On the surface, this sounds simple. However, further investigation REVEALS new questions that must be answered first:

1. WHAT *capabilities* and *levels of performance* can the EQUIPMENT Element provide with current technologies and resource constraints such as cost and schedule?
2. WHAT *skills* and *levels of performance* do members of the PERSONNEL Element currently possess or can be trained to perform?
3. WHAT *level of risk* are we *willing to accept* in the PERSONNEL and EQUIPMENT elements.

We can ultimately condense these questions into: *WHAT types of operations and tasks can the EQUIPMENT Element and PERSONNEL Element or combinations of the two perform best*? To answer these questions, we need to identify the key strengths of PERSONNEL and EQUIPMENT performance.

Key Strengths of Human Performance

Humans excel in a number of skills and mental strengths when contrasted with EQUIPMENT. In general, human performance exceeds EQUIPMENT performance in the following areas:

1. Value-based judgments and decisions
2. Priority selections
3. Resource allocations—over time
4. Impromptu tasks
5. Creative, nonrepetitive tasks
6. Sensitivity to painful conditions
7. Human communications
8. Smell and touch
9. Adaptive behavior

Author's Note 44.1 *Numbered items in the list above and below are for reference purposes and SHOULD NOT be interpreted as rank ordered list.*

Key Strengths of EQUIPMENT Performance

In contrast, EQUIPMENT, when designed correctly from a User's perspective, excels in other areas when compared with humans. In general, EQUIPMENT performance exceeds human performance in the following areas:

1. Processing, storing, and retrieving vast amounts of data.
2. Computing complex algorithms and trends in a short period of time.

528 Chapter 44 Human–System Integration

3. Transmitting and receiving large amounts of error-free data under time constraints.
4. Artificially controlling human performance under prescribed safety conditions—aircraft performance.
5. Sensing and analyzing *microscopic* scale variations in electrical, mechanical, optical, environmental, and chemical conditions.
6. High-speed, noncreative, repetitive tasks—such as mass production.
7. Controlling high-risk operations that pose safety and health threats to humans and the environment—such as steel mill, handling hazardous and toxic materials.
8. Leveraging human physical capabilities.
9. Measuring parameters—such as time, mass, and material composition.

Guidepost 44.1 *Given this comparison of strengths, how do SEs determine the appropriate allocation and mix of performance-based operations and tasks to the EQUIPMENT and PERSONNEL elements? First, let's identify the types of human–system interfaces that provide the basis for decision making.*

Human Interface Classes

The FAA's National Airspace System-*SE Manual* Table 4.8–9 identifies eight classes of human interfaces:

1. Functional interfaces
2. Information interfaces
3. Environmental interfaces
4. Operational interfaces
5. Organizational interfaces
6. Cooperation interfaces
7. Cognitive interfaces
8. Physical interfaces

Table 44.1 provides a description of each of these classes, performance dimension, and performance objectives.

Guidepost 44.2 *Given these classes of interfaces, SEs need to understand HOW these interfaces affect various human performance characteristics. So, we need to identify human factors that are affected by human–system interactions.*

Human Factors

HSI requires consideration of several human characteristics that influence SYSTEM and EQUIPMENT element design. Human Factors Engineering (HFE) identifies five key factors concerning human characteristics that have an impact on PERSONNEL–EQUIPMENT *interactions* that lead to system design considerations:

1. Anthropometric factors
2. Sensory factors
3. Cognitive factors

Table 44.1 FAA perspective on classes of human interfaces

ID	Human Interface Class/Scope	Performance Dimension	Performance Objective
44.1	**Functional Interfaces** For operations and maintenance—role of the human versus automation; functions and tasks; manning levels; skills and training	Task performance	Ability to perform tasks within time and accuracy constraints
44.2	**Information Interfaces** Information media, electronic or hardcopy; information characteristics, and the information itself	Information handling/processing performance	Ability to identify, obtain, integrate, understand, interpret, apply, and disseminate information
44.3	**Environmental Interfaces** Physical, psychological, and tactical environments	Performance under environmental stress	Ability to perform under adverse environmental stress, including heat and cold, vibration, clothing, illumination, reduced visibility, weather, constrained time, and psychological stress
44.4	**Operational Interfaces** Procedures, job aids, embedded or organic training, and on-line help	Sustained performance	Ability to maintain performance over time
44.5	**Organizational Interfaces** Job design, policies, lines of authority, management structure, and organizational infrastructure	Job performance	Ability to perform jobs, tasks, and functions within the management and organizational structure
44.6	**Cooperation Interfaces** Communications, interpersonal relations, team performance	Team performance	Ability to collectively achieve mission objectives
44.7	**Cognitive Interfaces** Cognitive aspects of human computer interfaces (HCI), situational awareness, decision making, information integration, short-term memory	Cognitive performance	Ability to perform cognitive operations (e.g., problem-solving, decision making, information integration, situational awareness)
44.8	**Physical Interfaces** Physical aspects of the system with which the human interacts (e.g., HCI, controls and displays, workstations, and facilities)	Operations and maintenance performance	Ability to perform operations and maintenance at workstations and work sites, and in facilities using controls, displays, equipment, and tools

Source: National Airspace System-*System Engineering Manual*, Section 4.8.3.3, Table 4.8–9.

4. Psychological factors
5. Physiological factors.

To better understand the scope each of these factors, Table 44.2 lists general human characteristics related to each of these factors.

PERSONNEL–EQUIPMENT Trade-Offs

When specification requirements are written, *avoid* specifying operational tasks to be performed by PERSONNEL unless there are *compelling reasons* supported by analysis or trade studies. As the SE Process Model executes through multiple levels of system design, trade-offs are made with *Decision Support Practices* (Chapters 50–52). Decision support is tasked to determine the best mix of PERSONNEL versus EQUIPMENT tasks. This may require analyses, prototypes, and simulations to ensure that overall SYSTEM performance is *optimal*.

Author's Note 44.2 *Since SYSTEM operations involve human operator and external interactions that are variable and have a degree of uncertainty, we use the term optimal versus optimum. Given these uncertainties, seldom will overall SYSTEM performance be optimum.*

Table 44.2 Common human characteristics associated with human factors

Human Factors	Human Characteristics
Anthropometric factors	• Human physical dimensions • Body posture • Repetitive motion • Physical interface
Sensory factors	• Hearing • Vision • Touch • Balance
Cognitive factors	• Mental ability • Skills • Decision making • Training requirements
Psychological factors	• Human needs • Attitudes • Expectations • Motivations
Physiological factors	• Human reactions to environments • Strength (lifts, grip, carrying, etc.) • Endurance

Source: DoD Human Factors Engineering Critical Process Assessment Tool (CPAT), Table 1, p. 7.

PERSONNEL–EQUIPMENT Interactions

Once the initial allocations are made, the next question is *HOW will the PERSONNEL and EQUIPMENT Elements interact with each other to accomplish interface objectives*. We refer to these PERSONNEL interactions as input/output (I/O) operations that include:

1. Audio-visual stimuli and cues.
2. *Tactile cues*—such as touch and vibratory cues.
3. Physical products and services—such as hardcopies and data files.

As this information is collected and processed, the EQUIPMENT Element produces various preprogrammed cues that, at a *minimum*, include:

1. Prompts to the operator(s) to react to a question or make a decision.
2. General health status.
3. Problem reporting.
4. Progress reporting in performing a task.

Since audio, visual, and vibratory cues are an integral part of human–system interfaces; let's scope what we mean by each term.

Auditory Cues. *Audio cues* consist of any type of *warnings*, *cautions*, and *alerts* that *notify* the operator of specific EQUIPMENT health and status conditions. Various tonal frequencies as well as sequences and patterns of tones are employed to symbolize SYSTEM conditions. *Auditory cues* are often employed to *alert* system operators, especially when they are not attentive or observing the visual cues.

Visual Cues. Visual cues consist of any optical *warnings*, *cautions*, or *normal indications* or *messages* to inform operators and maintainers of the current SYSTEM conditions, status, or health.

Vibratory Cues. Where *auditory* or *visual cues* are not desirable or preferable, *vibratory cues* may be employed to *alert* SYSTEM operators. *Vibratory cues* consist of devices that employ electromechanical mechanisms that *vibrate* on command. Examples include silenced cell phones and pagers.

Guidepost 44.3 *Given an understanding of types of human–system interaction cues, HOW are these cues communicated? This brings us to our next topic, system command and control (C^2) devices.*

SYSTEM Command and Control (C^2) Devices

PERSONNEL require mechanisms to *command* and *control* (C^2) SYSTEM operations and performance. We refer to these mechanisms as *I/O devices*. Let's identify some of the various types of I/O devices that serve as candidate solutions for human–system interations.

Data Entry Devices. *Data entry devices* consist of electromechanical-optical mechanisms, such as keyboards or touch panels that enable SYSTEM operators and maintainers to enter alphanumeric information.

Pointing Control Devices. *Pointing control devices* consist of mechanisms, such as a computer-based trackball, eyeball trackers, or mouse to enable SYSTEM operators and maintainers to point, manipulate, or maneuver data such as "drag and drop."

Mechanical Control Devices. *Mechanical control I/O devices* include mechanical tools that enable operators to *calibrate, align, control,* or *adjust* the SYSTEM configuration, operation, and performance.

Electronic Control Devices. *Electronic control I/O devices* consist of electronic or electromechanical mechanisms, such as remote controls, toggle or rotary switches, dials, and touch screen displays, configured to communicate operator controlled pointing position or displacement to specific data items.

Translational Displacement Control Devices. *Translational displacement control devices* such as joysticks and track balls employ electronics that *transform* or *translate* mechanical movements—by angular displacement, stress, or compression—into electronic signals that are used to control systems.

Sensory I/O Devices. *Sensory I/O devices* consist of mechanisms that sense the presence, degree, proximity, and strength of human interaction.

Audio I/O Devices. *Audio input devices* consist of electro-mechanical mechanisms that *translate* sound waves into inputs that are compatible with and recognized by the system—such as speech recognition. *Audio output devices* consist of electromechanical mechanisms—such as speakers and headphones—that *communicate* tones or messages to the system operator(s).

Generalized and Specialized I/O Solutions

In Chapter 12, *System Interfaces*, we introduced the concept of *generalized* and *specialized* interfaces. We apply the same approach to human–system interfaces.

PERSONNEL–EQUIPMENT interactions begin with a simple acknowledgement that a *logical entity relationship* or *association* exists between two interacting systems, a *generalized* solution. As the System Developer analyzes the interface and its requirements, candidate solutions are investigated—as *specialized* solutions. Consider the following example:

EXAMPLE 44.1

A System Developer has a contract to design a system that requires an operator to enter data. Thus, we have a *logical* association between the PERSONNEL and EQUIPMENT Elements—a *generalized* solution. As a *generalized* solution we have not specified HOW the operator inputs data into the system. The development team responsible for providing the interface explores several viable candidate solutions that include standard keyboards, touch screen displays, and so forth. Subsequently, the team selects a standard keyboard for the application—a *specialized* solution. However, further analysis reveals that the keyboard may be susceptible to dirt, dust, rain, and snow. So, the team decides to go with a ruggedized keyboard—which is, another level of *specialized* solution—capable of surviving in the system's OPERATING ENVIRONMENT.

Guidepost 44.4 *At this juncture we have introduced some of the fundamentals of PERSONNEL–EQUIPMENT interfaces. We now shift our focus to designing system interfaces that integrate PERSONNEL–EQUIPMENT interactions. This brings us to our next topic, Human System Integration (HSI).*

44.4 HUMAN SYSTEM INTEGRATION (HSI) ELEMENTS

Human Systems Integration (HSI) in system design requires consideration of several elements. Since human factors are *critical operational and technical issues*, especially in aerospace and defense systems design, a significant amount of the knowledge we have today originates from those industries.

MIL-HDBK-46855A (para. 5.1.2.1 HSI Elements) identifies seven HSI elements that consist of:

1. Human factors engineering (HFE)
2. Manpower
3. Personnel
4. Training
5. Safety
6. Health hazards
7. Human survivability

Each of these HSI elements includes various *areas of concern* that provide a basis for further investigation. Table 44.3 provides a listing of example *areas of concern* that require SE attention.

44.5 HUMAN FACTORS ENGINEERING (HFE)

Human–system interactions encompass more than simply checking to see if the EQUIPMENT and PERSONNEL Elements are assigned the RIGHT mix of tasks. There are other dimensions of the interaction that require consideration by a specialty discipline referred to as Human Factors Engineering (HFE).

When *human interactions* are a key element of system operations, technical decisions have to be made regarding the *man–machine interface* (MMI) between the PERSONNEL and EQUIPMENT System Elements. Depending on the system applications, these decisions often involve a single philosophy. Reiterating previous themes: *for a given set of operating conditions and constraints:*

Table 44.3 HSI element areas of concern

HSI element	Example Areas of Concern
Human factors engineering (HFE)	1. Unnecessarily stringent selection criteria for physical and mental capabilities. 2. Compatibility of design with anthropometric and biomedical criteria 3. Workload situational awareness, and human performance reliability 4. Human–system interface 5. Implications of mission and system performance requirements on human operator, maintainer, supporter. 6. Effects of design on skill, knowledge, and aptitudes requirements 7. Design driven human performance, reliability, effectiveness, efficiency, and safety performance requirements 8. Simplicity of operation, maintenance, and support 9. Costs of design-driven human error, inefficiency, or ineffectiveness

(continued)

Table 44.3 *continued*

HSI element	Example Areas of Concern
Manpower	1. Manpower requirements 2. Deployment considerations 3. Team and organizational structure 4. Operating strength 5. Manning concepts 6. Manpower policies
Personnel	1. Personnel selection and classification 2. Demographics 3. Accession rates 4. Attrition rates 5. Career progression and retention rates 6. Promotion flow 7. Personnel training and pipeline 8. Qualified personnel where and when needed 9. Projected user population/recruiting 10. Cognitive, physical, and educational profiles
Training	1. Training concepts and strategy 2. Training tasks and training development methods 3. Media, equipment, and facilities 4. Simulation 5. Operational tempo 6. Training system suitability, effectiveness, efficiency, and costs 7. Concurrency of system with trainers
Safety	1. Safety of design and procedures under deployed conditions 2. Human error 3. Total system reliability and fault reduction 4. Total system risk reduction

Health hazards	1. Health hazards induced by systems, environment, or task requirements
	2. Areas of special interest include (but not limited to):
	   a. Acoustics
	   b. Biological and chemical substances
	   c. Radiation
	   d. Oxygen deficiency and air pressure
	   e. Temperature extremes
	   f. Shock and vibration
	   g. Laser protection

Human survivability	1. Threat environment 2. Identification friend or foe 3. Potential damage to crew compartment and personnel 4. Camouflage/concealment 5. Protective equipment 6. Medical injury 7. Fatigue and stress

Source: MIL-HDBK-46855A, Table 1: HSI Elements and the Areas of Concern for Each, p. 20.

1. *Compared to EQUIPMENT, what tasks can the PERSONNEL element perform best?*
2. *Compared to PERSONNEL, what tasks can the EQUIPMENT element—such as machines—perform best?*
3. What controls should be allocated for humans to perform versus EQUIPMENT—for examples, *automated* versus *manual* controls?
4. What work task ergonomic factors should be considered in human-system interface designs.
5. *What are the performance effects of human-system interactions and outputs—such as PRODUCTS, BY-PRODUCTS, and SERVICES—on the environment, safety, and health of SUPPORT SYSTEMs and the general public?*

Answers to these questions require specialty skills referred to as Human Factors Engineering (HFE). DOD 5000.2-R (Section C5.2.3.5.9.1) describes HFE as "established to develop effective human–machine interfaces, and *minimize or eliminate system characteristics that require extensive cognitive, physical, or sensory skills; require excessive training or workload for intensive tasks; or result in frequent or critical errors or safety/health hazards.*"

This leads to the question: WHAT do HFEs investigate? The answer resides in understanding our next topic, HFE analyses.

HFE Analyses

HFE performs various analyses such as manpower, personnel, training, and safety/health hazards to ensure that *System Performance Specification (SPS)* requirements are met. Human Factors (HF) engineers employ various tools and methods to perform operational sequence evaluations, timeline and task analyses, and error analyses.

Since these decisions have an impact on the *SPS* and *item development specification* requirements, HFE should be an integral part of system and specification development activities beginning during the proposal phase. Failure to do so may have a *major impact* on contract technical, cost, and schedule delivery performance as well as *severe consequences* if *catastrophic failures* occur after deployment of the system, product, or service.

The DoD Human Factors Engineering Critical Process Assessment Tool (CPAT) (Section 1.1.3) identifies three analytical HFE techniques for application to human–system interface design decision making:

- **Operational Sequence Evaluations** "[D]escribe the flow of information and processes from mission initiation through mission completion. The results of these evaluations are then used to determine how decision–action sequences should be supported by the human–system interfaces."

- **Task Analysis** "[T]he study of task and activity flows and human characteristics that may be anticipated in a particular task. Task analysis is used to detect design risks associated with human capabilities, such as skill levels and skill types. Task analysis also provides data for man–machine trade-off studies. The results of a task analysis allow the system designer to make informed decisions about the optimal mix of automation and manual tasking."

- **Error Analysis** "[I]s used to identify possible system failure modes. Error analysis is often conducted as part of human–machine trade-off studies to reveal and reduce (or eliminate) human error during operation and maintenance of the system. The error analysis results eventually are integrated into reliability failure analyses to determine the system level effects of any failures."

Tests and Demonstrations "[A]re often necessary to identify mission critical operations and maintenance tasks, validate the results of the human factors related analyses, and verify that human factors design requirements have been met. These tests and demonstrations are used to identify mission critical operations and maintenance tasks. Therefore, they should be completed at the earliest time possible in the design development process."

Subjects of Concern to HFE

From a system design perspective, the scope Human Factors Engineering (HFE) applies knowledge of *human characteristics* and methods to ensure human performance is accorded proper consideration while achieving overall system performance. To do this, MIL-HDBK-46855A (Table II, p. 26) identifies nine system design *subjects of concern* to HFE:

1. Human perceptual/performance characteristics
2. Display and control design
3. Design of equipment, vehicles, and other systems
4. Workplace, crew station, and facilities design
5. Automation and human–machine integration
6. Environmental conditions
7. Work design
8. Health and safety
9. System evaluation

Each *area of interest* consists of a set of concerns that require particular attention by HFEs. Table 44.4 provides a summary listing.

44.6 HSI ISSUE AREAS

As evidenced by the preceding discussions, when human–system interfaces are designed, there are a number of design considerations that, if not properly addressed, can become issues areas. The FAA's National Airspace System-*System Engineering Manual*, for example, identifies 23 issue areas and associated concerns. Table 44.5 provides a listing of these areas and provides a brief description of design considerations for each issue area.

HFE Prototypes and Demonstrations

One of the best approaches to support HFE design decision-making is to simply prototype design areas that may have a *moderate* to *high level* of risk. *Spiral development* provides a good strategy for refining human–system interfaces with Users via rapid prototyping to drive out key requirements. Rapid prototyping includes cardboard model mockups, sample displays, and so forth.

What does prototyping accomplish? The DoD Human Factors Engineering CPAT (Section 1.1.3) offers several reasons. "Operator/maintainer interfaces should be prototyped to:

1. Develop or improve display/software and hardware interfaces.
2. Achieve a design that results in the required effectiveness of human performance during system operation and maintenance.
3. Develop a design that makes economical demands upon personnel resources, skills, training and costs."

Table 44.4 MIL-HDBK-46855 perspective on HFE subjects of concern

Item	Area of Interest	HFE Subjects of Concern
44.1	Human Perceptual/Performance Characteristics	1. Anatomy and physiology 2. Anthropometry and biomechanics 3. Sensory processes in vision, hearing, and other senses 4. Spatial awareness and perceptual organization 5. Attention, workload, and situational awareness 6. Learning and memory 7. Decision making and problem solving 8. Perceptual motor skills and motion analysis 9. Complex control processes 10. Performance speed and reliability 11. Language 12. Stress, fatigue, and other psychological and physiological states 13. Individual differences
44.2	Display and Control Design	1. Input devices and controls 2. Grouping and arrangement controls 3. Process control and system operation 4. Control/display integration 5. Visual displays 6. Audio, tactile, motion, and mixed-modality displays 7. Information presentation and communication 8. Human–computer interfaces
44.3	Design of Equipment, Vehicles, and Other Systems	1. Portable systems and equipment 2. Remote handling equipment 3. Command and control systems 4. Equipment labeling 5. Ground vehicles 6. Marine craft 7. Aircraft and aerospace vehicles 8. Manpower and crew size requirements 9. Maintainability 10. Reliability 11. Usability
44.4	Workplace, Crew Station, and Facilities Design	1. Console and workstation dimensions and layout 2. General workplace and building design 3. Design of non-work (service) facilities 4. Design of multi-building environments 5. Design of self-contained working/living environments
44.5	Automation and Human-Machine Integration	1. Allocation of functions between human and machine 2. Automation of human tasks 3. Aiding of operators, maintainers, and teams 4. Artificial intelligence 5. Virtual environments 6. Robotics
44.6	Environmental Conditions	1. Illumination 2. Noise 3. Vibration 4. Acceleration/deceleration

(continued)

Table 44.4 *continued*

Item	Area of Interest	HFE Subjects of Concern
		5. Whole-body velocity (motion)
		6. Temperature extremes and humidity
		7. Micro-gravity
		8. Underwater conditions
		9. Restrictive environments
		10. Time-zone shifts
44.7	Work Design	1. Task description, task analysis, and task allocation
		2. Job skills, structure, and organization
		3. Crew coordination and crew resource management
		4. Work duration, shift work, and sustained and continuous operations
		5. Job attitudes, job satisfaction, and morale
		6. Personnel selection and evaluation
		7. Training, instructional manuals, and job support
44.8	Health and Safety	1. Health hazard assessment
		2. Risk perception and risk management
		3. Biological, chemical, radiation, and other hazards
		4. Occupational safety and health
		5. Alarms, warnings, and alerts
		6. Preventive education and training
		7. Protective clothing and personal equipment
		8. Life support, ejection, escape, survival equipment
44.9	System Evaluation	1. Modeling and simulation
		2. Mock-ups, prototypes, and models
		3. Mannequins and fitting trials
		4. Information flow and processing
		5. Systems analysis
		6. Reliability, failure, and error analysis
		7. Operational effectiveness training
		8. Usability testing

Source: MIL-HDBK-46855A, Table II, *Subjects of Concern to HE*, pp. 26–27.

Guidepost 44.5 *At this juncture we have introduced primary information used to design the EQUIPMENT Element hardware and software items. We now shift our emphasis to the PERSONNEL Element. Obviously, we cannot engineer a system's human operator. However, we can identify specific mission tasks assigned to the system operators and maintainers and train them to perform those tasks in a safe, orderly, and proficient manner.*

44.7 HUMAN–SYSTEM TASKING

Interactions between humans and systems involve three types of System Element requirements allocations: EQUIPMENT Element, PERSONNEL Element, and PROCEDURAL DATA Element.

- EQUIPMENT Element requirements specify the capability to:
    1. Display or output data to system operators and external interfaces.
    2. Accept input data from system operators to control system behavior and sustain performance.

Table 44.5 FAA perspective on HSI issue areas

Item	Issue Area	Description
44.1	Allocation of Function	System design reflecting assignment of those roles/functions/tasks for which the human performs better, or assignment to the equipment that it performs better while maintaining the human's awareness of the operational situation.
44.2	Anthropometrics and Biomechanics	System design accommodation of personnel (e.g., from the 1st through the 99th percentile levels of human physical characteristics) represented in the user population.
44.3	Communications and Teamwork	System design considerations to enhance required user communication and teamwork.
44.4	Computer–Human Interface (CHI)	Standardization of CHI to access and use common functions employing similar and effective user dialogues, interfaces, and procedures.
44.5	Displays and Controls	Design and arrangement of displays and controls to be consistent with the operator's and maintainer's natural sequence of operational actions and provide easily understandable supporting information.
44.6	Documentation	Preparation of user documentation and technical manuals in a suitable format of information presentation, at the appropriate reading level, easily accessible, and with the required degree of technical sophistication and clarity.
44.7	Environment	Accommodation of environmental factors (including extremes) to which equipment is to be subjected and the effects of environmental factors on human–system performance.
44.8	Functional Design and Operational Suitability	Use of a human-centered design process to achieve usability objectives and compatibility of equipment design with operation and maintenance concepts and legacy systems.
44.9	Human Error	Examination of unsafe acts, contextual conditions, and supervisory and organization influences as causal factors contributing to degradation in human performance, and consideration of error tolerance, resistance, and recovery in system operation.
44.10	Information Presentation	Enhancement of operator and maintainer performance through use of effective and consistent labels, symbols, colors, terms, acronyms, abbreviations, formats, and data fields.
44.11	Information Requirements	Availability of information needed by the operator and maintainer for a specific task when it is needed and in the appropriate sequence.
44.12	Input/Output Devices	Design of input and output devices and methods that support performing a task quickly and accurately, especially critical tasks.
44.13	Knowledge, Skills and Abilities (KSA)	Measurement of the knowledge, skills, and abilities required to perform job-related tasks. Necessary to determine appropriate selection requirements for operators.
44.14	Procedures	Design of operational and maintenance procedures for simplicity and consistency with the desired human–system interface functions.
44.15	Safety and Health	Reduction/prevention of operator and maintainer exposure to safety and health hazards.

(continued)

Table 44.5 *continued*

Item	Issue Area	Description
44.16	Situation Awareness	Consideration of the ability to detect, understand, and project the current and future operational situations.
44.17	Skills and Tools	Considerations to minimize the need for unique operator or maintainer skills, abilities, or characteristics.
44.18	Staffing	Accommodation of constraints and opportunities on staffing levels and organizational structures.
44.19	Subjective Workload	The operator's or maintainer's perceived effort involved in managing the operational situation.
44.20	Task Load	Objective determination of the numbers and types of tasks that an operator performs.
44.21	Training	Consideration of the acquisition and decay of operator and maintainer skills in the system design and capability to train users easily, and design of the training regimen to result in effective training.
44.22	Visual/Auditory Alerts	Design of visual and auditory alerts (including error messages) to invoke the necessary operator and maintainer response to adverse and emergency situations.
44.23	Workspace	Adequacy of workspace for personnel and their tools and equipment, and sufficient space for movements and actions they perform during operational and maintenance tasks under normal, adverse, and emergency conditions.

Source: FAA's National Airspace System *System Engineering Manual*, Section 4.8.3.3, Table 4.8–10.

- PERSONNEL Element requirements specify tasks that deploy, operate, and support the system to:
  1. Monitor system information and outputs.
  2. Interpret system operations and health conditions.
  3. Control system behavior and performance.
  4. Perform corrective action to mitigate risk to the operators, EQUIPMENT, public, and the environment.
- PROCEDURAL DATA Element requirements instruct the operators or maintainers in HOW to:
  1. Safely and effectively deploy, operate, and maintain the system.
  2. Proceed in normal and emergency operations.
  3. Perform corrective actions or remediate the effects of system failures should an emergency occur.
  4. Understand system capability and performance limitations.

Since system operator and maintainer requirements are *task-based* actions, let's define the *attributes* of a task.

Defining Task Attributes

As you analyze most tasks, you will discover that they share a common set of attributes. These attributes enable human factors specialists to understand the *mission* and *scenario conditions* that bound the human–system interactions. MIL-STD-1908B identifies six attributes of a task:

Table 44.6 MIL-STD-1908B definitions of task attributes

Item	Task Attribute	Definition
44.1	Mission	What the system is supposed to accomplish (e.g., mission).
44.2	Scenario/Conditions	Categories of factors or constraints under which the system will be expected to operate and be maintained (e.g., day/night, all weather, all terrain operation).
44.3	Function	A broad category of activity performed by a system (e.g., transportation).
44.4	Job	The combination of all human performance required for operation and maintenance of one personnel position in a system (e.g., driver).
44.5	Duty	A set of operationally related tasks within a given job (e.g., driving, system servicing, communicating, target detection, self-protection, and operator maintenance).
44.6	Task	A composite of related activities (perceptions, decisions, and responses) performed for an immediate purpose, written in operator/maintainer language (e.g., change a tire).
44.7	Subtask	An activity (perceptions, decisions and responses) that fulfills a portion of the immediate purpose within the task (e.g., remove lug nuts).
44.8	Task Element	The smallest logically and reasonably definable unit of behavior required in completing a task or subtask, e.g., apply counterclockwise torque to the lug nuts with a lug wrench.

Source: MIL-HDBK-1908B, Task Analysis Definition para. 3.0, p. 32.

1. Mission
2. Function
3. Job
4. Duty
5. Task
6. Subtask

Table 44.6 defines each of these attributes.

44.8 GUIDING PRINCIPLES

In summary, the preceding discussions provide the basis with which to establish the guiding principles that govern human–system interfaces practices.

Principle 44.1 *Optimal* HSI design requires balancing requirements allocation in two areas:

1. WHAT the PERSONNEL Element does best.
2. WHAT the EQUIPMENT Element does best.

Principle 44.2 When designing HSIs, structure layouts and display information based on use case prioritized operator/practitioner needs, skill levels, and frequency of usage priorities.

Principle 44.3 AVOID operator decision-making data overloads: keep interface information simple and intuitive.

Principle 44.4 When presenting information, AVOID presentation of trivial data unless the operator(s) has CONSCIOUSLY requested it, assuming existence of the information is known.

Principle 44.5 When training PERSONNEL to operate a SYSTEM, employ the EDEV approach:

1. Explanation.
2. Demonstration.
3. Experimentation.
4. Verification.

Principle 44.6 Interface simplicity rules!

44.9 SUMMARY

As an overview discussion of *human–system interfaces*, we investigated the key SE considerations for allocating EQUIPMENT and PERSONNEL requirements based on the premise of WHAT the system can perform best versus WHAT humans perform best—the Human-in-the-Loop (HITL) system.

Our initial discussions focused on understanding human–system interactions. We discussed types of human–system interfaces—audio, visual, and vibratory—and provided examples of *command* and *control* (C^2) I/O devices used in Human-in-the-Loop (HITL) systems.

Based on an understanding of human–system interactions, we shifted our focus to the design aspect of Human System Integration (HSI). We identified key elements of HSI and their respective areas of concern that require Human Factors Engineering (HFE). We illustrated how HFE categorizes *human characteristics* into *human factors* that impact human–system performance.

We concluded our discussion PERSONNEL tasks and identified their attributes.

GENERAL EXERCISES

1. Answer each of the *What You Should Learn from This Chapter* questions identified in the *Introduction*.
2. Refer to the list of systems identified in Chapter 2. Based on a selection from the preceding chapter's General *Exercises* or a new system selection, apply your knowledge derived from this chapter's topical discussions. Specifically identify the following:
   (a) What types of I/O devices are employed as human–system interfaces?
   (b) Identify examples of the types of factors listed below from Table 44.2 that should be considered in the system's design:
      1. Anthropometric design factors
      2. Sensory design factors
      3. Cognitive design factors
      4. Psychological design factors
      5. Physiological design factors
   (c) Identify examples of human-system tasks that represent decisions the System Developer may have made regarding EQUIPMENT element versus PERSONNEL Element decision-making allocations.

ORGANIZATIONAL CENTRIC EXERCISES

1. Research your organization's command media for guidance and direction in implementing the *human–system interface practices* from an SE perspective.

(a) What requirements are levied on SE responsibilities for HF/HFE?
(b) What overall process is required and how do SEs contribute?
(c) What SE work products and quality records are required?
(d) What verification and validation activities are required?
2. Contact two contract programs within your organization.
(a) What human factors requirements were levied on the system design?
(b) Who are the Users/Operators of the system?
(c) What critical human–system interface decisions were made?
(d) What PERSONNEL Element requirements were levied on the system operators?
(e) What task analysis was performed and linked to specification requirements and system capabilities?

REFERENCES

DoD 5000.2-R. 2002. *Mandatory Procedures for Masor Defense Acquisition Programs (MDAPS) and Major Automation Information (MAIS) Acquisition Programs*, Washington, DC: Department of Defense (DoD).

DoD 1998 *Human Factors Engineering* Critical Process Assessment Tool (CPAT), Military Specifications and Standards Reform Program (MSSRP), Washington, DC: Department of Defense.

DoD 5000.59-M. 1998. *DoD Modeling and Simulation (M&S) Glossary*. Washington, DC: Department of Defense (DoD).

MIL-HDBK-46855A. 1999. *Human Engineering Process and Program*. Washington, DC: Department of Defense (DoD).

MIL-HDBK-470A. 1997. *Designing and Developing Maintainable Products and Systems*. Washington, DC: Department of Defense (DoD).

MIL-HDBK-1908B. 1999. *DoD Definitions of Human Factors Terms*. Washington, DC: Department of Defense (DoD).

MIL-STD-882D. 2000. DoD *Standard Practice for System Safety*. Washington, DC: Department of Defense (DoD).

Federal Aviation Administration (FAA), ASD-100 Architecture and System Engineering. 2003. *National Air Space System—Systems Engineering Manual*. Washington, DC: FAA.

ADDITIONAL READING

ANSI/HFS 100. 1988, Human Factors and Ergonomics Society, Santa Monica, CA. *American National Standard for Human Factors Engineering of Visual Display Terminal Workstations*.

MIL-HDBK-759C. 1998. DoD *Handbook for Human Engineering Design Guidelines*. Washington, DC: Department of Defense.

MIL-STD-1472F. 1999. *DoD Design Criteria Standard: Human Engineering*. Washington, DC: Department of Defense.

JACKO, JULIE A., and SEARS, ANDREW, eds. 2003. *The Human-Computer Interaction Handbook: Fundamentals, Evolving Technologies, and Emerging Applications*. Mahwah, NJ: Lawrence Erlbaum Associates.

MULLET, KEVIN, and SANO, DARRELL. 1994. *Designing Visual Interfaces, Communication Oriented Techniques* 1st ed. Englewood Cliffs, NJ: Prentice-Hall.

NOYES, JANET M, and BRANSBY, MATTHEW. 2002. *People in Control: Human Factors in Control Room Design*, Vol. 60. IEE Control Engineering series. Institute of Electrical and Electronic Engineers (IEEE) New York, NY.

SALVENDY, GAVRIEL. 2005. *Handbook of Human Factors and Ergonomics*, New York: Wiley.

SANDERS, MARK S. and MCCORMICK, ERNEST J. 1993. *Human Factors in Engineering and Design*. New York: McGraw-Hill.

TILLEY, ALVIN R. 2001. *The Measure of Man and Woman: Human Factors in Design*. New York: Wiley.

WICKENS, CHRISTOPHER; LEE, JOHN; LIU, YILI; GORDON-BECKER, SALLIE. 2003. *An Introduction to Human Factors Engineering*. Englewood Cliffs, NJ: Prentice-Hall.

WOODSON, WESLEY E., TILLMAN, BARRY, and TILLMAN, PEGGY. 1992. *Human Factors Design Handbook*. New York: McGraw-Hill.

Chapter 45

Engineering Standards, Frames of Reference, and Conventions

45.1 INTRODUCTION

As the system architecture evolves, *interactions* between the system and its OPERATING ENVIRONMENT and internal elements such as SUBSYSTEMS must be *compatible* and *interoperable*. This requires that both sides of the interface *adhere* to *standards*. For some systems, the interface may require new development; in other cases, *legacy interfaces* may already exist, thereby establishing the *baseline* for *compliance* for new development.

Whichever is the case, SEs that engineer these interactions for new system, products, or services must be *synchronized* in thought, process, and methods, and share a common system design perspective for accomplishing the interface. The mechanism for ensuring a common perspective resides in the establishment of *engineering standards*, *frames of reference*, and *conventions*.

Many systems have been developed that failed system integration or their missions due to simple, human errors in *communicating* and *interpreting* the mechanical, electrical, chemical, optical, software, and information that apply to both sides of the interface.

This chapter addresses the need for SEs to establish *engineering standards*, *frames of reference*, and *conventions* "up front" as one of the cornerstones of system design. Our discussion explores each of these topical areas and provides examples to illustrate the importance of synchronizing design mindsets to a *common*, *shared* viewpoint for designing compatible and interoperable interfaces.

What You Should Learn from This Chapter

1. What is an *engineering standard*?
2. What is a *convention*?
3. Why do we need engineering standards?
4. What are the key subject matter categories of engineering standards?
5. What are some types of *standards*?
6. What are some types of *conventions*?
7. What are some common *standards* and conventions *issues*?
8. Who establishes engineering standards?
9. What are *local standards and conventions*?

System Analysis, Design, and Development, by Charles S. Wasson
Copyright © 2006 by John Wiley & Sons, Inc.

10. What are some examples of local standards and conventions?
11. How is the degree of compliance with engineering standards *verified*?

Definitions of Key Terms

- **Compliance** The act of adhering to the letter of a requirement without exception.
- **Conformance** The act of *adapting* or *customizing* organizational *work products*, *processes*, and *methods* to meet the *spirit and intent* of a required action or objective.
- **Convention** A method established by external or internal standards for conveying how engineers are to interpret *system* or *entity* configurations, orientations, directions, or actions.
- **Coordinate System** A two- or three-dimensional axis frame of reference used to establish system configuration and orientation conventions, support physical analysis, and facilitate mathematical computations.
- **Dimension** A physical property inherent to an object that is independent of the system of measure used to quantify its magnitude.
- **Open Standards** "Widely accepted and supported standards set by recognized standards organizations or the market place. These standards support interoperability, portability, and scalability and are equally available to the general public at no cost or with a moderate license fee." (Source: Adapted from DSMC—*Glossary: Defense Acquisition Acronyms and of Terms*)
- **Standard** "A document that establishes engineering and technical requirements for products, processes, procedures, practices, and methods that have been decreed by authority or adopted by consensus." (Source: ANSI/EIA 632-1999, *Processes for Engineering a System*, p. 67)
- **Technical Standard** "A common and repeated use of rules, conditions, guidelines or characteristics for products or related processes and production methods. It includes the definition of terms, classification of components, delineation of procedures, specification of dimensions, materials, performance, designs, or operations. It includes measurement of quality and quantity as well as a description of fit and measurements." (Source: NASA SOW NPG 5600.2B)

45.2 ENGINEERING STANDARDS

Engineering standards provide a mechanism for organizations and industries to:

1. Establish consensus performance requirements for development of systems, products, and services.
2. Audit compliance of those deliverable work products.
3. Provide a framework for targeting improvements related to performance and safety.

Standards evolve from *lessons learned*, *best practices*, and *methods* within an organization and across industry domains. They:

1. Ensure product *compatibility* and *interoperability* and avoid the consequences of lessons learned.
2. Ensure *consistency*, *uniformity*, *precision*, and *accuracy* in materials, processes, and weights and measures.
3. Promote *modularity*, *interchangeability*, *compatibility*, and *interoperability*.

4. Ensure *safety* to the general public and environment.
5. Promote *ethical* business relationships.

Where a standard is employed as the basis for evaluating work related performance, the term PERFORMANCE STANDARD is employed.

Standard Normative and Informative Clauses

In general, *standards express* performance requirements via clauses. Standards clauses generally fall into two types of clauses: *normative* clauses and *informative clauses*.

- *Normative* clauses or requirements express *mandatory* criteria for compliance with the standard and include the word "shall" to indicate required performance.
- *Informative* clauses express information for *voluntary* compliance or to provide guidance/clarification for implementing the *normative requirements*.

Local organization engineering standards should clearly delineate *normative* and *informative* clauses.

Engineering Standards Authorities

National and international organizations establish standards. To posture themselves for this position, they must be *recognized* and *respected* within a given industry or business domain as the *authoritative* proponent or issuer of standard practices. Consult your contract, industry, and engineering discipline for specific standards that may be applicable to your business and contract.

Dimensional Properties and Systems of Units

Standards express two types of information that serve as the *frame of reference* for measurements: dimensional properties and systems of units.

- *Dimensional properties* represent inherent physical properties of an object such as mass, length, width, weight.
- *Systems of units* are standards of measure that form the basis for measuring the magnitudes of an object's dimensional properties.

Guidepost 45.1 *At this point we have established WHAT a standard is and WHAT it expresses. Now let's shift our attention to understanding WHAT types of information they communicate.*

Standards Subject Matter

Standards are employed to express performance requirements for a variety of applications such as documentation, processes, methods, materials, interfaces, frames of reference, weights and measures, domain transformations, demonstrations, and conventions. Of these items, *engineering weights and measures*, *conventions*, and *frames of reference* require specific emphasis, especially in creating consistency across a contract program.

45.3 ENGINEERING STANDARDS FOR WEIGHTS AND MEASURES

Perhaps the most fundamental concept of engineering is establishing a system of *weights* and *measures*. Technical expressions that describe the *form*, *fit*, and *function* of a system, product, or service are totally dependent on usage of standard units for *weights* and *measures*.

45.3 Engineering Standards for Weights and Measures

One of the most fundamental examples of establishing *weights* and *measures* is simply selecting and applying a system of units. There are two primary standard systems of units in use today:

1. International System of Units (SI).
2. British Engineering System (BES).

Let's contrast each of these systems.

International System of Units (SI)

The International System of Units (SI) was approved by the 11th General Conference on Weights and Measures (CGPM) in 1960. The CGPM adopted the SI designation from the French Le Système International d'Unites. The SI, which is sometimes referred to as the *metric* or *mks* for meter–kilogram–second system, is based on seven base units that are determined to be independent. Table 45.1 provides a listing of these *base units*.

British Engineering System (BES)

The British Engineering System (BES) consists of five base units listed in Table 45.2.

Scientific Notation

In addition to defining the system of units for measurement, we need to express the magnitudes associated with those units in a manner that is easy to read. We do this with scientific notation as illustrated in Table 45.3.

Data Accuracy and Precision

When data are measured or computed, it is critical for a program to establish a policy for data *accuracy* and *precision*. In addition to establishing mathematical models, transformations, and conver-

Table 45.1 Base units of the International System of Units (SI)

Base Quantity	Name	Symbol
Length	meter	m
Mass	kilogram	kg
Time	second	s
Electric current	ampere	A
Thermodynamic pressure	kelvin	K
Amount of substance	mole	mol
Luminous intensity	candela	cd

Source: Taylor, *The International System of Units*. Table 1, p. 9, 2001.

Table 45.2 Base units of the British Engineering System (BES)

Base Quantity	Name	Symbol
Length	foot	ft
Force	pound	lb
Time	second	s
Temperature	degree fahrenheit	°F
Luminous intensity	candle	

Table 45.3 Scientific notation symbology

Power	Prefix	Notation
10^{-9}	nano	n
10^{-6}	micro	μ
10^{-3}	milli	m
10^{-2}	centi	c
10^{-1}	deci	d
10^{3}	kilo	k
10^{6}	mega	M
10^{9}	giga	G
10^{12}	tetra	T

sions, the integrity of the chain of computations is totally dependent on the accuracy of the data that feed each one.

You should recall our earlier discussions of the "supply chain" (Figure 13.3). Humans, as developers of systems, have two roles as a MISSION SYSTEM and a SUPPORT SYSTEM. As a MISSION SYSTEM, we perform a SUPPORT SYSTEM role to the next person or organization performing their MISSION SYSTEM role.

To illustrate the importance of *data precision*, consider a simple definition of the mathematical symbol, *pi*. Are two digits of precision—namely 3.14—*necessary* and *sufficient* criteria for downstream computations? Four digits? Eight digits? You have to decide and establish the standard for the program.

Closing Point

As the technical lead for a program, SEs need to establish a program consensus regarding a standard system of units for expressing physical quantities. This requires three actions:

1. Establish *unit conversion tables* via project memorandums or reference to a standard.
2. Thoroughly *scrutinize* interface *compatibilities* and *interoperability* throughout the System Development Phase, especially at reviews.
3. DEMAND professional discipline and compliance.

45.4 COORDINATE SYSTEMS

The engineering of systems and products often require that an observer's *frame of reference* be established for characterizing the engineering mechanics and dynamics of the system as a body relative to other systems. Basic concepts for observer *frames of reference* have their origins in physics. So, *what is a frame of reference*?

A *frame of reference* is expressed as a *three-dimensional axis system* with the *origin* located at a designated point on or within a MISSION SYSTEM. The orientation or perspective of the 3-axis coordinate system depends on an observer's *eye point* location when viewing the MISSION SYSTEM. This brings up two key questions:

1. *How are integrated EQUIPMENT Element component displacements and relationships characterized in terms of a 3-axis frame of reference?*
2. *How do we express movements relative to the observer's X-, Y-, or Z-axes?*

The answer to the first question resides in establishing a *coordinate reference system*. The second question requires establishing a *convention* for application to the coordinate system. First, let's consider what is meant by *convention*.

Frame of Reference Conventions

A *convention* is a designation of orientation used to describe actions observed relative to the observer's *frame of reference*. The *point of origin* for the *frame of reference* resides at the observer's *eye point*.

Conventions enable SEs to express relationships relative to the observer using the observer's frame of reference such as:

1. Spatial position of an object relative to an observer's eye point positioned at the origin of a 3-axis frame of reference.
2. Direction of movement relative to observer's eye point.
3. Translational movements relative to each axis of the observer's frame of reference.

The Right-Hand Rule Convention

The *Right-Hand Rule* states that if you configure your right hand so that the thumb points UPWARD to represent the direction of an axis and the fingers are coiled about the axis, the direction of the fingers about the axis symbolizes a CLOCKWISE rotation in the direction of the thumb that is POSITIVE. A COUNTERCLOCKWISE direction is considered NEGATIVE.

The Right-Handed Cartesian Coordinate System

Building on the Right-Hand Rule, we can create *two-dimensional* and *three-dimensional* Cartesian coordinate systems to support system design needs. The challenge comes in establishing a *convention* that enables us to:

1. Communicate the configuration of the axes to each other relative to the observer's frame of reference.
2. Express relative motion about the principal axes of the *frame of reference*.

Establishing an Observer's Frame of Reference Coordinate System. The preceding discussions established the need to define a *coordinate system* as an observer's *frame of reference* to facilitate communications, characterize free-body movements, orientation, calibration, and alignment. Figure 45.1 provides an illustration. This system enables us to establish an observer's eye point viewing position at the origin of a 3-axis coordinate system.

1. The *X*-axis represents the observer's *direction of observation relative to the eye point*.
2. The *Z*-axis points UPWARD from the observer's *eye point as the origin*.
3. The *Y*-axis extends to the LEFT of the observer's *eye point* and is orthogonal to the *X*- and *Z*-axes.

Table 45.4 provides examples of conventions employed to characterize these actions.

This description simply establishes HOW the observer expresses observations about objects within the viewing space relative to the frame of reference. The question is: *HOW does the observer express motion about those axes?* This brings us to our next topic, establishing *YAW, PITCH, and ROLL conventions*.

550 Chapter 45 Engineering Standards, Frames of Reference, and Conventions

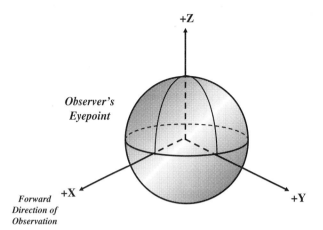

Figure 45.1 Right-Handed Cartesian Coordinate System as an Observer's Frame of Reference

Table 45.4 Examples of conventions applied to a 3-axis observer's frame of reference

Item	Observation	Directional Conventions
45.1	Object *spatial position* relative the observer's frame of reference	Left, right, up, down
45.2	Direction of *movement* relative to observer's frame of reference	Forward, backward, left, right, upward, and downward
45.3	*Translational movements* about the observer's 3-axis frame of reference	Rotations about an axis employing the Right-Hand Rule—e.g., yaw, pitch, and roll

YAW, PITCH, and ROLL Conventions. We can express translational motion about a 3-axis system by using the illustrations shown in Figure 45.2. Since the axes are different, we need to establish terms that enable us to differentiate motion. These terms are PITCH, YAW, and ROLL. For applications in which a frame of reference is allowed to rotate freely in any direction, we refer to it as a *free body axis system*. Table 45.5 lists descriptions of PITCH, YAW, and ROLL conventions applied a *free body axis system*.

Guidepost 45.2 *Our discussions up to this point provide generic descriptions of coordinate systems. Now let's explore some actual systems that implement these coordinate systems.*

World Coordinate System (WCS)

Systems, such as land surveying, aircraft, and military troops, track their geospatial positions based on displacement of the origin of the MISSION SYSTEM relative to an Earth-based coordinate system. To do this, they establish a 3-axis frame of reference using the Earth's center as the origin. We refer to this system as the World Coordinate System (WCS) as illustrated in Figure 45.3. Whereas navigators employed magnetic compasses and sextants to determine geographic position with some level of accuracy, we navigate using the Global Positioning System (GPS) satellites located in Earth orbit.

For the WCS:

45.4 Coordinate Systems

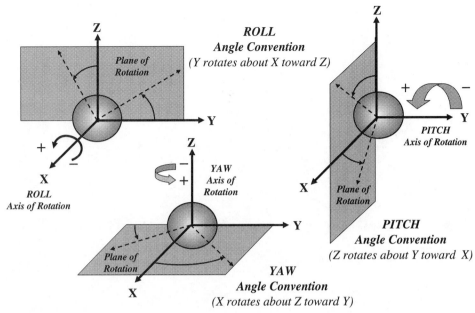

Figure 45.2 6 Degree of Freedom (DOF) Roll, Pitch, & Yaw Conventions

Table 45.5 ROLL, PITCH, and YAW conventions for a right-handed, body-axis, cartesian coordinate system relative to the origin

Parameter	Convention	Action Observed[a]
ROLL	*Positive* (+) ROLL	*Clockwise* angular rotation about the longitudinal X-axis (i.e., along the direction of forward travel) in which the Y-axis rotates toward the Z-axis
	Negative (−) ROLL	*Counterclockwise* angular rotation about the X-axis (Z into Y)
PITCH	*Positive* (+) PITCH	*Clockwise* angular rotation about the Y-axis in which the Z-axis rotates toward the X-axis
	Negative (−) PITCH	*Counterclockwise* angular rotation about the Y-axis (X into Z)
YAW	*Positive* (+) YAW	*Clockwise* angular rotation about the Z-axis in which the X-axis rotates toward the Y-axis
	Negative (−) YAW	*Counterclockwise* angular rotation about the Z-axis (Y into X)

[a]As viewed by an observer with their eye point located at the origin of a right-handed coordinate system.

1. The Z-axis extends from the center of the Earth (origin) through the North Pole.
2. The X-axis extends from the Earth's center (origin) through the Prime Meridian at the equator.
3. Finally, the Y-axis extends from the origin through 90 degrees East at the equator.

Engineering computations and simulations employ models that require certain assumptions about the Earth's characteristics such as Earth Centered, Earth Fixed (ECEF); Earth Centered Rotating (ECR); and Earth Centered Inertial (ECI) models.

552 Chapter 45 Engineering Standards, Frames of Reference, and Conventions

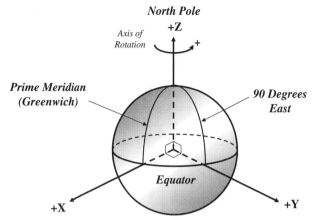

Figure 45.3 World Coordinate System (WCS) Application of the Right Hand Rule Rotation Convention

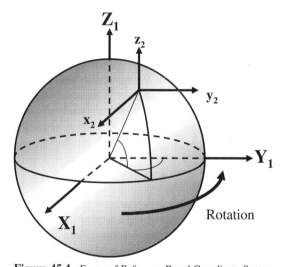

Figure 45.4 Frame of Reference Based Coordinate System

The WCS enables us to reference a specific point on the surface of the Earth. But, *HOW do air-based systems such as aircraft and spacecraft relate to the WCS?* Obviously, these systems employ on-board GPS technology. However, the systems are free bodies that are in motion relative to another body, the Earth, which is also in motion. *HOW do we express their heading and rotational velocities and accelerations relative to their frame of reference?*

Free Body Dynamics Relative to a Fixed Body

Complex systems often require dynamic characterizations of a free body relative to another body that is assumed to be fixed as illustrated in Figure 45.4. Where this is the case, select the coordinate system(s) to be applied to each body axis system. For this illustration, we establish an X_1, Y_1, Z_1 coordinate system to *represent* the Earth's frame of reference and an X_2, Y_2, Z_2 coordinate system to *represent* the orientation of a free body in space relative to the Earth.

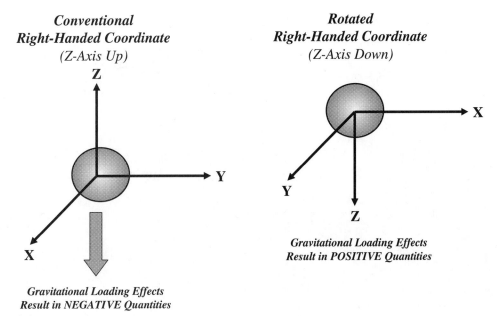

Figure 45.5 Engineering Statics and Mechanics Conventions

Six Degree of Freedom (6 DOF) Models. A free body in space has a relative position within the base frame of reference. Within that space its travels in any direction can be translated into motion relative to the *X*-, *Y*-, or *Z*-axes. It can also freely rotate about the axes of its own frame of reference. As such, we label this type of system as having *six degrees of freedom (6 DOF)*.

State Vectors. The 6 DOF discussion also provides the basis for the concept of *state vectors*. *State vectors* enable us to express the relative heading, velocities, and accelerations about the axes of a free body in space.

Applications of Coordinate Systems and Conventions to Engineering Mechanics

The preceding discussions illustrate the basic concept of standard right-handed Cartesian coordinate system as illustrated at the left side of Figure 45.5. For many applications this convention is acceptable. However, from an engineering *mechanics* perspective, the establishment of the *positive* Z-axis pointing upward means that analysis of gravity-based loading effects result in *negative* Z components.

For systems with this challenge, we can facilitate computation by rotating the 3-axis system so that the *positive* Z-axis points downward toward the center of the Earth. To illustrate this application, consider NASA's Space Shuttle example shown at the right side of Figure 45.6. Here the X-axis represents the *forward direction of travel*; the Y-axis extends from the center of gravity (CG) through the right wing.

Now consider the added complexity in which the Shuttle *maneuvers* in orbit to fly UPSIDE DOWN and BACKWARD during most of its mission to.

1. *Minimize* the thermal and radiation effects of the sun by turning the vehicle's belly toward the Sun and using the heat tiles as a solar shield.

554 Chapter 45 Engineering Standards, Frames of Reference, and Conventions

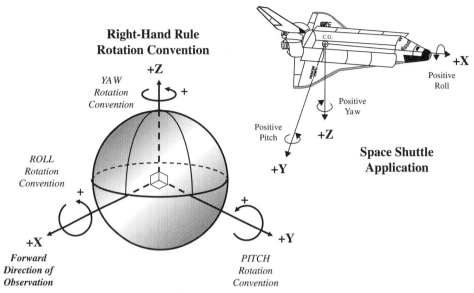

Figure 45.6 Space Shuttle Coordinate System Application
Source: http://liftoff.msfc.nasa.gov/academy/rocket_sci/shuttle/coord/orb_body.html

2. *Minimize* the impact areas of orbital debris and meteorites that may cross its path that could damage tiles on the vehicle's nose or leading edges of the wings.
3. *Shield* Cargo Bay equipment from impacts that could damage dangerous or fragile instruments.
4. *Minimize* the energy required to deploy payloads.
5. *Slow* the vehicle for reentry by firing small nozzles, which are located on the rear of the vehicle, in the direction of forward travel.

Guidepost 45.3 *The preceding discussions relate to a free body operating as a MISSION SYSTEM relative to another free body. Now let's shift our attention to focus on the MISSION SYSTEM as an integrated set of components that must interface as an integrated framework. This leads to the next question: How do we reference a system consisting of multiple components that are integrated to form the free body system?*

Dimensional Reference Coordinate Systems

One of the challenges in the engineering of systems is expressing *the relative position* and *orientation* of physically integrated components within the system. The objective is to ensure they interoperate in *form*, *fit*, and *function* and individually do not interfere with each other *unintentionally*, or have a *negative impact* on the performance of the system, its operators, and its facilities or mission objectives.

This challenge requires positioning integrated components in a virtual frame of reference or structure and attaching them at integration points (IPs) or nodes to ensure interoperability. Then, identifying additional attachment points such as *lift points* that enable external systems to lift or move the integrated system. This requires establishing a *dimensional coordinate system*.

NASA's Space Shuttle coordinate system shown in Figure 45.7 provides an excellent example of a *dimensional coordinate system*. Here, the Orbiter Vehicle (OV), External Tank (ET), and Solid

45.4 Coordinate Systems

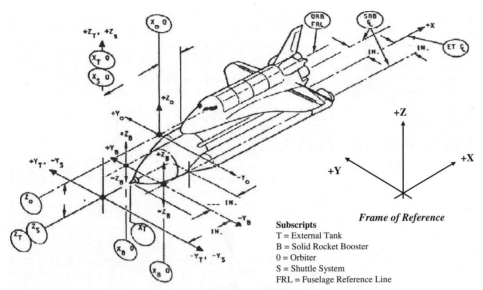

Figure 45.7 Space Shuttle Coordinate System and Dimensions
Source: *Report of the PRESIDENTIAL COMMISSION on the Space Shuttle Challenger Accident* Figure 1: Shuttle Coordinate Systems and Dimensions (http://history.nasa.gov/rogersrep/v3o378a.htm)

Rocket Boosters (SRBs) each have their own respective *frame of reference coordinate systems*. These systems are then referenced relative to an integrated vehicle coordinate system with the origin designated as X_-, Y_-, Z_-. This figure illustrates two key decision areas for SEs regarding dimensional reference coordinate systems.

First, note in this figure HOW the *principal X-axis* extends through the rear of the vehicle. Any point along this axis is considered *positive* by convention. Systems such as aircraft establish dimensional benchmarks or "stations" along the fuselage as a means of referencing component locations. Using a *positive X-axis* convention extending through the rear of the aircraft eliminates the need for *negative values* for dimensions or "stations."

Second, note the location of the origin in front of the ET nose. Aircraft designers select these virtual origins for a variety of engineering based rationale. One rationale, for example, includes accommodating free space forward of the nose for additional components that may be added later. In this case, a component attached to the nose would remain within the *positive X-axis* space. Other placement considerations include movements about the center of gravity (CG) and center of rotation.

Third, compare and contrast the coordinate system in Figure 45.6 with the one in Figure 45.7. This illustrates that a given system design may employ common frames of reference but different orientations, each intended to support specific needs. Figure 45.6 facilitates engineering mechanics computations for system modeling; Figure 45.7 facilitates *dimensional* space coordinate reference systems for product engineers and designers.

Angular Displacement Reference Systems

Some systems employ common components in *left-hand* and *right-hand* orientations that require the need for a convention. Referring to Figure 45.7, *how can the SRBs be uniquely identified*? NASA designates each one as illustrated in Figure 45.8.

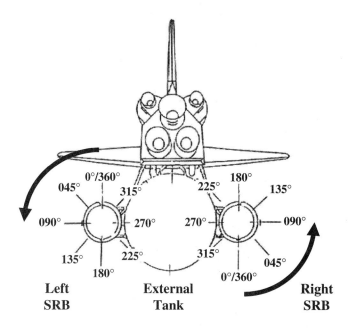

(1) View is Forward (Direction of Flight) or "Up" When Vehicle is on Launch Pad

(2) Angles Increase *Counterclockwise*

Figure 45.8 Angular Coordinate System for Solid Rocket Boosters/Motors
Source: *Report of the PRESIDENTIAL COMMISSION on the Space Shuttle Challenger Accident* Figure 26: Challenger Report http://history.nasa.gov/rogersrep/vlp69.htm

In this figure the observer's *eye point* origin is positioned to the rear of the Shuttle vehicle looking forward along the X-axis in the vehicle's forward direction of travel. Relative to the observer's *frame of reference*, the SRB on the left is designated as LEFT and the other as RIGHT.

Now observe that the LEFT and RIGHT SRBs each employ *an angular displacement system* to reference specific locations about their principal axis. Note the location of the respective 0°/360° reference marks when the SRBs are attached and integrated on opposite sides. Configuration orientation diagrams such as this are developed by SEs in collaboration with subject matter experts (SMEs) and are critical to design decision making.

45.5 OTHER EXAMPLES OF ORIENTATION CONVENTIONS

Briefly, other examples of *engineering conventions* include the following:

- When viewing some integrated circuit (IC) devices with the notch at the top and pins on the left and right sides, Pin 1 is located in the upper left corner; the remaining pins are numbered *counterclockwise* from Pin 1 with the last ID assigned to the pin in the upper right corner. If a notch is not present, Pin 1 is noted by a small dot printed on the device or impressed into the body of the chip next to the Pin 1 location.

- Doors are designated as *left-hand* or *right-hand* as viewed by an observer. When an observer approaches a door, *right-handed* doors have the knob on the RIGHT side, open toward the observer, and swing to their LEFT; *left-handed* doors have the knob on the LEFT side, open toward the observer, and swing to their RIGHT. Special considerations may have to be given to "inswing" or "outswing".

- *Mission Event Timelines (METs)* include events that must be referenced to a common point of reference in time. Thus designations such as T0, T1, T2, etc., or t_0, t_1, t_2, etc., are established.

The bottom line is: SEs must analyze their systems to identify areas that may require conventions for identification and interfacing components to avoid confusion and safety issues.

45.6 STANDARD ATMOSPHERES

Systems and products interact with other external systems in their OPERATING ENVIRONMENT under a variety of NATURAL and INDUCED ENVIRONMENT conditions. As part of the *problem space* and *solution space* definitions, SEs and others have to make ASSUMPTIONS that *characterize* and *bound* OPERATING ENVIRONMENT conditions.

Adding to the complexity of these assumptions is the fact that NATURAL ENVIRONMENT conditions vary throughout the day, month, year, and world. So, *how to we standardize to support informed OPERATING ENVIRONMENT decision making*? The answer resides in creating the Standard Atmosphere.

Scientists express the *Standard Atmosphere* via models that describe the interrelationships among *air temperature*, *density*, and *pressure* as a function of time of day, month, and geographical region location. Figure 45.9 serves as an example.

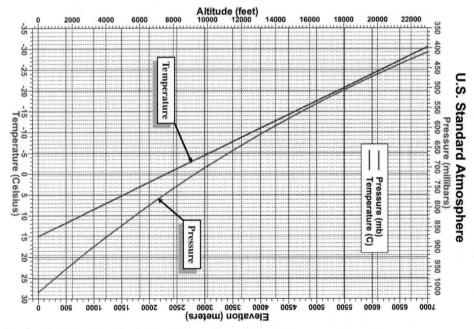

Figure 45.9 Graphical plot of Standard Atmosphere—Altitude, Pressure, and Temperature
Source: Climate Research Division, Scripps Institution of Oceanography, University of California, San Diego, CA based on *Smithsonian Meteorological Tables*, 6E, Smithsonian Institution, Washington, DC, 1966

Standard atmosphere models and tables for geographical locations are available from government organizations such as the US National Oceanographic and Atmospheric Administration (NOAA), and National Aeronautics and Space Administration (NASA).

45.7 APPLYING ENGINEERING STANDARDS AND CONVENTIONS TO PROGRAMS

As a lead on a contract development effort program, SEs must ensure that all applicable *standards and conventions* used on the program are well documented and *explicitly communicated* and *well understood* by all program personnel. So, *how are these documented*?

Standards and conventions are documented via specifications, plans, design documents, *Interface Requirements Specifications (IRS)*, *Interface Control Documents (ICD)*, and so forth. To understand some of the types of information that have to be communicated across these interfaces to achieve interoperability, consider the following examples:

EXAMPLE 45.1

- Weights and measures
- Coordinate system frame of reference
- Computational precision and accuracy
- Base number systems
- Units of conversion
- Calibration standards
- Engineering standards sources
- Documentation guidelines, standards, and conventions
- Data communications standards

Finally, establish an official list of program acronyms and definitions of key terms. Some people refer to this as "languaging" the program. Although this simple task seems trivial, it will save many hours of personnel thrashing about, each with different views of the terms. People need to share a common mindset of terminology.

45.8 ENGINEERING STANDARDS AND CONVENTIONS LESSONS LEARNED

Engineering *standards* and *conventions* involve a number of application and implementation lessons learned. Let's explore some of the more common lessons learned.

Lesson Learned 1: Tailoring Organizational Standards

Organizational standards are established to accommodate a wide variety of system applications. Since every system is different, the standards must include *tailoring instructions* guidance for applying the standard to a program.

Lesson Learned 2: Organizational Standards as Contract Requirements

When organizational standards are unavailable to Offerors for a formal *Request for Proposal (RFP)* solicitation or contract, the Acquirer must ensure that each Offeror is provided a copy or access to a copy during normal business hours.

Lesson Learned 3: Program Standards and Conventions Document

A common problem related to standards implementation is *failure* to *define* the standards to be used in developing system or item interfaces, coordinate transformations, and software coding. Therefore, SEs should establish, baseline, and release an *Engineering Standards and Conventions* document for application on each program.

Lesson Learned 4: Conflicts between Standards

Occasionally, *conflicts* occur within and between engineering standards. Standards organizations work to ensure that *conflicts* are *avoided*. If, as an Acquirer, you mandate that a system, product, or service meet specific standards, make sure specification requirements do not conflict.

Lesson Learned 5: Scope of Requirements

From an SE perspective, *standards* are often written for broader application to a variety of business domains. As an SE, your role is *collaborate* with subject matter experts (SMEs) to ensure that the explicit provisions of the standard that are applicable to a contract or system development effort are identified by paragraph number. This avoids *confusion* and unnecessary verification expenses ferreting out WHAT IS and IS NOT applicable.

Lesson Learned 6: Application of Standards Use

When proposing or developing new systems, Acquirer SEs should ensure that all standards referenced in procurement packages or specifications are the most current, approved, released version of the document. Likewise, System Developer SEs, via contract protocols, should request the Acquirer to CLARIFY broad references such as *"ANSI-STD-XXXX shall apply"* via specific requirements such as *"ANSI-STD-XXXX- (version) para. x.x.x shall apply."*

Lesson Learned 7: Assumptions About Conventions

One of the challenges SEs face is making *assumptions* about performance boundary conditions and conventions. Inevitably, people assume everyone understands to use the XYZ convention. *ALWAYS document the conventions in project memorandums to be used; leave nothing to chance!* Every technical review should include a technical description of conventions such as a graphic, where applicable, that ensures *consistency* and *completeness* of SE design activities.

Lesson Learned 8: Coordinate System Transformations

Our discussions included a focus on designating a coordinate system for a free body such as an air vehicle. Aircraft, especially military aircraft, serve as a platform for other payload systems such as sensors and missile systems. Just because the aircraft might use a *right-handed* Cartesian coordinate system with the Z-axis pointing downward *does not* mean interfacing components will also.

If a payload is developed for a specific aircraft, it makes sense to establish compatible coordinate systems. However, what happens if a decision is made to add a payload to an air vehicle that uses a different coordinate system. You have a challenge.

Where the coordinate systems of interfacing systems are different, developers perform *coordinate transformations* from one system to another. This requires additional processing resources and consumes valuable time: you may not have a choice. In any case, ALWAYS create a simple diagram that illustrates the coordinate systems used and standardize across the program WHAT methods will be employed to perform the coordinate transformations. Establish which interfacing system is required to perform the coordinate transformation in the respective specification.

Lesson Learned 9: Standard Terminology

Successful system design and development requires that everyone on the program have a *common* vision and mindset and communicate in a language that is *consistent* and *mutually understood* by everyone. This requires establishing *terminology* and *acronyms* for universal application throughout all documentation.

Suggestions to establish standard terminologies are usually met with disparaging remarks about how terms are *intuitively obvious* are to smart people. Then, when the program encounters major team-based work scope *issues* or *interpretations* caused by failures in application of terminology, everyone suddenly has one of those *"why didn't we think of this sooner"* responses. How do you avoid this?

The System Engineering and Integration Team (SEIT) must:

1. Establish and be the "keeper" of standard terms, definitions, and acronyms.
2. Make the list readily accessible via on-line network drives or Web sites.
3. Communicate updates to everyone on the program.
4. Ensure compliance across all documentation.

45.9 GUIDING PRINCIPLES

In summary, the preceding discussions provide the basis with which to establish the guiding principles that govern engineering standards and conventions practices.

Principle 45.1 Engineering standards and conventions are crucial for decision making, communications, and documentation. Neglect them and they will become high risk; nurture them and they will guide your path toward success.

Principle 45.2 Each interface should be specified in one and only one official document that is baselined and controlled by the interface owner.

45.10 SUMMARY

Our discussions of *engineering standards*, *frames of reference*, and *conventions* highlighted the need to establish corporate, national, and international standards for guiding the system development effort. We introduced the meanings of *normative* and *informative* clauses and how they were to be implemented.

We also highlighted the need to establish interface standards and conventions such as inertial frames of reference, English (BES) versus metrics (SI) systems, and coding guidelines to guide development work.

You may say, *"Look, we do not develop free body air vehicles. Why should we bother with considerations with coordinate systems?"* The examples discussed above were used to illustrate some of the complexities of coordinate systems. Similar concepts are employed for designing and using numerical control machines, aligning sensors and optical devices, land surveying, and so forth.

- Each program must establish and communicate a set of conventions for application to system configuration and interface applications, engineering design processes and methods, and work products.
- Each program must conduct periodic assessments of engineering standards compliance and review engineering standards and conventions compliance as a key element of all technical reviews.

GENERAL EXERCISES

1. Answer each of the *What You Should Learn from This Chapter* question identified in the *Introduction*.
2. Refer to the list of systems identified in Chapter 2. Based on a selection from the preceding chapter's General *Exercises* or a new system selection, apply your knowledge derived from this section's topical discussions. Specifically identify the following:
   (a) What frame of reference would you apply to the MISSION SYSTEM?
   (b) What conventions are applicable to this system?
   (c) What are some examples of standards that are applicable to the development of this system or product?

ORGANIZATIONAL CENTRIC EXERCISES

1. Investigate several contract programs within your organization.
   (a) Identify the engineering standards and conventions required by contract and others used voluntarily.
   (b) Identify what types of physical measurement systems are used.
   (c) If the system or product employs navigational systems or dimensional frame of reference, define the system that is used.
   (d) Report your findings and observations.
2. Research the library or Internet for examples of other frame of reference systems that are used and identify the MISSION SYSTEM where used.
3. For Item 2 above, identify historical instances where the lack of engineering standards and conventions or the misapplication contributed to system or product failures.

REFERENCES

ANSI/EIA 632-1999. 1999. *Processes for Engineering a System*. Electronic Industries Alliance (EIA). Arlington, VA.

Defense Systems Management College (DSMC). 2001. *Glossary: Defense Acquisition Acronyms and Terms*, 10th ed. Defense Acquisition University Press Ft. Belvoir, VA.

NPG 5600.2B NASA Procedures and Guidelines. 1997. *Statement of Work (SOW): Guidance for Writing Work Statements*. NASA, Office of Procurement. Washington, DC.

TAYLOR, BARRY N., ed. 2001. *The International System of Units (SI)*. National Institute of Standards and Technology (NIST) Special Publication 330. Washington, DC.

TAYLOR, BARRY N., ed. 1995. *Guide for the Use of the International System of Units*, National Institute of Standards and Technology (NIST) special publication 811, Washington, DC.

ADDITIONAL READING

American National Standards Institute (ANSI) standards.

American Society for Testing and Materials (ASTM) standards.

DoD 5000.59-M. 1998. *DoD Modeling and Simulation (M&S) Glossary*. Washington, DC: Department of Defense (DoD).

G-003A-1996. 1996. *AIAA Guide to Reference and Standard Atmosphere Models*. American Institute of Aeronautics and Astronautics (AIAA). Reston, VA.

IEEE Std 610.3-1989. 1989. *IEEE Standard Glossary of Modeling and Simulation Terminology*. Institute of Electrical and Electronic Engineers (IEEE), New York, NY.

IEEE Std 610.12-1990. 1990. *IEEE Standard Glossary of Modeling and Simulation Terminology*. Institute of Electrical and Electronic Engineers (IEEE), New York, NY.

National Space Transportation System (NSTS) 1988 News Reference Manual. Kennedy Space Center (KSC), FL.

National Oceanic and Atmospheric Administration (NOAA), National Aeronautics and Space Administration (NASA), and the US Air Force. 1976. *U.S. Standard Atmosphere*. Washington, DC: Government Printing Office.

1987. *Handbook of Chemistry and Physics*. Boca Raton, FL: CRC Press. 85th edition, 2004, David R. Lide (editor).

Chapter 46

System Design and Development Documentation

46.1 INTRODUCTION

When a program is proposed and implemented, there are several key questions SEs need to be able to answer.

1. *What is the minimum level of documentation required to design and develop a system?*
2. *What documentation is required to be delivered and when?*
3. *What other nondeliverable documentation do we need to produce?*
4. *What level of formality and detail are required?*
5. *WHAT design data are required to perform the job, WHO owns it, WHAT are the risks in obtaining it, and if we get it, is it accurate, valid, and under CM control?*

These questions have answers that range from simple to very challenging, especially for documentation owned by other organizations.

This section explores the types of documentation produced for most system design and development programs. Our discussions introduce key contracting terms such as CDRLs, DAL, and DCL and describe each one and its relationship to the overall system development contract. We identify four types of document classifications such as plans, specifications, design, and test documentation and provide a graphical schedule of their release points.

The discussions provide commonly used rules for developing SE and development documentation. We emphasize the importance and criticality of assessing documentation for export control of sensitive data and technology. Our final discussion topic identifies several types of documentation issues SEs need to be prepared to address.

What You Should Learn from This Chapter

1. What is a *Contract Data Requirements List (CDRL)*?
2. What is the nominal *sequencing* and release of SE documentation?
3. What is a *Subcontract Data Requirements List (SDRL)*?
4. What is a *Data Item Description (DID)*?
5. What is a *Data Accession List (DAL)*?
6. What is a *Data Criteria List (DCL)*?

7. What are *product data*?
8. What are *product description data*?
9. What are *product design data*?
10. What are *product support data*?
11. What is the difference between *approved data* and *released data*?
12. What mechanisms are used to officially release SE documentation?
13. How should data exchanges between organizations be performed?
14. Why is it important to be sensitive to *Export Control*?

Definitions of Key Terms

- **Authorized Access** A formal approval issued by an Acquirer or program organization signifying that an *internal* or *external* individual or organization with a NEED TO KNOW is *authorized* for *limited access rights* to specific types of data for a constrained period of time subject to handling and control procedures established by the contract or program.
- **Contract Data Requirements List (CDRL)** An attachment to a contract that identifies a list of documents to be delivered under the Terms and Conditions (Ts&Cs) of the contract. Each CDRL item should reference delivery instructions including: 1) when the documents are to be delivered; 2) outline, format, and media to be used; 3) to whom and in what quantities; 4) level of maturity such as outline, draft, and final; 5) requirements for corrective action; and 6) approvals.
- **Data Accession List (DAL)** "An index of data that may be available for request." (Source: DID DI-MGMT-81453, para. 3.1)
- **Design Criteria List (DCL)** A listing of design data that characterize the capabilities and performance of an external system or entity that serves as an interfacing element or model for emulation or simulation.
- **Engineering Release Record** "A record used to release configuration documentation." (Source: MIL-STD-973 [Canceled] *Configuration Management*, para. 3.38)
- **Released Data** "(1) Data that has been released after review and internal approvals. (2) Data that has been provided to others outside the originating group or team for use (as opposed to for comment)." (Source: ANSI/EIA 649-1998, para. 3.0, p. 6)
- **Subcontract Data Requirements List (SDRL)** A listing of data deliverables required by a subcontract. See CDRL above.
- **Technical Data** "Technical data is recorded information (regardless of the form or method of recording) of a scientific or technical nature (including computer software documentation) relating to supplies procured by an agency. Technical data does not include computer software or financial, administrative, cost or pricing, or management data or other information incidental to contract administration. . . ." (Source: Former MIL-STD-973 [Canceled], *Configuration Management*, para. 3.87)
- **Technical Data Package** "A technical description that is adequate to support acquisition of an item, including engineering, production, and logistic support. The technical description defines the design configuration and procedures required to ensure adequacy of item performance. It consists of all applicable technical data, such as engineering drawings, associated lists, product and process specifications and standards, performance requirements, quality assurance provisions, and packaging details." (Source: MIL-HBDK-59B [Canceled], p. 57)

- **Working Data** "Data that have not been reviewed or released; any data that are currently controlled solely by the originator including new versions of data that were released, submitted, or approved." (Source: ANSI/EIA-649-1998, para. 3.0 *Definitions*, p. 7)

Quality System and Engineering Data Records

System development, as an incremental decision-making process, requires that technical specifications, plans, design documentation, test procedures, and test results be documented for deliverables and decision making, and be archived for historical purposes. The integrity of the documentation data require verification and validation reviews to ensure *accuracy*, *preciseness*, *consistency*, and *completeness*.

For ISO 9001-based Quality Systems, each step of the documentation process produces *data records* that serve as *objective evidence* to demonstrate that you accomplished WHAT you planned to do. When you plan your technical program, plan for, create, and capture *data records*.

Based on this introduction, let's begin with a high-level perspective on documentation produced by a program.

46.2 SYSTEM DESIGN AND DEVELOPMENT DATA

System data consists of five basic types:

1. Contract deliverable data.
2. Subcontract, vendor, supplier required data.
3. Operation and support data.
4. Working data.
5. Local command media required data.
6. Personal engineering data records.

Let's explore each of these further.

Contract Deliverable Data

The Acquirer of a SYSTEM levies documentation requirements on a System Developer via Terms and Conditions (Ts&Cs) of the contract. Organizations issue contracts containing a *Contract Data Requirements List (CDRL)* which identify specific documents to be delivered throughout the contract.

CDRL Items. CDRL item numbers specify deliver instructions on a CDRL item form. Each CDRL Item form identifies the CDRL item number, documentation formats, and conditions for submittal, submittal dates, and distribution lists.

Data Item Descriptions (DIDs). Most contracts require CDRLs to be delivered in a specific format. The DoD, for example, employs *Data Item Descriptions (DIDs)* that provide preparation instructions regarding the deliverable document's outline and contents.

Subcontractor, Vendor, and Supplier Data

Acquirers (role) levy data requirements on contractors via the contract. These data apply to:

1. *Configuration items* (CIs) and *items* such as commercial off-the-shelf (COTS) items and nondevelopmental items (NDIs) procured under subcontract.

2. *Certificates of Compliance (C of C)* for purchased items.
3. Design and management reporting data to support SE design activities.

Subcontracts include a *Subcontract Data Requirements List (SDRL)* that identifies the set of data deliverables. In some cases, the System Developer may flow down CDRL requirements via the SDRL to subcontractors or extract specific CDRL requirements and incorporate into SDRLS to provide inputs to support System Developer CDRL documents.

Vendor and supplier deliverables and data are often procured as part of a purchase order, for example. The order may require the vendor to supply a *Certificate of Compliance* (C of C) certifying that the delivered item *meets* or *exceeds* the vendor or supplier's product specification requirements.

Operations and Support (O&S) Data

Operations and support data consist of the data required to *operate* and *maintain* the system, product, or service. Examples include technical manuals and *Standard Operating Practices and Procedures (SOPPs)*. O&S data for a system may be delivered as part of the System Developer's contract, support contract, or developed internally by the User.

Working Data

Working data represent data such as analyses, trade studies, modeling and simulation results, test data, technical decisions and rationale generated during the course of designing and developing the system. Unless these data are *explicitly* identified as CDRL items under the Ts&Cs of the contract, they are for *internal* usage by the System Developer's personnel.

Occasionally the Acquirer may request the opportunity to view these data on the System Developer's premises with the understanding they are not deliverables. This is accommodated via the Data *Accession List (DAL)* discussed later.

Local Command Media Data Requirements

Another type of system development data are *work products* required by *local command media*—the policies, processes, and procedures. Data required by local command media may include plans, briefings, drawings, wiring lists, analyses, reports, models, and simulations. Unless specified by the CDRL or SDRL, these data are considered to be non-deliverables.

You may ask: *If a document is not required by contract,* WHY *consume resources producing it*? In general, the answer is organizations that possess levels of capability in SE *know* and *understand* that specific data—such as plans, specifications, and test procedures—are crucial for success, regardless of CDRL or SDRL data requirements. CDRL and SDRL data requirements may be inadvertently *overlooked* during the formal solicitation process or, as is often the case, the Acquirer decides not to procure the data. If you determine that specific data items are missing during the formal *Request for Proposal (RFP)* solicitation process, confer with the proposal leader regarding how to address the missing items.

Personal Engineering Data Records

The final type of SE data includes personal data records as part of their normal tasking. Personal data records should be maintained in an engineering notebook or computer-based analogy. Personal data include plans, schedules, analyses, sketches, reports, meeting minutes, action items, tests conducted, and test results. These data types are summarized in technical reports and progress and status reports.

46.3 DESIGN CRITERIA LIST (DCL) AND DATA ACCESSION LIST (DAL)

Contracts and subcontracts often require two types of documentation as part of the CDRL or SDRL deliveries:

1. *Data Accession List (DAL)*
2. *Design Criteria List (DCL)*

Let's describe each of these documents.

Data Accession List (DAL)

The development of most systems and products involves two types of documentation: *deliverable data* and *nondeliverable data*. An Acquirer (role) *specifies* and *negotiates* the documentation to be delivered as part of the contract delivery work products. During the course of the contract, the System Developer or subcontractor may produce additional documentation that is not part of the contract but may be needed later by the Acquirer or User.

Where this is the case, Acquirer (role) contracts often require the System Developer (role) to prepare and maintain a *Data Accession List (DAL)* that lists all documentation produced under the contract for *deliverables* and *nondeliverables*. This enables the Acquirer to determine if additional data are available for procurement to support system maintenance. If so, the Acquirer may *negotiate* with the contractor or subcontractor and modify the original contract to procure that data.

Each contract or subcontract should include a requirement for a System Developer to provide a *Data Accession List (DAL)* as a CDRL/SDRL item to identify all CDRL and *nondeliverable* documentation produced under the contract for review and an opportunity to procure the data at a later date.

Design Criteria List (DCL)

Systems, *products*, and *services* are often required to be integrated into HIGHER ORDER systems. For physical systems, *interoperable* interfaces or models are CRUCIAL. For simulations and emulations, each model must provide the precise *form*, *fit*, and *function* of the simulated or emulated device.

In each of these cases, the physical system, model, simulation, or emulation must *comply* with specific *design criteria*. Source data *authentication*, *verification*, and *validation* are critical to ensure the integrity of the SE decision-making efforts.

How do you identify the design data that will be required to support the SE design effort? The process of procuring this data begins with a *Design Criteria List (DCL)*. The DCL is developed and evolves to identify specific design documents required to support the system development effort. Generally, SEs and Integrated Product Teams (IPTs) are responsible for submitting a detailed list of items and attributes—such as title, document ID—to the Data Manager for acquisition. On receipt, the Data Manager:

1. Forwards the documents to Configuration Management for processing and archival storage.
2. NOTIFIES the design data requestors regarding RECEIPT of the documents.

Each contract or subcontract should require publication of a *Design Criteria List (DCL)* by the System Developer identifying specific documentation sources.

Guidepost 46.1 *This concludes our overview on types of contract and subcontract SE data. Let's shift our focus to SE and Development Documentation Sequencing.*

46.4 SE AND DEVELOPMENT DOCUMENTATION SEQUENCING

One of the challenges for SEs is determining WHEN various documents should be *prepared*, *reviewed*, *approved*, *baselined*, and *released*. Although every contract and program requirements vary by Acquirer, there are some general schedules that can be used to prepare SE and development documentation. In general, we can categorize most SE documentation into four classes:

- Planning documentation
- Specification documentation
- System design documentation
- Test documentation

Planning Documentation

Figure 46.1 illustrates a basic schedule as general guidance for preparing and releasing various types of *technical plans*. These documents include:

1. Key *technical management plans (TMPs)* such as hardware development plans, software development plans, configuration and data management plans, and risk management plans.
2. Supporting technical plans such as system safety plans, manufacturing plans, and system support plans.
3. Test plans such as system integration, test, and verification plans and hardware and software test plans.

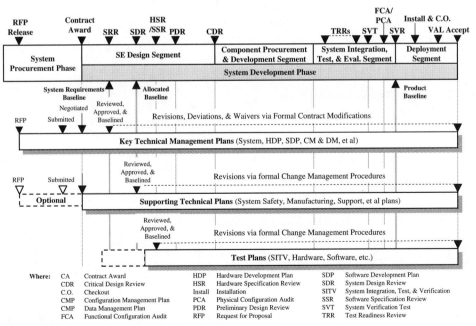

Figure 46.1 Planning Documentation Development

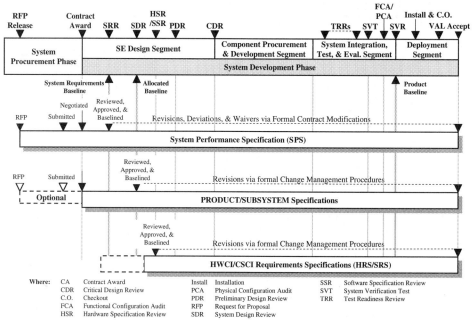

Figure 46.2 Specification Documentation Development

Specification Documentation

Figure 46.2 illustrates a basic schedule as general guidance for preparing and releasing various types of *specifications*. These documents include:

1. *System Performance Specification (SPS)*
2. *PRODUCT/SUBSYSTEM Development Specifications*
3. *HWCI/CSCI Requirements Specifications (HRS/SRS)*

System Design Documentation

Figure 46.3 illustrates a basic schedule as general guidance for preparing and releasing various types of *technical documentation*. These documents include:

1. *Concept of Operations (ConOps)*
2. *System/Segment Design Description (SSDD)*
3. *Interface Control Documents (ICDs)* and drawings for hardware items
4. *Interface Design Description (IDD)* for software items
5. *Software Design Description (SDD)*
6. *Database Design Description (DBDD)*

Test Documentation

Figure 46.4 illustrates a basic schedule as general guidance for preparing and releasing *test* documentation. These documents include various levels of test procedures and test quality records.

46.5 Documentation Levels of Formality

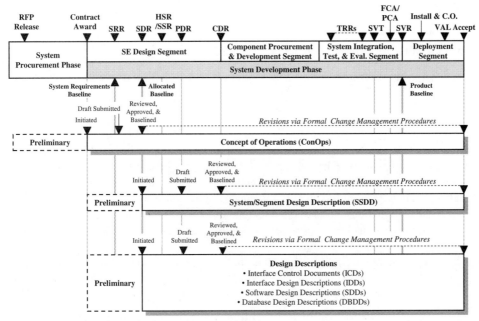

Figure 46.3 System Design Documentation Development

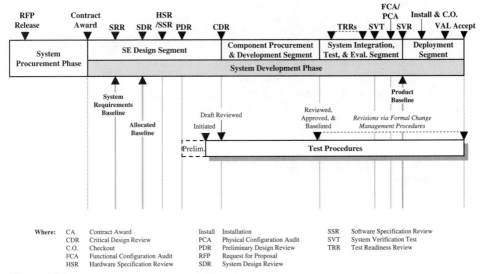

Figure 46.4 Test Procedures Documentation Development

46.5 DOCUMENTATION LEVELS OF FORMALITY

Most people despise preparing documentation. The general viewpoint is documentation is *non-value-added bureaucracy*. Engineers, in particular, rationalize that if they had wanted to *specialize* in documentation, they would have pursued it as a course of study in college. This viewpoint is contradictory to an engineering development environment that is so dependent on documented data maturity for decision making.

570 Chapter 46 System Design and Development Documentation

SE development documentation involves two key decisions:

1. WHAT must be documented?
2. To WHAT DEGREE do you document the details?

The previous discussions addressed WHAT you should document. Let's explore the second point further.

Level of Details

If you ask engineers to prepare a document, they lament about having an impossible task, regardless of timeframe to complete. The knowledge and maturity of a professional is reflected in the individual's ability to sift rapidly through a large amount of data, identify the key points, and articulate the results in summary form. In general, if a manager allows you one hour, both parties must recognize that at the end the hour, you get the one-hour version of the report. Eight hours gets the manager the eight-hour version—*an elaboration of details to key points expressed in the one-hour version.*

When you document plans, specifications, and reports, there are key points relevant to the subject that must be reflected in the document outline. These points, in turn, require various levels of details. The key points are:

1. Engineers need to learn to identify WHAT information (i.e., major points) is required to be communicated.
2. Apply common sense to scale the level of detail to fit the available time and resources.

Key Point *If you have addressed the critical points, anything else should be just supporting detail.* The ability to scale these *levels of detail* and still address the key points depends on personal *knowledge*, *experience* and seasoned *maturity*.

Guidepost 46.2 *The preceding discussions highlight some of the key SE documentation concepts. Our last topic addresses some of the issues related to release of SE documentation to authorized recipients and the media used to do so.*

46.6 EXPORT CONTROL OF SENSITIVE DATA AND TECHNOLOGY

Today the Internet provides a mechanism for immediate data communications and access between organizations in-country as well as internationally. As a result, the Internet offers tremendous opportunities for contract programs to exploit the technology by establishing Web sites that enable *authorized* organizations and individuals who are geographically dispersed to post and access technical program information. This environment, when *uncontrolled*, however, adds new dimensions and threats to information access.

People often confuse EXPORT CONTROL information with security classification systems. Although both are certainly interrelated, EXPORT CONTROL information may be *sensitive* but *unclassified*. Classified data handling and procedures are yet another issue.

Export Control of Technology

Many people erroneously believe that a technology or information export to a foreign national, organization, or country occurs when that information is physically transferred outside the country.

The fact is *technology transfer* also occurs in-country with foreign nationals, whether direct or via the Internet. *Technology transfer* is governed in the United States by EXPORT CONTROL laws and regulations such as the International Traffic and Arms Regulations (ITAR). So, *what does this mean to SE?*

If you:

1. Post EXPORT CONTROL technology or information to an *uncontrolled* Web site.
2. E-mail the information via the Internet.
3. Simply give that information to foreign nationals, organizations, or governments without being *licensed* and taking *reasonable measures* to restrict access to only authorized users, you have violated the US ITARS governance.

Warning! ALWAYS consult with your organization's EXPORT CONTROL Officer and security, legal, and contracts organizations before initiating an action that may violate EXPORT CONTROL or security regulations and procedures.

The US government and other countries have *explicit laws* and *regulations* that govern the *EXPORT of technology* and *data* to foreign nationals, organizations, and countries. The laws carry severe penalties for noncompliance.

46.7 SYSTEM DOCUMENTATION ISSUES

System documentation has a number of issues that SEs must address. Here are several issues that commonly challenge many programs:

Issue 1: Data validation and authentication
Issue 2: Posting acquirer and vendor documentation
Issue 3: Proprietary data and nondisclosure agreements
Issue 4: Vendor-owned data
Issue 5: Electronic signatures

Issue 1: Data Validation and Authentication

Since SE technical decision making is predicated on data integrity, the challenge is determining if *external* and *internal* data are valid and authentic. This includes DCL items for models and simulations, interfaces, and test data. Additionally, production contracts may require using SE documentation created by the original vendor. *So, how do you determine the validity and authenticity of the data?*

Data *validity* and *authenticity* require rigorous investigation, by inspection, sampled testing, Functional Configuration Audit (FCA), or Physical Configuration Audit (PCA). Data integrity is particularly troublesome when the original System Developer delivered the Product Baseline and transferred CM control to the User via the Acquirer. The challenge is: *Does the "As Maintained" system or product reflect the "As Designed, Built, Verified, and Validated" SE configurations?*

Issue 2: Posting Acquirer and Vendor Documentation

Many engineers *erroneously* believe that they can arbitrarily copy and post Acquirer and vendor documentation on a program Web site. Although Acquirer solicitation and other data may be posted for public access, DO NOT *copy* this material and POST it unless your organization has prior written authorization from the Acquirer or Vendor. Posting creates configuration management control, proprietary, data concurrency, and copyright issues. If you need to provide program personnel access to this data, simply provide links from the program Web site to the Acquirer or vendor Web site, assuming they approve. This *avoids* ownership, proprietary, concurrency, copyright, and other issues.

Issue 3: Proprietary Data and Non-Disclosure Agreements

Finally, before any EXCHANGES of data can occur between any organization and the Acquirer, User, subcontractors, and vendors, execute Proprietary Data Agreements and/or Nondisclosure Agreements between the parties.

Referral *ALWAYS consult with the Program Director and/or Technical Director as well as the contracts, legal, and EXPORT CONTROL organizations for the proper procedures for:*

1. *Communicating and exchanging data with external organizations.*
2. *Establishing proprietary data agreements and/or nondisclosure agreements.*

Issue 4: Vendor-Owned Data

When Acquirers procure a system, they are often willing to trade off documentation funds for system capabilities, especially when they have limited budgets. Later, if they decide to procure system upgrades that are dependent on these data, they *shift* the *burden* and *risk* of obtaining the data to the System Developer. If the original developer and new System Developer are competitors, this creates a very challenging problem for programs and SEs have to solve it.

The challenge is one of cost and schedule. Programs *inevitably* underestimate the amount of resources required to acquire documentation owned by other organizations. *Thoroughly investigate* these issues during the proposal phase and establish data exchange agreements, terms, and conditions prior to Contract Award. *As an Acquirer, if you wait until after Contract Award, guess WHO is in the POWER position to control the negotiation?* The supplier is and you can rest assured the *procrastination* will cost you *significantly*!

Issue 5: Electronic Signatures

When implementing an integrated data environment IDE, one of the key issues is establishing a *secure* method for SE documentation reviewers to electronically *approve* the documents. This requires that "electronic signature" standards, methods, and tools be established to ensure that only *authorized* approvers can review and release documentation for implementation.

46.8 GUIDING PRINCIPLES

In summary, the preceding discussions provide the basis with which to establish the guiding principles that govern system engineering and development documentation practices.

Principle 46.1 Establish information protection guidelines; train personnel to implement them.

Principle 46.2 Every document cover and interior page must be marked in accordance with program guidance including authorized recipients.

46.9 SUMMARY

This concludes our overview discussion of *System Engineering and Development Documentation Practices*. The checklists we described are examples of types of SE documentation that help Acquirer and System Developer roles develop their systems, products, and services.

GENERAL EXERCISES

1. Answer each of the *What You Should Learn from This Chapter* questions identified in the *Introduction*.
2. Refer to the list of systems identified in Chapter 2. Based on a selection from the preceding chapter's *General Exercises* or a new system selection, apply your knowledge derived from this chapter's topical discussions. Specifically identify the following:
   (a) If you were the Acquirer of the system, what types of CDRL documents would require for the system design and development effort and why?
   (b) Rank-order SE design and development documentation in terms of significance to the program.

ORGANIZATIONAL CENTRIC EXERCISES

1. Research your organization's command media.
   (a) What types of program documentation do the command media require for programs of all sizes?
   (b) What does the command media say about the level of detail and formality?
   (c) What process is to be used to approve and release the documents?
2. Contact a small, a medium, and a large organization.
   (a) What types of deliverable documentation does each of the contracts require?
   (b) When are these documents to be delivered?
   (c) Are there document outline requirements? If not, what does your organization provide in terms of standard outlines?
   (d) Interview program personnel. What deliverable documents were valuable? Which ones were not?
   (e) Create a timeline indicating when the contract requires delivery of the documents? Include major technical review events (Contract Award, SRR, SDR, etc.).
   (f) Do any of the contracts have export control requirements?

REFERENCES

ANSI/EIA 646-1998. 1998. EIA Standard: *National Consensus Standard for Configuration Management*. Electronic Industries Alliance. Arlington, VA.

DID DI-MGMT-81453. 1995. Data Accession List (DAL), Washington, DC: Department of Defense (DoD).

MIL-HDBK-59B (canceled). 1994. *Continuous Acquisition and Life Cycle Support (CALS) Implementation Guide*. Washington, DC: Department of Defense (DoD).

MIL-STD-973 (canceled). 2000. *Military Standard: Configuration Management*. Notice 4. Washington, DC: Department of Defense (DoD).

ADDITIONAL READING

MIL-HDBK-61. 1997. *Military Handbook: Configuration Management Guidance*, Washington, DC: Department of Defense (DoD).

DoD. *US Government International Traffic and Arms Regulations (ITAR)*. Washington, DC: Department of Defense (DoD).

Chapter 47

Analytical Decision Support

47.1 INTRODUCTION

Earlier we defined "engineering" as the application of *mathematical* and *scientific* principles. We then expanded the scope of this definition for SE to include the application of *analytical* principles. While you can debate for hours the implicit *nature* and relationship of "analysis" to applying *mathematical* and *scientific* principles, we expanded the scope of the definition and elevated the *analytical* aspects to address the hierarchical organization, decomposition, and bounding practices of SE.

The realities are that SEs must be capable of *analyzing*:

1. The User's OPERATING ENVIRONMENT, its constituent *opportunity space*, *problem space*, and *solution space* elements, and relevance to specific SYSTEM missions and applications.
2. A SYSTEM OF INTEREST (SOI), its capabilities, and performance.
3. Interactions between the SOI and its prescribed OPERATING ENVIRONMENT.

This chapter introduces *analytical decision support practices*. Our discussions provide insights into the analytical decision-making environment, factors *affecting* the decision-making process, technical reporting of analytical results, and its challenges and issues.

What You Should Learn from This Chapter

1. What is *analytical decision support*?
2. What are the *attributes* of a technical decision?
3. What is the *SEs role* in technical decision making?
4. What is *system performance evaluation and analysis*?
5. What *types* of engineering analyses are performed?
6. How are *engineering analyses* documented?
7. What *format* should be used for documenting engineering analyses?
8. What are SE *design and development rules* for analytical decision making?

Definitions of Key Terms

- **Analysis** A structured investigation or examination using a proven methodology.
- **Analysis Paralysis** A condition whereby an analyst becomes preoccupied or immersed in the details of an analysis while failing to recognize the marginal utility of continual investigation.

System Analysis, Design, and Development, by Charles S. Wasson
Copyright © 2006 by John Wiley & Sons, Inc.

- **Effectiveness Analysis** "An analytical approach used to determine how well a system performs in its intended utilization environment." (Source: Kossiakoff and Sweet, *System Engineering*, p. 448)
- **Sanity Check** "An approximate calculation or estimation for comparison with a result obtained from a more complete and complex process. The differences in value should be relatively small; if not, the results of the original process are suspect and further analysis is required." (Source: Kossiakoff and Sweet, *System Engineering*, p. 453)
- **Suboptimization** The preferential emphasis on the performance of a lower level entity at the expense of overall system performance.
- **System Optimization** The act of adjusting the performance of individual elements of a system to peak the maximum performance that can be achieved by the integrated set for a given set of boundary conditions and constraints.

47.2 WHAT IS ANALYTICAL DECISION SUPPORT?

Before we begin our discussions of *analytical decision support practices*, we need to first understand the context and anatomy of a technical decision. *Decision support* is a technical services response to a contract or task commitment to gather, analyze, clarify, investigate, recommend, and present fact-based, *objective evidence*. This enables decision makers to SELECT a proper (best) *course of action* from a set of *viable* alternatives bounded by specific constraints—cost, schedule, technical, technology, and support—and *acceptable* level of *risk*.

Analytical Decision Support Objective

The primary objective of *analytical decision support* is to respond to tasking or the need for technical analysis, demonstration, and data collection recommendations to support *informed* SE Process Model decision making.

Expected Outcome of Analytical Decision Support

Decision support *work products* are identified by task objectives. *Work products* and *quality records* include analyses, trade study reports (TSRs), and performance data. In support of these work products, *decision support* develops operational prototypes and proof of concept or technology demonstrations, models and simulations, and mock-ups to provide data for supporting the analysis.

From a technical decision making perspective, decisions are substantiated by the facts of the formal *work products* such as analyses and TSRs provided to the decision maker. The reality is that the decision may have *subconsciously* been made by the decision maker long BEFORE the delivery of the formal work products for approval. This brings us to our next topic, *attributes of technical decisions*.

47.3 ATTRIBUTES OF A TECHNICAL DECISION

Every decision has several attributes you need to understand to be able to properly respond to the task. The attributes you should understand are:

1. WHAT is the *central issue* or *problem* to be addressed?
2. WHAT is the *scope of the task* to be performed?
3. WHAT are the *boundary constraints* for the solution set?

4. What is the *degree of flexibility* in the constraints?
5. Is the *timing* of the decision crucial?
6. WHO is the *user* of the decision?
7. HOW will the decision be used?
8. WHAT *criteria* are to be used in making the decision?
9. WHAT *assumptions* must be made to accomplish the decision?
10. WHAT *accuracy* and *precision* is required for the decision?
11. HOW is the decision is to be *documented* and *delivered*?

Scope the Problem to be Solved

Decisions represent approval of solutions intended to lead to *actionable* tasks that will resolve a *critical operational or technical or issue (COI/CTI)*. The analyst begins with understanding what:

1. Problem is to be solved.
2. Question is to be answered.
3. Issue is to be resolved.

Therefore, begin with a CLEAR and SUCCINCT *problem statement*.

Referral For more information about writing problem statements, refer to Chapter 14 on *Understanding The Problem, Opportunity, and Solution Spaces concept*.

If you are tasked to solve a technical problem and are not provided a documented tasking statement, discuss it with the decision authority. *Active listening* enables analysts to *verify* their *understanding* of the tasking. Add corrections based on the discussion and return a courtesy copy to the decision maker. Then, when briefing the status of the task, ALWAYS include a restatement of the task so ALL reviewers have a *clear* understanding of the analysis you were tasked to perform.

Decision Boundary Condition Constraints and Flexibility

Technical decisions are bounded by cost, schedule, technology, and support constraints. In turn, the constraints must be reconciled with an *acceptable* level of risk. *Constraints* sometimes are also *flexible*. Talk with the decision maker and assess the amount of *flexibility* in the constraint. Document the constraints and *acceptable level of risk* as part of the task statement.

Criticality of Timing of the Decision

Timing of decisions is CRUCIAL, not only from the perspective of the decision maker but also that of the SE supporting the decision making. Be *sensitive* to the decision authority's schedule and the prevailing environment when the recommendations are presented. If the schedule is *impractical*, discuss it with the decision maker including level of risk.

Understand How the Decision Will Be Used and by Whom

Decisions often require approvals by multiple levels of organizational and customer stakeholder decision makers. Avoid wasted effort trying to solve the wrong problem. Tactfully *validate* the decision *problem statement* by consensus of the stakeholders.

Document the Criteria for Decision Making

Once the *problem statement* is documented and the boundary constraints for the decision are established, identify the *threshold criteria* that will be used to assess the success of the decision results. Obtain stakeholder concurrence with the decision criteria. Make corrections as necessary to clarify the criteria to avoid *misinterpretation* when the decision is presented for approval. If the decision criteria are not documented "up front," you may be subjected to the discretion of the decision maker to determine when the task is complete.

Identify the Accuracy and Precision of the Analysis

Every technical decision involves data that have a *level of accuracy* and *precision*. Determine "up front" what *accuracy* and *precision* will be required to support analytical results, and make sure these are clearly communicated and understood by everyone participating. One of the worst things analysts can do is discover *after the fact* that they need four-digit decimal data *precision* when they only measured and recorded two-digit data. Some data collection exercises may not be *repeatable* or *practical*. THINK and PLAN ahead: similar rules should be established for *rounding* data.

Author's Note 47.1 *As a reminder, two-digit precision data that require multiplication DO NOT yield four-digit precision results; the best you can have is two-digit result due to the source data precision.*

Identify How the Decision Is to Be Delivered

Decisions need a point of closure or delivery. Identify what *form* and *media* the decision is to be delivered: as a document, presentation, or both. In any case, make sure that your response is documented for the record via a cover letter or e-mail.

47.4 TYPES OF ENGINEERING ANALYSES

Engineering analyses cover a spectrum of disciplinary and specialty skills. The challenge for SEs is to understand:

1. WHAT analyses may be required.
2. At WHAT level of detail.
3. WHAT tools are best suited for various analytical applications.
4. WHAT level of formality is required for documenting the results.

To illustrate a few of the many analyses that might be conducted, here's a sample list.

- Mission operations and task analysis
- Environmental analysis
- Fault tree analysis (FTA)
- Finite element analysis (FEA)
- Mechanical analysis
- Electromagnetic interference (EMI)/electromagnetic countermeasures (EMC) analysis
- Optical analysis
- Reliability, availability, and maintainability (RAM) analysis
- Stress analysis

- Survivability analysis
- Vulnerability analysis
- Thermal analysis
- Timing analysis
- System latency analysis
- Life cycle cost analysis

Guidepost 47.1 *The application of various types of engineering analyses should focus on providing objective, fact-based data that support informed technical decision making. These results at all levels aggregate into overall system performance that forms the basis of our next topic, system performance evaluation and analysis.*

47.5 SYSTEM PERFORMANCE EVALUATION AND ANALYSIS

System performance evaluation and analysis is the investigation, study, and operational analysis of *actual* or *predicted* system performance relative to *planned* or *required* performance as documented in performance or *item development specifications*. The analysis process requires the planning, configuration, data collection, and post data analysis to thoroughly understand a system's performance.

System Performance Analysis Tools and Methods

System performance evaluation and *analysis* employs a number of decision aid tools and methods to collect data to support the analysis. These include models, simulations, prototypes, interviews, surveys, and test markets.

Optimizing System Performance

System components at every *level of abstraction* inherently have *statistical* variations in physical characteristics, reliability, and performance. Systems that involve humans involve statistical variability in knowledge and skill levels, and thus involve an element of *uncertainty*. The challenge question for SEs is: *WHAT combination of system configurations, conditions, human-machine tasks, and associated levels of performance optimize system performance?*

System optimization is a term relative to the *stakeholder*. *Optimization criteria* reflect the appropriate balance of cost, schedule, technical, technology, and support performance or combination thereof.

Author's Note 47.2 *We should note here that optimization is for the total system. Avoid a condition referred to as suboptimization unless there is a compelling reason.*

Suboptimization

Suboptimization is a condition that exists when one element of a system—the PRODUCT, SUBSYSTEM, ASSEMBLY, SUBASSEMBLY, or PART level—is *optimized* at the expense of overall system performance. During System Integration, Test, and Evaluation (SITE), system items at each level of abstraction may be optimized. Theoretically, if the item is designed correctly, *optimal* performance occurs at the planned midpoint of any adjustment ranges.

The underlying design philosophy here is that if the system is properly designed and component statistical variations are *validated*, only minor *adjustments* may be required for an output to be centered about some hypothetical *mean* value. If the variations have not been taken into account or design modifications have been made, the output may be "off-set" from the *mean* value but within its operating range when "optimized." Thus, at higher levels of integration, this *off-nominal* condition may impact overall system performance, especially if further adjustments beyond the components adjustment range are required.

The Danger of Analysis Paralysis

Analyses serve as a powerful tool for understanding, predicting, and communicating system performance. Analyses, however, cost money and consume valuable resources. The challenge question for SEs to consider is, *How GOOD is good enough? At what level or point in time does an analysis meet minimal sufficiency criteria to be considered valid for decision making?* Since engineers, by nature, tend to *immerse* themselves in analytics, we sometimes suffer from a condition referred to as "analysis paralysis." So, *what is analysis paralysis?*

Analysis paralysis is a condition where an analyst becomes preoccupied or *immersed* in the details of an analysis while failing to recognize the marginal utility of continual investigation. So, *HOW do SEs deal with this condition?*

First, you need to learn to recognize the signs of this condition in yourself as well as others. Although the condition varies with everyone, some are more prone than others. *Second*, aside from personality characteristics, the condition may be a response mechanism to the work environment, especially from paranoid, control freak managers who suffer from the condition themselves.

47.6 ENGINEERING ANALYSIS REPORTS

As a discipline requiring *integrity* in analytical, mathematical, and scientific data and computations to support downstream or lower level decision making, engineering documentation is often sloppy at best or simply nonexistent. One of the hallmarks of a professional discipline is an expectation to document recommendations supported by *factual, objective evidence* derived empirically or by observation.

Data that contribute to informed SE decisions are characterized by the *assumptions, boundary conditions*, and *constraints* surrounding the data collection. While most engineers competently consider relevant factors affecting a decision, the *tendency* is to *avoid* recording the results; they view paperwork as unnecessary, *bureaucratic documentation* that does not add value directly to the deliverable product. As a result, a professional, high-value analysis ends in mediocrity due to the analyst *lacking* personal initiative to perform the task correctly.

To better appreciate the professional discipline required to document analyses properly, consider a hypothetical visit to a physician:

EXAMPLE 47.1

You visit a medical doctor for a condition that requires several treatment appointments at three-month intervals for a year. The doctor performs a high-value diagnosis and prescribes the treatments but fails to record the medication and actions performed at each treatment event. At each subsequent treatment you and the doctor have to reconstruct *to the best of everyone's knowledge* the assumptions, dosages, and actions performed. Aside from the medical and legal implications, can you imagine the frustration, foggy memories, and "guesstimates" associated with these interactions. Engineering, as a professional discipline, is no different. Subsequent decision making is highly dependent on the documented assumptions and constraints of previous decisions.

The difference between mediocrity and high-quality *professional results* may be only a few minutes to simply document critical considerations that yielded the analytical result and recommendations presented. For SEs, this information should be recorded in an *engineering laboratory notebook* or on-line in a network-based *journal*.

47.7 ENGINEERING REPORT FORMAT

Where practical and appropriate, engineering analyses should be documented in formal technical reports. Contract or organizational command media sometimes specify the format of these reports. If you are expected to formally report the results of an analysis and do not have specific format requirements, consider the example outline below.

EXAMPLE 47.2

The following is an example of an outline that could be used to document a technical report.

1.0. INTRODUCTION

The introduction establishes the context and basis for the analysis. Opening statements identify the document, its context and usage in the program, as well as the program this analysis is being performed to support.

- **1.1.** Purpose
- **1.2.** Scope
- **1.3.** Objectives
- **1.4.** Analyst/Team Members
- **1.5.** Acronyms and Abbreviations
- **1.6.** Definitions of Key Terms

2.0. REFERENCED DOCUMENTS

This section lists the documents referenced in other sections of the document. Note the *operative* title "*Referenced Documents*" as opposed to "*Applicable Documents*."

3.0. EXECUTIVE SUMMARY

Summarize the results of the analysis such as *findings*, *observations*, *conclusions*, and *recommendations*: tell them the bottom line "up front." Then, if the reader desires to read about the details concerning HOW you arrived at those results, they can do so in subsequent sections.

4.0. CONDUCT OF THE ANALYSIS

Informed decision making is heavily dependent on *objective*, fact based data. As such, the conditions under which the analysis is performed must be established as a means of providing credibility for the results. Subsections include:

- **4.1.** Background
- **4.2.** Assumptions
- **4.3.** Methodology
- **4.4.** Data Collection
- **4.5.** Analytical Tools and Methods
- **4.6.** Versions and Configurations
- **4.7.** Statistical Analysis (if applicable)
- **4.8.** Analysis Results
- **4.9.** Observations

4.10. Precision and Accuracy
4.11. Graphical Plots
4.12. Sources

5.0. FINDINGS, OBSERVATIONS, AND CONCLUSIONS

As with any scientific study, it is important for the analyst to communicate:

- WHAT they found.
- WHAT they observed.
- WHAT *conclusions* they derived from the *findings* and *observations*. Subsections include:

   5.1. Findings
   5.2. Observations
   5.3. Conclusions

6.0. RECOMMENDATIONS

Based on the analyst's *findings*, *observations*, and *conclusions*, Section 6.0 provides a set of *prioritized* recommendations to decision makers concerning the objectives established by the analysis tasking.

APPENDICES

Appendices provide areas to present supporting documentation collected during the analysis or that illustrates how the author(s) arrived at their *findings*, *conclusions*, and *recommendations*.

Decision Documentation Formality

There are numerous ways to address the need to *balance* document decision making with time, resource, and formality constraints. Approaches to document critical decisions range from a single page of *informal*, handwritten notes to *highly* formal documents. Establish disciplinary standards for yourself and your organization related to documenting decisions. Then, scale the documentation formality according to task constraints. Regardless of the approach used, the documentation should CAPTURE the key attributes of a decision in sufficient detail to enable "downstream" understanding of the factors that resulted in the decision.

The *credibility* and *integrity* of an analysis often depends on who collected and analyzed the data. Analysis report appendixes provide a means of organizing and preserving any supporting vendor, test, simulation, or other data used by the analyst(s) to support the results. This is particularly important if, at a later date, *conditions* that served as the basis for the initial analysis task change, thereby creating a need to revisit the original analysis. Because of the changing conditions, some data may have to be regenerated; some may not. For those data that have not changed, the appendices *minimize* work on the new task analysis by *avoiding* the need to *recollect* or *regenerate* the data.

47.8 ANALYSIS LESSONS LEARNED

Once the performance analysis tasking and boundary conditions are established, the next step is to conduct the analysis. Let's explore some lessons learned you should consider in preparing to conduct the analysis.

Lesson 1: Establish a Decision Development Methodology

Decision paths tend to veer off-course midway through the decision development process. Establish a decision making methodology "up front" to serve as a *roadmap* for keeping the effort on track. When you establish the methodology "up front," you have the *visibility of clear*, unbiased THINKING unemcumbered by the adventures along the decision path. If you and your team are convinced you have a good methodology, that plan will serve as a compass heading. This is not to say that some conditions may warrant a change in methodology. Avoid changes unless there is a *compelling* reason to change.

Lesson 2: Acquire Analysis Resources

As with any task, success is partially driven by simply having the RIGHT resources in place when they are required. This includes:

1. Subject matter experts (SMEs)
2. Analytical tools
3. Access to personnel who may have relevant information concerning the analysis area
4. Analytical tools
5. Data that describe operating conditions and observations relevant to the analysis, and so forth.

Lesson 3: Document Assumptions and Caveats

Every decision involves some level of *assumptions* and/or *caveats*. *Document the assumptions in a clear, concise manner.* Make sure that the CAVEATS are documented on the same page as the decision (footnotes, etc.) or recommendations. If the decision *recommendations* are copied or *removed* from the document, the *caveats* will ALWAYS be in place. Otherwise, people may *intentionally* or *unintentionally* apply the decision or recommendations *out of context*.

Lesson 4: Date the Decision Documentation

Every page of a decision document should marked indicating the document title, revision level, date, page number, and classification level, if applicable. Using this approach, the reader can always determine if the version they possess is current. Additionally, if a single page is copied, the source is readily identifiable. Most people fail to perform this simple task. When multiple versions of a report, especially drafts, are distributed without dates, the de facto version is determined by WHERE the document is within a stack on someone's desk.

Lesson 5: State the Facts as Objective Evidence

Technical reports must be based on the latest, *factual* information from *credible* and *reliable* sources. Conjecture, hearsay, and personal opinions should be *avoided*. If requested, qualified opinions can be presented *informally* with the delivery of the report.

Lesson 6: Cite Only Credible and Reliable Sources

Technical decisions often *leverage* and *expand* on existing knowledge and research, published or verbal. If you use this information to support findings and conclusions, *explicitly* cite the source(s) in *explicit* detail. Avoid vague references such as *"read the [author's] report"* documented in an obscure publication published 10 years ago that may be inaccessible or only available to the author(s). If these sources are unavailable, quote passages with permission of the owner.

Lesson 7: REFERENCE Documents versus APPLICABLE Documents

Analyses often reference other documents and employ the terms APPLICABLE DOCUMENTS or REFERENCED DOCUMENTS. People *unknowingly* interchange the terms. Using conventional outline structures, Section 2.0 should be titled REFERENCED DOCUMENTS and list all sources cited in the text. Other source or related reading material *relevant* to the subject matter is cited in an ADDITIONAL READING section provided in the appendix.

Lesson 8: Cite Referenced Documents

When citing *referenced documents*, include the date and version containing data that serve as inputs to the decision. People often believe that if they reference a document by title they have satisfied analysis criteria. Technical decision making is only as good as the *credibility* and *integrity* of its sources of *objective, fact-based information*. Source documents may be revised over time. Do yourself and your team a favor: make sure that you *clearly* and *concisely* document the critical attributes of source documentation.

Lesson 9: Conduct SME Peer Reviews

Technical decisions are sometimes *dead on arrival (DOA)* due to poor assumptions, flawed decision criteria, and bad research. Plan for success by conducting an *informal* peer review by trusted and qualified colleagues—the subject matter experts (SMEs)—of the evolving decision document. Listen to their *challenges* and *concerns*. Are they highlighting *critical operational and technical issues (COIs/CTIs)* that remain to be *resolved*, or *overlooked* variables and solutions that are *obscured* by the analysis or research? We refer to this as "posturing for success" before the presentation.

Lesson 10: Prepare Findings, Conclusions, and Recommendations

There are a number of reasons as to WHY an analysis is conducted. In one case the technical decision maker may not possess current technical expertise or the ability to *internalize* and *assimilate* data for a *complex* problem. So they seek out those who do posses this capability such as consultants or organizations. In general, the analyst wants to know WHAT the subject matter experts (SMEs) who are closest to the problems, issues, and technology suggest as recommendations regarding the decision. Therefore, analyses should include *findings*, *recommendations*, and *recommendations*.

Based on the results of the analysis, the decision maker can choose to:

1. Ponder the *findings* and *conclusions* from their own perspective.
2. *Accept* or *reject* the recommendations as a means of arriving at an informed decision.

In any case, they need to know WHAT the subject matter experts (SMEs) have to offer regarding the decision.

47.9 GUIDING PRINCIPLES

In summary, the preceding discussions provide the basis with which to establish the guiding principles that govern analytical decision support practices.

Principle 47.1 Analysis results are only as VALID as their underlying assumptions, models, and methodology. Validate and preserve their integrity.

47.10 SUMMARY

Our discussion of *analysis decision support* provided data and recommendations to support the SE Process Model at all levels of abstraction. As an introductory discussion, analytical decision support employs various tools addressed in the sections that follow:

- Statistical influences on SE decision making
- System performance analysis, budgets, and safety margins
- System reliability, availability, and maintainability
- System modeling and simulation
- Trade studies: analysis of alternatives

GENERAL EXERCISES

1. Answer each of the *What You Should Learn from This Chapter* questions identified in the *Introduction*.
2. Refer to the list of systems identified in Chapter 2. Based on a selection from the preceding chapter's General *Exercises* or a new system selection, apply your knowledge derived from this chapter's topical discussions. If you were the project engineer or Lead SE:
   (a) What types of engineering analyses would you recommend?
   (b) How would you collect data to support those analyses?
   (c) Select one of the analyses. Write a simple analysis task statement based on the *attributes of a technical decision* discussed at the beginning of this section.

ORGANIZATIONAL CENTRIC EXERCISES

1. Research your organization's command media for guidance and direction concerning the implementation of *analytical decision support practices*.
   (a) What requirements are levied on programs and SEs concerning the conduct of analyses?
   (b) Does the organization have a standard methodology for conducting an analysis? If so, report your findings.
   (c) Does the organization have a standard format for documenting analyses? If so, report your findings.
2. Contact small, medium and large contract programs within your organization.
   (a) What analyses were performed on the program?
   (b) How were the analyses documented?
   (c) How was the analysis task communicated? Did the analysis report describe the objectives and scope of the analysis?
   (d) What *level of formality*—engineering notebook, informal report, or formal report—did technical decision makers levy on the analysis?
   (e) Were the analyses conducted without constraints or were they conducted to justify a predetermined decision?
   (f) What *challenges* or *issues* did the analysts encounter during the conduct of the analysis?
   (g) Based on the program's lessons learned, what recommendations do they offer as guidance for conducting analyses on future programs?

3. Select two analysis reports from different contract programs.
   (a) What is your assessment of each report?
   (b) Did the program apply the right level of formality in documenting the analysis?

REFERENCES

IEEE Std 610.12-1990. 1990. *IEEE Standard Glossary of Modeling and Simulation Terminology*. Institute of Electrical and Electronic Engineers (IEEE). New York, NY.

KOSSIAKOFF, ALEXANDER, and SWEET, WILLIAM N. 2003. *Systems Engineering Principles and Practice*, New York: Wiley-InterScience.

ADDITIONAL READING

ASD-100. 2004. National Airspace System *System Engineering Manual, ATO Operations Planning*. Washington, DC: Federal Aviation Administration (FAA).

Chapter 48

Statistical Influences on System Design

48.1 INTRODUCTION

For many engineers, system design evolves around abstract phrases such as "bound environmental data" and "receive data." The challenge is: *HOW do you quantify and bound the conditions for a specific parameter*? Then, *how does an SE determine conditions such as*:

1. Acceptable signal and noise (S/N) ratios?
2. Computational errors in processing the data?
3. Time variations required to process system data?

The reality is that the hypothetical boundary condition problems engineers studied in college aren't so ideal. Additionally, when a system or product is developed, multiple copies may produce varying degrees of responses to a set of controlled inputs. So, *how do SEs deal with the challenges of these uncertainties*?

Systems and products have varying degrees of stability, performance, and uncertainty that are *influenced* by their unique form, fit, and function *performance* characteristics. Depending on the price the User is willing to pay, we can improve and match material characteristics and processes used to produce the SE systems and products. If we analyze a system's or product's performance characteristics over a controlled range of inputs and conditions, we can *statistically* state the *variance* in terms of *standard deviation*.

This chapter provides an introductory overview of how statistical methods can be applied to system design to improve capability performance. As a prerequisite to this discussion, you should have basic familiarity with statistical methods and their applications.

What You Should Learn from This Chapter

1. How do you characterize random variations in system inputs sufficiently to bound the range?
2. How do SEs establish criteria for acceptable system inputs and outputs?
3. What is a design range?
4. How are upper and lower tolerance limits established for a design range?
5. How do SEs establish criteria for CAUTION and WARNING indicators?
6. What development methods can be employed to improve our understanding of the variability of engineering input data?

System Analysis, Design, and Development, by Charles S. Wasson
Copyright © 2006 by John Wiley & Sons, Inc.

7. What is circular error probability (CEP)?
8. What is meant by the degree of correlation?

Definitions of Key Terms

- **Circular Error Probability (CEP)** The Gaussian probability density function (normal distribution) referenced to a central point with concentric rings representing the standard deviations of data dispersion.
- **Cumulative Error** A measure of the total cumulative errors inherent within and created by a system or product when processing statistically variant inputs to produce a standard output or outcome.
- **Logarithmic Distribution (Lognormal)** An asymmetrical, graphical plot of the Poisson probability density function depicting the dispersion and frequency of independent data occurrences about a mean that is skewed from a median of the data distribution.
- **Normal Distribution** A graphical plot of the Gaussian probability density function depicting the symmetrical dispersion and frequency of independent data occurrences about a central mean.
- **Variance (Statistical)** "A measure of the degree of spread among a set of values; a measure of the tendency of individual values to vary from the mean value. It is computed by subtracting the mean value from each value, squaring each of these differences, summing these results, and dividing this sum by the number of values in order to obtain the arithmetic mean of these squares." (Source: DSMC T&E Mgt. Guide, *DoD Glossary of Test Terminology*, p. B-21)

48.2 UNDERSTANDING THE VARIABILITY OF THE ENGINEERING DATA

In an ideal world, engineering data are precisely linear or identically match *predictive* values with zero *error margins*. In the real world, however, variations in mass properties and characteristics; attenuation, propagation, and transmission delays; and human responses are among the uncertainties that must be accounted for in engineering calculations. In general, the data are *dispersed* about the *mean* of the *frequency distribution*.

Normal and Logarithmic Probability Density Functions

Statistically, we characterize the range dispersions about a central mean in terms of normal (Gaussian) and logarithmic (Poisson) *frequency distributions* as shown in Figure 48.1.

Normal and *logarithmic* frequency distributions can be used to mathematically *characterize* and *bound* engineering data related to *statistical process control* (SPC); *queuing* or *waiting line theory* for customer service and message traffic; production lines; maintenance and repair, pressure containment; temperature/humidity ranges; and so on.

Applying Statistical Distributions to Systems

In Chapter 3 *What Is A System* we characterized a system as having *desirable* and *undesirable* inputs and producing *desirable* outputs. The system can also produce *undesirable* outputs—be they electromagnetic, optical, chemical, thermal, or mechanical—that make the system *vulnerable* to adversaries or create self-induced feedback that *degrades* system performance. The challenge for SEs is bounding:

588   Chapter 48   Statistical Influences on System Design

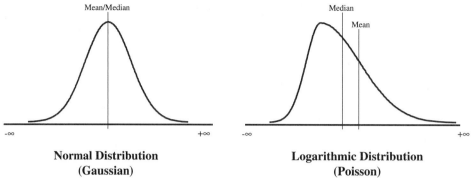

Figure 48.1 Basic Types of Statistical Frequency Distributions

1. The range of *desirable* or *acceptable inputs* and *conditions* from *undesirable* or *unacceptable inputs*.
2. The range of *desirable* or *acceptable outputs* and *conditions* from *undesirable* or *unacceptable outputs*.

Recall Figure 3.2 of our discussion of *system entity concepts* where we illustrate the challenge in SE Design decision making relative to *acceptable* and *unacceptable* inputs and outputs.

Design Input/Output Range Acceptability. Statistically we can bound and characterize the *range of acceptable inputs* and *outputs* using the frequency distributions. As a simple example, Figure 48.2 illustrates an example of a Normal Distribution that we can employ to characterize input/output variability.

In this illustration we employ a Normal Distribution with a *central mean*. Depending on bounding conditions imposed by the system application, SEs determine the acceptable design range that includes upper and lower limits relative to the *mean*.

Range of Acceptable System Performance. During normal system operations, system or product capabilities perform within an acceptable (Normal) Design Range. The challenge for SEs is determining WHAT the thresholds for alerting system operators and maintainers WHEN system performance is OFF nominal and begins to pose a risk or threat to the operators, EQUIPMENT, public, or environment. To better understand this point, let's examine it using Figure 48.2.

In the figure we have a Normal Distribution about a central mean and characterized by *four* types of operating ranges:

- **DESIGN Range** The range of engineering parameter values for a specific capability and conditions that *bound* the ACCEPTABLE upper and lower *tolerance* limits.
- **NORMAL Operating Range** The range of *acceptable* engineering parameter values for a specific capability within the design range that clearly indicates capability performance under a given set of conditions is operating as expected and does not pose a risk or threat to the operators, EQUIPMENT, general public, or environment.
- **CAUTIONARY Range** The range of engineering parameter values for a specific capability that clearly indicates capability performance under a given set of conditions is *beyond* or

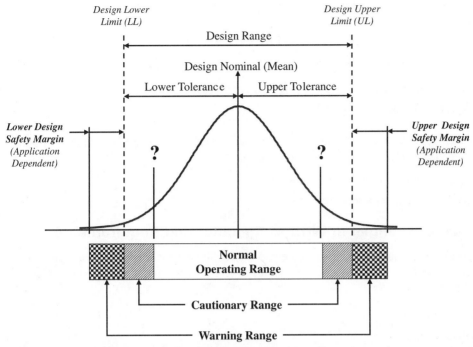

Figure 48.2 Application Dependent Normalized Range of Acceptable Operating Condition Limits

OUTSIDE the Normal Operating Range and potentially poses a *risk* or *threat* to the operators, EQUIPMENT, general public, or environment.

- **WARNING Range** The range of engineering parameter values for a specific capability and conditions that clearly poses a high level of *risk* or a *threat* to the operators, EQUIPMENT, public, or environment with *catastrophic consequences*.

This presents several decision-making challenges for SEs:

1. What is the *acceptable* Design Range that includes upper and lower Caution Ranges?
2. WHAT are the UPPER and LOWER limits and conditions of the *acceptable* Normal Operating Range?
3. WHAT are the *thresholds* and *conditions* for the WARNING Range?
4. WHAT upper and lower Design Safety Margins and conditions must be established for the system relative to the Normal Operating Range, Caution Range, and Warning Range?

These questions, which are application dependent, are typically difficult to answer. Also keep in mind that this graphic reflects a single *measure of performance* (MOP) for one system entity at a specific level of abstraction. The significance of this decision is *exacerbated* by the need to allocate the design range to lower level entities, which also have comparable performance distributions, ranges, and safety margins. Obviously, this poses a number of risks. For large, complex systems, *HOW do we deal with this challenge?*

There are several approaches for supporting the *design thresholds* and *conditions*.

First, you can model and simulate the system and employ Monte Carlo techniques to assess the *most likely* or *probable* outcomes for a given set of use case scenarios. *Second*, you can leverage modeling and simulation results and develop a prototype of the system for further analysis and evaluation. *Third*, you can employ spiral development to evolve a set of requirements over a set of sequential prototypes.

Now let's shift our focus to *understanding* how statistical methods apply to system development.

48.3 STATISTICAL METHOD APPLICATIONS TO SYSTEM DEVELOPMENT

Statistical methods are employed throughout the System Development Phase by various disciplines. For SEs, statistical challenges occur in two key areas:

1. *Bounding* specification requirements.
2. *Verifying* specification requirements.

Statistical Challenges in Writing Specification Requirements

During specification requirements development, Acquirer SEs are challenged to specify the *acceptable and unacceptable* ranges of inputs and outputs for performance-based specifications. Consider the following example:

EXAMPLE 48.1

Under specified operating conditions, the Sensor System shall have a probability of detection of 0.XX over a (magnitude) spectral frequency range.

Once the contract is awarded, System Developer SEs are challenged to *determine*, *allocate*, and *flow down* system *performance budgets* and *safety margins* requirements derived from higher level requirements such as in the example above. The challenge is analyzing the example above to *derive* requirements for PRODUCT, SUBSYSTEM, ASSEMBLY, and other levels. Consider the following example.

EXAMPLE 48.2

The (name) output shall have a ±3σ worst-case error of 0.XX for Input Parameter A distributions between 0.000 vdc to 10.000 vdc.

Statistical Challenges in Verifying Specification Requirements

Now let's suppose that an SE's mission is to verify the requirement stated in Example 48.2. For simplicity, let's assume that the sampled end points of Input data are 0.000 vdc and 10.000 vdc with a couple of points in between. We collect data measurements as a function of Input Data and plot them. Panel A of Figure 48.3 might be representative of thess data.

Applying statistical methods, we determine the *trend line* and ±3σ boundary conditions based on the requirement. Panels C and D of Figure 48.3 could represent these data. Then we superimpose the trend line and ±3σ boundaries and verify that all system performance data are within the *acceptable* range indicating the system *passed* (Panel D).

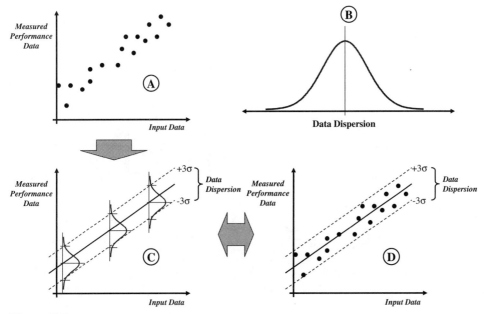

Figure 48.3 Understanding Engineering Data Dispersion

48.4 UNDERSTANDING TEST DATA DISPERSION

The preceding discussion *focuses on design decisions*. Let's explore how system analysts and SEs statistically deal with test data that *may be used as an input into SE design decision making or verifying that* a *System Performance Specification (SPS)* or *item development specification* requirements has been achieved.

Suppose that we conduct a test to measure system or entity performance over a range of input data as shown in Panel A of Figure 48.3. As illustrated, we have a number of data points *that have* a *positive* slope. This graphic has two important aspects:

1. Upward sloping *trend* of data
2. A dispersion of data along the trend line.

In this example, if we performed a Least Squares mathematical fit of the data, we could establish the slope and intercepts of the trend line using a simple $y = mx + b$ construct.

Using the trend line as a central *mean* for the data set as a function of Input Data (X-axis), we find that the corresponding Y data points are dispersed about the *mean* as illustrated in Panel C. Based on the *standard deviation* of the data set, we could say that there is 0.9973 probability that a given data point lies within the $\pm 3\sigma$ about the mean. Thus, Panel D depicts the results of projecting the $\pm 3\sigma$ lines along the trend line.

48.5 CUMULATIVE SYSTEM PERFORMANCE EFFECTS

Our discussions to this point focus on statistical distributions relative to a specific capability parameter. The question is: *HOW do these errors propagate throughout the system*? There are several factors that contribute to the error propagation:

592 Chapter 48 Statistical Influences on System Design

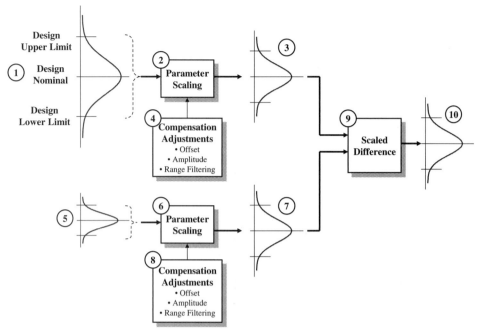

Figure 48.4 Understanding Cumulative Error Statistics

1. OPERATING ENVIRONMENT influences on system component properties.
2. Timing variations.
3. Computational precision and accuracy.
4. Drift or aliasing errors as a function of time.

From a total system perspective, we refer to this concept as *cumulative error*. Figure 48.4 provides an example.

Let's assume we have a simple system that computes the difference between two parameters A and B. If we examine the characteristics of parameters A and B, we find that each parameter has different data dispersions about its predicted *mean*.

Ultimately, if we intend to compute the difference between parameter A and parameter B, both parameters have to be *scaled* relative to some *normalized* value. Otherwise, we get an "apples and oranges" comparison. So, we scale each input and make any correctional offset adjustments. This simply solves the functional aspect of the computation. Now, *what about errors originating from the source values about a nominal mean plus all intervening scaling operations?* The answer is: SEs have to *account* for the *cumulative error distributions* related to errors and dispersions. Once the system is developed, integrated, and tested, SYSTEM Level *optimization* is used to correct for any *errors* and *dispersions*.

48.6 CIRCULAR ERROR PROBABILITY (CEP)

The preceding discussion focused on analyzing and allocating system performance within the system. The ultimate test for SE decision making comes from the actual field results. The question is: *How do cumulative error probabilities impact overall operational and system effectiveness?* Perhaps the best way to answer this question is a "bull's-eye target" analogy using Figure 48.5.

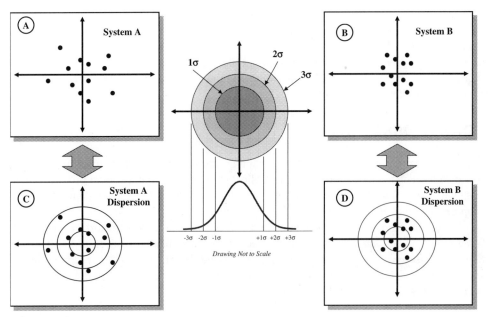

Figure 48.5 Circular Error Probability Example

Our discussions up to this point have focused on the *dispersion* of data along *linear* trend lines with a central mean. There are system applications whereby data are dispersed about a central point such as the "bull's eye" illustrated in Figure 48.5. In these cases the ±1σ, ±2σ, and ±3σ points lie on concentric circles aligned about a central mean located at the bull's eye. Applications of this type are generally target based such as munitions, firearms, and financial plans. Consider the following example:

EXAMPLE 48.3

Suppose that you conduct an evaluation of two competing rifle systems, System A and System B. We will assume statistical sampling methods are employed to determine a statistically valid sample size. Specification requirements state that 95% of the shots must be contained within circle with a diameter of X inches centered at the bull's eye.

Each system is placed in a test fixture and calibrated. When environmental conditions are acceptable, expert marksmen "live fire" the required number of rounds from each rifle. Marksmen are unaware of the manufacturer of each rifle. Miss distance firing results are shown in Panels A and B.

Using the theoretical crosshair as the origin, you superimpose the concentric lines about the bull's eye representing the ±1σ, ±2σ, and ±3σ points as illustrated in the center of the graphic. Panels C and D depict the results with miss distance as the deciding factor, System B is superior.

In this simple, ideal example we focused exclusively on *system effectiveness*, not *cost effectiveness*, which includes system effectiveness. The challenge is: *things are not always ideal and rifles are not identical in cost*. What do you do? The solution lies in the Cost as an Independent Variable (CAIV) and *trade study utility function* concepts discussed earlier in Figure 6.1. *What is the utility function of the field performance test results relative to cost and other factors?*

594 Chapter 48 Statistical Influences on System Design

If System A costs one-half as much as System B, *does the increased performance of System B substantiate the cost*? You may decide that the ±3σ point is the *minimum threshold requirement* for *system acceptability*. Thus, from a CAIV perspective, System A meets the specification *threshold requirement* and costs one-half as much, yielding the best value.

You can continue this analysis further by evaluating the utility of hitting the target on the first shot for a given set of time constraints, and so forth.

48.7 DATA CORRELATION

Engineering often requires developing mathematical algorithms that model best-fit approximations to real world data set characterizations. Data are collected to *validate* that a system produces high-quality data within *predictable* values. We refer to the degree of "fit" of the actual data to the standard or approximation as *data correlation*.

Data correlation is a measure of the degree to which *actual* data regress toward a central mean of *predicted* values. When *actual values* match predicted *values*, data correlation is 1.0. Thus, as data set variances *diverge* away from the mean trend line, the *degree of correlation* represented by r, the *correlation coefficient*, *diminishes* toward zero. To illustrate the concept of data correlation and convergence, Figure 48.6 provides examples.

Positive and Negative Correlation

Data correlation is characterized as *positive* or *negative* depending on the SLOPE of the line representing the mean of the data set over a range of input values. Panel A of Figure 48.6 represents a *positive* (slope) correlation; Panel B represents a *negative* (slope) correlation. This brings us to our next point, *convergence* or *regression* toward the mean.

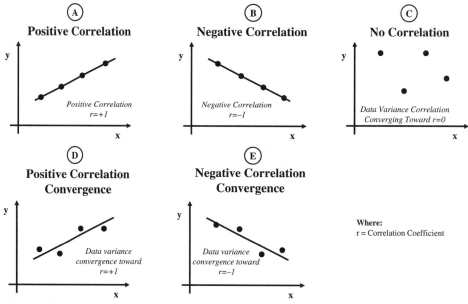

Figure 48.6 Understanding Data Correlation

Regression toward Convergence

Since engineering data are subject to variations in physical characteristics, *actual data* do not always perfectly match the *predicted values*. In an ideal situation we could state that the data *correlate* over a bounded range IF all of the values of the data set are perfectly aligned on the mean trend line as illustrated in Panels A and B of Figure 48.6.

In reality, data are typically dispersed along the trend line representing the mean values. Thus, we refer to the *convergence* or *data variance* toward the mean as the degree of *correlation*. As data sets *regress* toward a *central mean*, the *data variance* or *correlation* increases toward $r = +1$ or -1 as illustrated in Panels D and E. Data variances that *decrease* toward $r = 0$ indicate decreasing *convergence* or *low correlation*. Therefore, we characterize the relationship between data parameters as *positive* or *negative data variance convergence* or *correlation*.

48.8 SUMMARY

Our discussions of *statistical influences on system design practices* were predicated on a basic understanding of statistical methods and provided a high-level overview of key statistical concepts that influence SE design decisions.

We highlighted the importance of using statistical methods to define acceptable or desirable design ranges for input and output data. We also addressed the importance of establishing boundary conditions for NORMAL operating ranges, CAUTIONARY ranges, WARNING ranges, as well as establishing *safety margins*. Using the basic concepts as a foundation, we addressed the concept of *cumulative errors*, *circular error probabilities (CEP)*, and data correlation. We also addressed the need to bound *acceptable* or *desirable* system outputs that include products, by-products, and services.

Statistical data variances have significant influence on SE technical decisions such as system performance, budgets, and safety margins and operational and system effectiveness. What is important is that SEs:

1. Learn to *recognize* and *appreciate* engineering input/output data variances
2. Know WHEN and HOW to apply statistical methods to understand SYSTEM interactions with its OPERATING ENVIRONMENT.

GENERAL EXERCISES

1. Answer each of the *What You Should Learn from This Chapter* questions identified in the *Introduction*.
2. Refer to the list of systems identified in Chapter 2. Based on a selection from the preceding chapter's General *Exercises* or a new system, selection, apply your knowledge derived from this chapter's topical discussions. Specifically identify the following:
   (a) What inputs of the system can be represented by statistical distributions?
   (b) How would you translate those inputs into a set of input requirements?
   (c) Based on processing of those inputs, do errors accumulate and, if so, what is the impact?
   (d) How would you specify requirements to minimize the impacts of errors?

ORGANIZATIONAL CENTRIC EXERCISES

1. Contact a technical program in your organization. Research how the program SEs accommodated statistical variability for the following:
   (a) Acceptable data input and output ranges for system processing
   (b) External data and timing variability

2. For systems that require performance monitoring equipment such as gages, meters, audible warnings, and flashing lights, research how SEs determined threshold values for activating the notifications or indications.

REFERENCE

Defense Systems Management College (DSMC). 1998. DSMC *Test and Evaluation Management Guide*, 3rd ed. Defense Acquisition Press. Ft. Belvoir, VA.

ADDITIONAL READING

BLANCHARD, BENJAMIN S., and J. FABRYCKY, WOLTER. 1990. *Systems Engineering and Analysis*, 2nd ed. Englewood Cliff, NJ: Prentice-Hall.

LANGFORD, JOHN W. 1995. *Logistics: Principles and Applications*. New York: McGraw-Hill.

National Aeronautics and Space Administration (NASA). 1994. *Systems Engineering "Toolbox" for Design-Oriented Engineers*. NASA Reference Publication 1358. Washington, DC.

Chapter 49

System Performance Analysis, Budgets, and Safety Margins

49.1 INTRODUCTION

System effectiveness manifests itself via the cumulative performance results of the integrated set of System Elements at a specific instance in time. That performance ultimately determines *mission* and *system objectives* success—in some cases, *survival*.

When SEs allocate system performance, there is a tendency to think of those requirements as *static parameters*—for example, "*shall be +12.3 ± 0.10 vdc*." Aside from status switch settings or configuration parameters, seldom are parameters static or steady state.

From an SE perspective, SEs partition and organize requirements via a hierarchical framework. Take the example of static weight. We have a budget of 100 pounds to allocate equally to three components. Static parameters make the SE requirements allocation task a lot easier. This is not the case for many system requirements. *How do we establish values for system inputs that are subject to variations such as environmental conditions, time of day, time of year, signal properties, human error and other variables?*

System requirement parameters are often characterized by *statistical* value distributions—such as Normal (Gaussian), Binomial, and LogNormal (Poisson)—with frequencies and tendencies about a *mean value*. Using our static requirements example above, we can state that the voltage must be constrained to a range of +12.20 vdc (-3σ) to +12.40 vdc ($+3\sigma$) with a *mean* of +12.30 vdc for a prescribed set of operating conditions.

On the surface, this sounds very simple and straightforward. The challenge is: *How did SEs decide:*

1. That *the mean value needed to be +12.30 vdc*?
2. That *the variations could not exceed 0.10 vdc*?

This simple example illustrates one of the most challenging and perplexing aspects of System Engineering—*allocating dynamic parameters*.

Many times SEs simply do not have any *precedent* data. For example, consider human attempts to build bridges, develop and fly an aircraft, launch rockets and missiles, and land on the Moon and Mars. Analysis with a lot of *trial and error* data collection and observation may be all you have to establish initial estimates of these parameters.

There are a number of ways one can determine these values. Examples include:

System Analysis, Design, and Development, by Charles S. Wasson
Copyright © 2006 by John Wiley & Sons, Inc.

1. Educated guesses based on seasoned experience.
2. Theoretical and empirical trial and error analysis.
3. Modeling and simulation with increasing fidelity.
4. Prototyping demonstrations.

The challenge is being able to identify a reliable, low-risk, level of confidence method of determining values for statistically variant parameters.

This chapter describes how we allocate *System Performance Specification (SPS)* requirements to lower levels. We explore how *functional* and *nonfunctional performance* are analyzed and allocated. This requires building on previous practices such as *statistical influences on system design* discussed in the preceding chapter. We introduce the concepts of decomposing cycle time based performances into queue, process, and transport times. Finally, we conclude by illustrating how *performance budgets* and *safety margins* enable us to achieve SPS performance requirements.

What You Should Learn from This Chapter

1. What is *system performance analysis*?
2. What is a *cycle time*?
3. What is a *queue time*?
4. What is a *transport time*?
5. What is a *processing time*?
6. What is a *performance budget*?
7. How do you establish *performance budgets*?
8. What is a *safety margin*?

Definitions of Key Terms

- **"Design-to" MOP** A targeted *mean value* bounded by *minimum* and/or *maximum* threshold values levied on a system capability performance parameter to constrain decision making.
- **Performance Budget Allocation** A *minimum, maximum, or min-max* constraint that represents the absolute thresholds that bound a capability or performance characteristic.
- **Processing Time** The statistical *mean* time and tolerance that statistically characterizes the time interval between an input stimulus or cue event and the completion of processing of the input(s).
- **Queue Time** The statistical *mean* time and tolerance that characterizes the time interval between the *arrival* of an input for processing and the point where processing begins.
- **Safety Margin** A portion of an assigned capability or physical characteristic *measure of performance (MOP)* that is restricted from casual usage to cover instances in which the budgeted performance exceeds its allocated MOP.
- **System Latency** The time differential between a stimulus or cue events and a system response event. Some people refer to this as the *responsivity* of the system for a specific parameter.
- **Transport Time** The statistical *mean* time and tolerance that characterizes the time interval between transmission of an output and its receipt at the next processing task.

49.2 PERFORMANCE "DESIGN-TO" BUDGETS AND SAFETY MARGINS

Every functional capability or physical characteristic of a system or *item* must be bounded by *performance constraints*. This is very important in top-down/bottom-up/horizontal design whereby system functional capabilities are *decomposed*, *allocated*, and *flowed down* into multiple levels of design detail.

Achieving Measures of Performance (MOPs)

The mechanism for decomposing system performance into subsequent levels of detail is referred to as *performance budgets and margins*. In general, *performance budgets and margins* allow SEs to impose performance constraints on functional capabilities that include a *margin of safety*. Philosophically, if overall system performance must be controlled, so should the *contributing* entities at multiple *levels of abstraction*.

Performance constraints are further partitioned into: 1) *"design-to" measures of performance* (MOPs) and 2) *performance safety margins*.

Design-to MOPs

Design-to MOPs serve as the key mechanism for *allocating*, *flowing down*, and *communicating performance constraints* to lower levels system *items*. The actual allocation process is accomplished by a number of methods ranging from *equitable* shares to specific allocations based on *arbitrary* and *discretionary* decisions or decisions supported by design support analyses and trade studies.

Safety Margins

Safety margins accomplish two things. *First*, they provide a means to accommodate variations in tolerances, accuracies, and latencies in system responses plus errors in human judgment. *Second*, they provide a reserve for decision makers to trade off misappropriated performance inequities as a means of *optimizing* overall system performance.

Performance *safety margins* serve as *contingency reserves* to compensate for component variations or to accommodate *worst-case scenarios* that:

1. Could have been underestimated.
2. Potentially create safety risks and hazards.
3. Result from human errors in computational precision and accuracy.
4. Are due to physical variations in materials properties and components
5. Result from the "unknowns."

Every engineering discipline employs *rules of thumb* and *guidelines* for accommodating *safety margins*. Typically, safety margins might vary from 5% to 200% on average, depending on the application and risk.

There are limitations to the practicality of *safety margins* in terms of: 1) cost–benefits, 2) *probability* or *likelihood of occurrence*, 3) alternative actions, and 4) reasonable measures, among other things. In some cases, the implicit cost of increasing *safety margin* MOPs above a *practical* level can be offset by taking appropriate system or product safety precautions, safeguards, markings, and procedures that reduce the probability of occurrence.

600 Chapter 49 System Performance Analysis, Budgets, and Safety Margins

Warning! ALWAYS seek guidance from your program and technical management, disciplinary standards and practices, or local organization engineering command media to establish a consensus about safety margins for your program. When these are established prior to the Contract Award, document the authoritative basis for the selection and disseminate to all personnel or incorporate into program command media—as project memoranda or plans.

Safety margins, as the name implies, involve technical decision making to prevent potential safety hazards from occurring. Any potential safety hazards carry safety, financial, and legal liabilities. Establish safety margins that safeguard the SYSTEM and its operators, general public, property, and the environment.

Applying Design-to MOPs and Safety Margins

Figure 49.1 illustrates how *Design-To MOPs* and *safety margins* are established. In this figure the *measures of performance (MOPs)* for system capabilities and physical characteristics are given in generic terms or units. Note that the units can represent time, electrical power, mass properties, and so on.

Author's Note 49.1 *The example in Figure 49.1 shows the basic method of allocating performance budgets and safety margins. In reality, this highly iterative, time-consuming process often requires analyses, trade studies, modeling, simulation, prototyping, and negotiations to balance and optimize overall system performance.*

Let's assume the *SPS* specifies that Capability A have a performance constraint of 100 units. System designers decide to establish a 10% safety margin at all levels of the design. Therefore, they establish a *Design-To MOP* of 90 units and a safety margin of 10 units. The *Design-To MOP* of 90 units is allocated as follows: Capability A_1 is allocated an MOP of 40 units, Capability A_2 is allocated an MOP of 30 units, and Capability A_3 is allocated an MOP of 20 units.

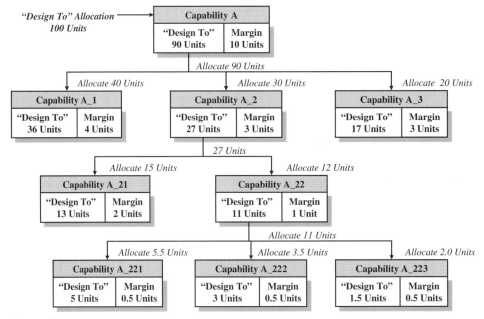

Figure 49.1 Performance Budget & Design Margin Allocations

Focusing on Capability A_2, the MOP of 30 units is partitioned into a *Design-To MOP* of 27 units and a *safety margin MOP* of 3 units. The resultant *Design-To MOP* of 27 units is then allocated to Capability A_21 and Capability A_22 in similar manner, and so forth.

Some SEs argue that once you establish the initial 10 unit *safety margin MOP* at the Capability A level, there is no need to establish *design safety margins* at lower levels of the capability— (capability A_3 safety margin, etc.). Observe how the second level has allocated an additional 10 units of margin to the Capability A-1, A-2, and A-3 budgets above and beyond the 10 units at the Capability A level. They emphasize that as long as all subordinate level capabilities meet their *Design-To MOP* performance budgets, the 10 unit MOP safety margin adequately covers situations where lower level performance exceeds allocated budgets.

They continue that imposing *safety margins* at lower levels unnecessarily CONSTRAINS critical resources and *increases* system cost due to the need for *higher* performance equipment. For some *nonsafety* critical, non–real-time performance applications, this may be true; every system and application is unique.

As an initial starting point, ALWAYS begin with multi-level safety margins. If you encounter difficulties meeting lower level performance allocation constraints, you should weigh options, benefits, costs, and risks. Since system *item* performance inevitably requires adjustment for *optimization*, ALWAYS establish design *safety margins* at every level and for every system *item* to ensure flexibility in the integrated performance in achieving SPS requirements. Once all levels of design are defined, rebalance the hierarchical structure of performance budgets and design safety margins as needed.

To illustrate how we might implement time-based performance budgets and margins allocations, let's explore another example.

EXAMPLE 49.1

To illustrate this concept, refer to Figure 49.2. The left side of the graphic portrays a hierarchical decomposition of a Capability A. The *SPS* requires that Capability A complete processing within a specified period of time such as 200 milliseconds. System designers designate the initiation of Capability A as *Event 1* and its completion as *Event 2*. A Mission Event Timeline (MET) depicting the two *event* constraints is shown in the top portion of the graphic. System Designers partition the Capability A time interval constraint into a *Design-To MOP* and a *safety margin MOP*. The Design-To MOP constraint is designated as *Event 1.4*.

Author's Note 49.2 *Initially MET Event 1.4 may not have this label. We have simply applied the Event 1.4 label to provide a degree of sequential consistency across the MET (1.1, 1.2, 1.3, etc.). Events 1.2 and 1.3 are actually established by lower level allocations for Capabilities A_1 and A_2.*

As shown at the left side, the Capability A requirement is analyzed and decomposed into Capability A_1 and Capability A_2 requirements. Thus, the *Design-To MOP* for Capability A is partitioned into a Capability A_1 time constraint and a Capability A_2 time constraint as MOPs. Likewise, Capability A_1 and Capability A_2 are decomposed into lower level requirements, each with its respective *Design-To MOP* and *safety margin*. The process continues to successively lower levels of system *items*.

Reconciling Performance Budget Allocations and Safety Margins. As design teams apply Design-To MOP allocations, *what happens if there are critical performance issues with the initial allocations?* Let's assume that Capability A_22 in Figure 49.2 was initially allocated 12 units. An initial analysis of Capability A_22 indicates that 13 units are required. *What should an SE do?*

602 Chapter 49 System Performance Analysis, Budgets, and Safety Margins

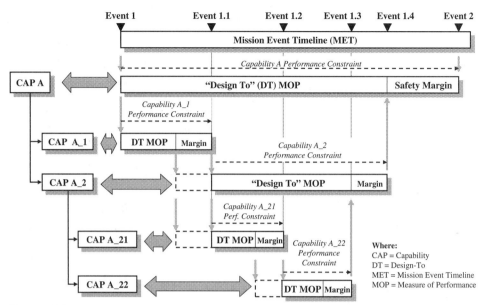

Figure 49.2 Performance Budgets & Safety Margins Application

Capability A_22 owner confers with the higher level Capability A_2 and peer level Capability A_21 owners. During the discussions the Capability A_21 owner indicates that Capability A_21 was allocated 15 units but only requires 14 units, which includes a *safety margin*. The group reaches a *consensus* to *reallocate* an MOP of 14 units to Capability A_21 and an MOP of 13 units to Capability A_22.

Final Thoughts about Performance Budgets and Margins. The process of allocating *performance budgets* and *safety margins* is a top-down/bottom-up/left-right/right-left negotiation process. Within human decision-making terms, the intent is to reconcile the inequities as a means of achieving and *optimizing* overall system performance. Without *negotiation* and *reconciliation*, you get a condition referred to as *suboptimization* of a single item, thereby *degrading* overall system performance.

Performance Budget and Safety Margin Ownership

A key question is: *WHO owns performance budgets and safety margins*? In general, the owner of the specification that contains the capabilities and physical characteristics that are allocated as performance budget MOPs and safety margins is the owner.

How are Performance Budgets and Margins Documented?

Performance budgets and safety margins are documented a number of ways, depending on program size, resources, and tools.

First, requirements allocations should be documented in a decision database or spreadsheet controlled by the Lead SE or System Engineering and Integration Team (SEIT). Requirements management tools based on relational databases provide a convenient mechanism to record the allocations. *Second*, as performance allocations, they should be *formally* documented and controlled as specific requirements flowed down to lower level specifications.

A relational database requirements management tools allow you to:

1. Document the allocation.
2. Flow the allocation down to lower level specifications with traceability linkages back to the higher level *parent* performance constraint.

49.3 ANALYZING SYSTEM PERFORMANCE

The preceding discussion introduced the basic concepts of *performance budgets* and *design safety margins*. Implementations of these *Design-To MOPs* are discussed in engineering textbooks such as electronics engineering and mechanical engineering. However, from an SE perspective, integrated electrical, mechanical, or optical systems have performance *variations* interfacing with similar EQUIPMENT and PERSONNEL within larger *structural* systems. The interactions among these systems and *levels of abstractions* require in-depth analysis to determine acceptable limits for performance variability. At all *levels of abstraction*, capabilities are typically event and/or task driven—that is, an external or time-based stimulus or cue activates or initiates a capability to action.

Understanding System Performance and Tasking

Overall, system performance represents the *integrated* performance of the System Elements—such as EQUIPMENT, PERSONNEL, and MISSION RESOURCES—that provide system capabilities, operations, and processes. As integrated elements, if the performance of any of these mission critical items—(PRODUCTS, SUBSYSTEMS, etc.) is *degraded*, so is the overall system performance, depending on the *robustness* of the system design. *Robust* designs often employ *redundant* hardware and/or software design implementations to *minimize* the effects of system performance *degradation* on achieving the mission and its objectives.

Referral For more information about redundant systems, refer to Chapter 50 on *Reliability, Availability, and Maintainability (RAM) Practices* and also Chapter 36 on *System Architecture Development Practices*.

From a systems perspective, SYSTEM capabilities, operations, and executable processes are response mechanisms to "tasking" assigned and initiated by peer or HIGHER ORDER Systems with authority. Thus, SYSTEM "tasking" requires the integrated set of *sequential* and *concurrent* capabilities to accomplish a desired performance-based *outcome* or *result*.

To illustrate the TASKING of capabilities, consider the simple graphic shown in Figure 49.3. Note the chain of sequential tasks, Tasks A through *n*. Each task has a finite duration bounded by a set of *Mission Event Timeline (MET)* performance parameters. The time period marked by the start of one task until the start of another is referred to as the *throughput* or *cycle time*. The *cycle time* parameter brings up an interesting point, especially in establishing a *convention* for decision making.

Establishing Cycle Time Conventions

When you establish *cycle times*, you need to define a convention that will be used consistently throughout your analyses. There are a couple of approaches as shown in Figure 49.4 Convention A defines *cycle time* as beginning with the START of Task A and the START of Task B. In contrast, Convention B defines *cycle time* as starting at the END of Task A and completing at the start of Task B. You can use either method. For discussion purposes, we will use Convention A.

604 Chapter 49 System Performance Analysis, Budgets, and Safety Margins

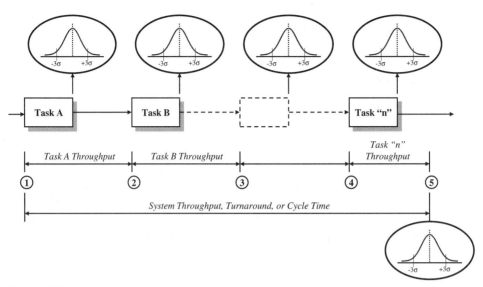

Figure 49.3 Mission Event Timeline (MET) Analysis

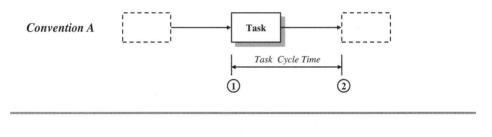

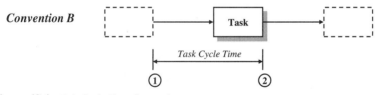

Figure 49.4 Task Cycle Time Convention

Referral For more information about *conventions*, refer to Chapter 45 on *Engineering Standards, Frames of Reference*, and *Conventions Practices*.

49.4 OPERATIONAL TASKING OF A SYSTEM CAPABILITY

Most tasks, whether performed by human operators or EQUIPMENT, involve three phases:

1. *Pre-Mission* Preparation or configuration/reconfiguration.
2. *Mission* Performance of the task.
3. *Post-mission* Delivery of the required results and any residual housecleaning in preparation for the next tasks.

49.4 Operational Tasking of a System Capability

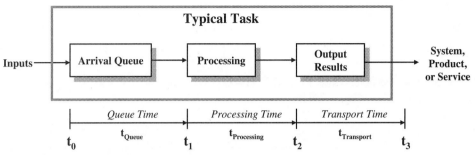

Figure 49.5 Anatomy of Typical Task Performance

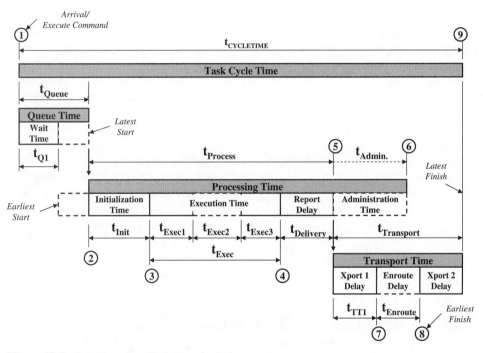

Figure 49.6 Task Throughput/Cycle Time Analysis

Communications intensive systems such as humans, computers, et al have a similar pattern that typically includes QUEUES—or waiting line theory. We illustrate this pattern in Figure 49.5.

In the illustration, a typical task provides three capabilities: 1) queue new arrivals, 2) perform processing, and 3) output results. Since new arrivals may *overwhelm* the processing function, queue new arrivals establish a *buffer* or *holding area* for first in–first out (FIFO) processing or some other *priority* processing algorithm.

Each of these capabilities is marked by its own *cycle time*: t_{Queue} = Queue Time, $t_{Process}$ = Processing Time, and $t_{Transport}$ = Transport Time. Figure 49.6 illustrates how we might decompose each of these capabilities into lower level time constraints for establishing budgets and safety margins.

Guidepost 49.1 *At this point we have identified and partitioned task performance into three phases: queue time, processing time, and transport time. Philosophically, this partitioning enables*

606 Chapter 49 System Performance Analysis, Budgets, and Safety Margins

us to decompose allocated task performance times into smaller, more manageable pieces. Beyond this point, however, tasks analysis becomes more complicated due to timing uncertainties. This brings us to our next topic, understanding statistical characteristics of tasking.

49.5 UNDERSTANDING STATISTICAL CHARACTERISTICS OF TASKING

All tasks involve some level of performance *uncertainty*. The *level of uncertainty* is created by the inherent reliability of the System Elements—PERSONNEL, MISSION RESOURCES, and so forth. In general, task performance and its *uncertainty* can be characterized with statistics using normal (Gaussian) (Normal), Binomial, and LogNormal (Poisson) frequency distributions. Based on measured performance over a large number of samples, we can assign a PROBABILITY that task or capability performance will complete processing within a minimum amount of time and should not exceed a specified amount of time. To see how this relates to system performance and allocated performance budgets and margins, refer to Figure 49.7.

Let's suppose a task that is initiated at Event 1 and must be completed by Event 2. We refer to this time as $t_{Allocation}$. To ensure a *margin of safety* as a contingency and for growth, we establish a performance safety margin, t_{Margin}. This leaves a remaining time, t_{Exec}, as the maximum budgeted performance.

Using the lower portion of the graphic, suppose that we perform an analysis and determine that the task, represented by the gray rectangular box, is expected to have a mean finish time, t_{Mean}. We also determine with a *level of probability* that task completion may vary about the *mean* by ±3σ designated as *early finish* and *late finish*. Therefore, we *equate* the latest finish to the maximum budgeted performance, t_{Budget}. This means that once initiated at Event 1, the task must complete and deliver the output or results to the next task no later than t_{Budget}. Based on the projected distri-

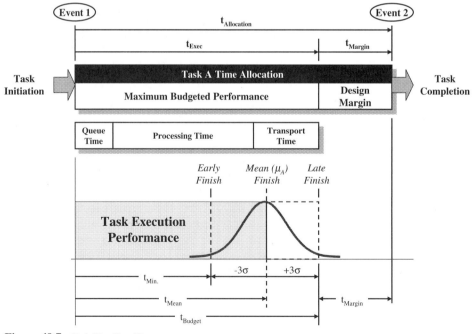

Figure 49.7 Task Timeline Elements

bution, we also expect the task to be completed no sooner than the -3σ point—the *early finish* point.

If we translate this analysis into the Requirements Domain Solution, we generate a requirements statement that captures the capability and its associated performance allocation. Let's suppose that Capability B requires that Capability A complete processing and transmit data within 250 milliseconds AFTER Event 1 occurs. Consider the following example of a specification requirement statement.

EXAMPLE 49.2

"When event 1 occurs, (Capability A) shall process the incoming data and transmit the results to (Capability B) within 250 milliseconds."

Now let's suppose that Capability B requires receipt of the data within a window of time between 240 milliseconds and 260 milliseconds. The requirement might read as follows:

EXAMPLE 49.3

"When event 1 occurs, (Capability A) shall process the incoming data and transmit the results to (Capability B) within 250 ± 10 milliseconds."

Guidepost 49.2 Our discussion emphasizes how overall system task performance can vary. To understand how this variation occurs, let's take it one step further and discuss it relative to the key task phases.

49.6 APPLYING STATISTICS TO MULTI-TASK SYSTEM PERFORMANCE

Our previous discussion focused on the *performance* of a *single capability*. If the *statistical* variations for a single capability are *aggregated* for a *multi-level* system, we can easily see how this impacts overall system performance. We can illustrate this by the example shown in Figure 49.8.

Let's assume we have an overall task called Perform Task A. The purpose of Task A is to perform a computation using *variable inputs*, I_1 and I_2, and produce a computed value as an output. The key point of our discussion here is to illustrate time-based *statistical* variances to complete processing.

EXAMPLE 49.4

Let's assume that Task A consists of two subtasks, Subtask A_1 and Subtask A_2. Subtask A_1 enters inputs I_1 and I_2. Each input, I_1 and I_2, has values that vary about a mean.

When Subtask A_1 is initiated, it produces a response within t_{A1Mean} that varies from t_{A1Min} to t_{A1Max}. The output of Subtask A_1 serves as an input to Subtask A_2. Subtask A_2 produces a response within t_{A2Mean} that may occur as early as t_{A2Min} or as late as t_{A2Max}.

If we investigate the overall performance of Task A, we find that Task A is computed within $t_{Compute}$ as indicated by the central mean. The overall Task A performance is determined by the statistical variance of the summation of Subtask A_1 and Subtask A_2 processing times.

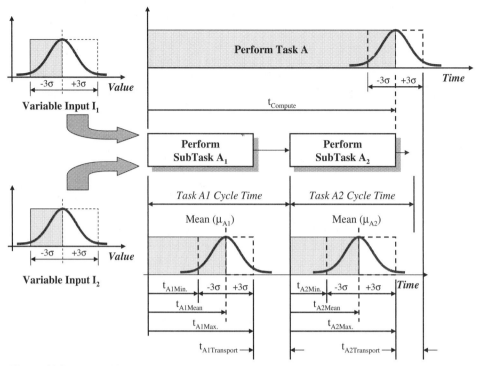

Figure 49.8 Task Timeline (MET) Statistical Analysis

Guidepost 49.3 *We have seen how statistical variations in inputs and processing affect system performance from a timing perspective. Similar methods are applied to the statistical variations of inputs 1 and 2 as independent variables over a range of values. The point of our discussion is to heighten your awareness of these variances. THINK about and CONSIDER statistical variability when allocating performance budgets and margins as well as analyzing data produced by the system to determine compliance with requirements.*

Referral For more information about statistical variability, refer to Chapter 48 on *Statistical Influences on System Design Practices*.

Given this understanding, let's return to a previous discussion about *applying statistical variations to phases of a task*.

49.7 APPLYING STATISTICAL VARIATIONS TO INTRA-TASK PHASES

In our earlier discussion of task phases—consisting of *queue time, capability performance time*, and *transport time*—we identified the various time segments within each phase. Let's examine how statistical influences affect those phases. Figure 49.9 serves as the focal point for our discussion.

The central part of the figure represents an overall task and its respective *queue time, processing time*, and *transport time* phases. Below each phase is a statistical *representation* of the execution time. The top portion of the chart illustrates the *aggregate* performance of the overall task execution.

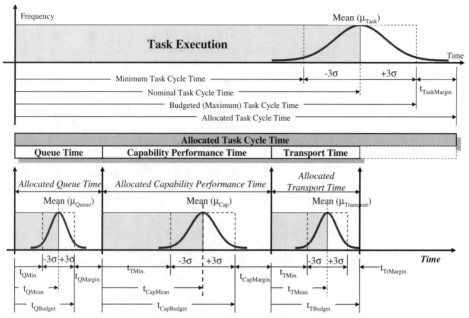

Figure 49.9 Task Timeline (MET) Statistical Analysis

How does this relate to SE? If a given capability or task is supposed to be completed within the allocated cycle time and you are designing a system—with queues, computational devices, and transmission lines—you need to *factor* in these times as *performance budgets* and *safety margins* to flow down to lower levels.

Applying Statistics to Overall System Performance

Our discussions to this point focused on the task and multi-tasking level. The ultimate challenge for SEs is: *HOW will the overall system perform*? Figure 49.10 illustrates the *effects* of statistical variability of the System Element Architecture and its OPERATING ENVIRONMENT.

49.8 MATHEMATICAL APPROXIMATION ALTERNATIVE

Our conceptual discussions of *statistical* system performance analysis were intended to highlight key considerations for establishing and allocating *performance budgets* and *margins* and analyzing data for system performance tuning. Most people do not have the time to perform the statistical analyses. For some applications, this may be acceptable and you should use the method appropriate for your application. There is an alternative method you might want to consider using, however.

Scheduling techniques such as the Program Evaluation and Review Technique (PERT) employ approximations that serve as analogs to a Gaussian (normal) distribution. The formula stated below is used:

$$\text{Expected or mean time} = \frac{t_a + 4t_b + t_c}{6} \quad (49.1)$$

where

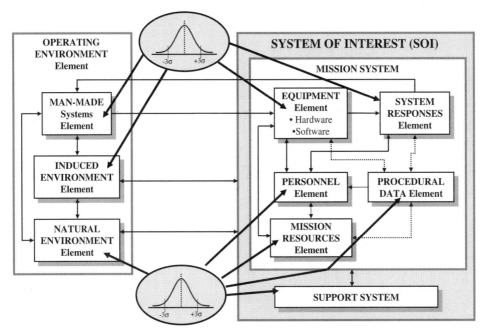

Figure 49.10 Statistical Variations in SYSTEM Level Performance

t_a = optimistic time

t_b = most likely time

t_c = pessimistic time

Do SEs Actually Perform This Type of Analysis?

Our discussion here highlights the theoretical viewpoint of task analysis. A question people often ask is: *Do people actually go to this level of detail?* In general, the answer is yes, especially in manufacturing and scheduling environments. In those environments statistical process control (SPC) is used to *minimize* process and material *variations* in the production of parts, and this *translates* into *cost reduction*.

Modeling and Simulation

If you develop a model of a system whereby each of the capabilities, operations, processes, and tasks is represented by a series of sequential and concurrent *elements* or feedback loops, you can apply statistics to the processing time associated with each of those elements. By analyzing how each of the input variables varies over value ranges bounded on the ±3σ points, you can determine the overall system performance relative to a *mean*.

Guidepost 49.4 *Our discussions highlighted some basic task-oriented methods that support a variety of systems engineering activities. These methods can be applied to Mission Event Timelines (METs), system capabilities and performance as a means of determining overall system performance. Through decomposition and allocation of overall requirements, developers can establish the appropriate performance budgets and safety margins for lower level system entities.*

49.9 REAL-TIME CONTROL AND FRAME-BASED SYSTEMS

Some systems operate as *real-time*, *closed loop*, feedback systems. Others are multi-tasking whereby they have to serve multiple processing tasks on a priority basis. Let's explore each of these types further.

Real-Time, Closed Loop Feedback Systems

Electronic, mechanical, and electromechanical systems include real-time, closed loop, feedback systems that condition or process input data and produce an output, which is sampled and summed with the input as negative feedback. Rather than feedback impulse functions to the input, filters may be required to dampen the system response. Otherwise, the system might overcompensate and go unstable while attempting to regain control. The challenge for SEs is determining and allocating performance for the optimal *feedback responses* to ensure system stability.

Frame-Based System Performance

Electronic systems often employ software to accomplish CYCLICAL data processing tasks using combinations of OPEN and CLOSED loop cycles. Systems of these types are referred to as *frame-based systems*.

Frame-based systems perform accomplish *multi-task* processing via time-based blocks of time such as 30 Hertz (Hz) or 60 Hz. Within each block, processing of multiple CONCURRENT tasks is accomplished by allocating a portion of each frame to a specific task, depending on *priorities*. For these cases, apply performance analysis to determine the appropriate mix of concurrent task processing times.

Author's Note 49.3 *One approach to frame-based system task scheduling is rate monotonic analysis (RMA). Research this topic further if frame-based systems apply to your business domain.*

49.10 SYSTEM PERFORMANCE OPTIMIZATION

System performance analysis provides a valuable tool for *modeling* and *prediction* the intended interactions of the SYSTEM in its OPERATING ENVIRONMENT. Underlying *assumptions* are validated when the SYSTEM is powered up and operated. The challenge for SEs becomes one of *optimizing* overall system performance to *compensate* for the variability of the embedded PRODUCTS, SUBSYSTEMS, ASSEMBLIES, SUBASSEMBLIES, and PARTS.

Minimum Conditions for System Optimization

When the system enters the System Integration, Test, and Evaluation (SITE) Phase, *deficiencies* often *consume* most of the SE's time. The challenge is getting the system to a *state of equilibrium* that can best be described as *nominal* and *acceptable* as defined by the *System Performance Specification (SPS)*.

Once the system is in a state of nominal operation with no outstanding deficiencies, there may be a need to tweak performance to achieve optimum performance. Let's reiterate the last point: you must correct all major deficiencies BEFORE you can attempt to optimize system performance in a specific area. *Exceptions* include minor items that are not system performance effecters. Consider the following example:

EXAMPLE 49.5

Suppose that you are developing a fuel-efficient vehicle and are attempting to optimize road performance via a test track. If the fuel flow has a deficiency, can you optimize overall system performance? Absolutely not! *Does having a taillight out impact fuel efficiency performance?* No, its not a contributing element to fuel flow. It may, however, impact safety.

Pareto-Based Performance Improvements

There are a number of ways to optimize system performance, some of which can be very time-consuming. For many systems, however, there is *limited* time to *optimize* performance prior to delivery, simply because the system development schedule has been consumed with correcting system deficiencies.

When you investigate options for WHERE to focus system performance *optimization* efforts, one approach employs the Pareto 80/20 rule. Under the 80/20 rule, 80% of system performance is attributable to 20% of the processing tasks. If you accept this analogy, the key is to identify the 20% processes and focus performance analysis efforts on *maximizing* or *minimizing* their impact. So, employ *diagnostic* tools to understand HOW each item is performing as well as interface latencies between items via networks, and so forth.

Today electronic instrumentation devices and software are available to track processing tasks that consume system resources and performance. Plan from the beginning of system development HOW these devices and software can be employed to identify and *prioritize* system processing task performance and *optimize* it.

System performance *optimization* begins on day 1 after Contract Award via:

1. System performance requirements allocations.
2. Plans for "test points" to monitor performance after the system has been fully integrated.

49.11 SYSTEM ANALYSIS REPORTING

As a disciplined professional, document the results of each an analysis. For engineering tasks that involve simple assessments, ALWAYS document the results, even informally, in an engineering notebook. For tasks that require a more formal, structured approach, you may be expected to deliver a formal report. A common question many SEs have is: *HOW do you structure an analytical report*?

First, you should consult your contract for specific requirements. If the contract does not require a specific a structure, consult your local command media. If your command media does not provide guidance, consider using the outline provided in Chapter 47 on *Analytical Decision Support Practices*.

49.12 GUIDING PRINCIPLES

In summary, the preceding discussions provide the basis with which to establish the guiding principles that govern system performance analysis, budgets, and safety margins practices.

Principle 49.1 Every SYSTEM/entity design must have performance budgets and margins that bound capabilities and provide a margin of safety to accommodate uncertainty in material, component, and operational performance.

Principle 49.2 Every measure of performance (MOP) has an element of development risk. Mitigate the risk by establishing one or more boundary condition thresholds to trigger risk mitigation actions.

Principle 49.3 When planning task processing, at a *minimum*, include considerations for *queue time*, *processing time*, and *transport time* in all computations.

Principle 49.4 System performance for a specific capability should only be *optimized* when all *performance effecters* are operating *normally* within their specification tolerances.

49.13 SUMMARY

Our discussions of *system performance analysis, budgets, and margins practices* provided an overview of key SE design considerations. We described the basic process and methods for:

1. Allocating measures of performance (MOPs) to lower levels
2. Investigating task-based processing relative to statistical variability
3. Publishing analysis results in a report using a recommended outline.

We also offered an approach for estimating task processing time durations and introduced the concept of rate monotonic analysis (RMA) for frame-based processing.

Finally, we showed from an SE perspective the performance variability of System Elements that must be factored into performance allocations.

- Each SPS or item specification measure of performance (MOP) should be partitioned into a "design-to" MOP and a safety margin MOP.
- Each program must provide guidance for establishing safety margins for all system elements.

GENERAL EXERCISES

1. Answer each of the *What You Should Learn from This Chapter* questions identified in the *Introduction*.
2. Refer to the list of systems identified in Chapter 2. Based on a selection from the preceding chapter's General *Exercises* or a new system selection, apply your knowledge derived from this chapter's topical discussions. Select an element of the system. Specifically identify the following:
   (a) If you had been the developer of the system, what guidance would you have provided for performance allocations, budgets, and safety margins? Give examples.
   (b) What types of budgets and margins would you recommend?

ORGANIZATIONAL CENTRIC EXERCISES

1. Contact an in-house program that designs *real time software intensive systems* for laboratory environments. Interview SE personnel regarding the following questions and prepare a report on your findings and observations.
   (a) How were the system tasking, task performance, and EQUIPMENT sized, timed, and optimized?
   (b) How did the program document this information?
   (c) Is rate monotonic analysis (RMA) applicable to this program and how is it applied?
2. Contact an in-house program that designs *real time products for deployment in external environments*—such as missiles or monitoring stations. Interview SE personnel regarding the following questions and prepare a report on your findings and observations.

(a) How were the system tasking, task performance, and EQUIPMENT sized, timed, and optimize.

(b) How did the program document this information?

(c) Is RMA applicable to this program and how is it applied?

3. Search the Internet for papers on rate monotonic analysis (RMA). Develop a *conceptual report* on how you would apply RMA to the following types of systems or products or others of your choosing. Include responses to inputs and output results.

   (a) Automobile engine controller

   (b) Multi-tasking computer system

4. Contact a small, medium, and a large contract program within you organization.

   (a) What guidance did they establish for design safety margins? Provide examples.

   (b) How were design-to parameters and safety margins allocated, documented, and tracked? Provide examples.

   (c) How was the guidance communicated to developer?

   (d) Were technical performance measures (TPMs) required?

   (e) How did the program link TPMs and performance budgets and safety margins?

ADDITIONAL READING

SHA, LUI and SATHAYE, SHIRISH. *Distributed System Design Using Generalized Rate Monotonic Theory*, Technical Report CMU/SEI-95-TR-011, Carneige-Mellon University, 1995.

KLEIN, MARX, RALYA, THOMAS; POLLAK, BILL; OBENZA, RAY; HARBOUR, MICHAEL GONZALE. 1993. A Practitioners' Handbook for Real-Time Analysis: Guide to Rate Monotonic Analysis for Real-Time Systems. Boston, MA: Kluwer Academic. Publishers, Norwell, MA.

Chapter 50

Reliability, Availability, and Maintainability

50.1 INTRODUCTION

The purpose of any system is to accomplish the intended User's missions and objectives as bounded by a prescribed OPERATING ENVIRONMENT. The focus of any mission is on successful completion of the objectives *efficiently* and *effectively* with *minimal* or no safety risk to its system's operators, the system, general public, and the environment.

What does it take to accomplish these objectives? Three key system requirements emerge.

- The system has to be immediately available *on-demand* when the User wants to conduct their missions.
- The system must be *reliable* and *dependable* to complete the mission safely.
- The system must be capable of being *maintained* and achieve turnaround requirements that support its operational readiness—or *availability*—to conduct new missions.

We refer to the three requirements as: 1) availability, 2) reliability, and 3) maintainability respectively.

As an SE you may be expected to:

1. Lead efforts to specify (Acquirer role), analyze and allocate (System Developer) RAM requirements for the *System Performance Specification (SPS)* and item development specifications.
2. Integrate engineering specialty disciplines into the RAM specification and analysis decision-making process.
3. Ensure those engineering specialties are integrated into the multi-level SYSTEM and item SE design and development activities.

What You Should Learn from This Chapter

1. What is RAM?
2. Why are reliability, availability, and maintainability addressed as a set?
3. Who is accountable for RAM?
4. What is reliability?
5. How is reliability determined?

System Analysis, Design, and Development, by Charles S. Wasson
Copyright © 2006 by John Wiley & Sons, Inc.

6. What is availability?
7. How is availability determined?
8. What is maintainability?
9. How is maintainability determined?
10. What is the RAM trade space?
11. How do you improve availability?
12. How do you improve reliability?
13. How do you improve maintainability?
14. When is RAM addressed in a program?

Definitions of Key Terms

- **Compensating Provision** "Actions that are available or can be taken by an operator to negate or mitigate the effect of a failure on a system." (Source: MIL-STD-1629A [canceled], para. 3.1.3)
- **Corrective Maintenance** "All actions performed as a result of failure, to restore an item to a specified condition. Corrective maintenance can include any or all of the following steps: Localization, Isolation, Disassembly, Interchange, Reassembly, Alignment and Checkout. (Source: MIL-HDBK-470A, Appendix G, *glossary*, p. G-3)
- **Delay Time** That element of downtime during which no maintenance is being accomplished on the item because of either supply or administrative delay." (Source: MIL-HDBK-470A, Appendix G, *Glossary*, p. G-3)
- **End Effect** "The consequence(s) a failure mode has on the operation, function, or status of the highest indenture level." (Source: MIL-STD-1629A [canceled], para. 3.1.13.3)
- **Failure** "The event, or inoperable state, in which any item or part of an item does not, or would not, perform as previously specified." (Source: MIL-HDBK-470A, Appendix G)
- **Failure Cause** "The physical or chemical processes, design defects, quality defects, part misapplication, or other processes which are the basic reason for failure or which initiate the physical process by which deterioration proceeds to failure." (Source: MIL-STD-1629A [canceled], para. 3.1.12)
- **Failure Effect** "The consequence(s) a failure mode has on the operation, function, or status of an item. Failure effects are typically classified as Local, Next Higher Level, and End Effect." (Source: MIL-HDBK-470A, Appendix G)
- **Failure Mode and Effects Analysis (FMEA)** "A procedure by which each potential failure mode in a product (system) is analyzed to determine the results or effects thereof on the product and to classify each potential failure mode according to its severity or risk probability number." (Source: MIL-HDBK-470A, Appendix G)
- **Latent Defects** Fault characteristics in an item's physical implementation due to design errors and flaws, component and material properties, imperfections, abnormalities, or impurities; poor work process and method practices due to poor quality assurance or lack of training.
- **Maintainability** Refer to the definition provided in Chapter 35 on *System To/For Design Objectives*.
- **Reliability** Refer to the definition provided in Chapter 35 on *System Design To/For Objectives*.

50.2 APPROACH TO THIS CHAPTER

RAM is a specialty engineering skill. Due to the SERIOUSNESS of safety, legal, financial, and political consequences resulting from the lack of application or misapplication of the specialty, QUALIFIED, COMPETENT, professional reliability engineers and maintainability engineers must perform RAM. With qualified credentials, these engineers are subject matter experts (SMEs) who understand and *know* HOW TO:

1. *Apply* RAM methodologies
2. *Interpret* the results and their subtleties for informed decision making.

Because competencies and experience are required to understand which formulas to apply as well as the risk of *misapplication*, you are hereby advised to employ the services of a CERTIFIED professional RAM SME with a PROVEN track record, either on your staff or as a consultant. DO NOT attempt to apply these practices unless you are properly TRAINED and CERTIFIED.

Your job, as an SE is to develop a working knowledge of the basic RAM concepts to enable you to:

1. Know WHEN to employ these services.
2. Gain a *level of confidence* in the individual(s) performing RAM.
3. Understand how to *apply* the results provided by the SMEs.

Our discussion is NOT intended to be an all-encompassing discourse on RAM. The intent is to *introduce* you to RAM concepts that provide a *foundation* for understanding and conversing about the subject matter.

So, rather than repeat what other authors have published in detail, we will approach RAM from an SE task perspective. As such, SEs need to understand two aspects of RAM: its fundamentals and its implementation. For detailed RAM specifics, research the following sources or later editions:

- Patrick D. T. O'Connor, *Practical Reliability Engineering* (Wiley, 3rd edition, 1992).
- Benjamin S. Blanchard and Wolter J. Fabrycky, *System Engineering and Analysis*, second edition (Wiley, 1990).

A prerequisite to RAM is a *basic* knowledge of mathematical frequency distributions and their application as best-fit characteristic models. Specifically, RAM employs these distributions to model component failure rates that can be integrated at higher levels to provide an estimate of overall system RAM.

Frequency Distribution Modeling

There are numerous mathematical models with characteristic curves and parameters that enable us to model a specific component, item, or system. Example distributions include normal, lognormal, binomial, exponential, and Weibull, distributions. For discussion purposes, there are three standard types of distributions that provide a *best fit* for most reliability applications: 1) normal or Gaussian distribution, 2) lognormal distribution, and 3) negative exponential distribution. Figure 50.1 illustrates the characteristic profiles of these distributions.

Let's begin our discussion of RAM with the *system reliability*.

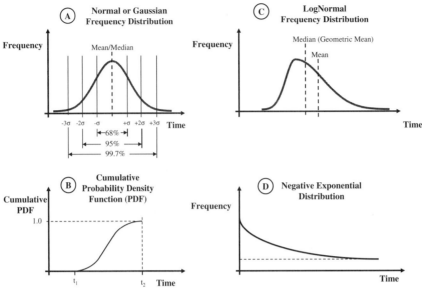

Figure 50.1 Types of Frequency Distributions

50.3 SYSTEM RELIABILITY

What is *system reliability*? Reliability is defined as the *probability* of successfully completing a bounded mission with *specified* objectives within a *specific* operating duration for a *prescribed* set of OPERATING ENVIRONMENT conditions, *use cases*, and *scenarios* without a failure. This leads to the question: *WHAT constitutes a failure?*

What Constitutes a Failure?

MIL-HDBK-470A (Appendix G, p. G-5) defines a *failure* as "*an event, or inoperable state, in which any item or part of an item does not, or would not, perform as previously specified.*" Taking a literal interpretation of this definition, a lightbulb in a car's trunk that becomes inoperable during the course of a day trip—the mission—*qualifies* as a failure. However, a trunk light failure at night may be important, especially if the driver has to retrieve a tire jack, tools, and space tire to replace a flat or damaged tire.

A *failure* is determined by:

1. Criteria that establish the *criticality* of system components to achieving the mission.
2. What degree of performance *degradation* is allowable and still meet performance standards.

We refer to these components as *mission critical items*. Consider the following example:

EXAMPLE 50.1

A car should have four tires and a spare in a qualified condition prior to its driver leaving on a trip. During travel if a flat occurs, the mission can continue, though with some level of elevated risk due to having to use the spare until the tire is repaired. Tires on a car irrefutably are *mission critical items*. The significance of a tire *failure* event depends on criteria such as the condition of the tires prior to the trip, trip distance of 1 mile or 1000 miles, and the circumstances required for emergency applications.

Reliability Philosophy Precepts

Reliability is founded on a multi-faceted philosophy that includes the following *precepts*:

1. Every EQUIPMENT element has a *probability of success* that is determined by its pre-mission operational health status, mission duration, and prescribed set of OPERATING ENVIRONMENT *conditions*, *uses cases*, and *scenarios*.
2. Systems have an *inherent failure rate* at delivery due to *latent defects* from *design errors* and *flaws*, *component/material properties*, and *workmanship processes and methods*.
3. The characterization and modeling of component RAM can be *approximated* with various mathematical distributions.
4. With proper attention, *latent defects* within a specific system are *eliminated* via *corrective maintenance actions* over time when identified.
5. By *eliminating* potential fail points, the system *failure rate* decreases, thereby improving the reliability, assuming the system is *deployed*, *operated*, and *supported* according to the System Developer's instructions.
6. At some point during a *mission* or *useful service life* of a SYSTEM, *failure rates* begin to increase due to the *combinational* and *cumulative effects* of component failure rates resulting from physical interactions and mass property *degradation* due to operating and environmental stresses.
7. These effects are *minimized* and the component's *useful service life* extended via a *proactive* program of system training, proper use, handling, and timely *preventive* and *corrective maintenance* actions at specified operating intervals.

To better understand these precepts, let's explore the service life profile of a fielded system.

System Service Life Profiles

Humans and human-made systems exhibit service life profiles characterized by a level of infant mortality early followed by a growth and stability and finally aging. Textbook discussions on reliability often introduce the concept of service life profiles via a model referred to as the Bathtub Curve. Conceptually, the Bath Tub Curve represents a plot of failure rates over the *active service life* of the equipment. The name is derived from the characteristic Bathtub Curve hazard rate profile as illustrated in Figure 50.2. We will delineate hazard rate versus failure rate later.

The Bathtub Curve consists of three distinct service life profile regions: 1) a period of *decreasing* hazard rates, 2) a period of *stabilized* hazard rates characterized by random failures, and 3) a period of *increasing* hazard rates. Each of these periods is characterized by *different* types of exponential distributions.

The three service life profile regions are often referred to by a variety of names. The period of *decreasing failures* is also referred to as the Infant Mortality, Burn-In, and Early Failure Period. The period of *stabilized failures* is often referred to as the Useful Service Life period. The period of *increasing failures* is sometimes referred to as the Wearout Period.

The Bathtub Curve is actually a *paradigm*, a model for thinking. Unfortunately, like most paradigms, the Bathtub has become ingrained as a "one size" mindset that it applies universally to all systems and components. *This is a fallacy!* Most experts agree that field data, though limited, suggest that the failure rate profiles vary significantly by component type. This is illustrated in Figure 50.2 by the dashed lines in the Period of Decreasing Failures and the Period of Increasing Failures.

Smith [1992] notes the origin of the Bathtub Curve dates back to the 1940's and 1950's with the embryonic stages of reliability engineering. Early attempts to characterize failure rates of elec-

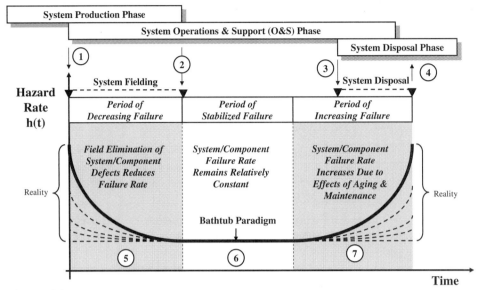

Figure 50.2 Bathtub Curve Paradigm Equipment Failure Profile

tronic components lead to the formulation of the Bathtub Curve. Examples of these components included vacuum tube and early semiconductor technologies that exhibited high failure rates.

Many subject matter experts (SMEs) question the validity of the Bathtub Curve in today's world. Due to the higher component reliabilities available today, most systems become *obsolete* and are *replaced* or *disposed* of long before their *useful service life* profile reaches the Period of Increasing Failure. Most computers will last for many years or decades. Yet, newer technologies drive the need to *upgrade* or *replace* computers every 2–3 years.

Smith [1992] observes that the Bathtub Curve may provide an appropriate profile for a few components. *However, the Bathtub Curve has been ASSUMED to be applicable to more components than is supported by actual field data measurements.* Large, statistically valid sample sizes are required to establish age-reliability characteristics of components. Often, large populations of data are difficult to obtain due to their proprietary nature, assuming they exist. Nelson [1990] also notes that the Bathtub Curve only represents a small percentage of his experiences.

Anecdotal evidence suggests that most reliability work in the US is based on the Period of Stabilized Failure, primarily due to the simplicity of dealing with the constant *hazard rate* from negative exponential distributions. Although any *decreasing* or *increasing* exponential distribution can be used to model the three failure rate regions, the Weibull distribution typically provides more flexibility in accurately shaping the characteristic profile.

Based on these observations, the validity of the Bathtub Curve may better serve as an instructional tool for describing "curve fitting" than as "one size fits all" paradigm for every type of electronic equipment.

Low Rate and Mass Production Systems. The discussion of the *decreasing failure period* of the Bathtub Curve does provide some key insights, especially in terms of systems planned for production. Figure 50.3 might represent the initial failure rate period for a *first article* system and set of OPERATING ENVIRONMENT conditions. For example, assume the *instantaneous failure rate* or *hazard rate* at POWER ON is f_0, which represents the mean of a statistically valid sample of *first article* systems. During the System Design Segment, f_0 is strictly an analytical estimate.

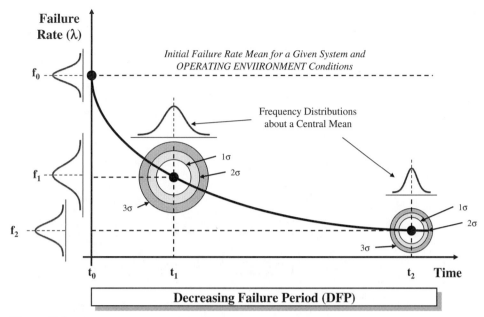

Figure 50.3 Electronic Equipment Failure Profile

Once we have a statistically valid sampling of *first article* systems, we can validate the distribution with *actual* field data. By correcting latent *defects* such as design *errors and flaws*, component material *defects*, and workmanship process and method *problems* for a specific serial number system, its failure rate should *decrease* thereby *improving* its reliability. The strategy is to posture the system design and manufacturing processes for a Low Rate Initial Production (LRIP) failure rate near f_2, not f_0.

From an infant mortality perspective, Figure 50.3 also illustrates the effects of a *diminishing* failure rate resulting from a component "burn-in." Based on this premise, some organizations establish a "burn-in" strategy to specific components for a prescribed number of hours prior to assembly to reduce infant mortality.

System/Component Mortality

Traditionally, complex systems were viewed as having a failure rate that decreased with time based on the frequency distribution of initial failures. These failures were referred to as *infant mortality* due to weak components. Therefore, the *Decreasing Failure Period (DFP)* was viewed as a burn-in period to eliminate these failures.

If the failure rate diminishes initially during the defect elimination phase, *why not subject the system/component to an operational burn-in period at the factory*? Some System Developer or Production organizations do this. They know that *mission critical* components must achieve a specified level of reliability and perform a burn-in. Some organizations driven by profitability may tend to rebuff this extra *burn-in* as follows:

1. Unnecessary due to the higher reliability of today's components.
2. Consumes investment resources with no immediate return.

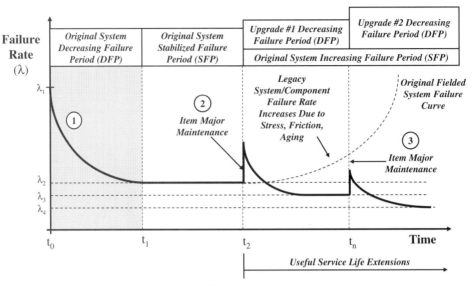

Figure 50.4 Mechanical Equipment Service Life Extension

3. Drives up the cost such that the product will no longer be competitive.
4. Performed by consumers (i.e., if the item "breaks" consumers will bring it back as "defective").

Characterizing the Failure Rate Profiles

The Bathtub Curve provides a graphical concept for R&M engineers to characterize each of the three failure regions with mathematical models. Initially, analytical models are developed to characterize a component, item, or system's reliability profile. Data from existing systems or prototypes may be used to collect data for validating the models. When the system is fielded, *actual* data should be available to validate the models.

The Reliability Function

The cornerstone of reliability is *success* over a specified mission duration and prescribed operating conditions. *The degree of success*, in turn, is dependent on the *degree of failure*, which is *probabilistic* and can be characterized and *estimated statistically*.

Using probability theory, we can state that the *probability of mission success*, $P_{Success}$, is a function of the *probability of failure*, $P_{Failure}$ as a function of elapsed time. Since the probability density function (PDF) is a normalized to 1.0, we can mathematically express this relationship as a function of time as:

$$P_{Success}(t) = 1.0 - P_{Failure}(t) \tag{50.1}$$

If the *degree of success* represents the *reliability* of an entity at a specific point in time for a prescribed set of OPERATING ENVIRONMENT conditions, we equate:

$$P_{Success}(t) \equiv \text{Reliability}, R(t) \tag{50.2}$$

Failures, as random *time* and *condition-based* functions, can occur throughout a mission. An item's *probability of failure*, $P_{Failure(t)}$, or mission UNRELIABILITY, is the represented by its *cumulative failure probability density function* (PDF), $F(t)$ at a specific point in time. Therefore, we can express Reliability (t) as follows:

$$R(t) \equiv 1 - P_{\text{Failure}(t)} \quad \text{for } t \geq 0 \quad (50.3)$$

We can state that $R(t)$ represents the probability that the system or entity will SURVIVE a planned mission of a specific duration under a prescribed set of operating conditions and elapsed time since the start of the mission without failure. We refer to this bounded condition as the system or entity's *reliability*, which is also referred to as the Survival Function.

Using Eq. 50.3, we can express reliability, $R(t)$, over the time interval from t_0, the installation time, to t_1, the current time, in terms of the Cumulative PDF, $F(t)$, as shown in Eq. 50.4.

$$R(t) \equiv 1 - F(t) = 1 - \int_{t0}^{t1} f(t) dt \quad (50.4)$$

Since the total area under the density function is 1.0, we can rewrite Eq. 50.4 over the time interval from t_1 to t_∞ as:

$$R(t) = \int_{t1}^{\infty} f(t) dt = \int_{t1}^{\infty} \lambda e^{-\lambda t} dt \quad \text{for } t > 0 \text{ and } \lambda > 0 \quad (50.5)$$

$$R(t) = e^{-\lambda t} \quad (50.6)$$

The Hazard Rate

Our discussion up to this point characterizes the reliability of a system in SURVIVING from t_0 to the current time, t_1. Given that the system has survived to t_1, the question is: *what rate of failure can be expected in the next Δt time as $\Delta t \to 0$*. We refer to the *rate of change* over this time increment as the *hazard rate*, $h(t)$; some refer to it as the *instantaneous failure rate*. Figure 50.5 Panel A, which represents a NEGATIVE exponential failure density distribution for a system of component, illustrates the $t_1 + \Delta t$ time increment. Based on this discussion, we define the hazard rate as follows:

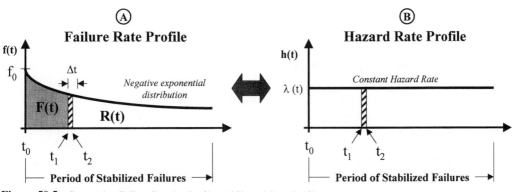

Figure 50.5 Contrasting Failure Density Profile and Hazard Rate Profile

- **Hazard Rate**—the *conditional* probability density that the system or entity will continue to perform its mission within specification limits without failure through the next time increment $t_1 + \Delta t$.

Since a hazard represents a condition that has a *likelihood of occurrence*, probability theory is used to characterize the $f(t)$ and $R(t)$ relationship over the next time increment. For systems or entities that can be characterized using the Exponential Failure Law (EFL), we can express the *hazard rate*, $h(t)$, in terms of its failure density, $f(t)$, and Reliability, $R(t)$, for a specific point in time. For NEGATIVE exponential distributions, $h(t)$ can be expressed as shown in Eq. 50.7:

$$h(t) = \frac{f(t)}{R(t)} \qquad \text{For } t > 0 \qquad (50.7)$$

Referral *For information concerning the derivation of the Hazard Function, refer to Billinton and Allan pp. 126–129.*

Eq. 50.7, which appears to be a typical time variant equation, offers a unique proportionality that is CONSTANT over time. Let's explore this point further.

For negative exponential failure density distributions:

$$f(t) = \lambda e^{-\lambda t} \qquad \text{For } t \geq 0 \text{ and } \lambda > 0 \qquad (50.8)$$

Substituting $f(t)$ from Eq. 50.8 and R(t) from Eq. 50.6 into Eq. 50.7 yields Eq. 50.9

$$h(t) = \frac{\lambda e^{-\lambda t}}{e^{-\lambda t}} \qquad \text{For } t \geq 0 \text{ and } \lambda > 0$$
$$h(t) = \lambda \qquad (50.9)$$

The result shown in Eq. 50.9 illustrates that the *hazard rate*, $h(t)$, for systems or components exhibiting a NEGATIVE exponential distribution, is a CONSTANT as illustrated in Figure 50.5 Panel B.

This point characterizes the Period of Stabilized Failures in Figure 50.2 representing the Useful Service Life of a system or component. Most reliability computations, for analytical simplicity and convenience, are based on a system or entity with negative exponential characteristics operating in this period.

Author's Note 50.2 *Remember—the Periods of Decreasing Failure, Stabilized Failure, and Increasing Failure are INDEPENDENT and characterized by DIFFERENT failure rate distributions. Observe that the Y-axis in Figure 50.2 is labeled Hazard Rate, NOT Failure Rate. These are different computations.*

Understanding Mean Time Between Failures (MTBF)

To better understand SYSTEM failure concepts, we shift our attention to the base element of reliability, PART level or component reliability. Figure 50.6 illustrates key points of our discussion.

Failure Frequency Distributions. Beginning with graph A, suppose we test the life span of 50 lightbulbs in a controlled laboratory environment with a specified set of OPERATING ENVIRONMENT conditions. We allow the lightbulbs to burn continuously until they fail. Thus, at the *time of failure* we record a failure event and the *elapsed* operating hours from the time the lights were powered ON until failure.

When we complete testing of all lightbulbs, we plot the results and observe a Normal or Gaussian distribution provides a best-fit characterization. We compute the *mean* of the distribution, μ. Reviewing the plot we observe that over the *population* of 50 light bulbs for a given set of OPERATING ENVIRONMENT conditions, we had our first bulb failure, a single instance, at t_1 indicat-

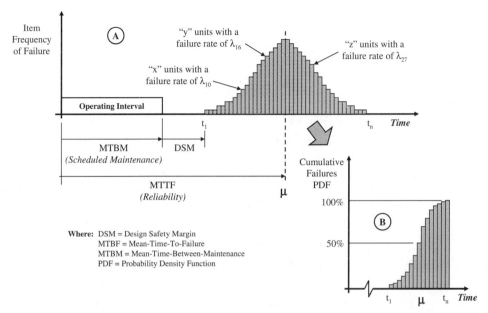

Figure 50.6 Maintenance Interval Concepts

ing a *failure rate* of λ_1. The last failure, a single instance, occurs at t_n indicating a *failure rate* of λ_n. Based on this analysis, we appropriately label the mean, μ, as the *Mean Time To Failure (MTTF)*.

So, *what is the point of this discussion*? It provides insights about an entity's failure frequency distribution. So, if we are asked, *how many failures can we expect at t_x*, we can point to Figure 50.5 at time t_x and observe that based on a sample testing of 50 lightbulbs, a specific number of failures can be expected to fail. In this regard, we refer to this as the instantaneous *failure rate* at time, t_x. Now suppose that we are asked WHAT percentage of the lightbulbs can be expected to fail by t_x. This brings us to our next topic, the *probability density function (PDF)* of *cumulative failure*.

Cumulative Failure Probability Density Function (PDF). Beginning with t_1 when the first bulb fails, we plot the *cumulative failures* over time until all have failed by t_n. Since the area under the Normal Distribution curve is normalized to 1.0, we can exploit the characteristic profile to derive probabilities. As a result, we plot the cumulative probability density function of the frequency distribution as illustrated in Panel B of Figure 50.6. Thus, we can say that at t_μ, there is a probability of 0.50 that 25 lightbulbs can be expected to fail between $t = 0$ and that instant in time.

MTBF, TTF, and MTTR Relationships. Engineers often interchange terms that have subtle but important differences. Such is the case with two reliability terms, Mean-Time-Between Failures (MTBF) and Mean-Time-To-Failure (MTTF).

MTTF, as the name implies, encompasses the elapsed time duration for an item from installation until failure. Once the item fails, there is a Mean-Time-To-Repair, (MTTR) to perform a corrective action. In contrast, MTBF is the summation of MTTF and MTTR as indicated in Eq. 50.10:

$$\text{MTBF} = \text{MTTF} + \text{MTTR} \tag{50.10}$$

The mean failure rate, μ, is the reciprocal of MTTF:

$$\mu \equiv \frac{1}{\text{MTTF}} \tag{50.11}$$

Since MTTF >> MTTR, we can say that:

$$\text{MTBF} \cong \text{MTTF}$$

Therefore:

$$\mu \cong \frac{1}{\text{MTBF}} \tag{50.12}$$

Modeling Reliability Configurations

Given a basic understanding in entity reliability, we now shift our attention to modeling configurations of entity reliabilities. Three basic constructs enable us to compute the reliability of networked entities. These constructs are: 1) *series*, 2) *parallel*, and 3) *series parallel*. Let's explore each of these individually.

Series Network Configuration Reliability. The first reliability network construct is a *series network configuration*. This configuration consists of two or more entities connected in SERIES as shown in Panel A of Figure 50.7. Mathematically, we express this relationship as follows:

$$R_{\text{Series}} = (R_1)(R_2)(R_3)\ldots(R_n) \tag{50.13}$$

where

- R_{Series} = overall reliability of the series network configuration
- R_1 = reliability of Item 1
- R_2 = reliability of Item 2
- $\ldots$
- R_n = reliability of Item n

Substituting Eq. (50.6) into Eq. (50.13) using a NEGATIVE *exponential distribution*:

$$R_{\text{Series}} = e^{-(\lambda_1 + \lambda_2 + \lambda_3 \ldots \lambda_n)(t)} \tag{50.14}$$

where:

- λ_n = unique failure rate of the nth item
- t = NOMINAL operating cycle duration

Parallel Network Configuration Reliability. The second reliability network construct is based on a set entities connected in *parallel* as illustrated in Panel B of Figure 50.7. We refer to this construct as a Parallel Reliability Network and express this relationship as follows:

$$R_{\text{Parallel}} = (R_1) + (R_2) - (R_1)(R_2) \quad \text{for one out of two redundancies} \tag{50.15}$$

50.3 System Reliability

(A) Series Entities Reliability Construct

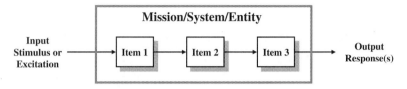

(B) Parallel Entities Reliability Construct

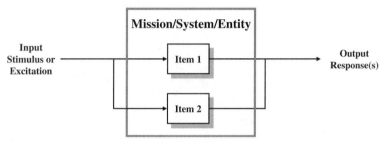

Figure 50.7 Examples of Series and Parallel Reliability Configuration Constructs

where:

R_{Parallel} = overall reliability of the *parallel* network configuration
R_1 = reliability of component 1
R_2 = reliability of component 2

Equation (50.15) can become unwieldy. However, we can *simplify* the equation by restructuring it in the form noted below:

$$R_{\text{Parallel}} = 1 - [(1 - R_1)(1 - R_2) \ldots (1 - R_n)] \quad \text{for one out of } n \text{ redundancies} \qquad (50.16)$$

Observe that each $(1 - R)$ term represents the respective item's *unreliability*.

While each additional entity interjects an additional component subject to failure, the parallel network configuration *increases* the overall reliability of the system to successfully complete its mission.

To illustrate this point further, consider the illustrations shown in Panels A to C of Figure 50.8. Note how the addition of redundant entities with identical reliabilities increases the overall reliability.

Series-Parallel Network Configuration Reliability. The third reliability network construct is configured with a combination of *series-parallel* branches as shown in Figure 50.9. We refer to this construct as a Series-Parallel Network. Computation of the reliability is a multi-step process:

Compute the reliability of PRODUCT A.

Step 1: Compute the reliability for the *series* network containing Subsystems A1 and A3.

Step 2: Compute the reliability for the *series* network containing Subsystems A2 and A4.

Step 3: Compute the reliability for PRODUCT A using the *parallel* network construct composed of the results from Step 1 (Subsystems A1–A3) and Step 2 (Subsystems A2–A4).

628 Chapter 50 Reliability, Availability, and Maintainability

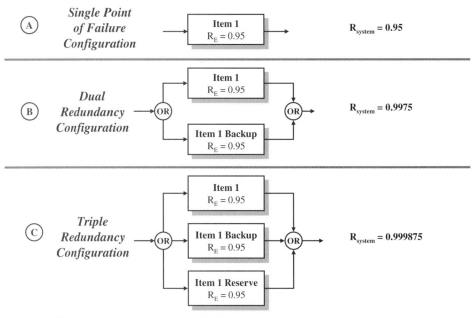

Figure 50.8 Improving System Reliability via Redundancy

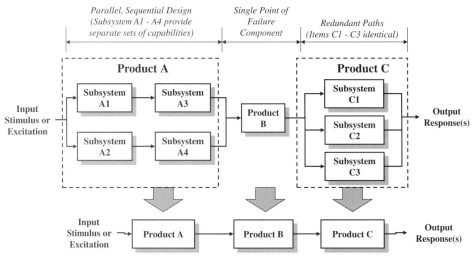

Figure 50.9 Example of a Series-Parallel Reliability Reliability Construxt

Compute the reliability of PRODUCT C.

Step 4: Compute the reliability for the *parallel* network containing Subsystems C1 and C2.

Step 5: Compute the reliability of the *parallel* network consisting of Subsystems C1 and C2 combination in *parallel* with C3.

Compute the reliability of the SYSTEM.

Step 6: Compute the reliability for the SYSTEM using the product of reliability *series* composed of PRODUCT A (Step 3), PRODUCT B, and PRODUCT C (Step 5).

Author's Note 50.3 *As illustrated by the series configuration, every item added to a system increases the system's probability of experiencing a failure (proportional to the failure rate). Adding redundant items increases the number of items and thus requires more repairs. However, insightful use of redundant items may substantially INCREASE the likelihood of a successful mission. However, increased maintenance is likely due to the added hardware.*

Reliability Predictions or Estimates?

Based on the preceding discussions, reliability is usually approached with a series of characteristic equations. This often leaves the casual observer with the *misperception* that reliability is an *exact science*. In fact, reliability is heavily based on probability *theory*. It employs statistical *methods* and *techniques* based on *approximations* and *assumptions* about a population of system entities. As a result, you will hear people comment about PREDICTING the reliability of a system. So, *if reliability theory is based on assumptions, observations, approximations, and probabilities, do you PREDICT or ESTIMATE the reliability of a system*?

If the basis for computation is *approximations*, the best you can *expect* to achieve is an *estimate*, not a *prediction*. The *estimate* represents the probability of a component, item, or system completing its mission based on a prescribed set of OPERATING ENVIRONMENT *conditions* and *assumptions* about system dynamics, statistics, knowledge, experience, and field data.

Since we tend to associate *predictions* with mythical wizards, soothsayers, and crystal balls, our discussions in this chapter will employ the term *estimate*. Combined with seasoned experience of a *qualified* R&M engineer, estimates can be as *cost effective* and *accurate* as in-depth analyses that require *assumptions* and consume extensive *resources* without necessarily *improving* the reliability of the product. Reliability estimates, coupled with "worst case" analyses, can be very effective.

Expressing Reliability Estimates

The preceding discussion brings us to a key question: HOW *do you express the reliability of your system*? If you ask a system designer, they might respond 0.91. There is a fundamental problem with this response. Remember, *reliability* is more than simply picking a number. Reliability estimates require four key elements:

1. *Probability* of successfully completing a defined mission.
2. *Bounded* mission duration.
3. *Elapsed* operating time since the start of the mission.
4. *Prescribed* set of OPERATING ENVIRONMENT conditions.

Therefore, when you simply respond that your system reliability is 0.91, you have *failed* to *conditionally* constrain the parameter in terms of the other three qualifying elements.

Reliability can be specified in several ways. The most common approach is to express reliability as a percentage. Consider the following example:

EXAMPLE 50.2

Electronics reliability can be stated as a *not-to-exceed* failure rate, typically as number/percentage of failures per hour or failures per million hours, in a given environment.

The problem of specifying in these *rate terms* is that *redundancies*, which can *improve* overall mission reliability, *increase* the potential rate of failure by adding to the parts count. When specified as a percentage, redundancies can easily be calculated that show the *mission reliability* improvement.

Reliabilities are sometimes stated in terms of a *Mean Time Between Failure (MTBF)*. However, instantaneous *failure rates* change with each hour of operation and operating condition. The MTBF is the *reciprocal* of the *failure rate* ONLY when using a NEGATIVE *exponential distribution*! In general, the *negative exponential distribution* applies ONLY to systems that have all *exponentially distributed* components, such as electronics, with NO redundancies. In keeping with the true definition of reliability and its four elements, it is strongly suggested that reliability be expressed as a PERCENTAGE instead of MTBF for highly complex systems. The same applies to PRODUCT and SUBSYSTEM allocation. Additionally, reliability engineers suggest avoiding the specification of reliability below the Subsystem Level. This may unnecessarily constrain the solution set that might otherwise improve overall reliability and reduce cost. Exceptions include externally procured components based on proper reliability allocations.

Guidepost 50.1 *Our discussions to this point focus on the reliability of systems to accomplish their missions of a specified duration and set of OPERATING ENVIRONMENT conditions. Each mission assumes that the system will begin operating at a given level of capability, performance, and physical condition. If properly operated and maintained in accordance with the prescribed PROCEDURAL DATA instructions, the User can expect a useful service life of XX hours, years, and so forth. Under these assumptions the system can expect a nominal failure rate that varies over the service life profile. So, how is the nominal failure rate achieved? This brings us to our next discussion topic, maintainability.*

50.4 SYSTEM MAINTAINABILITY

When systems fail, they impact the revenue stream or could adversely affect the operator and public. Corrective actions are required immediately to return the system to normal active service. The *efficiency* and *effectiveness* of these actions centers on the system's *maintainability*. Maintainability is a measure of a system, product, or service's capability to be returned from a failed state to a functioning state within time limits established by a performance standard.

Maintainability must be an integral part of the SE Design segment beginning with Contract Award. The starting point begins with formulation of the *maintenance concept*, our next topic.

The Importance of the Maintainability Concept

The cornerstone for planning system maintainability resides in the *maintenance concept*. The maintenance concept often begins as a basic strategy addressed conceptually in the system *Concept of Operations (ConOps)* document.

The ConOps document provides an initial starting point for documenting the maintenance concept unless there is a compelling need for a stand-alone document. As a conceptual strategy, the ConOps documents the WHO, WHAT, WHEN, WHERE, and HOW the system, product, or service will be maintained. For example, maintenance on complex systems may be performed by: 1) the User, 2) an internal support organization, 3) contractors, or 4) a combination of these.

Subsequently, a *System Operations and Support (O&S) Plan* or *Integrated Support Plan (ISP)* may be required or prepared to document the maintenance concept implementation details such as *types* of maintenance and *levels* of maintenance.

Types of Maintenance

Once a system is deployed to the field, there are two basic types of maintenance that are performed: *preventive maintenance* and *corrective maintenance*.

Preventive Maintenance. *Preventive Maintenance* refers to scheduled, *periodic*, maintenance actions performed on the system or product in accordance with the manufacturer's instructions. Note the operative term *scheduled*. The *intent* is to perform *proactive* maintenance in risk areas BEFORE they start to have an *impact* or result in *degraded* system performance as illustrated in Figure 50.5.

EXAMPLE 50.3

Automobile manufacturers recommend that you change oil at 3,000 mile or greater intervals.

In general, routine system maintenance elements include fluids, lubricants, and filters and other maintenance actions such as calibrations and alignments performed at intervals. Components, referred to as *line replaceable units* (LRUs), are replaced at longer intervals. In these cases, *corrective maintenance* actions may also be required in conjunction with the *preventive maintenance*.

EXAMPLE 50.4

Some automobile manufacturers recommend that you replace the engine timing chain at 90,000 mile intervals for specific types of cars to *avoid* a *serious* engine malfunction.

Because of the expense of repairs and unnecessary preventive maintenance, some complex systems include *condition-based maintenance* (CBM) systems. These systems *monitor* and *analyze* the status and health of engine fluids, such as motor oils, to determine the level of metal filings and other conditions that may indicate *premature wear* or *potential failure* condition. CBM systems are critical for systems such as aircraft engines applications.

Corrective Maintenance. *Corrective maintenance is referred to as unscheduled* maintenance. It represents those maintenance actions required to restore the system or product performance back to the manufacturer's performance specifications after failure. In general, *corrective maintenance* includes *removal* and *replacement* (R&R) of LRUs, recalibration, and realignment.

Levels of Maintenance

Organizations often employ a three-level maintenance model: 1) organizational maintenance, 2) intermediate maintenance, and 3) factory maintenance.

Field or Organizational Level Maintenance. *Field or organizational level maintenance* consists of *preventive* and *corrective* maintenance actions performed in the field by the User or their support organization. At a MINIMUM these actions entail *removal and replacement* (R&R) of an item with a new or repaired item. In some instances, the item may be repaired at this level with *common support equipment* (CSE) tools. The support organization may be *internal* to the User's organization or contractor support referred to as *contract logistics support* (CLS).

Intermediate Maintenance. *Intermediate maintenance* consists of *corrective maintenance* actions that require *removal* of an item by the User or CLS personnel and return to a central repair facility that may have specialized *peculiar support equipment* (PSE) to accomplish the maintenance action. When the repaired item is returned to the field, User or CLS personnel *reinstall*, *checkout*, and *verify* the item's operation.

Factory Maintenance. *Factory maintenance* actions require that an item be sent by the central repair facility back to the factory that has the specialized PSE and expertise required to determine the cause of failure. This may include site visits back to the User to interview personnel, review records, and inspect other items that may provide insights into the *circumstances* and *operating conditions* leading to the required maintenance action.

50.5 FAILURE REPORTING, ANALYSIS, AND CORRECTIVE ACTION SYSTEM (FRACAS)

When a component, item, or system is developed, Reliability and Maintainability Engineers employ SE designs and *Engineering Bills of Materials (EBOMs)* and vendor/manufacturer data to construct reliability and maintainability models to estimate system performance. These models provide insights related to the *premature* wear, thermal stress, *adequacy* of periodic inspections, and performance metrics tracking. These data provide early indications of *wear out* that may lead to failure. Once the component, item, or system is fielded, actual field data should be used to further *refine* and *validate* the models. So, *how can these data be obtained and tracked*?

The solution resides in a Failure Reporting, Analysis, and Corrective Action System (FRACAS). MIL-HDBK-470A (para. 4.5.1.1, p. 4–58) describes FRACAS as "*a closed-loop data reporting system for the purpose of systematically recording, analyzing, and resolving equipment reliability AND maintainability problems and failures.*"

Depending on the organizational plans for the system, the FRACAS provides a *history* of maintenance ACTIONS for each component and replacements with new components. This enables the organization to continue to cost effectively *extend* the *useful service life* and employment of the asset.

Author's Note 50.4 *When you review FRACAS data, make sure that you understand the context of the failure. Depending on the discipline instilled in maintenance personnel, errors such as ordering the incorrect part number and installing the incorrect part may sometimes be recorded as a failure initiating yet another maintenance action. In addition FRACAS as a maintainability task may not include maintenance time tracking.*

Maintainability Computations

Organizationally, the subject of *maintainability* often receives "lip service" and usually falls second in priority to *reliability*. Yet, operational support costs, which include maintenance, account for approximately 70% of system life cycle costs, especially for complex systems. As a result, post deployment support costs and their contributory factors should be a MAJOR concern EARLY in the SE Design Segment. The primary system design factors that contribute to support costs are: 1) *reliability*, which impacts the frequency of maintenance, and 2) *maintainability*, which impacts the amount of time required to perform *preventive* and *corrective* maintenance.

Once a system is fielded, the User must live with the consequences of SE design decision-making and its accountability or the lack thereof for the reliability and maintainability factors. Additionally, these considerations manifest themselves in another factor, *availability*, which represents the level of operational readiness of a system to perform its mission on demand for a given set of operating conditions. If *unavailable*, the system: 1) is NOT generating revenue and 2) even worse, is costing the organization funds for repairs.

Engineering textbooks often approach maintainability with *equation-itis*. Equation-itis, like analysis paralysis, is a condition created by a preoccupation with equations WITHOUT understanding the operational challenges Users have to address and the base assumptions to be established that lead to the need for equations. Maintainability is a classic topical example. To illustrate WHY we need to understand User challenges, let's briefly explore a SUPPORT SYSTEM scenario.

A system is either: 1) operating—performing a mission or supporting training, 2) in storage, 3) awaiting maintenance, 4) being maintained, or 5) awaiting return to active service. System maintenance is categorically referred to as Maintenance Down Time (MDT). Every hour spent in MDT equates to $$ of lost revenue as in the case of commercial systems such as production machinery, airlines, and so forth. To ensure that the enterprise can sustain schedules, extra systems may be procured, leased, rented, and so forth to maintain continuity of operations while one system is being maintained.

When a system failure occurs, you need a system of spare parts on-site along with skilled maintenance technicians who can perform the repair with the least amount of MDT. Therefore, you need some insight concerning: 1) the quantity of spare parts required for a given type of failure, 2) the quantity maintenance technicians required "full time" and "part time", 3) the quantity of maintenance workstations, if applicable, 4) the types and quantities of test equipment, 5) system and parts storage space allocations, 6) ordering systems, 7) logistical support systems, and so forth. This challenge is compounded by seasonal usage factors for some systems. All of this translates into $COST. To minimize the cost of maintaining an inventory of spare parts, many organizations arrange strategic partnerships or tier level suppliers to provide parts Just in Time (JIT), when required.

How do we minimize the cost of maintenance in the SE Design Segment? We can: 1) incorporate a Built-In Test (BIT) capability into major items to detect Line Replaceable Unit (LRU) failures, 2) perform corrective maintenance on mission critical items during a preventive maintenance cycle BEFORE they fail, 3) promote standardization and interchangeability of LRUs, 4) provide easy access, quick removal and installation of LRUs, and so forth. For example, some complex, mission critical systems are designed for *condition-based maintenance* (CBM).

As an Acquirer representing the User's contract and technical interests, your job as an SE may be to prepare *System Performance Specification* (SPS) requirements for mission reliability (MTBF), maintainability (MTTR), and operational availability (A_o) discussed in a later section. As a System Developer SE, you may be confronted with analyzing, allocating, and flowing these requirements down to configuration items and their LRUs. For example, *how do you design the system for a 30-minute MTTR?*

To answer this question, maintainability engineers employ a series of equations and analysis data to support informed SE Design. The strategy is that if we know the quantity of fielded systems and the state of maintenance of each system, failure rate data for system LRUs, and the prescribed preventive maintenance schedule, we can ESTIMATE the parameters of the SUPPORT SYSTEM such as quantities of spare parts, maintenance technicians and levels of training, work stations, ordering systems, and so forth.

The intent of the computation discussion that follows is to provide you with a *general working knowledge* of the maintainability terminology, metrics, and their interrelationships, not "plug and chug" equations. Given a working knowledge, you should be better prepared to communicate with maintainability engineers and understand the specification requirements they respond to as well as review the work products they produce in response to contract requirements.

Maintenance Down Time (MDT). When a system fails, Users want to know: *how long the system will be out of service*—DOWNTIME. In the commercial world, system *downtime* means an interruption to the revenue stream. In the military and medical fields, *downtime* can mean the difference in life or death situations.

The answer to the question resides in an Organizational Level metric referred to as Maintenance Down Time (MDT). MDT is a function of three key elements: 1) Mean Active Maintenance Time, 2) Logistic Delay Time, and 3) Administrative Delay Time. Eq. 50.17 illustrates the computation.

$$\text{MDT} = \overline{\text{M}} + \text{LDT} + \text{ADT} \qquad (50.17)$$

Where:

$\overline{M}$ = Mean Active Maintenance Time (Eq. 50.22)

LDT = Logistics Delay Time

ADT = Administrative Delay Time

The three MDT elements have two perspectives: System Developer versus Users.

- System Developers influence the Mean Active Maintenance Time ($\overline{M}$) via the System Design process and should be *familiar* with the LDT and ADT of the User's support system. However, the System Developer DOES NOT have *control* over LDT and ADT; those are typically User responsibility functions.
- Users EXERCISE degrees of control over the $\overline{M}$, LDT, and ADT, depending on their organizational and political authority.

We will see these subtleties emerge during our discussion of *system availability*. On the topic of availability, many of the maintenance parameters are key elements of availability equations.

Logistics Delay Time (LDT). *Logistics Delay Time* (LDT) is the actual time required to perform actions that support a corrective maintenance. Examples include: 1) research for replacement or equivalent parts, 2) procurement and tracking of parts orders, 3) delivery of parts from the supplier, 4) receiving inspection of parts, 5) routing to maintenance technician, 6) reordering of incorrect parts, and so forth. LDT is sometimes expressed in terms of *Mean Logistics Delay Time*, $\overline{M}_{LD}$.

Administrative Delay Time (ADT). *Administrative Delay Time* (ADT) consists of actual time spent on activities such as staffing the maintenance organization; training personnel; vacation, holiday, and sick leave; and so forth. ADT is sometimes expressed in terms of Mean Administrative Delay Time, $\overline{M}_{AD}$.

Preventive Maintenance (PM) Time. SEs often have to answer: *how much time is required to perform preventive maintenance actions on the system*? The solution resides in an Organizational level metric referred to a Preventive Maintenance (PM) Time. PM time represents the *actual* time required to perform routine maintenance such as inspections and replacement of fluids, filters, belts, and so forth.

The *preventive maintenance time* for a specific type of maintenance action is represented by M_{pt}. $\overline{M}_{pt}$, which represents the *mean* time for all preventive maintenance actions within the system, is computed as shown in Eq. 50.18.

$$\overline{M}_{pt} = \frac{\sum_{i=1}^{n}(f_{pt(i)})(M_{pt(i)})}{\sum f_{pt(i)}} \qquad (50.18)$$

Where:

$\overline{M}_{pt}$ = Mean Preventive Maintenance Time

$M_{pt(i)}$ = Time required to perform the ith preventive maintenance action

i = Sequential identifier for a specific type of preventive maintenance action

$f_{pt(i)}$ = Frequency of the ith preventive maintenance action

50.5 Failure Reporting, Analysis, and Corrective Action System (FRACAS)

Mean Corrective Maintenance Time ($\overline{M}_{ct}$). The *Mean Corrective Maintenance Time* ($\overline{M}_{ct}$), which is equivalent to *Mean Time to Repair* (MTTR), represents the actual time required at the Organizational Level to perform a repair action on a system to replace a *failed* or *damaged* LRU. MTTR, for example, is usually expressed in hours or fractions thereof. *What does the repair action include?* It includes the time required to:

1. Isolate the fault to a Line Replaceable Unit (LRU).
2. Quickly and easily replace the LRU.
3. Retest to assure that the total system is functioning at a specified or acceptable performance levels.

The *corrective maintenance time* for a specific type of corrective maintenance action is represented by M_{ct}. $\overline{M}_{ct}$, which represents the mean for all corrective maintenance actions within the system, is computed as shown in Eq 50.19.

$$\overline{M}_{ct} = \frac{\sum_{i=1}^{n} (\lambda_{(i)})(M_{ct(i)})}{\sum \lambda_{(i)}} \qquad (50.19)$$

Where:

$\overline{M}_{ct}$ = Mean Corrective Maintenance Time
$M_{ct(i)}$ = Time required to perform the ith corrective maintenance action
i = Sequential identifier for a specific type of corrective maintenance action
$\lambda_{(i)}$ = Frequency of the ith corrective maintenance action

A plot of the frequency of *corrective maintenance actions* is often assumed to follow a *LogNormal Distribution* as illustrated in Figure 50.1. That is, a few of the failures can be *isolated* and *repaired* rather quickly. Most of the corrective actions will require longer time periods and center about the *median* of the frequency distribution. The remainder of the corrective actions may be few in number but require a considerable number of hours to *correct*.

Failure Rate or Corrective Maintenance Frequency (λ). The Corrective Maintenance Frequency is represented by λ and is computed as shown in Eq. 50.20.

$$\lambda = \frac{1}{\text{MTTF}} \qquad (50.20)$$

Where:

MTTF = Mean Time To Failure

From Eq. 50.12, λ, which is referred to as the failure rate, is approximately equivalent to the MTBF.

Repair Rate. Some systems track the *repair rate*, μ, which is computed as illustrated in 50.21

$$\mu = \frac{1}{\overline{M}_{ct}} \qquad (50.21)$$

Where:

$\overline{M}_{ct}$ = Mean Corrective Maintenance Time (Eq. 50.19)

Mean Active Maintenance Time ($\overline{M}$). As the system design evolves, SEs need to know: *what is the mean time to accomplish a preventive maintenance action or a corrective maintenance action*? The solution resides in an Organizational Level metric referred to as the *Mean Active Maintenance Time* ($\overline{M}$). Eq. 50.22 illustrates the computation.

$$\overline{M} = \frac{f_{pt}(\overline{M}_{pt}) + (\lambda)(\overline{M}_{ct})}{f_{pt} + \lambda} \qquad (50.22)$$

Where:

- $\overline{M}$ = Mean Active Maintenance Time
- $\overline{M}_{pt}$ = Mean Preventive Maintenance Time (Eq. 50.18)
- f_{pt} = Frequency of Preventive Maintenance
- $\overline{M}_{ct}$ = Mean Corrective Maintenance Time (Eq. 50.19)
- λ = Failure rate or frequency of corrective maintenance (Eq. 50.20)

Mean Time Between Maintenance (MTBM). *Mean Time Between Maintenance* (MTBM) is the *mean operating interval* since the last maintenance action. MTBM is expressed in terms of elapsed operating hours for a specific set of operating conditions for the system based on the mean of all *preventive* ($\overline{M}_{pt}$) and *corrective* ($\overline{M}_{ct}$) maintenance actions. MTBM is computed as shown in Eq. 50.23.

$$\text{MTBM} = \frac{1}{\frac{1}{\overline{M}_{pt}} + \frac{1}{\overline{M}_{ct}}} = \frac{1}{f_{pt} + \frac{1}{\lambda}} \qquad (50.23)$$

Where:

- MTBM = Mean Time Between Maintenance
- $\overline{M}_{pt}$ = Mean Preventive Maintenance time (Eq. 50.18)
- $\overline{M}_{ct}$ = Mean Corrective Maintenance Time (Eq. 50.19)

Sources of Maintainability Data. The *validity* of maintenance computations is dependent on having good data sets derived from *actual* designs and field data. Potential sources of maintainability data include:

- Historical data from similar items
- Item design and/or manufacturing data
- Data recorded during item demonstration
- Field maintenance repair reports or FRACAS data.

Guidepost 50.2 *At this juncture we have addressed a few of the fundamentals of system reliability and maintainability. The critical question is: Will the system be operationally available to conduct its missions when tasked? You can have a system that is reliable and well maintained but be unavailable when needed to conduct missions. This leads us to our next topic, system availability.*

50.6 SYSTEM AVAILABILITY

Availability focuses on a SYSTEM's state of readiness to perform missions WHEN tasked. This requires two criteria:

1. Operational readiness in terms of completion of *preventive* and *corrective maintenance* actions within a specified timeframe.
2. The *ability* to reliably start and operate.

Therefore, a SYSTEM'S *availability* is a function of its design reliability and its maintainability.

Availability can drive the System Developer in a decision-making circle as they attempt to *balance* maintainability and reliability objectives. Consider the two cases at *opposite* extremes:

- SYSTEM A is a highly reliable SYSTEM. During development the Acquirer had a choice: improve reliability or improve maintainability. The Acquirer chooses to focus design resources on reliability. The rationale is: *"If the system is highly reliable, we don't need a lot of repairs, and therefore won't need a large maintenance organization."* So, when the system does need maintenance, it requires weeks for corrective maintenance actions due to the difficulty in accessing maintenance repair areas. Clearly, the highly reliable SYSTEM is *unavailable* for missions for long periods of time when maintenance is required.
- SYSTEM B has less reliability because the Acquirer had to make a choice similar to that for SYSTEM A above. The Acquirer reduces the reliability requirement and focuses on IMPROVING the maintainability. The rationale is: *"We have an outstanding maintenance organization that can repair anything."* As a result, SYSTEM B fails *continuously* and is prone to numerous *corrective maintenance actions*, thereby making it unavailable to perform most missions.

The point of this discussion is: *system design and development must include a practical availability requirement that enables the achievement of a balance between a system's reliability and maintainability and still meet mission and organizational objectives*. Close collaboration between the Acquirer and User is important during the System Procurement Phase and with the System Developer after Contract Award.

There are three types of availability: 1) *operational availability* (A_o), 2) *achieved availability* (A_a), and 3) *inherent availability* (A_i). Let's explore each of these concepts briefly.

Operational Availability (A_o)

Operational availability, which is symbolized by A_o, represents the probability that the system will operate in accordance with its specified performance and prescribed operating environment conditions when tasked to perform its mission. A_o includes maintenance delays and other factors independent of its design. An item's operational availability, A_o, is expressed mathematically as follows:

$$A_o = \frac{\text{MTBM}}{\text{MTBM} + \text{MDT}} \quad (50.24)$$

where:

A_o = Operational Availability
MDT = Maintenance Downtime (Eq. 50.17)
MTBM = Mean Time Between Maintenance (Eq. 50.23)

Observe that the MDT parameter includes both *preventive* (scheduled) and *corrective* (unscheduled) maintence times.

Achieved Availability (A_a)

System Developers typically have *no control* over the User's support system factors such as LDT and ADT. So, achieved availability (A_a) represents the level of availability under the control of the System Developer as constrained by the *System Performance Specification (SPS)*.

An item's Achieved Availability, A_a, is the ratio of the Mean Time Between Maintenance (MTBM) to the *summation* of MTBM and Mean Active Maintenance Time ($\bar{M}$):

$$A_a = \frac{\text{MTBM}}{\text{MTBM} + \bar{M}} \qquad (50.25)$$

where:

- A_a = Achieved Availability
- $\bar{M}$ = Mean Active Maintenance Time (Eq. 50.22)
- MTBM = Mean time Between Maintenance (Eq. 50.23)

Inherent Availability (A_i)

Finally, the third type of availability is *inherent availability*, A_i. Inherent availability assumes that the system operates within its specified performance limits and prescribed operating environment conditions with an ideal support system that performs only corrective maintence actions, $\bar{M}_{ct}$. Preventive maintenance actions, LDT, and ADT are excluded. An item's inherent availability, A_i, is the ratio of the Mean Time Between Failure (MTBF) to the summation of MTBF and Mean Corrective Maintenance Time ($\bar{M}_{ct}$). A_i is defined as follows:

$$A_i = \frac{\text{MTBF}}{\text{MTBF} + \bar{M}_{ct}} \qquad (50.26)$$

where:

- A_i = Inherent Availability
- $\bar{M}_{ct}$ = Mean Corrective Maintenance Time or MTTR (Eq. 50.19)
- MTBF = Mean Time Between Failure

Note the switch from MTBM used to compute A_o and A_a to MTBF for A_i.

The FAA s National Airspace System-*Systems Engineering Manual* (2003, para. 4.8.2.1.1.3) notes two key points about *Inherent Availability*.

- First, *"This availability is based solely on the MTBF and MTTR characteristics of the system or constituent piece and the level of redundancy, if any, provided. For systems or constituent pieces employing redundant elements, perfect recovery is assumed. Downtime occurs only if multiple failures within a common timeframe result in outages of the system or one or more of its pieces to the extent that the need for redundant resources exceed the level of redundancy provided. Inherent availability represents the maximum availability that the system or constituent piece is theoretically capable of achieving... if automatic recovery is 100 percent effective."*

- Secondly, inherent availability *"does not include the effects of scheduled downtime, shortages of spares, unavailability of service personnel, or poorly trained service personnel."*

Final Thoughts

In summary, we have briefly defined each of the three types of availability. From a contractual *System Performance Specification (SPS)* perspective, A_o obviously has a *major* significance to the

User versus A_a and A_i. So, *how does a System Developer deal with the LDT and ADT factors for which they have no control*?

First, the contract should *delineate* Acquirer and User *accountability* for supplying LDT and ADT *data* and/or *assumptions*. Second, the SE should consider creating a Logistical Support Integrated Product Team (IPT) that includes the LDT and ADT stakeholders to supply the data and participate in making recommendations via the formal contract protocol.

Assuming the IPT performs as expected, these stakeholders must assume *accountability* for their *contributions* to achieving the overall success of the system and verification of the appropriate SPS requirements. Blanchard (*SE Management*, p. 121) observes that there is *limited* value in *constraining* the design of EQUIPMENT to achieve a 30 minute $\bar{M}_{ct}$ if the *supply system* requires three months on average, consisting of LDT + ADT, to RESPOND with parts. A_o is certainly not achieved in this manner.

Guidepost 50.3 *The preceding discussions addressed some of the fundamentals of RAM. Given this foundation, we shift our attention to applying RAM to system design.*

50.7 APPLYING RAM TO SYSTEM DESIGN

The application of RAM to system design is one of the most challenging of the SE discipline. The challenge comes in *scoping*, *estimating*, and *quantifying* minimum parameter performance to meet User operational mission and system service life needs while balancing those needs with available User budgets and schedules.

One of the ways of meeting this challenge is to establish a RAM strategy. The strategy, which includes both Acquirer and System Developer role-based tasks, should be documented in the program's *Technical Management Plan (TMP)*. At a *minimum*, RAM tasks include the following:

Task 1: Establish User mission RAM requirements.
Task 2: Analyze and allocate RAM requirements.
Task 3: Model system RAM performance.
Task 4: Review RAM modeling results.
Task 5: Implement compensating provision actions.
Task 6: Implement a parts program (optional).
Task 7: Improve EQUIPMENT characteristic profiles.

Each of these tasks iterates with others from Contract Award through system delivery and acceptance.

Task 1: Establish User Mission RAM Requirements

The objective of Task 1 is to develop or contract the development of *objective*, *quantifiable* performance-based RAM requirements *necessary* and *sufficient* for procurement or development actions. This requires mission analysis of the item or system within its prescribed OPERATING ENVIRONMENT use cases and scenarios.

The challenge for the Acquirer (role) is to *collaborate* with the User to *understand* and *objectively*:

1. Specify the *minimum*, *quantifiable*, and *affordable* mission reliability that satisfies budget and schedule constraints.
2. Allocate the appropriate RAM requirements to the various System Elements.

Accomplishment of these two objectives is a *highly iterative* and time-consuming process.

When specifying RAM, SEs and Reliability and Maintainability (R&M) Engineers have to answer a challenging question: *WHAT level of RAM is the minimum required for a system or product to accomplish a mission of a specified duration in a prescribed OPERATING ENVIRONMENT?* This is the hard part for SEs and R&Ms that often requires extensive analysis, modeling, simulation, and collaboration with Users—the operators and maintainers.

Every system has its own unique applications, system *mission use cases* and *scenarios*, mission objectives, priorities, and values to the Users. So, the answer to this question resides in high-level SE concepts that must be *quantitatively* translated into *objective* capability and performance requirements.

Create the Initial Reliability and Availability Starting Points. As an *initial starting point*, you must identify a system *reliability* number that expresses the probability of mission success *required* of the item. You also need to *select* an *availability* number based on the criticality of the SYSTEM to support the mission. Finally, you need a *maintenance concept* that describes the philosophy of maintenance.

Author's Note 50.5 *For this first step most people are reluctant to pick a reasonable number. From an engineering perspective, you need an initial starting point and that is ALL it is. Unfortunately, there are those who criticize these efforts without recognizing that it may not be the "end game" reliability number; it is only a starting point . . . Nothing more, and nothing less.*

Establish Development Objectives and Constraints. One of the key User objectives is to minimize system or product life cycle cost. Inevitably, there are trade-offs at high levels between system development costs and support costs. Depending on the type of system, 60% to 70% of the *recurring* life cycle costs of many systems occur in the System Operations and Support Phase (O&S). Most of these costs are incurred by system *corrective* and *preventive* maintenance. (Refer to DSMC, *Acquisition Logistics Guide*, Fig. 13-1, p. 13–6, and Fig. 20-2, p. 20–3.). So, you may have to make trade-off decisions for every dollar you spend on design reliability. *WHAT is the cost avoidance in maintenance and support costs?* Let's explore this point further.

Using the illustrative example of Figure 50.10, let's *increase* the design reliability of a system or product—its system development cost. We should then *expect* a corresponding *decrease* in the system maintenance costs. If we sum the nonrecurring *system development cost* and *system maintenance costs* over the range of reliability, we emerge with a bowl-shaped curve representing *total life cycle cost*, or more appropriately the *total ownership cost* (TOC). Ideally, there is a Target Reliability Objective, R_A, and a Target LC Cost Objective, C_A, that represent the *minimum* TOC (A) for the User. These objectives, in combination with the core RAM criterion of identifying the reliability level, should form the basis for specifying RAM requirements.

Philosophically, the TOC *minimization* approach assumes that the User, by virtue of resources, has the *flexibility* to select the *optimal* level of reliability. However, as is typically the case, the User has limited resources for how much *reliability* they can afford "up front" as *system development costs*. Depending on the *operational need(s)* and *level of* urgency to procure the system, they may be confronted with selecting a more affordable reliability, R_B, based on system development cost limitations that result in a higher than planned system maintenance cost ($MAINT_B$) and a higher TOC_B.

For innovative organizations, this presents both an *opportunity* and a *challenge*. For organizations that do not perform SE from the standpoint of *analysis of alternative* (AoA) solutions and jump to a point solution, this may be a *missed* business opportunity. During the AoA activity new approaches may be discovered that enables the User to *minimize* TOC within the constraints of system development costs.

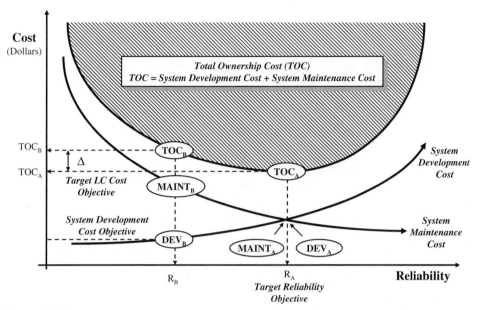

Figure 50.10 Notional Optimal Reliability

Task 2: Analyze and Allocate RAM Requirements

When potential Offerors receive the RAM requirements via a formal *Request for Proposal (RFP)* solicitation, conduct analyses to fully understand WHAT is required to satisfy the system or item's specification requirements. Therefore, the objective of Task 2 is to analyze RAM requirements specified in the RFP *System Requirements Document (SRD)* or contract *System Performance Specification (SPS)* requirements and allocate RAM requirements to PRODUCTs, SUBSYSTEMs, and so forth.

From an SE perspective, accomplishment of the mission requires that the MISSION SYSTEM and SUPPORT SYSTEM operate at a level of RAM sufficient for successful completion of the mission. Recall that a system, by definition, includes MISSION SYSTEM and SUPPORT SYSTEM elements: PERSONNEL, EQUIPMENT, PROCEDURAL DATA, MISSION RESOURCES, FACILITIES, and SYSTEM RESPONSES. Therefore, the overall *mission reliability* must be allocated to each of these system elements and subelements. On allocation, this requires the following considerations:

- PERSONNEL Element considerations include RAM related to human system interfaces, human factors and performance, training, and skills.
- EQUIPMENT Element considerations include SYSTEM, PRODUCT, SUBSYSTEM, ASSEMBLY, SUBASSEMBLY level RAM related to HARDWARE and SOFTWARE design and development.
- PROCEDURAL DATA Element considerations include RAM of organizational *Standard Operating Procedures and Practices (SOPPs)* and EQUIPMENT operating procedures required to safely conduct system missions, operate the EQUIPMENT, and return.
- MISSION RESOURCES Element considerations include RAM related to the timely delivery and quality of mission data, expendables, and consumables required to perform the mission and sustain mission operations.

- FACILITIES Element (SUPPORT SYSTEM) considerations include RAM related to MISSION SYSTEM interfaces, environmental conditions that impact storage "shelf life" and maintenance.
- SYSTEM RESPONSES Element considerations include *products*, *by-products*, and *services* RAM related to:
  1. The timely delivery of system outputs to accomplish mission objectives and *Mission Event Timelines (METs)*.
  2. "Fitness-for-use" quality requirements.
  3. Compliance to standards.

The resulting RAM allocations to the System Elements and subelements provide the basis for procurement of contractor PERSONNEL services or training of internal PERSONNEL, EQUIPMENT, FACILITIES, MISSION RESOURCES, and PROCEDURAL DATA. Our discussion in the remainder of this chapter will focus on the EQUIPMENT element RAM.

If a system is being developed using several item/configuration item (CI) suppliers, then it will be necessary to ALLOCATE a proportion of the total system reliability to each SUBSYSTEM. This enables each supplier to understand WHAT their allocated contribution to reliability is.

Since the final SYSTEM Level reliability will be the multiplicative product of all item reliabilities, individual components generally require a *higher reliability* requirement than the total system. However, there is chance of *misallocating* and *specifying* requirements too high for some items, resulting in substantial additional costs to the program. As a general rule, *avoid* allocating BELOW the highest possible levels of *abstraction*—the SUBSYSTEM—*unless* there is a requirement to do so, due to the potential added program costs.

Task 3: Model System RAM Performance

Once the initial RAM performance allocations are made to items, Task 3 determines if the design solution for each item *meets* or *exceeds* the allocated RAM requirements. The objective of Task 3 is to employ the simplest, *most* cost-effective modeling method to determine an item's compliance with the RAM requirement. If conditions require it, higher fidelity models may be required.

Develop Quick Look Reliability Models. Due to the need to make *assumptions* about configurations and component/item failure rates, the first step is to create a simple model of the item's RAM. Typically, component *failure rate* data and quantities are derived from each item's Engineering Bill of Materials (EBOM) and entered into a spreadsheet to arrive at a "ballpark" estimate. This is referred to as a *parts count estimate*. The estimate either does or does not meet the *minimum* RAM allocation requirements threshold. If the model is *insufficient*, more *complex* mathematical models that employ curve fitting reflecting the applications are employed.

The *parts count* estimate employs average stresses and environments. The *preferred* data source is *actual* test data collected from a statistically *valid* sample of test articles and ultimately system population test and operations data from your system. However, time and cost constraints usually *prevents* or *limits* accumulation of sufficient data to be of value, especially early in the program where FMEA is most economically effective. In most cases data can be *extracted* from handbooks such as

1. MIL-HDBK-217 *Reliability Prediction Of Electronic Equipment*.
2. NPRD-95 *Non-Electronic Parts Reliability Data (NPRD)*.

MIL-HDBK-217 provides a knowledgebase of reliability information such as general *failure rates* for various types of electronic components, quality factors, and environmental factors, temperature derating curves. Later in the development when the circuit parameters are defined, a *stress level* factor can be applied using actual design values to refine the estimate.

Employ Physical Models as the Basis for Reliability Estimates. On initial inspection, reliability modeling may appear to be a simple and straightforward procedure. SEs are tempted to employ *functional* models of the system as the basis for estimating reliability. However, recognize that the system functional model *may not* be an accurate reliability model. A model of the *physical* system provides the most *accurate representation*; the *fidelity* of the model depends on the application.

A *functional* model of the system is necessary and serves as the basis for the reliability model. However, in most cases the two models will NOT be identical. In simple cases, the functional model may be *adequate* and actually serve as the initial reliability model. The reliability model nevertheless must account for any system redundancies, and any peculiarities such as component design and properties that might have impact on completion of the mission. Functional models MAY NOT sufficiently reveal this.

Model Architectural Configuration Reliability. Reliability computations for any SYSTEM or entity are predicated on its physical architecture configuration at various levels of abstraction. At any *level of abstraction*, the SYSTEM/entity can be characterized by one of three types of constructs that are representative of various architectural network configurations: 1) *series*, 2) *parallel*, and 3) *series-parallel*.

Perform a FMEA/FMECA. Simply constructing reliability network models of system architectural elements provides some indication of system reliability. However, component *failure effects* range from *nonthreatening* to *catastrophic*. As a result, SEs need to understand HOW and in WHAT ways a system fails and WHAT the potential *ramifications* of that failure are to completing the mission.

Human-made systems should undergo safety analysis to assess the risks and potential adverse impacts to the system, general public, and the environment. The safety analysis involves conducting a *failure modes and effects analysis* (FMEA). The purpose of the FMEA is to understand HOW a system or product and its components might fail because of *misapplication*, *misuse*, or *abuse* by operators or Users, poor design, or a single point of failure.

One method for understanding HOW and in WHAT ways a system may *fail* is to create *fault trees*. Figure 50.11 provides an example of a simple remote controlled television system *fault tree*. Through FMEA, SEs employ *compensating provisions* such as redesign and procedural changes that enable cost-effective ways to mitigate the risks of failure mode effects.

For systems that require high levels of reliability such as spacecraft and medical equipment, the FMEA may be expanded to include a *criticality analysis* of specific component reliability and their effects. We refer to this as a *failure modes, effects, and criticality analysis* (FMECA).

The FMEA should recommend *cost effective*, corrective action solutions referred to as *compensating provisions* to the system elements—including PERSONNEL (skills/training), EQUIPMENT (design), and PROCEDURAL DATA (usage). The FMEA assesses design documentation such as functional block diagrams (FBDs), assembly drawings, schematics, Engineering Bill of Materials (EBOM), and fault trees to *identify* and *prioritize* areas that may be prone to failure and their associated level of impacts on the system.

Identify Reliability Critical Items (RCIs). The FMEA and FMECA should *analytically* identify reliability critical items (RCIs) that require specification, selection, and oversight. Consider the following example:

644 Chapter 50 Reliability, Availability, and Maintainability

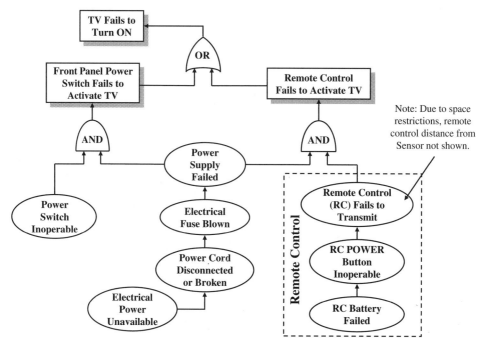

Figure 50.11 Remote Control TV-FTA Feature Example

EXAMPLE 50.5

Failures in gyroscopes and accelerometers in flight control systems for aircraft and sensors can result in major accuracy, safety, and political risks. Therefore, these components may be designated as RCIs.

RCI solutions include singling out these components, specifying higher reliability components, or implementing redundancy with LESS reliability components. Redundancy, however, compounds the system reliability and *increases* expense and maintenance because of *increased* parts counts.

Some systems require special considerations such as electrostatic discharge (ESD) during manufacture and testing to preclude *premature failures* due to *poor* procedures. Additionally, systems that go into storage for extended periods of time prior to usage may have components with a limited shelf life. Therefore, factor ESD and shelf life considerations into *reliability estimates* and *design requirements*.

Perform an Electronic Parts/Circuits Tolerance Analysis. Some systems may require investigation using tolerance analysis sometimes referred to as *worst-case analysis*. Electronic parts/circuits components are *derated* and assigned their *maximum* tolerance values that will have *maximum* effect on circuit output. The system is then evaluated to assure that it can still function under these *extreme* conditions. The analysis is then evaluated again at the other *extreme* end of the *maximum* tolerance values to assure that the system can still function. Since most of today's products are digital and have high reliabilities, the *marginal utility* of this analysis should be assessed on a program needs basis.

Eliminate Single-Point Failures in Mission Critical Items. Since failures in mission critical items can force an *aborted* mission, SEs and R&Ms often *specify* higher levels of reliability that drive up system costs. Yet, they ignore interfaces to the item that are also *single points of failure*

(SPFs) and may have a higher *probability* of failure. When searching for SPFs, investigate not only at the component reliability but also its interface implementation reliability.

Improve System Reliability through Redundancy. *Mission critical systems* often have *stringent* reliability requirements that are very *expensive* to achieve. Additionally, the *possibility* of a SPF may be too *risky* from a *safety* point of view. One approach to developing solutions to these types of challenges is through system design *redundancy*. In some cases elements with *lesser* reliabilities and costs can be combined to achieve higher performance reliability requirements but at the *expense* of *increased* parts counts and maintenance.

Author's Note 50.6 *Customers and specifications often mandate redundancy, without determining if there is a compelling need. Remember, redundancy is a design method option exercised to achieve a system reliability requirement or to eliminate a single point of failure of a mission critical item. Avoid specifying redundancy, which drives up costs, without first determining IF the current design solution is insufficient to meet the reliability requirements.*

Architectural Redundancy versus Component Redundancy. SEs sometimes convince themselves they have designed in system redundancy by simply adding redundant components. This *may* be a false perception. There is a difference between creating redundant *components* and design implementation *redundancy*. To see this, refer to Figure 50.12.

In the figure, we have a simple system that includes Item 1 and Item 1 Backup. Panel A shows the design implementation *redundancy* in signal flow by way of separate, independent interfaces to Item 1 and Item 1 Backup from a single input stimulus or excitation. The same is true with the output interface. Architecturally, if either interface fails and both Item 1 and Item 1 Backup are working properly, an output response is produced. Thus, we can *avoid* a *single point of failure (SPF)*.

In contrast, Panel B includes redundant elements Item 1 and Item 1 Backup. Here, there is a SINGLE interface entry to Item 1 and Item 1 Backup. Therefore, the interface becomes a *single*

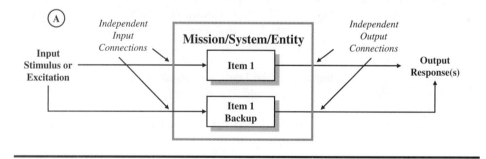

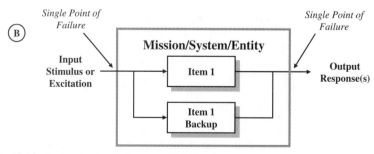

Figure 50.12 System Design Redundancy Considerations

point of failure such that IF it fails, Item 1 and Item 1 Backup are both *useless* as *redundant* architectural elements despite their redundant architectural elements.

Task 4: Review RAM Modeling Results

As the RAM models evolve, reviews should be conducted with the System Engineering and Integration Team (SEIT). The primary objective of the review is to perform a sanity check of the system RAM model results to:

1. *Validate* assumptions.
2. Identify failures modes and effects.
3. Characterize component reliabilities.
4. Identify *compensating provisions* for trade-offs and implementation.

Evaluation of the system RAM *results* and *recommendations* should be performed by *competent, qualified* SMEs.

Review Reliability Estimates. Reliability estimates should be evaluated at each major design review. As is typically the case, subject matter experts (SMEs) are seldom available to *scrutinize* reliability estimates, nor are any of the other attendees *inclined to listen* to debates over reliability prediction approaches. Because of the *criticality* of this topic, conduct a review of the reliability data and approaches PRIOR TO major program reviews. Then, report a summary of the results at the review.

Identify Compensating Provision Actions. As part of the review, *compensating provisions* should be identified to *mitigate* any risks related to achieving RAM requirements. Each *compensating provision* should be assigned a unique action item and tracked to *closure*.

Allocate RAM Resources Based on Risk. For some programs, there may only be a *limited* number of resources for reliability tasks. *How might you deal with this? First*, you should investigate further funding resources. *Second*, you may have to *allocate* those resources based on reliability *estimates* of the system that may have the *highest* risk.

How do you identify CANDIDATE *risk items?* The *risk items* might include: 1) newly developed items, 2) new technologies, 3) mission critical items, 4) critical interfaces, 5) thermal areas, 6) power conditioning, 7) pressurization systems, and 8) toxic failures.

Task 5: Implement Compensating Provision Actions

Once decisions are made concerning the RAM recommendations, the objective of Task 5 is to implement the FMEA/FMECA *compensating provisions*. These actions may require:

1. Trade-off and reallocation of RAM performance requirements at the SYSTEM, PRODUCT, SUBSYSTEM, ASSEMBLY, SUBASSEMBLY or PART levels.
2. Redesign of items.
3. Renegotiation of the contract/subcontract/task.

Task 6: Implement a Parts Program

Some aerospace and defense systems establish *standards* to ensure that *quality* components are used. Where this is the case, a parts program may be created to *establish* part specifications and standards and to *screen* incoming parts for *compliance*. Parts programs are very *costly* to implement and maintain. An alternative to a parts program may be to identify and manage *reliability critical items* (RCIs).

Task 7: Improve EQUIPMENT Characteristic Profiles

A common question SEs must face is: *HOW do we minimize the number of initial failures and prolong life expectancy of useful service regions?* You can:

1. Increase the reliability by using *lower* failure rate components, which *increases* system development cost.
2. *Conduct* environmental stress screening (ESS) selection of incoming components.
3. *Improve* system design practices.
4. *Improve* quality assurance/control to drive out inherent errors and latent defects prior to fielding.

The last point reinforces the need for a robust *testing program* for new system or product designs. The bottom line is: increase the useful *service life* by *engineering* the job correctly beginning with the proposal and certainly at Contract Award.

Concluding Point

RAM engineers CANNOT and SHOULD NOT perform these tasks in a vacuum. As specialty engineers, they provide critical decision support to the system developers in the "engineering of systems." Your job, as an SE, is to make sure a collaborative *engineering* environment exists with frequent reviews and communications between R&M engineers and Integrated Product Teams (IPTs). Assuming they are implemented properly, IPTs can provide such an environment.

50.8 RAM CHALLENGES

Implementation of RAM practices poses a number of challenges. Let's explore some of the key challenges.

Challenge 1: Defining and Classifying Failures

Despite all the complexities of curve fitting and creating mathematical equations to model RAM, one of the most *sensitive* issues is simply WHAT constitutes a failure. *Does it mean loss of a critical mission function?* Your task, as an SE, is to develop a consensus definition of a failure that is shared by Acquirer and System Developer teams.

Challenge 2: Failure to Document Assumptions and Data Sources

The validity of any engineering data typically requires:

1. Making *assumptions* about the mission/SYSTEM use cases and scenarios and OPERATING ENVIRONMENT conditions.
2. Identifying *credible* data sources.
3. Documenting any assumptions or observations.

Yet, few organizations *ingrain* this *discipline* in their engineers. Then, when the time comes to make critical *informed* decisions, the decision process is left in a quandary as to whether the R&M

Engineer DID or DID NOT consider all *relevant* factors. Time *exacerbates* the problem. Therefore, train engineers in your organization to document *assumptions* and *data sources* as part of a RAM analysis.

Challenge 3: Validating RAM Models

RAM analyses, assuming they are performed properly, are only as good as the models employed to generate data for the analyses. George E.P. Box (1979, p. 202) once remarked *"All models are wrong but some are useful."* From the beginning of a program, strive to validate the models used to generate decision-making data with *actual* vendor or *field* data.

Challenge 4: The Criticality of Scoping System Availability

Today, the government and private sectors are moving toward a contracting for services environment in which a contractor provides systems and services on a *fee-per-service* basis. For example, contracts will often state that the EQUIPMENT and services must be operationally *available* from 8 a.m. to 5 p.m. Monday through Friday and at other times under specified conditions. The Acquirer pays a fee for mission time—for system use within those time frames.

Here is something for a System Developer/Services Provider to consider. Depending on the *size* and *complexity* of the EQUIPMENT and services rendered, some missions MAY NOT require specific pieces of EQUIPMENT to be operationally available *simultaneously. Should the Services Provider be penalized? What about holiday occurrences between Monday and Friday?*

The challenge in developing the contract is to thoroughly understand all of the *use cases* and *scenarios* that may become *major showstoppers* for progress payments in performance of the contract and clarifying HOW all parties will accommodate them. Think SMARTLY before signing contracts that have *system availability* requirements.

Challenge 5: Quantifying the RAM of Software

We live in an Information Age where seemingly all systems, especially complex systems, are software intensive. Despite RAM curve fitting and the mathematical equations discussed, *HOW do you measure and model the RAM of software*? Unlike hardware components that can be tested in controlled laboratory conditions and maintained via repairs, HOW *do you model the RAM of software*?

This is perhaps one of the most perplexing areas of engineering. *What about the RAM of software used in basic computer applications versus mission critical software for the International Space Station (ISS), Space Shuttle, passenger aircraft, and medical equipment?* There are no easy or simple answers.

One solution may be to employ the services of *independent verification and validation* (IV&V) contractors. IV&V contractors perform services to review and test software designs for purposes of *identifying* and *eliminating design flaws, coding errors, and product defects*—and their services can be very costly, especially from a system development perspective. Depending on the legal and financial ramifications of system *abuse* or *misuse*, the *return on investment* (ROI) for IV&V activities may be cost effective. Contact SMEs in your industry to gain insights into how to specify software RAM.

50.9 GUIDING PRINCIPLES

In summary, the preceding discussions provide the basis to establish several guiding principles to govern *Reliability, Availability, and Maintainability* practices.

Principle 50.1 Express reliability in terms of its four key elements:

1. A *probability* of successfully completing a defined mission.
2. A *bounded* mission duration.
3. *Elapsed* operating time since the start of the mission.
4. A *prescribed* set of OPERATING ENVIRONMENT conditions.

Avoid using MTBF as the reliability requirement without bounding these conditions.

Principle 50.2 Components have service life profiles that may exhibit regions of decreasing, stable, and increasing failure rates, each with differing failure rate distributions.

Principle 50.3 Only systems or components that are characterized by *negative exponential distributions* have a constant hazard rate (Period of Stabilized Failures).

Principle 50.4 Avoid RAM analysis paralysis; couple RAM analysis with "worst case" analysis.

50.10 SUMMARY

Our discussion of RAM practices is intended to provide a basic understanding that will enable you to communicate with reliability engineers, logisticians, and others. There are several key points you should remember:

- RAM practices apply model-based mathematical and scientific principles to *estimate* reliability, availability, and maintainability to support SE design decision making.
- RAM estimates are only as *valid* as the *assumptions* and *inputs* used to generate the data from validated models.
- RAM models require best-fit selection to provide an *estimate* of probability related to mission, system, and item success.
- RAM practices involve *art*, *science*, and *sound judgment: art* from the standpoint of empirical knowledge, wisdom, and experience gleaned over an entire career, *science* from the application of mathematical and scientific principles, and *sound judgment* from being able to know and understand the *difference* between the *art* and the *science*.
- ALWAYS entrust the RAM estimates to a *qualified*, *professional*, Reliability and Maintainability (R&M) Engineer recognized as a subject matter expert (SME), either as a staff member or as a credible consultant with integrity. Remember, RAM involves critical areas that involve ethical, legal, and safety issues and their associated ramifications.

GENERAL EXERCISES

1. Answer each of the What You Should Learn from This Chapter questions identified in the Introduction.

ORGANIZATIONAL CENTRIC EXERCISES

1. Research your organization's command media to learn what *processes* and *methods* are to be employed when conducting RAM practices. Report your findings.
2. Contact several contract programs within your organization and research the requirements for RAM. Interview program personnel to determine the following:

(a) How did they allocate and document the requirements allocations to system elements?
(b) What lessons did the development team learn?
(c) How would they advise approaching RAM on future programs?
(d) What sources of RAM data and models did they use to develop RAM predictions?
(e) What type of FRACAS system did the User have in place?
(f) Did the development team use this information?
(g) How was availability computed for the system?

REFERENCES

ASD-100. 2004. *System Engineering Manual*, National Airspace System, Architecture and System Engineering, Washington, DC: Federal Aviation Administration (FAA).

BILLINTON, ROY, and ALLAN, RONALD N. 1987. *Reliability Evaluation of Engineering Systems: Concepts and Techniques*. New York: Plenum Press.

BLANCHARD, BENJAMIN S. 1998. *System Engineering Management*, 2nd ed. New York: Wiley.

BLANCHARD, BENJAMIN S., and FABRYCKY, WOLTER J. 1990. *Systems Engineering and Analysis*, 2nd ed. Engelwood Cliffs, NJ: Prentice-Hall.

BOX, GEORGE E.P. 1979. "Robustness is the Strategy of Scientific Model Building" in Robustness in Statistics. eds, R.L. Launer and G.N. Wilkinson, Academic Press, New York, NY.

Defense Systems Management College (DSMC). 2001. *DoD Glossary: Defense Acquisition Acronyms and Terms*, 3rd ed. Defense Acquisition University Press Ft. Belvoir, VA.

Defense Systems Management College (DSMC). 1997. *DSMC Acquisition Logistics Guide*, 3rd ed. Ft. Belvoir, VA: DSMC.

MIL-HDBK-470A. 1997. *DoD Handbook: Designing and Developing Maintainable Systems and Products*, Vol. 1. Washington, DC: Department of Defense (DoD).

MIL-STD-1629A (canceled). 1980. *Military Standard Procedures for Performing a Failure Modes, Effects, and Criticality Analysis*. Washington, DC: Department of Defense (DoD).

NELSON, WAYNE. 1990. *Accelerated Testing: Statistical Models, Test Plans, and Data Analyses*. New York: Wiley.

ROSSI, MICHAEL J. 1987. *NonOperating Reliability Databook* (aka, NONOP-1). Griffiss Air Force Base, NY: Reliability Analysis Center, Rome Air Development Center.

NPRD-95. *Non-electronic Parts Reliability Data (NPRD)*. 1995. Griffiss Air Force Base, NY: Reliability Analysis Center, Rome Air Development Center.

O'CONNOR, PATRICK D. 1995. *Practical Reliability Engineering*. New York: Wiley.

SMITH, ANTHONY M. 1992. *Reliability-Centered Maintenance*: New York: McGraw-Hill.

ADDITIONAL READING

HDBK-1120. 1993. *Failure Modes, Effects, and Criticality Analysis*, Rome, NY: Reliability Analysis Center (RAC).

HDBK-1140. 1991. *Fault Tree Analysis Application Guide*, Rome, NY: Reliability Analysis Center (RAC).

HDBK-1610. 1999. *Evaluating the Reliability of Commercial Off-the-Shelf (COTS) Items*, Rome, NY: Reliability Analysis Center (RAC).

HDBK-1190. 2002. *A Practical Guide to Developing Reliable Human-Machine Systems and Processes*, Rome, NY: Reliability Analysis Center (RAC).

HDBK-3180. 2004. *Operational Availability Handbook*, Rome, NY: Reliability Analysis Center (RAC).

LEITCH, R.D. 1988. *BASIC Reliability Engineering Analysis*. Stoneham, MA: Butterworth.

MIL-HDBK-0217. 1991. *Reliability Prediction of Electronic Equipment*. Washington, DC: Department of Defense (DoD).

MIL-HDBK-470A. 1997. *Designing and Developing Maintainable Products and Systems*. Washington, DC: Department of Defense (DoD).

MIL-HDBK-1908B. 1999. *DoD Definitions of Human Factors Terms*. Washington, DC: Department of Defense (DOD).

MIL-HDBK-2155. 1995. (Cancelled) *Failure Reporting, Analysis, and Corrective Action System*. Washington, DC: Department of Defense (DoD).

MIL-STD-721C. 1981. *Definition of Terms for Reliability and Maintainability*. Washington, DC: Department of Defense (DoD).

MIL-STD-882D. 2000. *DoD Standard Practice for System Safety*. Washington, DC: Department of Defense (DOD).

National Aeronautics and Space Administration (NASA). 1994. *Systems Engineering "Toolbox" for Design-Oriented Engineers*. NASA Reference Publication 1358. Washington, DC.

Chapter 51

System Modeling and Simulation

51.1 INTRODUCTION

Analytically, System Engineering requires several types of technical decision-making activities:

1. **Mission Analysis** Understanding the User's *problem space* to identify and bound a solution space that provides *operational utility, suitability, availability*, and *effectiveness*.
2. **Architecture Development** Hierarchical organization, decomposition, and bounding of operational *problem space* complexity into manageable levels of *solutions spaces*, each with a bounded set of requirements.
3. **Requirements Allocation** Informed appropriation and assignment of capabilities and quantifiable performance to each of the *solution spaces*.
4. **System Optimization** The evaluation and refinement of system performance to maximize *efficiency* and *effectiveness* in achieving solution space mission objectives.

Depending on the *size* and *complexity* of the system, most of these decisions require tools to facilitate the decision making. Because of the *complex* interacting parameters of the SYSTEM OF INTEREST (SOI) and its OPERATING ENVIRONMENT, humans are often unable of *internalize* solutions on a personal level. For this reason engineers as a group tend to *employ* and *exploit* decision aids such as models and simulations to gain insights into the system *interactions* for a prescribed set of operating scenarios and conditions. *Assimilation* and *synthesis* of this knowledge and interdependencies via *models* and *simulations* enable SEs to collectively make these decisions.

This chapter provides an introductory overview of how SEs employ models and simulations to implement the SE Process Model. Our discussions are not intended to instruct you in model or simulation development; numerous textbooks are available on this topic. Instead, we focus on the *application* of models and simulations to facilitate SE decision making.

We begin our discussion with an introduction to the *fundamentals* of models and simulations. We identify various types of models, define model fidelity, address the need to certify models, and describe the integration of models into a simulation. Then, we explore HOW SEs employ models and simulations to support technical decisions involving architecture evaluations, performance requirement allocations, and validating the performance.

What You Should Learn from This Section

1. What is a *model*?
2. What are the various *types of models*?
3. How are models employed in SE decision making?

System Analysis, Design, and Development, by Charles S. Wasson
Copyright © 2006 by John Wiley & Sons, Inc.

4. What is a *simulation*?
5. How are simulations employed in SE decision making?
6. What is a *mock-up*?
7. What is a *prototype*?
8. What is a *testbed*?
9. How is a *testbed* employed in SE decision making?

Definitions of Key Terms

- **Accreditation** "The formal certification that a model or simulation is acceptable for use for a specific purpose. Accreditation is conferred by the organization best positioned to make the judgment that the model or simulation in question is acceptable. That organization may be an operational user, the program office, or a contractor, depending upon the purposes intended." (DSMC *SE Fundamentals*, Section 13.4 *Verification, Validation, and Accreditation*; p. 120)

- **Certified Model** A formal designation by an officially recognized decision authority for validating the products and performance of a model.

- **Deterministic Model** "A model in which the results are determined through known relationships among the states and events, and in which a given input will always produce the same output; for example, a model depicting a known chemical reaction. Contrast with: stochastic model. (DIS *Glossary of M&S Terms*, and IEEE STD 610.3, (references (b) and (c))" (Source: DoD 5000.59-M *Modeling and Simulation (M&S) Glossary*, Part II, item 153, p. 102)

- **Event** "A change of object attribute value, an interaction between objects, an instantiation of a new object, or a deletion of an existing object that is associated with a particular point on the federation time axis. Each event contains a time stamp indicating when it is said to occur. (*High Level Architecture Glossary*, (reference (m))." (Source: DoD 5000.59-M *Modeling and Simulation (M&S) Glossary*, Part II, item 193, p. 107)

- **Fidelity** "The accuracy of the representation when compared to the real world. (DoD Publication 5000.59-P, (reference (g))." (Source: DoD 5000.59-M *Modeling and Simulation (M&S) Glossary*, Part II, item 218, p. 112)

- **Initial Condition** "The values assumed by the variables in a system, model, or simulation at the beginning of some specified duration of time. Contrast with: boundary condition; final condition. (DIS *Glossary of M&S Terms*, (reference (b))." (Source: DoD 5000.59-M *Modeling and Simulation (M&S) Glossary*, Part II, item 270, p. 123)

- **Initial State** "The values assumed by the state variables of a system, component, or simulation at the beginning of some specified duration of time. Contrast with: final state. (DIS *Glossary of M&S Terms*, (reference (b))." (Source: DoD 5000.59-M *Modeling and Simulation (M&S) Glossary*, Part II, item 271, p. 123)

- **Model** A virtual or physical representation of an entity for purposes of presenting, studying, and analyzing its characteristics such as appearance, behavior, or performance for a prescribed set of OPERATING ENVIRONMENT conditions and scenarios.

- **Model-Test-Model** "An integrated approach to using models and simulations in support of pre-test analysis and planning; conducting the actual test and collecting data; and post-test analysis of test results along with further validation of the models using the test data." (Source: DoD 5000.59-M *Modeling and Simulation (M&S) Glossary*, Part II, item 342, p. 137)

- **Monte Carlo Algorithm** "A statistical procedure that determines the occurrence of probabilistic events or values of probabilistic variables for deterministic models; e.g., making a random draw. (DSMC 1993–94 *Military Research Fellows Report*, (reference (k))." (Source: DoD 5000.59-M *Modeling and Simulation (M&S) Glossary*, Part II, item 345, p. 138)

- **Monte Carlo Method** "In modeling and simulation, any method that employs Monte Carlo simulation to determine estimates for unknown values in a deterministic problem. (DIS *Glossary of M&S Terms*, and IEEE STD 610.3, (references (b) and (c))." (Source: DoD 5000.59-M *Modeling and Simulation (M&S) Glossary*, Part II, item 346, p. 138)

- **Simulation Time** "A simulation's internal representation of time. Simulation time may accumulate faster, slower, or at the same pace as sidereal time. (DIS *Glossary of M&S Terms*, and IEEE STD 610.3, (references (b) and (c))." (Source: DoD 5000.59-M *Modeling and Simulation (M&S) Glossary*, Part II, item 473, p. 159)

- **Stimulate** "To provide input to a system in order to observe or evaluate the system's response." (Source: DSMC Simulation Based Acquisition: A New Approach—Dec. 1998)

- **Stimulation** "The use of simulations to provide an external stimulus to a system or subsystem." (Source: DSMC Simulation Based Acquisition: A New Approach—Dec. 1998)

- **Stochastic Process** "Any process dealing with events that develop in time or cannot be described precisely, except in terms of probability theory. (DSMC 1993–94 *Military Research Fellows Report*, (reference (k))." (Source: DoD 5000.59-M *Modeling and Simulation (M&S) Glossary*, Part II, item 493, p. 161)

- **Validated Model** An analytical model whose outputs and performance characteristics identically or closely match the products and performance of the physical system or device.

- **Validation (Model)** "The process of determining the manner and degree to which a model is an accurate representation of the real world from the perspective of the intended uses of the model, and of establishing the level of confidence that should be placed on this assessment." (DSMC *SE Fundamentals*, Section 13.4 *Verification, Validation, and Accreditation*; p. 120.)

- **Verification (Model)** "The process of determining that a model implementation accurately represents the developer's conceptual description and specifications that the model was designed to." (DSMC *SE Fundamentals*, Section 13.4 *Verification, Validation, and Accreditation*; p. 120.)

51.2 TECHNICAL DECISION-MAKING AIDS

SE decision making related to system performance allocations, performance budgets and safety margins, and design requires *decision support* to ensure that *informed, fact-based recommendations* are made. Ideally, we would prefer to have an *exact* representation of the system you are analyzing. In reality, *exact* representations do not exist until the system or product is designed, developed, verified, and validated.

There are, however, some decision aids SE can employ to provide degrees of *representations* of a system or product to facilitate technical decision making. These include models, prototypes, and mock-ups. The purpose of the decision aids is to create a *representation* of a system on a small, low-cost scale that can provide empirical *form, fit, or function data* to support design decision making on a much larger scale.

51.3 MODELS

Models generally are of two varieties: *deterministic* and *stochastic*.

Deterministic Models

Deterministic models are structured on known relationships that produce predictable, repeatable results. Consider the following example:

EXAMPLE 51.1

Each work period an employee receives a paycheck based on a formula that computes their gross salary—hours worked times hourly rate *less* any distributions for insurance, taxes, charitable contributions, and so forth.

Stochastic Models

Whereas deterministic models are based on precise relationships, *stochastic models* are structured using probability theory to process a set of random event occurrences. In general, *stochastic* models are constructed using data from statistically valid samples of a population that enable us to *infer* or *estimate* results about the population as a whole. Consider the following *theoretical* example:

EXAMPLE 51.2

A manufacturer produces 0.95 inch spacer parts in which three of the parts are assembled onto an axle. The axle is then installed into a *constrained* space of a larger ASSEMBLY. If we produced several thousand parts, we might discover the individual spacers *randomly* vary in size from 0.90 to 1.0 inch; no two parts are exactly the same. So, *what must the dimension of the constrained space be to accommodate dimensional variations of each spacer?* We construct a *stochastic* model that *randomly* selects dimensions for each of three spacers within their allowable ranges. Then, it computes the integrated set of dimensions to estimate the mean of a typical *stacked* configuration. Based on the results, a dimension of the *constrained* space is selected that factors in any additional considerations such degrees of looseness, if applicable.

The example above illustrates theoretically how a stochastic model could be employed. SEs often use a *worst-case analysis* in lieu of developing a model. For some applications this may be *acceptable*. However, suppose a *worst-case analysis* results in too much slack between the spools? What the SEs need to know is: *For randomly selected parts assembled into the configuration, based on frequency distributions of part dimensions and their standard deviations, WHAT is the estimated mean of any configuration?*

In summary, *stochastic models* enable us to *estimate* or draw *inferences* about system performance for highly complex situations. These situations involve random, uncontrollable events or inputs that have a frequency of occurrence under prescribed conditions. Based on the frequency distributions of sampled data, we can apply statistical methods that *infer* a most probable outcome for a specific set of conditions. Examples include environmental conditions and events, human reactions to publicity, and pharmaceutical drug medications.

Model Development

Analytical methods require a *frame of reference* to represent the characteristics of an entity. For most systems, a model is created using an observer's *frame of reference*, such as the *right-handed coordinate system*.

Analytically, *model development* is similar to *system development*. SE model developers should fully understand the *problem space* a model's *solution space* is intended to satisfy. Based on this understanding, design methodology requires that we first survey or research the marketplace to see if the model(s) we require has already been developed and is available. If available, we need to determine if it has *necessary* and *sufficient* technical detail to support our system or entity application. Conventional design wisdom (Principle 41.1) says that new models should *only* be developed *after* you have *exhausted* all other *alternatives* to locate an *existing* model.

Model Validation

Models are only as *valid* as the quality of its *behavioral* and *physical* performance characteristics to *replicate* the real world entity. We refer to the *quality* of a model in terms of its *fidelity*—meaning its degree of realism. So, the challenge for SEs is: *Even if we develop a model of a system or product, HOW do we gain a level of confidence that the model is valid and accurately and precisely represents the physical instance of an item and its interactions with a simulated real world OPERATING ENVIRONMENT?*

In general, when developing models, we attempt to represent physical reality with *simulated* or *scaled* reality. Our goal is to try to achieve *convergence* of the two within the *practicality* or *resource constraints*. So, *HOW do we achieve convergence*? We do this by collecting empirical *data* from actual physical systems, prototypes, or field tests. Then, we *validate* the model by comparing the actual field data with the simulated behavioral and physical characteristics. Finally, we refine the model until its results closely match those of the actual system. This leads to the question: *HOW do we get field data to validate a model for a system or product we are developing?*

There are a number ways of obtaining field data. We can:

1. Collect data using controlled laboratory experiments subjected to OPERATING ENVIRONMENT conditions and scenarios.

2. Install a similar component on a fielded system and collect measurement data for OPERATING ENVIRONMENT conditions and scenarios.

3. Instrument a field platform such as an aircraft with transducers and sensors to collect OPERATING ENVIRONMENT data.

Regardless of the method we use, a model is *calibrated*, *adjusted*, and *refined* until it is *validated* as an *accurate* and *precise* representation of the physical system or device. The model is then placed under formal configuration control. Finally, we may decide to have an independent decision authority or subject matter expert (SME) to *certify* the model, which brings us to our next topic.

Model Certification

The creation of a model is one thing; creating a *valid* model and getting it *certified* is yet another. Remember, SE technical decision making must be founded in *objective, fact-based* data that *accurately* represent real world situations and conditions. The same applies with models. So, *what is certification*?

IEEE 610.12–1990 defines *certification* as "*A written guarantee that a system or component complies with its specified requirements and is acceptable for operational use. For example, a written authorization that a computer system is secure and is permitted to operate in a defined environment.*" For many applications an independent authority validates a model by authenticating that model results identically match those obtained from measurements of an actual system operating under a specified set of OPERATING ENVIRONMENT conditions.

In general, one SE can demonstrate to a colleague, their manager, or a Quality Assurance (QA) representative that the data match. *Certification* comes later when a recognized decision authority within industry or governmental organization reviews the data validation results and officially issues a *Letter of Certification* declaring the model to be certified for use in specific applications and conditions.

Do you need certified models? This depends on you're the program's needs. *Certification*:

1. Is expensive establish and maintain.
2. Has an intrinsic value to the creator and marketplace.

Some models are used one time; others are used repeatedly and refined over several years. Since engineering decisions must be based on the integrity of data, models are generally *validated* but not necessarily *certified*.

Understanding Model Characteristics

Models are generally developed to satisfy specific needs of the analyst. Although models may appear to match two different analysis needs, they may not satisfy the requirements. Consider the following example:

EXAMPLE 51.3

Let's suppose Analyst A requires a sensor model to investigate a technical issue. Analyst A develops a *functional model* of Sensor XYZ to meet their needs of understanding the behavioral responses to external stimuli. Later, Analyst B in another organization researches the marketplace and learns that Analyst A has already developed a sensor model that may be available. However, Analyst B soon learns that the model describes the *functional behavior* of Sensor XYZ whereas Analyst B is interested in the *physical model* of Sensor XYZ. As a result, Analyst B either creates their own *physical model* of Sensor XYZ or translates the functional domain model into the physical domain.

Understanding Model Fidelity

One of the challenges of modeling and simulation is determining the type of model you need. Ideally, you would want a perfect model readily available so that you could provide simple inputs and conduct WHAT IF games with reliable results.

Due to the *complexities* and *practicalities* of modeling, among which are cost and schedule constraints, models are *estimates or approximations* of reality—termed *levels of fidelity*. For example, is a first-order approximation sufficient? Second-order? Third-order? etc. The question we have to answer is: *WHAT minimum level of fidelity do we need for a specific area of investigation?* Consider the following examples:

EXAMPLE 51.4

Hypothetically, a mechanical gear system has a *transfer function* that can be described mathematically as $y = 0.1x$ where x = input and y = output. You may find a simple analytical *math model* may be sufficient for some applications. In other applications the area of analytical investigation might require a physical model of each component within the gear train, including the frictional losses due to the loading effects on axle bearings.

EXAMPLE 51.5

Let's assume you are developing an aircraft simulator. The key questions are:

1. *Are computer-generated graphic displays of cockpit instruments with simulated moving needle instruments and touch screen switches sufficient, or do you need the actual working hardware used in the real cockpit to conduct training?*
2. *What level of fidelity in the instruments do you need to provide pilot trainees with the "look and feel" of flying the actual aircraft?*
3. *Is a static cockpit platform sufficient for training, or do you need a three-axis motion simulator to provide the level of fidelity in realistic training?*

The point of these examples is: SEs, in collaboration with analysts, the Acquirer, and Users, must be able to determine WHAT *levels of fidelity* are required and then be able to *specify* it. In the case of simulator training systems, various levels of fidelity may be acceptable. Where this is the case, create a matrix to specify the *level of fidelity* required for each physical item and include scoping definitions of each level of fidelity. To illustrate this point, consider the following example:

EXAMPLE 51.6

The *level of fidelity* required for some switches may indicate computer-generated images are sufficient. Touch screen displays that enable switch activation by touch may be acceptable to create the effects of flipping switches.

EXAMPLE 51.7

In other instances hand controls, brake pedals, and other mechanisms may require *actual* working devices that provide the tactile *look* and *feel* of devices of the actual system.

Specifying Model Fidelity

Understanding model fidelity is often a challenge. One of the objectives of modeling and simulation is being able to realistically model the real world. In the case of training simulators that require visual representations of the environment inside and outside the *simulated* vehicle, what *level of fidelity* in the terrain and trees and cultural features such as roads, bridges, and buildings is *necessary* and *sufficient* for training purposes?

- Are computer-generated images with synthetic texture sufficient for landscapes?
- Do you require photographic images with computer-generated texture?

The answer to these questions depends on trade-offs between resources available and the *positive* or *negative* impacts to training. Increasing the *level of fidelity* typically requires significantly more resources—such as data storage or computer processing performance. Concepts such as *cost as an independent variable* (CAIV) enable Acquirer decision makers to assess WHAT level of CAPABILITY can be achieved at WHAT cost.

51.4 SYSTEM SIMULATION

Models serve as "building block" *representations* or *approximations* of physical reality. When we *integrate* these models into an executable framework that enables us stimulate interactions and behavioral responses under controlled conditions, we create a simulation of a SYSTEM OF INTEREST (SOI).

As *analytical* models, simulations enable us to conduct WHAT IF exercises with each model or system. In this context the intent is for SEs to understand the *functional* or *physical* behavior and interactions of the system for a given set of OPERATING ENVIRONMENT scenarios and conditions.

Guidepost 51.1 *The preceding discussions provide the foundation for understanding models and simulations. We now shift our focus to understanding HOW SEs employ models and simulations to support analytical decision making as well as create deliverable products for Users.*

51.5 APPLICATION EXAMPLES OF MODELING AND SIMULATION

Modeling and simulation (M&S) are applied in a variety of ways by SEs to support technical decision making. SEs employ models and simulations for several types of applications:

Application 1: Simulation-based architecture selection
Application 2: Simulation-based architectural performance allocations
Application 3: Simulation-based acquisition (SBA)
Application 4: Test environment stimuli
Application 5: Simulation-based failure investigations
Application 6: Simulation based training
Application 7: Test bed environments for technical decision support

To better understand HOW SEs employ models and simulations, let's describe each type of application.

Application 1: Simulation-Based Architecture Selection

When you engineer systems, you should have a range of alternatives available to support informed selection of the best candidate to meet a set of prescribed OPERATING ENVIRONMENT scenarios and conditions. In practical terms, you cannot *afford* to develop every candidate architecture just to study it for purposes of selecting the best one. We can construct, however, *models* and *simulations* that represent *functional* or *physical* architectural configurations. To illustrate, consider the following example using Figure 51.1.

> **EXAMPLE 51.8**
>
> Let's suppose we have identified several promising Candidate Architectures 1 through *n* as illustrated on the left side of the diagram. We conduct a trade study analysis of alternatives (AoA) and determine that the complexities of selecting the RIGHT architecture for a given system application requires employment of models and simulations. Thus, we create Simulation 1 through Simulation *n* to provide the *analytical* basis for selecting the *preferred* architectural configuration.

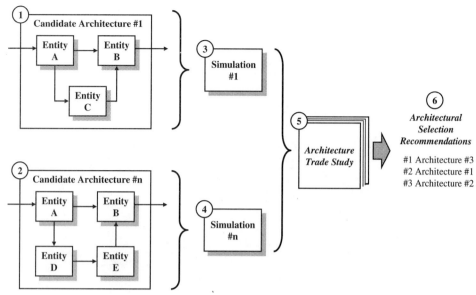

Figure 51.1 Simulation-Based Architecture Selection

We exercise the simulations over a variety of OPERATING ENVIRONMENT scenarios and conditions. Results are analyzed and compiled and documented in an *Architecture Trade Study*. The *Architecture Trade Study* rank orders the results as part of its recommendations. Based on a review of the *Architecture Trade Study*, SEs select an architecture. Once the architecture is selected, the simulation serves as the *framework* for evaluation and refining each simulated architectural entity at lower levels of abstraction.

Application 2: Simulation-Based Architectural Performance Allocations

Modeling and simulation are also employed to perform simulation-based performance allocations as illustrated in Figure 51.2. Consider the following example:

EXAMPLE 51.9

Suppose that Requirement A describes and bounds Capability A. Our initial analysis derives three subordinate capabilities, A1 through A3, that are specified and bounded by Requirements A1 through A3: The challenge is: *How do SEs allocate Capability A's performance to Capabilities A1 through A3?*

Let's assume that basic analysis provides us with an initial set of performance allocations that is "in the ballpark." However, the interactions among entities are complex and require *modeling* and *simulation to support performance allocation decision making*. We construct a model of the Capability A's architecture to investigate the performance relationships and interactions of Entities A1 through A3.

Next, we construct the Capability A simulation consisting of models, A1 through A3, *representing* subordinate Capabilities A1 through A3. Each supporting capability, A1 through A3, is modeled using the System Entity Capability Construct shown in Figure 22.1. The simulation is *exercised* for a variety of *stimuli, cues*, or *excitations* using Monte Carlo methods to understand the behavior of the interactions over a range of operating environment scenarios and conditions. The results of the interactions are captured in the system behavioral response characteristics.

660 Chapter 51 System Modeling and Simulation

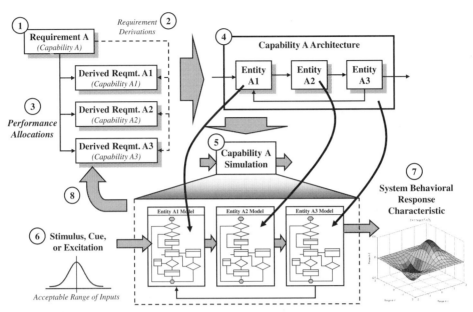

Figure 51.2 Simulation-Based Performance Allocations

After several iterations to *optimize* the interactions, SEs arrive at a final set of *performance allocations* that become the basis for requirements specifications for capability A. Is this perfect? No! Remember, this is a *human approximation* or *estimate*. Due to *variations* in physical components and the OPERATING ENVIRONMENT, the final simulations may still have to be *calibrated, aligned*, and *tweaked* for field operations based on actual field data. However, we initiated this process to *reduce* the *complexity* of the *solution space into more* manageable *pieces*. Thus, we arrived at a very close approximation to support requirements' allocations without having to go to the expense of developing the actual working hardware and software.

Application 3: Simulation-Based Acquisition (SBA)

Traditionally, when an Acquirer acquired a system or product, they had to wait until the System Developer delivered the final system for Operational Test and Evaluation (OT&E) or final acceptance. During OT&E the User or an Independent Test Agency (ITA) conducts field exercises to evaluate system or product performance under actual OPERATING ENVIRONMENT conditions. *Theoretically* there should be no surprises. *Why*?

1. The *System Performance Specification (SPS)* perfectly described and bounded the *well-defined solution space*.

2. The System Developer created the ideal physical solution that perfectly complies with the SPS.

In REALITY every system design solution has *compromises* due to the *constraints* imposed. Acquirers and User(s) of a system need a level of confidence "up front" that the system will perform as intended. *Why*? The cost of developing large complex systems, for example, and ensuring that they meet User validated operational needs is challenging.

One method for improving the chances of delivery success is *simulation-based acquisition (SBA). What is SBA?* In general, when the Acquirer releases a formal *Request for Proposal (RFP)* solicitation for a system or product, a requirement is included for each Offeror to deliver a *working* simulation model along with their technical proposal. The RFP stipulates criteria for meeting a

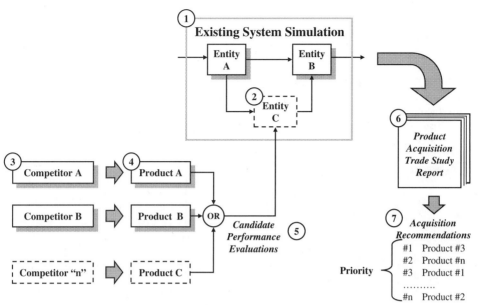

Figure 51.3 Simulation-Based Acquisition (SBA)

prescribed set of functionality, interface and performance requirements. To illustrate how SBA is applied, refer to Figure 51.3.

EXAMPLE 51.10

Let's suppose a User has an existing system and decides there is a need to replace a SUBSYSTEM such as a propulsion system. Additionally, an Existing System Simulation is presently used to investigate system performance issues.

The User selects an Acquirer to procure the SUBSYSTEM replacement. The Acquirer releases an RFP to a qualified set of Offerors, competitors A through *n*. In response to RFP requirements, each Offeror delivers a simulation of their *proposed* system or product to support the evaluation of their technical proposal.

On delivery, the Acquirer Source Selection Team evaluates each technical proposal using predefined proposal evaluation criteria. The Team also integrates the SUBSYSTEM simulation into the Existing System Simulation for further technical evaluation.

During source selection, the offeror's technical proposals and simulations are evaluated. Results of the evaluations are documented in a *Product Acquisition Trade Study*. The TSR provides a set of Acquisition Recommendations to the Source Selection Team (SST), which in turn makes Acquisition Recommendations to a Source Selection Decision Authority (SSDA).

Application 4: Test Environment Stimuli

System Integration, Test, and Evaluation (SITE) can be a very expensive element of system development, not only from its labor intensiveness but also the creation of the test environment interfaces to the *unit under test* (UUT). There are several approaches SEs can employ to test a UUT. The usual *system integration, test, and evaluation (SITE) options include:* 1) *stimulation*, 2) *emulation*, and 3) *simulation*. The *simulations* in this context are designed to reproduce external system interfaces to the (UUT). Refer to Figure 51.4.

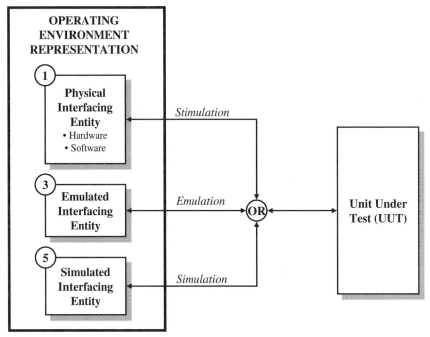

Figure 51.4 Stimulation, Emulation, and Simulation Testing Options

Application 5: Simulation-Based Failure Investigations

Large complex systems often require simulations that enable decision makers to explore different aspects of performance in employing the system or product in a prescribed OPERATING ENVIRONMENT.

Occasionally, these systems encounter an unanticipated *failure mode* that requires in-depth investigation. The question for SEs is: *What set of system/operator actions or conditions and use case scenarios contributed to the failure?* Was the root cause due to: 1) latent defects, design flaws, or errors, 2) reliability of components, 3) operational fatigue, 4) lack of proper maintenance, 5) misuse, abuse, or misapplication of the system from its intended application, or 6) an anomaly?

Due to safety and other issues, it may be advantageous to explore the *root cause* of the FAILURE using the existing simulation. The challenge for SEs is being able to:

1. Construct the chain of events leading to the failure.
2. Reliably replicate the problem on a predictable basis.

A decision could be made to use the simulation to explore the *probable cause* of the failure mode. Figure 51.5 illustrates how you might investigate the cause of failure.

Let's assume that a *System Failure Report* (1) documents the OPERATING ENVIRONMENT *scenarios* and *conditions* leading to a *failure event*. It includes a maintenance history record among the documents. Members of the failure analysis team *extract* the *Operating Conditions and Data* (2) from the report and incorporate *actual* data and into the Existing System Simulation (3). SEs perform analyses using Validated Field Data (4)—among which are the instrument data and a metallurgical analysis of components/residues—and they derive additional inputs and make valid assumptions as necessary.

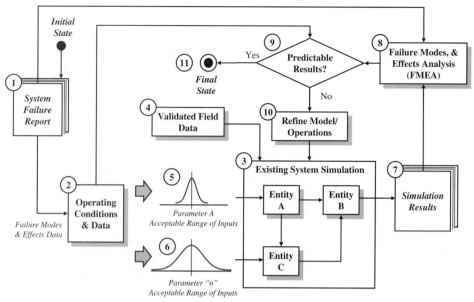

Figure 51.5 Simulation-Based Failure Mode Investigations

The failure analysis team explores all possible *actions* and rules out *probable causes* using Monte Carlo simulations and other methods. As with any failure mode investigation, the approach is based on the premise that all *scenarios* and *conditions* are suspect until they are *ruled out* by a process of fact-based *elimination*. Simulation Results (7) serve as inputs to a Failure Modes and Effects Analysis (FMEA) (8) that compares the results the scenarios and conditions identified in the *System Failure Report (1)*. If the results are not *predictable (9)*, the SEs continue to Refine the Model/Operations (10) until they are successful in duplicating the *root cause* on a predictable basis.

Application 6: Simulation-Based Training

Although simulations are used as analytical tools for technical decision making, they are also used to train system operators. Simulators are commonly used for air and ground vehicle training. Figure 51.6 provides an illustrative example.

For these applications, simulators are developed as deliverable instructional training devices to provide the *look* and *feel* of *actual* systems such as aircraft. As instructional training devices, these systems support all phases of training including: 1) briefing, 2) mission training, and 3) post-mission debriefing. From an SE perspective, these systems provide a Human-in-the Loop (HITL) training environment that includes:

1. *Briefing Stations* (3) support trainee briefs concerning the planned missions and mission scenarios.
2. *Instructor/Operator Stations (IOS)* (5) control the training scenario and environment.
3. *Target System Simulation* (1) simulates the physical system the trainee is being trained to operate.
4. *Visual Systems* (8) generate and display (9) (10) simulated OPERATING ENVIRONMENTS.
5. *Databases* (7) support visual system environments.
6. *Debrief Stations* (3) provide an instructional replay of the training mission and results.

664 Chapter 51 System Modeling and Simulation

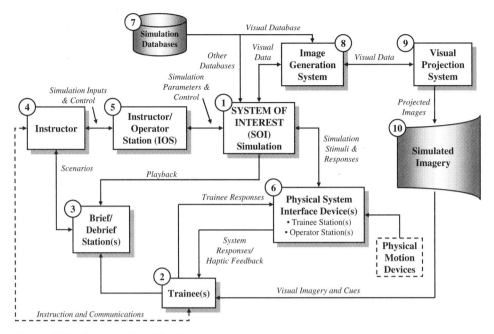

Figure 51.6 Simulation-Based Training

Training Simulator Implementation. In general, there are several types of training simulators:

- **Fixed Platform Simulators** Provide static implementation and use only visual system motion and cues to represent dynamic motion of the trainee.
- **Motion System Simulators** Employ one-, two-, or three-axis *six-degree-of-freedom* (6 DOF) motion platforms to provide an enhanced realism to a simulated training session.

One of the challenges of training simulation development is the cost related to hardware and software. Technology advances sometimes outpace the time required to develop and delivery new systems. Additionally, the capability to create an *immersive* training environment that *transcends* the *synthetic* and *physical* worlds is challenging.

One approach to these challenges is to develop a virtual reality simulator. *What is a virtual reality simulation?*

- **Virtual Reality Simulation** The employment of physical elements such as helmet visors and sensory gloves to psychologically *immerse* a subject in an audio, visual, and haptic feedback environment that creates the *perception* and *sensation* of physical reality.

Application 7: Test Bed Environments for Technical Decision Support

When we develop systems, we need *early feedback* on the downstream impacts of technical decisions. While methods such as breadboards, brassboards, rapid prototyping, and technical demonstrations enable us to reduce risk, the reality is that the effects of these decisions may not be known until the System Integration, Test, and Evaluation (SITE) Phase. Even worse, the *cost to correct* any *design flaws* or *errors* in these decisions or physical implementations increases significantly as a function of time after Contract Award.

From an engineering perspective, it would be desirable to *evolve* and *mature* models, or prototypes of a laboratory "working system," directly into the deliverable system. An approach such as this provides continuity of:

1. The evolving system design solution and its element interfaces.
2. Verification of those elements.

The question is: *HOW can we implement this approach?*
One method is to create a test bed. So, *WHAT is a Test Bed and WHY do you need one?*

Test Bed Development Environments. A *test bed* is an architectural *framework* and ENVIRONMENT that allows *simulated, emulated,* or *physical* components to be integrated as "working" representations of a physical SYSTEM or configuration item (CI) and be replaced by *actual* components as they become available. IEEE 610.12 (1990) describes a test bed as *"An environment containing the hardware, instrumentation, simulators, software tools, and other support elements needed to conduct a test."*

Test beds may reside in environmentally controlled laboratories and facilities, or they may be implemented on mobile platforms such as aircraft, ships, and ground vehicles. In general, a *test bed* serves as a mechanism that enables the virtual world of modeling and simulation to transition to the physical world over time.

Test Bed Implementation. A *test bed* is implemented with a central *framework* that *integrates* the system elements and controls the interactions as illustrated in Figure 51.7. Here, we have a Test Bed Executive Backbone (1) framework that consists of Interface Adapters (2), (5), (10) that serve as interfaces to *simulated* or *actual* physical elements, PRODUCTS A through C.

During the early stages of system development, PRODUCTS A, B, and C are MODELED and incorporated into simulations: Simulation A (4); Simulations B1 (7), B2 (9), B3 (8); and Simulation

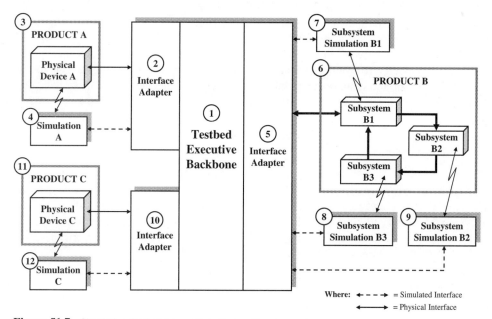

Figure 51.7 Simulation Testbed Approach to System Development

C (12). The objective is to investigate *critical operational or technical issues* (COIs/CTIs) and facilitate technical decision making. These initial simulations may be of LOW to MEDIUM *fidelity*. As the system design solution evolves, HIGHER *fidelity* models may be developed to replace the lower fidelity models, depending on specific requirements.

As PRODUCTS A, B, and C or their subelements are physically implemented as prototypes, breadboards, brassboards, and the like, the physical entities may replace simulations A through C as *plug-and-play* modules. Consider the following example:

EXAMPLE 51.11

During the development of PRODUCT B, SUBSYSTEMS B1 through B3 may be implemented as Simulation B1, B2, and B3. At some point in time SUBSYSTEM B2 is physically prototyped in the laboratory. Once the SUBSYSTEM B2 physical prototype reaches an *acceptable level* of *maturity*, Simulation B2 is *removed* and *replaced* by the SUBSYSTEM B2 prototype. Later, when the SUBSYSTEM B2 developer delivers the *verified* physical item, the SUBSYSTEM B2 prototype is replaced with the deliverable item.

In summary, a *test bed* provides a controlled framework with interface "stubs" that enable developers to integrate—"plug-and-play"—functional models, simulations, or emulations. As physical hardware (HWCI) and software configuration items (CSCIs) are *verified*, they replace the models, simulations, or emulations. Thus, over time the test bed evolves from an initial set of *functional* and *physical models* and simulation *representations* to a fully integrated and *verified* system.

Reasons That Drive the Need for a Test Bed. Throughout the System Development and the Operation and Support (O&S) phases of the *system/product life cycle*, SEs are confronted with several challenges that drive the need for using a *test bed*. Throughout this decision-making process, a mechanism is required that enables SEs to incrementally *build* a level of *confidence* in the evolving system architecture and design solution as well as to support field upgrades after deployment.

Under conventional system development, breadboards, brassboards, rapid prototypes, and technology demonstrations are used to investigate *COIs/CTIs*. Data collected from these decision aids are *translated* into design requirements—as mechanical drawings, electrical assembly drawings and schematics, and software design, for example.

The translation process is *prone* to human errors; however, integrated tool environments *minimize* the human translation errors but often suffer from format compatibility problems. Due to *discontinuities* in the design and component development workflow, the success of these decisions and implementation may not be known until the System Integration, Test, and Evaluation (SITE) Phase.

So, *how can a test bed overcome these problems*? There are several reasons why test beds can facilitate system development.

Reason 1: *Performance allocation–based decision making.* When we engineer and develop systems, *recursive* application of the SE Process Model requires *informed*, *fact-based* decision making at each level of abstraction using the most current data available. *Models* and *simulations* provide a means to *investigate* and *analyze* performance and system responses to OPERATING ENVIRONMENT scenarios for a given set of WHAT IF assumptions. The challenge is that models and simulations are ONLY as GOOD as the algorithmic *representations* used and *validated* based on *actual* field data measurements.

Reason 2: *Prototype development expense.* Working *prototypes* and *demonstrations* provide mechanisms to investigate a system's behavior and performance. However, full *pro-*

totypes for some systems may be too risky due to the MATURITY of the technology involved and expense, schedule, and security issues. The question is: *Do you have incur the expense of creating a prototype of an entire system just to study a part of it?* Consider the following example:

EXAMPLE 51.12

To study an aerodynamic problem, you may not need to physically build an entire aircraft. Model a "piece" of the problem for a given set of *boundary conditions*.

Reason 3: *System component delivery problems.* Despite insightful planning, programs often encounter late vendor deliveries. When this occurs SITE activities may severely impact contract schedules unless you have a good *risk mitigation plan* in place. SITE activities may become bottlenecked until a *critical* component is delivered. *Risk mitigation activities might* include some form of representation—simulation, emulation, or stimulation—of the missing component to enable SITE to continue to avoid interrupting the overall program schedule.

Reason 4: *New technologies.* Technology drives many decisions. The challenge SEs must answer is:

1. Is a technology as mature as its literature suggests.
2. Is this the RIGHT technology for this User's application and longer term needs.
3. Can the technology be seamlessly integrated with the other system components with minimal schedule impact.

So a test bed enables the integration, analysis, and evaluation of new technologies without *exposing* an existing system to *unnecessary* risk. For example, new engines for aircraft.

Reason 5: *Post deployment field support.* Some contracts require field support for a specific time frame following system delivery during the System Operations and Support (O&S) Phase. If the Users are planning a series of *upgrades* via *builds*, they have a choice:

1. Bear the cost of operating and maintaining a test article(s) of a fielded system for assessing incremental upgrades to a fielded configuration.
2. Maintain a test bed that allows the evaluation of configuration upgrades.

Depending on the type of system and its complexity, test beds can provide a lower cost solution.

Synthesizing the Challenges. In general, a *test bed* provides for *plug-and-play* simulations of a configuration items (CIs) or the actual physical component. Test beds are also useful for work arounds because they can minimize SITE schedule problems. They can be used to:

- Integrate early versions of an architectural configuration that is populated with simulated model *representations* (functional, physical, etc.) of configuration items (CIs).
- Establish a *plug-and-play* working test environment with prototype system components before an entire system is developed.
- Evaluate systems or configuration items (CIs) to be *represented* by *simulated* or *emulated* models that can be replaced by higher fidelity models and ultimately by the actual physical configuration item (PCI).

- Apply various *technologies* and alternative *architectural* and *design solutions* for configuration items (CIs).
- Assess *incremental* capability and performance upgrades to system field configurations.

Evolution of the Test Bed. *Test beds* evolve in a number of different ways. *Test beds* may be operated and maintained until the final deliverable system completes SITE. At that point *actual systems* serve as the basis for *incremental* or *evolutionary* development. Every system is different. So assess the cost–benefits of maintaining the test bed. All or portions of the *test bed* may be *dismantled*, depending on the development needs as well as the utility and expense of maintenance.

For some large complex systems, it may be impractical to conduct WHAT IF experiments on the ACTUAL *systems* in enclosed facilities due to:

1. Physical space requirements.
2. Environmental considerations.
3. Geographically dispersed development organizations.

In these cases it may be *practical* to keep a test bed *intact*. This, in combination with the capabilities of high-speed Internet access, may allow geographically dispersed development organizations to conduct work with a test bed without having to be physically colocated with the actual system.

51.6 MODELING AND SIMULATION CHALLENGES AND ISSUES

Although modeling and simulation offer great opportunities for SEs to *exploit* technology to understand the *problem* and *solution* spaces, there are also a number of challenges and issues. Let's explore some of these.

Challenge 1: Failure to Record Assumptions and Scenarios

Modeling and simulation requires establishing a base set of *assumptions*, *scenarios*, and *operating conditions*. Reporting modeling and simulation results without recording and noting this information in technical reports and briefings *diminishes* the *integrity* and *credibility* of the results.

Challenge 2: Improper Application of the Model

Before applying a model to a specific type of decision support task, the intended application of the model should be *verified*. There may be instances where models do not exist for the application. You may even be confronted with a model that has only a *degree of relevance* to the application. If this happens, you should take the relevancy into account and apply the results cautiously. The best approach may be to *adapt* the current model.

Challenge 3: Poor Understanding of Model Deficiencies and Flaws

Models and simulations generally evolve because an organization has an *operational need* to satisfy or resolve. Where the need to resolve *critical operational* or *technical issues* (COIs/CTIs) is immediate, the investigator may only model a segment of an application or "piece of the problem." Other Users with *different* needs may want to *modify* the model to satisfy their own "segment" needs. Before long, the model will evolve through a series of *undocumented* "patches," and then documentation *accuracy* and *configuration control* become critical issues.

To a potential user, such a model may have risks due to potential *deficiencies* or *shortcomings* relative to the User's application. Additionally, undiscovered *design flaws* and *errors* may exist because parts of the model have not been exercised. *Beware* of this problem. Thoroughly *investigate* the model before selecting it for usage. Locate the originator of the model, assuming they can be located or are available. ASK the developers WHAT you should know about the model's *performance, deficiencies,* and *flaws* that may be *untested* and *undocumented*.

Challenge 4: Model Portability

Models tend to get passed around, patched, and adapted. As a result, configuration and version control becomes a critical issue. Maintenance and configuration management of models and simulations and their associated documentation is very *expensive*. Unless an organization has a need to use a model for the long term, the item may go onto a shelf. While the *physics* and *logic* of the model may remain constant over time, the *execution* of the model on newer computer platforms may be *questionable*. This often necessitates *migration* of the model to a new computer system at a *significant* cost.

Challenge 5: Poor Model and Simulation Documentation

Models tend to be developed for *specific* rather than *general* applications. Since models and simulations are often nondeliverable items, documentation tends to get *low priority* and is often inadequate. Management decision making often follows a *"do we put $1.00 into making the M&S better or do we place $1.00 into documenting the product"* mindset. Unless the simulation is a deliverable, the view is that it is only for internal use and so *minimal* documentation is the strategy.

Challenge 6: Failure to Understand Model Fidelity

Every model and simulation has a *level of fidelity* that characterizes its performance and quality. Understand what *level of fidelity* you need, investigate the level of fidelity of the candidate model, and make a determination of utility of the model to meet your needs.

Challenge 7: Undocumented Features

Models or simulations developed as laboratory tools typically are not documented with the level of discipline and scrutiny of formal deliverables. For this reason a model or simulation may include *undocumented* "features" that the developer *forgot* to record because of the available time, budgets cuts, and the like. Therefore, you may think that you can easily *reuse* the model but *discover* that it contains *problem areas*. A *worst-case scenario* is believing and planning to use a model only to discover *deficiencies* when you are "too far down the stream" to pursue an alternative course of action.

51.7 GUIDING PRINCIPLES

In summary, the preceding discussions provide the basis with which to establish the guiding principles that govern modeling and simulation practices.

Principle 51.1   Model fidelity resides in the User's mind. HIGH fidelity to one person may be MEDIUM fidelity to a second person and LOW fidelity to a third person.

Principle 51.2   "All models are wrong but some are useful." [George E.P. Box (1979) p. 202]

51.8 SUMMARY

In our discussion of *modeling and simulation practices* we identified, defined, and addressed various types of models and simulations. We also addressed the implementation of test beds as evolutionary "bridges" that enable the virtual world of *modeling and simulation* to evolve to the *physical world.*

GENERAL EXERCISES

1. Answer each of the *What You Should Learn from This Chapter* questions identified in the *Introduction*.
2. Refer to the list of systems identified in Chapter 2. Based on a selection from the preceding chapter's General *Exercises* or a new system selection, apply your knowledge derived from this chapter's topical discussions. If you were the Acquirer of the system:
    (a) Are there *critical operational* and *technical issues* (COIs/CTIs) that drive the need to employ models and simulations to support system development? Identify the issues.
    (b) What elements of the system require modeling and simulation?
    (c) Would a test bed facilitate development of this system? HOW?
    (d) What requirements would you levy on a contractor in terms of documenting a model or simulation?
    (e) What strategy would you validate the model or simulation?
    (f) Could the models be employed as part of the deliverables operational system?
    (g) What types of system upgrades do you envision for the evolution of this system? How would a test bed facilitate evaluation of these upgrades?

ORGANIZATIONAL CENTRIC EXERCISES

1. Research your organization's command media concerning *modeling and simulation* practices.
    (a) What requirements and guidance are provided?
    (b) What requirements are imposed on documenting models and simulations?
2. How are *models and simulations* employed in your line of business and programs?
3. Contact small, medium, and large contract programs within your organization.
    (a) How do they employ models and simulations in their technical decision-making processes?
    (b) What types of models do they use?
    (c) How did the program employ models and simulations (in architectural studies, performance allocations, etc.)?
    (d) What experiences have they had in external model documentation or developing documentation for models developed internally?
    (e) What lessons learned in the employment and application of models and simulations do they suggest?
    (f) Do the programs employ test beds or use test beds of other external organizations?
    (g) Did the contract require delivery of any models or simulations used as *contract line items (CLINs)*? If so, what *Contract Data Requirements List (CDRL)* items were required, and when?

REFERENCES

DoD 5000.59-M. 1998. *DoD Modeling and Simulation (M&S) Glossary*. Washington, DC: Department of Defense. (DoD).

DSMC. 1998. *Simulation Based Acquisition: A New Approach*. Defense System Management College (DSMC) Press Ft. Belvoir, VA.

IEEE Std 610.12-1990. 1990. *IEEE Standard Glossary of Software Engineering Terminology.* New York: Institute of Electrical and Electronic Engineers (IEEE).

KOSSIAKOFF, ALEXANDER, and SWEET, WILLIAM N. 2003. *Systems Engineering Principles and Practice*. New York: Wiley-InterScience.

ADDITIONAL READING

FRANTZ, FREDERICK K. 1995. *A Taxonomy of Model Abstraction Techniques*. Computer Sciences Corporation. Proceedings of the 27th Winter Simulation Conference.

DSMC. 1998. *Simulation Based Acquisition: A New Approach*. Defense System Management College (DSMC) Press Ft. Belvoir, VA.

LEWIS, JACK W. 1994. *Modeling Engineering Systems*. Engineering Mentor series. Solana Beach, CA: HighText Publications.

Chapter 52

Trade Study Analysis of Alternatives

52.1 INTRODUCTION

The engineering and development of systems requires SEs to identify and work through a large range of *critical operational* and *technical issues (COIs/CTIs)*. These issues range from the miniscule to the complex, requiring in-depth analyses supported by models, simulations, and prototypes. Adding to the complexity, many of these decisions are interrelated. *How can SEs effectively work through these issues and keep the program on schedule?*

This section answers this question with a discussion of *trade study analysis of Alternatives (AoA)*. We:

1. Explore WHAT a trade study is and how it relates to a *trade space*.
2. Introduce a methodology for conducting a trade study.
3. Define the format for a *Trade Study Report (TSR)*.
4. Suggest recommendations for presenting trade study results.
5. Investigate challenges, issues, and risks related to performing trade studies.

We conclude with a discussion of trade study issues that SEs need to be prepared to address.

What You Should Learn from This Chapter

1. What is a *trade study*?
2. What are the *attributes* of a trade study?
3. How are trade studies conducted?
4. Who is *responsible* for conducting trade studies?
5. When are trade studies conducted?
6. Why do you need to do trade studies?
7. What is a *trade space*?
8. What *methodology* is used to perform a trade study?
9. How do you select *trade study decision factors/criteria and weights*?
10. What is a *utility function*?
11. What is a *sensitivity* analysis?

System Analysis, Design, and Development, by Charles S. Wasson
Copyright © 2006 by John Wiley & Sons, Inc.

12. What is the *work product* of a trade study?
13. How do you *document, report,* and *present trade study results*?
14. What are some of the issues and risks in conducting a trade study?

Definitions of Key Terms

- **Conclusion** A reasoned opinion derived from a preponderance of *fact-based findings* and other *objective evidence*.
- **Decision Criteria** Attributes of a decision factor. For example, if a decision factor is maintainability, *decision criteria* might include component modularity, interchangeability, accessibility, and test points.
- **Decision Factor** A key attribute of a system, as viewed by Users or stakeholders, that has a major influence on or contribution to a requirement, capability, *critical operational, or technical issue (COI/CTI)* being evaluated. Examples include elements of technical performance, cost, schedule, technology, and support.
- **Finding** A commonsense observation supported by in-depth analysis and distillation of facts and other objective data. One or more *findings* support a *conclusion*.
- **Recommendation** A logically *reasoned* plan or course of action to achieve a specific outcome or results based on a set of conclusions.
- **Sensitivity Analysis** "A procedure for testing the robustness of the results of trade-off analysis by examining the effect of varying assigned values of the decision criteria on the result of the analysis." (Source: Kossiakoff and Sweet, *System Engineering*, p. 453)
- **Trade Space** An *area of evaluation* or interest bounded by a *prescribed* set of *boundary constraints* that serve to scope the set of acceptable candidate alternatives, options, or choices for further trade study investigation and analysis.
- **Trade Study** "An objective evaluation of alternative requirements, architectures, design approaches, or solutions using identical ground rules and criteria." (Source: former MIL-STD-499)
- **Trade Study Report (TSR)** A document prepared by an individual or team that captures and presents key considerations—such as objectives, candidate options, and methodology—used to recommend a prioritized set of options or course of action to resolve a critical operational or technical issue.
- **Utility Function** A linear or nonlinear characteristic profile or value scale that represents the level of importance different stakeholders place on a system or entity attribute or capability relative to constraints established by a specification.
- **Utility Space** An area of interest bounded by *minimum* and/or *maximum* performance criteria established by a specification or analysis and a degree of utility within the performance range.
- **Viable Alternative** A candidate approach that is qualified for consideration based on its technical, cost, schedule, support, and risk level merits relative to decision boundary conditions.

Trade Study Semantics

Marketers express a variety of terms to Acquirers and Users that communicate lofty goals that SEs aspire to achieve. Terms include *best solution, optimal solution, preferred solution, solution of choice, ideal solution,* and so on. Rhetorically speaking:

- *HOW do we structure a course of action to know when we have achieved a "best solution"?*
- *WHAT is a "preferred" solution? Preferred by WHOM?*

These questions emphasize the importance of structuring a *course of action* that enables us to arrive at a consensus of what these terms mean. The mechanism for accomplishing this *course of action* is a *trade study*, which is an *analysis of alternatives* (AoA).

To better understand HOW trade studies establish a *course of action* to achieve lofty goals, let's begin by establishing the objectives of a trade study:

52.2 TRADE STUDY OBJECTIVES

The objectives of a *trade study* are to:

1. INVESTIGATE a *critical operational* or *technical issue* (COI/CTI).
2. IDENTIFY VIABLE candidate solutions.
3. EXPLORE the *fact-based* MERITS of candidate solutions relative to decision criteria derived from stakeholder requirements—via the contract, *Statement of Objectives (SOO)*, specification requirements, user interviews, cost, or schedules.
4. PRIORITIZE solution recommendations.

In general, COIs/CTIs are often too *complex* for most humans to *internalize* all of the technical details on a personal level. Adding to the complexity are the *interdependencies* among the COIs/CTIs. Proper analysis requires *assimilation* and *synthesis* of large *complex* data sets to arrive at a preferred approach that has relative value or merit to the stakeholders such as Users, Acquirer, and System Developer. The solution to this challenge is to conduct a trade study that consists of a structured *analysis of alternatives* (AoA).

Typical Trade Study Decision Areas

The hierarchical decomposition of a system into entities at *multiple levels of abstraction* and selection of physical components requires a multitude of technical and programmatic decisions. Many of these decisions are driven by the system design-to/for *objectives* and resource *constraints*.

Referral For more information about system development objectives, refer to Chapter 35 on *System Design To/For Objectives*.

If we *analyze* the *sequences* of many technical decisions, categories of trade study areas emerge across numerous system, product, or service domains. Although every system, product, or service is *unique* and has to be evaluated on its own merits, most system decisions can be characterized using Figure 52.1. Specifically, the large vertical box in the center of the graphic depicts the top-down *chain of decisions* common to most entities regardless of system *level of abstraction*.

Beginning at the top of the center box, the decision sequences include:

- Architecture trades
- Interface trades including human-machine interfaces
- Hardware/software (HW/SW) trades
- Commercial off-the-shelf (COTS)/nondevelopmental item (NDI)/new development trades
- HW/SW component composition trades
- HW/SW process and methods trades
- HW/SW integration and verification trades

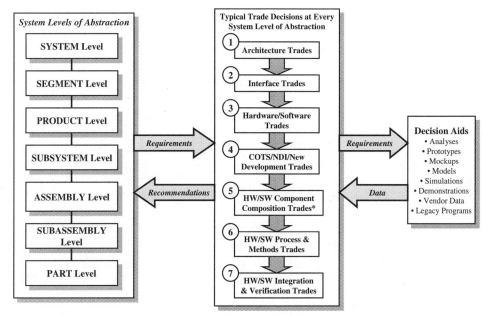

Figure 52.1 Typical Trade Study Decision Sequences

This *chain of decisions* applies to entities at every system level of abstraction—from SYSTEM, to PRODUCT, to SUBSYSTEM, and so forth, as illustrated by the left facing arrows. SEs employ *decision aids* to support these decisions, such as analyses, prototypes, mock-ups, models, simulations, technology demonstrations, vendor data, and their own experience, as illustrated by the box shown at the right-hand side. The question is: *HOW are the sequences of decisions accomplished?*

Trade Studies Address Critical Operational/ Technical Issues (COIs/CTIs)

The sequence of trade study decisions represents a basic "line of questioning" intended to facilitate the SE design solution of each entity.

1. *What type of architectural approach enables the USER to best leverage the required system, product, or service capabilities and levels of performance?*
2. *Given an architecture decision, what is the best approach to establish low risk, interoperable interfaces or interfaces to minimize susceptibility or vulnerability to external system threats?*
3. *How should we implement the architecture, interfaces, capabilities, and levels of performance? Equipment? Hardware? Software? Humans? Or a combination of these?*
4. *What development approach represents a solution that minimizes cost, schedule, and technical risk? COTS? NDI? Acquirer furnished equipment (AFE)? New development? A combination of COTS, NDI, AFE, and new development?*
5. *Given the development approach, what should the composition of the HWCI or CSCI be in terms of hardware components or software languages, as applicable?*
6. *For each HWCI, CSCI, or HWCI/CSCI component, what processes and methods should be employed to design and develop the entity?*
7. *Once the HWCI, CSCI, or HWCI/CSCI components are developed, how should they be integrated and verified to demonstrate full compliance?*

676 Chapter 52 Trade Study Analysis of Alternatives

We answer these questions through a series of technical decisions. A trade study, as an *analysis of alternatives* (AoA), provides a basis for comparative evaluation of available options based on a predefined set of decision criteria.

52.3 SEQUENCING TRADE STUDY DECISION DEPENDENCIES

Technical programs usually have a number of *COIs/CTIs* that must be resolved to enable progression to the next decision in the *chain of decisions*. If we analyze the sequences of these issues, we discover that the process of decision making resembles a tree structure over time. Thus, the branches of the structure represent decision dependencies as illustrated in Figure 52.2.

During the proposal phase of a program, the proposal team conducts preliminary trade studies to *rough out* key design *decisions* and *issues* that require more detailed attention after Contract Award (CA). These studies enable us to understand the COI or CTI to be resolved after CA. Additionally, thorough studies provide a *level of confidence* in the cost estimate, schedule, and risk—leading to an understanding of the *problem* and *solution spaces*.

Author's Note 52.1 *Depending on the type of program/contract, a trade study tree is often helpful to demonstrate to a customer that you have a logical decision path toward a timely system design solution.*

52.4 SYSTEM ARCHITECTURAL ELEMENT TRADE STUDIES

Once an entity's *problem* and *solution space(s)* are understood, one of the first tasks a team has to perform is to select an architecture. Let's suppose you are leading a team to develop a new type of vehicle. *What are the technical decisions that have to be made?* We establish a hierarchy of vehicle architecture elements as illustrated in Figure 52.3.

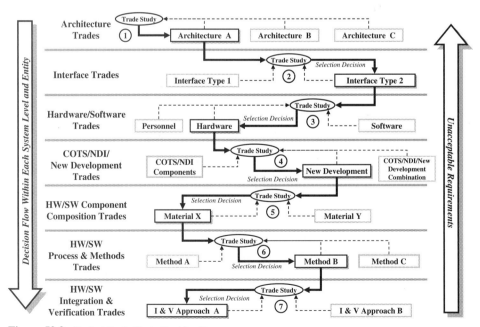

Figure 52.2 Typical Trade Study Decision Tree

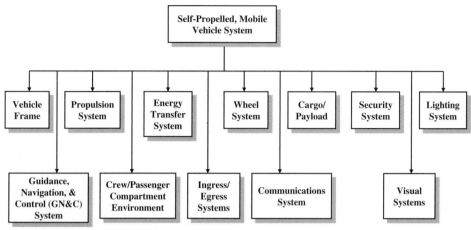

Figure 52.3 Mobile Vehicle Trade Study Areas Example

Each of these elements involves a *series* of technical decisions that form the basis for *subsequent*, lower level decisions. Additionally decisions made in one element as part of the SE process may have an *impact* on one or more other elements. Consider the following example:

EXAMPLE 52.1

Cargo/payload constraints influence decision factors and criterion used in the Propulsion System trades—involving technology and power; vehicle frame trades—involving size, strength, and materials; wheel system trades—involving type and braking; and other areas as well.

52.5 UNDERSTANDING THE PRESCRIBED TRADE SPACE

Despite the appearance that trade study efforts have the *freedom* to *explore* and *evaluate* options, there are often *limiting* constraints. These constraints bound the *area of study*, *investigation*, or *interest*. In effect the bounded area scopes what is referred to as the *trade space*.

The Trade Space

We illustrate the basic *trade space* by the diagram in Figure 52.4. Let's assume that the *System Performance Specification (SPS)* identifies specific *measures of performance (MOPs)* that can be *aggregated* into a *minimum acceptable* level of performance—by a *figure of merit (FOM)*—as noted by the vertical gray line. Marketing analyses or the Acquirer's proposal requirements indicate there is a *per unit cost ceiling* as illustrated by the horizontal line. If we focus on the area bounded by the *minimum acceptable performance* (vertical line), *per unit cost ceiling* (horizontal line), and cost–performance curve, the *bounded* area *represents* the *trade space*.

Now suppose that we conduct a trade study to evaluate candidate Solutions 1, 2, 3, and 4. We construct the cost–performance curve. To ensure a level of objectivity, we *normalize* the per unit cost ceiling to the Acquirer maximum requirement. We plot cost and relative performance of each of the candidate Solutions 1, 2, 3, and 4 on the *cost–performance* curve.

By *inspection* and *comparison* of plotted cost and technical performance relative to required performance:

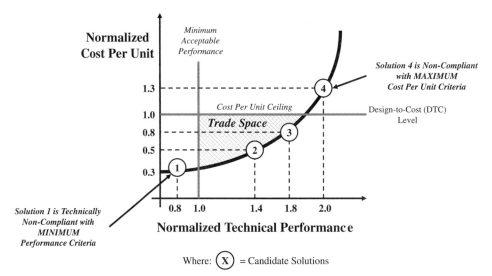

Figure 52.4 Candidate Trade Space Zone

- Solutions 1 and 4 fall outside the trade space.
- Solution 1 is *cost compliant* but *technically noncompliant*.
- Solution 4 is *technically compliant* but *cost noncompliant*.

When this occurs, the *Trade Study Report (TSR)* documents that Solutions 1 and 4 were considered and determined by analysis to be *noncompliant* with the trade space decision criteria and were *eliminated* from consideration.

Following elimination of Solutions 1 and 4, Solutions 2 and 3 undergo further analysis to thoroughly evaluate and score other considerations such as organizational risk.

Optimal Solution Selection

The previous discussion illustrates the basic concept of a two-dimensional *trade space*. A *trade space*, however, is *multidimensional*. For this reason it is more aptly described as a multidimensional *trade volume* that encompasses technical, life cycle cost, schedule, support, and risk decision factors.

We can illustrate the *trade volume* using the graphic shown in Figure 52.5. To keep the diagram simple, we constrain our discussion to a three-dimensional model representing the convolution of technical, cost, and schedule factors. Let's explore each dimension represented by the trade volume.

- **Performance–Schedule Trade Space** The graphic in the upper left-hand corner of the diagram *represents* the performance vs. schedule trade space. Identifier 1 marks the location of the selected performance versus schedule solution.
- **Performance–Cost Trade Space** The graphic in the upper right-hand corner includes *represents* the performance–cost trade space. Identifier 2 marks the location of the selected performance versus cost solution.
- **Cost–Schedule Trade Space** The graphic in the lower right-hand corner of the diagram *represents* the Cost–Schedule *trade space*. Identifier 3 marks the location of the selected cost versus schedule solution.

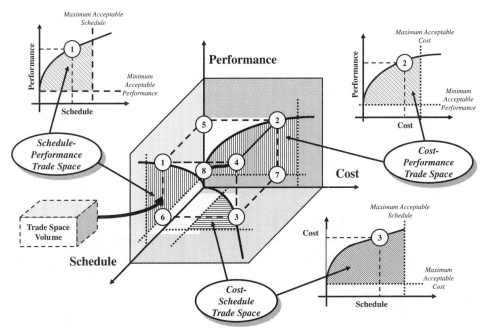

Figure 52.5 Trade Space Interdependencies

If we *convolve* these trade spaces and their boundary constraints into a three-dimensional model, the cube in the center of the diagram results.

The *optimal* solution selected is *represented* by the intersection of orthogonal lines in their respective planes. Conceptually, the *optimal* solution would lie on a curve that *represents* the *convolution* of the performance–schedule, performance–cost, and cost–schedule curves. Since each plane includes a restricted trade space, the integration of these planes into a three-dimensional model results in a *trade space volume*.

52.6 THE TRADE STUDY PROCESS

Trade studies consist of *highly iterative* steps to analyze the issues to be resolved into a set prioritized recommendations. Figure 52.6 *represents* a basic Trade Study Process and its process *steps*. Let's briefly examine the process through each of its steps.

Process Step 1: Define the trade study objective(s).
Process Step 2: Identify decision stakeholders.
Process Step 3: Identify trade study individual or team.
Process Step 4: Define the trade study decision factors/criteria.
Process Step 5: Charter the trade study team.
Process Step 6: Review the Trade Study Report (TSR)
Process Step 7: Select the preferred approach.
Process Step 8: Document the decision.

680 Chapter 52 Trade Study Analysis of Alternatives

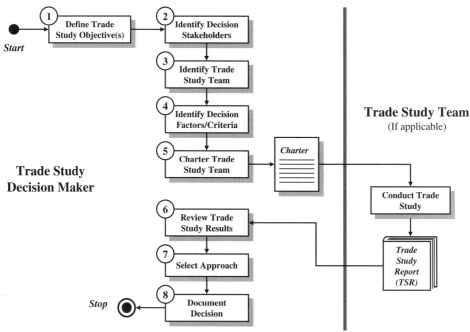

Figure 52.6 Trade Study Process

Guidepost 52.1 *Our discussion has identified the overall Trade Study Process. Now let's focus our attention on understanding the basic methodology that will be employed to conduct the trade study.*

52.7 ESTABLISHING THE TRADE STUDY METHODOLOGY

Objective technical and scientific investigations require a methodology for making decisions. The methodology facilitates the development of *strategy*, *course of action*, or "roadmap" of the planned technical approach to *investigate*, *analyze*, and *evaluate* the candidate solution approaches or options. Methodologies, especially proven ones, keep the study effort on track and prevent unnecessary excursions that *consume* resources and yield no productive results.

There are numerous ways of establishing the *trade study methodology*. Figure 52.7 provides an illustrative example:

Step 1: Understand the problem statement.
Step 2: Define the evaluation decision factors and criteria.
Step 3: Weight decision factors and criteria.
Step 4: Prepare utility function profiles.
Step 5: Identify candidate solutions.
Step 6: Analyze, evaluate, and score the candidate options.
Step 7: Perform sensitivity analysis.
Step 8: Prepare the Trade Study Report (TSR).
Step 9: Conduct peer/subject matter expert (SME) reviews.
Step 10: Present the TSR for approval.

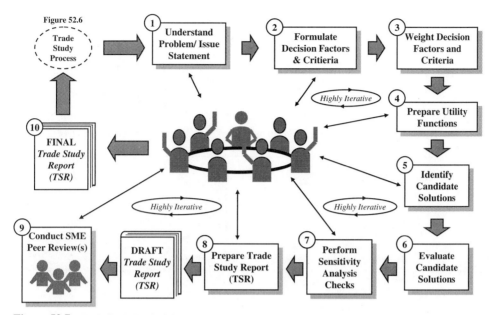

Figure 52.7 Trade Study Methodology

Guidepost 52.2 *At this point we have established the basic trade study methodology. On the surface the methodology is straightforward. However, HOW do we evaluate alternatives that have degrees of utility to the stakeholder? This brings us to a special topic, trade study functions.*

52.8 TRADE STUDY UTILITY FUNCTIONS

When scoring some decision factors and criteria, the natural *tendency* is to do so on a linear scale such as 1–5 or 1–10. This method *assumes* that the User's *value scale* is linear; in many cases it is *nonlinear*. In fact, some candidate options data have *levels of* utility. One way of addressing this issue is to employ *utility functions*.

Understanding the Utility Function and Space

The *trade space* allows us to sort out *acceptable* solutions that fall within the boundary constraints of the *trade space*. Note we used the term *acceptable* as in the context of satisfying a minimum/maximum threshold. The reality is some solutions are, by a *figure of merit (FOM)*, better than others. We need a means to express the *degree of utility* mathematically. Figure 52.8 provides examples of HOW Users might establish utility function profiles. To see this point better, consider the following example:

EXAMPLE 52.4

A User requires a vehicle with a *minimum* speed within a mission area of 50 miles per hour (mph) under specified operating environment conditions. Mission analysis, as validated by the User, indicates that 64.0 mph is the *maximum* speed required. Thus, we can state that the *minimum* utility to the User is 50 mph and the *maximum* utility is 64.0 mph.

682 Chapter 52 Trade Study Analysis of Alternatives

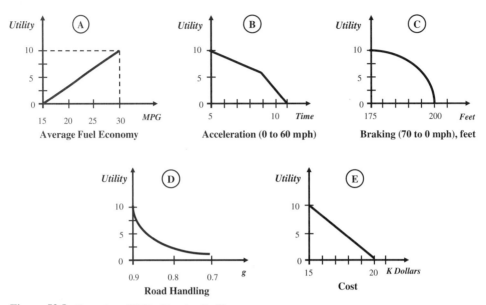

Figure 52.8 Examples of Utility Function Profiles
Source: Adapted from *NASA System Engineering "Toolbox" for Design-Oriented Engineers*, Figure 2-1 "Example utility functions"; p. 2-7.

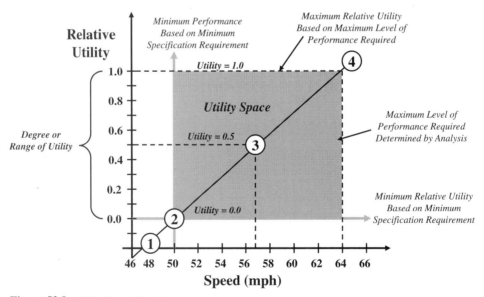

Figure 52.9 Utility Space Illustration

Assigning the Relative Utility Value Range. Since utility represents the value profile a User places on an attribute, we assign the *minimum* utility a value of 0.0 to represent the *minimum* performance requirement—which is 50 mph. We assign a utility value of 1.0 to represent the *maximum* requirement—which is 64.0 mph. The net result is the establishment of the *utility space* as indicated by the shaded area in Figure 52.9.

Determining Candidate Solution Utility. Once the *utility range* and *space* are established, the relative utility of candidate options can be evaluated. Suppose that we have four candidate vehicle solutions—1, 2, 3, and 4—to consider.

- Vehicle 1 has a *minimum* speed of 48 mph.
- Vehicle 2's *minimum* speed is 50 mph—the threshold specification requirement.
- Vehicle 3's *minimum* speed is 57 mph.
- Vehicle 4's *minimum* speed is 65 mph.

So we assign to each vehicle the following utility values relative to the minimum specification requirement:

1. Vehicle 1 = *unacceptable* and *noncompliant*
2. Vehicle 2 at 50 mph = utility value of 0, the *minimum* threshold
3. Vehicle 3 at 57 mph = utility value of 0.5
4. Vehicle 4 = exceeds the *maximum* threshold and therefore has a utility value of 1.0.

This approach creates several issues:

First, if Vehicle 1 has a *minimum* speed of 48 mph, does this mean that it has a utility value of <0.0 (i.e., *disutility*) or 0? The answer is no, because we assigned 0.0 to be the *minimum* specification requirement of 50 mph which vehicle 2 meets.

Second, if Vehicle 4 exceeds the *maximum* speed requirement, do we assign it a utility value of 1.0+ (i.e., >1.0), or do we *maximized* its utility at 1.0? The answer depends on whether vehicle 4 already exists or will be developed. You generally are not paid to *overdevelop* a system beyond its required capabilities—in this case, 64 mph.

Third, if we apply the utility value to the trade study scoring criteria (decision factor × weight × utility value), HOW do we deal with a system such as Vehicle 4 that has a utility value of 0.0 but meets the *minimum* specification requirement?

Utility Value Correction Approach 1

In the preceding example we started with good intentions—to find *value-based* decision factors via utility functions—but have created another problem. *How do we solve it?* There are a couple of solutions to *correct* this situation.

One approach is to simply establish a utility value of 1.0 to represent the *minimum* specification requirement. This presents an issue. In the example Vehicle 1 has a *minimum* speed of 48 mph under specified operating conditions. If a utility value of 1.0 represents the *minimum* performance requirement, Vehicle 1 will have a utility value of −0.2.

Simply applying this utility value infers *acceptance* as a *viable* option and allows it to continue to be evaluated in a trade study evaluation matrix. Our intention is to eliminate *noncompliant solutions*—which is to remove Vehicle 1 from consideration. Thus, if a solution is *unacceptable*, it should have a utility value of 0.0. This brings us to Approach 2.

Utility Value Correction Approach 2

Another utility correction approach that overcomes the problems of Correction Approach 1 involves a hybrid *digital* and an *analog* solution. Rather than IMMERSING ourselves in the mathematical concepts, let's simply THINK about what we are attempting to accomplish.

The reality is that either a candidate option satisfies a *minimum/maximum* requirement or it doesn't. The result is digital: 1 = meets requirement, and 0 = does not meet requirement.

This then leads to the question: *If an option meets the requirement, how well does it meet the requirement*—meaning an analog result? We can state that more concisely as:

$$\text{Utility value} = \text{Digital utility} + \text{Analog utility} \qquad (52.3)$$

where:

Digital utility (DU) = 1 or 0

Analog utility (AU) = 0.0 to 1.0 (variable scale)

In summary, *should you use utility functions in your trade studies*? The decision depends on a case-by-case basis. In general, the preceding discussion simply provides a means of *refining* and delineating the *degree of utility* for specific capabilities relative to a User's mission application.

52.9 SENSITIVITY ANALYSIS

Although trade studies are intended to yield THE best answer that stands out for a given set of decision criteria, many times they do not. Often the data for competing alternatives are clustered together. *How do you decluster the data to resolve this dilemma?*

Theoretically, we perform a *sensitivity analysis* of the data. The *sensitivity analysis* enables us to vary any one decision factor weight by some quantity such as 10% to *observe* the *effects* on the decision. However, this may or may not decluster the data. So, let's take another approach.

Better Sensitivity Analysis Approach

A better approach to differentiating *clustered* trade study data resides in *selection* of the decision factors and criteria. When we initially identified decision factors/criteria with the stakeholders, chances are there was a board range of factors. To keep things simple, assume we arbitrarily selected 6 criteria from a set of 10. Weight each criterion. As a result, the *competing* solutions became clustered.

The next logical step is to *factor* in the $n + 1$ criterion and *renormalize* the weights based on earlier ranking. Then, you continue to factor in additional criteria until the data decluster. Recognize that if the $n + 1$ has a relative weight of 1%, it may not significantly influence the results. This leaves two options:

Option A: Make a judgmental decision and pick a solution.

Option B: Establish a ground rule that the initial selection of decision criteria should not constitute more than 90% or 95% of the total points and scale the list to 100%. This effectively leaves 5% to 10% for the $n + 1$ or higher terms that may have a level of significance on the outcome. As each new item is added back, rescale the weights relative to their initial weights within the total set.

52.10 TRADE STUDY REPORTS (TSRs)

On completion of the trade study analysis, the next challenge is being able to articulate the results in the *Trade Study Report (TSR)*. The TSR serves as a *quality record* that documents accomplishment of the chartered or assigned task.

Why Document the Trade Study?

A common question is WHY document a trade study? There are several reasons:

First, your *Contract Data Requirements List (CDRL)* or organization command media may require that you to document trade studies.

Second, trade studies formally document key *decision criteria*, *assumptions*, and *constraints* of the trade study environment for posterity. Since SE, as a *highly iterative* process, requires decision making based on given conditions and constraints, those same conditions and constraints can change quickly or over time. Therefore, you or others could have to *revisit* previous trade study decisions to investigate HOW the changing *conditions* or *constraints* could have impacted the decision or *course of action* such that corrective actions should be initiated.

Third, as a professional, document key decisions and rationale as a matter of disciplinary practice.

Trade Study Documentation Formality

Trade studies are documented at various *levels of formality*. The level of formality ranges from simply recording the considerations and deliberations in a SE's engineering notebook to formally approved and published reports intended for wide distribution. ALWAYS check your contract, local organization command media, and/or program's Technical Management Plan (TMP) for explicit direction concerning the *level of formality* required. At a *minimum*, document the key facts of the trade study in a personal engineering notebook.

Preparing the TSR

There are numerous ways of preparing the TSR. First and foremost, ALWAYS consult your contract or organizational command media for guidance. If there are no specific outline requirements, consider using or tailoring the outline provided below:

1.0 INTRODUCTION
    1.1 Scope
    1.2 Authority
    1.3 Trade Study Team Members
    1.4 Acronyms and Abbreviations
    1.5 Definitions of Key Terms

2.0 APPLICBLE DOCUMENTS
    2.1 Acquirer (role) Documents
    2.2 System Developer (role) Documents
    2.3 Vendor Documents

3.0 EXECUTIVE SUMMARY
    3.1 Trade Study Objective(s)
    3.2 Trade Study Purpose
    3.3 Trade Space Boundaries
    3.4 Findings
    3.5 Conclusions
    3.6 Recommendations
    3.7 Other Viewpoints
    3.8 Selection Risks and Impacts

4.0 TRADE STUDY METHODOLOGY

    4.1 Step 1

    4.2 Step 2

    ⋮

    4.z Step z

5.0 DECISION CRITERIA, FACTORS, AND WEIGHTS

    5.1 Selection of Decision Factors and Criteria

    5.2 Selection of Weights for Decision Factors or Criteria

7.0 EVALUATION AND ANALYSIS OF ALTERNATIVES

    7.1 Option A

    7.2 Option B

    ⋮

    7.x Option x

8.0 FINDINGS AND CONCLUSIONS *(optional)*

9.0 RECOMMENDATIONS *(optional)*

APPENDIXES SUPPORTING DATA

    A Data Item 1

    B Data Item 2

    ⋮

Warning: Proprietary, Copyrighted, and Export Controlled Information Most vendor literature is *copyrighted* or deemed *proprietary*. So *avoid* reproducing and/or posting any copyrighted information unless you have the expressed, written permission from the owner/vendor to reproduce and distribute the material. ALWAYS establish proprietary data exchange agreements before accepting any proprietary vendor data.

As stated previously, some vendor data may be EXPORT controlled and subject to the US International Traffic and Arms Regulations (ITAR). So ALWAYS consult with your program, Contracts, legal, and Export Control organizations before disseminating technical information that may be subject to this constraint.

Proof Check the TSR Prior to Delivery

Some Trade Study Teams develop powerful and compelling trade studies only to have the effort falter due to *poor writing* and communications skills. When the TSR is prepared, thoroughly edit it to ensure *completeness* and *consistency*. Perform a spell check on the document. Then, have peers review the document to see that it is self-explanatory, consistent, and does not contain *errors*. Make sure the deliverable TSR reflects the *professionalism*, *effort*, and *quality* of effort contributed to the trade study.

Presenting the Trade Study Results

Once team members prepare and approve the *Trade Study Report (TSR)*, present the trade study results. There are a number of approaches for presenting TSR results. The approaches generally include delivery of the TSR as a document, briefings, or combination.

Trade Study Report (TSR) Submittal. For review, approval, and implementation, deliver the TSR directly to the *chartering decision authority* that commissioned the study. The TSR should always include a cover letter prepared by the Trade Study Team lead and reviewed for concurrence by the team.

TSRs can be delivered via the mail or by personal contact. It is advisable that the Trade Study Team lead or team, if applicable, *personally* deliver the TSR to the decision authority. This provides an opportunity to *briefly* discuss the contents and recommendations.

During the meeting the decision authority may *solicit* team recommendations for disseminating the TSR to stakeholders. If a meeting forum is selected to present a TSR briefing, the date, time, and location should be *coordinated* through *notification* to the stakeholders and Trade Study Team members.

Advance Review of the Trade Study Report. The *Trade Study decision authority*—such as the Technical Director, Project Engineer, or System Engineering and Integration Team (SEIT)—may request an *advance* review of the TSR by *stakeholders* prior to the TSR presentation and discussion. If a decision is expected at the meeting, advance review of the TSR enables the *stakeholders* to come prepared to:

1. Address any open questions or concerns
2. Make a decision concerning the recommendations.

TSR Briefings. TSR briefings to stakeholders can be *helpful* or a *hindrance*. They are helpful if additional clarification is required. Conversely, if the presenter does a poor job with presentation, the level of *confidence* in the TSR may be questioned. Therefore, BE PREPARED.

52.11 TRADE STUDY RISK AREAS

Trade studies, like most decisions, have a number of risk areas: let's explore a few examples.

Risk Area 1: Test Article Data Collection Failures

When *test articles* are on *loan* for technical evaluation, *failures* may occur and PRECLUDE completion of data collection within the allowable time frame. Because of the limited time for the trade study, replacement of the test article(s) may not be *practical* or *feasible*. *Plan for contingencies and mitigate their risks!*

Risk Area 2: Poor or Incorrect Assumptions

Formulation of candidate solutions often requires a set of dependencies and assumptions—such as availability of funding or technology. *Stakeholders* often challenge trade study results because *poor* or *incorrect assumptions* were made by the trade study team. Where *appropriate* and necessary, *discuss* and *validate* assumptions with the decision authority to preclude consuming resources developing a decision that was flawed due to *poor* or *incorrect assumptions* from the start.

Risk Area 3: Data Validity

Technical decision making must be accomplished with the latest most *complete*, *accurate*, and *precise* data available. Authenticate the *currency*, *accuracy*, and *precision* of all data as well as vendor *commitment* to stand behind the *integrity* of the data.

Risk Area 4: Selection Criteria Relevancy

Occasionally selection criteria that have little or no contribution to the selection focus and objective get onto the list of *Decision Factors* or *Criteria*. Scrutinize the *validity* of factors and selection criteria. Document the supporting *rationale*.

Risk Area 5: Overlooked Selection Criteria

Sometimes there are *critical operational* or *technical issues* (COIs/CTIs) attributes that do not make the Selection Criteria List. Selection criteria *checks and balances* should include verification of traceability of selection criteria to the COIs/CTIs to be resolved.

Risk Area 6: Failure to Get the User "Buy In"

Contrary to popular opinion, customer satisfaction does not begin at delivery of the system. The process begins at Contract Award. Toward this end, keep the User involved as much as practical in the technical decision-making process to provide the foundation for positive delivery satisfaction. When high-level trade studies are performed that have an impact on system capabilities, interfaces, and performance, *solicit* User *validation* of Selection Decision Factors and Criteria and their respective weights. Give the User some level of *ownership* of the system/product, starting at Contract Award.

Risk Area 7: Unproven or Invalid Methodology

Trade study success begins with a strong, robust strategy and methodology that will *withstand professional scrutiny*. Flaws in the methodology *influence* and *reduce* the *integrity* and *effectiveness* of the trade study. Solicit peer reviews by trusted colleagues to ensure that the trade study begins on the RIGHT track and yields results that will withstand professional scrutiny by the organization, Acquirer, User, and *professional* community, as applicable.

Risk 8: Scaling the Trade Study Task Activities to Resource Constraints

As with most SE tasks, you may not always have an *adequate* amount of time to perform trade studies. Yet, the results are *expected* to be *professionally* and *competently* accomplished.

Whatever time frame you have available, assuming it is *reasonable* and *practical*, the key *results* and *findings*, in general, need to be *comparable* whether you have one day or one week. If you have one day, the decision authority gets a one-day trade study and data; one month gets a month's level of analysis and data. So, *HOW do you deal with the time constraints*? The level of detail, research, analysis, and reporting may have to be *scaled* to the available time.

52.12 TRADE STUDY SUGGESTIONS

Although trade studies are intended to resolve *critical operational* or *technical* (COI/CTI) *issues*, one of their ironies is they sometimes create their own set of issues related to scope, context, conduct, and reporting. Here are a few suggestions to consider based on issues common to many trade studies:

Suggestion 1: Select the right methodology.
Suggestion 2: Adhere to the methodology.
Suggestion 3: Select the right decision factors, criteria, and weights.

Suggestion 4: Avoid predetermined trade study decisions.
Suggestion 5: Establish acceptable data collection methods.
Suggestion 6: Ensure data source integrity and credibility.
Suggestion 7: Reconcile inter-COI/CTI dependencies.
Suggestion 8: Accept/reject trade study report recommendations.
Suggestion 9: Document the trade study decision.
Suggestion 10: Create trade study report credibility and integrity.
Suggestion 11: Respect trade study dissenting technical opinions.
Suggestion 12: Maintain trade study reports.

Concluding Point

The perception of *clear-cut options* available for selection is sometimes *deceiving*. The reality is none of the options may be *acceptable*. There may even be instances where the trade study may lead to yet another option that is based on *combinations* of the options considered or options not considered. Remember, the trade study process is NOT intended to answer: Is it A, B, or C. The objective is to determine the *best solution* given a set of prescribed decision criteria. That includes other options that may not have been identified prior to the start of the trade study.

52.13 GUIDING PRINCIPLES

In summary, the preceding discussions provide the basis with which to establish the guiding principles that govern trade study practices.

Principle 52.1 An undocumented trade study is nothing more than personal opinion.

Principle 52.2 Trade studies are only as valid as their task constraints and underlying assumptions, methodology, and data collection. Document and preserve the integrity of each.

Principle 52.3 A trade study team without a methodology is prone to wander aimlessly.

Principle 52.4 Analyses investigate a specific condition, state, circumstances, or "cause-and-effect" relationships; trade studies analyze alternatives and propose prioritized recommendations.

Principle 52.5 (Wasson's Task Significance Principle) The amount of time allowed for task performance and completion is often inversely proportional to the task's level of significance to the deliverable system, User, Acquirer, or organization.

52.14 SUMMARY

During our discussion of trade study practices, we defined what a trade study is, discussed why trade studies are important and should be documented, and outlined the basic trade study process and methodology. We also addressed methods for documenting and presenting the TSR results.

GENERAL EXERCISES

1. Answer each of the *What You Should Learn from This Chapter* questions identified in the *Introduction*.
2. Refer to the list of systems identified in Chapter 2. Based on a selection from the preceding chapter's General *Exercises* or a new system, selection, apply your knowledge derived from this chapter's topical discussions. Specifically identify the following:
   (a) *Critical operational and technical issues (COIs/CTIs)* that had to be resolved.
   (b) Based on your own needs, what decision factors and criteria would you establish for each COI/CTI? How would you weight them?

ORGANIZATIONAL CENTRIC EXERCISES

1. Your management has decided to procure a specific computer for an application and wants you to develop a trade study justifying the selection and present the results to corporate headquarters. How would you approach the situation?
2. Contact two or three contract programs within you organization. Interview the Technical Director or Project Engineer and SEs regarding what approaches were used to perform trade studies. Identify the following for each program, report and discuss your findings with peers:
   (a) What methodology steps were used?
   (b) How were decision criteria determined?
   (c) How were decision criteria weights established?
   (d) How many candidates were evaluated?
   (e) Was a detailed analysis performed on each candidate or only on the final set of candidates?
   (f) What type of sensitivity analysis was employed to decluster candidates, if applicable?
   (g) What lessons did the program learn from the trade studies? What worked/didn't work?

REFERENCES

Kossiakoff, Alexander, and Sweet, William N. 2003. *Systems Engineering Principles and Practice.* New York: Wiley-InterScience.

MIL-STD-499B. 1994. (canceled draft). *Systems Engineering.* Washington, DC: Department of Defense (DoD).

National Aeronautics and Space Administration (NASA). 1994. *Systems Engineering "Toolbox" for Design-Oriented Engineers.* NASA Reference Publication 1358. Washington, DC.

ADDITIONAL READING

US Federal Aviation Administration (FAA), ASD-100 Architecture and System Engineering. 2003. *National Airspace System—Systems Engineering Manual.* Washington, DC: FAA.

Chapter 53

System Verification and Validation

53.1 INTRODUCTION

For most non–research and development (R&D) programs, there are two options to engineering and developing systems:

- Option 1 Employ the hobbyist approach based on the BUILD, TEST, FIX paradigm *"until we get it right"* philosophy.
- Option 2 Do the job RIGHT the first time.

Stockholders, corporations, and Acquirers, among many others, want to know "up front" that their money is going to be applied *efficiently* and *effectively* within cost and schedule constraints. From the corporation's perspective, this means *winning* contracts, *surviving*, *growing* its business, *increasing* shareholder value, and *achieving* a return on the investment (ROI).

In a highly competitive marketplace, organizations wrestle over every contract to answer many key questions. Consider the following examples:

1. *How do we assess and maintain the technical integrity of the evolving system design solution?*
2. *How do we avoid expensive "fixes" and "retrofits" in the field after system delivery due to latent defects?*
3. *How can we reduce the cost of maintenance due to correct latent defects?*
4. *How do we institute controls to protect our investment in the system development through reduction of defects and errors?*
5. *How can we reduce development cost, schedule, technical, technology, and support risks?*
6. *How do we validate that the specified system will meet the User's intended operational needs?*

So, *HOW do SEs get system development RIGHT the first time while satisfying these questions*? You institute a series of ongoing *verification* and *validation* activities throughout the System Development Phase. These activities are deployed at *staging* or *control points*—the major milestones and reviews. *Why is this necessary?* To ensure that the evolving Developmental Configuration progresses toward maturity with an *acceptable* level of risk to the stakeholders and is *compliant* with the *System Performance Specification (SPS)*, contract cost and schedule constraints, and ultimately satisfies the User's *validated* operational needs.

System Analysis, Design, and Development, by Charles S. Wasson
Copyright © 2006 by John Wiley & Sons, Inc.

Definitions of Key Terms

- **Analysis** (Verification Method) "Use of analytical data or simulations under defined conditions to show theoretical compliance. Used where testing to realistic conditions cannot be achieved or is not cost-effective." (Source: INCOSE SE Handbook Version 2.0, July 2000, para. 4.5.18 Verification Analysis, p. 275)

- **Certification** "Refers to verification against legal and/or industrial standards by an outside authority without direction to that authority as to how the requirements are to be verified. Typically used in commercial programs. For example, this method is used for CE certification in Europe, and UL certification in the US and Canada. Note that any requirement with a verification method of "certification" is eventually assigned one or more of the four verification methods (inspection, analysis, demonstration, or test)." (Source: INCOSE *SE Handbook* Version 2.0, July 2000, para. 4.5.18 Verification Analysis, p. 275)

- **Classification of Defects** "The enumeration of possible defects of the unit or product, classified according to their seriousness. Defects will normally be grouped into the classes of critical, major or minor: however, they may be grouped into other classes, or into subclasses within these classes." (Source: Former MIL-STD-973, para. 3.7)

- **Deficiency** "1. Operational need minus existing and planned capability. The degree of inability to successfully accomplish one or more mission tasks or functions required to accomplish a mission or its objectives. Deficiencies might arise from changing mission objectives, opposing threat systems, changes in the environment, obsolescence, or depreciation in current military assets. 2. In contract management—any part of a proposal that fails to satisfy the government's requirements." (Source: *DSMC Defense Acquisition Acronyms and Terms*, 10th edition, 2001.)

    Deficiencies consist of two types:
    1. "Conditions or characteristics in any item which are not in accordance with the item's current approved configuration documentation." (Source: Former MIL-STD-973 para. 3.28)
    2. "Inadequate (or erroneous) item configuration documentation, which has resulted, or may result, in units of the item that do not meet the requirements for the item." (Source: Former MIL-STD-973 para. 3.28)

- **Demonstration** (Verification Method) "A qualitative exhibition of functional performance, usually accomplished with no or minimal instrumentation." (Source: INCOSE SE *Handbook* Version 2.0, July 2000, para. 4.5.18 Verification Analysis, p. 275)

- **Deviation** Refer to the definition in Chapter 28 on *System Specification Practices*.

- **Discrepancy** A statement highlighting the variance between *what exists* and *minimum requirements* for standard process, documentation, or practice performance compliance.

- **Independent Verification and Validation (IV&V)** "Verification and validation performed by an organization that is technically, managerially, and financially independent of the development organization." (Source: IEEE 610.12-1990 *Standard Glossary of Software Engineering Terminology*)

- **Inspection** (Verification Method) "Visual examination of the item (hardware and software) and associated descriptive documentation which compares appropriate characteristics with predetermined standards to determine conformance to requirements without the use of special laboratory equipment or procedures." (Source: Adapted from *DSMC Glossary*: *Defense Acquisition Acronyms and Terms*)

- **Similarity** (Verification Method) The process of demonstrating, by traceability to source documentation, that a *previously* developed and verified SE design or item applied to a new program complies with the same requirements thereby eliminating the need for design level reverification.
- **Test** (Verification Method) The act of executing a formal or informal scripted procedure, measuring and recording the data and observations, and comparing to expected results for purposes of evaluating a system's response to a specified stimuli in a prescribed environment with a set of constraints and initial conditions.
- **Validation** The act of assessing the requirements, design, and development of a work product to assure that it will meet the User's operational needs and expectations at delivery.
- **Verification** "The process of evaluating a system or component to determine whether the products of a given development phase satisfy the conditions imposed at the start of that phase. Formal proof of program correctness." (Source: IEEE 610.12-1990 *Standard Glossary of Software Engineering Terminology*)
- **Verification and Validation (V&V)** "The process of determining whether the requirements for a system or component are complete and correct, the products of each development phase fulfill the requirements or conditions imposed by the previous phase, and the final system or component complies with specified requirements." (Source: IEEE 610.12-1990 *Standard Glossary of Software Engineering Terminology*)
- **Waiver** Refer to the definition in Chapter 28 on *System Specification Practices*.

53.2 SYSTEM VERIFICATION AND VALIDATION OVERVIEW

Verification and *validation* are intended to satisfy some very critical program needs and questions that serve as the basis for V&V objectives as illustrated in Table 53.1.

Based on this introduction, let's begin our first discussion topic by *correcting the V&V myth*.

Correcting the V&V Myth

The first thing you should understand is that V&V activities are performed throughout the system/product life cycle. V&V activities begin at Contract Award and follow through contract delivery *system acceptance* at the end of the System Development Phase.

Many people erroneously believe that V&V activities are only performed at the end of a program as part of system acceptance. Although formal V&V activities include acceptance *tests* and *field trials*, System Developers initiate V&V activities at the time of Contract Award and continue throughout all segments of the System Development Phase as shown in Figure 24.3. V&V is performed at all system *levels of abstraction* and on each entity within a level.

Under ISO 9000, technical plans state how multi-disciplined system engineering and development are to be accomplished, tasks to be performed, schedule, and work products to be produced. V & V activities employ these work products at various stages of completion to assess compliance of the evolving system design solution to technical plans, tasks, and specifications.

53.3 SYSTEM VERIFICATION PRACTICES

Verification encompasses all of the System Development Phase activities from Contract Award through system acceptance. This includes Developmental Test and Evaluation (DT&E) activities such as technology validation, manufacturing process proofing, quality assurance and acceptance, as well as the Operational Test and Evaluation (OT&E).

Table 53.1 V&V solutions to program challenges

Challenge	Program Challenge	V&V Objective
1	How do we preserve the technical integrity of the evolving system design solution?	Conduct periodic Requirements Traceability Audits (RTAs).
2	How do we avoid expensive "fixes" and "retrofits" in the field after system delivery?	Identify and correct deficiencies and defects "early" to avoid increased "downstream" development and operational costs and risks.
3	How do we ensure the specified system will meet the User's validated operational needs?	Coordinate and communicate with the User to ensure that expected system outcomes, requirements, and assumptions are valid.
4	How can we reduce technical, technology, and support risks?	Institute crosschecks of analyses, trade studies, demonstrations, models, and simulation results.
5	How do we protect our investment in the system development?	Conduct technical reviews and audits. Require periodic risk assessments. Conduct proof of concept or technology demonstrations

Importance of System Verification

System verification provides *incremental*, OBJECTIVE *evidence* that the evolving, multi-level system design solution, as captured by the Developmental Configuration, is progressing to maturity. COMPLIANCE *verification*, in turn, provides a level of confidence in meeting the planned capabilities and levels of performance.

Verification Tasks, Actions, and Activities

Verification tasks and actions, which apply to all facets of system development, include: 1) analyses, 2) design, 3) technical reviews, 4) procurement, 5) modeling and simulation, 6) technology demonstrations, 7) tests, 8) deployment, 9) operations, 10) support, and 11) disposal. Verification tasks enable technical programs to evaluate: risk assessments; people, product, and process capabilities; compliance with requirements; proof of concept, and so forth.

53.4 WHAT DO PROGRAMS VERIFY?

Most engineers use the term *verify* casually and interchangeably with *validation* without understanding the scope of its context. The key question is: *WHAT is being verified?* The answer resides in the System Development Phase segments of the system/product life cycle.

The key segments of the System Development Phase are illustrated in Figure 24.3 and include System Engineering Design, Component Procurement and Development, System Integration and Test, System Verification, Operational Test and Evaluation (OT&E), and Authentication of System Baselines. Let's investigate WHAT is being *verified* during the System Development Phase of the system/product life cycle.

System Design Segment Verification

During the SE Design Segment, multiple levels of the SE design solution are *verified*—via document reviews, technical reviews, prototypes and technology demonstrations, models and simulations, and requirements traceability—for compliance with the contract and specification requirements.

Component Procurement and Development Segment Verification

Component procurement and development *verification* occurs in two forms:

1. Receiving inspection of external vendor products such as components and raw materials based on procurement "fitness-for-use" criteria.
2. Internally produced or modified components.

Externally procured components and materials undergo receiving inspection verification that the item(s) comply with the procurement specifications. The verification may be accomplished by:

1. Random samples selected for analysis and test.
2. Inspection of *Certificates of Certification (CofCs)* certified by the vendor's quality assurance organization.
3. By 100% testing of each component.

Internally developed or modified components are subjected to INSPECTION, ANALYSIS, DEMONSTRATION, or TEST to verify that each component fully complies with its design requirements—technical drawings, and so forth.

In all cases, component or material deficiencies such as design flaws and substandard work quality are recorded as *discrepancy reports* (DRs) and dispositioned for corrective action, as appropriate.

System Integration, Test, and Evaluation (SITE) Segment Verification

During the System Integration, Test, and Evaluation (SITE) segment, each integrated system entity is *verified* for *compliance* to its respective *performance* or *item development specification* using pre-approved test procedures. If *noncompliances* are identified, a DR is documented and dispositioned for corrective action.

Operational Test and Evaluation (OT&E) Segment Verification

The Operational Test & Evaluation (OT&E) segment focuses on *validating* that the User's documented operational needs, as stated in the *Operational Requirements Document (ORD)*, have been met. However, system verification occurs during this segment to ensure that all system elements are in a *state of* readiness to perform *system validation*.

Authenticate System Baselines Segment Verification

When a system completes its System Verification Test (SVT) and Operational Test and Evaluation (OT&E), performance results from the *As Built*, *As Verified*, and *As Validated* system configurations are verified via a *functional configuration audit* (FCA) and a *physical configuration audit* (PCA), as applicable. The results of the FCA and PCA are formally authenticated in a System Verification Review (SVR).

53.5 MULTI-LEVEL APPLICATION OF VERIFICATION

Verification is performed at all levels of the system and every item within each level. This includes SYSTEM, PRODUCT, SUBSYSTEM, ASSEMBLY, SUBASSEMBLY, and PART levels.

If a commercial off-the-shelf (COTS) item, a nondevelopmental item (NDI), or a configuration item (CI) is procured according to design requirements—such as procurement specifications, control drawings, or vendor product specifications—you must *verify*, via some form of *objective evidence* or *quality record*, that the item fully complies with its requirements.

For internal and subcontracted *configuration item* (CI) development efforts, formal *acceptance test procedures (ATPs)* must be successfully completed, witnessed, and documented. COTS/NDI items generally include a *Certificate of Compliance (CofC)* from the vendor unless prior arrangements have been made.

System Verification Responsibilitys

System verification is performed *formally* and *informally* every day on every task for both *internal* and *external* customers. You should recall our discussion in Figure 13.3 about the *internal* and *external* customer "supply chains." So, *WHO is responsible for verification?* Anyone who produces a *work product* regardless of whether the "customer is internal or external in the workflow process."

Informal Verification Responsibilities. From the moment a new contract is signed until the system is delivered to the field, every task:

1. Accepts outputs from at least one or more predecessor tasks.
2. Performs value-added processing in accordance with established standards and practices.
3. Delivers the resulting *work product* to meet the needs and expectations of the next "downstream" task.

As each task is performed, the *accountable* individual or team *verifies* that:

1. The task inputs comply with "*fitness-for-use*" criteria.
2. Their personal *work products* meet specific requirements.

Therefore, verification activities incrementally build integrity into the *value chain* to ultimately deliver physical components and work products that *comply* with organizational and contract requirements.

Formal Verification Responsibilities. We noted in our *Introduction* to system V&V that verification activities occur throughout the system/product life cycle at strategic *staging* or *control points*. System Development Phase *critical staging* or *control points* are documented in the *Integrated Master Plan (IMP)* as *events, accomplishments that support events, and criteria that support accomplishments*. These include major technical reviews, technology demonstrations, document reviews, component inspections, and multi-level system acceptance tests. Each of these events is assigned to a responsible individual or IPT for completion accountability including V&V.

Verification Control Points

Verification is performed at various *staging* or *control points* in the System Development Process. These control points, which are both work product and contract event based, include document reviews, technical reviews, modeling and simulation, and prototyping and technology demonstrations. Let's explore each of these briefly.

Document Reviews. Document reviews serve as one of the earliest opportunities for verification tasks. Reviewers assess the *completeness*, *consistency*, and *traceability* of documentation relative to source requirements, risk, soundness of strategy, logic, and engineering approach. *Work product* examples include plans, specifications, *Concept of Operation* (ConOps), analyses and trade studies.

Technical Reviews. Technical reviews are conducted *informally* and *formally*. Reviews, which typically include the Acquirer and User, provide a forum to address, *resolve*, and *verify any critical operational* or *technical issue* (COI/CTI) resolution relative to contract requirements and direction.

Referral For more information about technical reviews, refer to Chapter 54 on *Technical Review Practices*.

Models and Simulations. During the early phases of the program, candidate architectures are evaluated for performance trade-offs, requirements allocation decisions are made, and system timing is determined. Models and simulations provide early verification insight into *critical operational and technical issues (COIs/CTIs)*, and reveal some *unknowns* as well as *how* the proposed system solution may perform in a simulated operating environment.

Referral For more information about *System Modeling and Simulation Practices*, refer to Chapter 51.

Prototypes and Technology Demonstrations. *Critical operational and technical issues (COIs/CTIs)* that cannot be *resolved* via models and simulations can often be investigated using *prototypes* and *demonstrations* as a risk reduction method of how a proposed solution might perform. Work product examples include test flights and test beds. During these demonstrations, the System Developer and Users should *verify* and *validate* system performance and risk prior to committing to a larger scale program.

Requirements Traceability. Developing a system with *integrity* requires that each system component at every system *level of abstraction* is *interoperable* and *traceable* back to a set of source requirements. Developers employ tools such as spreadsheets or requirements management tools to document traceability links.

Referral For more information about requirements traceability, refer to Chapter 31 on *Requirements Derivation, Allocation, Flow Down, and Traceability Practices*.

53.6 VERIFICATION METHODS

The process of *verifying* multi-level SE design compliance to the *System Performance Specification (SPS)* or *item development specification* (IDS) requires standard *verification* methods that are well defined and understood. Verification includes five commonly recognized methods: 1) INSPECTION, 2) ANALYSIS, 3) DEMONSTRATION, 4) TEST, and 5) CERTIFICATION. A sixth method, SIMILARITY, is permitted as a verification method in some business domains. The NASA *SE Handbook* (para. 6.6.1, p. 118) further suggests two additional verification methods: SIMULATION and VALIDATION OF RECORDS.

Let's explore each of these types of *verification methods*.

Verification by TEST

Test is used as a verification method to prove by measurement and results that a system or product complies with its specification or design requirements. Testing employs a prescribed set of OPERATING ENVIRONMENT conditions using test procedures approved by the Acquirer (role). Testing occurs in two forms: *functional* testing and *environmental* testing.

Testing can be very expensive and should only be used if *inspection, analysis*, and *demonstration* do not *individually* or *collectively* provide the *objective evidence* required to prove compliance.

Verification by ANALYSIS

Specific aspects of system performance that call for verification by ANALYSIS are documented in a formal technical report. In general, these analyses are placed under configuration control.

Verification by DEMONSTRATION

Verification by DEMONSTRATION is typically performed without instrumentation. The system or product is presented in various facets of operation for witnesses to OBSERVE and document the results. DEMONSTRATION is often used in field-based applications and operational scenarios involving, reliability, maintainability, human engineering, and final on-site acceptance following formal verification.

Verification by SIMILARITY

The NASA *SE Handbook* (para. 6.6.1, p. 119) describes verification by SIMILARITY as: *"[T]he process of assessing by review of prior acceptance data or hardware configuration and applications that the article is similar or identical in design and manufacturing process to another article that has previously been qualified to equivalent or more stringent specifications."*

Author's Note 53.1 *Remember, there are two contexts of system development: design verification and product verification. "Design verification" is a rigorous process that proves an SE design meets specification and design requirements. You ONLY verify a design once, unless you make changes to the As Designed, As Verified, and As Validated Product Baseline. Once the design is committed to production, "product verification"—physical realization of the design—is applied to prove that the physical instance of a product—the model and serial number—performs intended capabilities without errors or defects.* Therefore, verification by similarity requires that you present quality records of the design verification previously accomplished.

Verification by INSPECTION

The NASA *SE Handbook* (para. 6.6.1, p. 119) defines verification by INSPECTION as the *"[P]hysical evaluation of equipment and/or documentation to verify design features. Inspection is used to verify construction features, workmanship, and physical dimensions and condition (such as cleanliness, surface finish, and locking hardware)."* Some organization use verification by EXAMINATION instead of INSPECTION.

Verification by Simulation

The NASA *SE Handbook* (para. 6.6.1, p. 119) states verification *by SIMULATION is* "*[E]mployed when analytical methods may be impractical and physical EQUIPMENT System Element hardware and software have not been produced to provide actual data.*"

Verification by Validation of Records

The NASA *SE Handbook* (para. 6.6.1, p. 119) defines verification by VALIDATION of RECORDS as "*[T]he process of using manufacturing records at end-item acceptance to verify construction features and processes for flight hardware.*"

Verification Method Cost Factors

In general, the cost to perform these verification methods varies significantly. *Relative* costs by verification method are:

- Inspection (low cost)
- Similarity (low cost)
- Analysis (low to moderate cost)
- Demonstration (moderate cost)
- Test (moderate to high cost)
- Certification (moderate to high cost)

When proposing a new system, you should strive to identify the *method* with the *lowest cost* and *risk* that will provide *compelling*, *objective evidence* that a requirement has been satisfactorily *accomplished*. If you have a requirement that you can verify by ANALYSIS (low to moderate cost), WHY would you want to commit to TEST (moderate to high cost) as the *verification method* and drive up your proposal price and potentially risk losing the contract?

Referral For additional information about selection of specification requirements verification methods, refer to Chapter 33 on *Requirements Statement Development Practices*.

53.7 SYSTEM VALIDATION PRACTICES

Whereas system *verification* asks if we built the work product RIGHT, system *validation* answers the Acquirer question "*Did we acquire the RIGHT system to meet the User's validated operational needs?*"

Validation Methods

Validation is performed with a number of methods such as User interviews, prototyping, demonstration, qualification tests, test markets, and field trials. In general, *validation* consists of any method that enables the System Developer of a *system*, *product*, or *service* to establish a *level of confidence* that the evolving *work product* satisfies the User's *documented* operational need(s).

Author's Note 53.2 *Note the use of the term "documented"; this is very good practice. People tend to change their minds about WHAT they said or intended to say. To avoid any misinterpretation by either party, document the mutual understanding of User needs, including what criterion represent mission success at acceptance.*

Importance System Validation Responsibility

Validation employs User feedback mechanisms to keep the activities that produce work products *focused* on the key factors *critical* to User satisfaction.

The scope of validation encompasses more than a User function. Remember, the *System Performance Specification (SPS)* provides a basis to decompose a high-level *problem space* into lower levels of *solution spaces*, even within the System Developer's program organization. In this context the User (role) is the higher level team assigned the PRODUCT, SUBSYSTEM, ASSEMBLY, and SUBASSEMBLY *problem space*. As the lower level *solution space* "designs" evolve, higher level users must validate that the evolving item will satisfy needs.

EXAMPLE 53.1

A Control Station Integrated Product Team (IPT), which is assigned implementation responsibility for the *Control Station Performance Specification*, has a *problem space* to resolve. The *problem space* is decomposed into several *solution spaces*. Requirements for Computer Software Configuration Item (CSCI) 1 are allocated and documented in the CSCI 1's *Software Requirements Specification* and assigned to the CSCI 1's IPT. As CSCI 1's design is formulated, the IPT employs *iterative* rapid prototyping methods to generate sample operator displays and transitions for evaluation and feedback (i.e., *validation*). The Control Station IPT invites the Users, via Acquirer contract protocol, to participate in a demonstration to review and provide feedback as *validation* that the evolving display designs and transitions satisfy the User's needs. Results of the demonstration are documented as a *quality record* and used to support developer decision making.

Validation Responsibility

Everyone on the program on *every* task performs validation by making sure that each work product or stage of a work product's completion meets the *expected* needs of each task and customer in the *supply chain*.

EXAMPLE 53.2

An SE is tasked to produce an analysis. The SE performs *validation* by talking with the task source to:

1. Fully understand the *critical operational or technical issue (COI/CTI)* to be resolved.
2. Scope areas for investigation, objectives, and constraints.
3. Understand how the results are to be documented to ensure the *work product* will meet the task source's needs.

Validation continues after the *work product* (PRODUCT, SUBSYSTEM, etc.) is delivered until the User's intended operational need is satisfied.

53.8 SYSTEM PERFORMANCE VERIFICATION AND VALIDATION

Our discussion of system *verification* and *validation* up to this point has focused on the *process* and *mechanisms* of V&V implementation. Here, we explore some approaches that focus on *system performance verification* and *validation*. These approaches include:

- Acceptance tests
- Technical demonstrations

- Qualification tests
- System verification tests (SVTs)
- Field trails and test markets

Acceptance Tests

Acceptance tests are formal tests that establish the *technical* and *legal criteria* for acceptance of the system or product by the Acquirer for the User based on *compliance* with specification requirements. *Compliance* results provide a *prerequisite* for the Acquirer to formally *accept* the system.

In general, *acceptance tests* are conducted with a set of formally *approved* and *released* acceptance test procedures (ATPs) that have been agreed to by the Acquirer and User as applicable, and by the System Developer. System level acceptance tests:

1. Are derived from the *System Performance Specification (SPS)* Section 3.0 *Requirements*.
2. Apply *verification methods* identified in the Section 4.0 *Qualification Provisions* to verify accomplishment of each Section 3.0 *Requirement(s)*.

For some programs, the term *acceptance test* (AT) is synonymous with the System Verification Test (SVT). In other cases, an AT may utilize a subset of the SVT ATPs after system installation at a User site as a final verification prior to formal system acceptance. Acceptance tests, as a generic term, are also used by IPTs to demonstrate entity compliance to a higher level team such as a System Engineering and Integration Team (SEIT).

Acceptance tests may be *procedure-based* or *scenario-based*.

Procedure-Based ATPs. *Procedure-based* ATPs provide detailed, scripted instructions. Test operators are required to follow *prescribed* scripts to establish a specific test configuration, switch positions, and inputs to stimulate or enter into the *system*. Each test procedure identifies the *expected results* for verification against specification requirements.

Scenario-Based ATPs. *Scenario-based* ATPs provide general guidance in the form of operational scenarios. Detail actions such as switch settings and configuring the software required to operate the *system* are left to the system/test operator. ATP data sheets generally include a field for recording the *actual* measurements and observations. Relevant witnesses from the System Developer, Acquirer, and User organizations to the System Developer and Quality Assurance (QA) representatives authenticate the ATP results as quality records.

Technical Demonstrations

Technical demonstrations are sometimes performed as part of the Developmental Test and Evaluation (DT&E) procedures through the use of prototypes, technology demonstrations, and proof of concept demonstrations. These activities provide an excellent opportunity to *assess* and *evaluate* system performance and obtain more insight into system requirements and their refinements. The results of technology demonstrations sometimes ARE NOT the actual *test articles* but the set of requirements and data *derived* from the technical demonstration.

System Verification Test (SVT)

System Verification Tests (SVTs) are *formal* tests performed on systems and products at a designated facility and witnessed by the Acquirer and System Developer QA. A User representative(s) is typically invited by the Acquirer to participate. The purpose of the SVT is to *prove* that the system or product fully *complies* with and meets its *System Performance Specification (SPS)* requirements.

In preparation for the SVT, formal *acceptance test procedures (ATPs)* are developed, reviewed, and approved. Prior to the SVT, a test readiness review (TRR) may be conducted to determine the *state of readiness* of the *test article(s)* and supporting *test environment*—including EQUIPMENT, FACILITIES, and PERSONNEL Elements as well as processes, methods, and tools used prior to the test.

Referral For more information about the test readiness review and system verification tests, refer to Chapter 54 on *Technical Review Practices*.

Qualification Tests

When a system completes its formal SVT, system validation may be performed as a *qualification test*, depending on contract requirements. *Qualification tests*, or Formal Qualification Tests (FQTs), consist of two types:

1. Tests conducted under *actual* or *realistic* field conditions to demonstrate that the system design and its development (components, workmanship, etc.) meet contract requirements.

2. *Scenario-driven* tests conducted by an Independent Test Agency (ITA) for the User organization as part of the Operational Test and Evaluation (OT&E) activity.

During the OT&E *qualification test*, the System Developer is normally *precluded* from participating in the test. Generally, the System Developer is kept informed about the evolving results of OT&E. *Qualification tests* should be conducted by the User personnel to assess not only the *system* (e.g., EQUIPMENT Element performance) but also overall Human Systems Integration (HSI) effectiveness. *Environmental* or *qualification tests* are often followed by a Formal Qualification Review (FQR), which assesses the results of the *qualification tests*.

EXAMPLE 53.3

Qualification tests subject system and component test article(s) to HARSH environmental test conditions in a NATURAL ENVIRONMENT or controlled laboratory or field test environments. Conditions may include shock, vibration, electromagnetic interference (EMI), temperature, humidity, and salt spray to qualify the design for its intended application (in space, air, sea, land, etc.).

Field Trails and Test Markets

Commercial System Developers conduct formal *system*, *product*, and *service verification* testing via field trials in the marketplace. Systems, products, and services evolve through a series of design iterations based on feedback from Users in field trials or test markets. Ultimately, marketplace *supply* and *demand* determine the public's response in terms of system ACCEPTABILITY.

53.9 INDEPENDENT VERIFICATION AND VALIDATION (IV&V)

Due to technical issues such as risk, interoperability, safety, and health related to large complex, expensive systems, government organizations such as DoD and NASA may issue IV&V contracts to assess the work of System Developers during the System Development Phase. In this context, the IV&V contractor provides an independent assessment to the Acquirer that the provisions of the system development contract are being implemented properly. A *key objective* is to ensure that systems or products will have *acceptable risk* and proper considerations for the *safety* and *health* of the system operators and the general public.

Need for Independence

Potential hardware and software design *flaws, deficiencies*, or *errors* are sometimes *missed* due to limited personnel availability and skills, incorrect requirements, or changes in hardware or software platforms. *Critical flaws* can result in *cost overruns*, or even *catastrophic failure* that poses safety risks to the public and the NATURAL ENVIRONMENT. An independent observer can be an important ally to the Acquirer in preventing these problems.

Degree of Independence

A common question is: *How "independent" must an IV&V organization be?* IEEE 610-1990 describes an IV&V organization as "*technically, managerially, and financially independent of the development organization.*"

Benefits of IV&V

People often ask, "*Why should we go to the expense of performing IV&V, either by contract or by internal assessments? What's the return on investment (ROI)?*" There are several reasons; some are objective and others subjective. In general, IV&V:

1. Improves system or product safety.
2. Provides increased visibility into the System Development Process.
3. Identifies *nonessential* requirements and design features.
4. Assesses compliance between specification and performance.
5. Identifies potential risks areas.
6. Reduces the quantity of *latent defects* such as design flaws, errors, defective materials or components, and workmanship problems.
7. Reduces development, operations, and support costs.

When performed *constructively* and *competently*, IV&V can be of benefit to both the Acquirer and the System Developer. Depending on the role assigned by the Acquirer to the IV&V contractor, adding another player into the System Development Process may require the System Developer to plan for and obtain supplemental resources. The challenge for the System Developer may be dealing in an environment whereby the Acquirer *believes* the IV&V contractor is not "earning their keep" unless they FIND a lot of microscopic deficiencies, even if the *work products* are more than adequate *technically, professionally*, and *contractually*.

EXAMPLE 53.4

As an example, software is playing an increasing role in day to day. For each NASA mission or project to execute successfully, it is imperative that the software operates *safely* and within its designed parameters. Failure of a single piece of *mission critical software* within a NASA mission can potentially result in loss of life, dollars, and/or data. IV&V serves as a mechanism to underscore the importance of software safety and helps ensure safe and successful NASA missions.

53.10 COMMON SYSTEM V&V CHALLENGES AND ISSUES

System *verification* and *validation*, in general, involve a number of issues that SEs must be prepared to address. Let's examine a few of the challenges and issues.

Challenge 1:   Acquirer need for V&V.
Challenge 2:   Establishing the scope of V&V activities.
Challenge 3:   Developing confidence in the evolving work product.
Challenge 4:   System developer need for V&V.
Challenge 5:   V&V impacts on cost and schedule.
Challenge 6:   Degree of V&V team independence.
Challenge 7:   Incorporation of IV&V findings and recommendations.
Challenge 8:   Program personnel V&V skills and knowledge levels.
Challenge 9:   Model and simulation certification.
Challenge 10: Verified systems that fail system validation.

Challenge 1: Acquirer Need for V&V

From an Acquirer's perspective, you need a technical basis to determine if the deliverable system and all other aspects of the contract are progressing in accordance with the *performance* standards established by the contract. V&V activities provide a mechanism to the degree funding constraints allow. *The questions is: Do you allocate funding for V&V or target those funds to adding more system capabilities?*

Large, complex systems require disciplined approaches to ensuring that the system or product will satisfy its requirements. Accomplishment of this objective is achieveable without performing V&V during Developmental Test and Evaluation (DT&E) and Operational Test and Evaluation (OT&E). The question is: *Depending on the system application, are you willing to assume the risk of not performing V&V?*

Challenge 2: Establishing the Scope of V&V

The scope, breadth, and depth of the V&V effort are often *critical issues*. Since cost CONSTRAINS the extent of V&V, do you:

- *Perform V&V at high levels across the whole program?*
- *Focus V&V resources in specific areas that may be high risk? In any case, the program should make an assessment on a case-by-case basis.*

Contracts typically specify some form of V&V activities such as technical reviews and acceptance tests to demonstrate accomplishment of the SPS requirements. IV&V may be a different matter. If IV&V is not required, here's the challenge: *Are you prepared to deal with the consequences from the lack of V&V such as the cost-to-correct latent defects and design flaws*. The answer requires consideration of the *size* and *complexity* of system development and operational risks.

Challenge 3: Developing Confidence in the Evolving Work Product

As an Acquirer or System Developer, you must establish a system of *checks* and *balances* throughout the System Development Phase to build a *level of confidence* that the evolving Developmental

Configuration will satisfy the *System Performance Specification (SPS)* requirements and the User's validated operational needs.

Challenge 4: System Developer Need for V&V

From a System Developer's perspective, whether or not the contract requires V&V activities, you need a *mechanism* to demonstrate satisfactory achievement of the SPS requirements, accomplishment and closure of the contract requirements, and so forth. If V&V activities are not part of the contract, consider conducting some level of internal V&V activities.

Challenge 5: V&V Impacts on Cost and Schedule

Cost and schedule impacts are the *critical issues* in V&V implementation. If the V&V activity has been planned and scheduled into the *Master Program Schedule (MPS)*, *Integrated Master Plan (IMP)*, and *Integrated Master Schedule (IMS)*, cost and schedule impacts should not necessarily be a problem, assuming they were properly estimated, allocated, and managed.

The challenge here is to overcome *perceptions* that V&V, or its results, typically place programs behind schedule. Conducting a V&V activity, which is often viewed by SE Design Team members as *unnecessary* and *non–value added*, identifies potential risk areas that can become ROOT CAUSES for late program delivery. This translates into increased *costs*, impacts contract performance *incentives*, and subsequently impacts *profit*, depending on the type of contract.

Challenge 6: Degree of V&V Team Independence

One of the *critical issues* in V&V is WHO performs the V&V activity. Ideally, an independent verification and validation (IV&V) organization or contractor performs V&V. If this is impractical, an in-house team independent of the program may be a candidate to perform *verification*, depending on the application. The Acquirer, however, may offer a few comments about the *credibility* of this effort because of a potential organizational *conflict of interest*.

The criticality of team *independence* is driven by the fact that an IPT is too close to the evolving Developmental Configuration and may not be objective. You could say that the IPT has a *conflict of interest* (COI); designers SHOULD NOT evaluate their own work. Capable personnel from other IPTs can provide a comparable V&V function though the degree of *independence* organizationally may still be questionable. But, it may your only choice.

Challenge 7: Incorporation of IV&V Findings and Recommendations

If we assume the V&V activity produces *substantive* and *meaningful* results, consideration should be given to implementing the *findings* and *recommendations*. *If you are going to spend the money, you should be prepared to seriously consider implementing the results*. The underlying assumption here is that the IV&V findings are within the scope of your contract. If not, the System Developer should discuss and negotiate the issue with the Acquirer.

Granted, there are sometimes circumstances where the V&V Audit Team may not have knowledge of all the factors affecting a decision. So, the team should investigate and discuss questionable *areas* with the SEIT or IPT to *validate* any evidence that may lead to a finding. Likewise, the IPT should ensure that the V&V Audit Team has all the information they require to make well-founded assessments.

Challenge 8: Program Personnel V&V Skills and Knowledge Levels

One of the issues with V&V is that program personnel may lack common skills, experience, and insights into conducting V&V activities. As an SE, you need to be the "leveling" agent and make sure everyone is trained and skilled in V&V practices, especially from a decision-making perspective. The realities are you have to devise a way to build in a *minimum* set of V&V activities to ensure that the system meets the *minimizing* costs requirements and remains competitive. Good practices consist of establishing key V&V objectives, soliciting proposed solutions, and selecting a preferred approach.

Author's Note 53.3 *A worst-case program scenario becomes reality when executives with limited experience exploit their position and authority by making command decisions that jeopardize program success. Examples are shortcutting or eliminating the review process to save dollars, eliminating critical documents, and configuration management and quality assurance. Executives who perceive themselves, by virtue of power and authority, as subject matter experts (SMEs) don't necessarily place their careers on the line by doing this. They ALWAYS have a way of finding scapegoats in the technical ranks to place blame for poor contract performance resulting from their innovative cost-cutting shortcuts. Then, continue to be promoted to higher positions because of their "outstanding" leadership.*

Challenge 9: Model and Simulation Certification

Development of credible models and simulations may require *verification* by SMEs or SME organizations. *Certification* brings a level of *accountability* for maintaining and controlling the models and simulations in the form of expenses, distribution, licensing, proprietary issues, and documentation.

Referral For more information about *modeling and simulation*, refer to Chapter 51 on *System Modeling and Simulation Practices*.

Challenge 10: Verified Systems That Fail System Validation

One of the most critical points of SE is the integrity of the "value chain" from the identification and validation of User operational needs through system delivery. You can incorrectly:

1. And *inaccurately* identify, bound, and specify the *problem space(s)* and the *solution space(s)*.
2. Prepare the *Operational Requirements Document (ORD)*, RFP *System Requirements Document (SRD)*, *System Performance Specifications (SPS)*, *item development specifications (IDS)*, and design specification requirements.
3. Select the wrong COTS/NDI products.

Errors, omissions, and *defects* in the ORD to SRD to SPS decision-making process can result in a system that fully *complies* with its SPS but *fails* to meet the User-validated operational needs documented in the ORD. Therefore, system specification, design, and development process *decisions* and *integrity* are *critical issues* to the User, Acquirer, and System Developer.

53.11 IS V&V A HELP OR HINDRANCE?

People typically view V&V activities as *unnecessary* tasks that consume critical skills, cost, and schedule resources and do not produce a new product. Contrary to this shortsighted mindset, V&V activities result in a *higher quality product* and AVOID *costly rework*. In the end the program directors, technical directors, project engineers, SEs, and others who are *accountable* for technical program performance and customer acceptance must live with the consequences of their *decisions*.

Some people apply V&V and view it as a *pathway leading to success*; others view it as an unnecessary hindrance. Program performance tends to correlate with these two perspectives. The bottom line is: *invest in correcting design flaws, errors, discrepancies, and deficiencies "up front" OR pay significantly more at higher levels of integration.*

If you are *serious* about your reputation and career, you should ensure that some form of *checks* and *balances* are in place to verify that the system development effort will produce *systems, products*, and *services* that comply with contract requirements. V&V activities provide a means to accomplish this. *Does V&V guarantee success?* Absolutely not! Like most human activities, the quality of the V&V effort is *only as good as the people who perform the work, methods and tools used, and the resources allocated to the activity.*

53.12 GUIDING PRINCIPLES

In summary, the preceding discussions provide the basis with which to establish the guiding principles that govern system verification and validation practices.

Principle 53.1 Verification assesses compliance by asking the question: *Did we develop the work product RIGHT—in accordance with specification, design, or task requirements?*

Principle 53.2 Validation asks the question: *Did we acquire the RIGHT work product to satisfy the User's intended operational need?*

Principle 53.3 To avoid or minimize legal issues, ALWAYS perform system verification, preferably with an Acquirer representative as a witness.

Principle 53.4 ALWAYS have an Independent Test Agency (ITA) or organization perform validation System Developers have a conflict of interest.

Principle 53.5 Every SYSTEM/entity failure during system verification and validation has a cost. Explicitly document Acquirer and System Developer cost accountability before the contract is signed.

53.13 SUMMARY

During our discussion of *system V&V practices*, we defined V&V; its objectives; *HOW, WHEN*, and *WHERE* V&V is accomplished; *who* is accountable; and *methods* for conducting V&V activities. In support of this overview, chapter 54 on *Technical Review practices* address specific verification activities that support key decision control points.

We described the various verification methods: 1) inspection, 2) analysis, 3) demonstration, 4) test, 5) certification, and 6) *similarity*, if permissible. Each of the verification methods is supported by verification activities such as design reviews, prototypes, and audits. We also discussed independent verification and validation (IV&V) concepts, their implementation, and their benefits.

GENERAL EXERCISES

1. Answer each of the *What You Should Learn from This Chapter* questions identified in the *Introduction*.
2. Refer to the list of systems identified in Chapter 2. Based on a selection from the preceding chapter's General *Exercises* or a new system selection, apply your knowledge derived from this chapter's topical discussions. Specifically identify the following:
   (a) How would you approach *verification* and *validation* for the system, product, or service?
   (b) Do you believe there is a need for IV&V. Provide your rationale?
   (c) What aspects of the design and development must be verified by *inspection*, *analysis*, *demonstration*, and *test*?

ORGANIZATIONAL CENTRIC EXERCISES

1. Research your organization's command media for guidance and direction for performing verification and validation (V&V) activities.
   (a) Identify specific requirements for V&V.
   (b) When are V&V activities to be performed?
   (c) How are V&V results to be documented, briefed, and implemented?
   (d) Is IV&V required?
2. Contact a small, a medium, and a large contract program within your organization.
   (a) Do their respective contracts require any V&V activities?
   (b) How are the results to be documented?
   (c) Does the contract include an IV&V contractor?
   (d) What lessons learned from previous programs has the contract program learned, and how are the results being implemented?
3. For your business domain, which verification methods are typically required for most contracts or tasks?

REFERENCES

Defense Systems Management College (DSMC). 2001. *Glossary: Defense Acquisition Acronyms and Terms*, 10th ed. Defense Acquisition University Press. Ft. Belvoir, VA.

IEEE Std 610.12-1990. 1990. *IEEE Standard Glossary of Modeling and Simulation Terminology*. Institute of Electrical and Electronic Engineers (IEEE), New York, NY.

International Council on System Engineering (INCOSE). 2000. *INCOSE System Engineering Handbook*, Version 2.0. Seattle, WA.

MIL-STD-973 (canceled). 1992. *Configuration Management*. Washington, DC: Department of Defense (DoD).

DSMC. 1998. *Simulation Based Acquisition: A New Approach*. Defense System Management College (DSMC) Press. Ft. Belvoir, VA.

National Aeronautics and Space Administration (NASA). 1995. SP-610S *NASA Systems Engineering Handbook*. Washington, DC.

ADDITIONAL READING

Kossiakoff, Alexander, and Sweet, William N. 2003. *Systems Engineering Principles and Practice*. New York: Wiley-InterScience.

MIL-STD-490A (canceled). 1985. *Military Standard: Specification Practices*. Washington, DC: Department of Defense (DoD).

MIL-STD-961D. 1995. *DoD Standard Practice for Defense Specifications*. Washington, DC: Department of Defense (DoD).

Lewis, Robert O. 1992. *Independent Verification and Validation: A Life Cycle Engineering Process for Quality Software*. New York: Wiley.

Federal Aviation Administration (FAA), ASD-100 Architecture and System Engineering. 2004. *National Air Space System—Systems Engineering Manual*. Washington, DC: FAA.

Defense Systems Management College (DSMC). 1998. *Test and Evaluation Management Guide*, 3rd ed. Defense Acquisition Press. Ft. Belvoir, VA.

Chapter 54

Technical Reviews

54.1 INTRODUCTION

Successful system development requires periodic assessments of the status, progress, maturity, and risk of the evolving system design solution at *critical staging* or *control points*. These critical *staging* or *control points* serve as *gating mechanisms* and are intended to answer the following questions:

1. Is there agreement between the key stakeholders—the User, Acquirer, System Developer—concerning the requirements for the system, product, or service? Do the stakeholders share a common understanding and interpretation of the requirements? Have all requirements issues been resolved? Are the requirements in balance with the planned technical, technology cost, and schedule constraints and risks?
2. Does the System Developer have a SYSTEM level design solution selected from a set of viable candidate solutions that represent the best balance of technical technology, cost, schedule, and support performance and risks?
3. Have System Performance Specification (SPS) requirements been allocated via multi-level item development specifications to PRODUCTs, SUBSYSTEMs, hardware configuration items (HWCIs), computer software configuration items (CSCIs), and so forth?
4. Do we have a multi-level, preliminary design solution that is sufficient to commit to detailed design with minimal risk in terms of meeting technical, cost, and schedule performance requirements?
5. Do we have a multi-level, detailed design solution that is sufficient to commit to component procurement and development with minimal risk in terms of meeting technical, cost, and schedule performance requirements?
6. Are the procured and developed components ready for system integration, test, and evaluation (SITE) verification and validation?
7. Is the system or product ready to undergo formal system verification and acceptance by the Acquirer, as the User's contractual and technical?
8. Does the system meet the User's validated operational needs—its operational utility, suitability, and effectiveness?

As the questions above *evolve* with the system development, technical reviews enable the Acquirer to *verify* that the evolving Developmental Configuration complies with specification requirements and is progressing with acceptable risk toward delivery on-schedule and within budget.

System Analysis, Design, and Development, by Charles S. Wasson
Copyright © 2006 by John Wiley & Sons, Inc.

What You Should Learn from This Chapter

1. What is a *technical review*?
2. *Why* should you conduct a technical review?
3. Who is *responsible* for conducting technical reviews?
4. How are *technical reviews documented*?
5. What is the *relationship* between a technical review and contract phase?
6. What is an *Integrated Baseline Review (IBR)* and what is it intended to accomplish?
7. What is a *System Requirements Review (SRR)* and what is it intended to accomplish?
8. What is a *System Design Review (SDR)* and what is it intended to accomplish?
9. What is a *Hardware Specification Review (HSR)* and *Software Specification Review (SSR)* and what is it intended to accomplish?
10. What is a *Software Specification Review (SSR)* and what is it intended to accomplish?
11. What is a *Preliminary Design Review (PDR)* and what is it intended to accomplish?
12. What is a *Critical Design Review (CDR)* and what is it intended to accomplish?
13. What is a *Test Readiness Review (TRR)* and what is it intended to accomplish?
14. What is a *Production Readiness Review (PRR)* and what is it intended to accomplish?

Definitions of Key Terms

- **Conference Minutes** A *quality record* that serves as a written summary of a formal program event and documents the attendees, agenda, discussion topics, decisions, actions items, and handouts.
- **Critical Design Review (CDR)** A major technical program event conducted by a System Developer with the Acquirer and User to assess the progress, status, maturity, plans, and risks of each configuration item's (CI) detailed design solution. The event serves as a *critical staging point* for *authorizing* and *committing* resources for the Component Procurement and Development Segment of the System Development Phase.
- **Hardware/Software Specification Review (SSR)** An assessment of each unique HWCI or CSCI requirements specifications to determine their adequacy to authorize preliminary hardware design or preliminary software design and commit resources to support those activities.
- **In-Process Review (IPR)** An interim or incremental assessment of a document or design solution during its development to provide "early" stakeholder feedback for purposes of ensuring that the work product is progressing to maturity with acceptable risk within cost, schedule, and technical constraints. IPRs may be conducted internally or with the external participants and should result in a documented set of *conference minutes*.
- **Peer Review** A formal or informal review of an SE's or IPT's *work products* by knowledgeable subject matter experts (SMEs).
- **Preliminary Design Review (PDR)** A major technical program event conducted by a System Developer with the Acquirer and User to review a SYSTEM's HWCI or CSCI designs with intent to authorize and commit resources to detailed design.
- **Production Readiness Review (PRR)** "A formal assessment by system stakeholders to authenticate the current Product Baseline and the readiness of the Technical Data Package (TDP) for production." (Source: Former MIL-STD-1521)

- **Ready-to-Ship Review (RTSR)** A *formal* assessment by system stakeholders to determine the *readiness* of the system to be disassembled and shipped to a User's designated location.
- **System Design Review (SDR)** An assessment of the evolving system design solution to evaluate the completeness and maturity of the system architecture and its interfaces, identification of PRODUCTS/SUBSYSTEMS, allocation of SPS requirements to PRODUCTS/SUBSYSTEMS, and risks.
- **System Requirements Review (SRR)** An assessment of the conciseness, completeness, accuracy, reasonableness, and risk of the *System Performance Specification (SPS)* requirements to permit the development of a system with an intent to AVOID misinterpretations, inconsistencies, and errors, especially during *system verification, validation*, and *acceptance*.
- **System Verification Review (SVR)** A formal assessment by system stakeholders to authenticate the results of the functional configuration audit (FCA) and physical configuration audits (PCA) relative to *System Performance Specification (SPS)* requirements and verification methods.
- **Technical Reviews** A series of system engineering activities by which the technical progress on a project is assessed relative to its technical or contractual requirements. The reviews are conducted at logical transition points in the development effort to identify and correct problems resulting from the work completed thus far before the problems can disrupt or delay the technical progress. The reviews provide a method for the contractor and the Acquirer to determine that the development of a configuration item and its documentation have met contract requirements. (Source: Adapted from Former MIL-STD-973, para. 3.89)
- **Test Readiness Review (TRR)** An assessment of the maturity of all aspects of a multi-*item*/configuration item (CI) to determine: 1) its *state of readiness* to proceed with testing with a critical focus on environmental, safety, and health (ES&H) concerns and 2) authorize initiation of the tests.

54.2 TECHNICAL REVIEWS OVERVIEW

System development is highly dependent on progressive, efficient, technical decision making throughout the System Development Phase that involve the Acquirer and User. The *timing* of constructive technical assessment and feedback by stakeholders at *critical control* or *staging points* enables the System Development Team to evolve the system design solution to maturity. The mechanism for staging these assessments and making key technical decisions consist of a series of technical reviews.

Technical Review Objective

The objective of a technical review is to enable *key stakeholders* to assess the evolving system design solution at critical *staging* or *control points* to determine the *progress, status, maturity, integrity, plans*, and *risk* as a condition of a SYSTEM, configuration item (CI), or non-developmental item (NDI) for committing resources for the next segment or phase of program activities.

Categories of Technical Reviews

Technical reviews consist of *major reviews* conducted formally and *internal reviews* that tend to be less formal. In general, the *major reviews* are typically required by contract and involve the Acquirer and User *stakeholder* representatives as participants.

Formal Contract Technical Reviews

Contract reviews, which are referred to as program *events*, are formally conducted in accordance with the *terms and conditions* (Ts&Cs) of the contract. Generally, the contract identifies the reviews to be conducted and specifies guidance for preparing, conducting, and completing the review. In addition to the contract guidance that defines HOW the reviews are to be conducted, there is also a *protocol* associated with providing directions and guidance during the review, attendees to be invited and by whom, and so forth.

Traditional Contract Technical Reviews

Years ago contracts used technical reviews as a limited, but important, window for the customer (Acquirer) to look into the contractor's operations and assess HOW WELL the effort was progressing. Depending on contract size, complexity, and priority, the customer assigned an on-site representative at the contractor's facility. This individual's task was to monitor day-to-day operations and communicate to the "home organization" views on HOW WELL the contractor efforts were progressing. In general, the reviews enable customer project managers to ask themselves "*Does the System Developer's design solution review materials correlate with the progress depicted in prior phone conversations with the contractor—'glowing' contractor status and progress reports, and so forth?*"

Several weeks prior to a technical review, the System Developer prepared large documentation packages for distribution to the customer and User for review. During this review the contents of the documentation package were discussed over a period of several days in agonizing detail. As a result, technical reviews consumed large amounts of time, were costly, and provided "feedback" too late, which caused *rework* and *impacted* customer schedules. These issues, in conjunction with escalating contract costs and process inefficiencies, prompted the need for better status information faster and cost of rework by the US government under Acquisition Reform policies in the early 1990s.

Acquisition Reform and Integrated Process and Product Development (IPPD)

The need for acquisition reform, streamlining, process improvement, and reduced rework influenced a move toward Integrated Process and Product Development (IPPD) environments. IPPD environments, which include Acquirer as an integral part of the "team," provide on-site access to the details and nuances of the product development effort.

As a result of the Acquisition Reform initiatives in the DoD, for example, the technical reviews paradigm began to shift. Whereas technical reviews consumed several days agonizing over documentation details, the paradigm shifted to simply spending a few hours resolving *critical operational* and *technical issue* (COI/CTI) decisions. *Why?* If the User and Acquirer are participants in the IPPD team processes, either on-site or virtually, they should be *intimately* familiar with the design details. The only agenda topics remaining focused on resolution of issues between the acquirer and System Developer.

Finally, another aspect was the shift from *date-driven* to *event-driven* contract reviews.

Date-Driven versus Event-Driven Reviews

Technical reviews, specified by contract, are of two types: *date driven* or *event driven*.

- *Date-driven reviews* MANDATE that a review to be conducted X days after Contract Award (ACA)—at a specific calendar date.

- *Event-driven reviews* are conducted when the development efforts reach specific maturity levels, usually within a general time frame. The time frame may be specified as "within *XX* days/months After Contract Award (ACA)."

Guidepost 54.1 *Given this historical background to today's technical reviews, we now shift our focus to investigating the general conduct of technical reviews.*

54.3 TECHNICAL REVIEWS—GENERAL

Technical reviews are accomplished in accordance with the terms of the contract, subcontract, or associated agreements. Every contract should specify the WHO, WHAT, WHEN, WHERE, HOW, and WHY technical and programmatic reviews are to be accomplished. If not, DO NOT sign until all of the *terms and conditions (Ts&Cs)* of the technical reviews are clearly delineated, mutually understood, and agreed to by all parties.

Technical reviews are more than simply forums to make nice and orderly presentations. The reviews are an opportunity to brief the Acquirer and User "for the record" on WHAT progress has been made in maturing the system design solution since the last review. Reviews provide an opportunity for the Acquirer to validate what is documented in the monthly contract progress reports.

Checks and Balances Benefits

Technical reviews provide checks and balances for the System Developer, Acquirer, and User with inherent benefits for all. For the Acquirer and User, technical reviews provide the opportunity to:

1. Assess progress, maturity, and risks of the product development efforts.
2. Factor the results into Acquirer delivery schedules.
3. Express preferences and priorities.
4. Provide legal technical direction for the contract, depending on type.

For the System Developer or Service Provider, the technical reviews provide the opportunity to:

1. Clearly demonstrate product development maturity and progress.
2. Address and resolve critical operational and technical issues (COIs/CTIs).
3. Assess Acquirer priorities.
4. Obtain Acquirer agreement, if appropriate for the type of contract, to baseline system documentation to *scope* and *bound* future discussion and scoping of technical direction.

Type of Contract

In general, the type of contract determines HOW a review is accomplished and the degree of *influence* or *approval* the customer has over the information presented. In the case of *firm, fixed price (FFPs) contracts*, the System Developer conducts technical reviews to inform the Acquirer and User about progress to date, current status, and risks.

Depending on the type of the contract, the Acquirer may be limited in the *degree* to which they can approve/reject the System Developer's solution without contract modification. In contrast, *cost plus fixed fee (CPFF)* contracts typically enable the Acquirer to exert a large amount of *control* over the contractor's decision making and make adjustments in cost and schedule to *accommodate* changes in the contract's technical direction.

As is ALWAYS the case, understand your contract's T&Cs, and consult with your organization's Program Management Team, contracts, and legal organizations for specific guidance and expertise in these areas.

Review Entry Criteria

Contracts often establish entry and exit criteria for reviews. Because of space restrictions entry criteria are not shown for the review descriptions that follow. Entry criteria should be derived based on the checklist items. Refer to Tables 54.1 through 54.10.

Standard Review Work Products and Quality Records

The preceding discussions highlighted WHAT is to be accomplished at reviews. In a decision-making event it is important to document the review materials and results for *historical* and *reference* purposes. The following is a list of example *work products* and *quality records* that, at a *minimum*, should be produced for each of the major technical reviews.

EXAMPLE 54.1

- Conference agenda
- Attendees list
- Presentation materials
- Handouts (analyses, trade studies, modeling and simulation results, etc.)
- Conference notes
- Conference minutes
- Action items CLOSED/OPEN
- Other supporting documentation, as required

54.4 CONTRACT REVIEW REQUIREMENTS

Major technical reviews are conducted by the System Developer as a *contract event* in accordance with the *Master Program Schedule (MPS)*, *Integrated Master Plan (IMP)*, or *Integrated Master Schedule (IMS)*, as appropriate.

Technical Reviews Location

Technical reviews are conducted at locations specified in accordance with the *terms and conditions* (Ts&Cs) of the contract. Generally, the reviews are performed at the *item's* System Developer's facility due to close proximity to the documentation and actual hardware and software for demonstrations.

Author's Note 54.1 *Remember, "item" in the context above represents a SYSTEM, PRODUCT, SUBSYSTEM, and so forth. For example, if a subcontractor is developing a SUBSYSTEM, reviews are conducted at the subcontractor's facility and include invitations to the prime System Development contractor, who may elect to invite their customer, the Acquirer. Under contract protocols the Acquirer ALWAYS invites the User unless prior arrangements have been made with the System Developer.*

Orchestrating the Review

Major technical reviews are often referred to as *conferences*. Each *conference* consists of three phases of activities to ensure successful completion of the review: Pre-conference, Conference, and Postconference.

- *Preconference Phase* activities include coordination between the System Developer and Acquirer to set the date, time, and location of the review, as well as to set the agenda, invitees, special facility access, and arrangements such as security, parking, and protocols.
- *Conference Phase* activities include conducting the conference in accordance with the planned agenda, classification level, rules of conduct/engagement, recording of *conference notes*, and *action items*.
- *Postconference Phase* activities include resolving conference action items, preparing and approving the conference minutes, incorporating corrections into documentation, and establishing baselines, where applicable.

Contract Review Completion Exit Criteria

Some contracts require completion of technical events, such as reviews, as a prerequisite for obtaining contract progress payments. *Exit criteria* are employed to *explicitly* identify WHAT must be accomplished as a contract *condition* for Acquirer *acceptance*.

Some formal solicitations such as *Request for Proposals (RFPs)* require identification of *exit criteria* as a requirement. Where this is the case, the Offeror's proposal may become part of the contract. If contract progress payments to the System Developer are linked to program events, such as reviews, make sure that *exit criteria* are explicitly stated in language that DOES NOT require interpretation when time comes to contractually CLOSE the review and pay the contractor.

A Word of Caution

Review *entry* and *exit criteria* may appear simple on *initial inspection*. However, they can easily become MAJOR SHOWSTOPPERS due to misinterpretation(s), especially where contract progress payments are "on the line."

When you state that the software design is complete, be *very explicit* as to WHAT you mean by "software design complete." It bears emphasizing that the Acquirer interprets "software design complete" as EVERYTHING. THINK about these statements. Recognize the scope of your intent BEFORE you write them into a proposal and sign a contract, especially if progress payments are at risk. This includes the *Integrated Master Plan (IMP)* and associated dictionaries. *Why?* Simply stated, if you DO NOT follow the contract words, you don't get PAID! And guess who negotiated the contract words? Your organization did!

Posting and Distribution of Technical Review Materials

Today's contracting environment often involves contractor teams across the country and around globe to integrate their efforts via collaborative development and review environments. For technical reviews, World Wide Web (WWW) based reviews provide opportunities to perform IPRs without the expense of travel or disruption of work. However, keep in mind that this media also presents major data security issues related to proprietary data, copyright law, security classification, and export control.

Warning! The US International Traffic and Arms Regulations (ITAR) govern the EXPORT of critical technologies and data by any media including the Internet. ALWAYS refer to your Legal and Contracts organizations as well as the EXPORT CONTROL Officer for guidance in these areas.

Technical Review Contract Direction

The conduct of contract *technical reviews* is often viewed as a causal event. Beware! Despite the suave coolness of the System Developer's *"we're glad you are here"* and Acquirer's *"we're glad to be here"* rhetoric, the activity carries some very SERIOUS *political* and *legal* implications that demand your full attention and awareness.

Remember, ONLY the Acquirer's Contracting Officer (ACO) is officially *authorized* to provide contract direction to the System Developer regarding the technical review. The Acquirer's Program Manager provides the program and technical guidance to the ACO. As a result, you will often hear an Acquirer Program Manager make introductory remarks at the beginning of the review and begin with a disclaimer: *"We are here to listen... anything we say or ask does not infer or should be construed as technical direction... only the Contracting Officer can provide contractual direction."*

Technical Issue Resolution

One of the greatest challenges of technical reviews is making sure all SHOWSTOPPER *critical operational* and *technical issues (COIs/CTIs)* are resolved to the mutual *satisfaction* of the Acquirer and System Developer organizations.

COIs/CTIs often have far-reaching impacts that range from achieving the mission objectives to simply printing out a report. COIs *impact* one of more of the integrated set of System Elements—EQUIPMENT, PERSONNEL, FACILITIES, and so forth. COI impacts span the spectrum from *System Performance Specification (SPS)* requirements to PART level design requirements, and vice versa. Therefore, it is imperative that these issues be *recognized* and *resolved* at the technical reviews to *avoid* schedule impacts.

Programs often refuse to acknowledge or publicize *COIs/CTIs*. Ironically, these issues are often ignored until the program finally has to confront them. Humans have a natural propensity to *pretend* and *believe* that problems—such as COIs/CTIs—will simply *"just go away."*

Sometimes this happens; however, in most cases this is a *panacea*. Although there are reasonable and practical bounds to issue resolution, sooner or later you will learn to confront these issues "early on." Historically, you either *pay the price now* or *pay an even GREATER price* farther downstream to resolve an outstanding issue, assuming the set of *potential* solutions are *viable* at that point in time in terms of cost and schedule performance.

The realities are Users, Acquirers, and System Developers *procrastinate* on resolving *critical operational* and *technical issues*. Organizations spend millions of dollars every year publicizing *how* they have cut back expenses by reducing the number of pencils and paper people use. Yet, these expenses are often insignificant when compared to the dollars wasted procrastinating to resolve *critical technical issues and inefficiencies*. There ARE NO easy solutions or "quick fixes" other than to say that all parties—Users, Acquirers, System Developers, subcontractors among these—collectively need to do a better job confronting these challenges. So, *how does resolution of critical operational or technical issues (COIs/CTIs) relate to technical reviews*?

Technical reviews, in a generic sense, provide the forum for time constrained "open technical debate" to resolve any outstanding or lingering critical technical issues. Granted, some people "debate" just to hear themselves talk. The context here is an environment in which the participants remove their organizational "hats and badges" and focus all energies on alternative paths to bring an issue to *consensual closure* to the mutual satisfaction of all parties. *Tough to do?* You bet! *Are there alternatives?* Yes, there is the "programmatic approach" whereby the program managers for the Acquirer or System Developer may *dictate* a solution to meet schedule. This is not a desired path *technically* or *programmatically*. If you don't like the "programmatic direction approach," ENSURE the "technical approach" works.

Obviously, you do not want to consume valuable time and resources in a technical review "debating" *COIs and CTIs*, unless they surface *unexpectedly* from the *unknown*. If COIs/CTIs are *known* prior to a review, special arrangements should be made before a review to schedule some form of "working group" meeting of all parties. The session must be constrained with the understanding that group must bring the issue to closure and present the recommendations at the major review.

Final Point

One of the most *overlooked* aspects of technical reviews is the *psychology* of managing customer PERCEPTIONS. From a System Developer's perspective, the reviews are opportunities to manage customer *expectations*. Likewise, Acquirers and Users formulate *perceptions* during the review as well as throughout contract performance. Perceptions of System Developer contract performance influence *future interactions*, whether it be future business, follow-on business, or negotiable items of the contract. While the primary focus on a technical review is the current contract or task, the User and Acquirer may be subconsciously asking themselves, *"Do we want to do business with this System Developer again?"*

Although not discussed openly, the reviews enable Acquirers to *validate* levels of *confidence* in their mind that they made the RIGHT choice in selecting your organization to perform on this contract as opposed to your competition. Remember, successful contract performance reflects on Acquirers and HOW their customers, the User and executive management, view your organization.

54.5 MAJOR TECHNICAL REVIEWS

Technical review agendas consist of two types of discussion topics:

1. Programmatic topics
2. Technical topics unique to the type of review

In general, more than 90% of the review time should be devoted to technical agenda items.

Standard Technical Review Items

Standard technical review topics include the following examples:

EXAMPLE 54.2

- Review updates to technical plans, approaches, or procedures.
- Review risk mitigation plans and approaches.
- Review of current schedule status and progress.
- Review *Contract Data Requirements List (CDRL)* item status.
- Review *Contract Line Item Number (CLIN)* status.
- Review results of the requirements traceability audit (RTA).
- Provide authority, if applicable to the contract, to proceed and commit resources to the next system development segment—the preliminary design, detailed design, and so forth.

These topics serve as standard agenda topics for every review and are presented in summary form as part of the introductory remarks. In addition to the standard review topics, review unique topics are provided in the discussions of each review that follow.

Common Types of Technical Reviews

Common types of major SYSTEM level technical reviews include:

- Integrated Baseline Review (IBR)
- System Requirements Review (SRR)
- System Design Review (SDR)
- Software Specification Reviews (SSRs)
- Hardware Specification Reviews (HSRs)
- Preliminary Design Review (PDR)
- Critical Design Review (CDR)
- Test Readiness Reviews (TRRs)
- System Verification Review (SVR)
- Production Readiness Review (PRR)

The sequencing of these reviews is illustrated in Figure 54.1.

Author's Note 54.2 *For programs that employ the "Waterfall" development strategy, these reviews occur only once. For other types of development strategies—such as the Incremental or Spiral Model—these reviews may be repeated for each instance and iteration of the system or product development cycle.*

Referral For more information concerning development strategies, please refer to Chapter 24 on *System Development Workflow Strategy Practices*.

Each of the SYSTEM level technical reviews is supported by PRODUCT, SUBSYSTEM, and ASSEMBLY level assessments that culminate in a baseline for each *item* or configuration item (CI). Because of the expense of travel, some reviews may be conducted *virtually* via audio teleconferences, video teleconferences (VTCs), or on site, depending on *item* maturity. In some instances,

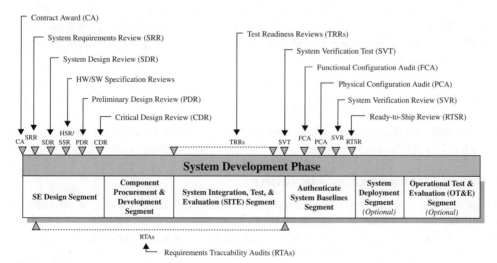

Figure 54.1 Technical Review Sequencing

several PRODUCT, SUBSYSTEM, or ASSEMBLY level reviews may be conducted sequentially during the same time frame at a location, typically the System Developer's facility. Consider the following example:

EXAMPLE 54.3

Successful completion of SUBSYSTEM CDRs serve as *entry criteria* to a *system level* CDR, which then determines if the system design is mature enough to proceed with the Component Procurement and Development Segment of the System Development Phase.

Integrated Baseline Review (IBR)

The Integrated Baseline Review (IBR) is typically the first review conducted for a program. The objective of an IBR is "to conduct a joint review by Acquirer and contractor program managers and their technical staff personnel, following Contract Award, to confirm that:

1. The contractor's Performance Measurement Baseline (PMB)—technical requirements and cost/schedule constraints—covers the entire scope of work.
2. The work is *realistically* and *accurately* scheduled, that the proper amount and mix of resources have been assigned to tasks.
3. Proper objective indicators have been selected for measurement of task accomplishment." (Source: Adapted from the BMDO 5004-H *IBR Glossary*)

The IBR consists of an integrated assessment of all key contract technical, cost, and schedule elements. As such, the IBR seeks to assess:

1. WHAT work is to be performed—for example, *Contract Statement of Work (CSOW), Contract Work Breakdown Structure (CWBS), Integrated Master Plan (IMP), System Performance Specification (SPS), Contract Data Requirements List (CDRL), Contract Line Item Numbers (CLINs)*, and organizational command media.
2. WHEN the work is to be performed—for example, *Master Program Schedule (MPS)* and *Integrated Master Schedule (IMS)*.
3. WHO is accountable for performing the work—for example, program organization and IPT charters.
4. HOW the work will be resourced and controlled—for example, control accounts and work packages linked to the *Contract Work Breakdown Statement (CWBS)*.

The *primary* IBR objectives are accomplished by completion of *supporting objectives* or *exit criteria*. Table 54.1 provides an example listing of these supporting objectives/exit criteria and expected decisions.

On successful completion of the IBR, control accounts and work packages are activated, and work is initiated for the SE Design Segment of the System Development Phase.

System Requirements Review (SRR)

The System Requirements Review (SRR) is typically the first opportunity for technical representatives from the User, Acquirer, and System Developer organizations to get together in a common forum to review, interpret, clarify, and correct, if appropriate, the system requirements.

The primary objectives of the SRR are to:

Table 54.1 Example IBR supporting objectives/exit criteria and expected decisions

Item	Example IBR Objectives/Exit Criteria	Decision Expected
1	Assess the *adequacy, completeness, consistency*, and risk of the System Performance Specification (SPS).	• Action item(s) or • Concurrence/approval
2	Assess the *adequacy, completeness, consistency*, and *risk* of the contract and program schedule elements: • Master Program Schedule (MPS) • Integrated Master Plan (IMP) • Integrated Master Schedule (IMS)	• Action item(s) or • Concurrence/approval
3	Assess the *adequacy, completeness, consistency*, and *risk* of the contract and program cost elements: • Contract Statement of Work (CSOW) • Contract Work Breakdown Structure (CWBS) • Contract Data Requirements List (CDRL) items • Contract Line Item Numbers (CLINs) • Control accounts • Work packages	• Action item(s) or • Concurrence/approval
4	Establish the Performance Measurement Baseline (PMB): • Technical performance baseline elements • Schedule performance baseline elements • Cost performance baseline elements	• Action item(s) or • Concurrence/approval

1. *Review, clarify, correct,* and *baseline* the set of SPS requirements to ensure a common *interpretation* and *understanding* among decision makers.
2. Establish the System Requirements Baseline.

The *primary* SRR objectives are accomplished by completion of *supporting objectives* or *exit criteria*. Table 54.2 provides an example listing of these supporting objectives/exit criteria and expected decisions.

System Design Review (SDR)

The System Design Review (SDR) follows the SRR and is conducted in accordance with the Ts&Cs of the contract CSOW, SDR entry criteria, and program schedule. Following the SRR, the System Developer matures the system design solution derived from the System Requirements Baseline. At a *minimum*, the solution includes development and maturation of the system architecture, system interface requirements, *Concept of Operations (ConOps)*, and preliminary allocations of SPS requirements to the PRODUCT or SUBSYSTEM levels of abstraction.

The *primary* objective of the SDR is to establish the SYSTEM level design solution and Allocated Baseline as part of the *evolving* Developmental Configuration. The *primary* SDR objective is accomplished by completion of *supporting objectives* or *exit criteria*. Table 54.3 provides a listing of these supporting objectives/exit criteria and expected decisions.

On successful completion of the SDR, emphasis shifts to formulating and maturing HWCI unique and CSCI unique requirements specifications (HRS/SRS).

Table 54.2 Example SRR supporting objectives/exit criteria and expected decisions

Item	Example SRR Objectives/Exit Criteria	Decision Expected
1	Verify that the User's *problem space* and solution(s) space(s) are properly UNDERSTOOD and BOUNDED—as what IS/IS NOT part of the *solution space*.	• Action item(s) or • Concurrence/approval
2	Verify that the requirements *completely*, *consistently*, *accurately*, and *concisely* articulate and bound the User's *solution space(s)* for the SYSTEM in a manner that AVOIDS misinterpretation by multiple reviewers.	• Action item(s) or • Concurrence/approval
3	Verify that any ambiguous, overlapping, incomplete, inconsistent, or unbounded requirements are ELIMINATED.	• Action item(s) or • Concurrence/approval
4	Evaluate the quality—as bounded, measurable, testable, and verifiable—of the requirements in terms of required capabilities and performance.	• Action item(s) or • Concurrence/approval
5	Determine if all stakeholder requirements have been adequately addressed subject to contract cost schedule constraints.	• Action item(s) or • Concurrence/approval
6	Verify that each SPS Section 3.X requirement has at least one or more Section 4.X verification methods.	• Action item(s) or • Concurrence/approval
7	Verify that the Section 4.0 Qualification Provision verification methods *represent* the least cost, schedule, and technical risk approach to proving each requirement's compliance?	• Action item(s) or • Concurrence/approval
8	Where appropriate, obtain consensus for *System Performance Specification (SPS)* requirements, interpretations and clarifications, modifications, etc. subject to Acquirer's Contracting Officer (ACO) approval.	• Action item(s) or • Concurrence/approval
9	Where applicable, obtain authorization to establish the System Requirements Baseline.	• Action item(s) or • Concurrence/approval

Hardware/Software Specification Reviews (HSRs/SSRs)

Once the PRODUCT or SUBSYSTEM Level requirements allocations have been established as the Allocated Baseline at the PDR, the architectures for PRODUCT or SUBSYSTEM level solutions are developed and matured. Each solution evolves from analyses of allocated requirements.

Trade studies are conducted to select a preferred PRODUCT or SUBSYSTEM architecture consisting of items, HWCIs, and CSCIs from a set of viable candidate solutions. PRODUCT or SUBSYSTEM *development specification* requirements are next allocated to the items, HWCIs, and CSCIs, as applicable. HWCI and CSCI allocated and derived requirements are documented respectively in HWCI unique PRELIMINARY *Hardware Requirements Specifications (HRSs)* and CSCI unique PRELIMINARY *Software Requirements Specifications (SRSs)*. The culmination of this activity results is an HSR or SSR, as applicable.

In addition to the standard review items, the *primary objective* of the HSR/SSR is to establish an HWCI's or CSCI's requirements specification (HRS/SRS) baseline for its Developmental Configuration to provide a basis for peer level and lower level decision making. The *primary* HSR/SSR objective is accomplished by completion of *supporting objectives* or *exit criteria*. Table 54.4 provides an example listing of these supporting objectives/exit criteria and expected decisions.

Table 54.3 Example SDR supporting objectives/exit criteria and decisions expected

Item	Example SDR Objectives/Exit Criteria	Decision Expected
1	Assess the progress, status, maturity, and risk of the SYSTEM level design solution—architecture, interfaces, etc.	• Action item(s) or • Concurrence/approval
2	Review the preliminary Concept of Operations (ConOps).	• Action item(s) or • Concurrence/approval
3	Review *Mission Event Timeline (MET)* allocations.	• Action item(s) or • Concurrence/approval
4	Review and approve, if appropriate, SPS requirements allocations to PRODUCTs or SUBSYSTEMS and other components.	• Action item(s) or • Concurrence/approval
5	Review any supporting analyses and trade studies relevant to SDR decision making.	• Action item(s) or • Concurrence/approval
6	Review *preliminary* PRODUCT or SUBSYSTEM level *item development specifications*.	• Action item(s) or • Concurrence/approval
7	Resolve any critical *operational* or *technical* issues (COIs/CTIs) related to system capabilities, performance, interfaces, safety, and design criteria—the data.	• Mutual resolution • Closure
8	Review the system life cycle cost analysis.	• Action item(s) or • Concurrence/approval
9	Establish the Allocated Baseline plus any corrective actions required as criteria for acceptance.	• Action item(s) or • Concurrence/approval

On successful completion of each HSR and SSR for their respective HWCI or CSCI, emphasis shifts to formulating and maturing the PRELIMINARY SYSTEM design. This includes development of HWCI and CSCI level designs that respond to their respective HRS and SRS.

Preliminary Design Review (PDR)

The Preliminary Design Review (PDR) represents the fourth major technical review in the design of a SYSTEM, PRODUCT, SUBSYSTEM, ASSEMBLY, and so forth. The review is conducted as a program event to assess the *adequacy, completeness, and risk* of the evolving system design solution down to the HWCI and CSCI design levels.

Following the HSR/SSR, the architectural solution of each HWCI and CSCI evolves. Analyses and trade studies are performed to select a preferred HWCI or CSCI architecture that represents the best solution as determined by a set of pre-defined evaluation criteria. As each HWCI and CSCI solution reaches a level of maturity, a PDR is conducted for each one and culminates in a SYSTEM Level PDR.

In addition to the standard review objectives, the *primary objective* of the PDR is to review and concur/approve the PRELIMINARY SYSTEM/item design solution down to HWCI/CSCI architecture levels. The *primary* PDR objective is accomplished by completion of *supporting objectives* or *exit criteria*. Table 54.5 provides an example listing of these supporting objectives/exit criteria and expected decisions.

On successful completion of the PDR, emphasis shifts to formulating and maturing HWCI-unique and CSCI-unique detailed designs for presentation at the Critical Design Review (CDR).

Table 54.4 Example HSR/SSR objectives/exit criteria and decisions expected

Item	Example HSR/SSR Objectives/Exit Criteria	Decision Expected
1	Assess the *adequacy* and *completeness* of requirements allocations and traceability of an HWCI or CSCI to its *higher level item development specification*.	• Action items or • Concurrence/approval
2	Resolve any *critical operational* or *technical issues* (COIs/CTIs) related to HWCI or CSCI capabilities, performance, interfaces, and design criteria—the data.	• Action items or • Concurrence/approval
3	Establish the criteria and corrective actions required to baseline the HWCI or CSCI requirements.	• Action items or • Concurrence/approval
4	Review each HWCI/CSCI, including use cases, inputs, processing capabilities, and outputs.	• Action items or • Concurrence/approval
5	Review HWCI/CSCI performance requirements, including those for execution time, storage requirements, and similar constraints.	• Action items or • Concurrence/approval
6	Review *control flow* and *data flow interactions* between each of the hardware and software capabilities that comprise the HWCI/CSCI.	• Action items or • Concurrence/approval
7	Review interface requirements between the HWCI/CSCI and all other configuration items both internal and external to the HWCI/CSCI.	• Action items or • Concurrence/approval
8	Review *qualification* or *verification* requirements that identify applicable levels and methods of testing for the software requirements that comprise the HWCI/CSCI.	• Action items or • Concurrence/approval
9	Review any special delivery requirements for the HWCI/CSCI.	• Action items or • Concurrence/approval
10	Review quality factor requirements: correctness, reliability, efficiency, integrity, usability, maintainability, testability, flexibility, portability, reusability, safety and interoperability.	• Action items or • Concurrence/approval
11	Review mission requirements of the system and its associated operational and support environments related to the HWCI/CSCI.	• Action items or • Concurrence/approval
12	Review HWCI/CSCI functions and characteristics within the overall system.	• Action items or • Concurrence/approval
13	Identify any HSR/SSR corrective actions required to establish the HWCI/CSCI Requirements baseline.	• Action items or • Concurrence/approval

Critical Design Review (CDR)

The Critical Design Review (CDR) is the fifth major technical review in the design of a SYSTEM, PRODUCT, SUBSYSTEM, ASSEMBLY, and so forth. The review is conducted as a program event to assess the *adequacy*, *completeness*, and *risk* of the evolving system design solution down to the HWCI, ASSEMBLY, and PART levels and CSCI computer software component (CSC) and unit (CSU) levels.

In addition to the standard review objectives, the primary objectives of the system level CDR are to:

Table 54.5 Example PDR objectives/exit criteria and expected decisions

Item	Example PDR Objectives/Exit Criteria	Decision Expected
1	Briefly review any updates to the SYSTEM level architecture.	None
2	Briefly review any updates to the PRODUCT or SUBSYSTEM architecture.	None
3	Review HWCI design solutions: • HWCI requirements allocations • HWCI specification requirements and traceability • HWCI use cases • HWCI theory of operations • HWCI architecture • HWCI support of system phases, modes, and states of operation • HWCI performance budgets and margins • HWCI technical performance measures (TPMs) • HWCI analyses and trade studies • HWCI-to-HWCI interoperability • HWCI critical technical issues (CTIs) • HWCI critical technology issues • HWCI-to-CSCI(s) integration	• Action items or • Concurrence/approval
4	Review CSCI design solutions. • CSCI requirements allocations • CSCI specification requirements and traceability • CSCI use cases • CSCI theory of operations • CSCI architecture • CSCI support of SYSTEM level phases, modes, and states of operation • CSCI analyses and trade studies • CSCI performance budgets and margins • CSCI-to-CSCI interoperability • CSCI-to-HWCI integration • CSCI critical technical issues (CTIs) • CSCI critical technology issues	• Action items or • Concurrence/approval
5	Review specialty engineering considerations such as: • Human factors engineering (HFE) • Logistics • Reliability • Availability • Maintainability • Supportability • Producibility • Environmental • Training • Vulnerability • Safety • Survivability • Susceptibility	• Action items or • Concurrence/approval
6	Review hardware/software/operator integration issues	• Action items or • Concurrence/approval

1. Review and concur/approve the SYSTEM/CI design solution.
2. Make a decision to *authorize* and *commit* resources to the Component Procurement and Development Segment of the System Development Phase.

The *primary* CDR objectives are accomplished by completion of *supporting objectives* or *exit criteria*. Table 54.6 provides an example listing of these supporting objectives/exit criteria and expected decisions.

On successful completion of the CDR, emphasis shifts to procurement and development of physical components that implement the detailed design requirements. This includes new development and acquisition of commercial off-the-shelf (COTS) and nondevelopmental items (NDIs).

Table 54.6 Example CDR objectives/exit criteria and expected decisions

Item	Example CDR Objectives/Exit Criteria	Decision Expected
1	Determine if the detailed design satisfies the performance and engineering requirements specified in an *item's development specification*.	• Action items or • Concurrence/approval
2	Assess the detailed design *compatibility* and *interoperability* internal to the item and externally to other system elements: • EQUIPMENT (hardware and software) • FACILITIES • PERSONNEL • MISSION RESOURCES • PROCEDURAL DATA • SYSTEM RESPONSES—behavior, products, by-products, or services.	• Action items or • Concurrence/approval
3	Assess item compliance with allocated technical *performance budgets and safety margins*.	• Action items or • Concurrence/approval
4	Assess *specialty engineering* considerations such as: • Reliability, availability, and maintainability (RAM) • Producibility • Logistics • Survivability • Vulnerability • Producibility • Survivability • Safety • Environmental • Susceptibility • Human factors engineering (HFE)	• Action items or • Concurrence/approval
5	Assess any detailed analyses, trade studies, modeling and simulation, or demonstration results that support decision making.	• Action items or • Concurrence/approval
6	Assess the adequacy of verification test plans for each "item."	• Action items or • Concurrence/approval
7	Review *preliminary* test procedures.	• Action items or • Concurrence/approval
8	Freeze the *Developmental Configuration*.	• Action items or • Concurrence/approval

Author's Note 54.3 *The timing of the CDR may be incompatible with long lead item procurement required to meet contract deliveries. Where this is the case, the System Developer may have to assume a risk by procuring long lead items early recognizing the Acquirer has not approved the CDR design.*

Test Readiness Reviews (TRRs)

At each level of abstraction—PART, SUBASSEMBLY, ASSEMBLY, or SUBSYSTEM—some systems require Test Readiness Reviews (TRRs) to be conducted. TRRs range from major program events for large complex systems to simple team coordination meetings among development team members.

The *primary* objectives of a TRR are to:

1. Assess the *readiness* and *risks* of the *test article*, environment, and team to conduct a test or series of tests.
2. Ensure that all test roles are identified, assigned to personnel, and allocated responsibilities.
3. Authorize initiation of test activities.

The *primary* TRR objectives are accomplished by completion of *supporting objectives* or *exit criteria*. Table 54.7 provides an example listing of these supporting objectives/exit criteria and expected decisions.

On successful completion of a TRR, an authority to proceed with specific tests is granted.

Table 54.7 Example TRR objectives/exit criteria and expected decisions

Item	Example TRR Objectives/Exit Criteria	Decision Expected
1	Assess the readiness of the *test article* to undergo testing—destructive or nondestructive.	• Action item(s) or • Concurrence/approval
2	Coordinate and assess the readiness of all test article interfaces and resources.	• Action item(s) or • Concurrence/approval
3	Verify test plans and procedures are approved and communicated.	• Action item(s) or • Concurrence/approval
4	Identify and resolve all critical technical, test, statutory, and regulatory issues are resolved.	• Action item(s) or • Concurrence/approval
5	Assess the readiness of the test environment to support the test.	
6	Verify all safety, health, and environmental concerns are resolved and adequate EMERGENCY processes, services, and equipment are in place to support all aspects of the test.	• Action item(s) or • Concurrence/approval
7	Verify all lower level *Discrepancy Report (DR) corrective actions* have been completed and *verified* for HWCIs and CSCIs.	• Action item(s) or • Concurrence/approval
8	Verify that the "*As Built*" test article identically matches its "*As Designed*" documentation.	• Action item(s) or • Concurrence/approval
9	Coordinate responsibilities for test conduct, measurement, and reporting	• Action item(s) or • Concurrence/approval
10	Designate range safety officers (RSOs) and security personnel, at appropriate.	• Action item(s) or • Concurrence/approval
11	Obtain *authority-to-proceed* with specific tests.	• Action item(s) or • Concurrence/approval

System Verification Review (SVR)

When system verification testing is completed, a System Verification Review (SVR) is conducted. The *primary* objectives of the SVR are:

1. *Authenticate* the results of System Verification Test (SVT), including the FCA and PCA results.
2. Establish the Product Baseline. Accomplishment of the primary SVR objectives is supported by secondary objectives and exit criteria that culminate in key decisions. Table 54.8 provides example objectives/exit criteria and expected decisions.

The *primary* SVR objectives are accomplished by completion of *supporting objectives* or *exit criteria*. Table 54.8 provides an example listing of these supporting objectives/exit criteria and expected decisions.

On successful completion of the SVR, the workflow progresses to a Ready-to-Ship Review (RTSR) decision.

Ready to Ship Review (RTSR)

Following the SVR, a Ready to Ship Review (RTSR) is conducted. The primary objective of the RTSR is to determine the system's state of *readiness* for *disassembly* and *deployment* to the designated delivery site.

The *primary* RTSR objectives are accomplished by completion of *supporting objectives* or *exit criteria*. Table 54.9 provides an example listing of these supporting objectives/exit criteria and expected decisions.

On successful completion of the RTSR, the system is disassembled, packaged, crated, and shipped to the designated deployment or job site in accordance with the contract or direction from the ACO.

Production Readiness Review (PRR)

For systems planned for production, a Production Readiness Review (PRR) is conducted shortly after production contract award. The primary objectives of the PRR are to:

Table 54.8 Example SVR objectives/exit criteria and expected decisions

Objective	Example SVR Objective/Exit Criteria	Decision Expected
1	Audit and certify the results of the Functional Configuration Audit (FCA).	• Action item(s) or • Concurrence/approval
2	Audit and certify the results of the Physical Configuration Audit (PCA).	• Action item(s) or • Concurrence/approval
3	Identify any outstanding inconsistencies, latent defects such as design errors, deficiencies, and flaws.	• Action item(s) or • Concurrence/approval
4	Verify that all approved *engineering change proposals (ECPs)* and *discrepancy reports (DRs)* have been incorporated and verified.	• Action item(s) or • Concurrence/approval
5	Authorization to establish the Product Baseline for the *As Specified*, *As Designed*, *As Built*, and "*As Verified*" configurations.	• Action item(s) or • Concurrence/approval

Table 54.9 Example RTSR objectives/exit criteria and expected decisions

Item	Example RTSR Objectives/Exit Criteria	Decision Expected
1	Verify that all compliance test data have been collected, documented, and certified.	• Action item(s) or • Concurrence/approval
2	Verify that all cabling and equipment items are inventoried and properly identified.	• Action item(s) or • Concurrence/approval
3	Verify that anything related to the configuration installation is documented.	• Action item(s) or • Concurrence/approval
4	Assess verification data completeness prior to system disassembly, packing, and shipping.	• Action item(s) or • Concurrence/approval
5	Assess storage or deployment site facility readiness to accept system delivery for installation and integration: • Environmental conditions • Interfaces • "Gate keeper" decision authority approvals	• Action item(s) or • Concurrence/approval
6	Verify coordination of enroute system transportation support including: • Licenses and permits • Route selection • Security • Resources	• Action item(s) or • Concurrence/approval

1. Authenticate the production baseline configuration.
2. Make a production "go-ahead" decision to commit to Low-Rate Initial Production (LRIP) or Full-Scale Production (FSP).

Author's Note 54.4 A critical issue for any production contract concerns the *currency* of the "*As-Maintained*" configuration documentation. That is, *does the documentation identically match the fielded system configuration(s)?* This question MUST be resolved before committing to production.

The *primary* TRR objectives are accomplished by completion of *supporting* objectives or exit criteria. Table 54.10 provides a listing of these *supporting* objectives/exit criteria and expected decisions.

On successful completion of the PRR, emphasis shifts to Low-Rate Initial Production (LRIP). Subsequent PRRs address readiness for full-scale production (FSP).

54.6 IN-PROCESS REVIEWS (IPRs)

In-Process Reviews (IPRs), which represent another class of technical reviews, occur in two forms:

1. As a System Developer's internal assessment of its evolving work products.
2. As an Acquirer's readiness assessment for conducting a major technical review for a contract.

Let's describe each of these types.

Table 54.10 Example PRR review objectives/exit criteria and decisions expected

Item	Example PRR Objectives/Exit Criteria	Decision Expected
1	Authenticate the *integrity*, *adequacy*, and *completeness* of the current Product Baseline and place it under configuration control.	• Action item(s) or • Concurrence/approval
2	Verify that design improvements—the approved engineering change proposals (ECPs)—to facilitate production have been incorporated, verified, and validated.	• Action item(s) or • Concurrence/approval
3	Resolve any vendor, production, materials, or process issues.	• Action item(s) or • Concurrence/approval
4	Make a production "go-ahead" decision and determine the scope of production.	• Action item(s) or • Concurrence/approval
5	Establish the production baseline.	• Action item(s) or • Concurrence/approval

System Developer IPRs

System Developer personnel conduct IPRs for their evolving work products with internal *stakeholders*. IPRs review plans, specifications, designs, and test procedures. Where program organizations are based on Integrated Product Teams (IPTs), the Acquirer and User may have standing invitations to participate in the reviews.

Although *IPRs* can be formal, they tend to be less structured and informal. Regardless of the level of formality, IPRs should be documented via *conference minutes* and action item assignments.

Contract IPRs

Major technical reviews can be very costly, especially if conducted *prematurely*. Some contracts require the System Developer to conduct a *Contract IPR* as a readiness assessment for conducting the formal technical review. *Contract IPRs* are conducted 30 to 60 days prior to a major technical review to assess the readiness to "go ahead" with SDR, PDR, CDR, and so forth.

54.7 GUIDING PRINCIPLES

In summary, the preceding discussions provide the basis with which to establish the guiding principles that govern technical review practices.

Principle 54.1 (Under contract protocol) Only the Acquirer's Contracting Officer (ACO) is officially authorized to issue technical direction to the System Developer in response to technical review comments, conference minutes, action items, contract modifications, and acceptance of contract documents.

Principle 54.2 Conduct technical reviews at critical control or staging points in system design maturity; no sooner, no later.

Principle 54.3 In most contract environments, interpret Acquirer personnel comments as personal opinion, NOT formal contract technical direction.

Principle 54.4 (Under contract protocol) System Developers invite the Acquirer to program events; the Acquirer extends invitations to the User unless the Acquirer has made other arrangements with the System Developer.

Principle 54.5 Technical reviews assess the status, progress, maturity, compliance, and risk of the evolving system design solution in meeting its contract and specification requirements.

Principle 54.6 Technical review agendas, attendees, discussions, decisions, and action items are documented via conference minutes, and reviewed, approved, and released via contract protocol.

Principle 54.7 Contract clauses are sometimes open to multiple interpretations by different organizations. THINK smartly about and AVOID ambiguous contract technical review entry and exit criteria language.

Principle 54.8 Conference minutes document attendees, agenda, discussion topics, meeting, and action items. Perform the task well and obtain Acquirer/User acceptance via the ACO.

54.8 SUMMARY

In our discussion of technical reviews we introduced the various types of reviews and their occurrence as critical control or staging points in a system's development. For each review we identified the key objectives and referenced checklists for conducting the review. We also highlighted the importance of conducting maturity-based event reviews for each stage of development. Finally, we provided guiding principles to consider when conducting the review.

GENERAL EXERCISES

1. Answer each of the *What You Should Learn from This Chapter* questions identified in the *Introduction*.

ORGANIZATION CENTRIC EXERCISES

1. Research your organization's command media.
   (a) What requirements and guidance is established for conducting formal technical reviews?
   (b) What guidance is provided concerning In-Process Reviews (IPRs)?
   (c) What requirements are levied for documenting the review conference minutes and action items?
   (d) What requirements are levied for disciplinary participants in the reviews?
2. Contact several contract programs in your organization ranging in size from small to large.
   (a) What type of contract—firm-fixed price (FFP) or cost plus incentive fee (CPIF)—does the program have?
   (b) Based on contract type, what degree of approvals or concurrence does the Acquirer have oversight of technical review decisions and baselines?
   (c) What technical review requirements are imposed by the contract?
   (d) How do the reviews differ from those specified in this section?
   (e) Where were the technical review requirements specified?
   (f) How are results of the reviews to be documented?
   (g) What criteria are established as entry and exit criteria for the reviews?

(h) Are PDRs, CDRs, and TRRs at the PRODUCT, SUBSYSTEM, and other levels sequenced relative to the SYSTEM level PDRs, CDRs, and TRRs?

3. Contact an organization that develops commercial products.

   (a) Research how the organization conducts commercial technical reviews.
   (b) Develop a mapping between the two types of approaches to technical reviews.
   (c) Contrast the two approaches and document your views about the advantages and disadvantages of each?

REFERENCES

BMDO 5004-H. 1998. *Integrated Baseline Review (IBR) Team Handbook*, Version 2.0. Ballistic Missile Defense Organization (BMDO). Washington, DC.

Defense Systems Management College (DSMC). 1998. DSMC *Test and Evaluation Management Guide*, 3rd ed. Defense Acquisition University Press Ft. Belvoir, VA.

MIL-STD-973 (canceled). 1992. *Military Standard: Configuration Management*. Washington, DC: Department of Defense (DoD).

MIL-STD-1521B (canceled). 1992. *Military Standard: Technical Reviews and audits for Systems, Equipments, and Computer Software*. Washington, DC: Department of Defense (DoD).

ADDITIONAL READING

Federal Aviation Administration (FAA), ASD-100 Architecture and System Engineering. 2004. *National Air Space System—Systems Engineering Manual*. Washington, DC: FAA.

Chapter 55

System Integration, Test, and Evaluation

55.1 INTRODUCTION

As each system component completes the Component Procurement and Development Segment of the System Development Process, it is ready for System Integration, Test, and Evaluation (SITE). Planning for SITE begins during the SE Design Segment and continues through system delivery and acceptance.

This chapter introduces *SITE practices* for implementing the right side of the V-Model illustrated in Figure 25.5. Our discussions address WHAT SITE is and HOW it is conducted. We begin with a discussion of the fundamentals of SITE. Next, we explore HOW SITE planning is performed and describe the test organization.

Given a basic understanding of SITE, we introduce key tasks that capture HOW developers incrementally test and verify compliance of multi-level test articles in accordance with their *performance* or *development specifications*. We also explore some of the challenges of test data collection and management. We conclude with a discussion of common integration and test issues.

What You Should Learn in This Chapter

1. What *is system integration, test, and evaluation (SITE)?*
2. *When* is SITE conducted?
3. What is the *relationship* of SITE to DT&E and OT&E?
4. What is SITE the expected to accomplish SITE?
5. What are the *work products of SITE*?
6. What are the *roles and responsibilities for SITE* accountability?
7. What are the *types* of test plans?
8. What is a *Test and Evaluation Master Plan (TEMP)*?
9. Who *owns* and *prepares* the TEMP?
10. What is a System Integration and Verification Plan (SIVP)?
11. Who owns and prepares the SIVP?
12. Differentiate the *contexts* of the TEMP versus the SIVP.
13. What is a *Test and Evaluation Working Group (TEWG)*?

System Analysis, Design, and Development, by Charles S. Wasson
Copyright © 2006 by John Wiley & Sons, Inc.

14. Why do you need *test logs*?
15. What is *regression testing*?
16. What is a *discrepancy report (DR)*?
17. When is a DR prepared?
18. How to *prioritize* SITE activities and record SITE deficiencies and issues?
19. What are *Test Readiness Reviews (TRRs)*?
20. What are the SITE phases and supporting tasks for an item?
21. What are *SITE observations, defects, deficiencies, and anomalies*?
22. How do you prepare verification *test reports (TRs)*?
23. What is the importance of *approving and archiving test data*?
24. How do you *certify* test data?
25. What are some *common* SITE challenges and issues?

SITE discussions employ several terms that require definitions. Let's begin by defining these terms.

Definitions of Key Terms

- **Acceptance Testing** "Formal testing conducted to determine whether or not a system satisfies its acceptance criteria and to enable the customer to determine whether or not to accept the system." (Source: IEEE 610.12-1990)

- **Acceptance Test Procedures (ATPs)** Formal procedures that verify the successful achievement of one or more specification requirements in accordance with the verification method(s) stated for the requirement(s).

- **Alpha Testing** "System testing performed at the developer's site, often by employees of the customer." (Source: Kossiakoff and Sweet, *System Engineering*, p. 445)

- **Anomaly** An unexplainable event or observation that cannot be replicated based on current knowledge or facts.

- **Automatic Test Equipment (ATE)** "Equipment that is designed to automatically conduct analysis of functional or static parameters and to evaluate the degree of UUT (Unit Under Test) performance degradation; and may be used to perform fault isolation of UUT malfunctions. The decision making, control, or evaluative functions are conducted with minimum reliance on human intervention and usually done under computer control." (Source: MIL-HDBK-470A, Appendix G—Glossary, p. G-2)

- **Beta Testing** "System testing performed at a customer's site without the developer's presence, and reported to the developer." (Source: Kossiakoff and Sweet, *System Engineering*, p. 445)

- **Destructive Tests** Tests that result in stressing a test article to failure beyond repair via destruction or loss. *Destructive tests* usually destroy, damage, or impair a test article's form, structure, capabilities, or performance beyond refurbishment at a practical and economically feasible level.
- **Formal Testing** "Testing conducted in accordance with test plans and procedures that have been reviewed and approved by a customer, user, or designated level of management." (Source: IEEE 610.12-1990 *Standard Glossary of Software Engineering Terminology*)
- **Nondestructive Tests** Tests that subject a test article to a prescribed set of input conditions and operating environment to demonstrate compliance to requirements. Nondestructive tests do not destroy, damage, or impair a test article's appearance, form, structure, capabilities, or performance other than minor refurbishment.

EXAMPLE 55.1

Nondestructive tests include qualification tests concerning temperature, humidity, shock, vibration, and so forth.

- **Qualification Test** "Simulates defined operational environmental conditions with a predetermined safety factor, the results indicating whether a given design can perform its function within the simulated operational environment of a system." (Source: *DSMC Simulation Based Acquisition: A New Approach*, Dec. 1998)
- **Regression Testing** "A selected set of previously conducted tests after a system change to reconfirm their validity." (Source: Kossiakoff and Sweet, *System Engineering*, p. 452)
- **Test** The act of subjecting, measuring, and evaluating a SYSTEM or entity's responses to a prescribed and controlled set of OPERATING ENVIRONMENT conditions, verification methods, and stimuli and comparing the results to a set of specified capability and performance requirements.
- **Test and Evaluation (T&E)** The *informal* or *semiformal* act of evaluating the behavioral response and reaction time performance of a system entity within a prescribed OPERATING ENVIRONMENT to a controlled set of stimuli for purposes of assessing entity functionality and eliminate defects. T&E establishes that an article is free of *latent defects* or *deficiencies*, as allowed, and is ready for formal verification. T&E should include a "dry run" of the formal verification *acceptance test procedures* (ATPs) prior to formal verification. T&E activities are generally *informal* contractor activities and may or may not be observed by the Acquirer.
- **Test and Evaluation Working Group (TEWG)** A team consisting of User, Acquirer, System Developer, Subcontractor, and vendor personnel stakeholders formed to plan, coordinate, implement, monitor, analyze, and evaluate test results.
- **Test Article** An initial unit of a SYSTEM or PRODUCT or one randomly extracted from a production lot to be used for conducting *nondestructive* or *destructive* tests.
- **Test Case** One instance of a series of use case scenario-based tests that employ combinations of test inputs and conditions to verify an item's ability to ACCEPT/REJECT ranges of inputs, perform value-added processing and to produce only ACCEPTABLE performance-based outcomes or results.
- **Test Configuration** A controlled architectural framework capable of representing an item's OPERATING ENVIRONMENT conditions—NATURAL, INDUCED, or HUMAN-MADE

SYSTEMS—via simulation, stimulation, or emulation to verify that the *test article* satisfies a specific requirement or set of requirements.

- **Test Coverage** "The degree to which a given test or set of tests addresses all specified requirements for a given system or component." (Source: IEEE 610.12-1990)
- **Test Criteria** "Standards by which test results and outcome are judged." (Source: DSMC, *Test & Evaluation Management Guide*, Appendix B—*Glossary of Test Terminology*)
- **Test Environment** The set of System Elements (EQUIPMENT, PERSONNEL, FACILITIES, PROCEDURAL DATA, MISSION RESOURCES, simulated NATURAL and INDUCED ENVIRONMENTs, etc.) configured to represent a test article's OPERATING ENVIRONMENT conditions.
- **Test and Measurement Equipment** "The peculiar or unique testing and measurement equipment which allows an operator or maintenance function to evaluate operational conditions of a system or equipment by performing specific diagnostics, screening or quality assurance effort at an organizational, intermediate, or depot level of equipment support." (Source: MIL-HDBK-881, Appendix H, para. H.3.6.1)

EXAMPLE 55.2

Examples include: "... test measurement and diagnostic equipment, precision measuring equipment, automatic test equipment, manual test equipment, automatic test systems, test program sets, appropriate interconnect devices, automated load modules, taps, and related software, firmware and support hardware (power supply equipment, etc.) used at all levels of maintenance." (Source: MIL-HDBK-881, Appendix H, para. H.3.6.1)

- **Test Incident Report** "A document that describes an event that occurred during testing which requires further investigation." (Source: IEEE 610.12-1990)
- **Test Instrumentation** "Test instrumentation is scientific, automated data processing equipment (ADPE), or technical equipment used to measure, sense, record, transmit, process or display data during tests, evaluations or examination of materiel, training concepts or tactical doctrine. Audio-visual is included as instrumentation when used to support Acquirer testing." (Source: Adapted from DSMC, *Test & Evaluation Management Guide*, Appendix B, *Glossary of Test Terminology*)
- **Test Range** An indoor or outdoor FACILITY that provides a safe and secure area for evaluating a SYSTEM or entity's capabilities and performance under near NATURAL ENVIRONMENT or simulated conditions.
- **Test Repeatability** "An attribute of a test, indicating that the same results are produced each time the test is conducted." (Source: IEEE 610.12-1990)
- **Test Resources** "A collective term that encompasses all elements necessary to plan, conduct and collect/analyze data from a test event or program." (Source: DSMC, *Glossary: Defense Acquisition Acronyms and Terms*)
- **Testing** "The process of operating a system or component under specified conditions, observing or recording the results, and making an evaluation of some aspect of the system or component." (Source: IEEE 610.12-1990)
- **Transient Error** An "Error that occurs once, or at unpredictable intervals." (Source: IEEE 610.12-1990)

55.2 SITE FUNDAMENTALS

To better understand the SITE tasks discussed in this section, let's introduce some of the fundamentals of that provide the foundation for SITE.

What Is SITE?

System integration, test, and evaluation (SITE) is the sequential, bottoms-up process of:

1. Incremtially interfacing previously *verified* system *items* and *configuration items (CIs)*—consisting of PARTS, SUBASSEMBLIES, ASSEMBLIES, SUBSYSTEMS and PRODUCTS—at *Integration Points (IPs)*, beginning with the lowest level.
2. Conducting *functional* and qualification tests of the integrated test article to verify all capabilities comply with specification and design requirements.
3. Evaluating the test results for compliance and optimizing test article performance.

Completion of SITE at each Integration Point (IP) is marked by a formal verification test, which may or may not be performed formal *acceptance test procedures* (ATPs). On completion of each test, we conduct compliance assessments based on *performance* or *development specification* verification requirements and methods. Each test *proves* the *test article's* capability to perform over the *prescribed* operating range of inputs and environmental conditions. For some programs, lower level SITE verification tests may be semiformal, *informal,* and *unwitnessed.* In other cases, the tests:

1. Are formal.
2. Employ approved *acceptance test procedures* (ATPs) derived from specification requirements and verification methods.
3. Are witnessed by all stakeholders including the Acquirer and User.

Depending on the contract, the Acquirer and User, at the request of the Acquirer, are invited to "witness" the SITE and verification activities.

Author's Note 55.1 *As a reminder, the System Developer is accountable for notifying the Acquirer regarding* SITE *in accordance with the terms and conditions of the contract. Under conventional contracting protocol, the Acquirer notifies the User unless the* Acquirer has made special arrangements with the System Developer. *Conversely, when the Acquirer and User have open invitations to observe and witness SITE activities, as a professional courtesy to the System Developer, the Acquirer should inform the System Developer's Program Director in advance WHO, HOW MANY people, and WHAT organization(s) will participate.*

SITE Objective

The objective of SITE is to subject a *test article* of a SYSTEM, item, or configuration item (CI) to a range of test cases, input stimuli and/or cues, and conditions representative of a prescribed OPERATING ENVIRONMENT so that the achievement of each capability requirement and its performance can be verified.

Engineers often have *misperceptions* of WHAT is to be accomplished by SITE. The ERRONEOUS view is that a *system* or *entity* is subjected to a set of conditions *bounded* by *minimum/maximum* performance requirement *thresholds.* The fallacy in this view is: *HOW does the item perform when subjected to conditions* BEYOND *these limits?* Obviously, you could test an item

738 Chapter 55 System Integration, Test, and Evaluation

beyond its physical limits such as environmental conditions, electrical overload, or shorts, but *what about improper data formats, magnitudes that are under/over range?* If the item is designed properly, it should accommodate these conditions without failure.

The objective should be to *exercise* and *assess* the item's capability to cope with *acceptable* and *unacceptable* input conditions. Likewise, for those input conditions, produce only acceptable SYSTEM RESPONSES such as *behavior*, *products*, *by-products*, and *services*.

SITE Conduct Sequencing

SITE, which is part of Developmental Test and Evaluation (DT&E), is conducted following the Component Procurement and Development Segment of the System Development Phase as illustrated in Figure 55.1. Throughout the SITE Segment, the System Developer conducts Test Readiness Reviews (TRRs) as *entry criteria* for various tests. SITE activities culminate in a System Verification Review (SVR).

What Do SITE Activities Prove?

We can create nice words about SITE such as verify *compliance* with specification requirements, but *WHAT do SITE activities really accomplish*. In very simple terms, SITE answers five key questions:

1. Can the SYSTEM/entity design *interoperate* with external systems in its OPERATING ENVIRONMENT?
2. Does the SYSTEM/entity predictably and responsively *function* as planned?
3. Can the SYSTEM/entity materials and components *survive* the *stresses* of the prescribed OPERATING ENVIRONMENT conditions for the duration of the mission/operating cycle limits?
4. Is the *quality of workmanship* and *construction* sufficient to ensure the integrity of component interfaces?
5. As applicable, can the SYSTEM/entity be easily maintained?

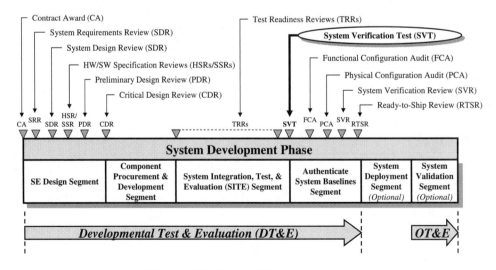

Where: OT&E = Operational Test & Evaluation

Figure 55.1 System Verification Test (SVT)

Engineering Model, First Article, Test Article, and Unit-Under-Test (UUT) Semantics

SITE terminology includes terms such as *engineering model, first article, test article,* and *unit under test (UUT).* Let's explore each term briefly.

Engineering Model(s). An engineering model, which may or may not be a contract deliverable, serves as an initial prototype. The model is used in collecting data to *validate* system models and simulations or in demonstrating technologies or proofs of concept. From a *spiral development* perspective, one or more engineering models may be developed in *iterative* sequences as *risk mitigation* devices over a period of time to identify requirements from which the Developmental Configuration design will be used to produce the initial *First Article(s).*

First Article(s). By the term *first article* we mean the initial units of the approved Developmental Configuration that are available for verification testing and subsequent delivery in accordance with the *terms and conditions* (Ts&Cs) of the contract. The term is sometimes a misnomer since several *first articles* may be produced from the Developmental Configuration. Collectively, the set of devices are referred to as *first articles*. Each initial set of *first articles* is subjected to various *nondestructive* and *destructive* tests. On successful completion of *system* verification *testing* of the *first article(s)*, the resulting Developmental Configuration is used to establish the Product Baseline.

Test Article and UUTs. The term *test article* is actually a generic term that represents any item used for test purposes. So, a *first-article* system progressing through its development is referred to as a *test article* during SITE and as a *unit under test (UUT)* when integrated into the test configuration for verification testing. Production articles may be *test articles* in the sense that only functional tests are performed.

Key Elements of SITE

Planning and implementation of SITE requires an understanding of its key elements—the PERSONNEL, EQUIPMENT, FACILITIES, and other system elements—and HOW to orchestrate these elements during the test. We represent these System Elements via the system block diagram (SBD) shown in Figure 55.2.

The *test article* serves as the SYSTEM OF INTEREST (SOI) for SITE. Surrounding the test article is the test environment, which consists of the test operator, test procedures, test log, test equipment and tools, controlled test environment and design documentation, all contained within the test facility.

Author's Note 55.2 *Note how the Controlled Test Environment is abstracted as an entity rather than shown as surrounding the test article. Analytically, both methods are acceptable. Abstracting the Controlled Operating Environment as a box with interfaces explicitly reminds you of the need to identify and specify the interfaces. If we encompassed the test article within the Controlled Operating Environment (i.e., a box within a box), the interface relationships would be less explicit and more difficult to identify.*

Guiding Philosophy of SITE

The guiding philosophy of SITE can be summarized in a few words: KEEP IT SIMPLE! *Simplicity* means establish an item *compliant* with its specification and then *incrementally* integrate other items one at a time.

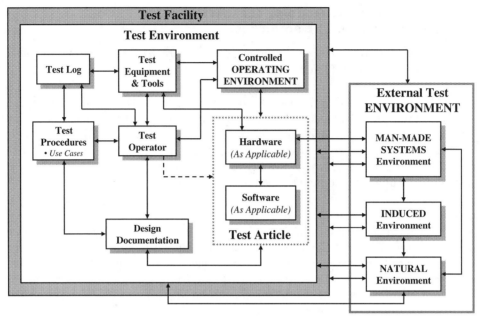

Figure 55.2 SITE Entity Relationships SBD Example

People often have the erroneous viewpoint that they can lash all PRODUCTs/SUBSYSTEMs into a gigantic test configuration and begin testing! If you do this, you may and probably will be confronted with *n* items, each with scores of UNPROVEN capabilities *interacting simultaneously*! If something does not work properly, HOW will you ever sort out WHAT is causing the problem? You will soon discover that you need to disassemble everything and start with a simple *proven* and *verified* item, integrate another item with it, and incrementally build a system. Remember, SITE *root wisdom* says integrate one *unproven* and *unverified* item at a time! At least you will know WHEN the problem entered the test article configuration and thereby can *simplify trouble shooting* and *fault isolation*. For that item, *disable* as many capabilities as you can and *incrementally* test and enable one new capability at a time.

Preparing Items for SITE

During the SE Design Segment of the System Development Phase, we *partitioned* and *decomposed* the *solution space* into successively lower levels of detail down to the PART level *item*. During Component Procurement and Development, each *item* is procured/fabricated, coded, assembled, and tested. Incrementally, we subject the *item* to *inspections* and *verification* to ensure that it complies with its *design requirements*—drawings, source codes, and so forth. Since each item serves as a "building block" for constructing or assembling the SYSTEM, item verification evaluates its *form*, *fit*, and *function*.

Once you understand the importance of the *form*, *fit*, *function* at the PART level, SITE is simply testing the *form*, *fit*, and *function* of physical *items* at higher *levels of* Integration Points (IPs) as a functioning system with a given set of capabilities. System integration is actually a *verification* exercise whereby: 1) components are *integrated*, 2) *interfaces* are verified, and 3) *interoperability* is demonstrated or tested.

Types of Testing

When organizations discuss SITE activities, you will often hear terms such as functional, environmental/qualification, destructive, and nondestructive testing. Let's clarify the context of each of these terms.

Functional and Environmental/Qualification Testing. In general, SITE activities perform two categories of testing: *functional testing* and environmental/*qualification testing*.

- *Functional testing* simply means that the components, when interconnected and integrated, perform actions and interoperate as planned with no errors under ambient conditions.
- *Environmental/qualification testing* is the next higher level and focuses on HOW WELL the *item* performs in its PRESCRIBED OPERATING ENVIRONMENT conditions. Qualification testing includes tests performed as part of Developmental Test and Evaluation (DT&E) *verification* activities and, if applicable, during Operational Test and Evaluation (OT&E) *validation* activities. Environmental/qualification testing involves environmental conditions such as temperature, humidity, shock, vibration, and Electromagnetic Interference (EMI).

Nondestructive and Destructive Testing. During *functional* and environmental/*qualification* testing, *test articles* may be subjected to a wide range of tests that may *destroy*, *blemish*, or *alter* the *test article's* appearance, structural integrity, capability, or performance. We refer to these as *destructive* tests. In contrast, *nondestructive* tests may not harm the item, which may be *refurbished* for delivery, assuming it is *permissible* by contract and safety practices.

During the System Production Phase of a program, *production sample test articles* may be randomly selected for a test program to assess the quality of materials, workquality, system capabilities and performance, shelf-life degradation, and so forth.

Author's Note 55.3 *ALWAYS consult your contract and program, contracts, and legal organizations for guidance concerning test article delivery, recovery, or refurbishment.*

Creating the Test Article's OPERATING ENVIRONMENT

When we perform SITE, we subject the test article to scenarios and conditions that represent its OPERATING ENVIRONMENT. This requires creating the HIGHER ORDER SYSTEMS and PHYSICAL ENVIRONMENT Elements, among these HUMAN-MADE SYSTEMS, the INDUCED ENVIRONMENT, and the NATURAL ENVIRONMENT. *How do we do this?*

There are several options for creating the OPERATING ENVIRONMENT as illustrated in Figure 51.5. We can *simulate*, *stimulate* or *emulate* entities within the environment or employ combinations of these options.

- **Simulate** To create a virtual model interface that *represents* the performance-based behavioral responses and characteristics of an external entity.
- **Stimulate** To create an interface using actual EQUIPMENT or a test set that has identical physical performance characteristics of the external entity.
- **Emulate** To create an interface that *identically mimics* the actual operations, processing sequences, and performance characteristics of an external system entity.

In each of these cases, the objective is to create an interface that is so *representative* or *realistic* of the external system that the test article is unable to discern its difference from reality.

SITE Conduct Procedures

Two types of procedures govern SITE conduct: *standard operating practices and procedures (SOPPs)* and *acceptance test procedures (ATPs)*.

Standard Operating Practices and Procedures (SOPPs). *Standard operating practices and procedures (SOPPs)* are organizational command media that apply to test conduct in test facilities and ranges. SOPPs focus on the *safe* and *proper* handling of EQUIPMENT, human and environmental safety, laboratory/test range procedures, security procedures, emergency procedures, as well as the *prevention* of *hazards*.

Acceptance Test Procedures (ATPs). *Acceptance test procedures* are derived from each Section 3.0 *System Performance Specification (SPS)* or *item developmental specification* (IDS) requirement using the prescribed Section 4.0 Verification Methods. Since a key objective is to *minimize* test costs, some Section 3.0 requirements may be verified simultaneously with a single test. ATPs should reflect this approach and *note* the requirements being *verified*.

Who Performs Tests?

Questions often emerge regarding WHO should perform the PART, SUBASSEMBLY, ASSEMBLY, SUBSYSTEM, PRODUCT, and SYSTEM level tests. This question has two aspects: testing personal work products and qualifying test personnel.

Testing Personal Work Products. Testing ranges from *informal* to *highly formal*. As a matter of practice, system designers SHOULD NOT test their own designs, provided that someone else is capable. The *underlying* philosophy of this principle is that it this represents a potential conflict of interest in objectively checking off your own work. In most organizations SE designers develop and perform informal testing of their work products, typically with *peer level* review *scrutiny*. Some organizations create *independent* test teams and assign them to perform testing at all levels of integration. Some contracts employ IV&V teams to perform testing, especially of software.

Qualifying Test Personnel. Testing requires knowledge, discipline, observational skills, adherence to safety practices, integrity, accuracy, and precision in reporting test results. In general, CONTRACT testing is not for amateurs; it requires training and experience, and often *certification*. Therefore, organizations should establish in-house command media policies that only testers who have been trained and *certified* for a defined period of time perform tests.

Simultaneous Testing of Multiple Requirements

Testing can rapidly become very expensive and consume valuable schedule resources. Remember, you only employ TEST where INSPECTION, ANALYSIS, and DEMONSTRATION verification methods are *insufficient* to demonstrate full compliance with a requirement.

There are ways to *efficiently* and *effectively* perform TESTs to reduce costs and schedule. Some people believe that you conduct one test for *each* requirement. In general, analyze a *performance* or *development specification* to identify dependencies and sets of requirements that can be tested *simultaneously* where TEST is the required verification method.

What Is Regression Testing?

SITE can be very expensive. As a result, you want to *minimize* the amount of *rework* and *retest*. The question is: *WHEN you discover a latent defect, design flaw, deficiency, or error that requires corrective action*, WHAT *tests impacted by the failure must be repeated before you can resume testing at the test sequence where the failure occurred?* The answer resides in *regression testing*.

During *regression testing* only those aspects of the component design that have been impacted by a *corrective action* are *re-verified*. *Form, fit, and function* checks of the UUT that were successfully completed and not affected by any corrective actions are not re-verified.

Discrepancy Reports (DRs)

Inevitably, *discrepancies* will appear between *actual* test data and bounded *expected* results specified in the SPS or *item development specification*. We refer to the occurrence of a discrepancy as a *test event*.

When test events such as failures occur, a *Discrepancy Report (DR)* is recorded. The Test Director should define WHAT constitutes a *test event* in the SIVP and event criteria for recording a DR. At a *minimum*, DRs document:

1. The test event, date, time.
2. Conditions and prior sequence of steps preceding a test event.
3. Test article identification.
4. Test configuration.
5. Reference documents and versions.
6. Specific document item requirement and expected results.
7. Results observed and recorded.
8. DR author and witnesses or observers.
9. Degree of significance requiring a level of urgency for corrective action.

DRs have levels of significance that affect the test schedule. They involve safety issues, data integrity issues, isolated tests that may not affect other tests, and cosmetic blemishes in the test article. As standard practice, establish a priority system to facilitate disposition of DRs. An example is provided in Table 55.1.

Table 55.1 Example test event DR classification system

Priority	Event	Description
1	Emergency condition	All testing must be TERMINATED IMMEDIATELY due to imminent DANGER to the Test Operators, test articles, or test facility.
2	Test component or configuration failure	Testing must be HALTED until a *corrective action* is performed. *Corrective* action may require redesign or replacement of a failed component.
3	Test failure	Testing can continue if the failure does not diminish the integrity of remaining tests; however, the test article requires corrective action reverification prior to integration at the next higher level.
4	Cosmetic blemish	Testing is permitted to continue, but corrective, action must be performed prior to system acceptance.

SITE Work Products

SITE *work products* that serve as *objective evidence* for a single item, regardless of level of abstraction, include:

1. A set of dated and signed entries in the Test Log that identify and describe:
   a. *Test Team*—the name of the responsible engineering team and lead.
   b. *Test Article*—what "parent" test article was integrated from what version of lower level test articles.
   c. *Test(s) conducted.*
   d. *Test results*—where recorded and stored.
   e. *Problems, conditions, or anomalies*—encountered and identified.
2. *Discrepancy report (DR).*
3. *Hardware trouble reports (HTRs)* documented and logged.
4. *Software change requests (CRs)* documented and logged.
5. State of readiness of the test article for scheduling formal verification.

Guidepost 55.1 *This completes our overview of SITE fundamentals. We now shift our attention to planning for SITE.*

55.3 PLANNING FOR SITE

SITE success begins with insightful planning to identify the test objectives; roles; responsibilities, and authorities; tasking, resources, facilities; and schedule. Testing, in general, involves two types of test plans:

1. The *Test and Evaluation Master Plan (TEMP).*
2. The *System Integration and Verification Plan (SIVP).*

The Test and Evaluation Master Plan (TEMP)

In general, the TEMP is a User's document that expresses HOW the User or an Independent Test Agency (ITA) representing the User's interests plans to validate the system, product, or service. From a User's perspective, development of a new system raises *critical operational and technical issues (COIs/CTIs)* that may become SHOWSTOPPERS to *validating* satisfaction of an organization's operational need. So, the scope of the TEMP covers the Operational Test and Evaluation (OT&E) period and establishes objectives to verify resolution of COIs/CTIs.

The TEMP is structured to answer a basic question: *Does the system, product, or services, as delivered, satisfy the User's validated operational needs—in terms of problem space and solution space?* Answering this question requires *formulation* of a set of *scenario*-driven test objectives—namely *use cases and scenarios.*

The System Integration and Verification Plan (SIVP)

The SIVP is written by the System Developer and expresses their approach for integrating and testing the SYSTEM or PRODUCT. The scope of the SIVP, which is contract dependent, covers the Developmental Test and Evaluation (DT&E) period from Contract Award through the formal System Verification Test (SVT), typically at the System Developer's facility.

The SIVP identifies objectives, organizational roles and responsibilities, tasks, resource requirements, strategy for sequencing testing activities, and schedules. Depending on contract requirements, the SIVP may include delivery, installation, and checkout at a User's designated job site.

Developing the System Integration and Test Strategy

The strength of a system integration and test program requires "up front" THINKING to ensure that the vertical integration occurs *just in time* (JIT) in the proper sequences. Therefore, the first step is to establish a strong *system integration and test strategy*.

One method is to construct a *System Integration and Test Concept* graphically or by means of an integration decision tree. The test concept should reflect:

1. WHAT component integration dependencies are critical?
2. WHO is responsible and accountable for the integration?
3. WHEN and in WHAT sequence they will be integrated?
4. WHERE is the integration to be performed?
5. HOW the components will be integrated?

The SITE process may require a single facility such as a laboratory or multiple facilities within the same geographic area, or integration across various geographical locations.

Destructive Test Sequence Planning

During the final stages of SITE Developmental Test and Evaluation (DT&E), several *test articles* may be required. The challenge for SEs is: *HOW and in WHAT sequence do we conduct* non-destructive *tests to collect data to verify design compliance prior to conducting destructive test that may destroy or damage the test article?* THINK through these sequences carefully.

Guidepost 55.2 *Once the site plans are in place, the next step requires establishing the test organization.*

55.4 ESTABLISHING THE TEST ORGANIZATION

One of the first steps following approval of the SIVP is establishing the test organization and assignment of roles, responsibilities, and authorities. Key roles include Test Director, Lab Manager, Tester, Test Safety Officer or Range Safety Officer (RSO), Quality Assurance (QA), Security Representative, and Acquirer/User Test Representative.

Test Director Role

The Test Director is a member of the System Developer's program and serves as the key decision authority for testing. Since SITE activities involve interpretation of specification statement language and the need to access test ports and test points to collect data for *test article* compliance verification, the Test Director role should be assigned EARLY and be a key participant in System Design Segment reviews. At a *minimum*, the primary Test Director responsibilities are:

1. Develop and implement the SIVP.
2. Chair the Test and Evaluation Working Group (TEWG), if applicable.

3. Plan, coordinate, and synchronize test team task assignments, resources, and communications.
4. Exercise authoritative control of the test configuration and environment.
5. Identify, assess, and mitigate test risks.
6. Review and approve test conduct rules and test procedures.
7. Account for personnel environmental, safety, and health (ES&H).
8. Train test personnel.
9. Prioritize and disposition of *discrepancy reports (DRs)*.
10. Verify DR corrective actions.
11. Accomplish contract test requirements.
12. Preserve test data and results.
13. Conduct failure investigations.
14. Coordinate with Acquirer/User test personnel regarding disposition of DRs and test issues.

Lab Manager Role

The Lab Manager is a member of the System Developer's program and supports the Test Director. At a *minimum*, the primary Lab Manager responsibilities are:

1. Implement the test configuration and environment.
2. Acquire of test tools and equipment.
3. Create the laboratory notebook.
4. Support test operator training.

Test Safety Officer or Range Safety Officer Role

Since testing often involves *unproven* designs and test configurations, safety is a very *critical issue*, not only for test personnel but also for the test article and facilities. Therefore, every program should designate a Safety Officer.

In general, there are two types of test safety officers: SITE Test Officer and Range Safety Officer (RSO).

- The Test Safety Officer is a member of the System Developer's organization and supports the Test Director and Lab Manager.
- The Range Safety Officer is a member of a test range.

In some cases, Range Safety Officers (RSOs) have authority to destruct test articles should they become *unstable* and *uncontrollable* during a test or mission and pose a threat to personnel, facilities, and/or the public.

Tester Role

As a general rule, system, product, or service developers should not test their own designs; it is simply a *conflict of interest*. However, at lower levels of abstraction, programs often *lack* the resources to adequately train testers. So, System Developers often perform their own informal testing. For some contracts Independent Verification and Validation (IV&V) Teams, *internal* or *external* to the program or organization, may perform the testing.

Regardless of WHO performs the *tester role*, test operators must be trained in HOW to *safely* perform the test, record and document results, and deal with anomalies. Some organizations formally train personnel and refer to them as *certified test operators* (CTOs).

Quality Assurance (QA) Representative

At a *minimum*, the System Developer's Quality Assurance Representative (QAR) is responsible for ensuring *compliance* with contract requirements, organizational and program command media, the SIVP, and ATPs for software-intensive system development efforts, a Software Quality Assurance (SQA) representative is assigned to the program.

Security Representative

At a *minimum*, the System Developer's Security Representative, if applicable, is responsible for the assuring *compliance* with contract security requirements, organizational and program command media, the programs security plan and ATPs.

Acquirer Test Representative

Throughout SITE, test issues surface that require an Acquirer decision. Additionally, some Acquirers represent several organizations, many with conflicting opinions. This presents a challenge for System Developers. One solution is for the Acquirer Program Manager to designate an individual to serve as an on-site representative at the System Developer's facility and provide a single voice representing all Acquirer viewpoints. Primary responsibilities are to:

1. Serve as THE single point of contact for ALL Acquirer and User technical interests and communications.
2. Work with the Test Director to resolve any *critical operational or technical test issues (COIs/CTIs)* that affect Acquirer-User interests.
3. Where applicable by contract, collaborate with the Test Director to assign priorities to *discrepancy reports (DRs)*.
4. Where appropriate, review and coordinate approval of *acceptance test procedures (ATPs)*.
5. Where appropriate, provide a single set of ATP comments that represent a consensus of the Acquirer-User organizations.
6. Witness and approve ATP results.

55.5 DEVELOPING ATPs

In general, ATPs provide the *scripts* to verify compliance with *SPS* or *item development specification* requirements. In *Chapters 13 through 17 System Mission Concepts,* we discussed that an SPS or item's development specification (IDS) is derived from *use cases* and *scenarios* based on HOW the User envisions using the system, product, or service. In general, the ATPs script HOW TO demonstrate that the SYSTEM or *item* provides a specified set of capabilities for a given phase and mode of operation. To a test strategy, we create *test cases* that verify these capabilities as shown in Table 55.2.

Types of ATPs

ATPs are generally of two types: *procedure-based* and *scenario-based* (*ATs*).

Table 55.2 Derivation of test cases

Phase of Operation	Mode of Operation	Use Cases (UCs)	Use Case Scenarios	Required Operational Capability	Test Cases (TCs)
Pre-mission phase	Mode 1	UC 1.0		ROC 1.0	TC 1.0
			Scenario 1.1	ROC 1.1	TC 1._
			...		TC 1._
			Scenario 1.n	ROC 1.n	TC 1._
			...		TC 1._
	Mode 2	UC 2.0		ROC 2.0	TC 2.0
			Scenario 2.1	ROC 2.1	TC 2._
			...		TC 2._
			Scenario 2.n	ROC 2.n	TC 2._
			...		TC 2._
		UC 3.0		ROC 3.0	TC 3.0
			Scenario 3.1	ROC 3.1	TC 3._
			...		TC 3._
			Scenario 3.n	ROC 3.n	TC 3._
			...		TC 3._
Mission phase	(Modes)	(Fill-in)	(Fill-in)	(Fill-in)	(Fill-in)
Post-mission phase	(Modes)	(Fill-in)	(Fill-in)	(Fill-in)	(Fill-in)

Note: A variety of TCs are used to test the SYSTEM/entity's inputs/outputs over *acceptable* and *unacceptable* ranges.

Procedure-Based ATPs. *Procedure-based ATPs* are highly detailed test procedures that describe test configurations, environmental controls, test input switchology, expected results and behavior, among other details, with *prescribed* script of *sequences* and *approvals*. Consider the example shown in Table 55.3 for a secure Web site log-on test by an *authorized* user.

Scenario-Based ATPs. *Scenario-based ATPs* are generally performed during system *validation* either in a controlled facility or in a prescribed field environment. Since *system validation* is intended to evaluate a system's capabilities to meet a User's *validated* operational needs, those needs are often *scenario based*.

Scenario-based ATPs employ objectives or missions as key drivers for the test. Thus, the ATP tends to involve very high level statements that describe the operational mission scenario to be accomplished, objective(s), expected outcome(s), and performance. In general, scenario-based ATPs defer to the Test Operator to determine which *sequences* of "switches and buttons" to use based on operational familiarity with the *test article*.

Table 55.3 Example procedure-based acceptance test (AT) form

Test Step	Test Operator Action to be Performed	Expected Results	Measured, Displayed, or Observed Results	Pass/Fail Results	Operator Initials and Date	QA
Step 1	Using the left mouse button, click on the Web site link.	Web browser is launched to selected Web site.	*Web site appears*	*Pass*	*JD 4/18/XX*	QA 205 *KW 4/18/XX*
Step 2	Using the left mouse button, click on the "Logon" button.	Logon access dialogue box opens up.	*As expected*	*Pass*	*JD 4/18/XX*	QA 205 *KW 4/18/XX*
Step 3	Position the cursor within the User Name field of the dialogue box.	Fixed cursor blinks in field.	*As expected*	*Pass*	*JD 4/18/XX*	QA 205 *KW 4/18/XX*
Step 4	Enter user ID (max. of 10 characters) and click SUBMIT	Field displayed.	*User ID entered*	*Pass*	*JD 4/18/XX*	QA 205 *KW 4/18/XX*

55.6 PERFORMING SITE TASKS

SITE, as with any system, consists of three phases: pre-testing phase, testing phase, and a post-testing phase. Each phase consists of a series of tasks for integrating, testing, evaluating, and verifying the design of an item's or configuration item (CI). Remember, every system is unique. The discussions that follow represent generic test tasks that apply to every level of abstraction. These tasks are *highly interactive* and may *cycle* numerous times, especially in the testing phase.

Task 1.0: Perform Pre-test Activities

- **Task 1.1** Configure the test environment.
- **Task 1.2** Prepare and instrument the test article(s) for SITE.
- **Task 1.3** Integrate the test article into the test environment.
- **Task 1.4** Perform a test readiness inspection and assessment.

Task 2.0: Test and Evaluate Test Article Performance

- **Task 2.1** Perform informal testing.
- **Task 2.2** Evaluate informal test results.
- **Task 2.3** Optimize design and test article performance.
- **Task 2.4** Prepare test article for formal verification testing.
- **Task 2.5** Perform a "dry run" test to check out ATP.

750   Chapter 55   System Integration, Test, and Evaluation

Task 3.0: Verify Test Article Performance Compliance

- **Task 3.1** Conduct a test readiness review (TRR).
- **Task 3.2** Formally verify the test article.

Task 4.0: Perform Post-test Follow-up Actions

- **Task 4.1** Prepare item *verification test reports* (VTRs).
- **Task 4.2** Archive test data.
- **Task 4.3** Implement all DR corrective actions.
- **Task 4.4** Refurbish/recondition test article(s) for delivery, if permissible.

Resolving Discrepancy Reports (DRs)

When a test failure occurs and a discrepancy report (DR) is documented, a determination has to be made as to the significance of the problem on the test article and test plan as well as isolation of the problem source. While the general tendency is to focus on the test article due to its *unproven* design, the *source* of the problem can *originate* from any one of the test environment elements shown in Figure 55.2 such as a test operator or test procedure error; test configuration or test environment problem, or combinations of these. From these contributing elements we can construct a fault isolation tree such as the one shown in Figure 55.3. Our purpose during an investigation such as this is to *assume* everything is *suspect* and *logically* rule out elements by a process of elimination.

Once the DR and the conditions surrounding the failure are understood, our first decision is to determine if the problem originated *external* or *internal* to the test facility, as applicable. For those problems originating within the facility, decide if this is a test operator, test article, test configuration, Test environment, or test measurement problem.

Since the test procedure is the orchestration mechanism, start with it and its test configuration. *Is the test configuration correct? Was the test environment controlled at all times without any dis-*

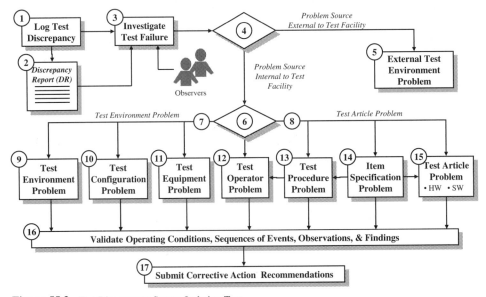

Figure 55.3   Test Discrepancy Source Isolation Tree

continuities? Did the operator perform the steps correctly in the proper sequence without bypassing any? Is the test procedure flawed? Does it contain errors? Was a dry run conducted with the test procedure prior to verify its logic and steps? If so, the test article may be suspect.

Whereas people tend to rush to judgment and take corrective action, VALIDATE the problem source by reconstructing the test configuration, operating conditions, sequence of events, test procedures, and observations as documented in the test log, DR, and test personnel interviews.

If the problem originated with the test article, retrace the development of the item in reverse order to its system development workflow. *Was the item built correctly per its design requirements? Was it properly inspected and verified? If so, was there a problem due to component, material, process, or workmanship defect(s) or in the verification of the item?* If so, determine if the test article will have to be *reworked, scrapped, reprocured, or retested.* If not, the design or specification may be suspect.

Audit the design. *Is it fully compliant with its specification? Does the design have an inherent flaw? Were there errors in translating specification requirements into the design documentation? Was the specification requirement misinterpreted?* If so, *redesign* to correct the flaw or error will have to be performed. If not, since the specification establishes the *compliance* thresholds for verification testing, you may have to consider: 1) revising the specification and 2) reallocating performance budgets and margins. Based on your findings, recommend, obtain approval, and implement the corrective actions. Then, perform *regression* testing based on the where the last validate test unaffected by the failure was completed.

55.7 COMMON INTEGRATION AND TEST CHALLENGES AND ISSUES

SITE practices often involve a number of challenges and issues for SEs. Let's explore some of the more common ones.

Challenge 1: SITE Data Integrity

Deficiencies in establishing the test environment, poor test assumptions, improperly trained and skilled test operators, and an uncontrollable test environment *compromise* the integrity of engineering test results. Ensuring the integrity of test data and results is *crucial* for downstream decision making involving formal acceptance, certification, and accreditation of the system.

Warning! *Purposeful actions to DISTORT or MISREPRESENT test data are a violation of professional and business ethics. Such acts are subject to SERIOUS criminal penalties that are punishable under federal or other statutes or regulations.*

Challenge 2: Biased or Aliased SITE Data Measurements

When instrumentation such as measuring devices are connected or "piggybacked" to "test points", the resulting impact can *bias* or *alias* test data and/or degrade system performance. Test data capture should be not degrade system performance. Thoroughly analyze the impact of *potential* effects of test device bias or alias on system performance BEFORE instrumenting a test article. *Investigate to see if some data may be derived implicitly from other data.* Decide:

1. How critical the data is needed.
2. If there are alternative data collection mechanism or methods.
3. Whether the data "value" to be gained is worth the technical, cost, and schedule risk.

Challenge 3: Preserving and Archiving Test Data

The end technical goal of SITE and *system verification* is to establish that a system, product, or service fully complies with its *System Performance Specification (SPS)*. The *validity* and *integrity* of the *compliance decision* resides in the formal *acceptance test procedure (ATP)* results used to record *objective evidence*. Therefore, ALL test data recorded during a formal ATP must be *preserved* by *archiving* in a permanent, safe, secure, and limited access facility. Witnessed or authenticated *test data* may be required to support:

1. A *functional configuration audit* (FCA) and a *physical configuration audit* (PCA) prior to system delivery and formal acceptance by the Acquirer for the User.
2. Analyses of system *failures* or *problems* in the field.
3. Legal claims.

Most contracts have requirements and organizations have policies that govern the storage and retention of contract data, typically several years after the completion of a contract.

Challenge 4: Test Data Authentication

When formal test data are recorded, the validity of the data should be authenticated, depending on end usage. *Authentication* occurs in a number of ways. Generally, the authentication is performed by an Independent Test Agency (ITA) or individual within the Quality Assurance (QA) organization that is *trained* and *authorized* to authenticate test data in accordance with *prescribed* policies and procedures. *Authentication* may also be required by higher level bonded, external organizations. At a *minimum*, *authentication criteria* include a witnessed affirmation of the following:

1. Test article and test environment configuration
2. Test operator qualifications and methods
3. Test assumptions and operating conditions
4. Test events and occurrences
5. Accomplishment of expected results
6. Pass/fail decision
7. Test discrepancies

Challenge 5: Dealing with One Test Article and Multiple Integrators and Testers

Because of the expense of developing large complex systems, multiple integrators may be required to work sequentially in shifts to meet development schedules. This potentially presents *problems* when integrators on the next shift waste time *uninstalling undocumented* "patches" to a build from a previous shift. Therefore, each SITE work shift should begin with a joint coordination meeting of persons going *off shift* and coming *on shift*. The purpose of the meeting is to make sure everyone *communicates* and *understands* the current configuration "build" that transpired during the previous shift.

Challenge 6: Deviations and Waivers

When a system or item fails to meet its performance, development, and/or design requirements, the *item* is tagged as *noncompliant*. For hardware, a *nonconformance report (NCR)* documents the discrepancy and *dispositions* it for corrective action by a Material Review Board (MRB). For software, a software developer submits a *Software Change Request (SCR)* to a Software Configuration Control Board (SCCB) for approval. *Noncompliances* are sometimes resolved by issuing a *deviation* or *waiver*, rework, or scrap without requiring a CCB action.

Challenge 7: Equipment and Tool Calibration and Certification

The *credibility* and *integrity* of a V&V effort is often dependent on:

1. *Facilities* and *equipment* to establish a controlled environment for system modeling, simulation, and testing.
2. Tools used to make precision adjustments in system/product functions and outputs.
3. Tools used to *measure* and *record* the system's environment, inputs, and outputs.
4. Tools used to *analyze* the system responses based on I/O data.

All of these factors:

1. Require *calibration* or *certification* to standards for weights, measures, and conversion factors.
2. Must be traceable to national/international standards.

Therefore, avoid *rework* and make sure that V&V activities have technical *credibility* and *integrity*. Begin with a firm foundation by ensuring that all *calibration* and *certification* is traceable to source standards.

Challenge 8: Insufficient Time Allocations for SITE

Perhaps one of the most SERIOUS challenges is making time allocations for SITE activities due to poor program planning and implementation. When organizations bid on system development contracts, executives bid an *aggressive* schedule based on some *clever* strategy to *win* the contract. This occurs despite the fact the organization may have had a long-standing history of poor contract performance such as cost/schedule overruns.

Technically, the more testing you perform, the more *latent defects* you can discover due to *design deficiencies*, *flaws*, or *errors*. While every system is different, on average, you should spend at least 40% of the schedule performing SITE. Most contract programs get behind schedule and compress that 40% into 10%. As a result, they rush SITE activities and inadequately test the system to "check the box" for completing schedule tasks.

There are four primary reasons for this:

1. Bidding *aggressive*, *unrealistic* schedules to win the contract.
2. Allowing the program to get off schedule, beginning with Contract Award, due to a lack of understanding of the *problem* and *solution spaces* or poor data delivery performance by external organizations.
3. Rushing *immature* designs into component procurement and development and finishing them during SITE.
4. Compressing component procurement and development to "check the box" and boldly PROCLAIM that SITE was entered "on time."
5. Assigning program management that understands meeting schedules and making profits but DOES NOT have a clue about the *magnitude* of the technical *problem* to be solved nor HOW TO *orchestrate* successful contract implementation and completion.

Challenge 9: Discrepancy Reporting Obstacles to SITE

One of the challenges of site is staying on schedule while dealing with *test discrepancies*. Establish a DR priority system to delineate those DRs from those that do not affect personal or equipment safety and do not jeopardize higher level test results. Establish *go/no-go* DR criteria to proceed to the next level of SITE.

Challenge 10: DR Implementation Priorities

Humans, by nature, like to work on "fun" things and the easy tasks. So, when DR corrective actions must be implemented, developers tend to work on those DRs that give them instant *credit* for completion. As a result, the challenging DRs get pushed to the bottom of the stack. Then, progress report metrics proudly proclaim the large quantity of DR corrective actions completed. In an earned value sense, you have a condition in which 50% of the DRs are completed the first month. Sounds great! Lots of productivity! Wrong! What we have is 50% of the quantity of DRs representing only 10% of the required work to be performed have been completed. Therefore, program technical management, who also contribute to this decision making, need to do the following:

1. Prioritize and schedule DRs for implementation.
2. Allocate resources based on those priorities.
3. Measure earned value progress based on relative importance or value of DRs.

In this manner, you tackle the hard problems first. Executive management and the Acquirer need to understand and commit their full support to the approach as *"the RIGHT thing to do!"* As *stakeholders*, they need to be participants in the prioritization process.

Challenge 11: Paragraph versus Singular Requirements

In addition to the SE Design Process challenges, the consequences of paragraph-based specification requirements arise during SITE. The realities of using a test to demonstrate several dependent or related requirements scattered in paragraphs throughout a specification can create many problems. THINK SMARTLY "up front" when specifications are written and create *singular* requirements statements that can be easily checked off as completed.

Referral For more information about singular requirements statements, refer to Chapter 33 on *Requirements Statement Development Practices*.

Challenge 12: Credit for Requirements Verified

The issue of *paragraph* versus *singular* requirements also presents challenges during verification. The challenge is paragraph-based requirements cannot be checked off as *verified* until ALL of the requirements in the paragraph have been verified. Otherwise, say the paragraph has 10 embedded requirements and you have completed nine. Guess WHAT! You are stuck without a verification check mark until the tenth requirement is *verified. Create singular requirement statements!*

Challenge 13: Refurbishing/Reconditioning Test Articles

System Verification Test (SVT) *articles*, especially expensive ones, may be *acceptable* for delivery under specified contract conditions—such as after refurbishment and touch-up. Establish acceptance *criteria* BEFORE the contact is signed concerning HOW SVT articles will be dispositioned AFTER the completion of system testing. Consult your contract and your program, legal, or contracts organization for guidance in this area.

Challenge 14: Calibration and Alignment of Test Equipment

Testing is very expensive and resource intensive. When the SVT is conducted, most programs are already behind schedule. During SITE, if it is determined that a *test article* is *misaligned* or your test equipment is *out of calibration*, you have test data integrity issues to resolve.

Do yourself a favor. Make sure ALL test equipment and tools are certified to be *calibrated* and *aligned* BEFORE you conduct formal tests. Since calibration certifications have expiration dates, *plan ahead* and have a contingency plan to *replace* test equipment items with calibration due to expire during the SITE. Tag all equipment and tools with *expired* calibration notices that are highly visible; *lock up* the expired equipment until *calibrated*.

Challenge 15: Test "Hooks"

Test hooks provide a means to capture data measurements such as test points, and software data measurements. Plan for these "hooks" during the SE Design Segment and make sure they do not *bias* or *alias* the *accuracy* of hardware measurements or *degrade* software performance. Identify and visibly *tag* each one for test reporting and easy *removal* later. Then, when test article verification is completed, make sure all test hooks are *removed*, unless they are required for higher level integration tests.

Challenge 16: Anomalies

Anomalies can and do occur during formal SITE. Make sure that test operator personnel and equipment properly *log* the *occurrence* and configuration and event *SEQUENCES* when anomalies occur to serve as a basis to *initiate* your investigation.

Anomalies are particularly troublesome on large complex systems. Sometimes you can *isolate* anomalies by luck; other times they are *elusive*, so you find them later by accident. In any case, when anomalies occur, record the *sequence of events* and *conditions* preceding the event. What appears to be an anomaly as a single event may have patterns of *recurrences* over time. Tracking anomaly records over time may provide clues that are traceable to a specific *root* or *probable cause*.

Challenge 17: Technical Conflict and Issue Resolution

Technical conflicts and issues can and do arise during formal SITE between the Acquirer's Test Representative and the System Developer, particularly over interpretations of readings or data. First, make sure that the test procedures are *explicitly stated* in a manner that avoids *multiple* interpretations. Where test areas may be questionable, the TEWG should establish prior to or during the Test Requirements Review (TRR) HOW conflicts will be managed. Establish a *conflict and issue resolution process* between the Acquirer (role) and System Developer (role) prior to formal testing. Document it in the System Integration and Verification Plan (SIVP).

Challenge 18: Creating the Real World Scenarios

During SITE planning, there is a tendency to "check-off" individual requirements via individual tests during verification. From a requirements compliance perspective, this has to be accomplished. However, verifying each requirement as a *discrete* test may not be desirable. There are two reasons: 1) cost and 2) real world.

First, just because a specification has separately stated requirements does not mean that you cannot conduct a single test that verifies a multiple number of requirements to minimize costs. This assumes the test is representative of system usage rather than a random combination of unrelated capabilities.

Secondly, most systems consist of multiple capabilities operating simultaneously. In some cases, interactions between capabilities may conflict. This can be a problem, especially if you discover later after individual specification requirement tests that indicated the system is compliant with the requirements. This point emphasizes the need for *use case* and *scenario based* tests and test cases to exercise and stress combinations of system/entity capabilities to expose potential interaction conflicts while verifying one or more specification requirements.

55.8 GUIDING PRINCIPLES

In summary, the preceding discussions provide the basis with which to establish the guiding principles that govern SITE practices.

Principle 5.1 Every Acquirer must have one Test Representative who serves as THE single voice of authority representing the Acquirer-User community concerning SITE decisions.

Principle 5.2 Specification compliance requires presentation of work products as objective evidence of satisfactory accomplishment of each requirement. *Apply the rule: if it isn't documented, it didn't occur.*

Principle 5.3 (Under contract protocol) System Developers invite Acquirer representatives to observe and/or witness a TEST; Acquirers extend invitations to Users unless prior arrangements with the System Developer have been made.

Principle 5.4 Every discrepancy has a level of significance and importance to the System Developer and Acquirer. Develop a consensus of priorities and allocate resources accordingly.

55.9 SUMMARY

Our discussion of SITE as one of the *verification* and *validation* (V&V) practices explored the key activities of Developmental Test and Evaluation (DT&E) under controlled laboratory conditions. *Work products* and *quality records* from these activities provide the contractual foundation for determining if a system or product fully complies with its *System Performance Specification (SPS)*. Data collected during SITE enables SEs to:

1. Develop confidence in the integrity of the Developmental Configuration.
2. Support the *functional configuration audit* (FCA) and *physical configuration audit* (PCA).
3. Answer the key verification question: *Did we build the system or product RIGHT—meaning in compliance with the SPS?*

GENERAL EXERCISES

1. Answer each of the *What You Should Learn from This Chapter* questions identified in the *Introduction*.
2. Refer to the list of systems identified in Chapter 2. Based on a selection from the preceding chapter's General *Exercises* or a new system selection, apply your knowledge derived from this chapter's topical discussions. Specifically identify the following:
   (a) Describe the primary test configurations for testing the system.
   (b) Given the size and complexity of the system, recommend a test organization and provide rationale for the role-based selections.
   (c) If you were the Acquirer of this system, would you require procedure-based ATPs or scenario-based ATPs? Provide supporting rationale.
   (d) What special considerations are required for testing this system—such as OPERATING ENVIRONMENT, tools, and equipment?

(e) Identify the basic integration and test strategy of steps you would specify in the SIVP.

(f) What types of DRs do you believe would require the Acquirer's Test Representative to review and approve? Provide examples.

ORGANIZATIONAL CENTRIC EXERCISES

1. Research your organization's command media for SITE policies and procedures.
    (a) Identify the command media requirements by source:
        1. Test planning
        2. Test procedures
        3. Test operator training
        4. Equipment alignment and calibration
        5. Test results reporting format and approvals
    (b) Document your findings and report your results.
2. Contact two or three contract programs within your organization.
    (a) What type of test plan, test procedures, and reporting requirements—such as the Contract Data Requirements List (CDRL)—are established by contract?
    (b) Does the Acquirer have approval of the test plan and procedures?
    (c) When are the test plan and procedures due for review and approval?
        1. How is the program implementing: 1) test logs and 2) test discrepancies (TDs), including TD tracking, emergency approvals, and TD corrective actions?
        2. Are TDs classified in terms of corrective action urgency? If so, what levels and criteria are used?
        3. Are all TDs required to be resolved prior to system delivery?
        4. What testing does the contract require during site installation for system acceptance?

REFERENCES

DSMC. 1998. *Simulation Based Acquisition: A New Approach*. Defense System Management College (DSMC) Press Ft. Belvoir, VA.

Defense Systems Management College (DSMC). 2001. *Glossary: Defense Acquisition Acronyms and Terms*, 10th ed. Defense Acquisition University Press Ft. Belvoir, VA.

IEEE Std 610.12-1990. 1990. *IEEE Standard Glossary of Modeling and Simulation Terminology*. Institute of Electrical and Electronic Engineers (IEEE). New York, NY.

KOSSIAKOFF, ALEXANDER, and SWEET, WILLIAM N. 2003. *Systems Engineering Principles and Practice*. New York: Wiley-InterScience.

MIL-HDBK-470A. 1997. *Designing and Developing Maintainable Products and Systems*. Washington, DC. Department of Defense (DoD).

MIL-HDBK-881. 1998. *Military Handbook—Work Breakdown Structure*, Washington, DC: Department of Defense (DoD).

MIL-STD-480B (canceled). 1988. *Configuration Control—Engineering Changes, Deviations, and Waivers*. Washington, DC: Department of Defense (DoD).

Defense Systems Management College (DSMC). 1998. *Test and Evaluation Management Guide*, 3rd ed. Defense Acquisition University Press Ft. Belvoir, VA.

ADDITIONAL READING

MIL-STD-0810. 2002. *DOD Test Method Standard for Environmental Engineering Considerations and Laboratory Tests*. Washington, DC: Department of Defense (DoD).

Chapter 56

System Deployment

56.1 INTRODUCTION

Most systems, products, or services require *deployment* or *distribution* by their System Developer or supplier to a User's designated field site. During the deployment, the system may be *subjected to* numerous types of OPERATING ENVIRONMENT threats and *conditions* such as temperature, humidity, shock, vibration, electromagnetic interference (EMI), electrostatic discharge (ESD), salt spray, wind, rail, sleet, and snow.

System deployment involves more than physically deploying the system or product. The system, at a *minimum* and as *applicable*, may also require:

1. Storage in and/or interface with temporary, interim, or permanent warehouse or support facilities.
2. Assembly, installation, and integration into the User's Level 0/Tier 0 system.
3. Support training of operators and maintainers.
4. Calibration and alignment.
5. Reverification.

To accommodate these challenges, system designs and components must be sufficiently *robust* to survive in these conditions, either in an *operational* or *nonoperational state*. For a system design to accommodate these challenges, the *System Performance Specification (SPS)* must *define* and *bound* the required operational capabilities and performance to SATISFY these conditions.

This chapter is intended to enhance your awareness of key considerations that must be factored into specifying *system*, *product*, or *service* requirements. During our discussion we will investigate the key concepts of system transportation, and operational site activation.

What You Should Learn from This Chapter

1. What is the *objective* of system deployment?
2. What is *site development*?
3. What types of considerations go into developing a site for a system?
4. What is *operational site activation*?
5. Compare and contrast *system deployment* versus *system distribution*.
6. What is a *site survey*?
7. Who conducts *site surveys*?

System Analysis, Design, and Development, by Charles S. Wasson
Copyright © 2006 by John Wiley & Sons, Inc.

8. How should you approach conducting a site survey?
9. How should a site survey be conducted?
10. What are some *considerations* that need to go into specifying systems for deployment?
11. What is *system installation and checkout*?
12. What are some *methods* to mitigate risk during system deployment?
13. Why is *environmental, safety, and health* (ES&H) a critical issue during system deployment?
14. What are some common system deployment issues?

Definitions of Key Terms

- **Deployment** The assignment, tasking, and physical relocation of a system, product, or service to a new active duty station for purposes of conducting organizational missions.
- **Disposal (Waste)** "The discharge, deposit, injection, dumping, spilling, leaking, or placing of any solid waste or hazardous waste into or on any land or water. The act is such that the solid waste or hazardous waste, or any constituent there of, may enter the environment or be emitted into the air or discharged into any waters, including ground water (40 CFR section 260.10)." (Source: AR 200-1 *Glossary*, p. 37)
- **Operational Site Activation** "The real estate, construction, conversion, utilities, and equipment to provide all facilities required to house, service, and launch prime mission equipment at the organizational and intermediate level." (Source: MIL-HDBK-881, Section H.3.8)
- **Site Installation** The process of unpacking, erecting, assembling, aligning, and calibrating a system and installing it into a facility, if applicable.
- **Site Selection** The process of identifying candidate sites to serve as the location for a system deployment and making the final selection that balances operational needs with environmental, historical, cultural, political, and religious constraints or customs.
- **Site Survey** A pre-arranged tour of a potential deployment site to understand the physical context and terrain; natural, historical, political, and cultural environments; and issues related to developing it to accommodate a system.

Objectives of System Deployment

The objective of system deployment is to safely and securely *relocate* or *reposition* a system or product from one field site to another using the most *efficient* methods available with the lowest *acceptable* technical, operational, cost, and schedule impacts and risk.

To accomplish a deployment for most mobile systems, we decompose this objective into several supporting objectives:

1. Prepare the *system* for shipment, including disassembly, inventory, packaging of components, and crating.
2. Coordinate the land, sea, air, or space based transportation.
3. Transport the device to the new location.
4. Store or install the system or product at the new site.
5. Install, erect, assemble, align, calibrate, checkout, and verify capabilities and performance.

System Deployment Contexts

System deployment has three contexts:

1. *First Article(s) Deployment* The relocation of *first article* systems to a test location range during the System Development Phase to perform operational test and evaluation (OT&E) activities.
2. *Production Distribution Deployment* The relocation of *production systems* via distribution systems to User sites or consumer accessible sites.
3. *Physical System Redeployment* The relocation of the *physical deployed* system during the System Operations and Support Phase (O&S) of the System/Product Life Cycle.

First Article Deployment. *First article(s) deployment* tends to be a very intensive exercise. Due to the cost of some systems, limited quantity, and length of time required to build a system with long lead time items, system deployment requires very close scrutiny. Time and/or resources may prohibit building another system, especially if it is inadvertently destroyed or damaged beyond repair during deployment.

First article deployment for commercial systems typically includes a lot of promotional fanfare and publicity. Referred to as a *rollout*, this event represents a key milestone toward the physical realization of the end deliverable system, product, or service. Depending on the system, *first article* deployment may involve moving a *first article* system test facilities or test ranges for completion of Developmental Test and Evaluation (DT&E) or initiation of Operational Test and Evaluation (OT&E).

System Certification Some systems require *certification* before they are permitted to be deployed for the System Operations and Support (O&S) phase. Examples are commercial aircraft *airworthiness certification*, *sea trials* for ships and submarines, weights and measures for businesses, calibration of test instrumentation, and so forth.

Since the *certification* process can take several months or years, perform advance planning *early* to *avoid* any *showstopper* events that *adversely* impact program costs and schedules.

Production Distribution System Deployment. Once the system or product is verified and validated, it is ready for the System Production Phase to be initiated. Production engineering efforts focus on the reduction of *nonrecurring* engineering cost and risk to the system, product, or service. This includes *innovating* packaging and delivery solutions to *deploy* large or mass quantities via distribution systems to the Users.

Whereas *first article* deployment tends to focus on protecting the *engineering model* or *prototype* system or product while in transit, these distribution systems have to *efficiently* move multiple packages via pallets and other instruments. Therefore, the System Developer must *factor* in design features that *facilitate* production and logistical distribution of systems and products, such as tracking bar coding and packaging for environmental conditions.

56.2 SE ROLES AND RESPONSIBILITIES DURING DEPLOYMENT

The major SEs activities related to *system deployment* occur during the System Procurement Phase and early SE Design Segment of the System Development Phase prior to the System Requirements Review (SRR). These activities include mission and system analysis, establishing site selection criteria, conducting site surveys, conducting trade-offs, deriving system requirements, and identifying system design and construction constraints.

During system deployment, the level of SE support varies depending on the situation. In general, some systems do not require any SE support. This is because appropriate *risk mitigation plans* (RMPs) are already in place. SEs should be available *on call* to immediately respond to any *emergency* or *crisis* situation. If possible, SEs should actively participate in the deployment and oversee some or all of the support operations.

Applying SE to Deployment

One of the preferred ways to identify SE requirements for deployment applications is to use a system block diagram (SBD). The SBD depicts the System Elements (EQUIPMENT, PERSONNEL, etc.) and their interfaces during the System Deployment Phase. Specific system engineering considerations include physical interfaces between the system being deployed and its transportation system as well as the OPERATING ENVIRONMENT encountered during the deployment.

System Deployment Interfaces

System deployment requires mechanical interfaces and measurement devices between the transportation device or container and the SYSTEM being deployed. This includes *electronic* temperature, humidity, shock, and vibration sensors to assess the *health* and *status* of the deployed SYSTEM and record worst case transients.

OPERATING ENVIRONMENT interfaces establish the conditions of the transfer of the system during the deployment. These may involve *shock* and *vibration, temperature, humidity, wind conditions, and toxic or hazardous materials*. Additional special considerations may include environmentally controlled transport containers to maintain or protect against cooling, heating, or humidity.

Each of these interfaces represents potential types of requirements for the *System Performance Specifications (SPS)*. These requirements must be identified during the System Procurement Phase of the System/Product Life Cycle of a program to ensure that the SE design of the deployed system fully accommodates these considerations.

Author's Note 56.1 *Unless there are compelling reasons, the System Developer does not need to be instructed in HOW to deploy the system for delivery or during the System Operations and Support (O&S) Phase. Instead, the required operational interface capabilities and performance should be bounded to allow the System Developer the flexibility to select the optimal mix of deployment method(s).*

Deployment Modes of Transportation Selection

Transportation by land, sea, air, space or combinations of these options is the way most systems, products, and services get from Point A to Point B. Each mode of transportation should be *investigated* in a trade study analysis of alternative modes that includes cost, schedule, efficiency, and timing considerations.

56.3 SELECTION AND DEVELOPMENT OF OPERATIONAL LOCATION

Preparations for system or product deployment to a field site require that the location be *selected, developed*, and *activated*. On delivery of the system, the site must be operationally ready to accept the system or product for system *installation* and *integration* into a higher level system (i.e., Level 0/Tier 0).

Development of the deployment site to support the system depends on the mission. Some systems require temporary *storage* until they are ready to be moved to a permanent location. Others require *assembly*, *installation*, and *integration* into a higher level system without disruption to existing facility operations. Some facilities provide high bays with cranes to accommodate system assembly, installation, and integration. Other facilities may require you to provide your own rental or leased equipment such as cranes and transport vehicles.

Whatever the plan is for the facility, SEs are tasked to *select*, *develop*, and *activate* the field site. These activities include site surveys, site selection, site requirements derivation, facility engineering or site planning, site preparation, and system deployment *safety* and *security*.

Special Site Selection Considerations

In our discussion of the OPERATING ENVIRONMENT architecture in Chapter 11, we noted that external HUMAN-MADE SYSTEMS include *historical*, *ethnic*, and *cultural* systems that must be *considered* and *preserved* when deploying a system. The same is true for NATURAL ENVIRONMENT ecosystems such as wetlands, rivers, and habitat.

The Need for Site Surveys

On-site surveys reveal significant information about doorways sizes, blocked entrances, entrance corridors with hairpin switchbacks, considerations of 60 Hz versus 50 Hz versus 400 Hz electrical power, 110 vac versus 230 vac, and so on, that drive SPS requirements. So, research and carry facility documentation with you. Visually *observe* the facility and *measure* it, if required, and *validate* the currency of the documentation.

Author's Note 56.2 *Site surveys are crucial for validating User documentation for decision making and observing obstacles. Organizational source documentation of fielded MISSION SYSTEMS and SUPPORT SYSTEMS tends to be lax; drawings are often out of date and do not reflect current configurations of EQUIPMENT and FACILITIES. ALWAYS make it a point to visit the location of the deployment, preferably AFTER a thorough review of site documentation. If impractical, you had better have an innovative, cost plus fixed fee (CPFF) contract or another contract that provides the flexibility to cover your costs at the Acquirer-User's expense for the UNKNOWN risks.*

Given this backdrop, let's address how site surveys are planned and conducted.

Planning and Conducting Site Surveys

Site surveys provide a key opportunity for a Site Survey Team to observe how the User envisions operating and maintaining a system, product, or service. Many sites may not be developed or postured to accommodate a new system. However, *legacy systems* typically exist that are *comparable* and provide an invaluable opportunity to explore the site and determine design options. These options may pertain to constrained spaces that limit hands-on access, height restrictions, crawl spaces, lighting, environmental control, communications, and so on. For system upgrades, the Site Survey Team can explore various options for installation.

Site surveys are also valuable means of learning about the environmental and operational challenges. Generally, the *site surveys* consist of a preparatory phase during which NATURAL ENVIRONMENT information about geographical, geologic, and regional life characteristics are collected and analyzed to properly *understand* the potential environmental issues.

Site surveys are more than surveying the landscape; the landscape includes environmental, historical, and cultural artifacts and sensitivities. As such, site survey activities include developing a list of *critical operational technical issues (COIs/CTIs)* to be resolved. For existing facilities, the *site surveys* also provide insights concerning the physical state of the existing facility, as well as COIs/CTIs related to *modifying* the building or *integrating* the new system while *minimizing* interruptions to the organization's workflow.

Site Survey Decision Criteria. *Site selection* often involves conducting a trade study. This requires *identifying* and *weighting* decision criteria based on the User values and priorities. Collaborate with the User via the Acquirer contract protocol to establish the decision criteria. Each criterion requires identifying HOW and from WHOM the data will be collected while on site or as follow-up data requests via the Acquirer.

Decision criteria include two types of data: *quantitative* and *qualitative*.

- *Quantitative* For example, the facility operates on 220 vac, 3-phase, 60 Hz power.
- *Qualitative* For example, WHAT problems you encountered with the existing system that you would like to AVOID when installing, operating, and supporting the new system.

Obviously, we prefer all data to be *quantitative*. However, *qualitative* data may express HOW the User truly *feels* about an *existing* system or the *agony* of installing it. Therefore, structure *open-ended* questions in a manner that encourages the User to provide elaborate answers—such as, *given Options A or B, WHICH would you prefer and WHY?*

Site Survey Data Collection. When identifying the list of site data to be collected, *prioritize* questions to accommodate time restrictions for site personnel interviews. There is a *tendency* to prepare site survey forms, send them out for responses, and then analyze the data. While this can be helpful in some cases, potential respondents today do not have time to fill out surveys.

One method for gaining advance information may come from alternative sources such satellite or aerial photos, assuming they are current, and teleconferences with the site personnel. So, when conducting teleconferences, ask *open-ended* and clarification questions that encourage the participants to answer freely rather than asking closed-ended questions that draw *yes* or *no* answers.

Site Survey Advance Coordination. One of the most fundamental rules of *site surveys* is advance coordination via Acquirer contract protocol. Security clearances, transportation, use of cameras and tape recorders, if permitted, are useful forms of documentation.

Author's Note 56.3 *ALWAYS confer with site officials prior to the visit as to what media are permitted on site for data collection and any data approvals required before leaving. Some organizations require data requests to be submitted during the visit with subsequent internal approvals and delivery following the visit.*

Conducting the Site Visit. During the site visit, *observe* and *ask* about everything related to the system or product's deployment, installation, integration, operation, and support including processes and procedures. LEAVE NOTHING TO CHANCE! Before you leave, request some time to assemble your team and reconcile notes. *THINK about WHAT you saw and DIDN'T see that you EXPECTED or would have EXPECTED to see and WHY NOT.* If necessary, follow up on these questions before you leave the site.

Selecting the Deployment Site

System deployment sites vary as a function of system mission. Consider the following examples:

> **EXAMPLE 56.1**

A computer system for an organization may have a *limited* number of deployment sites within an office building, even within the room of the existing computer system being replaced.

> **EXAMPLE 56.2**

The NASA Surveyor and Apollo programs had to select landing sites on the Moon.

> **EXAMPLE 56.3**

A business may require immediate access to a multi-modal facility that provides land, sea, and air transportation.

> **EXAMPLE 56.4**

An astronomical observatory may require a remote, unpopulated location far from the effects of atmospheric haze and scattering of city lights.

> **EXAMPLE 56.5**

Nuclear power plants require water resources—a MISSION RESOURCES Element—for cooling towers.

Site selection requires establishing *boundary constraints* that impact site selection. These include environmental threats to wildlife habitat, drinking water aquifers, preservation of historical and cultural sites, and political and religious sensitivities.

Site FACILITY Engineering Planning and Development

Successful system deployment at operational sites *begins* with the Site Activation Concept. The basic idea is to identify HOW and WHERE the User considers deploying the SYSTEM at the site, either *permanently* or *temporarily*.

In the *Introduction* of this book, we stated that system success *BEGINS with successful ENDINGS*. This requires deriving all the hierarchical tasks and activities that contribute to achieving success before any insightful planning, preparation, and enroute coordination of *events, accomplishments*, and *criteria* identified in the *Integrated Master Plan (IMP)* can take place. Such is particularly the case for facility engineering.

Modifying Existing Facilities. Preparation for *operational site activation* begins with establishing the FACILITY interface requirements to accommodate the new system. The *mechanism* for identifying and specifying these requirements is the *facility interface specification (FIS)*. The FIS *specifies* and *bounds* the boundary envelope conditions, capabilities and performance requirements to ensure that all facility interfaces are capable, compatible, and interoperable with the new SYSTEM.

Developing the Operational Site Activation Plan. The *operation site activation plan* describes the organization, roles and responsibilities, tasks, resources, and schedule required to develop and activate a new facility or to modify an existing facility. One of the key objectives of the plan is to *describe* HOW the system will be installed, aligned, calibrated, and integrated, as applicable, into higher level systems *without interrupting* normal operations, if relevant.

Site Development Planning Approvals. Approval of *operational site activation plans* sometimes requires several months or even years. At issue are things such as statutory and regulatory compliance, environmental impact considerations, and presence of historical artifacts. As an SE, AVOID the notion that all you have to do is *simply write the plan* and have it approved in a few days. When you prepare the plan, *employ* the services of a subject matter expert (SME) to make sure that all key tasks in activities are properly identified and *comply* with federal, state, and local statutes, regulations, and ordinances. Identify:

1. WHO are the key decision makers?
2. WHAT types of documentation are required to successfully complete the exercise?
3. WHEN must documentation be submitted for approval?
4. WHERE and to WHOM should documentation approval requests be submitted?
5. HOW long is the typical documentation approval cycle?

Other key considerations include zoning restrictions, permits or licenses, certifications, deviations, and waivers.

Site Preparation and Development

Site preparation includes all those activities required before the site can be developed to *accept* the deployed system. This may include grading the land, building temporary bridges, and installating primary power utility, sewer and drain lines.

Site Inspections. When the site has been prepared, the next step is to conduct *site* inspections. *Site inspections* may consist of on-site compliance assessments:

1. By the *stakeholders* to ensure that everything is in place to accept the new system.
2. By local and Acquirer representative authorities to verify *compliance* to statutory and regulatory constraints.

Author's Note 56.4 *Site inspections sometimes involve closed mindedness that has to be recognized and reconciled. As humans, we typically enter into site inspections from the perspective of viewing WHAT is currently in place. This is true especially for compliance checks. The critical issue, however, may be determining WHAT is NOT in place or compliant.*

Remember, local authorities verify whether site capabilities comply with local and statutory requirements. They do not, however, make a determination as to whether the site has ALL of the capabilities and people required to successfully support your system once it is be deployed. This is the SE's job! Therefore, make sure that all System Elements to be deployed are in place. Incorporate these requirements to the facility interface specification (FIS).

Enroute Modifications. Some system deployment efforts require *temporary modifications* to enroute roads and bridges, utility power poles and lines, signal light relocation, traffic rerouting, and road load bearing. Consider the following example:

> **EXAMPLE 56.6**
>
> An existing home is to be physically *relocated* elsewhere within a town. In preparation for the move, coordination is required to *temporarily move* utility lines and reroute traffic. In addition to the necessary permits and licenses, law enforcement officers will need to redirect traffic and utility crews will need to reinstate utility lines, including telephone, power, and signal lights cables.

System Deployment Safety and Security. The deployment of a new system from one location to another should be performed as *expeditiously*, *efficiently*, and *effectively* as practical. The intent is to *safely* and *securely* transport the system with *minimal* impact to it, the public, and the environment. Safety and security planning considerations include *protecting* the system to be deployed, the personnel who perform the deployment, and the support equipment.

System engineering requirement considerations should include *modularity* of the equipment. This means the *removal* of computer hard drives containing *sensitive* data that require special handling and *protection*. The same is true with *hazardous* materials such as flammable liquids, toxic chemicals, explosives, munitions, and ordinances. In these cases special equipment and tools may be needed to ensure their *safe* and *secure* transport by courier or by security teams. Additionally, environmental, safety, and health (ESH) *Material Safety Data Sheets (MSDS)* should accompany the deployment.

56.4 ENVIRONMENTAL CONSIDERATIONS DURING DEPLOYMENT

Environmental concerns have a major impact on all facets of system analysis, design, and development as well as on all phases of the system/product life cycle.

Statutory and Regulatory Requirements

Statutory and regulatory requirements concerning environmental protection and transportation of hazardous materials (HAZMAT) are *mandated* by local, state, federal, and international organizations. These regulations are intended to *protect* the cultural, historical, religious, and political environment and the public. SEs have *significant* challenges in ensuring that new systems and products are properly specified, developed, deployed, operated, supported, and fully *comply* with statutory and regulatory requirements. Consider the following example:

> **EXAMPLE 56.7**
>
> The US *National Environmental Policy Act (NEPA)* of 1969 and Environmental Protection Agency (EPA) establishes requirements on system deployment, operations, and support that impact the natural environment. In many cases System Developers, Acquirers, and Users are required to submit advance documentation such as *Environmental Impact Statements (EIS)* and other types of documents for approval prior to implementation.

Author's Note 56.5 *ALWAYS consult the provisions of your contract as well as with your contracts; legal; and environmental, safety, and health (ESH) organizations for guidance in complying with the appropriate statutory and regulatory environmental requirements.*

Environmental Mitigation Plans

Despite meticulous planning, environmental emergencies can and do occur during the deployment of a system. Develop risk mitigation plans and coordinate resources along the transportation route to *clean up* and *remediate* any environmental spills or catastrophes. Transportation vehicles, systems, and shipping containers should fully *comply* with all applicable federal and state statutory regulatory laws for labeling and handling. Organizations such as the Environmental Protection Agency (EPA), et al may require submittal of *environmental documentation* for certain types of programs. Investigate how the US *National Environmental Policy Act (NEPA)* and other legislation applies to your system's development and deployment.

Environmental Safety and Health (ES&H)

Environmental safety and health (ES&H) is a *critical issue* during system development and deployment. The objective is to *safely* and *securely* relocate a SYSTEM without impacting the system's capabilities and performance or endangering the health of the public or NATURAL ENVIRONMENT, nor that of the deployment team. ALWAYS investigate the requirements to ensure that the ES&H concerns are properly addressed in the design of the equipment as well as the means of transportation. ISO 14000 serves as the international standard used to assess and certify organizational *environmental management* processes and procedures.

Occupational Safety and Health Administration (OSHA) 29 CFR 1910 is the occupational safety and health standard in the US.

Environmental Reclamation

Environmental resources are *extremely* fragile. Today, great effort is being made to preserve the natural environment for future generations to enjoy. Therefore, during the transport of a system to a new job site, the risk of spills and emissions into the atmosphere should be *minimized* and *mitigated* to a level *acceptable* by law or *eliminated*.

56.5 SYSTEM INSTALLATION, INTEGRATION, AND CHECKOUT

Once the field site is prepared to accept the system, the next step is to install, integrate, and check out the system if the system's mission is at this facility. This requires conducting various levels of system installation and checkout tests.

Installation and Checkout Plan Activities

Installation and checkout activities cover a sequence of activities, organizational roles and responsibilities, and tasks before the newly deployed system can be located at a specific job site. SYSTEM requirements that are unique to on-site system installation and integration must be identified by *analysis* and incorporated into the *System Performance Specification (SPS)* PRIOR TO the Contract Award.

User Training

When a new system is ready to be deployed, a key task is to train Users to deploy, install, and check out the system. Generally, a system development contract will require the System Developer to train Users PRIOR TO *disassembly* for deployment from the System Developer's or system integration facility. The training sessions should prepare Users to *safely* and *properly* operate and to support Operational Test and Evaluation (OT&E) during the final portions of the System Development Phase.

"Shadow" Operations

Installation and checkout of new systems may require integration into higher level systems. The integration may involve the new system as an additional element or as a *replacement* for an existing or legacy system. Whichever is the case, system integration often involves *critical operational and technical issues (COIs/CTIs)*, especially from the standpoint of *interoperability* and security.

Depending on the criticality of the system, some Acquirers and Users require the new system to operate in a "shadow" mode to *validate* system responses to external stimuli while the existing system remains in place as the primary operating element. Upon completion of the evaluation, assessment, and certification, the new system may be brought "on line" as an Initial Operational Capability (IOC). *Incremental* capabilities may be added until Full Operational Capability (FOC) is achieved. To illustrate the *criticality* and importance of this type of *deployment* and *integration*, consider the following example:

EXAMPLE 56.8

Financial organizations such as banks depend on highly integrated, audited, certified systems that validate the integrity of the overall system. Consider the *magnitude* and *significance* of decisions related to integrating either new or replacement software systems to ensure *interoperability* without *degrading* system performance or compromising the *integrity* of the system or its security.

56.6 SYSTEM DEPLOYMENT ENGINEERING CONSIDERATIONS

System deployment engineering requires SEs to consider two key areas of system deployment:

- Operational transport considerations
- Environmental considerations

Table 56.1 provides system design areas for assessment.

Systems such as construction equipment cranes, bulldozers, the Space Shuttle, and military equipment are designed for multiple redeployments. In most cases the system is loaded onto a transport vehicle such as truck, aircraft, train, or ship.

Once loaded, the primary restrictions for travel include compatible tiedowns, load weights, size limitations, bridge underpass height, and shock/vibration suppression. Some systems may require shipping in specialized containers that are *environmentally controlled* for temperature, humidity, and protection from salt spray. *How do engineers accommodate these restrictions and AVOID surprises at deployment?*

The deployment engineering solution requires a structured analysis approach based on the System Operations Model discussed in Chapter 18. Perform operational and task analysis by sequencing the chain of events required to move the system from Point A to Point B. This includes *cost, performance,* and *risk trade-offs* for competing land, sea, and air *modes of transportation* and transport *mechanisms* such as truck, aircraft, ship, and train.

Once the *modes of transportation issues are* resolved, develop system interface diagrams for each phase of deployment. *Specify* and *bound* system requirements and incorporate them into the *System Performance Specification (SPS)* or system design.

Guidepost 56.1 *Our discussion has focused on deploying a MISSION SYSTEM and designing it to be compatible with an existing system performing a SUPPORT SYSTEM role. Now, let's switch the context and consider WHAT mission capabilities a SUPPORT SYSTEM requires.*

Table 56.1 System deployment engineering considerations

System Deployment Consideration	System Design
Operational transport considerations	• Tie downs, hold downs, and safety chains • Lift points • Land grades • Emergency stopping • Auxiliary power • Vehicle markings • Mock dry runs with simulated equipment • System security • Maintenance • Bridge, highway, and street load restrictions • Bridge and power line height restrictions.
Environmental considerations	• Shock and vibration • Saltwater and spray • Temperature and humidity control • Electrical fields and discharges (grounding) • Flying debris • Altitude and atmospheric pressure changes • Environmental instrumentation • Hazardous materials

Support Equipment

System design involves more than creating interfaces between a MISSION SYSTEM and its SUPPORT SYSTEM. Support equipment such as tools, sensors, and diagnostic equipment is often required to establish the interfaces for travel. Remember, support equipment includes:

1. *Common Support Equipment (CSE)* includes hammers, screwdrivers, and other hand tools that are applicable to most systems.
2. *Peculiar Support Equipment (PSE)* includes specialty tools and devices that are unique to a specific system.

56.7 COMMON SYSTEM DEPLOYMENT CHALLENGES

Depending on the risks associated with the system, deployment route, and site, *deployment* activities generate a lot of excitement and challenges. Let's investigate a few.

Challenge 1: Risk Mitigation Planning

When systems or products are deployed, the natural *tendency* is to *assume* you have selected the best approach for deployment. However, political *conditions* in foreign countries, such as labor strikes, can disrupt system deployment and force you to reconsider alternative methods. Develop risk mitigation plans that accommodate all or the most likely scenarios.

Challenge 2: Conduct a Mock Deployment

For large, complex systems and that require *special handling* considerations, a *mock deployment* exercise may be appropriate. A mock deployment involves some form of prototype or model having

Table 56.2 System deployment design and development rules

ID	Title	System Deployment Design and Development Rule
56.1	System deployment requirements and constraints	Bound and specify the set of system deployment requirements and constraints for every system, product, or service in the *System Performance Specification (SPS)*.
56.2	Deployment conditions	When planning deployment of a system or product, at a *minimum*, consider the following factors: 1. Deployment routes 2. System safety 3. System security 4. Logistical support operations 5. Training 6. Licenses, certifications, and permits 7. System/product stowage and protection during shipment 8. Environmental conditions (weather and road, etc.)
56.3	Deployment facilities	For systems or products that require facility interfaces, specify and bound interface requirements via a *facility interface specification (FIS)* or equivalent.
56.4	Site activation	Prepare a *site activation plan* for every system, product, or service for review and approval by the User and support facility.
56.5	System design	When developing a system, product, or service factor, at a *minimum*, factor in considerations to protect a system during deployment operations and conditions to minimize the effects of physical harm such as appearance, form, fit, or function.
56.6	Deployment interfaces	Verify that a system, product, or service design is *compatible* and *interoperable*, if necessary, with the deployment mechanism used to deploy the system to its designated field site.
56.7	Stakeholder decision-making participation	When planning deployment of a system or product, include considerations by those stakeholder organizations that permit system deployment through their jurisdictions—bridge heights above roads and weights permitted; barge, truck, aircraft payload restrictions; hazardous material passage through public areas, and aircraft landing constraints.

comparable form, fit, and weight to the actual system being fictionally transported to an approved site. The exercise debugs the deployment process and facilitates identification of unanticipated scenarios and events for the processes, methods, and tasks of deployment.

Challenge 3: Stakeholders Participation in Decision Making

System deployment often involves large numbers of geographically dispersed *stakeholders*. Therefore, *stakeholders* should be *actively* involved in the decision-making process, from the early planning stage but subject to contract type limitations. So, *what happens if you do not include these stakeholders*?

Depending on the situation, a stakeholder could become a SHOWSTOPPER and significantly impact system deployment schedules and costs. Do yourself and your organization a favor. Understand the deployment, site selection and development, and system installation and integration decision-making chain. This is key to ensuring success when the time comes to deploy the system.

AVOID a "downstream" SHOWSTOPPER situation simply because you and your organization chose to IGNORE some odd suggestions and views during the System Development Phase.

Challenge 4: System Security During Deployment

Some systems require a purposeful deployment using low *profile* or *visibility* methods to *minimize* publicity, depending on the *sensitivity* and *security* of the situation.

Challenge 5: Shock and Vibration

Professional systems and products such as instruments can be very sensitive to *shock* and *vibration*. Plan ahead for these critical *design factors*. Make sure that the system in a stand-alone mode is adequately designed to offset any *shock* or *vibration* conditions that occur during deployment. This includes establishing appropriate *design safety margins*. If appropriate, the SPS should specify shock and vibration requirements.

Challenge 6: Hazardous Material (HAZMAT) Spillage

When transporting EQUIPMENT systems and products that contain various fluids, hazardous material (HAZMAT) spillage is always a major concern, particularly for wildlife estuaries, rivers, and streams, but also underground water acquifers. Perform insightful ES&H planning and coordination to ensure that SUPPORT SYSTEMS—such as PERSONNEL, PROCEDURAL DATA, and EQUIPMENT—are readily available to enable the SUPPORT SYSTEM to rapidly respond to a hazardous event.

Challenge 7: Workforce Expertise Availability

Service contracts are often bid on the assumption of employing local resources to perform system operations, support, and maintenance. DO NOT ASSUME ANYTHING. Unless you are certain that these resources are *dependable* and will *commit* to the contract, you may be at risk. ALWAYS have a contingency plan!

56.8 GUIDING PRINCIPLES

In summary, the preceding discussions provide the basis with which to establish the guiding principles that govern system deployment practices.

Principle 56.1 Specify and bound a system's deployment mode(s) of transportation, distribution methods, and constraints; otherwise, the system's *form*, *fit*, and *function* may be incompatible with its deployment delivery system.

Principle 56.2 Site survey data quality begins with advance coordination and insightful planning. Obtain what you need on the first trip; you may not be permitted for a second visit.

56.9 SUMMARY

During our discussion of *system deployment*, we highlighted the importance for SEs to thoroughly understand all of the system deployment issues and ensure that requirements are properly addressed in the contract and *System Performance Specifications (SPS)*. When the system is being deployed, SEs should be involved to ensure that the system interfaces to the appropriate means of transportation and facilities.

GENERAL EXERCISES

1. Answer each of the *What You Should Learn from This Chapter* questions identified in the *Introduction*.
2. Refer to the list of systems identified in Chapter 2. Based on a selection from the preceding chapter's General *Exercises* or a new system selection, apply your knowledge derived from this chapter's topical discussions. Specifically identify the following:
3. Research the following topics and investigate how the system was deployed—packaging, handling, shipping, and transportation (PHS&T)—during integration and on system delivery to its field site:
   (a) International Space Station (ISS)
   (b) Hubble Space Telescope (HST)
   (c) Construction crane
   (d) Portable electronic device

ORGANIZATIONAL EXERCISES

1. Research your organizational command media for guidance concerning system deployment operations.
   (a) What requirements are imposed on system deployment requirements and design considerations?
   (b) What work products or quality records are required?
2. Contact several contract programs within your organization.
   (a) What system deployment requirements are stated in the contract and *System Performance Specification (SPS)*?
   (b) What capabilities and features are incorporated into the system design to comply with these requirements?
   (c) Does the program have a system deployment plan? If so, what types of routes, means of transportation, licenses, and permits are planned?
   (d) Did the program have requirements for measurement of deployment conditions—such as shock, vibration, temperature, and humidity?
   (e) How were countermeasures to temperature, humidity, shock, vibration, ESD, and salt spray accommodated in the design or packaging?

REFERENCES

AR 200-1. 1997. Army Regulation: *Environmental Protection and Enhancement.* Washington, DC: Department of Defense (DoD).

MIL-HDBK-881. 1998. *Military Handbook—Work Breakdown Structure.* Washington, DC: Department of Defense.

ADDITIONAL READING

ISO 14000 Series. *Environmental Management.* International Organization for Standardization (ISO). Geneva, Sweden.

MIL-STD-882D. 2000. *System Safety.* Washington, DC: Department of Defense (DoD).

OSHA 29 CFR 1910, (on-line always current date) *Occupational Safety and Health Standards*, Occupational Safety and Health Administration (OSHA), Washington, DC: Government Printing Office.

Public Law 91-190. 1969. *National Environmental Policy Act (NEPA).* Washington, DC: Government Printing Office.

Chapter 57

System Operations and Support

57.1 INTRODUCTION

As the preceding chapters have illustrated, the *engineering of systems* is not complete when the Acquirer formally accepts a new system for the User. In terms of SE *level of effort* to support the *development* of the system, product, or service this is true; however, the *system performance assessment* activities *continue throughout the system's active duty service*. Thus, when the system acceptance is complete, *the System Operations and Support (O&S) phase* begins. Depending on the type of system, product, or service, SE *technical* and *analytical* expertise may be required to monitor, track, and analyze system performance in the field.

Our discussion of system performance returns to where we started, to the System Element Architecture that forms the basis for HOW a system, product, or service is intended to operate. Since the SYSTEM OF INTEREST (SOI) is composed of one or more MISSION SYSTEMS that are supported by one or more SUPPORT SYSTEMS, we will approach the discussion from those two perspectives.

Our discussions explore system performance related to the MISSION SYSTEM and the SUPPORT SYSTEM. This includes:

1. Elimination of *latent defects* such as design *flaws*, *errors*, or safety issues.
2. Identification and removal of *defective* materials or components.
3. Optimization of system performance.

These activities also represent the BEGINNING of collecting requirements for:

1. Procuring follow-on systems, products, or services.
2. Upgrading capabilities of the existing system.
3. Refining current system performance.

Author's Note 57.1 This chapter has a twofold purpose. *First, it provides the basis for assessing the operational utility, suitability, availability, and effectiveness of new systems to ensure that they integrate and achieve levels of performance that satisfy organizational mission and system application objectives. Second, as we saw in Chapter 7 in our discussion of the System/Product Life Cycle, gaps evolve in a system, product, or service's capability to meet organizational objectives or external threats. In this chapter we learn about legacy and new system performance and need for developing operational capability requirements for the next generation of system, product, or service.*

System Analysis, Design, and Development, by Charles S. Wasson
Copyright © 2006 by John Wiley & Sons, Inc.

Chapter 57 System Operations and Support

What You Should Learn from This Chapter

1. What are the *primary objectives* for system operation and support (O&S)?
2. What are *key areas* for monitoring and analyzing *SYSTEM level performance*?
3. What are *key questions* for assessing *system element performance*?
4. What are common *methods* for assessing SYSTEM and element performance?
5. What are the key SE focus areas for *current system performance*?
6. What are the key SE focus areas for *planning next generation systems*?

Definitions of Key Terms

- **Equipment, Powered Ground (PGE)** "An assembly of mechanical components including an internal combustion engine or motor, gas turbine, or steam turbine engine mounted as a single unit on an integral base or chassis. Equipment may pump gases, liquids, or solids; or produce compressed, cooled, refrigerated or heater air; or generate electricity and oxygen. Examples of this equipment: portable cleaners, filters, hydraulic test stands, pumps and welders, air compressors, air conditioners. Term applies primarily to aeronautical systems." (Source: MIL-HDBK-1908, *Definitions*, para. 3.0, p. 20)

- **Field Maintenance** "That maintenance authorized and performed by designated maintenance activities in direct support of using organizations. It is normally limited to replacement of unserviceable parts, subassemblies, or assemblies." (Source: MIL-HDBK-1908, *Definitions*, para. 3.0, p. 15)

- **Human Performance** "A measure of human functions and action in a specified environment, reflecting the ability of actual users and maintainers to meet the system's performance standards, including reliability and maintainability, under the conditions in which the system will be employed." (Source: MIL-HDBK-1908, *Definitions*, para. 3.0, p. 18)

- **Line Replaceable Unit (LRU)** Refer to definition in Chapter 42 on *System Configuration Identification Practices*.

- **Logistic Support Analysis (LSA)** "The selective application of scientific and engineering efforts undertaken during the acquisition process, as part of the system engineering and design processes to assist in complying with supportability and other Integrated Logistics Support (ILS) objectives. ILS is a management process to facilitate development and integration of logistics support elements to acquire, field and support a system. These elements include: design, maintenance planning, manpower and personnel, supply support, support equipment, training, packaging and handling, transport, standardization and interoperability." (Source: MIL-HDBK-1908, *Definitions*, para. 3.0, p. 20)

- **Maintenance** "All actions necessary for retaining material in (or restoring it to) a serviceable condition. Maintenance includes servicing, repair, modification, modernization, overhaul, inspection, condition determination, corrosion control, and initial provisioning of support items." (Source: MIL-HDBK-1908, *Definitions*, para. 3.0, p. 21)

- **Problem Report** A formal document of a problem with or failure of MISSION SYSTEM or SUPPORT SYSTEM hardware—such as the EQUIPMENT System Element.

- **Provisioning** "The process of determining and acquiring the range and quantity (depth) of spares and repair parts, and support and test equipment required to operate and maintain an end item of material for an initial period of service. Usually refers to first outfitting of a ship, unit, or system." (Source: DSMC, *Glossary of Defense Acquisition Acronyms and Terms*)

- **Repair** "A procedure which reduces but does not completely eliminate a nonconformance resulting from production, and which has been reviewed and concurred in by the Material Review Board (MRB) and approved for use by the Acquirer. The purpose of repair is to reduce the effect of the nonconformance. Repair is distinguished from rework in that the characteristic after repair still does not completely conform to the applicable specifications, drawings or contract requirements." (Source: Adapted from Former MIL-STD-480, Section 3.1, *Definitions*, para. 3.1.58, p. 13)
- **Repairability** "The probability that a failed system will be restored to operable condition within a specified active repair time." (Source: DSMC, *Glossary of Defense Acquisition Acronyms and Terms*)
- **Repairable Item** "An item of a durable nature which has been determined by the application of engineering, economic, and other factors to be the type of item feasible for restoration to a serviceable condition through regular repair procedures." (Source: DSMC, *Glossary of Defense Acquisition Acronyms and Terms*)
- **Replacement Item** "An item which is replaceable with another item, but which may differ physically from the original item in that the installation of the replacement item requires operations such as drilling, reaming, cutting, filing, shimming, etc., in addition to the normal application and methods of attachment." (Source: Former MIL-STD-480, Section 3.1, *Definitions*, para. 3.1.45.2, p. 11)
- **Support Equipment** "All equipment required to perform the support function, except that which is an integral part of the mission equipment. SE includes tools, test equipment, automatic test equipment (ATE) (when the ATE is accomplishing a support function), organizational, intermediate, and related computer programs and software. It does not include any of the equipment required to perform mission operations functions." (Source: MIL-HDBK-1908, *Definitions*, para. 3.0, p. 30)
- **Common Support Equipment (CSE)** Refer to Chapter 10 on *SYSTEM OF INTEREST Architecture*.
- **Peculiar Support Equipment (PSE)** Refer to Chapter 10 on *SYSTEM OF INTEREST Architecture*.
- **Test, Measurement, and Diagnostic Equipment (TMDE)** Refer Chapter 10 on *SYSTEM OF INTEREST Architecture*.
- **Turn Around Time** "Time required to return an item to use between missions or after removed from use." (Source: DSMC, *Glossary of Defense Acquisition Acronyms and Terms*)

57.2 SYSTEM ENGINEERING OPERATIONS and SUPPORT (O&S) OBJECTIVES

Once a system is fielded, SEs should continue to be a vital element of the program. Specifically, system performance, efficiency, and effectiveness should be monitored continuously. This requires SE expertise to address the following objectives:

1. Monitor and analyze the *OPERATIONAL UTILITY, SUITABILITY, AVAILABILITY,* and *EFFECTIVENESS of system applications, capabilities, performance*, and the *service life* of the newly deployed system—including products and services—relative to its intended mission(s) in its OPERATING ENVIRONMENT.
2. Identify and CORRECT latent *defects*—such as design flaws and errors, faulty or defective components, and system *deficiencies*.

3. Maintain AWARENESS of the "gap"—in the *problem space*—that evolves over time between this existing system, product, or service and comparable competitor and adversarial capabilities.
4. Accumulate and evolve requirements for a new system or capability or upgrade to the existing system to fill the *solution space(s)* and eliminate the *problem space*.
5. Propose *interim* operational solutions—plans and tactics—to fill the "gap" until a replacement capability is established.
6. Maintain system configuration baselines.
7. Maintain MISSION SYSTEM and SUPPORT SYSTEM concurrency.

Let's explore each of these objectives further.

57.3 MONITOR AND ANALYZE SYSTEM PERFORMANCE

When new systems or capabilities are conceived, SEs are challenged to *translate* the operational needs that prompted the development into Requirements, Operations, Logical, and Physical Domain Solutions. System Developers apply *system verification practices* to answer the question: *Are we building the system RIGHT—in compliance with the specification requirements?* Acquirers or Users, or their Independent Test Agency (ITA), apply *System Validation Practices* to answer the question: *Did we procure the RIGHT system?*

System specification and development activities *represent* the best efforts of humans to *translate*, *quantify*, and *achieve* technical and operational expectations and thresholds derived from abstract visions. The ultimate question for Users to answer is: *Given our organizational objectives and budgetary constraints, did we procure the RIGHT system, product, or service to fulfill our needs*?

This question may appear to conflict with the *validation* purpose of Operational Test and Evaluation (OT&E). OT&E is intended to provide Acquirer and User pre-system *delivery* and *acceptance* answers to this question. However, OT&E activities have a finite length—of days, weeks, or months—under *tightly controlled* and *monitored* conditions representing or simulating the actual OPERATING ENVIRONMENT.

The challenge for the ITA is to *avoid* aliasing the answers due to the controlled environment. The underlying question is: *Will Users perform differently if they know they are being observed and monitored versus on their own*? The true answer resides with the User after the system, product, or service has "stood the test of time." Consider the following example:

EXAMPLE 57.1

During the course of normal operations in the respective OPERATING ENVIRONMENTS, construction, agricultural, military, medical, and other types of EQUIPMENT are subjected to various levels of *use*, *misapplication*, *misuse*, *abuse*, *maintenance*, or the lack thereof. During system development, time and resource constraints *limit* contract-based OT&E of these systems. Typically, a full OT&E of these systems is *impractical* and *limited* to a representative *sampling* of *most probable* or *likely* use cases and scenarios over a few weeks or months. In contrast, commercial product test marketing of some systems, products, or services over a period of time with large sample sizes may provide better insights as to whether the Users *love* or *loathe* a product.

Ultimately, the User is the SOLE entity that can answer four key questions that formed the basis for our System Design and Development Practices:

1. *Does the system add value to the User and provide the RIGHT capabilities to accomplish the User's organizational missions and objectives—OPERATIONAL UTILITY?*
2. *Does the system integrate and interoperate successfully within the User's "system of systems"—OPERATIONAL SUITABILITY?*
3. *Is the system operationally available when called upon by the User to successfully perform its missions—OPERATIONAL AVAILABILITY?*
4. *How well does the system support User missions as exemplified in achievement of mission objectives—OPERATIONAL EFFECTIVENESS?*

Answers to these questions require monitoring system performance. If your mission as an SE is to MONITOR overall system performance, *how would you approach this task?*

The answer resides in WHERE we began with the System Element Architecture template shown in Figure 10.1. The template serves as a simplistic starting point model of the SYSTEM. Since the model represents the integrated set of capabilities required to achieve organizational mission objectives, we allocate SPS requirements to each of the *system elements*—EQUIPMENT, PERSONNEL, FACILTIES, and so forth. Given this analytical framework, SEs need to ask two questions:

1. *HOW WELL is the interated System Element Architecture Model performing—system operational performance?*
2. Physically, *HOW WELL is each System Element and their set of physical components performing—system element performance?*

Let's explore each of these points further.

SYSTEM Operational Performance Monitoring

Real-time system *performance monitoring* should occur throughout the *pre-mission, mission, and postmission* phases of operation as paced by the Mission Event Timeline (MET). Depending on the system and mission application, the achievement of system objectives, in combination with the MET, provides the BASIS for operational system performance evaluation.

In general, answering HOW WELL the SYSTEM is performing depends on:

1. WHO you interview—the *stakeholders*.
2. WHAT their objectives are.

Consider the following example from the perspectives of the System Owner and the System Developer:

EXAMPLE 57.2

From the SYSTEM Owner's perspective, example questions might include:

1. *Are we achieving mission objectives?*
2. *Are we meeting our projected financial cost of ownership targets?*
3. *Are our maintenance costs in line with projections?*
4. *Is the per unit cost at or below what we projected?*

From the User's perspective, example questions might include:

1. *Is the system achieving its performance contribution thresholds established by the SPS?*
2. *Are there capability and performance areas that need to be improved?*

3. Is the SYSTEM responsive to our mission needs?
4. Does the SYSTEM exhibit any instabilities, latent defects, or deficiencies that require corrective action?

Author's Note 57.2 *Note the first item, SPS "performance contribution thresholds." Users often complain that a system does not "live up to their expectations." Several key questions:*

1. *Were those "expectations" documented as requirements in the SPS?*
2. *Did the Acquirer, as the user's contract and technical representative, VERIFY and ACCEPT the SYSTEM as meeting the SPS requirements?*
3. *Using system acceptance as a point of reference, have User expectations changed?*

Remember, aside from normal performance degradation from use, misuse, misapplication, lack of proper maintenance, and OPERATING ENVIRONMENT threats, a SYSTEM, as an inanimate object, does not change. People and organizations do. So, if the system IS NOT meeting User expectations, is it the SYSTEM or the operators/organization?

These are a few of the types of SYSTEM level questions to investigate. Once you have a good understanding of critical SYSTEM performance areas, the next question is: *WHAT System Elements are primary and secondary performance effecters that drive these results.*

System Element Performance Monitoring

Based on SYSTEM operational performance results, the key question for SEs to answer is: *what are the System Element contributions.* System Element performance areas include:

1. EQUIPMENT element
2. PERSONNEL element
3. MISSION RESOURCES element
4. PROCEDURAL DATA element
5. SYSTEM RESPONSES element—behavior, products, by-products, and services
6. SUPPORT SYSTEM element
   (a) Decision support operations
   (b) System maintenance operations
   (c) Manpower and personnel operations
   (d) Supply support operations
   (e) Training and training support operations
   (f) Technical data operations
   (g) Computer resources support operations
   (h) Facilities operations
   (i) Packaging, handling, storage, and transportation (PHST) operations
   (j) Publications support operations

Each of these System Element areas maps to the integrated set of physical architecture components that contribute to System Element and overall performance.

Guidepost 57.1 *At this juncture we have established WHAT needs to be monitored. We now shift our focus to HOW system performance monitoring is accomplished.*

57.4 PERFORMANCE MONITORING METHODS

The System Element performance monitoring presents several challenges.

First, User/Maintainer organization SEs are *accountable* for answering these questions. If they do not have SEs on staff, they may task support contractors to collect data and make recommendations.

Second, Offeror organizations intending to successfully propose *next-generation systems* or *upgrades* to legacy systems have to GET SMART in system performance areas and *critical operational* and *technical issues*, (COIs/CTIs) in a very short period of time. As a general rule, Offerors track and demonstrate performance in these areas over several years before they qualify themselves as competent suppliers. The competitive advantage *resides* with the *incumbent* contractor unless the User decides to change. Often your chances of success are diminished if you wait to start answering these questions when the DRAFT *Request for Proposal (RFP)* solicitation is *released*; it is simply *impractical*.

So, for those organizations that prepare for and posture themselves for *success*, *how do they obtain the data*? Basic data collection methods include, if accessible:

1. Personal interviews with User and maintainer personnel.
2. Postmission data analysis—such as *problem reports (PRs)*, mission debriefings, and post-action reports.
3. Visual inspections—such as on-site surveys and SYSTEM checkout lists.
4. Analysis of *preventive* and *corrective maintenance* records—such as a Failure Reporting and Corrective Action System (FRACAS), if available.
5. Observation of SYSTEM and PERSONNEL in action.

Although these methods look good on paper, they are ONLY as GOOD as the "corporate memories" of the User and maintainer organizations. Data retention following a mission drops *significantly* over several hours and days. Reality tends to become *embellished* over time. So, every event during the pre-mission, mission, and postmission phases of operation becomes a *critical staging point* for *after action* and *follow-up reporting*. This requires three actions:

1. Establishing record-keeping systems.
2. Ingrain professional discipline in PERSONNEL to record a mission or maintenance event.
3. Thoroughly document the sequence of actions leading up to a mission or maintenance event.

Depending on the consequences of the event, failure boards may be convened that investigate the WHO, WHAT, WHEN, WHERE, WHY, and WHY NOT of emergency and catastrophic events.

People, in general, and SEs, in particular, often lack proper training in reporting malfunctions or events. Most people, by nature, do not like to document things. So-called event reporting tools are not User friendly and typically perform poorly. For these reasons organizations should train personnel and assess their performance from day 1 of employment concerning:

1. WHAT data records are to be maintained and in what media.
2. WHY the data are required.
3. WHEN data are to be collected.
4. WHO the Users are.
5. HOW the User(s) apply the data to improving system performance

Final Thought

The preceding discussions are intended to illustrate HOW TO THINK about system performance monitoring. Earlier we described a condition referred to as *analysis paralysis*. These areas are prime example.

As a reality check, SEs need to ask themselves the question: *If we simply asked Users to identify and prioritize 3 to 5 system areas for improvement, would we glean as much knowledge and intelligence as analyzing warehouses full of data?* The answer depends. If you have to have *objective evidence* in hand to rationalize a decision, the answer is yes. Alternatively, establishing database reporting systems that can be queried provide an alternative. IF you choose to take the shortcut and only identify the 3 to 5 areas for improvement, you may *overlook* what seems to be a miniscule topic that may become tomorrow's headlines. Choose the method WISELY!

57.5 WHY ANALYZE SYSTEM PERFORMANCE DATA?

You may ask: *WHY do SEs need to analyze system performance data?* If the system works properly, *WHAT do you expect to gain from the exercise? WHAT is the return on investment (ROI)?* These are good questions. Actually there are at least several primary reasons *WHY* you need to analyze system performance data for some systems:

Reason 1: To benchmark nominal system performance.
Reason 2: To track and identify system performance trends.
Reason 3: To improve operator training, skills, and proficiency.
Reason 4: To correlate mission events with system performance.
Reason 5: To support gap analysis.
Reason 6: To validate models and simulations.
Reason 7: To evaluate human performance.

Let's explore each of these reasons further.

Reason 1: To Benchmark Nominal System Performance

Establish statistical performance *benchmarks* via *baselines*, where applicable, for WHAT constitutes actual nominal *system performance*. For example, your car's gas mileage has a statistical mean of 30 miles per gallon as measured over its first 20,000 miles.

Remember, the original System Performance Specification (SPS) was effectively a "Design-To limits" set of requirements based on human estimates of required performance. *Verification* simply proved that the *physical* deliverable system or product performed within *acceptable* boundary limits and conditions. Every HUMAN-MADE system has its own unique idiosyncrasies that require *awareness* and *understanding* whether it is performing *nominally* or *drifting in/out* of specification. Consider the following example:

EXAMPLE 57.3

If a hypothetical performance requirement is 100 ± 10, you NEED TO KNOW that System X's nominal performance is 90 and System Y's nominal performance is 100. Sometimes this is important; sometimes not. The borderline "90" system could stay at that level throughout its life while the perfect "100" system could drift out of tolerance and require continual maintenance.

Reason 2: To Track and Identify System Performance Trends

Use the nominal *performance baselines* as a benchmark comparison for system performance *degradation* and *trends* over its service life to ensure *preventive* and *corrective maintenance* occur at the proper time and are performed correctly. For example, suppose your car at 30,000 miles is only averaging 25 miles per gallon.

Reason 3: To Improve Operator Training and Skills

Determine IF and HOW systems are being unnecessarily *stressed*, *misused*, *abused*, or *misapplied* by particular operators.

Reason 4: To Correlate Mission Events with System Performance

Correlate *system mission events* and *operator observations* with recorded system responses and performance data. Ask yourself: *Are we observing a problem area or symptom of a problem that has a root case traceable to latent defects*?

Reason 5: To Support Gap Analysis

Collect *objective evidence* of existing system capabilities and performance to support "gap" analysis between the current system and projected competitive or adversarial system performance.

Reason 6: To Validate Models and Simulations

VALIDATE laboratory models and simulations against *actual* system performance to support future mission planning or assess proposed capability or performance upgrades.

Reason 7: To Evaluate Human Performance

Personnel such as infantry, pilots, and NASA astronauts are *subjected* to operating environments that can *overstress* human performance. Thus, human operator performance within the context of the overall MISSION SYSTEM performance must be well understood to ensure that training *corrects* or *enhances* operator performance for future missions.

57.6 SE FOCUS AREAS DURING O&S

During the System Operations and Support (O&S) Phase, there are a number of SE focus areas that represent *initial starting points* for tracking and assessing system, product, or service performance. As stated in the *Introduction*, this chapter has two contexts:

1. *Sustaining* and *improving* current system performance.
2. *Planning* for next generation systems.

Assessment of current system performance includes several focus areas:

Focus Area 1:   Correct latent defects.
Focus Area 2:   Improve Human–System Integration (HSI) performance.
Focus Area 3:   Maintain MISSION SYSTEM–training device concurrency.
Focus Area 4:   Maintain system baselines.

Focus Area 1: Correct Latent Defects

Systems, products, and services have degrees of perfection as viewed by the User and System Developer. System Developers employ *design verification* and *validation* practices to discover any *latent defects* and *deficiencies* early in the System Development Phase when *corrective actions* are less costly. Despite the best of human attempts to perfect systems, some latent *defects* and *deficiencies* remain hidden until someone discovers the problem during the System O&S Phase—hopefully WITHOUT *adverse* or *catastrophic* EFFECTS.

New systems, especially large complex systems, inevitably have *latent defects* and *deficiencies*. Sometimes these are *minor*; other times they are *major*. From a program and SE perspective, the *critical operational issue* (*COI*) is to ensure that the delivered EQUIPMENT and its operators are able to achieve their mission objectives without subjecting them to injury, damage, or threats that jeopardize the mission. *Software-intensive systems* are especially prone to *latent defects* that are not discovered during system verification testing.

Some *defects* or *deficiencies* not discovered until a system is fielded. Sometimes they are highly obvious, and sometimes they are only detected over a period of time. The discovery may occur directly or indirectly during analysis of large amounts of data. Therefore, monitor *Problem Reports* (*PRs*) closely to determine if there are *latent defects* that need to be corrected and, if so, the *degree of urgency* in correcting them.

Focus Area 2: Improve Human–System Integration (HSI) Performance

Our discussions up to this point focused on improving EQUIPMENT Element performance. As discussed earlier in this section, EQUIPMENT is just one of several system element performance effecters that contribute to overall system performance. Measurable system performance may also be achieved by improving the PERSONNEL Element performance aspects without having to procure new EQUIPMENT solutions. *How do we do this?*

EQUIPMENT system performance is often limited by system operator and maintainer skills, proficiency, and performance. System operators and maintainers require training to improve their knowledge, skills, and proficiency in understanding the limitations of the EQUIPMENT and HOW to properly apply the EQUIPMENT.

The *mechanisms* for improving human performance includes operator and maintainer selection, classroom training, field experience operating the system, and sometimes LUCK. Experienced system operators and maintainers are often responsible for training new students. The training, however, is dependent on the availability of training aides and devices that provide the students the *look*, *feel*, and *decision-making* environment that enables them to become proficient.

The question for SEs is: *HOW do we specify system capabilities that enable instructors to train and evaluate student performance on the EQUIPMENT?* Training sessions need to employ devices that *immerse* the student in the types of operational and decision-making environments they will actually confront in the OPERATING ENVIRONMENT. This presents several challenges for SEs.

Training Challenge for SEs. Most engineers, by virtue of education and lack of system development training, focus immediately on hardware and software design details. If you ask most to write a specification, seldom will you find anyone who SIMPLY asks the question: *Does this system have a training requirement? If so, WHAT capabilities must be incorporated into the system to support training?*

The solution to this challenge is to conduct on-site visits with the User stakeholders—the system operators, maintainers, and instructors—and determine what training requirements will be required for the system. Remember, the system requirements listed in the *Operational Require-*

ments Document (*ORD*) should include training requirements. Some of those requirements are allocated to training media, some to instructors, and some to the EQUIPMENT Element items that are to be procured. In some cases, the System Developer's contract may specify delivery of *training systems* to support the EQUIPMENT Element.

For most HUMAN-MADE systems, the general thinking is to train system operators to use a system or product after delivery. However, there are cases, especially in *single-use systems*, where the actual mission is the one and only real world training session. Consider the following example:

EXAMPLE 57.4

During the Apollo program, astronauts trained with laboratory simulators, physical landing device simulators, and neutral buoyancy tanks that provided the LOOK and FEEL of a zero gravity landing on the Moon. However, all training had to be accomplished in a normal Earth gravity environment. THINK about the training challenge! The actual landing of the manned Apollo *Lunar Lander*, as an *unprecedented* system, on the Moon's surface *represented* the first time all elements of a single-use system were operated simultaneously in its prescribed zero gravity OPERATING ENVIRONMENT.

So, *HOW do we create training aids to support training*? The answer is by way of *training devices*.

Training Devices. The cost, complexity, availability, and risk of using *actual* systems for training purposes often limits training. One mechanism for achieving similar results is to develop simulators that enable students to learn in a *realistic, immersive* environment without harming themselves, the public, instructors, or the EQUIPMENT. Training devices provide this capability. Training devices range from simple desktop computer models to highly complex aircraft simulators located on multi-axis motion platforms as illustrated in Figure 51.6.

In general, most systems have *training task lists* (*TTLs*) that identify specific mission tasks for assessing the knowledge, skills, and proficiency of trainee operators. Instructional System Development (ISD) personnel analyze these tasks and determine what types of training aides and media are required—whether to use classroom instruction, laboratory training devices or simulations, or actual EQUIPMENT systems. As a result of the ISD analyses, TTL items are allocated to one or more types of training. Therefore, training devices are required to provide capabilities to support all or some of the TTL items.

Training Device Specification Challenges. Training devices present a couple of key challenges to SEs.

Challenge 1: Degree of Realism. The need to *create* and *immerse* a student in a realistic environment sounds pretty simple. Wrong! The challenge for SEs is: HOW do we specify a system that is: 1) *realistic*, 2) *representative of the real world*, and 3) *provides effective training?* THINK about it. Do you specify that the simulated world should look and feel like the real world?

This challenge drives SEs to develop *levels of fidelity*—or the *degree of realism* representative of the physical world. Typically, attribute descriptors such as *low*, *medium*, and *high* are applied to levels of fidelity. However, *define low, medium, and high*? Medium fidelity to one person may be *low* fidelity to someone else. Does a training device require components—such as switches, dials, and gauges—that identically match the MISSION SYSTEM or are photographic models acceptable? SEs must *bound* and *define* these terms to avoid confusion and misinterpretation.

Challenge 2: Simulator Concurrency. Another challenge for SEs is specifying a simulator that identically matches the actual MISSION SYSTEM—e.g. aircraft, spacecraft, nuclear power plant, etc.—in *look, feel*, and *performance*. If the simulator is not *concurrent* with the actual system, you may have a *negative* training condition. If part of the actual system—e.g. vehicle—does not operate due to a lack of maintenance, etc., that training device capability should also be inoperable. This includes: switches, indicators lamps, displays, audio, etc.

Challenge 3: Scoring Capability Requirements. As with any training device, the question is: *HOW do we measure levels of student skills and proficiency?* This area represents a highly debated topic. The challenge for SEs is: *HOW do you specify scoring capability requirements?*

A word of caution: some SEs have inflated egos and *erroneously* believe they can specify training capabilities for this area. AVOID this notion; *recognize* the limitations of your expertise! Talk with the Instructional System Development (ISD) subject matter experts (SMEs) who are the professionals and understand WHAT constitutes *effective training* capabilities. *Collaborate* with the ISD SMEs to identify and specify the training capability requirements.

Challenge 4: Operational Task List Training. Large complex EQUIPMENT systems such as ground vehicles, aircraft, and spacecraft involve Training Task Lists (TTLs) that must be performed as part of normal system operations as well as in response to malfunctions.

Each of these operational tasks employs various system *capabilities* and *levels of performance* that are controlled by the Human-in-the-Loop (HITL) or by automated decision-making systems. When developing training devices, recognize that the simulator must provide the capability to train system operators in using the TTLs. Some TTL items may or may not be applicable or trainable except on the actual equipment. TTL items allocated to the simulator should be documented in the *System Performance Specification* (SPS).

The challenge for SEs is to specify the simulation device(s) with capabilities that enable effective training students in applying the respective TTLs in the simulated OPERATING ENVIRONMENT.

Challenge 5: Levels of Training Capabilities. Human performance improvement entails training in several skills areas. These include basic training, advanced training, and remedial training.

1. *Basic Training* Fundamental instruction complemented with hands-on experience to achieve a level of competence in basic system capabilities and levels of performance.
2. *Advance Training* Specialized training in mission scenario environments that challenge the limitations of the human and machine and ensure a level of proficiency in achieving mission objectives.
3. *Remedial/Refresher Training* Retraining in skills and proficiency areas that may be deficient.

Challenge 6: Other Human Performance Indicators. Earlier we highlighted the dependence of many engineers on actual EQUIPMENT data and noted that simple *inspection* by *examination*—such as for oil leaks, hydraulic leaks, or fatigue cracks—provides valuable insights into system performance. The same is true with human performance.

When systems are designed, vast energies are expended designing them *idealistically* for people who are in the 95th percentile of established anthropometric standards. For many applications, these design rules work. However, after a system is deployed to the field, SEs need to take a *field trip* to *observe* the system's operators and maintainers in action. Startling discoveries sometimes arise.

System operators and maintainers are *clever* and sometimes *show* the developers a *better way of doing things*. All this is dependent on HOW WELL the Users *like* or *loathe* the system or product. Also, observe the ergonomics of the work environment. *Are there work habits or indications that can degrade their performance?*

Focus Area 3: Maintain MISSION SYSTEM-Training Device Concurrency

Training devices such as models, simulations, and simulators must remain in *synchronized* lock step with the MISSION SYSTEM they simulate such as ground vehicles, aircraft, ships, nuclear reactors, robotics surgeries and so forth.

Typically, the MISSION SYSTEM developer is different from the *training device(s)* developer. Where concurrency is a *critical operational or technical issue* (*COI/CTI*), Users and the Acquirers must ensure that contracts are in place to support concurrency requirements to promote communications among the organizations.

Focus Area 4: Maintain System Baselines

One of the challenges of fielded systems is failure to keep the *As-Maintained* configuration baseline current. This commonly occurs when budgets are reduced or priorities are focused on other activities. Unfortunately, some organizations have a view that if an activity doesn't help the bottom line or accomplish a mission, it must not be worth investing resources. So, MISSION SYSTEM Users and System Developers ask: "*Do we invest money in the actual system to get more capability or in maintaining system baselines in lock step?*" Generally, the capability argument wins. As a result, there may be a discrepancy between the physical MISSION SYSTEM and its baselines.

Once a system is *developed, delivered*, and *accepted*, a key question is: *WHO maintains the product baseline for fielded systems?* Systems often go through a series of *major upgrades* over their useful *service life*. Each improvement or upgrade may be performed internally or externally by a new System Developer contract. The challenge for User and Acquirer to ultimately answer is: *HOW do we assure the developer that the baseline configuration documentation identically matches the improved or upgraded, As-Maintained system?* As a User SE, you will be expected to answer this question and vouch for the integrity of the *As-Maintained* Product Baseline.

Maintaining the *currency* of the Product Baseline is important not only for the existing fielded systems but also for future production runs. Some systems are fielded in low quantities as *trial runs* to assess consumer feedback in the marketplace. After the *initial* trials, production contracts may be released for large quantities. Some markets may saturate quickly or, if your organization is lucky, enjoy system production over several years. These production runs may involve a single production contractor or multiple contractors; the baseline integrity challenges, however, are the same.

57.7 PLANNING FOR NEXT-GENERATION SYSTEMS

Planning of next generation systems includes several focus areas:

Focus Area 5: Perform and maintain a "gap" analysis.
Focus Area 6: Bound and partition the problem and solution spaces.
Focus Area 7: Formulate and develop new capability requirements.

Focus Area 5: Perform and Maintain a "Gap" Analysis

As MISSION SYSTEM and SUPPORT SYSTEM data are collected and analyzed, a repository of knowledge is established. So, *what do you do with all this data*? Despite its title, this could be simply a single page for management with a brief rationale depicting the current gap and gap projections over time.

The OPERATING ENVIRONMENT element of most systems is often *highly competitive* and may range from *benign* or *adversarial*. Depending on the overall organizational mission, competitors and adversaries will continually *improve* and *upgrade* their system capabilities. Regardless of your organization's operating domain, the marketplace, military environment, space environment, and medical environment are *dynamic* and *continually change* in response to MISSION SYSTEM changes.

For some systems *survival* or *convenience* means changing, either by necessity due to component or *obsolescence* accordance cost of maintenance, in accord with marketplace trends and demands. Thus, operational capability and performance gaps *emerge* between your existing system and those of competitors and adversaries. This in turn forces investment to improve existing system performance or develop new *systems*, *products*, and *services*.

Focus Area 6: Bound and Partition the Problem and Solution Spaces

The gap analysis provides an assessment of current capabilities and performance relative to projected organizational needs as well as competitor and adversarial capability projections. Analysis of the "gaps" may reveal one or more potential *problem spaces*, each with its own degree of *urgency* for fulfillment. Each *problem space* in turn must be partitioned into one or more *solution spaces*.

Focus Area 7: Formulate and Develop New Capability Requirements

Over time, the capability "gap" between the existing system and projected needs widens. At some point in time a determination will be made to initiate actions to *improve* or *upgrade* existing system performance or to develop a new system. Thus, the evolving *problem* and *solution space* boundaries—the Requirements Domain—will have to be captured in terms of required operational capabilities and costs. As an SE, you may be assigned responsibility to collect and *quantify* these requirements.

Referral For more information regarding system capabilities, refer to Chapter 21 on *System Operational Capability Derivation and Allocation*.

57.8 GUIDING PRINCIPLES

In summary, the preceding discussions provide the basis with which to establish the guiding principles that govern System Operations and Support (O&S) practices.

Principle 57.1 SE does not end with the System Development Phase; it addresses system performance assessments throughout the System O&S Phase.

Principle 57.2 Ensure that the As-Maintained configuration baseline is current with the fielded system at all times.

57.9 SUMMARY

Our discussion of *system O&S practices* was intended to familiarize SEs with key areas that require definition prior to the formal solicitation effort that leads to new system development. We emphasized the two contexts of O&S:

1. Assessing existing system performance.
2. Planning for capability improvement upgrades and next-generation system requirements.

GENERAL EXERCISES

1. Answer each of the *What You Should Learn from This Chapter* questions identified in the *Introduction*.
2. Refer to the list of systems identified in Chapter 2. Based on a selection from the preceding chapter's General *Exercises* or a new system selection, apply your knowledge derived from this chapter's topical discussions. Assume that you have been contracted to develop requirements for the next-generation system.
   (a) What are the SYSTEM level questions you would pose to the owners and Users concerning the performance of current operations?
   (b) Using the System Elements as the frame of reference, what questions would you ask regarding the performance of each element?

ORGANIZATIONAL CENTRIC EXERCISES

1. Contact programs and functional organizations within your facility that procure systems for integration into contract deliverables or installation within the facility.
   (a) What experiences did they have with installation and integration of the new system?
   (b) What types of *latent defects* or *deficiencies* were discovered after system delivery?
   (c) Can they show metrics profiles concerning the frequency of finding defects over time?
   (d) Did the supplier provide training for the users?
   (e) For each of the System Elements, how would they grade the supplier's performance?
   (f) What lessons learned do they recommend avoiding in future procurements?

REFERENCES

Defense Systems Management College (DSMC). 2001. *Glossary: Defense Acquisition Acronyms and Terms*, 10th ed. Defense Acquisition University Press. Fr. Belvoir, VA.

MIL-STD-480B (canceled). 1988. *Military Standard Configuration Control: Engineering Changes, Deviations, and Waivers*. Washingtons, DC: Department of Defense (DoD).

MIL-HDBK-1908B. 1999. *DoD Definitions of Human Factors Terms*. Washington, DC: Department of Defense (DoD).

ADDITIONAL READING

AR 200-1. 1997. *Environmental Quality: Environmental Protection and Enhancement*, Army Regulation, Department of the Army. Washington, DC: Government Printing Office.

MIL-HDBK-470A. 1997. *Designing and Developing Maintainable Products and Systems*. Washington, DC: Department of Defense (DoD).

MIL-HDBK-881. 1998. *Military Handbook—Work Breakdown Structure*. Washington, DC: Department of Defense (DoD).

MIL-STD-961D. 1995. *DoD Standard Practice for Defense Specifications*. Washington, DC: Department of Defense (DoD).

Epilogue

The preceding chapters presented the key *concepts*, *principles*, and *practices* of *System Analysis, Design, and Development*. Our purpose has been to provide insights that enable system analysts and SEs to *bridge the gap* between a User's *abstract vision* of a system, product, or service into the *physical realization* that will satisfy their organizational or personal needs and objectives.

Our discussions highlighted key topics and methodologies that apply to the multi-disciplinary "engineering of systems." As a structured *problem solving/solution development* methodology, these *concepts* and *practices* enable us to develop systems ranging from *simple* to *highly complex*. The flexibility of the methodology facilitates application to any type of system, regardless of business domain. These systems may range from institutional/organizational systems such as a banking system, school, hospital system, et al. to the *"engineering of systems"* such as aircraft, ships, Space Shuttle, or International Space Station (ISS).

Equipped with a this new knowledge and insights for *thinking* about, *organizing*, *analyzing*, *designing*, and *developing* systems, you are now ready to embark on applying and tailoring what you have learned. With practical application, these practices should enable you to create a framework that can be tailored to best fit the needs of your *business domain* and *line of business*. With seasoned experience, you are better prepared to lead collaborative engineering development efforts to make informed decisions without having to take a *quantum leap* of faith to a point solution that may or may not fulfill the User's operational needs.

With these points in mind, I extend best wishes to you in your quest for *success* through the application of the concepts, principles, and practices of *System Analysis, Design, and Development*.

Charles S. Wasson

System Analysis, Design, and Development, by Charles S. Wasson
Copyright © 2006 by John Wiley & Sons, Inc.

Index

Note:
Boldface page numbers represent definitions
Def. = definition

Abstraction
  Defined **67**
  Hierarchical levels of 414
  System levels of 76, 80–82
Acceptability
  System 46
Accountability
  Technical 128
Aggregation (Composition) 7
Agreements
  Non-disclosure/Proprietary 572
Acquirer
  Contracting Officer (ACO) 313, 493, 505, 717, 730
  Furnished Equipment (AFE) 495, 675
  Furnished Property (AFP) **481**, 492, 493, 495
Actions
  Allowable/Prohibited 202
  Corrective 265, 270
Acceptance
  Tests 701
  Testing (def.) **734**
Acceptance Test Procedures (ATPs)
  Application 269, 271, 696
  Defined **734**
  Derivation 742
  Developing 747
  Procedure-based ATPs 701, 748
  Scenario-based ATPs 701, 748
  Types of 747
Accreditation 652
Accuracy and Precision 547, 577, 581
Actor (See UML)
Adaptation
  Defined **147**
  System 156
Affordability 49

Allocation(s)
  Performance budget 460, **598**, 601
  Requirements 221, 225, **358**, 365, 641–642
Alternative(s)
  Analysis of (AoA) 674
  Viable (def.) **673**
American Board of Engineering and Technology (ABET) 23
Analysis
  Aggregation (Composition) 7
  Conclusions 583
  Defined **574**
  Effectiveness 575
  Electromagnetic Compatibility (EMC) 577
  Electromagnetic Interference (EMI) 577
  Engineering Reports 579, 612
  Error 535
  Fault tree 577
  Findings 583
  Finite element 577
  Gap 126, 781, 786
  HFE task 526, 535, 603–611
  Human Factors Engineering (HFE) 535
  Lessons learned 581
  Logistic Support (LSA) (def.) **774**
  Mission 442
  Mission task 164
  of Alternatives (AoA) 672, 674
  Paralysis 574, 579, 632
  Part/circuits tolerance analysis 644
  Rate Monotonic (RMA) 611
  Recommendations 583
  Reliability 641
  Resources 582
  Sensitivity 673, 684
  Specification 327
  Statistical 580
  System operational capability 237
  System performance 603
  System Requirements (SRA) 329
  Task **526**, 535

System Analysis, Design, and Development, by Charles S. Wasson
Copyright © 2006 by John Wiley & Sons, Inc.

789

790   Index

Timeline 160
Transaction (def.) **452**
Types of 577–578
Use case 174
Worst case 644, 654
ANALYSIS (Verification Method)
  See Verification Methods
Anomaly
  Defined **734**
  Challenges 755
Anthropometry
  Defined **525**
Anthropometrics
  Defined **525**
Architect
  System 411, 412
Architecture(al)
  Architecting 411
  Attributes 414
  Behavioral Domain Solution 453
  Centralized 411, 418–420
  Client server 420
  Concerns **411**, 413
  Configuration redundancy 422
  Decentralized 411, 420
  Defined **411**
  Description (AD) **411**, 413
  Development 651
  Entities 414
  ES&H considerations 425
  Fault tolerant 421, **469**
  Fire detection and protection 425
  Flexibility **469**
  Formulation 412
  IEEE 1471–2000 413
  Item boundaries 500
  Logical 74, 415, 454
  Network 420
  Open system 411, **469**
  Operational 415, **439**
  Operations Domain Solution 440
  Physical 74, 415, **466**
  Physical attributes 469
  Physical Domain Solution 467
  Power considerations 425
  Redundancy 422
  Representation 414
  Requirements 415
  Requirements Domain Solution 415, 432
  Scalability **469**
  Selection 425
  Semantics 79
  Stakeholder concerns 413, 417
  SOI system elements 87

System security considerations 427
Trade studies 676
Transparency **469**
View **411**, 413, 414
Viewpoint **411**, 413
Assessment
  Situational (def.) **122**
Assumptions
  and caveats 582
  Failure to document 647
  Poor or incorrect 687
  Specification 311
Atmosphere
  Standard 577
Attributes
  Defined **28**
  Physical / Functional (def.) **28**
  System 30–33
Audits
  Functional Configuration (FCA) 56, **252**, 255, 256, 260, 571, 696, 728, 752
  Physical Configuration (PCA) 56, **253**, 255, 256, 260, 571, 696, 728, 752
  Requirements Traceability (RTA) 254, 718
Authorized Access
  Defined **563**
Availability
  Achieved (A_a) 537
  Defined **401**
  Design for 405
  Interface 521
  Inherent (A_i) 637
  Operational (A_o) 50, 637, 775
  System 636

Baseline Concept Description (BCD) 447
Baseline(s)
  Allocated (def.) **490**
  Configuration 502
  Control and maintenance 306
  CSCI **490**
  Defined **490**
  Functional 504
  Management (def.) **490**
  Product 504
  Production 504
  Requirements (def.) **491**
  Specification 355
Bathtub Curve
  Failure profile 620–621
Behavioral Domain Solution
  Architecture 453
  Challenges 461
  COIs / CTIs 461

Defined **276**
Dependencies 453
Development 451
Introduced 243–245
Key elements 452
Methodology 454
Need for 453
Objective 452
Optimization 282, 461
Responsibility 453
SE Process Model 281
Sequencing 453
V&V 461
Work products 463
Benchmark
System performance 780
Bill of Materials (BOM)
Engineering (EBOM) 82, 642, 643
Boards
Configuration Control Board (CCB) 432, 440, 453, 752
Material Review (MRB) 752
Software CCB (SCCB) 432, 453, 440, 752
Boehm, Dr. Barry 294, 296
Box, George E. P. 648, 669
British Engineering System (BES) 547
Budgets
Performance 460
Performance allocation 598–599
Safety margin 460
Build
Configuration 297–298
Business
Operational cycles 214
By-Products
Acceptable / Unacceptable 345

Calibration and Alignment
Equipment and tools 753, 754
Capability
Anatomy of 229
Automated or semi-automated (def.) **230**
Bounding 230
Construct 232
Defined **27**
Derivation 220, 434
Equating to modes 203
Force multiplier 142–143
Full operational (FOC) 62, 125, 140, 297, 414, 768
Gap 138
Initial operational (IOC) 62, 125, 140, 297, 414, 768
Level of urgency 125

Processing 171
Required Operational (ROC) 219
Reporting 235
Resource utilization 238
System operational 217
Case
Test (def.) **735**
Use (def.) **168**
Cautions and Warnings
User 219
Classification of Defects
Defined **692**
Certification
Defined **692**
Model & simulation 655, 706
System 760
Test equipment and tool 753
Characteristics
Aesthetic 35
Defined **29**
General operating 34
Operating or behavioral 34
Physical 34
Required Operational 382
Required Technical 382
System 34–35
Command Media
System design and documentation 565
Commercial-Off-the-Shelf (COTS)
COIs/CTIs 675
Defined **481**
Driving issues 484
Items 483
Physical items 469
Product (def.) **481**
Selection questions 484–487
Trade Studies 675
Verification 696
Compatibility 150
Defined **147**
Versus interoperability 150
Compensating Provision
Actions 646
Defined **616**
Use case 173
Completeness 414
Complexity
Reduction of 71
Compliance
Certificate of (C of C) 483, 565, 695, 696, 698
Consequences of non-compliance 155
Defined **545**
Specification 337–338

Command and Control (C²)
  HSI 469
  Input/Output (I/O) devices 531–532
  Interactions 153
Common Support Equipment (CSE)
  See Equipment
Comply
  Defined **147**
Component
  Defined **466**
  Design options 482
  External acquisition 493
  Origins 493
  Selection methods 483
Computer Software Configuration Item (CSCI)
  Computer SW Component (CSC) 269
  Computer SW Unit (CSU) 269
  Defined **490**
  Implementation 493
Concept
  Defined **4**
Concept of Operations (ConOps)
  Defined **178**
  Description 179
  Operations Domain Solution 446
  System Operations Model 179
Concerns
  Architectural 413, 414
Conclusion(s)
  Analysis report 581
  Prerequisite 36
  Trade Study Report (TSR) (def.) **673**
Conditions
  Abnormal operating 233
  Environmental 170
  Initial operating 36
  Specification of operating Environment 312
  System operating 321
Conference
  Agenda 716
  Minutes **711**, 716
  Action Items 716, 721–730
Configuration
  As Allocated **503**
  As Built 260, **503**
  As Designed 56, 260, 309, **503**
  As Maintained **503**, 785
  As Produced **503**
  As Specified **503**
  As Validated 56, **503**
  As Verified 56, 260, 309, **503**
  Baselines 502
  Defined **490**
  Developmental 61, 252, 502, 503
  Effectivity 335, 491, 497
  Identification responsibility 497
  Item selection 496
  Management (def.) **491**
  Redundancy 422
  Semantics 492, 494–495
Configuration Control Board (CCB)
  Behavioral Domain Solution 453
  Deviations and waivers 752
  Operations Domain Solution 440
  Requirements baselines 432
  SEIT function 309
Configuration Item (CI)
  Applications 223, 493
  Computer Software (CSCI)
  CSCI boundaries 496, 500
  Hardware Configuration (HWCI) 490, **491**, 493–494, 495, 515, 666
  Defined **491**
  Ownership 499
  Logical (LCI) (def.) **491**
  Ownership and control 499
  Physical (PCI) (def.) **491**
  Identification Responsibility 497
  Selection 496
Conflict of Interest
  V & V 705
Conform
  Defined **147**
Conformance
  Defined **545**
  Specification 337–338
Consumables
  Application 232, 406
  MISSION RESOURCES 91–92
Constraints
  Boundary condition 576
  Design and construction 217, 223
  Operational 160, 170
Construct(s)
  Behavioral interactions 153–155
  Behavioral responses model 148
  Phases and modes of operation 200
  Physical environment architecture 101
  System Entity 21
  System architecture 68–70, 99
  System capability 231
  System element architecture 87
Contract Data Requirement List (CDRL)
  Defined **563**
  Description 564

Discussed 313, 717
Technical reviews 718
Contract Line Item Numbers (CLINs)
Technical reviews 718
Contract Logistics Support (CLS)
Implementation 630–631
Contract/Subcontract
Statement of Objectives (SOO) 78, 126, 253, 254, 273, 279, 317, 327, 352, 674
Statement of Work (CSOW) 230, 313, 305, 354, 384
Work Breakdown Structure (CWBS)
CWBS Dictionary 95
Technical direction 717, 730
Terms and Conditions (Ts & Cs) 61, 78, 257, 267, 305, 331, 467, 493, 714
Type 714, 762
References 78, 95, 310, 447, 468, 501, 502
Control or Staging Points
Defined **178**
Developmental configuration 504
Mission 162
Verification 697
Control
Real-Time 611
Convention(s)
Angular displacement 555–556
Assumptions 559
Body axis 551
Control flow 209–210
Cycle Time 603–604
Data flow 209
Decomposition 10
Defined **545**
Lesson learned 558
Observer's frame of reference 549
PITCH, YAW, ROLL convention 550, 551–554
Right Hand Rule (RHR) 549
Semantics frame of reference 79
Coordinate System
Angular displacement 555–556
Defined **545**
Dimensional reference 554
Discussion 548–556
Frame of reference 548
Observer's eyepoint 549
Right Hand Cartesian 549
Transformation 559
World Coordinate System (WCS) 550
Correction
Error 592
Corrective
Action (def.) **265**

Maintenance (def.) **616**
Maintenance Frequency 635
Correlation
Positive / negative data 594
Cost
Effectiveness 52
Life Cycle 23, 52, 53
Design to (DTC) 403
Total Ownership Cost (TOC) 33, 470, 640
Cost as an Independent Variable (CAIV) 53, 162, 593, 657
Countermeasure
Application 152
Defined **147**
Counter-Counter Measure (CCM)
Application 152
Defined **147**
Criteria
Decision 673
Fitness for use 131, 362
Requirement validation 384
Critical Issues
Defined **391**
Critical Operational/Technical Issues (COIs/CTIs)
COI (def.) **391**
CTI (def.) **734**
Domain solutions 244
Evaluation of alternatives 296
Integrated Process and Product Development (IPPD) 713
Modeling and simulation 697
SME peer reviews 583
Specification issues and tracking 336–337
System deployment 768
System Design Process 267
System Design V & V strategy 259
System O & S 779
Technical review resolution 717–718
Trade studies 672, 674–676
Training device concurrency 785
Validation requirements 319
Cues
Auditory 531
Behavioral responses model 148
External inputs 92, 172
Modal transition 197
Vibratory 531
Visual 531
Use case 172
Customer
Needs, wants, can afford, willing to pay 49, 57, 360–361, 369

Perceptions 57, 125, 718
Satisfaction 485
Cycles
Embedded or nested 214

Data
Acquirer and vendor 571
Accuracy and precision 547
Archival preservation 752
Authentication 571, 752
Convergence 595
Correlation 594
Defined **563**
Deliverable 564
Design and development 564
Dispersion 591
Export control See ITAR
Flow 209–210
Integrity 751
Item Description (DID) 564
Maintainability 635–636
Operations & Support (O&S) 565
Personal engineering records 565
Proprietary 572
Regression 595
Released (def.) **563**
Source credibility 582
Sources—failure to document 647
Subcontractor, vendor, supplier 564
System performance 780
Technical (def.) **563**
Validation 571
Validity 687
Variability 587
Vendor 572
Working data **563**, 565
Data Accession List (DAL)
Application 565
Defined **563**
Description 566
Data Accession List (DAL)
Defined **563**
Description 566
Data Correlation
Convergence/divergence 595
Positive/negative 594
Regression 595
Data Flow Convention 209–210
Decision(s)
Criteria 577, 673
Factor 673
Make vs Buy 473, 481
Technical attributes 575–577

Decision Making
Aids 653
Delivery of analysis results 577
Documentation format 581
HSI 469
Stakeholder participation 770
Team-based 12–13
Timing 576
Decision Support
Process 257
SE Process Model 282
Decomposition
Decimal-based 11
Hierarchical 10
Rules 84
Tag-based 11
Defects
Classification of (def.) **692**
Correction of 270, 705
Design errors / flaws 62, 120, 233, 258, 270, 328, 753
Latent 120, 233, 258, 328, **616**, 669, 703, 705, 753, 775, 782
Workmanship 62
Deficiency
Defined **692**
Discussions 258, 270, 669, 703, 775, 782
Demonstration
Proof of concept 262
Proof of principle 262
Technical 701–702
DEMONSTRATION (Verification Method)
See Verification Methods
Deployment
Challenges 769–771
Decision making 770
Defined **759**
Design for 406
Engineering considerations 768
Environmental considerations 766, 769
Hazardous Materials (HAZMAT) 771
Interfaces 761
Mock 769
Modes of transportation 761
Objectives 759
Production distribution 760
Roles and responsibilities 760–761
Safety and security 766
SE application 761
SE roles 760
Security 766, 771
Shock and vibration 771
Site selection 763
Workforce availability 771

Index   795

Description
    Architectural (AD) 411, 413
    Baseline Concept (BCD) 447
    Interface Design 462
    System/Segment Design Description 462
Design
    Cautionary range 588
    Defects, deficiencies, errors, flaws See Defects
    Normal operating range 588
    Rationale document 330
    Warning range 589
Design and Construction Constraints
    Defined **217**
    Operational capability matrix 223
    Specification of 310, 311
Design Criteria List (DCL)
    Defined **563**
    Description 566
Design To/For Objectives
    Availability 401, 405
    Comfort 404
    Cost (DTC) 403
    Deployment 406
    Disposal 407
    Effectiveness 407
    Efficiency 401, 407
    Growth and expansion 405
    Integration, test, & evaluation 407
    Interoperability 401, 404
    Lethality 406
    Maintainability 401, 407
    Maneuverability 405
    Mission support 406
    Mobility 404
    Portability 401, 405
    Producibility 402, 405
    Reconfigurability 402, 407
    Reliability 402, 405
    Safety 402, 408
    Security and protection 408
    Single use / multi-use applications 404
    Survivability 406
    Training 406
    Transportability 403, 404
    Usability 403, 404
    Verification 407
    Vulnerability 403, 406
    Value (DTV) 403
Development
    Models 290
    Specifications 303
Developmental Test & Evaluation (DT&E)
    Defined **252**

    Description 261
    Risk mitigation 261
    System development overview 61
    System Integration, Test, & Evaluation (SITE) 738, 745
    System verification 704
Development Configuration
    Defined **252**
    System Development Process 256
Development Strategy
    Evolutionary 291, 293–294
    Incremental 291, 297–298
    Grand Design 291
    Spiral 291, 294–296
    System 292
    System versus component 299
    Waterfall 291, 292–293
Deviation
    Defined **303**
    Applications 752
Diagram Types
    Architecture Block (ABD) 414
    Functional Flow Block (FFBD) 414
    Interaction 9
    N2 (N × N) 417, 455, 456, 472
    State 190
    System Block (SBD) 6, 414, 458, 468, 739
    Sequence See UML
    Use Case See UML
Dictionary
    CWBS 95, 501
    System Operations 179, 184, 221
Dimension—System of Units
    Defined **545**
Discrepancy(ies)
    Defined **692**
    Classification of 743
    Report (DR) **265**, 743, 750
    Reporting obstacles 753
Disposal (Waste)
    Defined **759**
Distribution
    Exponential 624
    Logarithmic (LogNormal) **587**, 617
    Negative Distribution 617, 618
    Normal (Gaussian) **587**, 617
    Weibull 620
Documents
    Applicable versus referenced 580, 583
Documentation
    Acquirer and vendor 571
    Assumptions and sources 697
    Dates 582

Electronic signatures 572
Issues 571
Planning 567
Lessons learned 582–583
Levels of formality 569
Level of detail 570
Organizational command media 565
Performance budgets 602
Safety margins 602
Sequencing 567
Specification 568
System design 568
Test 568
Verification 697

Domain
HIGHER ORDER SYSTEMS 99–100
PHYSICAL ENVIRONMENT 100–101
Solution (def.) **276**

DoD Handbooks/Standards
DoD 5000.2-R 535
MIL-HDBK-217 642
MIL-HDBK-470 632
MIL-HDBK-881 736
MIL-HDBK-46855 533–534, 536–538
MIL-STD-822 525
MIL-STD-1908 540–541

Dynamics
Free body 552
Mission 36
State vectors 553

Effect(s)
Cumulative performance 592
Electromagnetic environment (E^3) 120
End (def.) **616**

Effectivity
Based specifications 335, 498
Configuration 497–498

Effectiveness
Cost 52
Design for 407
Measures of (MOEs) 47, 51
Operational 47, 51, 775
System 48, 53

Efficiency
Defined **401**
Design for 407
Organizational 128

Electrostatic Discharge (ESD) 644

Emulate
Defined **741**

Engineering
Analysis reports 579

Change Proposals (ECPs) 728
Defined **23**
Human Factors (HFE) 533
Mechanics 553
Models 739
Professional certification 475
Release Records (ERRs) (def.) **563**
Specialty 474
System (SE) (def.) **24**
Technical report format 580–581

Entity
Defined **67**

Entity Relationships
Defined **67**
Logical 9, 68, 71–73
Logical Configuration Item (LCI) 476
Operational capability 218
Organizational 129
Peer-to-peer 111, 154
Phases, modes, and states 191
Physical 9, 68, 73–74
Physical Configuration Item (PCI) 476

Environment(al)
Atmospheric 103
Biospheric 104
Conditions 170
Cosmospheric 102
Defined **68**
Geospheric 103
Global 102
Hazardous materials (HAZMAT) 766, 771
Hydrospheric 103
Induced 101–106
Local 103
HUMAN-MADE SYSTEMS 100, 104–105
NATURAL 102
Open Systems (OSE) 411
Operating 68
Reclamation 767
Remediation 186
Stress screening 647
System Threat 151

Environmental, Safety, and Health (ES&H)
Issues 767
Material Safety Data Sheets (MSDS) 766
NEPA (1969) 766
Physical Domain Solution considerations 475
Site selection considerations 746

Equilibrium—System
State of 36

Equipment
Alignment 754
Automated Test (ATE) (def.) **734**

Calibration 754
Common Support (CSE) 89, 631, 769, 775
Peculiar Support (PSE) 89, 769, 775
Powered Ground (PGE) (def.) **774**
Support & Handing 90
Support Equipment (def.) **775**
Test and Measurement **736**, 775
Test, Measurement, and Diagnostic (TMDE) 89
Versus human strengths 527–528
EQUIPMENT System Element
Description 88
Performance affecters 157
System architecture element 87
Ergonomics
Defined **525**
Error
Analysis 535
Correction 424
Cumulative **587**, 592
Latent defects 258
Transient 736
Evaluations
Operational Sequence 535
Event
Defined **652**
System performance 781
Triggering 190
Unscheduled or unanticipated 445
Evidence
Objective 582
Evolutionary Development Model
Strategy (def.) **291**
Description 293–294
EXAMINATION Verification Method
See Verification Methods
Exception Handling 232
Expendables
MISSION RESOURCES 91–92
References 232, 406
Export Control
See ITAR

FACILITY System Element
Description 94–95
Performance affecters 157
System architecture element 87
Failure
Cause (def.) **616**
Classification 647
Defined **616**
Effect (def.) **616**
Frequency distribution 624

Internal 165
Qualification 618
Rate 635
Period of Decreasing Failures 619–621
Period of Increasing Failures 619–621
Period of Stabilized Failures 619–620
Root cause 662
Service life profiles 619–620
Single point of (SPF) 419, 423, 644, 645
Failure Modes and Effects Analysis (FMEA)
Defined **616**
Interface failure mitigation 120
System mission analysis 165
Fault tolerant architectures 421
Behavioral Domain Solution 461
Physical Domain Solution 474
Reliability, Availability, and Maintainability (RAM) 643–644
Simulation-based failure investigations 663
Failure Reporting and Corrective Action System (FRACAS)
Defined **631**
Implementation 779
Fidelity
Model (def.) 652
Application 656
Figure of Merit (FOM)
Defined **391**
Performance 677
Finding
Analytical Report 581
Trade study (def.) **673**
Firmware 494
First Article
Defined **252**
Deployment 760
Description 739
Reliability 621
Force Multipliers
Capability 142–143
Form, Fit, and Function
Defined **28**
Discussions 28, 310, 740
Product 21
System 18
Frame of Reference 549
Conventions 549
Observer's 77, 549, 654
OPERATING ENVIRONMENT 106
Right Handed Cartesian 549
Right Hand Rule (RHR) 549
Semantics 77

Frequency
   Corrective Maintenance 634
   Distributions 624
Firmware 494
Function
   Cumulative failure PDF 625
   Defined **28**
   Hazard rate 623–624
   Operations (def.) **440**
   Reliability 623
   Response 149, 171
   Survival 623
   Transfer (def.) **149**
   Transfer response 171
   Utility **673**, 681–684
Functionality
   Out of the Box 481

Gaps
   Operational capability 448
   Problem-solution space 142
Generalization (UML) 8
Grand Design
   Development Strategy (def.) 291
Growth and expansion
   Design for 405

Haptic(s)
   Defined **525**
HARDWARE System Element
   Common Support Equipment (CSE) 89
   Description 89–90
   Peculiar support equipment (PSE) 89
   Support and handling equipment 90
   System architecture element 87
   Test, Measurement, and Diagnostic Equipment (TMDE) 89–90
HARDWARE Configuration Item (HWCI)
   Defined **491**
   Discussions 270, 493–494, 495
HARDWARE System Element
   Composition and description 89
Hazard Rate
   Function 623
Health Hazards
   Environmental, Safety, & Health See ESH
   HIS element 533–534
HIGHER ORDER Systems Domain
   Defined **100**
   Human Context 100
   Physical Environment 100
Human-System Integration (HSI)
   Areas of concern 533, 534
   Elements 533
   Issue areas 539–540
   Performance improvement 782
   Subjects of concern 536–538
Human
   Characteristics 530
   Factors (def.) **525**
   Factors 528, 530
   Factors Engineering (HFE) 533–536
   Performance strengths 527
   Interface classes 528–529
   In the Loop (HITL) 526, 542, 663, 784
   Issue areas 536, 539, 540
   Machine interfaces 111
   Subjects of concern 536–538
   Survivability 534
   System Integration (HSI) 533
   System tasking 538, 540–541
   Versus EQUIPMENT strengths 527–528
Human Factors
   Anthropometric factors 530
   Cognitive factors 530
   Defined **525**
   Psychological factors 530
   Physiological factors 530
   Prototypes and demonstrations 536
   Sensory factors 530
HUMAN-MADE SYSTEMS Element 100
   Business 105
   Cultural 105
   Defined **100**
   Description 104–105
   Educational 105
   Entity relationships 101
   Government 105
   Historical or heritage 104–105
   OPERATING ENVIRONMENT domain element 99
   Transportation 105
   Urban 105
Human Performance
   Defined **525**
   Evaluation 774, 781
   Improvements 782
   Indicators 784

IEEE
   Standard 1471–2000 412–413
Incremental Development Model
   Description 297–298
   Strategy (def.) **291**
Independent Test Agency (ITA)
   Defined **253**
   OT & E 253, 262
   Qualification tests 702

System O & S 776
System validation 56
Test data authentication 752
TEMP implementation 744
V & V implementation 707
Independent Verification and Validation (IV & V)
Benefits of 703
Defined **692**
Findings and recommendations 705–706
INDUCED ENVIRONMENT System Element
Defined **101**
Description 106
Entity relationships 101
OPERATING ENVIRONMENT Architecture element 99
System Element 101
Threat sources 151
INSPECTION (Verification Method)
See Verification Methods
Inputs
Acceptable 22, 345, 588
External 148, 170
Unacceptable 22, 345, 588
Input/Output (I/O) Devices
Audio I/O 532
Data entry 531
Electrical control 532
Mechanical control 532
Pointing control 531
Translation displacement control 532
Sensory I/O 532
Integrated Product & Process Development (IPPD) 713
Issues
Specification 336
Technical resolution 717
Integration
Rules of 83–84
Integration Point (IP)
Defined **266**
System Integration, Test, & Evaluation (SITE) 737
System of Systems (SoS) 77
Instructional System Development (ISD)
Personnel training 784
Interchangeability
Defined **111**
Interface(s)
Access 118
Acoustical 116
Active 114
Attributes 513
Availability 521

Biological 117
Challenges 519–522
Chemical 117
Command and Control (C^2) 531–532
Commitments 520
Compatibility 150, 520
Control (def.) **111**
Control Document (ICD) 516–518
Control Working Groups (ICWGs) 516
Coupling 508
Dedicated 117
Defined **111**
Definition and control challenges 519
Design Description (IDD) 462, 516
Design documentation 516, 518
Design solution 512
Development challenges 519
Devices (def.) **111**
Electrical 116
Electronic data 117–118
Environmental 146, 509
Failure 120
Failure mitigation & prevention 120
Generalized 510–511, 532
Human—Machine 111
Human—System 469, 528–529
Interactions matrix 509
Integrity 521
Interoperability 150, 520
Latency 120
Logical 115
Maintainability 521
Man-Machine Interface (MMI) (def.) **525**
Methodology (def.) **511**
Methodology—design 116
Natural Environment 117
Mechanical 116
Objectives (purpose) 112
Optical 116
Ownership (def.) **111**
Ownership and control 508, 515
Passive 114
Physical 115
Point-to-point 111
Reliability 521
Requirements Specification (IRS) 514
Specialized 510, 511, 532
Specification 514–515
Standardized 117, 519
Types of 114
User-Computer (UCI) (def.) **526**
Vulnerability 120, 521

International System (SI) of Units
  Metric (mks) system 547
International Traffic & Arms Regulations (ITAR)
  Export control 570–571, 572, 686, 716
Item
  Commercial-Off-the-Shelf (COTS) 481, 483, 675
  Configuration (CI) 491, 493–496
  Defined **491**
  Logical Configuration (LCI) 491
  Non-Developmental (NDI) 481, 483
  Physical Configuration (PCI) 491
  Repairable (def.) **775**
  Replacement (def.) **775**
  Reliability Critical (RCI) 643
Interaction(s)
  Analysis methodology 156
  Architectural 414
  Diagrams 9, 458
  Hierarchical (def.) **68**
  Levels of 156
  Peer Level (def.) **111**
  Peer-to-peer 154
  PERSONNEL-EQUIPMENT 531
  Point-to-point 111
  Strategic 156
  System 459
  System element 95
  System of Interest (SOI) 509
  Tactical 156
Interoperability
  Defined **111**
  Design for 404
  Interface 150
  Versus compatibility 150
Investment
  Return on (ROI) 780

Key Performance Parameters (KPPs)
  Defined **391**

Latency
  System 598
Lethality
  Design for 406
Levels of Abstraction
  Architectural 414
  ASSEMBLY 81, 84
  PART 82, 84
  PRODUCT 81, 84
  SEGMENT 81, 84
  SUBASSEMBLY 82, 84
  SUBSYSTEM 81, 84
  SYSTEM 79, 81

Life Cycle Phases
  Identified 60
  Life cycles within life cycles 63
  System Definition Phase 60
  System Development Phase 61–62
  System Disposal Phase 62–63
  System O & S Phase 62
  System Procurement Phase 61
  System Production Phase 62
Life Cycle(s)
  Cost 23, 52, 53
  Evolutionary 65
  System/product 60
  Within life cycles 63
Life Cycle Phase(s)
  Life Cycle 60
  System Definition 60, 253
  System Development 60, 254–257
  System Disposal 60
  System O & S 60
  System Procurement 60, 253
  System Production 60
Line Replaceable Unit (LRU)
  Defined **491**
  Corrective maintenance 631
Logical Entity Relationship
  Application 71–73
  Defined **68**
Logistic(s)
  Delay Time (LDT) 633
  Support Analysis (LSA) (def.) **774**

Make-Buy-Modify Decision
  Defined **481**
  Introduced 473
Maintainability
  Computations 632
  Concept 630
  Data sources 636
  Defined **401**
  Design for 407
  Interface 521
  System 630
Maintenance
  Administrative Delay Time (ADT) 634
  Condition-based (CBM) 631, 633
  Corrective (def.) 616
  Corrective implementation 630
  Defined **774**
  Downtime (MDT) 633
  Factory 631
  Field or organizational 631, **774**
  Preventive 214, 631

Index

System operations 184
Types of 630
Malfunction 233, 234
Manpower and Personnel
  HSI element 534
  SUPPORT SYSTEM operations 93
Man-Machine Interfaces (MMIs)
  Computer-Human Interfaces (def.) **539**
  Defined **525**
  Human—machine (def.) **111**
  User-Computer Interfaces (def.) **526**
Maneuverability
  Design for 405
Material Review Board (MRB)
  Corrective action disposition 752
Matrix
  I/F Interaction 509
  Logical interactions 455
  N × N 417, 456
  Requirements Verification (RVM) 375, 376, 378–379, 514
  System capability **217**, 220, 223
Mean
  Active Maintenance Time (MAMT) 636
  Corrective Maintenance Time (CMT) 635
  Time Between Failures (MTBF) 624, 630
  Time Between Maintenance (MTBM) 636
  Time To Repair (MTTR) 635, 638
Measurement
  Biased or aliased 751
  Technical Performance (TPM) 392
  TPM challenges 397
  TPM reporting 398
Measures of Effectiveness (MOE)
  Defined 47
  Mission gap analysis 126
  Description 51
Measures of Performance (MOP)
  Defined 47
  Design to **598**, 599
  Organizational 124
  Specification requirements 323
  TPM plots 396
  Trade space 677
Measures of Suitability (MOS)
  Application 51–52
  Defined 47
Methodology
  Behavioral Domain Solution 459–461
  Component selection 483
  COTS selection 483
  Interface definition 116
  Interface design 511

Logical-physical architecture 73, 454
Mission definition 160–165
Operating environment requirements 106
Operations Domain Solution 441–447
Physical Domain Solution 470–475
Requirements derivation 362–365
Requirements Domain Solution 432–436
SE Process Model 277
System interactions 156
Trade Study 680
Metrics (Refer to)
  Administrative Delay Time (ADT)
  Availability
  Corrective Maintenance (CMT)
  Figure of Merit (FOM)
  Measures of Suitability (MOSs)
  Measures of Performance (MOPs)
  Measures of effectiveness (MOEs)
  Logistics Delay Time (LDT)
  Maintenance Down Time (MDT)
  Mean Active Maintenance Time (MAMT)
  Mean Corrective Maintenance Time (CMT)
  Mean Time Between Failures (MTBF)
  Mean Time Between Maintenance (MTBM)
  Mean Time To Failure (MTTF)
  Mean Time To Repair (MTTR)
  Preventive Maintenance (PMT) 633
  Technical Performance Measures (TPMs)
Mission
  Analysis 159, 651
  Defined **159**
  Definition methodology 160–165
  Event Timeline (MET) 444
  Needs Statement (MNS) 160
  Objectives 127, 161
  Reliability 160, 161, 628
  Strategy 162
  Sustainment 164
  Unreliability 622
Mission Critical
  Component 618, 621, 644
  System (def.) 159
Mission Phase of Operation
  Objective 192
Mission Support
  Design for 406
MISSION SYSTEM
  Configuration 500
  Introduced 40
  Modal operations and interactions 195–198
  Organization roles and missions 130–133
  Personnel Roles 88
  Roles 40, 88

System architecture element 69, 87
Supply chain 130–132
System elements 88
MISSION RESOURCES System Element
Description 91–92
Performance affecters 157
System architecture element 87
Mobility
Design for 404
Mode(s) of Operation
ABNORMAL operations 199, 233–234, 320
ANALYSIS 199
CALIBRATE/ALIGN 199
CATASTROPHIC operations 321
CONFIGURE 199
Construct 195
Defined **190**
Derivation 194
Description 194
EMERGENCY operations 320
Equating to capabilities 203
MAINTENANCE 199
NORMAL operations 199, 320
Operations Domain Solution 445
of transportation 761
OFF 199
POWER-DOWN 199
POWER-UP/INITIALIZE 199
SAFING 199
TRAINING 199
Modal
Transitions 195–198
Triggering events 196
Model
6 Degree of Freedom (DOF) 551, 553
Application examples 658–668
Assumptions and scenarios 668
Behavioral response 148–149
Business operations 213
Certified (def.) **652**
Certification 655–656, 706
Closed loop 153
Defined **652**
Deficiencies and flaws 668
Deterministic (def.) **654**
Development 654
Documentation 669
Earth-Centered, Earth-Fixed (ECEF) 551
Earth-Centered Inertial (ECI) 551
Earth-Centered Rotating (ECR) 551
Engineering 739
Fidelity 652, 656–657
Hostile encounter 155
Improper application 668

Issue arbitration 155
Open loop 153
Portability 669
Problem-solving / solution Development 277
SE Process 275–289
Status and health request 154
Stochastic model 654
System operations 179–186
Test environment stimulus 662
Undocumented features 669
V-Model (def.) **256**, 272
Validated (def.) **652**
Validation (def.) **653**
Model-Test-Model 652
Modeling and Simulation (M&S)
Accreditation (def.) **652**
Architectural configuration 643, 658
Architecture selection example 658–659
Certification 655, 706
Certified (def.) **652**
Challenges and Issues 668–669
Characteristics 656
Defined **652**
Deterministic **652**, 654
Documentation 669
Event (def.) **652**
Examples 658–668
Failure investigations 662–663
Fidelity (def.) **652**
Frequency distribution 617
Initial conditions (def.) **652**
Initial state (def.) **652**
Monte Carlo algorithm (def.) **653**
Monte Carlo method (def.) **653**
Performance allocations 659
Physical 643
Quick look reliability 642
Performance 642
Reliability network 626
Simulation time 653
Simulation-Based Acquisition (SBA) 660
SOI Operations 208
Stimulation 653
Stochastic process **653**, 654
System performance analysis 610
Test beds 664–668
Time (def.) **653**
Training 663–664
Validated (def.) **652**
Validation (def.) **653**
Validation description 655
Validation reasons 781
Verification (definition) 653

Verification (description) 697
Virtual reality (def.) 664
Modes and states 445
  Physical Domain Solution 474
MOE, MOS, MOP, TPM
  Relationships 392
Moments of Truth 126
  Application x, xii
Monte Carlo
  Algorithm / Method (def.) **653**
  Techniques 589

National Aeronautics and Space Administration (NASA)
  Apollo Program 165, 457
  International Space Station (ISS) 161
  Space Shuttle 61, 198, 212, 235
NATURAL ENVIRONMENT System Element
  Classes 102
  Defined **100**
  Composition 102–104
  Entity relationships 101
  OPERATING ENVIRONMENT architecture element 99
  Threat sources 151
Need
  Operational 55
  Prioritized 126
  Specification of 311
Non-Developmental Items (NDIs)
  Certificate of compliance (C of C) 696
  Defined **481**
  Description 483
  Driving issues 484
  Realities 469
  Selection questions 484–487
Notation
  Decimal-based 11–12
  Scientific 547–548
  Tag-based 11–12
Notes
  Specification 311
Notebook
  System Design 330
NPRD-95 642

Operations and Support (O & S)
  Focus areas 781–785
  Phase objectives 775–776
Object
  Defined **68**
  Management Group (OMG) 6
Objectives
  Mission 127, 161

Operations and Support (O&S) 775
Phase 192, 444
System 127
Open
  Standards 545
  Systems Environment (OSE) 411
OPERATING ENVIRONMENT Domain
  Architecture 97
  Scope 98
  System architecture 69
Operation(s)
  ABNORMAL 233, 320
  Abnormal recovery 234
  CATASTROPHIC 321
  Concept of Operations (ConOps) 179
  Computer resources support 93
  Decision support 93
  Defined **5**
  Dictionary 221
  EMERGENCY 320
  Function (def.) **440**
  Manpower and personnel 93
  Mapping to phases 219
  Mode of 190
  NORMAL 320
  Packaging, Handling, Storage, & Transportation (PHS&T) 93
  Phase objectives 192
  Publications support 93
  Recovery 234
  Shadow 768
  State of 190
  System 179
  System maintenance 93
  Supply support 93
  Technical data 93
  Training and training support 93
Operations and Support (O & S)
  Objectives 775
Operational
  Architecture (def.) **439**
  Asset 439
  Availability 50, 57, 636–637, 775
  Capability bounding 230
  Concept Description (OCD) 126, 180, 198
  Control flow 209–210
  Cycles within cycles 214
  Data flow 209–210
  Domain Solution 439
  Effectiveness 47, 51, 57, 390, 775
  Need 55
  Need priorities 126
  Performance monitoring 777–780

Requirement document (ORD) 126, 253, 319, 706
Scenario 167
Site activation 759
States versus physical states 200–202
Suitability 48, 50, 57, 390, 775
Task sequence evaluations 535
Utility 57, 390, 775

Operational Requirements Document (ORD)
System development 253, 255
System O & S 783

Operational Test & Evaluation (OT&E)
Defined **253**
Risk mitigation 261–262
System Development Phase 61
System performance monitoring 776
System validation 56
Validation requirements 319
V & V activity 704

Operating Condition(s)
ABNORMAL operations 233
CATASTROPHIC operations 321
Categories 321
EMERGENCY operations 233
Initial 36
NORMAL operations 233
Prerequisite 36
Specification 321

OPERATING CONSTRAINTS System Element
Defined **100**
OPERATING ENVIRONMENT architecture element 99
System use case 170

Operations Domain Solution
Baseline Concept Description (BCD) 447
Challenges 448
Critical Operational Issues (COIs) 447
Critical Technical Issues (CTIs) 447
Defined **276**
Dependencies 440
Development 440
Introduced 243–245
Key elements 440
Methodology 441
Objective 440
Optimization 282
Responsibility 440
SE Process Model implementation 280–281
Sequencing 440
Verification and validation 447
Workflow sequences 443
Work products 448

Opportunity
Location-based 137
Space (def.) **136**
SE Process implementation 278, 279
Targets of (TOO) 123, 133
Time-based 137
Types of 136
Understanding 279

Optimization
Behavioral Domain Solution 282, 461
Operations Domain Solution 282
Physical Domain Solution 282, 475, 477
System 575, 578, 611
System performance 461, 578

ORGANIZATION System Element
Defined **100**
OPERATING ENVIRONMENT architecture element 99

Organizational
Accountability 128
Command media 565
Entity relationships 129
Roles and missions 130–133
System Elements (OSEs) 124

Out-of-the-Box Functionality 481

Outputs
Acceptable 21, 345, 738
Unacceptable 21, 345, 738

Outsourcing
Defined **481**

Packaging, Shipping, Handling & Transportation (PHS&T)
Support operations 93, 94

Paradigm(s)
Bathtub 619–620
Defined **122**
Requirements derivation 362

Parameters
Key Performance (KPPs) 391
Technical (Performance) (TPPs) 391

Parts Program
Reliability 646

Peculiar Support Equipment (PSE)
See Equipment

PERSONNEL System Element
Description 88
Equipment interactions 531
Equipment trade-offs 530
HSI element 532
Performance affecters 157
Roles—MISSION SYSTEM 88

Roles—SUPPORT SYSTEM 88
System architecture element 87
Performance
  Analysis tools 578
  Defined **28**
  Expected or mean approximation 609
  Evaluation 776
  Human-System Integration (HSI) 782
  Human 774, 781
  Improvements 612
  Level of 28
  Measures of (MOP) 47
  Monitoring 777–781
  Objective 29
  Organization System Element (OSE) 124
  Pareto ranked 612
  Personnel 88
  Processing time 605
  Safety margins 460, 474
  Subjective 29, 34
  Trends 781
  Verification and validation (V&V) 701–702
Performance Budgets
  Assignment 474
  Allocations 460, **598**, 601–602
  Documentation 602
  Ownership 474, 602
Phases of Operation
  Allocation of mission operations 192
  Defined **111**
  Mission 192
  Objectives 192, 444
  of Flight 193
  Pre-Mission 192
  Post-Mission 192
  Entity relationships 191
Phases, Modes and States
  Entity relationships (ERs) 191, 200, 202
PHS&T
  Description 93, 94
  Implementation 778
Physical Architecture
  Attributes 469
  Baseline 475
  Configuration states 473
  Development 467
  Formulation 470
Physical Configuration
  Audits (PCA) 56, **253**, 255, 256, 260, 571, 696, 729, 752
  States 200–202

Physical Entity Relationships
  Application 73–74
  Defined **68**
PHYSICAL ENVIRONMENT
  Domain 100
  Levels of abstraction 101
  System elements 102
Physical Domain Solution
  Challenges 476
  COIs/CTIs 469, 475
  Defined **276**
  Dependencies 467
  Development 465
  ESH Compliance 475
  Introduced 243–245
  Key elements 466
  Methodology 470
  Modes and states of operation 474
  Objectives 466
  Optimization 282, 475–477
  Performance budgets and margins 474
  Professional certification 475
  Requirements allocation 471
  Responsibility 467
  SE Process Model implementation 281
  Verification and Validation (V&V) 475
  Work Products 477
Plan
  Environmental mitigation 766
  Integrated Support (ISP) 630
  Integrated Master (IMP) 697, 705, 764
  Operations and Support (O&S) 630
  Operational site activation 764–765
  Strategic 122
  System Integration & Verification 744, 755
  System O & S 630
  Tactical 123
  Technical management 567, 639
  Test & Evaluation Master (TEMP) 319, 744
Planning
  Facility engineering 764
  Destructive test sequence 745
  Site survey 762
  Strategic 123–124
  System deployment risk mitigation 769
  Tactical 124
Point
  Critical staging or control 162, 178, 504, 697
  Integration (IP) 77, 266, 737
  of delivery 160
  of origination or departure 160, 193
  of termination or destination 160, 193
  Point-to-point (def.) **111**

Portability
Defined **401**
Challenges 669
Post-Mission Phase of Operation
Objective 192
Power, Balance of 37
Practice
Best or preferred 5, 311
Defined **5**
SE specification
Precedented
Systems 21, 266, 332, 457
Pre-Planned Product Improvements (P3I)
Releases/upgrades 93, 293
Principle
Defined **4**
Probability
Circular Error (CEP) (def.) **587**, 592–594
Density Function (PDF) 587, 622, 625
of occurrence (use case) 172
Problem
Reporting 744, 774
Statement 136, 139, 576
Problem Space
Boundaries 138, 279
Control 139
Defined **136**
Degree of Urgency 140
Dynamics 137
Forecasting 138
Partitioning 140–141, 278
Representation 141
Semantics communications 78
SE Process 278
Solving 71, 137
Space 136, 279, 433, 786
PROCEDURAL DATA System Element
Composition 91
Performance affecters 157
System architecture element 87
Process(ing)
Defined **4**
Decision Support 257
SE 277
Stochastic 653
System Design 267
Technical Management 257
Time **598**, 605
Trade study 679
Producer—Supplier
Relationships 132
Producibility
Defined **402**
Design for 405

Product
Defined **21**
Production Systems
Deployment 760
Low Rate Initial (LRIP) 621, 729
Mass or full scale 620–621, 729
Profile
Equipment characteristics 647
Failure rate 622
Service life 619
Product Structure
Defined **77**
Tree 467–468
Pre-Mission Phase of Operation
Objective 192
Preservation, Packaging, & Delivery
Specification of 312
Producer-supplier
Relationships 131–132
Procedures
Operating 91
Standard Operating Procedures and Practices (SOPPs) 91, 196, 565, 742
Proof of Concept/Principle
Demonstration 262
Properties
Defined **28**
Dimensional 546
System 27
Prototypes & Demos
Design verification support 697
Human Factors Engineering (HFE) 536
Technical demonstrations 701
Provisioning 774

Quality
Function Deployment (QFD) 408
System 564
Quality Record(s)
Defined **253**
SE Process 283
Queue
Time 598

Rate
Failure 633
Hazard 624
Repair 635
Recommendations
Analytical Report 581
Defined **673**
Reconfigurability
Defined **402**
Design for 407

Records
  Engineering Release (ERR) 563
  Personal engineering data 565
  Quality (QR) 253
Redundancy
  Architectural configuration 422, 645
  Cold or standby 422
  Component 645
  Compensating provisions 173, 646
  Data link 424
  Improving reliability 644
  K-out-of-n systems 423
  Like 423
  Operational 422
  Processing 423
  Service Request 424
  Unlike 423
Refurbishment/Reconditioning
  Test article 754
Regression
  Data 595
  Testing 742
Regulations See ITAR
Reliability
  Bathtub curve 619–620
  Challenges 647
  Compensating actions 173, 616, 646
  Critical items (RCIs) 643
  Defined **402**
  Design for 405
  Estimates 643
  Estimates versus predictions 629
  Function 623
  Hazard rate 624
  Interface 521
  Mission (def.) 160
  Mission discussion 160–162, 629, 639
  Models 642–643
  Parallel network configuration 626
  Parts/circuits tolerance analysis 644
  Parts program 646
  Period of Decreasing Failures 619, 620
  Period of Increasing Failures 619
  Period of Stabilized Failures 619
  Precepts 619
  Resource allocations 646
  Series network configuration 626
  Series-Parallel network configuration 627
  Service life profiles 619
  Single point of failure (SPF) 645
  Software 648
  Survival function 623

  System/component mortality 621
  Unreliability 622
Repair
  Defined **775**
Repairability
  Defined **775**
Replenishment
  Expendables and consumables 183
Report
  After Action/Follow-Up 779
  Analysis conclusions 581, 583
  Discrepancy (DR) **265**, 743, 750, 753
  Engineering analysis 580
  Findings 581, 583, 686
  Hardware Trouble (HTR) 744
  Problem **774**, 779, 782
  Recommendations 581, 583, 686
  Test Incident 736
  Trade Study 673, 678, 684–687
Report(ing)
  System Analysis 612
  System capability 235
Request for Proposal (RFP)
  Failure to research references 333
  Issues and clarifications 336
  Interface specification requirements 514
  SE Design 267
  Specification analysis 327
  System performance history 779
  System requirements 317
  System verification 55
Requirement(s)
  ABNORMAL operations 320
  Allocation 221, 225, **358**, 365, 641–642, 651
  Baseline (def.) **491**
  Capability 318
  CATASTROPHIC operations 320
  Categories 318
  Common problems 323
  Compliance 337
  Compound 382
  Conflicting 230, 323–324
  Conformance 337
  Deficiencies 323–324, 331–335
  Defined **316**
  Derivation 35, **358**, 359–365
  Derivation methodology 362–365
  Derivation paradigm 362
  Development 340
  Development guidelines 381–383
  Duplicated 230, 324
  Elicitation (def.) **316**, 317

808   Index

EMERGENCY operations 320
Flow Down 359, 365
Good 371
Hierarchy 322
Interfaces 318
Issues and Clarifications 435
Key characteristics 371–372
Leaf level (def.) **358**
Management tools 379
Meet 338
Minimization 385
Misplaced 323–324
Missing 230, 323–324
Modes of operation 320
MOEs, MOSs, MOPs 385
Need for 305
Non-functional **316**, 318
NORMAL operations 320
Objective **316**, 319
Official (formalization) 372–373
Operational 316, 318
Operational characteristics 382
Optimal 386–387
Origin of 65–66
Outcome-based 372
Performance (def.) **316**
PERSONNEL 312
Performance characteristics 365
Phases of operation 320
Primitive 374
Priorities 319
Reliability, Availability, Maintainability (RAM) 640, 642
SE Process 359
Singular 754
Source or originating **303**, 317
Stakeholder **316**, 317
Stakeholder objectives 317
Statement challenges 590
Statement development 373–375
SUPPORT SYSTEM 312
TBDs and TBSs 349
Technical characteristics 382
Testing 359, 383
Threshold **316**, 319
Time 160
Traceability 128, **359**, 365, 367–368, 462
Traceability audits 718
Traceability verification 697
Types of 317
Use case-based 371
Use of "all, and/or, etc." 381–383
Use of "shall" and "will" 381

Validation 258, **316**, 319, **359**
Validation criteria 258, 383
Verification **316**, 319, **359**, 754
Verification Matrix (RVM) 375, 378–379, 514
Verification method selection 375–378
Visual configuration (graphics) 364
Writing versus developing 230
Requirements Domain Solution
Architecture 432
Baseline 436
Challenges 436–437
Critical Operational Issues (COIs) 435
Critical Technical Issues (CTIs) 435
Defined **276**
Dependencies 431
Development 431
Introduced 243–245
Key elements 431
Methodology 432
Objective 431
Responsibility 432
SE Process Model implementation 279
Sequencing 431
V&V 436
Work Products 437
RESOURCES System Element
Defined **100**
OPERATING ENVIRONMENT domain element 99
Responses
Behavioral 92, 150
Return on Investment (ROI)
Discussions 16, 18, 23, 41, 46, 48, 52, 128, 132, 161, 691, 780
Review(s)
Categories 712
Common types of 719
Conference minutes 711, 716
Contract requirements 713, 715
Critical Design (CDR) 255, 259, 267, 287, 504, **711**, 726
Date driven 713
Entry/exit criteria 715, 716
Event driven 713
Formal Qualification (FQR) 504
Hardware/Software Specification (SSR) 504, **711**, 722–724
In-Process Review (IPR) 711, 729–730
Items 718
Integrated Baseline Review (IBR) 720
Location 716
Objective 712
Peer reviews 583, **711**

Preliminary Design (PDR) **711**, 723, 725
Presentation / handout materials 259, 504, 711, 716, 723, 725
Production Readiness (PRR) 504, **711**, 728–730
Quality records (QRs) 715
Ready-to-Ship Review (RTSR) **712**, 728–729
Software Specification (SSR) 259
System Design (SDR) 259, 504, **712**, 721, 723
System Requirements (SRR) 331, 332, 336, 352, 504, **712**, 720–721
System Verification (SVR) 56, 504, **712**, 728
Technical direction 717
Technical reviews **711**, 727
Test Readiness (TRR) 504, **712**, 727, 755
Verification 697
Work products 715
Risk 711
COIs/CTIs 461
Design 259
Mission 164
Mitigation (DT&E and OT&E) 261–262
Reducing 481
System 164
Specification reuse 332
Trade studies 687–688
Roles
MISSION SYSTEM 40
SUPPORT SYSTEM 40
ROLES AND MISSIONS System Element
Defined **100**
OPERATING ENVIRONMENT domain element 99
Rules
Operational capability analysis 237
System decomposition/integration 81–83

Safety
Critical (def.) **525**
Design for 408
HSI issue area 539
HSI subject of concern 538
System (def.) **402**
Safety Margin
Application 599–602
Defined **598**
Sanity Check 575
SE Process Model 277
Application 285
Behavioral Domain Solution 281
Characteristics 283
Decision support 282
Defined **276**
Entry criteria 277

Iterative characteristic **276**, 283
Methodology 277
Objective 277
Operations Domain Solution 280
Physical Domain Solution 281
Quality Records (QRs) 283
Recursive characteristic **276**, 285
Requirements Domain Solution 279
Specification development 350
Work products 283
Scenario
Consequences 171, 173
Operational (def.) **167**
Use case (def.) **168**
What if 219
Schedule
Master Program (MPS) 705
Integrated Master (IMS) 417, 705
Security
Communications 32
Deployment 766
Design for 408
System (def.) **402**
Types of 32
Semantics
Frame of reference 77
Levels of abstraction 76
Opportunity versus problem 137
Sensitivity Analysis
Defined **673**
Description 684
Services
Acceptable 345
Unacceptable 345
Serviceability
Defined **402**
Similarity
See Verification Methods
Simplicity
Defined **402**
Simulate
Defined **741**
Simulation
Discussion 658
Failure investigation 662
System 658
Time (def.) **653**
Virtual reality (def.) **664**
Simulator
Concurrency 783, 785
Fixed/motion platforms 664
Single-use applications
Design for 404

Site—Deployment
  Data collection 763
  Installation (def.) **759**
  Operational Site Activation (def.) **759**
  Preparation and development (def.) **765**
  Selection (def.) **759**
  Survey (def.) **759**, 762–763
  Visits 763
Software Configuration Control Board (SCCB)
  Behavioral Domain Solution 453
  Deviations and waivers 752
  Operations Domain Solution 440
  Requirements baselines 432
SOFTWARE System Element
  Change Requests (CRs) 744
  Description 90
  System architecture element 87
Solution(s)
  Behavioral Domain 451
  Candidate 143
  Domain (def.) **276**
  Generalized 406, 532
  Operations Domain 439
  Physical Domain 465
  Requirements Domain 430
  Specialized 406, 532
Solution Domains
  Introduced 244–245
  Defined **276**
  Sequencing 244
Solution Space
  Bounding 279
  Defined **136**
  Partitioning 278
  Requirements Domain 433
  SE Process 278
  System O & S 786
Space
  Opportunity 136, 279
  Problem **136**–141, 278, 279, 433, 786
  Solution **136**, 141–142, 278, 279, 433, 786
Specification(s)
  Acquirer perspective 328
  Ambiguity 333
  Analyzing 327
  Applicable documents 333
  Baseline control 306
  Broad references 333
  Change management 324, 325
  Common deficiencies 323, 331
  Compliance 337
  Conflicting requirements 323–324
  Conformance 337
  Content 310
  Defined **303**
  Design and construction constraints 311
  Detail (def.) **303**
  Developer perspective 328
  Development (def.) **303**
  Development checklist 354
  Development paradigm 349–352
  Development sequencing 313
  Deviation (def.) **303**, 752
  Effectivity-based 335, 498
  Evolution 313
  Facility Interface (FIS) 764, 770
  Feasibility 307
  General attributes 306–307
  General standard outline 306, 332, 342–343
  Generalized structure 312
  Graphical analysis 329
  Hardware Requirements (HRS) 496
  Hierarchical analysis 329
  Interface (def.) **303**
  Interface Requirements (IRS) 514–515
  Issues and concerns 336
  Key elements 310–312
  Language 307, **371**
  Misplaced requirements 323–324
  Missing requirements 323–324
  Model-based structured analysis 346–349
  Model fidelity 657
  Modeling & simulation analysis 329
  Notes & assumptions 311
  Objectives of 304
  Over/under 334
  Ownership and accountability 306, 309, 318, 333
  Packaging 312
  Performance (def.) **303**
  Performance criteria 312
  PERSONNEL requirements 312
  References versus applicable docs. 333
  References 333, 335
  Requirements categories 318–319
  Requirements duplication 324
  Requirements structure 371
  Requirements vs. CSOW tasks 305
  Reviews 341, 355
  SE Practices 311
  SE Process Model application 350
  Sections 352–353
  Sequencing 313
  Software Requirements (SRS) 496
  SUPPORT SYSTEM Requirements 311
  Tailoring (def.) **303**
  TBDs and TBSs 349

Technology analysis 329
Traceability 307
 Traceability to CWBS 310
 Types of 307, 308
 Use case relationship 175
 Verification method 311
 Waivers (def.) **303**
 Writers versus developers 335
Specification Development
 Architecture-based approach **341**
 Feature-based approach **341**, 343–344
 Performance-based approach **341**, 345–346
 Reuse-based approach **341**, 344–345
 Reuse risks 332
Specification Tree
 Defined **303**
 Implementation 307, 309, 498
 Ownership and control 309
 Tree alignment with CIs 498
Spiral Development Model
 Description 294–296
 Strategy (def.) 291
Staging or Control Points 162, 178, 504
Stakeholder
 Defined **39**
 Requirements 317
 Roles 42–45
Standard(s)
 Application 559
 Atmosphere 557
 Authorities 546
 Conflicts 559
 Defined **545**
 Deviation 586
 Informative clauses 546
 Lessons learned 558
 Normative clauses 546
 Open (def.) **545**
 Specification references 311
 Subject matter 546
 Technical (def.) **545**
 Weights and measures 546
Standard Operating Practices & Procedures
 (SOPPs) 91, 196, 565, 641, 742
State(s)
 Defined **190**
 Diagram 190
 Entity relationships 202
 Final 10, 170
 Initial 10, 36, 170
 Machine 190
 of equilibrium 36

State Machine Defined 190
 of operation **190**, 199
 Operational 200
 Physical configuration 201, 473
 Physical versus operational 200–202
Statement
 Mission Needs (MNS) 126, 160, 253
 of Objectives (SOO) See Contract SOO
 Problem **136**, 576
Statics 36
Statistical Data
 Application 590
 Dispersion 591
Statistical Distributions
 Normal (Gaussian) 587
 Logarithmic (LogNormal) 587
 Variance 587
Statistical Process Control (SPC) 587
Stimulate
 Defined **741**
Stimulation
 Defined **653**
Stimuli
 Behavioral response model 148
 External inputs 92, 172
 Modal transition dependency 197
 Test environment 661
 Use case 172
Strategic
 Plan **122**
 Planning 123
 Threats 147
Strategy
 Evolutionary development 291
 Grand Design 291
 Incremental development 291
 Mission 162
 SE design 269
 System Integration, Test, & Evaluation 268, 745
 Spiral development 291
 System development V & V 253–261
 System development workflow 251
 System Design Process 266
 System vs. component development 299
 Waterfall development 291
Strengths
 Human performance 527
 Machine performance 527–528
**Strengths, Weaknesses, Opportunities, & Threats
 (SWOT)** 44, 122, 126, 133
Subcontract Data Requirements List (SDRL)
 Defined **563**
 References 565

Suboptimization
  Defined **575**
  Description 578
Subphases
  of Operation 191, 193
Subsystems
  Mission specific 500–502
  Infrastructure 500–502
Suitability
  Measures of (MOS) 51
  Operational 48, 50, 775
Supply Chain
  Producer—supplier 130–132
Support
  and handling equipment 90
  Contract Logistics (CLS) 631
  Design for mission 406
  Equipment 775
  Supply 94
  System 92–94
SUPPORT SYSTEM
  Computer resources support operations 94, 778
  Decision support operations 93, 778
  Elements 87–88
  Facility operations 778
  Introduced 40
  Maintenance operations 93, 778
  Manpower and personnel operations 93–94, 778
  Organizational roles and missions 130–133
  Performance affecters 157
  Personnel roles 88
  PHS&T operations 94, 778
  Publications support operations 94, 778
  Roles 40
  Supply chain 131
  Supply support operations 94, 778
  System architecture element 69, 87
  System elements 88
  Technical data operations 94, 778
  Training and training support operations 94, 778
  Types of operations 93–94
Supportability
  Defined **402**
Survivability
  Defined **402**
  Design for 406
Susceptibility
  Defined **402**
Sustainability
  Defined **402**
Symptom Solving 137

System(s)
  Abuse 168, 173
  Acceptability challenges 48
  Adaptation 147
  Affordability 49
  Allowable actions 202
  Analytical representation 22
  Application/misapplication 168, 173
  Application types 211–212
  Attributes 30–33
  Availability 636
  Balance of power 37
  Benign 164
  Bounding 98
  Business 105
  Capabilities matrix 217
  Capability construct 231
  Certification 760
  Characteristics 34–35
  Checkout 767
  Complexity reduction 71
  Concept of operations (ConOps) 446
  Context 20
  Cooperative 164
  Cost 481
  Cultural 105
  Deactivation 183
  Decomposition rules 81–83
  Dedicated use 211
  Defined **18**
  Deployment 758
  Educational 105
  Effectiveness **48**, 53
  Element 68, 70
  End User/User 39
  Engagement 147
  Equilibrium 36
  Fault tolerant 421–422
  Feasibility 49
  Frame-based 611
  Friendly or cooperative 164
  Government 105
  Historical or heritage 104
  Hostile or adversarial 164
  Initial operating conditions 316
  Installation 767
  Integration 767
  Integration rules 81–83
  Introduced 70
  Latency 598
  Legacy (def.) **481**
  Levels of abstraction 76
  Life cycle phases 60
  Maintainability 629

Index **813**

Malfunction 234
MISSION 40
Mission critical 159
Mission dynamics 36
Objectives 127
of Interest (SOI) 40, 68
of Systems (SoS) 74, 77
of Units 546
Operational capability (def.) **217**
Operations (def.) **179**
Operations dictionary (def.) **179**, 184, 221
Operations Model (def.) **179**
Optimization 575, 578, 611, 651
Performance optimization 461, 578
Phase-out 183
Precedented 179, 266, 457
Prohibited actions 202
Production deployment 760
Quality 564
Real-time 611
Reliability 618
Repairability 775
Resources 100
Responses 92
Roles 40
Safety and security **402**, 766
Shadow operations 768
Simulation 658
Software 90
Software intensive **412**, 782
Sphere of influence 30, 135
Stabilization 37
Stakeholders 39, 42–45
Statics 36
State of equilibrium 36
Suboptimization 578
SUPPORT 40
Tactics 152
Threat 147
Transportation 105
Types of 20
Unanticipated/unscheduled events 445
Unprecedented 21, 179, 266, 457
Urban 105
Use/misuse 168, 173
Use cases 166–167, 444, 446
Verification and Validation (V & V) 54
Validation 56–57, 260, 699
Verification 55–56, 260, 694
Weights & Measures 546–547
System Applications
General use 211
Multi-purpose/use 211, 404
Reusable 211–212, 214
Single-use 211–212, 214
Types of system 211–212
System Documentation
Design 568
Export control Refer to ITAR
Integration and Test 568–569
Issues 571–572
Plans 567
Posting Acquirer / vendor data 571
Sanity check 575
Specifications 568
System Engineering
Defined **24**
System Engineering and Integration Team (SEIT)
Performance budgets and margins 602
RAM reviews 646
Specification development 349
Specification ownership 309, 349, 366
V & V 701
System Elements
Classes 70
Defined **68**
EQUIPMENT 70, 88–90, 641
FACILITIES 70, 94–95, 641
HARDWARE 89
INDUCED ENVIRONMENT 101, 106
Interactions 87, 91, 95
Introduced 70
Higher Order Domain 69
HUMAN-MADE SYSTEMS 100, 104–105
MISSION RESOURCES 70, 91–92, 641
NATURAL ENVIRONMENT 102–104
PERSONNEL 70, 88, 641
Performance 610
Physical Environment Domain 69
PROCEDURAL DATA 70, 91, 641
OPERATING CONSTRAINTS 70, 100
ORGANIZATION 70, 100, 124
RESOURCES 70, 100
ROLES AND MISSIONS 70, 100
SOFTWARE 90
System 68, 70
SYSTEM RESPONSES 70, 92, 641
System Integration, Test, & Evaluation (SITE)
Activities 738
Challenges & issues 751
Data integrity 751
Defined **737**
Design for 407, 737

814 Index

Deviations and waivers 752
Elements of 739
Fundamentals 737
Guiding philosophy 739
Objective 737–738
Planning 744, 762
Preparation 740
Procedures 742
Process strategy 745
Selection 759, 762–764
Tasks 749
Task sequencing 738
Time allocations 753
Work products 744
SITE Roles
Acquirer test representative 746
Lab manager 746
QA representative 746
Security representative 746
Test director 745
Test safety officer 746
Tester 746
System of Interest (SOI)
Architecture construct 87
Architectural system elements 87–95
MISSION SYSTEM role 40, 87
Specifying 311
SUPPORT SYSTEM role 40, 87
System Operations
Defined **179**
Dictionary (def.) **179**
Dictionary description 184, 221
Model (def.) **179**
Model description 180–186
System Performance
Acceptability 588
Affecters 157
Analysis 603
Benchmark 780
Cumulative effects 591
Evaluation 776–780
Monitoring 778
V & V 701–702
System Performance Specification (SPS)
Defined **303**
Discussions 78, 307
System Requirements Document (SRD) 225, 254, 317, 327, 352
SYSTEM RESPONSES Element 92
Description 92
Performance affecters 157
System architecture element 87
SysML 6

Tactics
System 152
Tactical
Plan (def.) **123**
Planning 124
Threats 148
Tailoring
Specification (def.) **303**
Targets of Opportunity (TOO) 123, 126, 133
Task(ing)
Analysis **526**, 535
Attributes 540–541
Defined **5**
Human-System tasking 538, 540–541
Mission 604
Modeling and simulation 610
Multi 607
Post-Mission 604
Pre-Mission 604
Order (def.) **160**
Statistical characteristics 606
System 603
Time approximation estimation 609
Taxonomy
Defined **68**
Technical/Technology
Demonstrations 701–702
Technical Data Package (TDP)
Defined **563**
Technical Performance Measurement (TPM)
Challenges 397–398
Defined **392**
Plotting 396
TPM selection 397
Technical Performance Parameters (TPPs)
Defined **391**
Test & Evaluation (T&E)
Defined **735**
Developmental (DT&E) 252, 261–262, 738, 745
Operational (OT&E) 253, 262, 738, 776
Working Group (TEWG) 735, 755
Test(ing)
Acceptance 734
Alignment 754
Alpha (def.) **734**
Alignment 754
Anomaly **734**, 755
Article **735**
Beta (def.) **734**
Case (def.) **735**
Challenges and issues 751–755
Configuration (def.) **735**
Conflicts 755

Index **815**

Coverage (def.) **735**
Criteria (def.) **735**
Data 751, 752
Daily Operational Readiness (DORT) 233
Data correlation 594–595
Data dispersion 591
Defined **735**, 736
Destructive 735, 741, 745
Discrepancy Reports (DRs) 743, 750–751
Environment (def.) **735**, 741
Equipment and tools 753
Equipment calibration 753, 754
Field trails 702
Formal (def.) **735**
Functional 741
Incident report (def.) **736**
Instrumentation (def.) **736**
Independent Test Agency (ITA) 56, 61, 253, 262, 702, 707, 752, 776
Issue resolution 755
Markets 702
and measurement equipment 736
Multiple requirements 742
Non-destructive **735**, 741
Operating environment 740, 741
Organization 745
Personnel qualification 742
Personnel roles 745–747
Qualification 702, 735, 741
Range (def.) **736**
Readiness Review (TRR) **727**
Regression (def.) **735**, 742–743
Repeatability (def.) **736**
Responsibility 742
Resources (def.) **736**
Scenarios 755
Sequence planning 745
System Verification (SVT) 56, 61, 260, 269, 313, 702
Transient error 736
Types of 741
Unit Under (UUT) 739
Verification method (def.) **693**
Work products 742
Test Article
Defined **735**
Discussions 687, 739, 752
Failure 687
Hooks 755
Refurbishment / reconditioning 754
Unit Under Test (UUT) 739
Utilization 752

TEST (Verification Method)
See Verification Methods
Testability
Defined **402**
Testbed
Application 664–668
Environment 664
Threat(s)
Alliances 151
Behavior 152
Countermeasures 152
Counter-countermeasures (CCM) 152
Encounters 152
Environment 151–152
Sources 151
Strategic (def.) **147**
System 147
Tactical (def.) **148**
Types of 151
Time
Administrative Delay (ADT) 633
Corrective Maintenance (CMT) 634
Delay (def.) **616**
Estimate approximation (nominal) 609
Logistics Delay (LDT) 634
Maintenance Down (MDT) 633
Mean Active Maintenance (MAMT) 636
Preventive Maintenance (PMT) 633
Process(ing) **598**, 605
Queue 598, **605**
Requirements 160
Simulation (def.) 653
Transport **598**, 605
Turn around (def.) **775**
Timeline
Analysis (def.) **160**
Event-based 170
Mission Event (MET) 129–130, 163, 279, 444, 460
Tool
Defined **21**
Requirements management 379
Traceability
Audits 254, 718
CWBS 463
Specification requirements 128, 307, 462
Trade Space
Cost—schedule 678
Defined **673**
Performance—cost 678
Performance—schedule 678
Utility 673

816 Index

Trade Study
  Analysis of alternatives (AoA) 672
  COIs/CTIs 675
  Conclusion (def.) **673**
  Conclusions 686
  Decision areas 674
  Decision criteria 673
  Decision factors 673, 680–681
  Defined **673**
  Dependency sequencing 676
  Findings (def.) **673**
  Findings 686
  Methodology 680, 688
  Objectives 674
  Outline 685–686
  PERSONNEL-EQUIPMENT Trade-Offs 530
  Process 679
  Recommendation (def.) **673**
  Recommendations 686
  Report 673
  Resource constraints 688
  Risk areas 687
  Selection criteria 688
  Semantics 674
  Sensitivity analysis (def.) **673**
  Suggestions 688
  Trade space (def.) **673**
  Report (TSR) 575, **673**, 684–687
  Utility function **681**, 681–684
  Utility space (def.) **673**
  Viable alternative (def.) **673**
Trade(s)
  Cost—schedule 678
  Performance—cost 678
  Performance—schedule 678
  Space 54, 673, 677–679
Training
  Challenges 783–784
  Design for 406
  Devices 783
  Device concurrency 785
  Operator 781, 782
  Scoring 784
  Task Lists (TTL) 784
  Levels of 784
  HSI element 533–534
Transaction
  Analysis (def.) **452**
  Defined **452**
Transitions
  Triggering event based 195–196
  Modal 197–198

Transportability
  Defined **403**
  Design for 404
Tree
  Specification 303, 309–310
  Product structure 467–468
Triggering
  Event 190
  Transition 195–196
UML™
  Activation box 10
  Activity 10
  Actor 11, **174**
  Decision block 10, 11
  Event 10
  Final State 10, 11
  Fork 11
  Interaction diagram 9, 458
  Initial State 10
  Lifeline 174
  Join 11
  References 6, 8, 172, 177, 453–463, 470
  Sequence diagrams 10, **167**, 172–173, 463
  Swimlane 174
  Synchronization bar 10, 11
  Use case diagram **168**, 172
Units
  British Engineering System (BES) 547
  International System (SI) of 547
  Metric System (mks) 547
  System of 546
Unprecedented
  Systems 21, 266, 457
Usability
  Defined **403**
  Design for 404
Use Case(s)
  Analysis 174
  Attributes 169–173
  Consequences 171
  Defined **168**
  Event-based timeline 170
  Diagram (def.) **168**
  Operational use cases 512
  Optimal quantity 175
  Priorities 170
  Preceding circumstances 170
  Problem of occurrence 172
  Resources 170
  Scenario **168**, 444
  Specification requirements 175
  System 167, 444, 446

Index **817**

System interactions 156
Thread 366
User/End User
   Defined **39**
   Moments of truth ix–x, 126
   System 77
   Training 767
Utility
   Bounding value ranges 682
   Function **673**, 681–684
   Operational 50, 775
   Space (def.) **673**

Validation
   Defined **693**
   Methods 700
   Model 655
   Requirements 258
   Responsibility 700
   System 56–57, 699–700
   V-Model 266
Value
   Design to (DTV) 403
Variance (Statistical)
   Defined **587**
   Deviation, waiver, departure **491**, 693, 752
Verification
   Acceptance Test Procedures (ATPs) 742
   Acceptance Test(ing) 701, 734
   Authenticate System Baseline 695
   Simulation (Method) 699
   Component procurement and development 695
   Control points 697
   Defined **693**
   Design for 407
   Methods 269, 698
   Modeling and simulation 697
   Multi-level application 696
   OT&E 695
   Requirements 375–378
   Requirements traceability 697
   Responsibility 696
   SITE 695
   System 55–56
   System Design 695
   System Verification Test (SVT) 56, 61, 260, 269, 313, 702
Verification & Validation (V&V)
   Authenticate System Baseline 260–261
   Behavioral Domain Solution 461
   Benefits 694, 707
   Certification 706

   Challenges and issues 704–706
   Classification of defects 692
   Component Procurement/Dev. 259
   Defined **693**
   Independent 692
   Introduced 54
   Myth 693
   Operations Domain Solution 447
   Physical Domain Solution 475
   Practices 694, 699–700
   Requirements Domain Solution 436
   Responsibility 700
   SE Design strategy 259
   Solution to program challenges 694
   System 54, 260, 691
   System IT&E strategy 259
   SPS strategy 258
   System performance 701
V & V Strategy
   Authenticate System Baselines 260–261
   Component Procurement and Development 259
   SE Design 259
   System Development Phase 254–255
   System Integration, Test, & Evaluation (SITE) 259–260
   System Performance Specification (SPS) 258
   System Procurement Phase 253–254
   Validate system 260
Verification Method(s)
   ANALYSIS (def.) **692**
   ANALYSIS method description 698
   ANALYSIS method selection 377
   Cost factors 699
   DEMONSTRATION (def.) **692**
   DEMONSTRATION method 698
   DEMONSTRATION selection 377
   INSPECTION / EXAMINATION **693**
   INSPECTION method description 699
   INSPECTION method selection 377
   Method selection process 376
   TEST (def.) **693**
   TEST method description 698
   TEST method selection 377
   Similarity 377, **693**, 698
   Validation of records 699
View
   Architectural 411, **413**
   Logical/Functional 415
   Operational architecture 415
   Physical architecture 415
   Requirements architecture 415

Viewpoint
  Architectural 411, 413
Vulnerability
  Defined **403**
  Design for 406
  Interface 521
  System success 97

Waiver
  Defined **303**
  Variance, deviation, waiver, departure **491**, 693, 752
Wasson's
  Law 335
  Task Significance Principle 689
Waterfall Development Model
  Application 267
  Development strategy (def.) **291**
  Description 292

Waypoint
  Defined **160**
  Mission timeline 162
Weibull Distribution 620
Weights and Measures 546–547
Work Product(s)
  Behavioral Domain Solution 463
  Physical Domain Solution 477
  Operations Domain Solution 448
  Requirements Domain Solution 437
  Review 715
  SE Process 283
  System Integration, Test, & Evaluation 744
  Testing 742
Workflow
  System development 251
Working Groups
  Interface Control (ICWG) 516
  Test & Evaluation (TEWG) 735